Lecture Notes in Computer Science 5305

Commenced Publication in 1973
Founding and Former Series Editors:
Gerhard Goos, Juris Hartmanis, and Jan van Leeuwen

Editorial Board

David Forsyth   Philip Torr
Andrew Zisserman (Eds.)

Computer Vision – ECCV 2008

10th European Conference on Computer Vision
Marseille, France, October 12-18, 2008
Proceedings, Part IV

 Springer

Volume Editors

David Forsyth
University of Illinois at Urbana-Champaign, Computer Science Department
3310 Siebel Hall, Urbana, IL 61801, USA
E-mail: daf@cs.uiuc.edu

Philip Torr
Oxford Brookes University, Department of Computing
Wheatley, Oxford OX33 1HX, UK
E-mail: philiptorr@brookes.ac.uk

Andrew Zisserman
University of Oxford, Department of Engineering Science
Parks Road, Oxford OX1 3PJ, UK
E-mail: az@robots.ox.ac.uk

Library of Congress Control Number: 2008936989

CR Subject Classification (1998): I.4, I.2.10, I.5.4, I.5, I.7.5

LNCS Sublibrary: SL 6 – Image Processing, Computer Vision, Pattern Recognition, and Graphics

ISSN   0302-9743
ISBN-10   3-540-88692-3 Springer Berlin Heidelberg New York
ISBN-13   978-3-540-88692-1 Springer Berlin Heidelberg New York

Springer is a part of Springer Science+Business Media

springer.com

© Springer-Verlag Berlin Heidelberg 2008
Printed in Germany

Typesetting: Camera-ready by author, data conversion by Scientific Publishing Services, Chennai, India
Printed on acid-free paper   SPIN: 12553631   06/3180   5 4 3 2 1 0

Preface

Welcome to the 2008 European Conference on Computer Vision. These proceedings are the result of a great deal of hard work by many people. To produce them, a total of 871 papers were reviewed. Forty were selected for oral presentation and 203 were selected for poster presentation, yielding acceptance rates of 4.6% for oral, 23.3% for poster, and 27.9% in total.

We applied three principles. First, since we had a strong group of Area Chairs, the final decisions to accept or reject a paper rested with the Area Chair, who would be informed by reviews and could act only in consensus with another Area Chair. Second, we felt that authors were entitled to a summary that explained how the Area Chair reached a decision for a paper. Third, we were very careful to avoid conflicts of interest.

Each paper was assigned to an Area Chair by the Program Chairs, and each Area Chair received a pool of about 25 papers. The Area Chairs then identified and ranked appropriate reviewers for each paper in their pool, and a constrained optimization allocated three reviewers to each paper. We are very proud that every paper received at least three reviews.

At this point, authors were able to respond to reviews. The Area Chairs then needed to reach a decision. We used a series of procedures to ensure careful review and to avoid conflicts of interest. Program Chairs did not submit papers. The Area Chairs were divided into three groups so that no Area Chair in the group was in conflict with any paper assigned to any Area Chair in the group. Each Area Chair had a "buddy" in their group. Before the Area Chairs met, they read papers and reviews, contacted reviewers to get reactions to submissions and occasionally asked for improved or additional reviews, and prepared a rough summary statement for each of the papers in their pool.

At the Area Chair meeting, groups met separately so that Area Chairs could reach a consensus with their buddies, and make initial oral/poster decisions. We met jointly so that we could review the rough program, and made final oral/poster decisions in groups. In the separate meetings, there were no conflicts. In the joint meeting, any Area Chairs with conflicts left the room when relevant papers were discussed. Decisions were published on the last day of the Area Chair meeting.

There are three more somber topics to report. First, the Program Chairs had to deal with several double submissions. Referees or Area Chairs identified potential double submissions, we checked to see if these papers met the criteria published in the call for papers, and if they did, we rejected the papers and did not make reviews available. Second, two submissions to ECCV 2008 contained open plagiarism of published works. We will pass details of these attempts to journal editors and conference chairs to make further plagiarism by the responsible parties more difficult. Third, by analysis of server logs we discovered that there had been a successful attempt to download all submissions shortly after the deadline. We warned all authors that this had happened to ward off dangers to intellectual property rights, and to minimize the chances that an attempt at plagiarism would be successful. We were able to identify the responsible party, discussed this matter with their institutional management, and believe we resolved the issue as well as we could have. Still, it is important to be aware that no security or software system is completely safe, and papers can leak from conference submission.

We felt the review process worked well, and recommend it to the community. The process would not have worked without the efforts of many people. We thank Lyndsey Pickup, who managed the software system, author queries, Area Chair queries and general correspondence (most people associated with the conference will have exchanged e-mails with her at some point). We thank Simon Baker, Ramin Zabih and especially Jiří Matas for their wise advice on how to organize and run these meetings; the process we have described is largely their model from CVPR 2007. We thank Jiří Matas and Dan Večerka, for extensive help with, and support of, the software system. We thank C. J. Taylor for the 3-from-5 optimization code. We thank the reviewers for their hard work. We thank the Area Chairs for their very hard work, and for the time and attention each gave to reading papers, reviews and summaries, and writing summaries.

We thank the Organization Chairs Peter Sturm and Edmond Boyer, and the General Chair, Jean Ponce, for their help and support and their sharing of the load. Finally, we thank Nathalie Abiola, Nasser Bacha, Jacques Beigbeder, Jerome Bertsch, Joëlle Isnard and Ludovic Ricardou of ENS for administrative support during the Area Chair meeting, and Danièle Herzog and Laetitia Libralato of INRIA Rhône-Alpes for administrative support after the meeting.

August 2008

Andrew Zisserman
David Forsyth
Philip Torr

Organization

Conference Chair

Jean Ponce	Ecole Normale Supérieure, France

Honorary Chair

Jan Koenderink	EEMCS, Delft University of Technology, The Netherlands

Program Chairs

David Forsyth	University of Illinois, USA
Philip Torr	Oxford Brookes University, UK
Andrew Zisserman	University of Oxford, UK

Organization Chairs

Edmond Boyer	LJK/UJF/INRIA Grenoble–Rhône-Alpes, France
Peter Sturm	INRIA Grenoble–Rhône-Alpes, France

Specialized Chairs

Frédéric Jurie	Workshops	Université de Caen, France
Frédéric Devernay	Demos	INRIA Grenoble–Rhône-Alpes, France
Edmond Boyer	Video Proc.	LJK/UJF/INRIA Grenoble–Rhône-Alpes, France
James Crowley	Video Proc.	INPG, France
Nikos Paragios	Tutorials	Ecole Centrale, France
Emmanuel Prados	Tutorials	INRIA Grenoble–Rhône-Alpes, France
Christophe Garcia	Industrial Liaison	France Telecom Research, France
Théo Papadopoulo	Industrial Liaison	INRIA Sophia, France
Jiří Matas	Conference Software	CTU Prague, Czech Republic
Dan Večerka	Conference Software	CTU Prague, Czech Republic

Program Chair Support

Lyndsey Pickup	University of Oxford, UK

Administration

Danile Herzog   INRIA Grenoble–Rhône-Alpes, France
Laetitia Libralato   INRIA Grenoble–Rhône-Alpes, France

Conference Website

Elisabeth Beaujard   INRIA Grenoble–Rhône-Alpes, France
Amaël Delaunoy   INRIA Grenoble–Rhône-Alpes, France
Mauricio Diaz   INRIA Grenoble–Rhône-Alpes, France
Benjamin Petit   INRIA Grenoble–Rhône-Alpes, France

Printed Materials

Ingrid Mattioni   INRIA Grenoble–Rhône-Alpes, France
Vanessa Peregrin   INRIA Grenoble–Rhône-Alpes, France
Isabelle Rey   INRIA Grenoble–Rhône-Alpes, France

Area Chairs

Horst Bischof   Graz University of Technology, Austria
Michael Black   Brown University, USA
Andrew Blake   Microsoft Research Cambridge, UK
Stefan Carlsson   NADA/KTH, Sweden
Tim Cootes   University of Manchester, UK
Alyosha Efros   CMU, USA
Jan-Olof Eklund   KTH, Sweden
Mark Everingham   University of Leeds, UK
Pedro Felzenszwalb   University of Chicago, USA
Richard Hartley   Australian National University, Australia
Martial Hebert   CMU, USA
Aaron Hertzmann   University of Toronto, Canada
Dan Huttenlocher   Cornell University, USA
Michael Isard   Microsoft Research Silicon Valley, USA
Aleš Leonardis   University of Ljubljana, Slovenia
David Lowe   University of British Columbia, Canada
Jiří Matas   CTU Prague, Czech Republic
Joe Mundy   Brown University, USA
David Nistér   Microsoft Live Labs/Microsoft Research, USA
Tomáš Pajdla   CTU Prague, Czech Republic
Patrick Pérez   IRISA/INRIA Rennes, France
Marc Pollefeys   ETH Zürich, Switzerland
Ian Reid   University of Oxford, UK
Cordelia Schmid   INRIA Grenoble–Rhône-Alpes, France
Bernt Schiele   Darmstadt University of Technology, Germany
Christoph Schnörr   University of Mannheim, Germany
Steve Seitz   University of Washington, USA

James Coughlan
David Crandall
Daniel Cremers
Antonio Criminisi
David Cristinacce
Gabriela Csurka
Navneet Dalal
Kristin Dana
Kostas Daniilidis
Larry Davis
Andrew Davison
Nando de Freitas
Daniel DeMenthon
David Demirdjian
Joachim Denzler
Michel Dhome
Sven Dickinson
Gianfranco Doretto
Gyuri Dorko
Pinar Duygulu Sahin
Charles Dyer
James Elder
Irfan Essa
Andras Ferencz
Rob Fergus
Vittorio Ferrari
Sanja Fidler
Mario Figueiredo
Graham Finlayson
Robert Fisher
François Fleuret
Wolfgang Förstner
Charless Fowlkes
Jan-Michael Frahm
Friedrich Fraundorfer
Bill Freeman
Brendan Frey
Andrea Frome
Pascal Fua
Yasutaka Furukawa
Daniel Gatica-Perez
Dariu Gavrila
James Gee
Guido Gerig
Theo Gevers

Christopher Geyer
Michael Goesele
Dan Goldman
Shaogang Gong
Leo Grady
Kristen Grauman
Eric Grimson
Fred Hamprecht
Edwin Hancock
Allen Hanson
James Hays
Carlos Hernández
Anders Heyden
Adrian Hilton
David Hogg
Derek Hoiem
Alex Holub
Anthony Hoogs
Daniel Huber
Alexander Ihler
Michal Irani
Hiroshi Ishikawa
David Jacobs
Bernd Jähne
Hervé Jégou
Ian Jermyn
Nebojsa Jojic
Michael Jones
Frédéric Jurie
Timor Kadir
Fredrik Kahl
Amit Kale
Kenichi Kanatani
Sing Bing Kang
Robert Kaucic
Qifa Ke
Renaud Keriven
Charles Kervrann
Ron Kikinis
Benjamin Kimia
Ron Kimmel
Josef Kittler
Hedvig Kjellström
Leif Kobbelt
Pushmeet Kohli

Esther Koller-Meier
Vladimir Kolmogorov
Nikos Komodakis
Kurt Konolige
Jana Košecká
Zuzana Kukelova
Sanjiv Kumar
Kyros Kutulakos
Ivan Laptev
Longin Jan Latecki
Svetlana Lazebnik
Erik Learned-Miller
Yann Lecun
Bastian Leibe
Vincent Lepetit
Thomas Leung
Anat Levin
Fei-Fei Li
Hongdong Li
Stephen Lin
Jim Little
Ce Liu
Yanxi Liu
Brian Lovell
Simon Lucey
John Maccormick
Petros Maragos
Aleix Martinez
Iain Matthews
Wojciech Matusik
Bruce Maxwell
Stephen Maybank
Stephen McKenna
Peter Meer
Etienne Mémin
Dimitris Metaxas
Branislav Mičušík
Krystian Mikolajczyk
Anurag Mittal
Theo Moons
Greg Mori
Pawan Mudigonda
David Murray
Srinivasa Narasimhan
Randal Nelson

Ram Nevatia
Jean-Marc Odobez
Björn Ommer
Nikos Paragios
Vladimir Pavlovic
Shmuel Peleg
Marcello Pelillo
Pietro Perona
Maria Petrou
Vladimir Petrovic
Jonathon Phillips
Matti Pietikäinen
Axel Pinz
Robert Pless
Tom Pock
Fatih Porikli
Simon Prince
Long Quan
Ravi Ramamoorthi
Deva Ramanan
Anand Rangarajan
Ramesh Raskar
Xiaofeng Ren
Jens Rittscher
Rómer Rosales
Bodo Rosenhahn
Peter Roth
Stefan Roth
Volker Roth
Carsten Rother
Fred Rothganger
Daniel Rueckert
Dimitris Samaras

Radim Šára
Eric Saund
Silvio Savarese
Daniel Scharstein
Yoav Schechner
Kondrad Schindler
Stan Sclaroff
Mubarak Shah
Gregory Shakhnarovich
Eli Shechtman
Jianbo Shi
Kaleem Siddiqi
Leonid Sigal
Sudipta Sinha
Josef Sivic
Cristian Sminchişescu
Anuj Srivastava
Drew Steedly
Gideon Stein
Björn Stenger
Christoph Strecha
Erik Sudderth
Josephine Sullivan
David Suter
Tomáš Svoboda
Hai Tao
Marshall Tappen
Demetri Terzopoulos
Carlo Tomasi
Fernando Torre
Lorenzo Torresani
Emanuele Trucco
David Tschumperlé

John Tsotsos
Peter Tu
Matthew Turk
Oncel Tuzel
Carole Twining
Ranjith Unnikrishnan
Raquel Urtasun
Joost Van de Weijer
Manik Varma
Nuno Vasconcelos
Olga Veksler
Jakob Verbeek
Luminita Vese
Thomas Vetter
René Vidal
George Vogiatzis
Daphna Weinshall
Michael Werman
Tomáš Werner
Richard Wildes
Lior Wolf
Ying Wu
Eric Xing
Yaser Yacoob
Ruigang Yang
Stella Yu
Lihi Zelnik-Manor
Richard Zemel
Li Zhang
S. Zhou
Song-Chun Zhu
Todd Zickler
Lawrence Zitnick

Additional Reviewers

Lourdes Agapito
Daniel Alexander
Elli Angelopoulou
Alexandru Balan
Adrian Barbu
Nick Barnes
João Barreto
Marian Bartlett
Herbert Bay

Ross Beveridge
V. Bhagavatula
Edwin Bonilla
Aeron Buchanan
Michael Burl
Tiberio Caetano
Octavia Camps
Sharat Chandran
François Chaumette

Yixin Chen
Dmitry Chetverikov
Sharat Chikkerur
Albert Chung
Nicholas Costen
Gabriela Oana Cula
Goksel Dedeoglu
Hervé Delingette
Michael Donoser

Mark Drew
Zoran Duric
Wolfgang Einhauser
Aly Farag
Beat Fasel
Raanan Fattal
Paolo Favaro
Rogerio Feris
Cornelia Fermüller
James Ferryman
David Forsyth
Jean-Sébastien Franco
Mario Fritz
Andrea Fusiello
Meirav Galun
Bogdan Georgescu
A. Georghiades
Georgy Gimel'farb
Roland Goecke
Toon Goedeme
Jacob Goldberger
Luis Goncalves
Venu Govindaraju
Helmut Grabner
Michael Grabner
Hayit Greenspan
Etienne Grossmann
Richard Harvey
Sam Hasinoff
Horst Haussecker
Jesse Hoey
Slobodan Ilic
Omar Javed
Qiang Ji
Jiaya Jia
Hailin Jin
Ioannis Kakadiaris
Joni-K. Kämäräinen
George Kamberov
Yan Ke
Andreas Klaus
Georg Klein
Reinhard Koch
Mathias Kolsch
Andreas Koschan
Christoph Lampert

Mike Langer
Georg Langs
Neil Lawrence
Sang Lee
Boudewijn Lelieveldt
Marc Levoy
Michael Lindenbaum
Chengjun Liu
Qingshan Liu
Manolis Lourakis
Ameesh Makadia
Ezio Malis
R. Manmatha
David Martin
Daniel Martinec
Yasuyuki Matsushita
Helmut Mayer
Christopher Mei
Paulo Mendonça
Majid Mirmehdi
Philippos Mordohai
Pierre Moreels
P.J. Narayanan
Nassir Navab
Jan Neumann
Juan Carlos Niebles
Ko Nishino
Thomas O'Donnell
Takayuki Okatani
Kenji Okuma
Margarita Osadchy
Mustafa Ozuysal
Sharath Pankanti
Sylvain Paris
James Philbin
Jean-Philippe Pons
Emmanuel Prados
Zhen Qian
Ariadna Quattoni
Ali Rahimi
Ashish Raj
Visvanathan Ramesh
Christopher Rasmussen
Tammy Riklin-Raviv
Charles Rosenberg
Arun Ross

Michael Ross
Szymon Rusinkiewicz
Bryan Russell
Sudeep Sarkar
Yoichi Sato
Ashutosh Saxena
Florian Schroff
Stephen Se
Nicu Sebe
Hans-Peter Seidel
Steve Seitz
Thomas Serre
Alexander Shekhovtsov
Ilan Shimshoni
Michal Sofka
Jan Solem
Gerald Sommer
Jian Sun
Rahul Swaminathan
Hugues Talbot
Chi-Keung Tang
Xiaoou Tang
C.J. Taylor
Jean-Philippe Thiran
David Tolliver
Yanghai Tsin
Zhuowen Tu
Vaibhav Vaish
Anton van den Hengel
Bram Van Ginneken
Dirk Vandermeulen
Alessandro Verri
Hongcheng Wang
Jue Wang
Yizhou Wang
Gregory Welch
Ming-Hsuan Yang
Caspi Yaron
Jieping Ye
Alper Yilmaz
Christopher Zach
Hongyuan Zha
Cha Zhang
Jerry Zhu
Lilla Zollei

Sponsoring Institutions

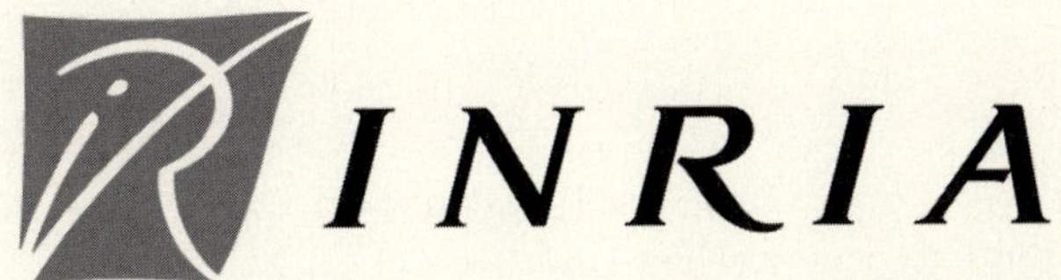

Région

Provence-Alpes-Côte d'Azur

Deutsche Telekom
Laboratories

Microsoft®
Research

EADS

TOSHIBA
Leading Innovation >>>

 Springer

Table of Contents – Part IV

Active Reconstruction

Image Segmentation in the Presence of Shadows and Highlights

Eduard Vazquez, Joost van de Weijer, and Ramon Baldrich

Computer Vision Center \ Dpt. Ciencies de la Computacio. Edifici O Universitat Autonoma de Barcelona. 08193 Cerdanyola del Valles, Barcelona, Spain

Abstract. The segmentation method proposed in this paper is based on the observation that a single physical reflectance can have many different image values. We call the set of all these values a dominant colour. These variations are caused by shadows, shading and highlights and due to varying object geometry. The main idea is that dominant colours trace connected ridges in the chromatic histogram. To capture them, we propose a new Ridge based Distribution Analysis (RAD) to find the set of ridges representative of the dominant colour. First, a multilocal crease-ness technique followed by a ridge extraction algorithm is proposed. Afterwards, a flooding procedure is performed to find the dominant colours in the histogram. Qualitative results illustrate the ability of our method to obtain excellent results in the presence of shadow and highlight edges. Quantitative results obtained on the Berkeley data set show that our method outperforms state-of-the-art segmentation methods at low computational cost.

1   Introduction

Image segmentation is a computer vision process consisting in the partition of an image into a set of non-overlapped regions. A robust and efficient segmentation is required as a preprocessing step in several computer vision tasks such as object recognition or tracking. On real images the varying shapes of the objects provoke several effects related with the illumination such as shadows, shading and highlights. These effects, are one of the main difficulties that have to be solved to obtain a correct segmentation.

There exist several different methods covering a broad spectrum of points of view. The work presented by Skarbek and Koschan [1], draws the basis of the current classifications of segmentation methods. Some other comprehensive surveys of colour segmentation techniques are presented in [2] and [3], where a similar schema is followed. From these works segmentation methods are divided in four main categories: feature-based, image-based, physics-based and hybrid approaches. Feature-based approaches are focused on the photometric information of an image represented on its histogram [4],[5]. Image-based approaches exploit the spatial information of the colour in an image, named *spatial coherence* [6]. Physics-based methods use physics and psychophysics information to

D. Forsyth, P. Torr, and A. Zisserman (Eds.): ECCV 2008, Part IV, LNCS 5305, pp. 1–14, 2008.

perform the segmentation. Finally, hybrid techniques combine methods of the previous categories.

This paper introduces a method that exploits exclusively the photometric information of an image on its histogram. Therefore, it belongs to the category of feature-based segmentation methods. This category can be further split in three main categories, i.e., histogram thresholding, clustering and fuzzy clustering. Histogram thresholding techniques assume that there exist a threshold value that isolates all pixels representative of an object in a scene. This basic concept is exploited in several ways as explained in [7]. Clustering techniques, also named hard clustering, perform a partition of the feature space under different criteria such a distance measure as k-means or ISODATA, probabilistic/statistical approaches, such as Mean Shift [8], or the spectral analysis of the data [9], based on the Karhunen-Loeve transformation. Fuzzy clustering includes methods such as fuzzy k-means, Gath-Geva clustering, or mixture models [10], [11] which are a way to look for areas of high density. The most related technique with the one introduced in this paper, is the Mean shift, which will be commented and compared with our method in section 4. Each technique has its own advantages and drawbacks. There is a difficulty shared between all these methods, i.e., their behaviour in the presence of shadows, shading, and highlights. Furthermore, the work presented by Martin *et al.* in [12], points out the existence of strong edges related with this physical effects in an image that are not considered in a human segmentation. These edges are detected for both image and feature based methods.

Our approach. Our approach to colour image segmentation is based on the insight that the distributions formed by a single-colored object have a physically determined shape in colour histogram-space. We model an image as being generated by a set of dominant colours (DC), where each dominant colour is described by a distribution in histogram-space. Each DC is related to a semantic object in the image. For example, in Figure 1 we distinguish between four different DC's, namely: red for the pepper, green and brown for the branch and black for the background.

A DC generates many image values due to geometrical and photometric variations. Our main aim is to find a good representation of the topologies which DC's are likely to form in histogram space. For this purpose, consider the distribution of a single DC as described by the dichromatic reflection model [13]:

$$\mathbf{f}\left(\mathbf{x}\right) = m^{b}\left(\mathbf{x}\right)\mathbf{c}^{b} + m^{i}\left(\mathbf{x}\right)\mathbf{c}^{i} \qquad (1)$$

in which $\mathbf{f} = \{R, G, B\}$, $\mathbf{c}^{b}$ is the body reflectance, $\mathbf{c}^{i}$ the surface reflectance, m^{b} and m^{i} are geometry dependent scalars representing the magnitude of body and surface reflectance. Bold notation is used to indicate vectors. For one DC we expect both $\mathbf{c}^{b}$ and $\mathbf{c}^{i}$ to be almost constant, whereas $m^{b}\left(\mathbf{x}\right)$ and $m^{i}\left(\mathbf{x}\right)$ are expected to vary significantly.

The two parts of the dichromatic reflectance model are clearly visible in the histogram of Figure 1b. Firstly, due to the shading variations the distribution of the red pepper traces an elongated shape in histogram-space. Secondly, the surface reflectance forms a branch which points in the direction of the reflected illuminant. In conclusion, the distribution of a single DC forms a ridge-like structure in histogram space.

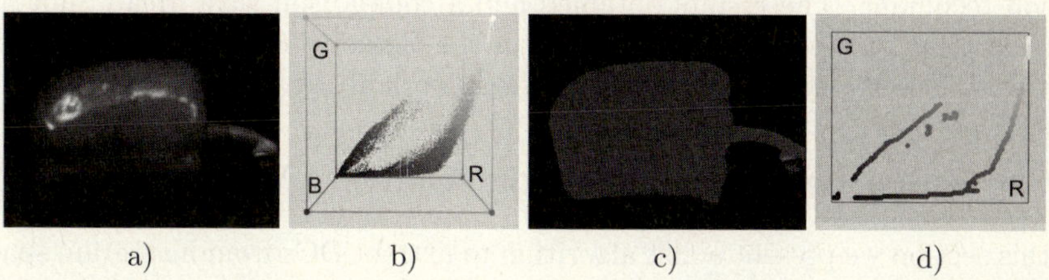

Fig. 1. (a) An image from [14] and (b) its histogram. The effects of shading and highlights are clearly visible in the red colours of the histogram. (c) Segmented images using RAD. (d) Ridges found with RAD. Note that the three branches of the red pepper are correctly connected in a single ridge.

To illustrate the difficulty of extracting the distributions of DCs consider Figure 2c, which contains a patch of the horse image. The 2D Red-Green histogram of the patch is depicted in Figure 2d to see the number of occurrences of each chromatic combination. This is done for explanation purposes. In this 2D histogram it can be clearly seen that the density of the geometric term $m_b(\mathbf{x})$ varies significantly, and the distribution is broken in two parts. However, we have an important clue that the two distributions belong to the same DC: the orientation of the two distribution is similar, which means they have a similar $\mathbf{c}^b$. We exploit this feature in the ridge extraction algorithm by connecting neighboring distributions with similar orientation.

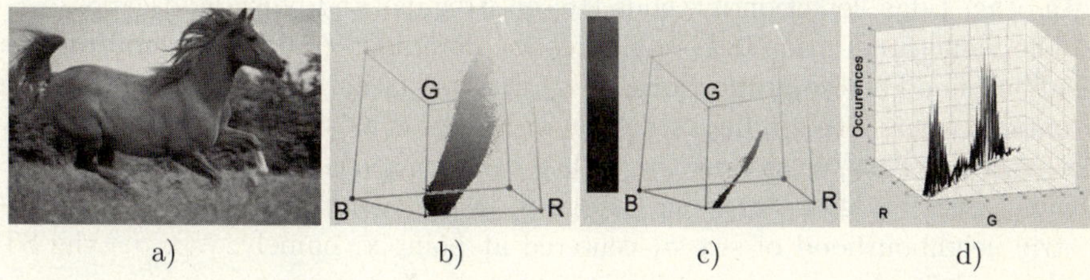

Fig. 2. (a) An image and (b) its 3D RGB histogram. (c) A patch of a) and its RGB histogram. (d) 2D histogram of c) to illustrate the discontinuities appearing on a DC.

In literature several methods have explicitly used the dichromatic model to obtain image segmentation, e.g. [15]. A drawback of such methods is however that for many images Eq. 1 does only approximately model the data. This can be caused by many reasons, such as non-linear acquisition systems, clipped highlights, and image compression. In this article we use Eq. 1 only to conclude that objects described by this equation will trace connected ridges in histogram space. This makes the method more robust to deviations from the dichromatic model.

This paper is organized as follows: in section 2 RAD is presented as a feature space analysis method. Afterwards, in section 3 RAD is introduced as a segmentation technique. The results obtained and a comparison with Mean Shift and various other state-of-the-art methods on the Berkeley dataset is presented in section 4. Finally, conclusions of the current work are given in section 5.

2   A Ridge Based Distribution Analysis Method (RAD)

In this section we present a fast algorithm to extract DCs from histogram space. Here we propose a method to find dominant structures (DS) for a d-dimensional feature space. In the context of this paper the dominant colours are the dominant structures of the 3D chromatic histogram. The proposed method is divided in two main steps. First, we propose a method to extract ridges as a representative of a DS. Afterwards a flooding process is performed to find the DSs from its ridges.

2.1   First Step: Ridge Extraction

To extract a DS descriptor we need to find those points containing the most meaningful information of a DS, *i.e.*, the ridge of a DS. We propose to apply a multilocal creaseness algorithm to find the best ridge point candidates. This operator avoids to split up ridges due to irregularities on the distribution, mainly caused by the discrete nature of the data. Afterwards, we apply a ridge extraction algorithm to find the descriptor.

Multilocal Creaseness: Finding Candidates and Enhancing Connectivity. In order to deal with this commonly heavily jagged DS (see Figure 2d) , we propose to apply the MLSEC-ST operator introduced by Lopez *et al.* in [16] to enhance ridge points. This method is used due to its good performance compared with other ridge detection methods [16] on irregular and noisy landscapes.

The Structure Tensor (ST) computes the dominant gradient orientation in a neighbourhood of size proportional to σ_d. Basically, this calculus enhances those situations where either a big attraction or repulsion exists in the gradient direction vectors. Thus, it assigns the higher values when a ridge or valley occurs. Given a distribution $\Omega(\mathbf{x})$, (the histogram in the current context), and a symmetric neighbourhood of size σ_i centered at point $\mathbf{x}$, namely, $N(\mathbf{x}, \sigma_i)$ the ST field S is defined as:

$$S(\mathbf{x}, \sigma) = N(\mathbf{x}, \sigma_\mathbf{i}) * (\nabla\mathbf{\Omega}(\mathbf{x}, \sigma_\mathbf{d}) \cdot \nabla\mathbf{\Omega^t}(\mathbf{x}, \sigma_\mathbf{d})) \tag{2}$$

where $\sigma = \{\sigma_\mathbf{i}, \sigma_\mathbf{d}\}$, and the calculus of the gradient vector field $\nabla\Omega(\mathbf{x}, \sigma_d)$ has been done with a Gaussian Kernel with standard deviation σ_d.

If $w(\mathbf{x}, \sigma)$ is the eigenvector corresponding to the largest eigenvalue of $S(\mathbf{x}, \sigma)$, then, the dominant gradient orientation $\overline{w}(\mathbf{x}, \sigma)$ in a neighbourhood of size proportional to σ_i centered at $\mathbf{x}$ is:

$$\overline{w}(\mathbf{x}, \sigma) = \mathbf{sign}(\mathbf{w^t}(\mathbf{x}, \sigma) \cdot \nabla^\mathbf{t}\mathbf{\Omega}(\mathbf{x}, \sigma_\mathbf{d}))\mathbf{w}(\mathbf{x}, \sigma) \tag{3}$$

The creaseness measure of $\Omega(\mathbf{x})$ for a given point $\mathbf{x}$, named $k(\mathbf{x}, \sigma)$, is computed with the divergence between the dominant gradient orientation and the normal vectors, namely n_k, on the r-connected neighbourhood of size proportional to σ_i. That is:

$$k(\mathbf{x}, \sigma) = -\mathbf{Div}(\overline{\mathbf{w}}(\mathbf{x}, \sigma)) = -\frac{\mathbf{d}}{\mathbf{r}} \sum_{\mathbf{k=1}}^{\mathbf{r}} \overline{\mathbf{w}}^{\mathbf{t}}(\mathbf{k}, \sigma) \cdot \mathbf{n_k} \tag{4}$$

where d is the dimension of $\Omega(\mathbf{x})$. The creaseness representation of $\Omega(\mathbf{x})$ will be referred hereafter as Ω^σ.

As an example, Figure 3a shows the opponent colour 2D histogram of 3g. Its creaseness values are showed in 3b. There are three enhanced areas which corresponds with the three dominant colours of the original image. They appear as three mountains in 3b, clearly separated by two valleys. Note that higher creaseness values have a larger probability to become a ridge point.

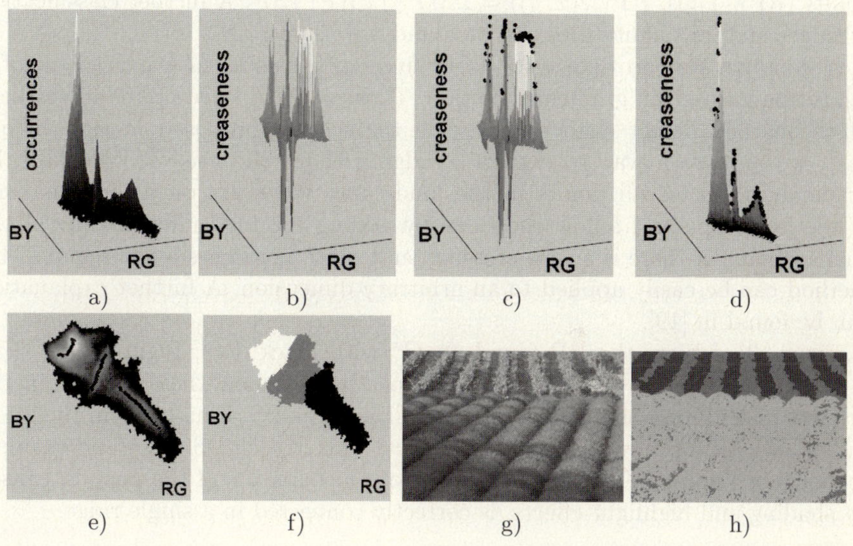

Fig. 3. A graphical example of the whole process. (a) Opponent Red-Green and Blue-Yellow histogram $\Omega(\mathbf{x})$ of g). (b) Creaseness representation of a). (c) Ridges found in b). (d)Ridges fitted on original distribution. (e) Top-view of d). (f)Dominant structures of a). (g) Original image. (h)Segmented image.

Ridge Detection. In the previous section we have detected a set of candidate ridge points. In this section we discard superfluous points. As a result only those points necessary to maintain the connectivity of a DS remain. These points form the ridges of Ω^σ.

We classify ridge points in three categories. First, Transitional Ridge Points (TRP): when there is a local maximum in a single direction. Second, Saddle

Points (SP): when there is a local maximum in one direction and a local minimum in another one. Third, Local Maximum Points (LMP). Formally, let $\Omega(x, y)$ be a continuous 2D surface and $\nabla\Omega(x, y)$ be the gradient vector of the function $\Omega(x, y)$. We define ω_1 and ω_2 as the unit eigenvectors of the Hessian matrix and λ_1 and λ_2 its corresponding eigenvalues with $\mid \lambda_1 \mid \leq \mid \lambda_2 \mid$. Then, for the 2D case:

$$LMP(\Omega(x,y)) = \{(x,y)|(\|\nabla\Omega(x,y)\| = 0), \lambda_1 < 0, \lambda_2 < 0\} \tag{5}$$

$$TRP(f(x,y)) = \{(x,y)|\|\nabla\Omega(x,y)\| \neq 0, \lambda_1 < 0, \nabla\Omega(x,y) \cdot \omega_1 = 0,$$
$$\|\nabla\Omega(x,y)\| \neq 0, \lambda_2 < 0, \nabla\Omega(x,y) \cdot \omega_2 = 0, \tag{6}$$
$$\|\nabla\Omega(x,y)\| = 0, \lambda_1 < 0, \lambda_2 = 0\}$$

$$SP(f(x,y)) = \{(x,y)|\|\nabla\Omega(x,y)\| = 0, \lambda_1 \cdot \lambda_2 < 0\} \tag{7}$$

This definition can be extended for an arbitrary dimension using the combinatorial of the eigenvalues. Hereafter we will refer to these three categories as *ridge points* (RP). Thus, $RP(\Omega(x,y)) = LMP \bigcup TRP \bigcup SP$. A further classification of ridges and its singularities can be found in [17] and [18].

A common way to detect RP is to find zero-crossing in the gradient of a landscape for a given gradient direction. Thus, we need to compute all gradient directions and detect changes following the schema proposed in [18]. In our case, we propose a way to extract a ridge without the need to calculate the gradient values for all points in the landscape. We begin on a local maxima of the landscape and follow the ridge by adding the higher neighbours of the current point, if there is a zero-crossing on it, until it reaches a flat region. This method can be easily applied to an arbitrary dimension. A further explanation can be found in [19].

Figure 3c depicts the RP found on Ω^σ with black dots. Figures 3d,e show a 3D view and a 2D projection view respectively of how these RPs fit in the original distribution as a representative of the three DS. Finally, from the set of RPs of a distribution we can perform the calculus of each DS. A second example is shown in Figure 1. The complicated colour distribution of the pepper, caused by shading and highlight effects, is correctly connected in a single ridge.

2.2   Second Step: DS Calculus from Its RPs

In this final step we find the DS belonging to each ridge found. From topological point of view, it implies finding the portion of landscape represented by each ridge. These portions of landscape are named *catchments basins*. Vincent and Soille [20] define a catchment basin associated with a local minimum M as the set of pixels p of Ω^σ such that a water drop falling at p flows down along the relief, following a certain descending path called the downstream of p, and eventually reaches M. In our case, M are the set of RPs found and then, DSs are found using the algorithm proposed in [20] applied on the inverse Ω^σ distribution. The proposed algorithm, is not based on the gradient vectors of a landscape [21] but on the idea of *immersion* which is more stable and reduces over-segmentation.

Basically, the flooding process begins on the local minima and, iteratively, the landscape sinks on the water. Those points where the water coming from different local minima join, compose the watershed lines. To avoid potential problems with irregularities [16], we force the flooding process to begin at the same time in all DS descriptors, on the smoothed $\Omega(\mathbf{x})$ distribution with a Gaussian kernel of standard deviation σ_d (already computed on the ST calculus). Then, we define RAD as the operator returning the set of DS of Ω^σ using RPs as marks:

$$RAD(\Omega(x)) = W(\Omega^\sigma, RP(\Omega^\sigma)) \tag{8}$$

Following this procedure, Figure 3f depicts the 2D projection of the DSs found on 3a.

3 Colour Image Segmentation Using RAD

Once RAD has been applied we need to assign a representative colour to each DS found. Thus, let $DS_n = \{\mathbf{x_1}, ..., \mathbf{x_r}\}$ be the nth DS of $\Omega(\mathbf{x})$, and $\Omega(\mathbf{x_i})$ the function returning the number of occurrences of $\mathbf{x_i}$ in Ω. Then, the dominant colour of DS_n, namely, $DC(DS_n)$ will be the mass center of $\Omega(DS_n)$:

$$DC(DS_n) = \frac{\sum_{i=1}^{r} \mathbf{x_i} \cdot \mathbf{\Omega}(\mathbf{x_i})}{\sum_{i=1}^{r} \Omega(\mathbf{x_i})} \tag{9}$$

The segmented image will have as many colours as the number of DSs found. Figure, 3h shows the segmentation obtained with RAD from 3g. This segmentation has been performed in the opponent colour histogram. Although RAD can be applied to any chromatic representation of an image such as CIE, RGB, Ohta spaces or 2-dimensional ones such as Opponent or normalized RGB.

4 Results and Performance Evaluation

In the experiments we qualitatively and quantitatively evaluate the proposed segmentation method. Firstly, RAD is compared with Mean Shift (MS) [8], [22]. MS has been chosen because it is widely used, has a public available version, the EDISON one [23] and it has demonstrated its good performance [24]. Additionally, Mean Shift is a feature space analysis technique, as well as RAD, and yields a segmentation in a rather reasonable time, in opposition to other set of methods such as the Graph-Based approaches [25] , (with the exception of the efficient graph-based segmentation method introduced in [26]). Secondly, our method is compared on the Berkeley data set against a set of state-of-the-art segmentation methods.

The MS method [22], consists of finding the modes of the underlying probability function of a distribution. The method finds the Mean Shift vectors in the histogram of an image that point to the direction of higher density. All values of the histogram attracted by one mode compound the basis of attraction of it. In a second step, the modes which are near of a given threshold are joined

8   E. Vazquez, J. van de Weijer, and R. Baldrich

in one unique mode. Finally, the basis of attraction of these modes will compose a dominant colour of the image. Mean Shift has two basic parameters to adapt the segmentation to an specific problem, namely, h_s, which controls a smoothing process, and h_r related with the size of the kernel used to determine the modes and its basis of attraction. To test the method, we have selected the set parameters $(h_s, h_r) = \{(7,3), (7,15), (7,19), (7,23), (13,7)(13,19), (17,23)\}$ given in [24] and [5]. The average times for this set of parameters, expressed in seconds, are 3.17, 4.15, 3.99, 4.07, 9.72, 9.69, 13.96 respectively. Nevertheless, these parameters do not cover the complete spectrum of possibilities of the MS. Here we want to compare RAD and MS from a soft oversegmentation to a soft undersegmentation. Hence, in order to reach an undersegmentation with MS, we add the following parameter settings $(h_s, h_r) = \{(20,25), (25,30), (30,35)\}$. For these settings, the average times are 18.05, 24.95 and 33.09 respectively.

The parameters used for RAD based segmentation are $(\sigma_d, \sigma_i) = \{$ (0.8,0.05), (0.8,0.5), (0.8,1), (0.8,1.5), (1.5,0.05), (1.5,0.5), (1.5,1.5), (2.5,0.05), (2.5,0.5), (2.5,1.5) $\}$. These parameters vary from a soft oversegmentation to an undersegmentation, and have been selected experimentally. The average times for RAD are 6.04, 5.99, 6.11, 6.36, 6.11, 5.75, 6.44, 5.86, 5.74 and 6.35. These average times, point out the fact that RAD is not dependent of the parameters used. In conclusion, whereas the execution time of Mean Shift increases significantly with increasing spatial scale, the execution time of RAD remains constant from an oversegmentation to an undersegmentation.

The experiments has been performed on the publicly available Berkeley image segmentation dataset and benchmark [12]. We use the Global Constancy Error (GCE) as an error measure. This measure was also proposed in [12] and takes care of the refinement between different segmentations. For a given pixel p_i, consider the segments (sets of connected pixels), S_1 from the benchmark and S_2

Fig. 4. Examples of segmentation. Original image. Columns from 2 to 5: segmentation for RAD on RGB with $(\sigma_d, \sigma_i) = \{(0.8,0.05), (1.5,0.05), (2.5,0.05), (2.5,1.5)\}$. Last column: human segmentation.

Image segmentation with non-local shape and color priors has attracted a lot of interest in the last years. As discussed above, most approaches use either local continuous optimization [27,7,24,9] or iterated minimization alternating graph cut and search over non-local parameter space [25,6,16]. Unfortunately, both groups of methods are prone to getting stuck in poor local minima. Global-optimization algorithms have also been suggested [12,26] . In particular, simultaneous work [10] presented a framework that also utilizes branch-and-bound ideas (paired with continuous optimization in their case). While all these global optimization methods are based on elegant ideas, the variety of shapes, invariances, and cues that each of them can handle is limited compared to our method.

Finally, our framework may be related to branch-and-bound search methods in computer vision (e.g. [1,21]). In particular, it should be noted that the way our framework handles shape priors is related to previous approaches like [14] that used tree search over shape hierarchies. However, neither of those approaches accomplish pixel-wise image segmentation.

3   Optimization Framework

In this section, we discuss our global energy optimization framework for obtaining image segmentations under non-local priors[1]. In the next sections, we detail how it can be used for the segmentation with non-local shape priors (Section 4) and non-local intensity/color priors (Section 5).

3.1   Energy Formulation

Firstly, we introduce notation and give the general form of the energy that can be optimized in our framework. Below, we consider the pixel-wise segmentation of the image. We denote the pixel set as $\mathcal{V}$ and use letters p and q to denote individual pixels. We also denote the set of edges connecting adjacent pixels as $\mathcal{E}$ and refer to individual edges as to the pairs of pixels (e.g. p,q). In our experiments, the set of edges consisted of all 8-connected pixel pairs in the raster.

The segmentation of the image is given by its $0-1$ labeling $\mathbf{x} \in 2^{\mathcal{V}}$, where individual pixel labels x_p take the values 1 for the pixels classified as the foreground and 0 for the pixels classified as the background. Finally, we denote the non-local parameter as ω and allow it to vary over a discrete, possibly very large, set Ω. The general form of the energy function that can be handled within our framework is then given by:

$$E(\mathbf{x},\omega) = C(\omega) + \sum_{p \in \mathcal{V}} F^p(\omega) \cdot x_p + \sum_{p \in \mathcal{V}} B^p(\omega) \cdot (1 - x_p) + \sum_{p,q \in \mathcal{E}} P^{pq}(\omega) \cdot |x_p - x_q| . \quad (1)$$

Here, $C(\omega)$ is a constant potential, which does not depend directly on the segmentation $\mathbf{x}$; $F^p(\omega)$ and $B^p(\omega)$ are the unary potentials defining the cost for

---

[1] The C++ code for this framework is available at the webpage of the first author.

assigning the pixel p to the foreground and to the background respectively; $P^{pq}(\omega)$ is the pairwise potential defining the cost of assigning adjacent pixels p and q to different segments. In our experiments, the pairwise potentials were taken non-negative to ensure the tractability of $E(\mathbf{x}, \omega)$ as the function of $\mathbf{x}$ for graph cut optimization [18].

All potentials in our framework depend on the non-local parameter $\omega \in \Omega$. In general, we assume that Ω is a discrete set, which may be large (e.g. millions of elements) and should have some structure (although, it need not be linearly or partially ordered). For the segmentation with shape priors, Ω will correspond to the product space of various poses and deformations of the template, while for the segmentation with color priors Ω will correspond to the set of parametric color distributions.

3.2   Lower Bound

Our approach optimizes the energy (1) exactly, finding its global minimum using branch-and-bound tree search [8], which utilizes the lower bound on (1) derived as follows:

$$
\min_{x \in 2^{\mathcal{V}}, \omega \in \Omega} E(\mathbf{x}, \omega) = \min_{x \in 2^{\mathcal{V}}} \min_{\omega \in \Omega} \left[C(\omega) + \sum_{p \in \mathcal{V}} F^p(\omega) \cdot x_p + \sum_{p \in \mathcal{V}} B^p(\omega) \cdot (1 - x_p) + \right.
$$

$$
\left. \sum_{p,q \in \mathcal{E}} P^{pq}(\omega) \cdot |x_p - x_q| \right] \geq \min_{\mathbf{x} \in 2^{\mathcal{V}}} \left[\min_{\omega \in \Omega} C(\omega) + \sum_{p \in \mathcal{V}} \min_{\omega \in \Omega} F^p(\omega) \cdot x_p + \right.
$$

$$
\left. \sum_{p \in \mathcal{V}} \min_{\omega \in \Omega} B^p(\omega) \cdot (1 - x_p) + \sum_{p,q \in \mathcal{E}} \min_{\omega \in \Omega} P^{pq}(\omega) \cdot |x_p - x_q| \right] =
$$

$$
\min_{x \in 2^{\mathcal{V}}} \left[C_\Omega + \sum_{p \in \mathcal{V}} F^p_\Omega \cdot x_p + \sum_{p \in \mathcal{V}} B^p_\Omega \cdot (1 - x_p) + \sum_{p,q \in \mathcal{E}} P^{pq}_\Omega \cdot |x_p - x_q| \right] = L(\Omega) . \quad (2)
$$

Here, C_Ω, F^p_Ω, B^p_Ω, P^{pq}_Ω denote the minima of $C(\omega)$, $F^p(\omega)$, $B^p(\omega)$, $P^{pq}(\omega)$ over $\omega \in \Omega$ referred below as *aggregated potentials*. $L(\Omega)$ denotes the derived lower bound for $E(\mathbf{x}, \omega)$ over $2^{\mathcal{V}} \otimes \Omega$. The inequality in (2) is essentially the Jensen inequality for the minimum operation.

The proposed lower bound possesses three properties crucial to the Branch-and-Mincut framework:

Monotonicity. For the nested domains of non-local parameters $\Omega_1 \subset \Omega_2$ the inequality $L(\Omega_1) \geq L(\Omega_2)$ holds (the proof is given in [23]).

Computability. The key property of the derived lower bound is the ease of its evaluation. Indeed, this bound equals the minimum of a submodular quadratic pseudo-boolean function. Such function can be realized on a network graph such that each configuration of the binary variables is in one-to-one correspondence with an st-cut of the graph having the weight equal to the value of the function (plus a constant C_Ω) [2,11,18]. The minimal st-cut corresponding to the minimum of $L(\Omega)$ then can be computed in a low-polynomial of $|\mathcal{V}|$ time e.g. with the popular algorithm [5].

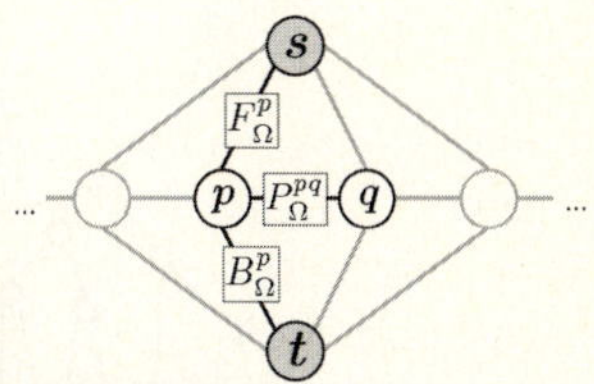

The fragment of the network graph realizing $L(\Omega)$ (edge weights shown in boxes). *(see e.g.[18] for details)*

Tightness. For a singleton Ω the bound is *tight*: $L(\{\omega\}) = \min_{x \in 2^\mathcal{V}} E(\mathbf{x}, \omega)$. In such case, the minimal st-cut also yields the segmentation $\mathbf{x}$ optimal for this ω ($x_p = 0$ *iff* the respective vertex belongs to the s-component of the cut).

Note, that the fact that the lower bound (2) may be evaluated via st-mincut gives rise to a whole family of looser, but cheaper, lower bounds. Indeed, the minimal cut on a network graph is often found by pushing *flows* until the flow becomes maximal (and equal to the weight of the mincut) [5]. Thus, the sequence of intermediate flows provides a sequence of the increasing lower bounds on (1) converging to the bound (2) (**flow bounds**). If some upper bound on the minimum value is imposed, the process may be terminated earlier without computing the full maxflow/mincut. This happens when the new flow bound exceeds the given upper bound. In this case it may be concluded that the value of the global minimum is greater than the imposed upper bound.

3.3   Branch-and-Bound Optimization

Finding the global minimum of (1) is, in general, a very difficult problem. Indeed, since the potentials can depend arbitrarily on the non-local parameter spanning arbitrary discrete set Ω, in the worst-case any optimization has to search exhaustively over Ω. In practice, however, any segmentation problem has some specifically-structured space Ω. This structure can be efficiently exploited by the branch-and-bound search detailed below.

We assume that the discrete domain Ω can be hierarchically clustered and the binary tree of its subregions $T_\Omega = \{\Omega = \Omega_0, \Omega_1, \ldots \Omega_N\}$ can be constructed (binarity of the tree is not essential). Each non-leaf node corresponding to the subregion Ω_k then has two children corresponding to the subregions $\Omega_{ch1(k)}$ and $\Omega_{ch2(k)}$ such that $\Omega_{ch1(k)} \subset \Omega_k$, $\Omega_{ch2(k)} \subset \Omega_k$. Here, $ch1(\cdot)$ and $ch2(\cdot)$ map the index of the node to the indices of its children. Also, leaf nodes of the tree are in one-to-one correspondence with singleton subsets $\Omega_l = \{\omega_t\}$.

Given such tree, the global minimum of (1) can be efficiently found using the *best-first* branch-and-bound search [8]. This algorithm propagates a *front* of nodes in the top-down direction (Fig. 1). During the search, the front contains a set of tree nodes, such that each top-down path from the root to a leaf contains

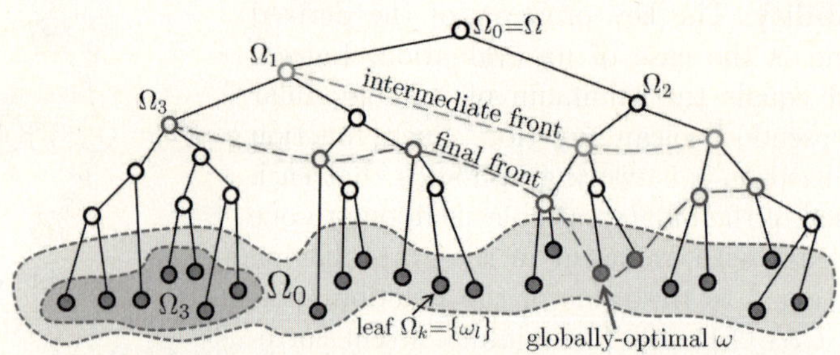

Fig. 1. *Best-first branch-and-bound* optimization on the tree of nested regions finds the globally-optimal ω by the top-down propagation of the *active front* (see text for details). At the moment when the lowest lower bound of the front is observed at leaf node, the process terminates with the global minimum found without traversing the whole tree.

exactly one active vertex. In the beginning, the front contains the tree root Ω_0. At each step the active node with the smallest lower bound (2) is removed from the active front, while two of its children are added to the active front (by monotonicity property they have higher or equal lower bounds). Thus, an active front moves towards the leaves making local steps that increase the lowest lower bound of all active nodes. Note, that at each moment, this lowest lower bound of the front constitutes a lower bound on the global optimum of (1) over the whole domain.

At some moment of time, the active node with the smallest lower bound turns out to be a leaf $\{\omega'\}$. Let $\mathbf{x}'$ be the optimal segmentation for ω' (found via minimum st-cut). Then, $E(x', \omega') = L(\omega')$ (tightness property) is by assumption the lowest bound of the front and hence a lower bound on the global optimum over the whole domain. Consequently, (x', ω') is a global minimum of (1) and the search terminates without traversing the whole tree. In our experiments, the number of the traversed nodes was typically very small (two-three orders of magnitude smaller then the size of the full tree). Therefore, the algorithm performed global optimization much faster than exhaustive search over Ω.

In order to further accelerate the search, we exploit the coherency between the mincut problems solved at different nodes. Indeed, the maximum flow as well as auxiliary structures such as shortest path trees computed for one graph may be "reused" in order to accelerate the computation of the minimal st-cut on another similar graph [3,17]. For some applications, this trick may give an order of magnitude speed-up for the evaluation of lower bounds.

In addition to the best-first branch-and-bound search we also tried the *depth-first* branch-and-bound [8]. When problem-specific heuristics are available that give good initial solutions, this variant may lead to moderate (up to a factor of 2) time savings. Interestingly, the depth-first variant of the search, which maintains upper bounds on the global optimum, may benefit significantly from the use of flow bounds discussed above. Nevertheless, we stick with the best-first branch-and-bound for the final experiments due to its generality (no need for initialization heuristics).

where S denotes the foreground segment, and $I(p)$ is a grayscale image. The first two terms measure the length of the boundary and the area, the third and the forth terms are the integrals over the fore- and background of the difference between image intensity and the two intensity values c^f and c^b, which correspond to the average intensities of the respective regions. Traditionally, this functional is optimized using level set framework converging to one of its local minima.

Below, we show that the discretized version of this functional can be optimized globally within our framework. Indeed, the discrete version of (6) can be written as (using notation as before):

$$E(\mathbf{x}, (c^f, c^b)) = \sum_{p,q \in \mathcal{E}} \frac{\mu}{|p-q|} \cdot |x_p - x_q| +$$

$$\sum_{p \in \mathcal{V}} \left(\nu + \lambda_1 (I(p) - c^f)^2 \right) \cdot x_p + \sum_{p \in \mathcal{V}} \lambda_2 \left(I(p) - c^b \right)^2 \cdot (1 - x_p) . \tag{7}$$

Here, the first term approximates the first term of (6) (the accuracy of the approximation depends on the size of the pixel neighborhood [4]), and the last two terms express the last three terms of (6) in a discrete setting.

The functional (7) clearly has the form (1) with non-local parameter $\omega = \{c^f, c^b\}$. Discretizing intensities c^f and c^b into 255 levels and building a quad-tree over their joint domain, we can apply our framework to find the global minima of (6). An example of a global minimum of (7) is shown to the right (this 183x162 image was segmented in 3 seconds, the proportion of the tree traversed was 1:115). More examples are given in [23].

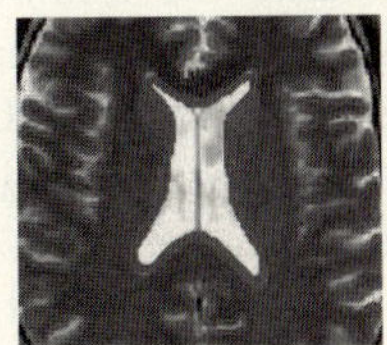

5.2 Segmenting Color Images: GrabCut functional

In [25], the *GrabCut* framework for the interactive color image segmentation based on Gaussian mixtures was proposed. In GrabCut, the segmentation is driven by the following energy:

$$E_{\mathrm{GrabCut}}(\mathbf{x}, (GM^f, GM^b)) = \sum_{p \in V} - \log(\mathrm{P}(K_p | GM^f)) \cdot x_p +$$

$$+ \sum_{p \in V} - \log(\mathrm{P}(K_p | GM^b)) \cdot (1 - x_p) + \sum_{p,q \in \mathcal{E}} \frac{\lambda_1 + \lambda_2 \cdot e^{-\frac{||K_p - K_q||^2}{\beta}}}{|p-q|} \cdot |x_p - x_q| . \tag{8}$$

Here, GM^f and GM^b are Gaussian mixtures in RGB color space and the first two terms of the energy measure how well these mixtures explain colors K_p of pixels attributed to fore- and background respectively. The third term is the contrast sensitive edge term, ensuring that the segmentation boundary is compact and tends to stick to color region boundaries in the image. In addition to this energy, the user provides supervision in the form of a bounding rectangle and brush strokes, specifying which parts of the image should be attributed to the foreground and to the background.

The original method [25] minimizes the energy within EM-style process, alternating between (i) the minimization of (8) over $\mathbf{x}$ given GM^f and GM^b and (ii) refitting the mixtures GM^f and GM^b given $\mathbf{x}$. Despite the use of the global graph cut optimization within the segmentation update step, the whole process yields only a local minimum of (8). In [25], the segmentation is initialized to the provided bounding box and then typically shrinks to one of the local minima.

The energy (8) has the form (1) and therefore can be optimized within Branch-and-Mincut framework, provided that the space of non-local parameters (which in this case is the joint space of the Gaussian mixtures for the foreground and for the background) is discretized and the tree of the subregions is built. In this scenario, however, the dense discretization of the non-local parameter space is infeasible (if the mixtures contain n Gaussians then the space is described by $20n - 2$ continuous parameters). It is possible, nevertheless, to choose a much smaller discrete subset Ω that is still likely to contain a good approximation to the globally-optimal mixtures.

To construct such Ω, we fit a mixture of $M = 8$ Gaussians $G_1, G_2, ...G_M$ with the support areas $a_1, a_2, ...a_M$ to the *whole* image. The support area a_i here counts the number of pixels p such as $\forall j \ \mathrm{P}(K_p|G_i) \geq \mathrm{P}(K_p|G_j)$. We assume that the components are ordered such that the support areas decrease ($a_i > a_{i+1}$). Then, the Gaussian mixtures we consider are defined by the binary vector $\beta = \{\beta_1, \beta_2 ... \beta_M\} \in \{0,1\}^M$ specifying which Gaussians should be included into the mixture: $\mathrm{P}(K | GM(\beta)) = \sum_i \beta_i a_i \mathrm{P}(K|G_i) / \sum_i \beta_i a_i$.

The overall set Ω is then defined as $\{0,1\}^{2M}$, where odd bits correspond to the foreground mixture vector β^f and even bits correspond to the background mixture vector β^b. Vectors with all even bits and/or all odd bits equal to zero do not correspond to meaningful mixtures and are therefore assigned an infinite cost. The hierarchy tree is naturally defined by the bit-ordering (the first bit corresponding to subdivision into the first two branches etc.).

Depending on the image and the value of M, the solutions found by Branch-and-Mincut framework may have larger or smaller energy (8) than the solutions found by the original EM-style method [25]. This is because Branch-and-Mincut here finds the global optimum over the subset of the domain of (8) while [25] searches locally but within the continuous domain. However, for all 15 images in our experiments, improving Branch-and-Mincut solutions with a few EM-style iterations [25] gave lower energy than the original solution of [25]. In most cases, these additional iterations simply refit the Gaussians properly and change very few pixels near boundary (see Fig. 4).

In terms of performance, for $M = 8$ the segmentation takes on average a few dozen seconds (10s and 40s for the images in Fig. 4) for 300x225 image. The proportion of the tree traversed by an active front is one to several hundred (1:963 and 1:283 for the images in Fig. 4).

Image+input   GrabCut[25](−618)   Branch&Mincut(−624)   Combined(−628)

Image+input   GrabCut[25](−593)   Branch&Mincut(−584)   Combined(−607)

Fig. 4. Being initialized with the user-provided bounding rectangle (shown in green in the first column) as suggested in [25], EM-style process [25] converges to a local minimum (the second column). Branch-and-Mincut result (the third column) escapes that local minimum and after EM-style improvement lead to the solution with much smaller energy and better segmentation accuracy (the forth column). Energy values are shown in brackets.

This experiment suggests the usefulness of Branch-and-Mincut framework as a mean of obtaining good initial point for local methods, when the domain space is too large for an exact branch-and-bound search.

6   Conclusion

The Branch-and-Mincut framework presented in this paper finds global optima of a wide class of energies dependent on the image segmentation mask and non-local parameters. The joint use of branch-and-bound and graph cut allows efficient traversal of the solution space. The developed framework is useful within a variety of image segmentation scenarios, including segmentation with non-local shape priors and non-local color/intensity priors.

Future work includes the extension of Branch-and-Mincut to other problems, such as simultaneous stitching and registration of images, as well as deriving analogous branch-and-bound frameworks for combinatorial methods other than binary graph cut, such as minimum ratio cycles and multilabel MRF inference.

Acknowledgements

We would like to acknowledge discussions and feedback from Vladimir Kolmogorov and Pushmeet Kohli. Vladimir has also kindly made several modifications of his code of [5] that allowed to reuse network flows more efficiently.

References

1. Agarwal, S., Chandaker, M., Kahl, F., Kriegman, D., Belongie, S.: Practical Global Optimization for Multiview Geometry. In: Leonardis, A., Bischof, H., Pinz, A. (eds.) ECCV 2006. LNCS, vol. 3951. Springer, Heidelberg (2006)
2. Boros, E., Hammer, P.: Pseudo-boolean optimization. Discrete Applied Mathematics 123(1-3) (2002)
3. Boykov, Y., Jolly, M.-P.: Interactive Graph Cuts for Optimal Boundary and Region Segmentation of Objects in N-D Images. In: ICCV 2001 (2001)
4. Boykov, Y., Kolmogorov, V.: Computing Geodesics and Minimal Surfaces via Graph Cuts. In: ICCV 2003 (2003)
5. Boykov, Y., Kolmogorov, V.: An Experimental Comparison of Min-Cut/Max-Flow Algorithms for Energy Minimization in Vision. PAMI 26(9) (2004)
6. Bray, M., Kohli, P., Torr, P.: PoseCut: Simultaneous Segmentation and 3D Pose Estimation of Humans Using Dynamic Graph-Cuts. In: Leonardis, A., Bischof, H., Pinz, A. (eds.) ECCV 2006. LNCS, vol. 3952. Springer, Heidelberg (2006)
7. Chan, T., Vese, L.: Active contours without edges. Trans. Image Process 10(2) (2001)
8. Clausen, J.: Branch and Bound Algorithms - Principles and Examples. Parallel Computing in Optimization (1997)
9. Cremers, D., Osher, S., Soatto, S.: Kernel Density Estimation and Intrinsic Alignment for Shape Priors in Level Set Segmentation. IJCV 69(3) (2006)
10. Cremers, D., Schmidt, F., Barthel, F.: Shape Priors in Variational Image Segmentation: Convexity, Lipschitz Continuity and Globally Optimal Solutions. In: CVPR 2008 (2008)
11. Greig, D., Porteous, B., Seheult, A.: Exact maximum a posteriori estimation for binary images. Journal of the Royal Statistical Society 51(2) (1989)
12. Felzenszwalb, P.: Representation and Detection of Deformable Shapes. PAMI 27(2) (2005)
13. Freedman, D., Zhang, T.: Interactive Graph Cut Based Segmentation with Shape Priors. In: CVPR 2005 (2005)
14. Gavrila, D., Philomin, V.: Real-Time Object Detection for "Smart" Vehicles. In: ICCV 1999 (1999)
15. Huang, R., Pavlovic, V., Metaxas, D.: A graphical model framework for coupling MRFs and deformable models. In: CVPR 2004 (2004)
16. Kim, J., Zabih, R.: A Segmentation Algorithm for Contrast-Enhanced Images. ICCV 2003 (2003)
17. Kohli, P., Torr, P.: Effciently Solving Dynamic Markov Random Fields Using Graph Cuts. In: ICCV 2005 (2005)
18. Kolmogorov, V., Zabih, R.: What Energy Functions Can Be Minimized via Graph Cuts. In: Heyden, A., Sparr, G., Nielsen, M., Johansen, P. (eds.) ECCV 2002. LNCS, vol. 2352. Springer, Heidelberg (2002)
19. Kolmogorov, V., Boykov, Y., Rother, C.: Applications of Parametric Maxflow in Computer Vision. In: ICCV 2007 (2007)
20. Pawan Kumar, M., Torr, P., Zisserman, A.: OBJ CUT. In: CVPR 2005 (2005)
21. Lampert, C., Blaschko, M., Hofman, T.: Beyond Sliding Windows: Object Localization by Efficient Subwindow Search. In: CVPR 2008 (2008)
22. Leibe, B., Leonardis, A., Schiele, B.: Robust Object Detection with Interleaved Categorization and Segmentation. IJCV 77(3) (2008)

23. Lempitsky, V., Blake, A., Rother, C.: Image Segmentation by Branch-and-Mincut. Microsoft Technical Report MSR-TR-2008-100 (July 2008)
24. Leventon, M., Grimson, E., Faugeras, O.: Statistical Shape Influence in Geodesic Active Contours. In: CVPR 2000 (2000)
25. Rother, C., Kolmogorov, V., Blake, A.: "GrabCut": interactive foreground extraction using iterated graph cuts. ACM Trans. Graph. 23(3) (2004)
26. Schoenemann, T., Cremers, D.: Globally Optimal Image Segmentation with an Elastic Shape Prior. In: ICCV 2007 (2007)
27. Wang, Y., Staib, L.: Boundary Finding with Correspondence Using Statistical Shape Models. In: CVPR 1998 (1998)

What Is a Good Image Segment?
A Unified Approach to Segment Extraction

Shai Bagon, Oren Boiman, and Michal Irani*

Weizmann Institute of Science, Rehovot, Israel

Abstract. There is a huge diversity of definitions of "visually meaningful" image segments, ranging from simple uniformly colored segments, textured segments, through symmetric patterns, and up to complex semantically meaningful objects. This diversity has led to a wide range of different approaches for image segmentation. In this paper we present a single unified framework for addressing this problem – "Segmentation by Composition". We define a good image segment as one which can be easily composed using its own pieces, but is difficult to compose using pieces from other parts of the image. This non-parametric approach captures a large diversity of segment types, yet requires no pre-definition or modelling of segment types, nor prior training. Based on this definition, we develop a segment extraction algorithm – i.e., given a single point-of-interest, provide the "best" image segment containing that point. This induces a figure-ground image segmentation, which applies to a range of different segmentation tasks: single image segmentation, simultaneous co-segmentation of several images, and class-based segmentations.

1   Introduction

One of the most fundamental vision tasks is image segmentation; the attempt to group image pixels into visually meaningful segments. However, the notion of a "visually meaningful" image segment is quite complex. There is a huge diversity in possible definitions of what is a good image segment, as illustrated in Fig. 1. In the simplest case, a uniform colored region may be a good image segment (e.g., the flower in Fig. 1.a). In other cases, a good segment might be a textured region (Fig. 1.b, 1.c) or semantically meaningful layers composed of disconnected regions (Fig. 1.c) and all the way to complex objects (Fig. 1.e, 1.f).

The diversity in segment types has led to a wide range of approaches for image segmentation: Algorithms for extracting *uniformly colored* regions (e.g., [1,2]), algorithms for extracting *textured* regions (e.g., [3,4]), algorithm for extracting regions with a distinct empirical color distribution (e.g., [5,6,7]). Some algorithms employ symmetry cues for image segmentation (e.g., [8]), while others use high-level semantic cues provided by object classes (i.e., class-based segmentation, see [9,10,11]). Some algorithms are unsupervised (e.g., [2]), while others require user interaction (e.g., [7]). There are also variants in the segmentation

---

* Author names are ordered alphabetically due to equal contribution.

D. Forsyth, P. Torr, and A. Zisserman (Eds.): ECCV 2008, Part IV, LNCS 5305, pp. 30–44, 2008.

(a)   (b)   (c)   (d)   (e)   (f)

Fig. 1. What is a good image segment? *Examples of visually meaningful image segments. These vary from uniformly colored segments (a) through textured segments (b)-(c), symmetric segments (d), to semantically meaningful segments (e)-(f). These results were provided by our single unified framework.*

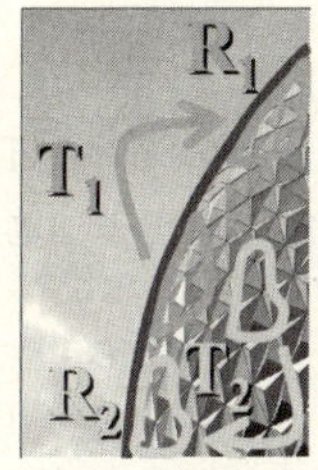

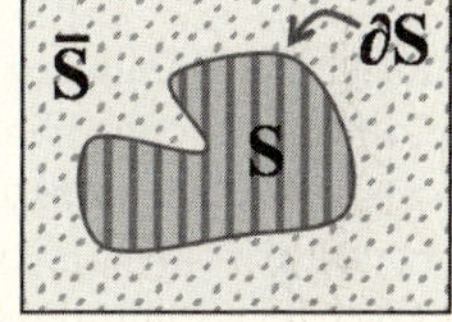

Fig. 3. Notations:
$Seg = \{S, \overline{S}, \partial S\}$ denotes a figure-ground segmentation. S is the foreground segment, $\overline{S}$ (its compliment) is the background, and ∂S is the boundary of the segment.

Fig. 2. Segmentation by composition: *A good segment S (e.g., the butterfly or the dome) can be easily composed of other regions in the segment. Regions R_1, R_2 are composed from other corresponding regions in S (using transformations T_1, T_2 respectively).*

tasks, ranging from segmentation of a single input image, through simultaneous segmentation of a pair of images ("Cosegmentation" [12]) or multiple images.

The large diversity of image segment types has increased the urge to devise a unified segmentation approach. Tu et al. [13] provided such a unified probabilistic framework, which enables to "plug-in" a wide variety of parametric models capturing different segment types. While their framework elegantly unifies these parametric models, it is restricted to a *predefined* set of segment types, and each specific object/segment type (e.g., faces, text, texture etc.) requires its own explicit parametric model. Moreover, adding a new parametric model to this framework requires a significant and careful algorithm re-design.

In this paper we propose a single unified approach to define and extract visually meaningful image segments, *without any explicit modelling*. Our approach defines a "good image segment" as one which is "easy to compose" (like a puzzle) using its own parts, yet it is difficult to compose it from other parts of the image (see Fig. 2). We formulate our "Segmentation-by-Composition" approach, using a unified non-parametric score for segment quality. Our unified score captures a wide range of segment types: uniformly colored segments, through textured segments, and even complex objects. We further present a simple interactive segment extraction algorithm, which optimizes our score – i.e., given a *single* point marked by the user, the algorithm extracts the "best" image segment containing that point. This in turn induces a figure-ground segmentation of the image. We provide results demonstrating the applicability of our score and algorithm to a diversity of segment types and segmentation tasks. The rest of this paper is organized as follows: In Sec. 2 we explain the basic concept behind our "Segmentation-by-Composition" approach for evaluating the visual quality of image segments. Sec. 3 provides the theoretical formulation of our unified segment quality score. We continue to describe our figure-ground segmentation algorithm in Sec. 4. Experimental results are provided in Sec. 5.

2   Basic Concept – "Segmentation By Composition"

Examining the image segments of Fig. 1, we note that good segments of significantly different types share a common property: Given any point within a good image segment, it is easy to compose ("describe") its surrounding region using other chunks of the same segment (like a 'jigsaw puzzle'), whereas it is difficult to compose it using chunks from the remaining parts of the image. This is trivially true for uniformly colored and textured segments (Fig. 1.a, 1.b, 1.c), since each portion of the segment (e.g., the dome) can be easily synthesized using other portions of the same segment (the dome), but difficult to compose using chunks from the remaining parts of the image (the sky). The same property carries to more complex structured segments, such as the compound puffins segment in Fig. 1.f. The surrounding region of each point in the puffin segment is easy to "describe" using portions of other puffins. The existence of several puffins in the image provides 'visual evidence' that the co-occurrence of different parts (orange beak, black neck, white body, etc.) is not coincidental, and all belong to a single compound segment. Similarly, one half of a complex symmetric object (e.g., the butterfly of Fig. 1.d, the man of Fig. 1.e) can be easily composed using its other half, providing visual evidence that these parts go together. Moreover, the simpler the segment composition (i.e., the larger the puzzle pieces), the higher the evidence that all these parts form together a single segment. Thus, the entire man of Fig. 1.e forms a better single segment than his pants or shirt alone.

The ease of describing (composing) an image in terms of pieces of another image was defined by [14], and used there in the context of image similarity. The pieces used for composition are *structured image regions* (as opposed to unstructured 'bags'/distributions of pointwise features/descriptors, e.g., as in [5,7]). Those structured regions, of *arbitrary shape and size*, can undergo a global geometric transformation (e.g., translation, rotation, scaling) with additional small local non-rigid deformations. We employ the composition framework of [14] for the purpose of image segmentation. We define a *"good image segment"* S as one that is easy to compose (non-trivially) using its own pieces, while difficult to

in case of one-pixel sized R_i). Using these observations, $Score\,(S)$ of (5) reduces to the empirical *KL divergence* between the region distributions of S and $\overline{S}$:

$$Score\,(S) = DL\left(S|\overline{S}\right) - DL\left(S|S\right) = |S| \cdot \left(\hat{H}\left(S,\overline{S}\right) - \hat{H}\left(S\right)\right) = |S| \cdot KL\left(S,\overline{S}\right)$$

In the case of single-pixel-sized regions R_i, this reduces to the KL divergence between the color distributions of S and $\overline{S}$.

A similar derivation can be applied to the general case of composing S from arbitrarily shaped regions R_i. In that case, $p\,(R_i|Ref)$ of (1) is the frequency of regions $R_i \subset S$ in $Ref = S$ or in $Ref = \overline{S}$ (estimated non-parametrically using region composition). This gives rise to an interpretation of the description length $DL\,(S|Ref)$ as a *Shannon entropy measure*, and our segment quality score $Score\,(S)$ of (5) can be interpreted as a *KL divergence* between the statistical distributions of regions (of arbitrary shape and size) in S and in $\overline{S}$.

Note that in the degenerate case when the regions $R_i \subset S$ are one-pixel sized, our framework reduces to a formulation closely related to that of GrabCut [7] (i.e., figure-ground segmentation into segments of distinct color distributions). However, our general formulation employs regions of arbitrary shapes and sizes, giving rise to figure-ground segmentation with distinct region distributions. This is essential when S and $\overline{S}$ share similar color distributions (first order statistics), and vary only in their structural patterns (i.e., higher order statistics). Such an example can be found in Fig. 5 which compares our results to that of GrabCut.

3.4 The Geometric Transformations T

The family of geometric transformations T applied to regions R in the composition process (Eq. 1) determines the degree of complexity of segments that can be handled by our approach. For instance, if we restrict T to pure translations, then a segment S may be composed by shuffling and combining pieces from Ref. Introducing scaling/rotation/affine transformations enables more complex compositions (e.g., compose a small object from a large one, etc.) Further including reflection transformations enables composing one half of a symmetric object/pattern from its other half. Note that different regions $R_i \subset S$ are 'generated' from Ref using different transformations T_i. Combining several types of transformations can give rise to composition of very complex objects S from their own sub-regions (e.g., partially symmetric object as in Fig. 10.b).

4 Figure-Ground Segmentation Algorithm

In this section we outline our figure-ground segmentation algorithm, which optimizes $Score\,(Seg)$ of (7). The goal of figure-ground segmentation is to extract an object of interest (the "foreground") from the remaining parts of the image (the "background"). In general, when the image contains multiple objects, a user input is required to specify the "foreground" object of interest.

| Input image | Our results
Init+ recovered S | Results of GrabCut [7]
Init bounding box   recovered S |

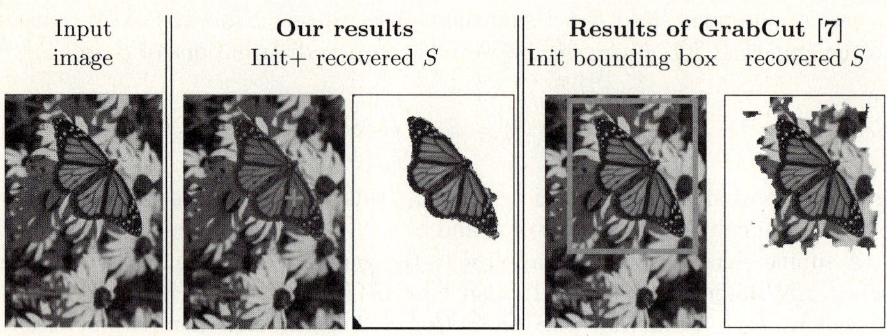

Fig. 5. Our result vs. GrabCut [7]. *GrabCut fails to segment the butterfly (foreground) due to the similar colors of the flowers in the background. Using composition with arbitrarily shaped regions, our algorithm accurately segments the butterfly. We used the GrabCut implementation of* `www.cs.cmu.edu/~mohitg/segmentation.htm`

Different figure-ground segmentation algorithms require different amounts of user-input to specify the foreground object, whether in the form of foreground/background scribbles (e.g., [6]), or a bounding-box containing the foreground object (e.g., [7]). In contrast, our figure-ground segmentation algorithm requires a *minimal amount* of user input – a *single* user-marked point on the foreground segment/object of interest. Our algorithm proceeds to *extract the "best" possible image segment containing that point.* In other words, the algorithm recovers a figure-ground segmentation $Seg = \left(S, \overline{S}, \partial S\right)$ s.t. S contains the user-marked point, and Seg maximizes the segmentation score of (7). Fig. 6 shows how different user-selected points-of-interest extract different objects of interest from the image (inducing different figure-ground segmentations Seg).

A figure-ground segmentation can be described by assigning a label l_i to every pixel i in the image, where $l_i = 1 \; \forall i \in S$, and $l_i = -1 \; \forall i \in \overline{S}$. We can rewrite $Score\,(Seg)$ of (7) in terms of these labels:

$$Score\,(Seg) = \sum_{i \in I} l_i \cdot \left(\text{PES}\,(i|S) - \text{PES}\,\left(i|\overline{S}\right)\right) + \frac{1}{2} \sum_{(i,j) \in \mathcal{N}} |l_i - l_j| \cdot \log Pr\,(Edge_{i,j})$$

(8)

where $\mathcal{N}$ is the set of all pairs of neighboring pixels. Maximizing (8) is equivalent to an energy minimization formulation which can be optimized using a MinCut algorithm [17], where $\left(\text{PES}\,(i|S) - \text{PES}\,\left(i|\overline{S}\right)\right)$ form the data term, and $\log Pr\,(Edge_{i,j})$ is the "smoothness" term. However, the data term has a complicated dependency on the segmentation into $S, \overline{S}$, via the terms $\text{PES}\,(i|S)$ and $\text{PES}\,\left(i|\overline{S}\right)$. This prevents straightforward application of MinCut. To overcome this problem, we employ EM-like iterations, i.e., alternating between estimating the data term and maximizing $Score\,(Seg)$ using MinCut (see Sec. 4.1).

In our current implementation the "smoothness" term $Pr\,(Edge_{i,j})$, is computed based on the edge probabilities of [16], which incorporates texture, luminance and color cues. The computation of $\text{PES}\,(i|Ref)$ for every pixel i (where

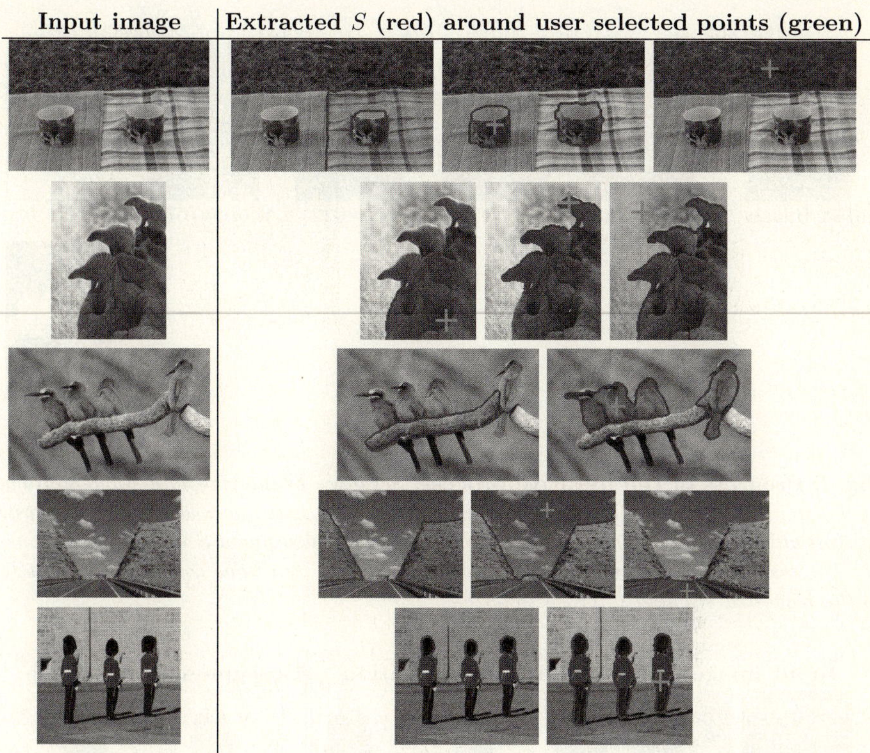

Fig. 6. Different input points result in different foreground segments

Ref is either S or $\overline{S}$) involves finding a 'maximal region' R surrounding i which has similar regions elsewhere in Ref, i.e., a region R that maximizes (4). An image region R (of any shape or size) is represented by a *dense and structured* 'ensemble of patch descriptors' using a star-graph model. When searching for a similar region, we search for a similar ensemble of patches (similar both in their patch descriptors, as well as in their relative geometric positions), up to a global transformation T (Sec. 3.4) and small local non-rigid deformations (see [18]). We find these 'maximal regions' R using the efficient region-growing algorithm of [18,14]: Starting with a small surrounding region around a pixel i, we search for similar such small regions in Ref. These few matched regions form *seeds* for the region growing algorithm. The initial region around i with its matching seed regions are simultaneously grown (in a greedy fashion) to find maximal matching regions (to maximize PES$(i|Ref)$). For more details see [18,14].

4.1 Iterative Optimization

Initialization. The input to our segment extraction algorithm is an image and a single user-marked point of interest q. We use the region composition procedure to generate maximal regions for points in the vicinity of q. We keep only the maximal regions that contain q and have high evidence (i.e., PES) scores. The

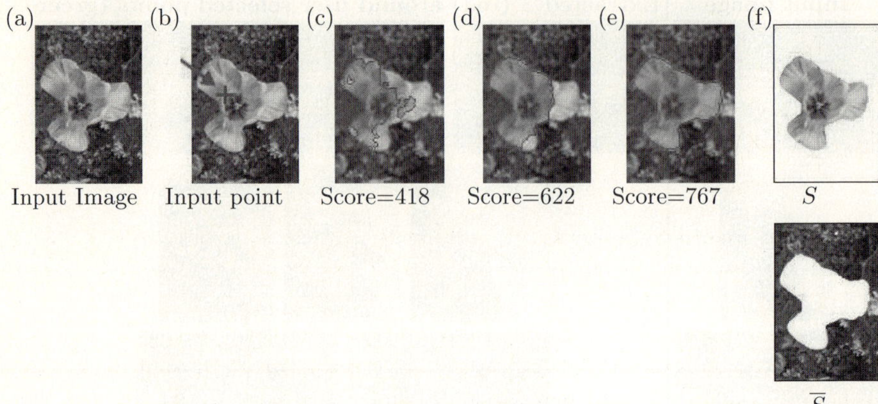

Fig. 7. Progress of the iterative process: *Sequence of intermediate segmentations of the iterative process. (a) The input image. (b) The user marked point-of-interest. (c) Initialization of S. (d) S after 22 iterations. (e) Final segment S after 48 iterations. (f) The resulting figure-ground segments, S and $\overline{S}$. The iterations converged accurately to the requested segment after 48 iterations.*

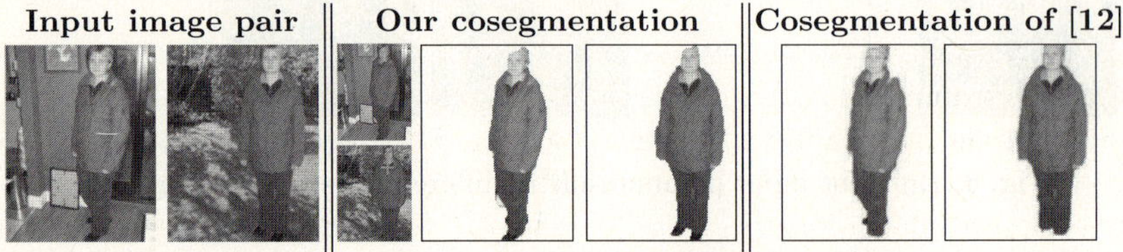

Fig. 8. Cosegmentation of image pair: *Comparing our result to that of [12]*

union of these regions, along with their corresponding reference regions, is used as a crude initialization, S_0, of the segment S (see Fig. 7.c for an example).

Iterations. Our optimization algorithm employs EM-like iterations: In each iteration we first fix the current segmentation $Seg = (S, \overline{S}, \partial S)$ and compute the data term by re-estimating $\text{PES}(i|S)$ and $\text{PES}(i|\overline{S})$. Then, we fix the data term and maximize $Score(Seg)$ using MinCut [17] on (8). This process is iterated until convergence (i.e., when $Score(Seg)$ ceases to improve). The iterative process is quite robust – even a crude initialization suffices for proper convergence. For computational efficiency, in each iteration t we recompute $\text{PES}(i|Ref)$ and relabel pixels only for pixels i within a narrow working band around the current boundary ∂S_t. The segment boundary recovered in the next iteration, ∂S_{t+1}, is restricted to pass inside that working band. The size of the working band is $\sim 10\%$ of the image width, which restricts the computational complexity, yet enables significant updates of the segment boundary in each iteration.

Input image   4 class images   Init point+recovered *Seg*

Fig. 9. Class-based Segmentation: *Segmenting a complex horse image (left) using 4 unsegmented example images of horses*

During the iterative process, similar regions may have conflicting labels. Due to the EM-like iterations, such regions may simultaneously flip their labels, and fail to converge (since each such region provides "evidence" for the other to flip its label). Therefore, in each iteration, we perform two types of steps successively: (i) an "expansion" step, in which only background pixels in $\overline{S_t}$ are allowed to flip their label to foreground pixels. (ii) a "shrinking" step, in which only foreground pixels in S_t are allowed to flip their label to background pixels. Fig. 7 shows a few steps in the iterative process, from initialization to convergence.

4.2 Integrating Several Descriptor Types

The composition process computes similarity of image regions, using local descriptors densely computed within the regions. To allow for flexibility, our framework integrates several descriptor-types, each handles a different aspect of similarity between image points (e.g., color, texture). Thus, several descriptor types can collaborate to describe a complex segment (e.g., in a "multi-person" segment, the color descriptor is dominant in the face regions, while the shape descriptor may be more dominant in other parts of the body). Although descriptor types are very different , the 'savings' in description length obtained by each descriptor type are all in the same units (i.e., bits). Therefore, we can integrate different descriptor-types by simply adding their savings. A descriptor type that is useful for describing a region will increase the savings in description length, while non-useful descriptor types will save nothing. We used the following descriptor types: (1) **SIFT** (2) **Color**: based on a color histogram (3) **Texture**: based on a texton histogram (4) **Shape**: An extension of Shape Context descriptor of Belongie et al. (5) The **Self Similarity** descriptor of Shechtman and Irani.

5 Results

We applied our segment extraction algorithm to a variety of segment types and segmentation tasks, using images from several segmentation databases [19,20,7]. In each case, a single point-of-interest was marked (a green cross in the figures). The algorithm extracted the "best" image segment containing that point

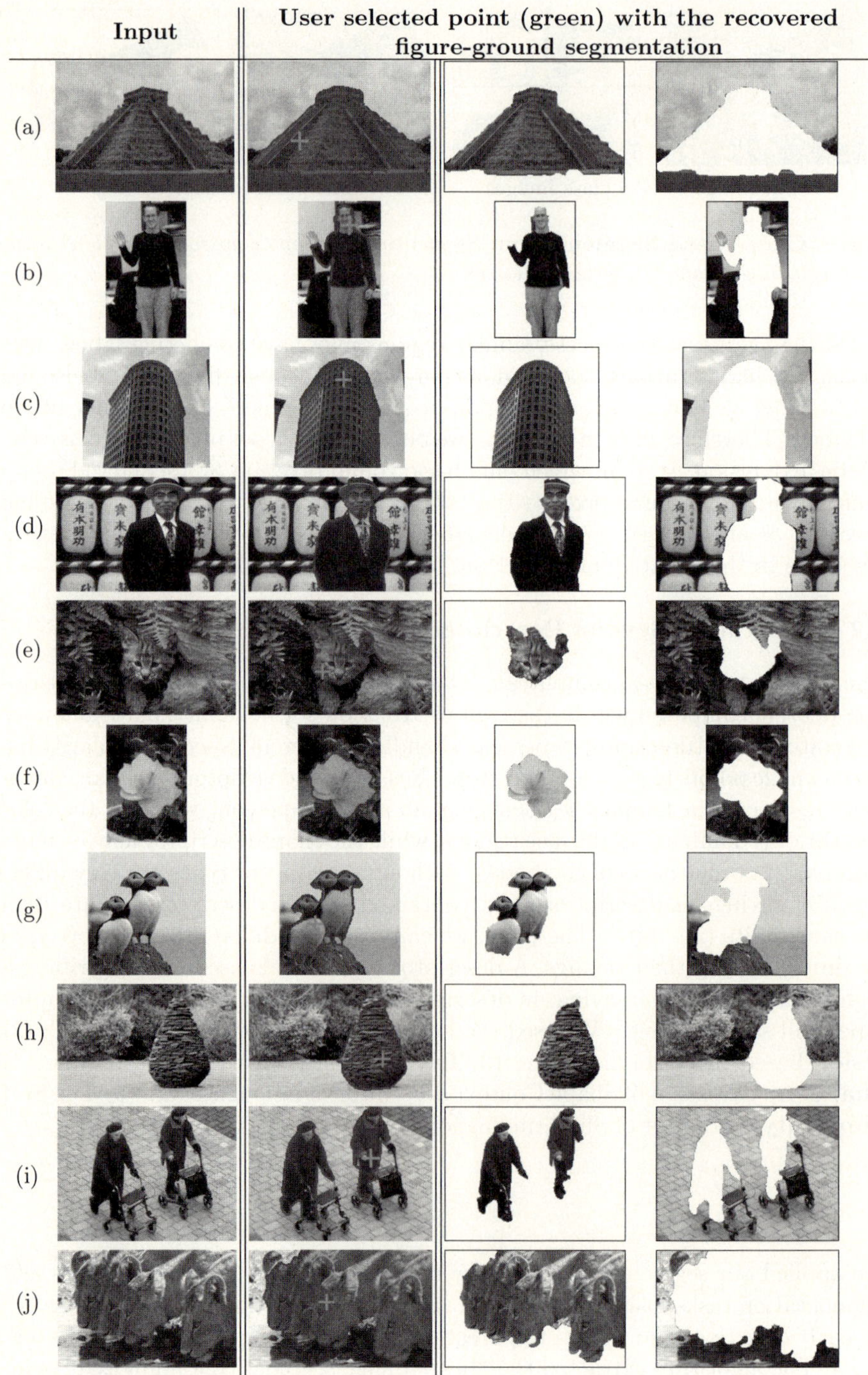

Fig. 10. Examples of figure-ground segmentations

(highlighted in red). Higher resolution images and many more results can be found in `www.wisdom.weizmann.ac.il/~vision/GoodSegment.html`.

Single-Image Segmentation. Fig. 10 demonstrates the capability of our approach to handle a variety of *different* segments types: uniformly colored segments (Fig. 10.f), complex textured segments (Fig. 10.h), complex symmetric objects (e.g., the butterfly in Fig. 5, the Man in Fig. 1.e). More complex objects can also be segmented (e.g., a non-symmetric person Fig. 10.b, or the puffins Fig. 10.g), resulting from combinations of different types of transformations T_i for different regions R_i within the segment, and different types of descriptors.

We further evaluated our algorithm on the benchmark database of [19], which consists of 100 images depicting a single object in front of a background, with ground-truth human segmentation. The total F-measure score of our algorithm was 0.87 ± 0.01 ($F = \frac{2 \cdot Recall \cdot Precision}{Recall + Precision}$), which is state-of-the-art on this database.

"Cosegmentation". We applied our segmentation algorithm *with no modifications* to a simultaneous co-segmentation of an image pair – the algorithm input is simply the concatenated image pair. The common object in the images is extracted as a single compound segment (Fig. 8, shows a comparison to [12]).

Class-Based Segmentation. Our algorithm can perform class-based segmentation given *unsegmented* example images of an object class. In this case, we append the example images to the reference $Ref = S$ of the foreground segment S. Thus the object segment can be composed using other parts in the segment as well as from parts in the example images. This process requires no pre-segmentation and no prior learning stage. Fig. 9 shows an example of extracting a complex horse segment using 4 unsegmented examples of horse images.

References

1. Comaniciu, D., Meer, P.: Mean shift: a robust approach toward feature space analysis. PAMI (2002)
2. Shi, J., Malik, J.: Normalized cuts and image segmentation. PAMI (2000)
3. Malik, J., Belongie, S., Shi, J., Leung, T.K.: Textons, contours and regions: Cue integration in image segmentation. In: ICCV (1999)
4. Galun, M., Sharon, E., Basri, R., Brandt, A.: Texture segmentation by multiscale aggregation of filter responses and shape elements. In: ICCV (2003)
5. Kadir, T., Brady, M.: Unsupervised non-parametric region segmentation using level sets. In: ICCV (2003)
6. Li, Y., Sun, J., Tang, C.K., Shum, H.Y.: Lazy snapping. ACM TOG (2004)
7. Rother, C., Kolmogorov, V., Blake, A.: "grabcut": Interactive foreground extraction using iterated graph cuts. In: SIGGRAPH (2004)
8. Riklin-Raviv, T., Kiryati, N., Sochen, N.: Segmentation by level sets and symmetry. In: CVPR (2006)
9. Borenstein, E., Ullman, S.: Class-specific, top-down segmentation. In: Heyden, A., Sparr, G., Nielsen, M., Johansen, P. (eds.) ECCV 2002. LNCS, vol. 2351. Springer, Heidelberg (2002)

10. Leibe, B., Schiele, B.: Interleaved object categorization and segmentation. In: BMVC (2003)
11. Levin, A., Weiss, Y.: Learning to combine bottom-up and top-down segmentation. In: Leonardis, A., Bischof, H., Pinz, A. (eds.) ECCV 2006. LNCS, vol. 3954. Springer, Heidelberg (2006)
12. Rother, C., Minka, T., Blake, A., Kolmogorov, V.: Cosegmentation of image pairs by histogram matching - incorporating a global constraint into mrfs. In: CVPR (2006)
13. Tu, Z., Chen, X., Yuille, A.L., Zhu, S.C.: Image parsing: Unifying segmentation, detection, and recognition. IJCV (2005)
14. Boiman, O., Irani, M.: Similarity by composition. In: NIPS (2006)
15. Cover, T.M., Thomas, J.A.: Elements of information theory. Wiley, Chichester (1991)
16. Martin, D.R., Fowlkes, C.C., Malik, J.: Learning to detect natural image boundaries using local brightness, color, and texture cues. PAMI (2004)
17. Boykov, Y., Veksler, O., Zabih, R.: Efficient approximate energy minimization via graph cuts. PAMI (2001)
18. Boiman, O., Irani, M.: Detecting irregularities in images and in video. IJCV (2007)
19. Alpert, S., Galun, M., Basri, R., Brandt, A.: Image segmentation by probabilistic bottom-up aggregation and cue integration. In: CVPR (2007)
20. Martin, D., Fowlkes, C., Tal, D., Malik, J.: A database of human segmented natural images and its application to evaluating segmentation algorithms and measuring ecological statistics. In: ICCV (2001)

Light-Efficient Photography

Samuel W. Hasinoff and Kiriakos N. Kutulakos[*]

Dept. of Computer Science, University of Toronto
{hasinoff,kyros}@cs.toronto.edu

Abstract. We consider the problem of imaging a scene with a given depth of field at a given exposure level in the shortest amount of time possible. We show that by (1) collecting a sequence of photos and (2) controlling the aperture, focus and exposure time of each photo individually, we can span the given depth of field in less total time than it takes to expose a single narrower-aperture photo. Using this as a starting point, we obtain two key results. First, for lenses with continuously-variable apertures, we derive a closed-form solution for the *globally optimal* capture sequence, *i.e.*, that collects light from the specified depth of field in the most efficient way possible. Second, for lenses with discrete apertures, we derive an integer programming problem whose solution is the optimal sequence. Our results are applicable to off-the-shelf cameras and typical photography conditions, and advocate the use of dense, wide-aperture photo sequences as a light-efficient alternative to single-shot, narrow-aperture photography.

1 Introduction

Two of the most important choices when taking a photo are the photo's exposure level and its depth of field. Ideally, these choices will result in a photo whose subject is free of noise or pixel saturation [1,2], and appears in-focus. These choices, however, come with a severe time constraint: in order to take a photo that has both a specific exposure level and a specific depth of field, we must expose the camera's sensor for a length of time dictated by the optics of the lens. Moreover, the larger the depth of field, the longer we must wait for the sensor to reach the chosen exposure level. In practice, this makes it impossible to efficiently take sharp and well-exposed photos of a poorly-illuminated subject that spans a wide range of distances from the camera. To get a good exposure level, we must compromise something – accepting either a smaller depth of field (incurring defocus blur [3,4,5,6]) or a longer exposure (incurring motion blur [7,8,9]).

In this paper we seek to overcome the time constraint imposed by lens optics, by capturing a sequence of photos rather than just one. We show that if the aperture, exposure time, and focus setting of each photo is selected appropriately, we can span a given depth of field with a given exposure level *in less total time than it takes to expose a single photo* (Fig. 1). This novel observation is based on a simple fact: even though wide apertures have a narrow depth of field (DOF), they are much more efficient than narrow apertures in gathering light from within their depth of field. Hence, even though

---

* This work was supported in part by the Natural Sciences and Engineering Research Council of Canada under the RGPIN program and by an Ontario Premier's Research Excellence Award.

D. Forsyth, P. Torr, and A. Zisserman (Eds.): ECCV 2008, Part IV, LNCS 5305, pp. 45–59, 2008.

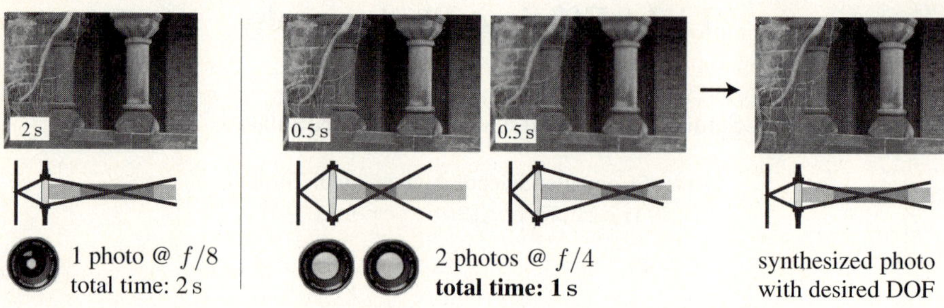

Fig. 1. *Left:* Traditional single-shot photography. The desired depth of field is shaded (red). *Right:* Light-efficient photography. Two wide-aperture photos span the same DOF as a single-shot narrow-aperture photo. Each wide-aperture photo requires $1/4$ the time to reach the exposure level of the single-shot photo, resulting in a $2\times$ net speedup for the total exposure time.

it is not possible to span a wide DOF with a single wide-aperture photo, it is possible to span it with several of them, and to do so very efficiently.

Using this observation as a starting point, we develop a general theory of *light-efficient photography* that addresses four questions: (1) under what conditions is capturing photo sequences with "synthetic" DOFs more efficient than single-shot photography? (2) How can we characterize the set of sequences that are *globally optimal* for a given DOF and exposure level, *i.e.* whose total exposure time is the shortest possible? (3) How can we compute such sequences automatically for a specific camera, depth of field, and exposure level? (4) Finally, how do we convert the captured sequence into a single photo with the specified depth of field and exposure level?

Little is known about how to gather light efficiently from a specified DOF. Research on computational photography has not investigated the light-gathering ability of existing methods, and has not considered the problem of optimizing exposure time for a desired DOF and exposure level. For example, even though there has been great interest in manipulating a camera's DOF through optical [10,11,12,13] or computational [5,14,15,16,17,18,2] means, current approaches do so without regard to exposure time – they simply assume that the shutter remains open as long as necessary to reach the desired exposure level. This assumption is also used for high-dynamic range photography [19,2], where the shutter must remain open for long periods in order to capture low-radiance regions in a scene. In contrast, here we capture photos with camera settings that are carefully chosen to minimize total exposure time for the desired DOF and exposure level.

Since shorter total exposure times reduce motion blur, our work can be thought of as complementary to recent *synthetic shutter* approaches whose goal is to reduce such blur. Instead of controlling aperture and focus, these techniques divide a given exposure interval into several shorter ones, with the same total exposure (*e.g.*, n photos, each with $1/n$ the exposure time [9]; two photos, one with long and one with short exposure [8]; or one photo where the shutter opens and closes intermittently during the exposure [7]). These techniques do not increase light-efficiency but can be readily combined with our work, to confer the advantages of both methods.

Moreover, our approach can be thought of as complementary to work on light field cameras [17,18,13], which are based on an orthogonal tradeoff between resolution and directional sampling. Compared to regular wide-aperture photography, these designs do not have the ability to extend the DOF when their reduced resolution is taken into account. Along similar lines, wavefront coding [11] exploits special optics to extend the DOF with no change in exposure time by using another orthogonal tradeoff – accepting lower signal-to-noise ratio for higher frequencies.

The final step in light-efficient photography involves merging the captured photos to create a new one (Fig. 1). As such, our work is related to the well-known technique of extended-depth-of-field imaging, which has found wide use in microscopy [18] and macro photography [20,17].

Our work offers four contributions over the state of the art. First, we develop a theory that leads to provably-efficient light-gathering strategies, and applies both to off-the-shelf cameras and to advanced camera designs [7,9] under typical photography conditions. Second, from a practical standpoint, our analysis shows that the optimal (or near-optimal) strategies are very simple: for example, in the continuous case, a strategy using the widest-possible aperture for all photos is either globally optimal or it is very close to it (in a quantifiable sense). Third, our experiments with real scenes suggest that it is possible to compute good-quality synthesized photos using readily-available algorithms. Fourth, we show that despite requiring less total exposure time than a single narrow-aperture shot, light-efficient photography provides more information about the scene (*i.e.*, depth) and allows post-capture control of aperture and focus.

2 The Exposure Time *vs.* Depth of Field Tradeoff

The *exposure level* of a photo is the total radiant energy integrated by the camera's entire sensor while the shutter is open. The exposure level can influence significantly the quality of a captured photo because when there is no saturation or thermal noise, a pixel's signal-to-noise ratio (SNR) always increases with higher exposure levels [1]. For this reason, most modern cameras can automate the task of choosing an exposure level that provides high SNR for most pixels and causes little or no saturation.

Lens-based camera systems provide only two ways to control exposure level – the diameter of their aperture and the exposure time. We assume that all light passing through the aperture will reach the sensor plane, and that the average irradiance measured over this aperture is independent of the aperture's diameter. In this case, the exposure level L is equal to

$$L = \tau D^2 \,, \tag{1}$$

where τ is exposure time, D is the effective aperture diameter, and the units of L are chosen appropriately.

Now suppose that we have chosen a desired exposure level L^*. How can we capture a photo at this exposure level? Equation (1) suggests that there are only two general strategies for doing this – either choose a long exposure time and a small aperture diameter, or choose a large aperture diameter and a short exposure time. Unfortunately, both strategies have important side-effects: increasing exposure time can introduce motion blur when we photograph moving scenes [8,9]; opening the lens aperture, on the

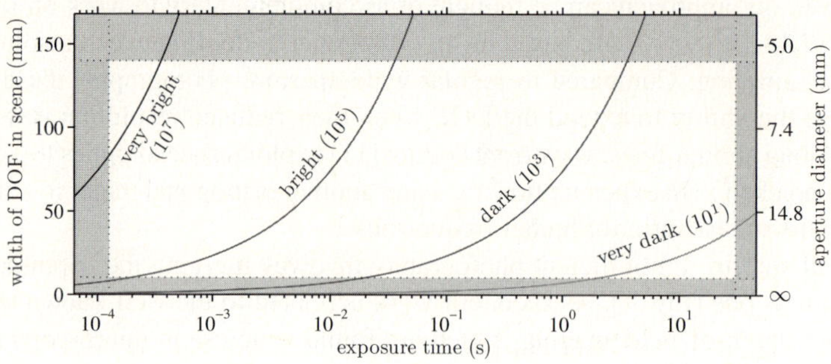

Fig. 2. Each curve represents all pairs (τ, D) for which $\tau D^2 = L^*$ in a specific scene. Shaded zones correspond to pairs outside the camera limits (valid settings were $\tau \in [1/8000\,\text{s}, 30\,\text{s}]$ and $D \in [f/16, f/1.2]$ with $f = 85\,\text{mm}$). Also shown is the DOF corresponding to each diameter D. The maximum acceptable blur was set to $c = 25\,\mu\text{m}$, or about 3 pixels in our camera. Different curves represent scenes with different average radiance (relative units shown in brackets).

other hand, affects the photo's *depth of field (DOF)*, *i.e.*, the range of distances where scene points do not appear out of focus. These side-effects lead to an important tradeoff between a photo's exposure time and its depth of field (Fig. 2):

> **Exposure Time *vs*. Depth of field Tradeoff:** *We can either achieve a desired exposure level L^* with short exposure times and a narrow DOF, or with long exposure times and a wide DOF.*

In practice, the exposure time *vs*. DOF tradeoff limits the range of scenes that can be photographed at a given exposure level (Fig. 2). This range depends on scene radiance, the physical limits of the camera (*i.e.*, range of possible apertures and shutter speeds), as well as subjective factors (*i.e.*, acceptable levels of motion blur and defocus blur).

Our goal is to "break" this tradeoff by seeking novel photo acquisition strategies that capture a given depth of field at the desired exposure level L^* much faster than traditional optics would predict. We briefly describe below the basic geometry and relations governing a photo's depth of field, as they are particularly important for our analysis.

2.1   Depth of Field Geometry

We assume that focus and defocus obey the standard thin lens model [3,21]. This model relates three positive quantities (Eq. (A) in Table 1): the focus setting v, defined as the distance from the sensor plane to the lens; the distance d from the lens to the in-focus scene plane; and the focal length f, representing the "focusing power" of the lens.

Apart from the idealized pinhole, all apertures induce spatially-varying amounts of defocus for points in the scene (Fig. 3a). If the lens focus setting is v, all points at distance d from the lens will be in-focus. A scene point at distance $d' \neq d$, however, will be defocused: its image will be a circle on the sensor plane whose diameter b is called the *blur diameter*. For any given distance d, the thin-lens model tells us exactly what focus setting we should use to bring the plane at distance d into focus, and what the blur diameter will be for points away from this plane (Eqs. (B) and (C), respectively).

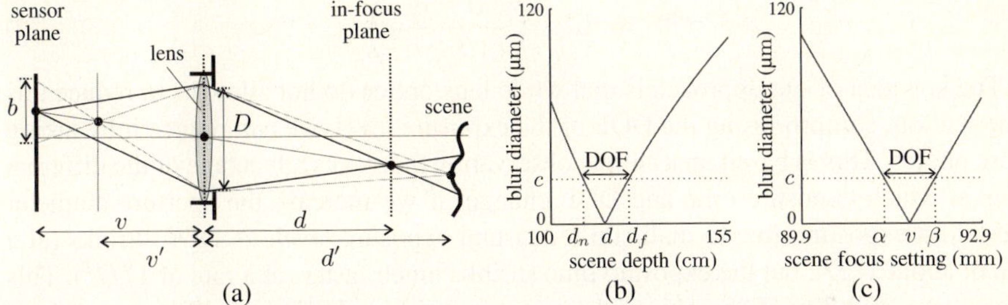

Fig. 3. (a) Blur geometry for a thin lens. (b) Blur diameter as a function of distance to a scene point. The plot is for a lens with $f = 85\,$mm, focused at $117\,$cm with an aperture diameter of $5.31\,$mm (*i.e.*, an $f/16$ aperture in photography terminology). (c) Blur diameter and DOF represented in the space of focus settings.

Table 1. Eqs. (A)–(F): Basic equations governing focus and DOFs for the thin-lens model

(A) Thin lens law	(B) Focus for distance d	(C) Blur diameter for distance d'	(D) Aper. diam. for DOF $[\alpha, \beta]$	(E) Focus for DOF $[\alpha, \beta]$	(F) DOF for aper. diam. D, focus v
$\dfrac{1}{v} + \dfrac{1}{d} = \dfrac{1}{f}$	$v = \dfrac{fd}{d-f}$	$b = D\dfrac{f\lvert d' - d\rvert}{d'(d-f)}$	$D = c\dfrac{\beta + \alpha}{\beta - \alpha}$	$v = \dfrac{2\alpha\beta}{\alpha + \beta}$	$\alpha, \beta = \dfrac{Dv}{D \pm c}$

For a given aperture and focus setting, the *depth of field* is the interval of distances in the scene whose blur diameter is below a maximum acceptable size c (Fig. 3b).

Since every distance in the scene corresponds to a unique focus setting (Eq. (B)), every DOF can also be expressed as an interval $[\alpha, \beta]$ in the space of focus settings. This alternate DOF representation gives us especially simple relations for the aperture and focus setting that produce a given DOF (Eqs. (D) and (E)) and, conversely, for the DOF produced by a given aperture and focus setting (Eq. (F)). We adopt this DOF representation for the rest of the paper (Fig. 3c).

A key property of the depth of field is that it shrinks when the aperture diameter increases: from Eq. (C) it follows that for a given out-of-focus distance, larger apertures always produce larger blur diameters. This equation is the root cause of the exposure time *vs.* depth of field tradeoff.

3   The Synthetic DOF Advantage

Suppose that we want to capture a single photo with a specific exposure level L^* *and* a specific depth of field $[\alpha, \beta]$. How quickly can we capture this photo? The basic DOF geometry of Sect. 2.1 tells us we have no choice: there is only one aperture diameter that can span the given depth of field (Eq. (D)), and only one exposure time that can achieve a given exposure level with that diameter (Eq. (1)). This exposure time is[1]

---

[1] The apertures and exposure times of real cameras span finite intervals and, in many cases, take discrete values. Hence, in practice, Eq. (2) holds only approximately.

$$\tau^{one} = L^* \cdot \left(\frac{\beta - \alpha}{c\,(\beta + \alpha)} \right)^2 . \tag{2}$$

The key idea of our approach is that while lens optics do not allow us to reduce this time without compromising the DOF or the exposure level, we *can* reduce it by taking more photos. This is based on a simple observation that takes advantage of the different rates at which exposure time and DOF change: if we increase the aperture diameter and adjust exposure time to maintain a constant exposure level, its DOF shrinks (at a rate of about $1/D$), but the exposure time shrinks much faster (at a rate of $1/D^2$). This opens the possibility of "breaking" the exposure time *vs.* DOF tradeoff by capturing a sequence of photos that jointly span the DOF in less total time than τ^{one} (Fig. 1).

Our goal is to study this idea in its full generality, by finding capture strategies that are provably time-optimal. We therefore start from first principles, by formally defining the notion of a *capture sequence* and of its *synthetic depth of field*:

Definition 1 (Photo Tuple). *A tuple $\langle\, D,\ \tau,\ v\, \rangle$ that specifies a photo's aperture diameter, exposure time, and focus setting, respectively.*

Definition 2 (Capture Sequence). *A finite ordered sequence of photo tuples.*

Definition 3 (Synthetic Depth of Field). *The union of DOFs of all photo tuples in a capture sequence.*

We will use two efficiency measures: the *total exposure time* of a sequence is the sum of the exposure times of all its photos; the *total capture time*, on the other hand, is the actual time it takes to capture the photos with a specific camera. This time is equal to the total exposure time, plus any overhead caused by camera internals (computational and mechanical). We now consider the following general problem:

> **Light-Efficient Photography:** *Given a set $\mathcal{D}$ of available aperture diameters, construct a capture sequence such that: (1) its synthetic DOF is equal to $[\alpha, \beta]$; (2) all its photos have exposure level L^*; (3) the total exposure time (or capture time) is smaller than τ^{one}; and (4) this time is a global minimum over all finite capture sequences.*

Intuitively, whenever such a capture sequence exists, it can be thought of as being optimally more efficient than single-shot photography in gathering light. Below we analyze three instances of the light-efficient photography problem. In all cases, we assume that the exposure level L^*, depth of field $[\alpha, \beta]$, and aperture set $\mathcal{D}$ are known and fixed.

Noise Properties. All photos we consider have similar noise, because most noise sources (photon, sensor, and quantization noise) depend only on exposure level, which we hold constant. The only exception is thermal noise, which increases with exposure time [1], and so will be lower for light-efficient sequences with shorter exposures.

4   Theory of Light-Efficient Photography

4.1   Continuously-Variable Aperture Diameters

Many manual-focus SLR lenses allow their aperture diameter to vary continuously within some interval $\mathcal{D} = [D_{min},\ D_{max}]$. In this case, we prove that the optimal capture sequence has an especially simple form – it is unique, it uses the same aperture diameter for all tuples, and this diameter is either the maximum possible or a diameter close to that maximum.

More specifically, consider the following special class of capture sequences:

Definition 4 (Sequences with Sequential DOFs). *A capture sequence has sequential DOFs if for every pair of adjacent photo tuples, the right endpoint of the first tuple's DOF is the left endpoint of the second.*

The following theorem states that the solution to the light-efficient photography problem is a specific sequence from this class:

Theorem 1 (Optimal Capture Sequence for Continuous Apertures). *(1) If the DOF endpoints satisfy $\beta < (7 + 4\sqrt{3})\alpha$, the sequence that globally minimizes total exposure time is a sequence with sequential DOFs whose tuples all have the same aperture. (2) Define $D(k)$ and n as follows:*

$$D(k) = c\,\frac{\sqrt[k]{\beta} + \sqrt[k]{\alpha}}{\sqrt[k]{\beta} - \sqrt[k]{\alpha}}\;, \qquad n = \left\lfloor \frac{\log\frac{\alpha}{\beta}}{\log\left(\frac{D_{max}-c}{D_{max}+c}\right)} \right\rfloor . \tag{3}$$

The aperture diameter D^* and length n^* of the optimal sequence is given by

$$D^* = \begin{cases} D(n) & \text{if } \frac{D(n)}{D_{max}} > \sqrt{\frac{n}{n+1}} \\ D_{max} & \text{otherwise.} \end{cases} \qquad n^* = \begin{cases} n & \text{if } \frac{D(n)}{D_{max}} > \sqrt{\frac{n}{n+1}} \\ n+1 & \text{otherwise.} \end{cases} . \tag{4}$$

Theorem 1 specifies the optimal sequence indirectly, via a "recipe" for calculating the optimal length and the optimal aperture diameter (Eqs. (3) and (4)). Informally, this calculation involves three steps. The first step defines the quantity $D(k)$; in our proof of Theorem 1 (see Appendix A), we show that this quantity represents the only aperture diameter that can be used to "tile" the interval $[\alpha, \beta]$ with exactly k photo tuples of the same aperture. The second step defines the quantity n; in our proof, we show that this represents the largest number of photos we can use to tile the interval $[\alpha, \beta]$ with photo tuples of the same aperture. The third step involves choosing between two "candidates" for the optimal solution – one with n tuples and one with $n + 1$.

Theorem 1 makes explicit the somewhat counter-intuitive fact that the most light-efficient way to span a given DOF $[\alpha, \beta]$ is to use images whose DOFs are very narrow. This fact applies broadly, because Theorem 1's inequality condition for α and β is satisfied for all lenses for consumer photography that we are aware of (*e.g.*, see [22]).[2] See Fig. 4 for an application of this theorem to a practical example.

Note that Theorem 1 specifies the number of tuples in the optimal sequence and their aperture diameter, but does not specify their exposure times or focus settings. The following lemma shows that specifying those quantities is not necessary because they are determined uniquely. Importantly, Lemma 1 gives us a recursive formula for computing the exposure time and focus setting of each tuple in the sequence:

---

[2] To violate the condition, the minimum focusing distance must be under $1.077f$, measured from the lens center.

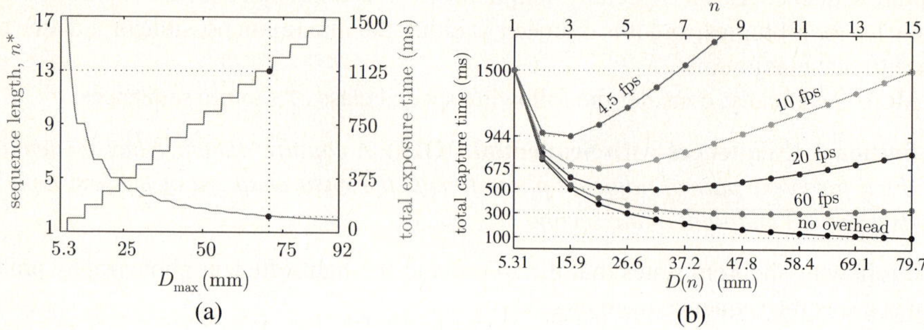

Fig. 4. (a) Optimal light-efficient photography of a "dark" subject spanning a DOF of [110 cm, 124 cm], using an $f = 85$ mm lens with a continuously-variable aperture. In this example, we can use a $f/16$ aperture (5.3 mm diameter) to cover the DOF with a single photo, which requires a 1.5 s exposure to obtain the desired exposure level. The plot illustrates the optimal sequences when the aperture diameter is restricted to a range $[f/16, D_{\max}]$: for each value of the maximum aperture, $D_{\max}$, Theorem 1 gives a unique optimal sequence. The graph shows the number of images n^* (red) and total exposure time (green) of this sequence. As $D_{\max}$ increases, the total exposure time of the optimal sequence falls dramatically: for lenses with an $f/1.2$ maximum aperture (71 mm), synthetic DOFs confer a $13\times$ speedup over single-shot photography for the same exposure level. (b) The effect of camera overhead for various frame-per-second (fps) rates. Each point represents the total capture time of a sequence that spans the DOF and whose photos all use the diameter $D(n)$ indicated. Even though overhead reduces the efficiency of long sequences, synthetic DOFs are faster than single-shot photography even for low fps rates.

Lemma 1 (Construction of Sequences with Sequential DOFs). *Given a left DOF endpoint α, every ordered sequence $D_1, \ldots, D_n$ of aperture diameters defines a unique capture sequence with sequential DOFs whose n tuples are*

$$\left\langle D_i,\ \frac{L^*}{D_i^{\,2}},\ \frac{D_i + c}{D_i}\,\alpha_i \right\rangle,\ \ i = 1, \ldots, n\ , \tag{5}$$

with α_i given by the following recursive relation:

$$\alpha_i = \begin{cases} \alpha & \text{if } i = 1\ , \\ \frac{D_i + c}{D_i - c}\,\alpha_{i-1} & \text{otherwise.} \end{cases} \tag{6}$$

4.2   Discrete Aperture Diameters

Modern auto-focus lenses often restrict the aperture diameter to a discrete set of choices, $\mathcal{D} = \{D_1, \ldots, D_m\}$. These diameters form a geometric progression, spaced so that the aperture area doubles every two or three steps. Unlike the continuous case, the optimal capture sequence is not unique and may contain several distinct aperture diameters. To find an optimal sequence, we reduce the problem to integer linear programming [23]:

Theorem 2 (Optimal Capture Sequence for Discrete Apertures). *There exists an optimal capture sequence with sequential DOFs whose tuples have a non-decreasing*

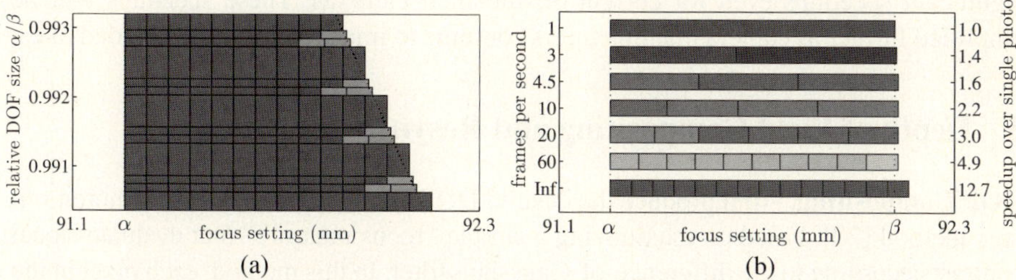

Fig. 5. Optimal light-efficient photography with discrete apertures, shown for a Canon EF85mm 1.2L lens (23 apertures, illustrated in different colors). (a) For a depth of field whose left endpoint is α, we show optimal capture sequences for a range of relative DOF sizes $\frac{\alpha}{\beta}$. These sequences can be read horizontally, with subintervals corresponding to the apertures determined by Theorem 2. Note that when the DOF is large, the optimal sequence approximates the continuous case. The diagonal dotted line indicates the DOF to be spanned. (b) Visualizing the optimal capture sequence as a function of the camera overhead for the DOF $[\alpha, \beta]$. Note that with higher overhead, the optimal sequence involves fewer photos with larger DOFs (*i.e.*, smaller apertures).

sequence of aperture diameters. Moreover, if n_i is the number of times diameter D_i appears in the sequence, the multiplicities $n_1, \ldots, n_m$ satisfy the integer program

$$\text{minimize} \quad \sum_{i=1}^{m} n_i \, \frac{L^*}{D_i^2} \tag{7}$$

$$\text{subject to} \quad \sum_{i=1}^{m} n_i \log \frac{D_i - c}{D_i + c} \;\leq\; \log \frac{\alpha}{\beta} \tag{8}$$

$$n_i \geq 0 \;\; \text{and integer} \,. \tag{9}$$

See [24] for a proof. As with Theorem 1, Theorem 2 does not specify the focus settings in the optimal capture sequence. We use Lemma 1 for this purpose, which explicitly constructs it from the apertures and their multiplicities.

While it is not possible to obtain a closed-form expression for the optimal sequence, solving the integer program for any desired DOF is straightforward. We use a simple branch-and-bound method based on successive relaxations to linear programming [23]. Moreover, since the optimal sequence depends only on the relative DOF size $\frac{\alpha}{\beta}$, we pre-compute it for all possible DOFs and store the results in a lookup table (Fig. 5a).

4.3 Discrete Aperture Diameters Plus Overhead

Our treatment of discrete apertures generalizes easily to account for camera overhead. We model overhead as a per-shot constant, τ^{over}, that expresses the minimum delay between the time that the shutter closes and the time it is ready to open again for the next photo. To find the optimal sequence, we modify the objective function of Theorem 2 so that it measures total capture time rather than total exposure time:

$$\text{minimize} \quad \sum_{i=1}^{m} n_i \left[\tau^{over} + \frac{L^*}{D_i^2} \right] \,. \tag{10}$$

Clearly, a non-negligible overhead penalizes long capture sequences and reduces the synthetic DOF advantage. Despite this, Fig. 5b shows that synthetic DOFs offer significant speedups even for current off-the-shelf cameras. These speedups will be amplified further as camera manufacturers continue to improve frame-per-second rates.

5   Depth of Field Compositing and Resynthesis

DOF Compositing. To reproduce the desired DOF, we use a variant of the Photomontage method [20], based on maximizing a simple "focus measure" that evaluates local contrast according to the difference-of-Gaussians filter. In this method, each pixel in the composite has a label that indicates the input photo for which the pixel is in-focus. These labels are optimized with a Markov random field network that is biased toward piece-wise smoothness. The resulting composite is a blend of the input photos, performed in the gradient-domain to reduce artifacts at label boundaries.

3D Reconstruction. The DOF compositing operation produces a coarse depth map as an intermediate step. This is because labels correspond to input photos, and each input photo defines an in-focus depth according to the focus setting with which it was captured. We found this depth map to be sufficient for good-quality resynthesis, although a more sophisticated depth-from-defocus analysis is also possible [6].

Synthesizing Photos for Novel Focus Settings and Aperture Diameters. To synthesize novel photos, we generalize DOF compositing and take advantage of the different levels of defocus throughout the capture sequence. We proceed in four basic steps. First, given a specific focus and aperture setting, we use Eq. (C) and the coarse depth map to assign a blur diameter to each pixel in the final composite. Second, we use Eq. (C) again to determine, for each pixel in the composite, the input photo whose blur diameter that corresponds to the pixel's depth matches most closely.[3] Third, for each depth layer, we synthesize a photo under the assumption that the entire scene is at that depth, and is observed with the novel focus and aperture setting. To do this, we use the blur diameter for this depth to define an interpolation between two of the input photos. We currently interpolate using simple linear cross-fading, which we found to be adequate when the DOF is sampled densely enough (*i.e.*, with 5 or more images). Fourth, we generate the final composite by merging all these synthesized images into one photo using the same gradient-domain blending as in DOF compositing, with the same depth labels.

6   Experimental Results

Figure 6 shows results and timings for two experiments, performed with two different cameras – a high-end digital SLR and a compact digital camera (see [24] for more results and videos). All photos were captured at the same exposure level for each experiment. In each case, we captured (1) a narrow-aperture photo and (2) the optimal capture sequence for the equivalent DOF and the particular camera. To compensate for the distortions that occur with changes in focus setting, we align the photos according

---

[3] Note each blur diameter is consistent with two depths (Fig. 3b). We resolve the ambiguity by choosing the matching input photo whose focus setting is closest to the synthetic focus setting.

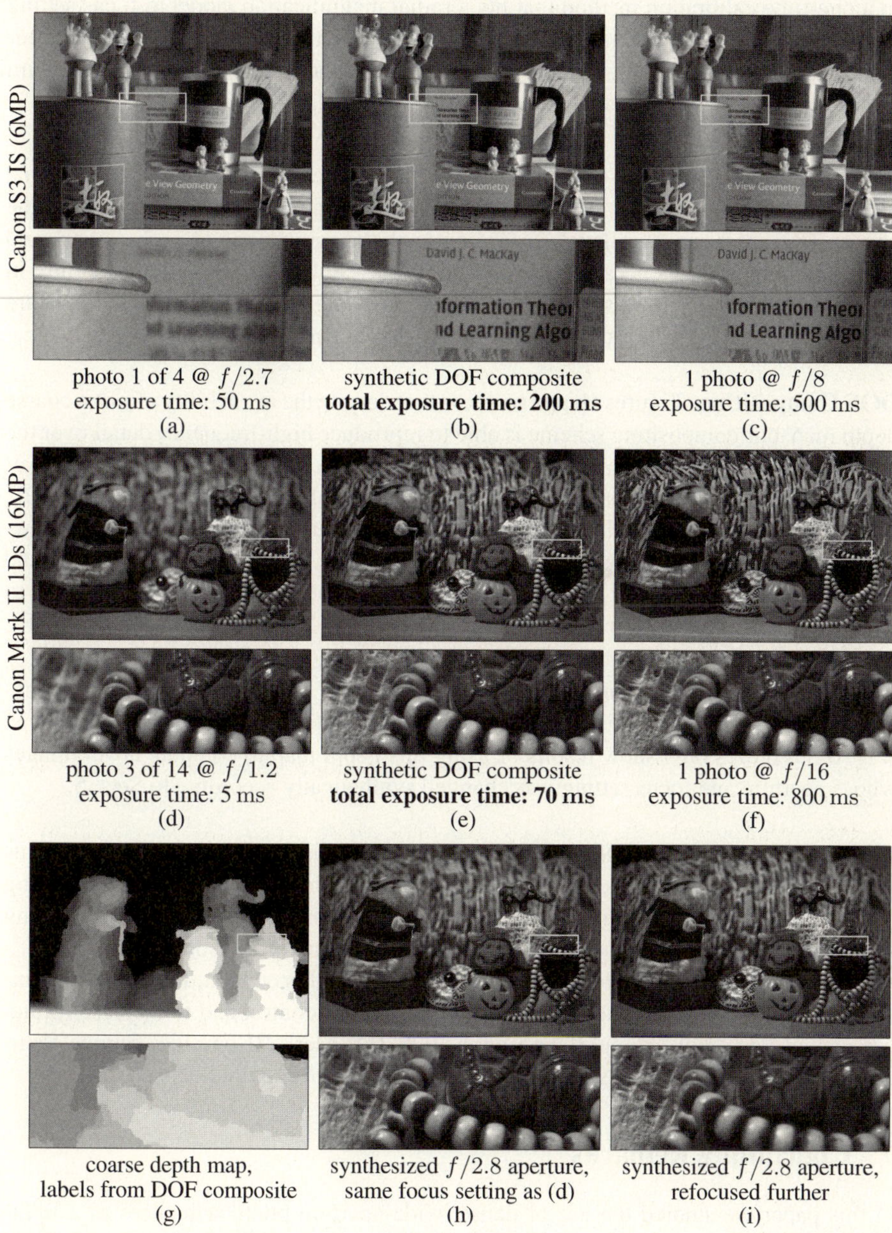

photo 1 of 4 @ $f/2.7$
exposure time: 50 ms
(a)

synthetic DOF composite
total exposure time: 200 ms
(b)

1 photo @ $f/8$
exposure time: 500 ms
(c)

photo 3 of 14 @ $f/1.2$
exposure time: 5 ms
(d)

synthetic DOF composite
total exposure time: 70 ms
(e)

1 photo @ $f/16$
exposure time: 800 ms
(f)

coarse depth map,
labels from DOF composite
(g)

synthesized $f/2.8$ aperture,
same focus setting as (d)
(h)

synthesized $f/2.8$ aperture,
refocused further
(i)

Fig. 6. Light-efficient photography timings and synthesis, for several real scenes, captured using a compact digital camera and a digital SLR. (a,d) Sample wide-aperture photo from the synthetic DOF sequence. (b,e) DOF composites synthesized from this sequence. (c,f) Narrow-aperture photos spanning an equivalent DOF, but with much longer exposure time. (g) Coarse depth map, computed from the labeling we used to compute (e). (h) Synthetically changing aperture size, focused at the same setting as (d). (i) Synthetically changing the focus setting as well.

to a one-time calibration method that fits a radial magnification model to focus setting [25]. To determine the maximum acceptable blur diameter c for each camera, we evaluated focus using a resolution chart. The values we found, 5 μm (1.4 pixels) and 25 μm (3.5 pixels) respectively, agree with standard values [21].

Timing Comparisons and Optimal Capture Sequences. To determine the optimal capture sequences, we assumed zero camera overhead and applied Theorem 2 for the chosen DOF and exposure level, according to the specifications of each camera and lens. The optimal sequences involved spanning the DOF using the largest aperture in both cases. As Fig. 6 shows, these sequences led to significant speedups in exposure time – $2.5\times$ and $11.9\times$ for the compact digital camera and digital SLR, respectively.

DOF Compositing. Figures 6b and 6e show that despite the availability of just a coarse depth map, our compositing scheme is able to reproduce high-frequency detail over the whole DOF without noticeable artifacts, even in the vicinity of depth discontinuities. Note that while the synthesized photos satisfy our goal of spanning a specific DOF, objects outside that DOF will appear more defocused than in the corresponding narrow-aperture photo (*e.g.*, see the background in Figs. 6e–f). While increased background defocus may be desirable (*e.g.*, for portrait or macro photography), it is also possible to capture sequences of photos to reproduce arbitrary levels of defocus outside the DOF.

Depth Maps and DOF Compositing. Despite being more efficient to capture, sequences with synthetic DOFs provide 3D shape information at no extra acquisition cost (Fig. 6g). Figures 6h–i show results of using this depth map to compute novel images whose aperture and focus setting was changed synthetically according to Sect. 5.

Implementation Details. Neither of our cameras provide the ability to control focus remotely. For our compact camera we used modified firmware that enables scripting [26], while for our SLR we used a computer-controlled motor to drive the focusing ring mechanically. Both methods incur high overhead and limit us to about 1 fps.

While light-efficient photography is not practical in this context, it will become increasingly so, as newer cameras begin to provide focus control and to increase frame-per-second rates. For example, the Canon EOS-1Ds Mark III provides remote focus control for all Canon EF lenses, and the Casio EX-F1 can capture 60 fps at 6MP.

7 Concluding Remarks

In this paper we studied the use of dense, wide-aperture photo sequences as a light-efficient alternative to single-shot, narrow-aperture photography. While our emphasis has been on the underlying theory, we believe our method has great practical potential.

We are currently investigating several extensions to the basic approach. These include designing light-efficient strategies (1) for spanning arbitrary defocus profiles, rather than just the DOF; (2) improving efficiency by taking advantage of the camera's auto-focus sensor; and (3) operating under a highly-restricted time-budget, for which it becomes important to weigh the tradeoff between noise and defocus.

References

1. Healey, G.E., Kondepudy, R.: Radiometric CCD camera calibration and noise estimation. TPAMI 16(3), 267–276 (1994)
2. Hasinoff, S.W., Kutulakos, K.N.: A layer-based restoration framework for variable-aperture photography. In: Proc. ICCV (2007)
3. Pentland, A.P.: A new sense for depth of field. TPAMI 9(4), 523–531 (1987)
4. Krotkov, E.: Focusing. IJCV 1(3), 223–237 (1987)
5. Hiura, S., Matsuyama, T.: Depth measurement by the multi-focus camera. In: CVPR, pp. 953–959 (1998)
6. Watanabe, M., Nayar, S.K.: Rational filters for passive depth from defocus. IJCV 27(3), 203–225 (1998)
7. Raskar, R., Agrawal, A., Tumblin, J.: Coded exposure photography: motion deblurring using fluttered shutter. In: SIGGRAPH, pp. 795–804 (2006)
8. Yuan, L., Sun, J., Quan, L., Shum, H.Y.: Image deblurring with blurred/noisy image pairs. In: SIGGRAPH (2007)
9. Telleen, J., Sullivan, A., Yee, J., Gunawardane, P., Wang, O., Collins, I., Davis, J.: Synthetic shutter speed imaging. In: Proc. Eurographics, pp. 591–598 (2007)
10. Farid, H., Simoncelli, E.P.: Range estimation by optical differentiation. JOSA A 15(7), 1777–1786 (1998)
11. Cathey, W.T., Dowski, E.R.: New paradigm for imaging systems. Applied Optics 41(29), 6080–6092 (2002)
12. Levin, A., Fergus, R., Durand, F., Freeman, W.T.: Image and depth from a conventional camera with a coded aperture. In: SIGGRAPH (2007)
13. Veeraraghavan, A., Raskar, R., Agrawal, A., Mohan, A., Tumblin, J.: Dappled photography: Mask enhanced cameras for heterodyned light fields and coded aperture refocusing. In: SIGGRAPH (2007)
14. Aizawa, K., Kodama, K., Kubota, A.: Producing object-based special effects by fusing multiple differently focused images. In: TCSVT 10(2) (2000)
15. Chaudhuri, S.: Defocus morphing in real aperture images. JOSA A 22(11), 2357–2365 (2005)
16. Hasinoff, S.W., Kutulakos, K.N.: Confocal stereo. In: Leonardis, A., Bischof, H., Pinz, A. (eds.) ECCV 2006. LNCS, vol. 3951, pp. 620–634. Springer, Heidelberg (2006)
17. Ng, R.: Fourier slice photography. In: SIGGRAPH, pp. 735–744 (2005)
18. Levoy, M., Ng, R., Adams, A., Footer, M., Horowitz, M.: Light field microscopy. In: SIGGRAPH, pp. 924–934 (2006)
19. Debevec, P., Malik, J.: Recovering high dynamic range radiance maps from photographs. In: SIGGRAPH, pp. 369–378 (1997)
20. Agarwala, A., Dontcheva, M., Agrawala, M., Drucker, S., Colburn, A., Curless, B., Salesin, D., Cohen, M.: Interactive digital photomontage. In: SIGGRAPH, pp. 294–302 (2004)
21. Smith, W.J.: Modern Optical Engineering, 3rd edn. McGraw-Hill, New York (2000)
22. Canon lens chart, `http://www.usa.canon.com/app/pdf/lens/`
23. Nocedal, J., Wright, S.J.: Numerical Optimization. Springer, Heidelberg (1999)
24. `http://www.cs.toronto.edu/~kyros/research/lightefficient/`
25. Willson, R., Shafer, S.: What is the center of the image? JOSA A 11(11), 2946–2955 (1994)
26. CHDK, `http://chdk.wikia.com/`

A   Proof of Theorem 1

Theorem 1 follows as a consequence of Lemma 1 and four additional lemmas. We first state Lemmas 2–5 below and then prove a subset of them, along with a proof sketch the theorem. All missing proofs can be found in [24].

Lemma 2 (Efficiency of Sequential DOFs). *For every sequence $\mathcal{S}$, there is a sequence $\mathcal{S}'$ with sequential DOFs that spans the same synthetic DOF and whose total exposure time is no larger.*

Lemma 3 (Permutation of Sequential DOFs). *Given the left endpoint, α, every permutation of $D_1, \ldots, D_n$ defines a capture sequence with sequential DOFs that has the same synthetic DOF and the same total exposure time.*

Lemma 4 (Optimality of Maximizing the Number of Photos). *Among all sequences with up to n tuples whose synthetic DOF is $[\alpha, \beta]$, the sequence that minimizes total exposure time has exactly n of them.*

Lemma 5 (Optimality of Equal-Aperture Sequences). *If $\beta < (7 + 4\sqrt{3})\alpha$, then among all capture sequences with n tuples whose synthetic DOF is $[\alpha, \beta]$, the sequence that minimizes total exposure time uses the same aperture for all tuples. Furthermore, this aperture is equal to*

$$D(n) \;=\; c\,\frac{\sqrt[n]{\beta} + \sqrt[n]{\alpha}}{\sqrt[n]{\beta} - \sqrt[n]{\alpha}} \;. \tag{11}$$

Proof of Lemma 1. We proceed inductively, by defining photo tuples whose DOFs "tile" the interval $[\alpha, \beta]$ from left to right. For the base case, the left endpoint of the first tuple's DOF must be $\alpha_1 = \alpha$. Now consider the i-th tuple. Equation (D) implies that the left endpoint α_i and the aperture diameter D_i determine the DOF's right endpoint uniquely:

$$\beta_i \;=\; \frac{D_i + c}{D_i - c}\,\alpha_i \;. \tag{12}$$

The tuple's focus setting in Eq. (5) now follows by applying Eq. (E) to the interval $[\alpha_i, \beta_i]$. Finally, since the DOFs of tuple i and $i+1$ are sequential, we have $\alpha_{i+1} = \beta_i$.
□

Proof of Lemma 4. From Lemma 2 it follows that among all sequences up to length n whose DOF is $[\alpha, \beta]$, there is a sequence $\mathcal{S}^*$ with minimum total exposure time whose tuples have sequential DOFs. Furthermore, Lemmas 1 and 3 imply that this capture sequence is fully determined by a sequence of n' aperture settings, $D_1 \leq D_2 \leq \cdots \leq D_{n'}$, for some $n' \leq n$. These settings partition the interval $[\alpha, \beta]$ into n' sub-intervals, whose endpoints are given by Eq. (6):

$$\alpha = \alpha_1 < \overbrace{\alpha_2 < \cdots < \alpha_{n'}}^{\text{determined by } \mathcal{S}^*} < \beta_{n'} = \beta \;. \tag{13}$$

It therefore suffices to show that placing $n' - 1$ points in $[\alpha, \beta]$ is most efficient when $n' = n$. To do this, we show that splitting a sub-interval always produces a more efficient capture sequence.

Consider the case $n = 2$, where the sub-interval to be split is actually equal to $[\alpha, \beta]$. Let $x \in [\alpha, \beta]$ be a splitting point. The exposure time for the sub-intervals $[\alpha, x]$ and $[x, \beta]$ can be obtained by combining Eqs. (D) and (1):

$$\tau(x) \;=\; \frac{L}{c^2}\left(\frac{x-\alpha}{x+\alpha}\right)^2 + \frac{L}{c^2}\left(\frac{\beta-x}{\beta+x}\right)^2 , \tag{14}$$

Differentiating Eq. (14) and evaluating it for $x = \alpha$ we obtain

$$\left.\frac{d\tau}{dx}\right|_{x=\alpha} \;=\; -\frac{4L}{c^2}\frac{(\beta-\alpha)\,\beta}{(\beta+\alpha)^3} \;<\; 0 . \tag{15}$$

Similarly, it is possible to show that $\frac{d\tau}{dx}$ is positive for $x = \beta$. Since $\tau(x)$ is continuous in $[\alpha, \beta]$, it follows that the minimum of $\tau(x)$ occurs strictly inside the interval. Hence, splitting the interval always reduces total exposure time. The general case for n intervals follows by induction. $\qquad\square$

Proof Sketch of Theorem 1. We proceed in four steps. First, we consider sequences whose synthetic DOF is equal to $[\alpha, \beta]$. From Lemmas 4 and 5 it follows that the most efficient sequence, $\mathcal{S}'$, among this set has diameter and length given by Eq. (3). Second, we show that sequences with a larger synthetic DOF that are potentially more efficient can have at most one more tuple. Third, we show that the most efficient of these sequences, $\mathcal{S}''$, uses a single diameter equal to $D_{\max}$. Finally, the decision rule in Eq. (4) follows by comparing the total exposure times of $\mathcal{S}'$ and $\mathcal{S}''$.

Flexible Depth of Field Photography[*]

Hajime Nagahara[1,2], Sujit Kuthirummal[2],
Changyin Zhou[2], and Shree K. Nayar[2]

[1] Osaka University
[2] Columbia University

Abstract. The range of scene depths that appear focused in an image is
known as the depth of field (DOF). Conventional cameras are limited by
a fundamental trade-off between depth of field and signal-to-noise ratio
(SNR). For a dark scene, the aperture of the lens must be opened up to
maintain SNR, which causes the DOF to reduce. Also, today's cameras
have DOFs that correspond to a single slab that is perpendicular to the
optical axis. In this paper, we present an imaging system that enables
one to control the DOF in new and powerful ways. Our approach is to
vary the position and/or orientation of the image detector, *during* the
integration time of a single photograph. Even when the detector motion
is very small (tens of microns), a large range of scene depths (several
meters) is captured both in and out of focus.

Our prototype camera uses a micro-actuator to translate the detector along the optical axis during image integration. Using this device,
we demonstrate three applications of flexible DOF. First, we describe
extended DOF, where a large depth range is captured with a very wide
aperture (low noise) but with nearly depth-independent defocus blur. Applying deconvolution to a captured image gives an image with extended
DOF and yet high SNR. Next, we show the capture of images with discontinuous DOFs. For instance, near and far objects can be imaged with
sharpness while objects in between are severely blurred. Finally, we show
that our camera can capture images with tilted DOFs (Scheimpflug imaging) without tilting the image detector. We believe flexible DOF imaging
can open a new creative dimension in photography and lead to new capabilities in scientific imaging, vision, and graphics.

1   Depth of Field

The depth of field (DOF) of an imaging system is the range of scene depths that
appear focused in an image. In virtually all applications of imaging, ranging
from consumer photography to optical microscopy, it is desirable to control the
DOF. Of particular interest is the ability to capture scenes with very large DOFs.
DOF can be increased by making the aperture smaller. However, this reduces the
amount of light received by the detector, resulting in greater image noise (lower

[*] Parts of this work were supported by grants from the National Science Foundation
(IIS-04-12759) and the Office of Naval Research (N00014-08-1-0329 and N00014-06-
1-0032.)

D. Forsyth, P. Torr, and A. Zisserman (Eds.): ECCV 2008, Part IV, LNCS 5305, pp. 60–73, 2008.

SNR). This trade-off gets worse with increase in spatial resolution (decrease in pixel size). As pixels get smaller, DOF decreases since the defocus blur occupies a greater number of pixels. At the same time, each pixel receives less light and hence SNR falls as well. This trade-off between DOF and SNR is one of the fundamental, long-standing limitations of imaging.

In a conventional camera, for any location of the image detector, there is one scene plane – the focal plane – that is perfectly focused. In this paper, we propose varying the position and/or orientation of the image detector *during* the integration time of a photograph. As a result, the focal plane is swept through a volume of the scene causing all points within it to come into and go out of focus, while the detector collects photons.

We demonstrate that such an imaging system enables one to control the DOF in new and powerful ways:

- **Extended Depth of Field.** Consider the case where a detector with a global shutter (all pixels are exposed simultaneously and for the same duration) is moved with *uniform speed* during image integration. Then, each scene point is captured under a continuous range of focus settings, including perfect focus. We analyze the resulting defocus blur kernel and show that it is nearly constant over the range of depths that the focal plane sweeps through during detector motion. Consequently, irrespective of the complexity of the scene, the captured image can be deconvolved with a single, known blur kernel to recover an image with significantly greater DOF. This approach is similar in spirit to Hausler's work in microscopy [1]. He showed that the DOF of an optical microscope can be enhanced by moving a specimen of depth range d, a distance $2d$ along the optical axis of the microscope, while filming the specimen. The defocus of the resulting captured image is similar over the entire depth range of the specimen. However, this approach of moving the scene with respect to the imaging system is practical only in microscopy and not suitable for general scenes. More importantly, Hausler's derivation assumes that defocus blur varies linearly with scene depth which is true only for the small distances involved in microscopy.

- **Discontinuous Depth of Field.** A conventional camera's DOF is a single fronto-parallel slab located around the focal plane. We show that by moving a global-shutter detector *non-uniformly*, we can capture images that are focused for certain specified scene depths, but defocused for in-between scene regions. Consider a scene that includes a person in the foreground, a landscape in the background, and a dirty window in between the two. By focusing the detector on the nearby person for some duration and the far away landscape for the rest of the integration time, we get an image in which both appear fairly well-focused, while the dirty window is blurred out and hence optically erased.

- **Tilted Depth of Field.** Most cameras can only focus on a fronto-parallel plane. An exception is the view camera configuration [2,3], where the image detector is tilted with respect to the lens. When this is done, the focal plane is tilted according to the well-known Scheimpflug condition [4]. We show that by *uniformly* translating an image detector with a rolling electronic shutter (different rows are exposed at different time intervals but for the same duration), we emulate a tilted image detector. As a result, we capture an image with a tilted focal plane. Furthermore, by translating the image detector *non-uniformly* (varying speed), we can emulate a non-planar image detector. This allows us to focus on curved surfaces in the scene.

An important feature of our approach is that the focal plane of the camera can be swept through a large range of scene depths with a very small translation of the image detector. For instance, with a 12.5 mm focal length lens, to sweep the focal plane from a distance of 450 mm from the lens to infinity, the detector has to be translated only about 360 microns. Since a detector only weighs a few milligrams, a variety of micro-actuators (solenoids, piezoelectric stacks, ultrasonic transducers, DC motors) can be used to move it over the required distance within very short integration times (less than a millisecond if required). Note that such micro-actuators are already used in most consumer cameras for focus and aperture control and for lens stabilization. We present several results that demonstrate the flexibility of our system to control DOF in unusual ways. We believe our approach can open up a new creative dimension in photography and lead to new capabilities in scientific imaging, computer vision, and computer graphics.

2   Related Work

A promising approach to extended DOF imaging is wavefront coding, where phase plates placed at the aperture of the lens cause scene objects within a certain depth range to be defocused in the same way [5,6,7]. Thus, by deconvolving the captured image with a single blur kernel, one can obtain an all-focused image. In this case, the effective DOF is determined by the phase plate used and is fixed. On the other hand, in our system, the DOF can be chosen by controlling the motion of the detector. Our approach has greater flexibility as it can even be used to achieve discontinuous or tilted DOFs.

Recently, Levin et al. [8] and Veeraraghavan et al. [9] have used masks at the lens aperture to control the properties of the defocus blur kernel. From a single captured photograph, they aim to estimate the structure of the scene and then use the corresponding depth-dependent blur kernels to deconvolve the image and get an all-focused image. However, they assume simple layered scenes and their depth recovery is not robust. In contrast, our approach is not geared towards depth recovery, but can significantly extend DOF irrespective of scene complexity. Also, the masks used in both these previous works attenuate some of the light entering the lens, while our system operates with a clear and wide aperture. All-focused images can also be computed from an image captured using integral photography [10,11,12]. However, since these cameras make spatio-angular resolution trade-offs to capture 4D lightfields in a single image, the computed images have much lower spatial resolutions when compared to our approach.

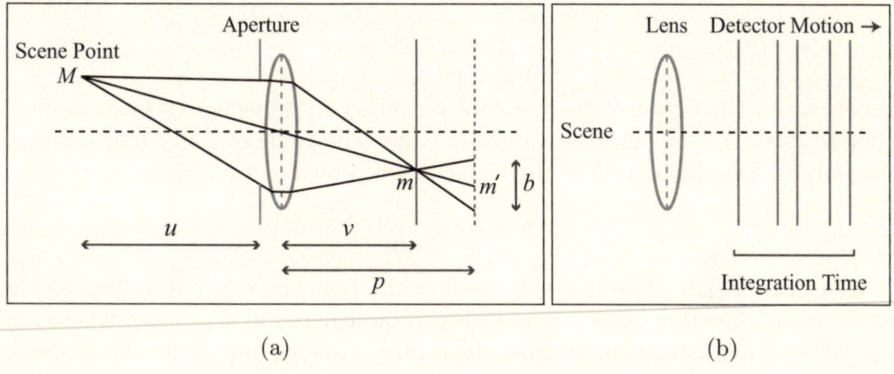

(a)   (b)

Fig. 1. (a) A scene point M, at a distance u from the lens, is imaged in perfect focus by a detector at a distance v from the lens. If the detector is shifted to a distance p from the lens, M is imaged as a blurred circle with diameter b centered around m'. (b) Our flexible DOF camera translates the detector along the optical axis during the integration time of an image. By controlling the starting position, speed, and acceleration of the detector, we can manipulate the DOF in powerful ways.

A related approach is to capture many images to form a focal stack [13,14,15]. An all-in-focus image as well as scene depth can be computed from a focal stack. However, the need to acquire multiple images increases the total capture time making the method suitable for only quasi-static scenes. An alternative is to use very small exposures for the individual images. However, in addition to the practical problems involved in reading out the many images quickly, this approach would result in under-exposed and noisy images that are unsuitable for depth recovery. Our approach does not recover scene depth, but can produce an all-in-focus photograph from a single, well-exposed image.

There is similar parallel work on moving the detector during image integration [16]. However, their focus is on handling motion blur, for which they propose to move the detector *perpendicular* to the optical axis. Some previous works have also varied the orientation or location of the image detector. Krishnan and Ahuja [3] tilt the detector and capture a panoramic image sequence, from which they compute an all-focused panorama and a depth map. For video super-resolution, Ben-Ezra et al. [17] capture a video sequence by instantaneously shifting the detector within the image plane, in between the integration periods of successive video frames.

Recently, it has been shown that a detector with a rolling shutter can be used to estimate the pose and velocity of a fast moving object [18]. We show how such a detector can be used to focus on tilted scene planes.

3   Camera with Programmable Depth of Field

Consider Figure 1(a), where the detector is at a distance v from a lens with focal length f and an aperture of diameter a. A scene point M is imaged in perfect focus at m, if its distance u from the lens satisfies the Gaussian lens law:

$$\frac{1}{f} = \frac{1}{u} + \frac{1}{v}.\tag{1}$$

As shown in the figure, if the detector is shifted to a distance p from the lens (dotted line), M is imaged as a blurred circle (the circle of confusion) centered around m'. The diameter b of this circle is given by

$$b = \frac{a}{v}|(v - p)|.\tag{2}$$

The distribution of light energy within the blur circle is referred to as the point spread function (PSF). The PSF can be denoted as $P(r, u, p)$, where r is the distance of an image point from the center m' of the blur circle. An idealized model for characterizing the PSF is the pillbox function:

$$P(r, u, p) = \frac{4}{\pi b^2}\Pi(\frac{r}{b}),\tag{3}$$

where, $\Pi(x)$ is the rectangle function, which has a value 1, if $|x| < 1/2$ and 0 otherwise. In the presence of optical aberrations, the PSF deviates from the pillbox function and is then often modeled as a Gaussian function:

$$P(r, u, p) = \frac{2}{\pi(gb)^2}\exp(-\frac{2r^2}{(gb)^2}),\tag{4}$$

where g is a constant.

We now analyze the effect of moving the detector during an image's integration time. For simplicity, consider the case where the detector is translated along the optical axis, as in Figure 1(b). Let $p(t)$ denote the detector's distance from the lens as a function of time. Then the aggregate PSF for a scene point at a distance u from the lens, referred to as the *integrated PSF* (IPSF), is given by

$$IP(r, u) = \int_0^T P(r, u, p(t))\ dt,\tag{5}$$

where T is the total integration time. By programming the detector motion $p(t)$– its starting position, speed, and acceleration – we can change the properties of the resulting IPSF. This corresponds to sweeping the focal plane through the scene in different ways. The above analysis only considers the translation of the detector along the optical axis (as implemented in our prototype camera). However, this analysis can be easily extended to more general detector motions, where both its position and orientation are varied during image integration.

Figure 2(a) shows our flexible DOF camera. It consists of a 1/3" Sony CCD (with 1024x768 pixels) mounted on a Physik Instrumente M-111.1DG transla-tion stage. This stage has a DC motor actuator that can translate the detector through a 15 mm range at a top speed of 2.7 mm/sec and can position it with an accuracy of 0.05 microns. The translation direction is along the optical axis of the lens. The CCD shown has a global shutter and was used to implement ex-tended DOF and discontinuous DOF. For realizing tilted DOFs, we used a 1/2.5" Micron CMOS detector (with 2592x1944 pixels) which has a rolling shutter.

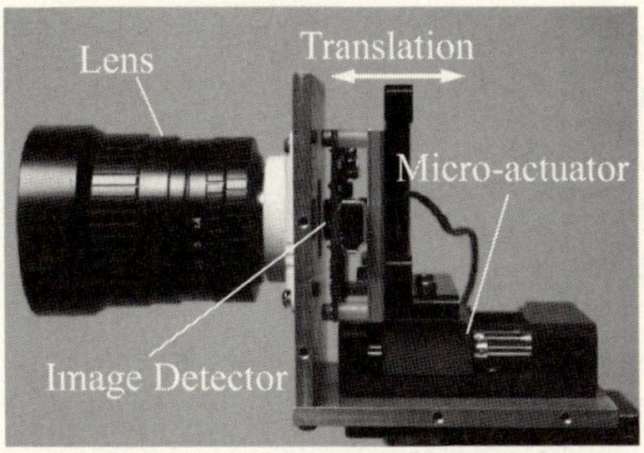

<table>
<tr><td rowspan="3"></td><td>Lens
Focal
Length</td><td>Scene
Depth
Range</td><td>Required
Detector
Translation</td><td>Maximum
Change in
Image Position</td></tr>
</table>

Lens Focal Length	Scene Depth Range	Required Detector Translation	Maximum Change in Image Position
	1m - ∞	81.7 μm	4.5 pixels
9.0mm	.5m - ∞	164.9 μm	5.0 pixels
	.2m - 0.5m	259.1 μm	7.2 pixels
	1m - ∞	158.2 μm	3.6 pixels
12.5mm	.5m - ∞	320.5 μm	5.6 pixels
	.2m - 0.5m	512.8 μm	8.5 pixels

(a) (b)

Fig. 2. (a) Prototype system with flexible DOF. (b) Translation of the detector required for sweeping the focal plane through different scene depth ranges. The maximum change in the image position of a scene point that results from this translation, when a 1024x768 pixel detector is used, is also shown.

The table in Figure 2(b) shows detector translations (third column) required to sweep the focal plane through various depth ranges (second column), using lenses with two different focal lengths (first column). As we can see, the detector has to be moved by very small distances to sweep very large depth ranges. Using commercially available micro-actuators, such translations are easily achieved within typical image integration times (a few milliseconds to a few seconds).

It must be noted that when the detector is translated, the magnification of the imaging system changes. The fourth column of the table in Figure 2(b) lists the maximum change in the image position of a scene point for different translations of a 1024x768 pixel detector. For the detector motions we require, these changes in magnification are very small. This does result in the images not being perspectively correct, but the distortions are imperceptible. More importantly, the IPSFs are not significantly affected by such a magnification change, since a scene point will be in high focus only for a small fraction of this change and will be highly blurred over the rest of it. We verify this in the next section.

4 Extended Depth of Field (EDOF)

In this section, we show that we can capture scenes with EDOF by translating a detector with a global shutter at a constant speed during image integration. We first show that the IPSF for an EDOF camera is nearly invariant to scene depth for all depths swept by the focal plane. As a result, we can deconvolve the captured image with the IPSF to obtain an image with EDOF and high SNR.

4.1 Depth Invariance of IPSF

Consider a detector translating along the optical axis with constant speed s, i.e., $p(t) = p(0) + st$. If we assume that the PSF of the lens can be modeled using the pillbox function in Equation 3, the IPSF in Equation 5 simplifies to

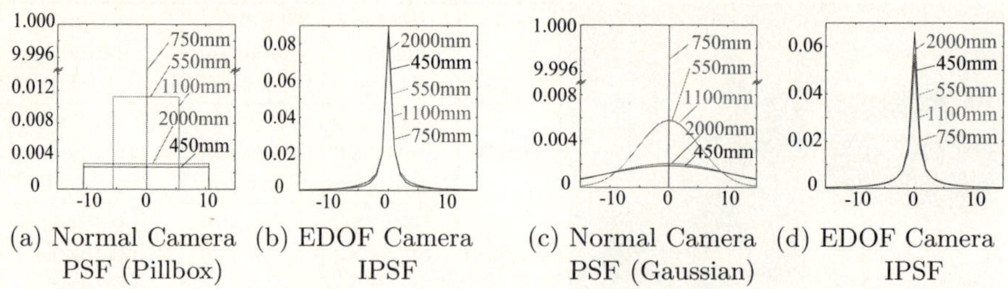

(a) Normal Camera   (b) EDOF Camera   (c) Normal Camera   (d) EDOF Camera
   PSF (Pillbox)   IPSF   PSF (Gaussian)   IPSF

Fig. 3. Simulated (a,c) normal camera PSFs and (b,d) EDOF camera IPSFs, obtained using pillbox and Gaussian lens PSF models for 5 scene depths. Note that the IPSFs are almost invariant to scene depth.

$$IP(r, u) = \frac{uf}{(u - f)\pi asT} \left(\frac{\lambda_0 + \lambda_T}{r} - \frac{2\lambda_0}{b(0)} - \frac{2\lambda_T}{b(T)} \right), \tag{6}$$

where, $b(t)$ is the blur circle diameter at time t, and $\lambda_t = 1$ if $b(t) \geq 2r$ and 0 otherwise. On the other hand, if we use the Gaussian function in Equation 4 for the lens PSF, we get

$$IP(r, u) = \frac{uf}{(u - f)\sqrt{2\pi} rasT} \left(\text{erfc} \left(\frac{r}{\sqrt{2}gb(0)} \right) + \text{erfc} \left(\frac{r}{\sqrt{2}gb(T)} \right) \right). \tag{7}$$

Figures 3(a) and (c) show 1D profiles of a normal camera's PSFs for 5 scene points with depths between 450 and 2000 mm from a lens with focal length $f = 12.5$ mm and $f/\# = 1.4$, computed using Equations 3 and 4 (with $g = 1$), respectively. In this simulation, the normal camera was focused at a distance of 750 mm. Figures 3(b) and (d) show the corresponding IPSFs of an EDOF camera with the same lens, $p(0) = 12.5$ mm, $s = 1$ mm/sec, and $T = 360$ msec, computed using Equations 6 and 7, respectively. As expected, the normal camera's PSF varies dramatically with scene depth. In contrast, the IPSFs of the EDOF camera derived using both pillbox and Gaussian PSF models look almost identical for all 5 scene depths, i.e., *the IPSFs are depth invariant.*

To verify this empirical observation, we measured a normal camera's PSFs and the EDOF camera's IPSFs for several scene depths, by capturing images of small dots placed at different depths. Both cameras have $f = 12.5$ mm, $f/\# = 1.4$, and $T = 360$ msec. The detector motion parameters for the EDOF camera are $p(0) = 12.5$ mm and $s = 1$ mm/sec. The first column of Figure 4 shows the measured PSF at the center pixel of the normal camera for 5 different scene depths; the camera was focused at a distance of 750 mm. (Note that the scale of the plot in the center row is 50 times that of the other plots.) Columns 2-4 of the figure show the IPSFs of the EDOF camera for 5 different scene depths and 3 different image locations. We can see that, while the normal camera's PSFs vary widely with scene depth, the EDOF camera's IPSFs appear almost invariant to both spatial location and scene depth. This also validates our claim that the small magnification changes that arise due to detector motion (discussed in Section 3) do not have a significant impact on the IPSFs.

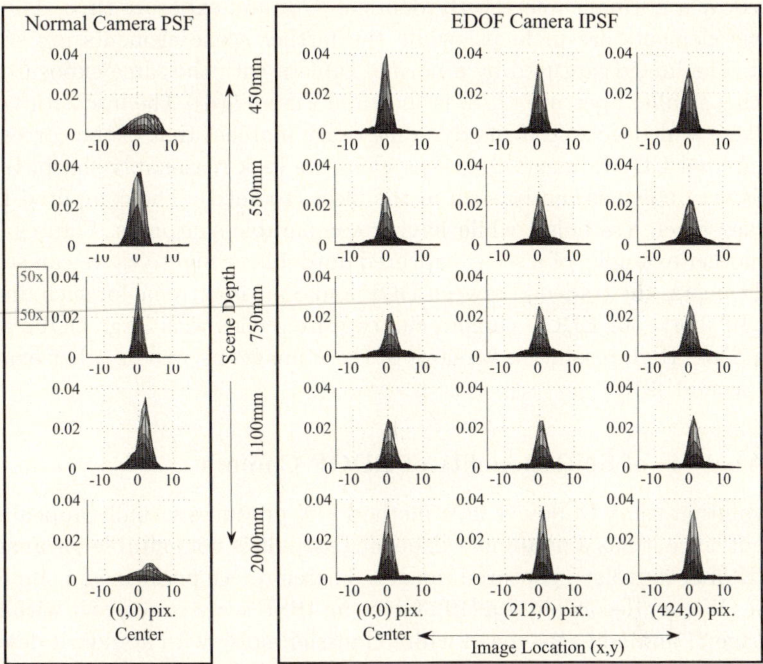

Fig. 4. (Left column) The measured PSF of a normal camera shown for 5 different scene depths. Note that the scale of the plot in the center row is 50 times that of the other plots. (Right columns) The measured IPSF of our EDOF camera shown for different scene depths (vertical axis) and image locations (horizontal axis). The EDOF camera's IPSFs are almost invariant to scene depth and image location.

4.2 Computing EDOF Images Using Deconvolution

Since the EDOF camera's IPSF is invariant to scene depth and image location, we can deconvolve a captured image with a single IPSF to get an image with greater DOF. A number of techniques have been proposed for deconvolution, Richardson-Lucy and Wiener [19] being two popular ones. For our results, we have used the approach of Dabov et al. [20], which combines Wiener deconvolution and block-based denoising. In all our experiments, we used the IPSF shown in the first row and second column of Figure 4 for deconvolution.

Figure 5(a) shows an image captured by our EDOF camera. It was captured with a 12.5 mm Fujinon lens with $f/1.4$ and 0.36 second exposure. Notice that the captured image looks slightly blurry, but high frequencies of all scene elements are captured. This scene spans a depth range of approximately 450 mm to 2000 mm – 10 times larger than the DOF of a normal camera with identical lens settings. Figure 5(b) shows the EDOF image computed from the captured image, in which the entire scene appears focused[1]. Figure 5(c) shows the image

---

[1] Mild ringing artifacts in the computed EDOF images are due to deconvolution.

captured by a normal camera with identical $f/\#$ and exposure time. The nearest scene elements are in focus, while the farther scene elements are severely blurred. The image captured by a normal camera with the same exposure time, but with a smaller aperture ($f/8$) is shown in Figure 5(d). The intensities of this image were scaled up so that its dynamic range matches that of the corresponding computed EDOF image. All scene elements look reasonably sharp, but the image is very noisy as can be seen in the inset (zoomed). The computed EDOF image has much less noise, while having comparable sharpness. Figures 5(e-h) show another example, of a scene captured outdoors at night. As we can see, in a normal camera, the tradeoff between DOF and SNR is extreme for such dimly lit scenes. In short, our EDOF camera can capture scenes with large DOFs as well as high SNR. High resolution versions of these images as well as other examples can be seen at [21].

4.3   Analysis of SNR Benefits of EDOF Camera

Deconvolution using Dabov et al.'s method [20] produces visually appealing results, but since it has a non-linear denoising step, it is not suitable for analyzing the SNR of deconvolved captured images. Therefore, we performed a simulation that uses Wiener deconvolution [19]. Given an IPSF k, we convolve it with a natural image I, and add zero-mean white Gaussian noise with standard deviation σ. The resulting image is then deconvolved with k to get the EDOF image $\hat{I}$. The standard deviation $\hat{\sigma}$ of $(I - \hat{I})$ is a measure of the noise in the deconvolution result when the captured image has noise σ.

The degree to which deconvolution amplifies noise depends on how much the high frequencies are attenuated by the IPSF. This, in turn, depends on the distance through which the detector moves during image integration – as the distance increases, so does the attenuation of high frequencies. This is illustrated in Figure 6(a), which shows (in red) the MTF (magnitude of the Fourier transform) for a simulated IPSF k_1, derived using the pillbox lens PSF model. In this case, we use the same detector translation (and other parameters) as in our EDOF experiments (Section 4.2). The MTF of the IPSF k_2 obtained when the detector translation is halved (keeping the mid-point of the translation the same) is also shown (in blue). As expected, k_2 attenuates the high frequencies less than k_1.

We analyzed the SNR benefits for these two IPSFs for different noise levels in the captured image. The table in Figure 6(b) shows the noise produced by a normal camera for different aperture sizes, given the noise level for the largest aperture, $f/1.4$. (Image brightness is assumed to lie between 0 and 1.) The last two rows show the effective noise levels for EDOF cameras with IPSFs k_1 and k_2, respectively. The last column of the table shows the effective DOFs realized; the normal camera is assumed to be focused at a scene distance that corresponds to the center position of the detector motion. One can see that, as the noise level in the captured image increases, the SNR benefits of EDOF cameras increase. As an example, if the noise of a normal camera at $f/1.4$ is 0.01, then the EDOF camera with IPSF k_1 has the SNR of a normal camera with $f/2.8$, but produces the DOF of a normal camera with $f/8$.

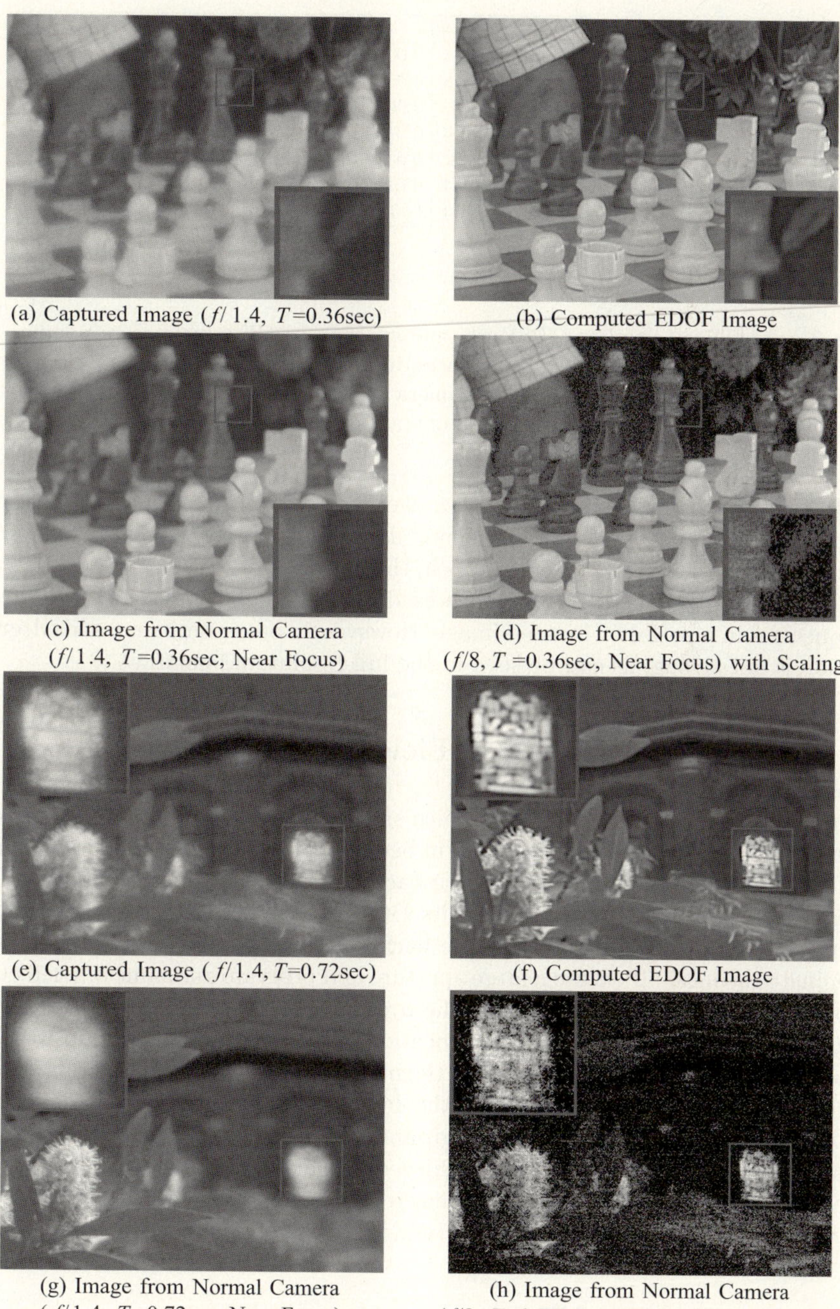

Fig. 5. (a,e) Images captured by the EDOF camera. (b,f) EDOF images computed from images in (a) and (e), respectively. Note that the entire scene appears focused. (c,g) Images captured by a normal camera with identical settings, with the nearest object in focus. (d,h) Images captured by a normal camera at $f/8$.

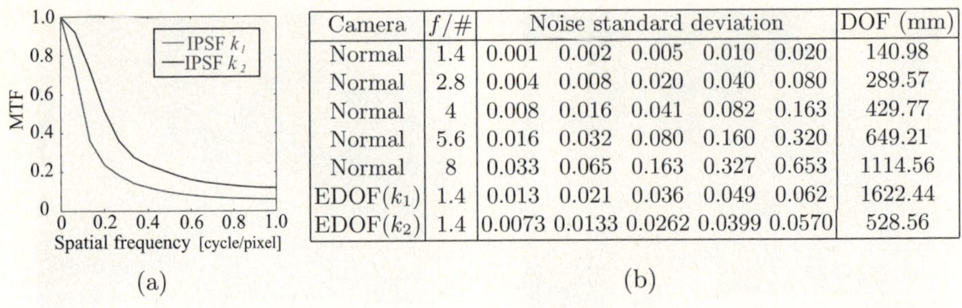

(a)

Camera	$f/\#$	Noise standard deviation					DOF (mm)
Normal	1.4	0.001	0.002	0.005	0.010	0.020	140.98
Normal	2.8	0.004	0.008	0.020	0.040	0.080	289.57
Normal	4	0.008	0.016	0.041	0.082	0.163	429.77
Normal	5.6	0.016	0.032	0.080	0.160	0.320	649.21
Normal	8	0.033	0.065	0.163	0.327	0.653	1114.56
EDOF(k_1)	1.4	0.013	0.021	0.036	0.049	0.062	1622.44
EDOF(k_2)	1.4	0.0073	0.0133	0.0262	0.0399	0.0570	528.56

(b)

Fig. 6. (a) MTFs of simulated IPSFs, k_1 and k_2, of an EDOF camera corresponding to the detector traveling two different distances during image integration. (b) Comparison of effective noise and DOF of a normal camera and a EDOF camera with IPSFs k_1 and k_2. The image noise of a normal camera operating at $f/1.4$ is assumed to be given.

In the above analysis, the SNR was averaged over all frequencies. However, it must be noted that SNR is frequency dependent - SNR is greater for lower frequencies than for higher frequencies in the deconvolved EDOF images. Hence, high frequencies in an EDOF image would be degraded, compared to the high frequencies in a perfectly focused image. However, in our experiments this degradation is not strong, as can be seen in the full resolution images at [21].

5   Discontinuous Depth of Field

Consider the image in Figure 7(a), which shows two toys (cow and hen) in front of a scenic backdrop with a wire mesh in between. A normal camera with a small DOF can capture either the toys or the backdrop in focus, while eliminating the mesh via defocusing. However, since its DOF is a single continuous volume, it cannot capture both the toys and the backdrop in focus and at the same time eliminate the mesh. If we use a large aperture and program our camera's detector motion such that it first focuses on the toys for a part of the integration time, and then moves quickly to another location to focus on the backdrop for the remaining integration time, we obtain the image in Figure 7(b). While this image includes some blurring, it captures the high frequencies in two disconnected DOFs - the foreground and the background - but almost completely eliminates the wire mesh in between. This is achieved without any post-processing. Note that we are not limited to two disconnected DOFs; by pausing the detector at several locations during image integration, more complex DOFs can be realized.

6   Tilted Depth of Field

Normal cameras can focus on only fronto-parallel scene planes. On the other hand, view cameras [2,3] can be made to focus on tilted scene planes by adjusting the orientation of the lens with respect to the detector. We show that our flexible

(a) Image from Normal Camera ($f/11$)   (b) Image from Our Camera ($f/1.4$)

Fig. 7. (a) An image captured by a normal camera with a large DOF. (b) An image captured by our flexible DOF camera, where the toy cow and hen in the foreground and the landscape in the background appear focused, while the wire mesh in between is optically erased via defocusing.

(a) Image from Normal Camera   (b) Image from our Camera
($f/1.4$, T=0.03sec)   ($f/1.4$, T=0.03sec)

Fig. 8. (a) An image captured by a normal camera of a table top inclined at 53° with respect to the lens plane. (b) An image captured by our flexible DOF camera, where the DOF is tilted by 53°. The entire table top (with the newspaper and keys) appears focused. Observe that the top of the mug is defocused, but the bottom appears focused, illustrating that the focal plane is aligned with the table top. Three scene regions of both the images are shown at a higher resolution to highlight the defocus effects.

DOF camera can be programmed to focus on tilted scene planes by simply translating (as in the previous applications) a detector with a rolling electronic shutter. A large fraction of CMOS detectors are of this type – while all pixels have the same integration time, successive rows of pixels are exposed with a slight time lag. When such a detector is translated with uniform speed s, during the frame read out time T of an image, we emulate a tilted image detector. If this tilted detector makes an angle θ with the lens plane, then the focal plane in the scene makes an angle ϕ with the lens plane, where θ and ϕ are related by the well-known Scheimpflug condition [4]:

$$\theta = \tan^{-1}(\frac{sT}{H}) \quad \text{and,} \quad \phi = \tan^{-1}\left(\frac{2f\tan(\theta)}{2p(0) + H\tan(\theta) - 2f}\right). \qquad (8)$$

Here, H is the height of the detector. Therefore, by controlling the speed s of the detector, we can vary the tilt angle of the image detector, and hence the tilt of the focal plane and its associated DOF.

Figure 8 shows a scene where the dominant scene plane – a table top with a newspaper, keys and a mug on it – is inclined at an angle of approximately 53° with the lens plane. As a result, a normal camera is unable to focus on the entire plane, as seen from Figure 8(a). By translating a rolling-shutter detector (1/2.5" CMOS sensor with a 70msec exposure lag between the first and last row of pixels) at 2.7 mm/sec, we emulate a detector tilt of 2.6°. This enables us to achieve the desired DOF tilt of 53° (from Equation 8) and capture the table top (with the newspaper and keys) in focus, as shown in Figure 8(b). Observe that the top of the mug is not in focus, but the bottom appears focused, illustrating the fact that the DOF is tilted to be aligned with the table top. It is interesting to note that, by translating the detector with varying speed, we can emulate non-planar detectors, that can focus on curved scene surfaces.

7   Discussion

In this paper we have proposed a camera with a flexible DOF. DOF is manipulated in various ways by changing the position of the detector during image integration. We have shown how such a system can capture arbitrarily complex scenes with extended DOF and high SNR. We have also shown that we can create DOFs that span multiple disconnected volumes. In addition, we have demonstrated that our camera can focus on tilted scene planes. All of these functionalities are achieved by simply controlling the motion of the detector during the exposure of a single image.

While computing images with extended DOF, we have not explicitly modeled occlusions at depth discontinuities or motion blur caused by object/camera motion. Due to defocus blur, images points that lie close to occlusion boundaries can receive light from scene points at very different depths. However, since the IPSF of the EDOF camera is nearly depth invariant, the aggregate IPSF for such an image point can be expected to be similar to the IPSF of points far from occlusion boundaries. With respect to motion blur, we have not observed any visible artifacts in EDOF images computed for scenes with typical object motion (see Figure 5). However, motion blur due to high-speed objects can be expected to cause problems. In this case, a single pixel sees multiple objects with possibly different depths. It is possible that neither of the objects are imaged in perfect focus during detector translation. This scenario is an interesting one that warrants further study.

In addition to the DOF manipulations shown in this paper, we have (a) captured extended DOF video by moving the detector forward one frame, backward the next, and so on (the IPSF is invariant to the direction of motion), (b) captured scenes with non-planar DOFs, and (c) exploited the camera's focusing mechanism to capture extended DOF by manually rotating a SLR camera lens' focus ring during image integration. For lack of space, we have not included these results here; they can be seen at [21].

References

1. Hausler, G.: A Method to Increase the Depth of Focus by Two Step Image Processing. Optics Communications, 38–42 (1972)
2. Merklinger, H.: Focusing the View Camera (1996)
3. Krishnan, A., Ahuja, N.: Range estimation from focus using a non-frontal imaging camera. IJCV, 169–185 (1996)
4. Scheimpflug, T.: Improved Method and Apparatus for the Systematic Alteration or Distortion of Plane Pictures and Images by Means of Lenses and Mirrors for Photography and for other purposes. GB Patent (1904)
5. Dowski, E.R., Cathey, W.T.: Extended Depth of Field Through Wavefront Coding. Applied Optics, 1859–1866 (1995)
6. George, N., Chi, W.: Extended depth of field using a logarithmic asphere. Journal of Optics A: Pure and Applied Optics, 157–163 (2003)
7. Castro, A., Ojeda-Castaneda, J.: Asymmetric Phase Masks for Extended Depth of Field. Applied Optics, 3474–3479 (2004)
8. Levin, A., Fergus, R., Durand, F., Freeman, B.: Image and depth from a conventional camera with a coded aperture. SIGGRAPH (2007)
9. Veeraraghavan, A., Raskar, R., Agrawal, A., Mohan, A., Tumblin, J.: Dappled photography: mask enhanced cameras for heterodyned light fields and coded aperture. SIGGRAPH (2007)
10. Adelson, E., Wang, J.: Single lens stereo with a plenoptic camera. IEEE Transactions on Pattern Analysis and Machine Intelligence, 99–106 (1992)
11. Ng, R., Levoy, M., Brdif, M., Duval, G., Horowitz, M., Hanrahan, P.: Light field photography with a hand-held plenoptic camera. Technical Report Stanford University (2005)
12. Georgiev, T., Zheng, C., Curless, B., Salesin, D., Nayar, S.K., Intwala, C.: Spatio-angular resolution tradeoff in integral photography. In: Eurographics Symposium on Rendering, pp. 263–272 (2006)
13. Darrell, T., Wohn, K.: Pyramid based depth from focus. CVPR, 504–509 (1988)
14. Nayar, S.K.: Shape from Focus System. CVPR, 302–308 (1992)
15. Subbarao, M., Choi, T.: Accurate Recovery of Three-Dimensional Shape from Image Focus. PAMI, 266–274 (1995)
16. Levin, A., Sand, P., Cho, T.S., Durand, F., Freeman, W.T.: Motion-Invarient Photography. SIGGRAPH, ACM Transaction on Graphics (2008)
17. Ben-Ezra, M., Zomet, A., Nayar, S.: Jitter Camera: High Resolution Video from a Low Resolution Detector. CVPR, 135–142 (2004)
18. Ait-Aider, O., Andreff, N., Lavest, J.M., Martinet, P.: Simultaneous Object Pose and Velocity Computation Using a Single View from a Rolling Shutter Camera. In: Leonardis, A., Bischof, H., Pinz, A. (eds.) ECCV 2006. LNCS, vol. 3952, pp. 56–68. Springer, Heidelberg (2006)
19. Jansson, P.A.: Deconvolution of Images and Spectra. Academic Press, London (1997)
20. Dabov, K., Foi, A., Katkovnik, V., Egiazarian, K.: Image restoration by sparse 3D transform-domain collaborative filtering. SPIE Electronic Imaging (2008)
21. `www.cs.columbia.edu/CAVE/projects/flexible_dof`

Priors for Large Photo Collections and What They Reveal about Cameras

Sujit Kuthirummal[1], Aseem Agarwala[2],
Dan B Goldman[2], and Shree K. Nayar[1]

[1] Columbia University
[2] Adobe Systems, Inc.

Abstract. A large photo collection downloaded from the internet spans a wide range of scenes, cameras, and photographers. In this paper we introduce several novel priors for statistics of such large photo collections that are independent of these factors. We then propose that properties of these factors can be recovered by examining the deviation between these statistical priors and the statistics of a slice of the overall photo collection that holds one factor constant. Specifically, we recover the radiometric properties of a particular camera model by collecting numerous images captured by it, and examining the deviation of this collection's statistics from that of a broader photo collection whose camera-specific effects have been removed. We show that using this approach we can recover both a camera model's non-linear response function and the spatially-varying vignetting of the camera's different lens settings. All this is achieved using publicly available photographs, without requiring images captured under controlled conditions or physical access to the cameras. We also apply this concept to identify bad pixels on the detectors of specific camera instances. We conclude with a discussion of future applications of this general approach to other common computer vision problems.

1 Introduction

Large publicly-available photo collections such as Flickr have recently spawned new applications such as Photo Tourism [1] and Internet Stereo [2]. They have also been exploited for filling in holes in images [3], inserting objects into scenes [4], and object recognition [5]. These research efforts have demonstrated the power of using large photo collections to develop novel applications as well as to solve hard computer vision problems.

In this paper, we examine the statistics of such a large photo collection and develop priors that are independent of the factors that influence any one photograph: the scene, the camera, and the photographer. Statistical priors for single images have already been used for a wide range of computer vision tasks [6,7,8,9,10,11]. We argue that priors on the statistics of photo collections have the potential to be similarly powerful, since the statistics of a slice of the photo collection that holds one factor constant should yield information as to how that factor distorts the priors. We investigate this approach to recover

D. Forsyth, P. Torr, and A. Zisserman (Eds.): ECCV 2008, Part IV, LNCS 5305, pp. 74–87, 2008.

camera properties. We first compute statistical priors from a photo collection with camera-specific effects removed; that is, we use known camera calibration profiles to remove radiometric distortion from a photo collection. As a result, that collection becomes camera-independent. Then, we describe and experimentally validate priors for (a) the spatial distribution of average image luminances and (b) the joint histogram of irradiances at neighboring pixels. Next, we compute these same statistics for a camera-model-specific photo collection whose images have *not* had their distortion removed. We can then recover that camera model's radiometric properties – its non-linear response function and the spatially-varying vignetting for different lens settings – by minimizing the deviation of these statistics from the camera-independent priors. We also show how the same concept can be used to identify bad pixels on the detectors of specific camera instances.

Our approach to recovering properties of specific camera models assumes that all instances of a model have the same properties. This is a reasonable assumption to make for point and shoot cameras [12] since they do not have the variability that arises from attaching different lenses to SLR camera bodies. Hence, in this paper, we restrict ourselves to only point-and-shoot cameras. Also, the camera model properties we recover are aggregate estimates over many instances of the model; for most applications, these estimates are more than adequate. Thus, our approach provides an attractive alternative to traditional camera calibration methods which are typically tedious. Also, since our approach can be used to recover a camera's properties using existing pictures, it provides a convenient means to create a database of camera properties. Such a database would be similar in spirit to the databases available with commercial products like DxO [13] and PTLens [14], but with the important advantage that the cost of creating it would be effectively zero — there would be no need to buy the cameras and manually calibrate them. A photo-sharing website could use our approach to leverage its growing image collection to continually update and add to its database of profiles, and allow users to either undistort their images or make photometrically-correct edits. More importantly, our results demonstrate that the statistics of large photo collections contain significant information about scenes, cameras, and photographers, and our work represents a first step towards extracting and exploiting that information.

2   Related Work

A number of image priors have been proposed to describe the statistics of individual photographs, such as the sparsity of outputs of band-pass filters (e.g. derivative filters) [6,7], biases in the distribution of gradient orientations [8,9], and $1/f$ fall-off of the amplitude spectrum [10,11]. These priors have been exploited for applications such as deriving intrinsic images from image sequences [15], super-resolution and image demosaicing [16], removing effects of camera shake [17], and classifying images as belonging to different scene categories [9,18]. We focus on the aggregate statistics of large photo collections, which tend to have less variability than the statistics of a single image. We thus propose two new priors for aggregate statistics of large photo collections and describe how they can be exploited to recover radiometric properties of cameras.

The most popular method for estimating the camera response function involves taking multiple registered images of a static scene with varying camera exposures [19,20]. Grossberg and Nayar [21] relax the need for spatial correspondences by using histograms of images at different exposures. If the exposure cannot be varied, but can be locked, the response can be estimated by capturing multiple registered images of a static scene illuminated by different combinations of light sources [22]. All these methods require significant user effort and physical access to the camera. Farid [23] assumes that the response function has the form of a gamma curve and estimates it from a single image. However, in practice response functions can differ significantly from gamma curves. Lin et al. [24] also estimate the response from a single image by exploiting intensity statistics at edges. Their results depend on the kinds of edges detected, and their method employs a non-linear optimization which needs multiple initial guesses for robustness. In contrast, we automatically and robustly estimate the response function using numerous existing photographs.

Vignetting can be estimated by imaging a uniformly illuminated flat textureless Lambertian surface, and comparing the intensity of every pixel with that of the center pixel (which is assumed to have no vignetting) [25,26]. Unfortunately, realizing such capture conditions is difficult. One approach is to use a device called an "integrating sphere," but this specialized hardware is expensive. Stumpfel et al. [27] capture many images of a known illuminant at different locations in the image and fit a polynomial to the measured irradiances. The same principle has been used to estimate vignetting from overlapping images of an arbitrary scene [28,29,30] using measured irradiances of the same scene point at different image locations. All these methods require the user to acquire new images under controlled conditions. Some of the above approaches [28,29] can be used to simultaneously estimate the vignetting and the response function of a camera, but there are typically ambiguities in recovering this information. Since we recover both properties independently, we do not have any ambiguities. Recently, Zheng et al. [31] have proposed estimating vignetting from a single image by assuming that a vignette-corrected image will yield an image segmentation with larger segments. Their optimization algorithm, which consists of many alternating image segmentation and vignetting estimation steps, is highly non-linear and hence is likely to have local minima issues. In contrast, we estimate vignetting linearly and efficiently.

During manufacturing, bad pixels are typically identified by exposing image detectors to uniform illuminations. However, some pixels develop defects later and it is difficult for consumers to create uniform environments to detect them. Dudas et al. [32] detect such pixels by analyzing a set of images in a Bayesian framework. However, they only show simulation results. We propose a simple technique that is able to detect bad pixels, albeit using many images.

3   Aggregate Statistics of Photo Collections

We now describe how we collect various internet photo collections and how we use them to form and experimentally validate two statistical priors that are independent of specific scenes, cameras, and photographers. We form image collections by downloading images from Flickr. Flickr supports searching for images from a particular camera model; we chose five popular models and downloaded thousands of images for each. We also manually calibrated these cameras using HDRShop [19] for response functions and an integrating sphere for vignetting (across different lens settings). To validate our approach, we then used the collection of one camera model – Canon S1IS – as a training set to undistort its corresponding downloaded images and form camera-independent priors;[1] the other camera models and their downloaded images were used to test our hypotheses.

Internet photo collections can contain outliers that corrupt our aggregate statistics. For example, images captured with flash, edited in Photoshop, or cropped would add distortion beyond the radiometric properties that we are recovering. Fortunately, EXIF tags allow us to cull most outliers; we remove flash images, images with certain Software fields, portrait-mode images, and images that are not full resolution. Our resultant collections contain about 40,000 images per camera model, which we then group using lens settings since camera properties vary with aperture and focal length. We would like to point out that since Flickr does not support searching for all images with a particular camera-lens setting, there are configurations for which we could not collect enough photographs to compute robust statistics. However, as we will show, for configurations with sufficient photographs, our approach gives uniformly good results.

3.1   Spatial Distribution of Average Image Luminances

Torralba and Oliva [18] and artist Jason Salavon (salavon.com) have made an interesting observation: the average of a set of photographs of similar scenes is not spatially stationary, but has a certain structure to it. So we ask: does the average photograph obtained by pixel-by-pixel averaging of many photographs captured with the same lens setting have a particular structure? To investigate this question we computed the average of the log-luminance of the photographs in the undistorted training set photo collection with the same lens setting. Figures 1 (a,b) show the average log-luminance of two groups of photographs captured with the same focal length, but different f-number. One can see that we have averaged out particular scenes, but the average image is not uniform. This is illustrated in Figures 1 (c) and (d) which show the contrast enhanced versions of the images in Figures 1 (a) and (b), respectively. We can immediately make two interesting observations. (i) The average images have a vertical gradient as can also be seen in Figure 1 (e) which shows log-luminances along a column of the

---

[1] We assume that undistortion is enough to make an image collection camera-independent for the purpose of training priors. While this may not be true in all cases, we have experimentally verified that our priors are accurate across all five camera models (from four different manufacturers) that we used in our experiments.

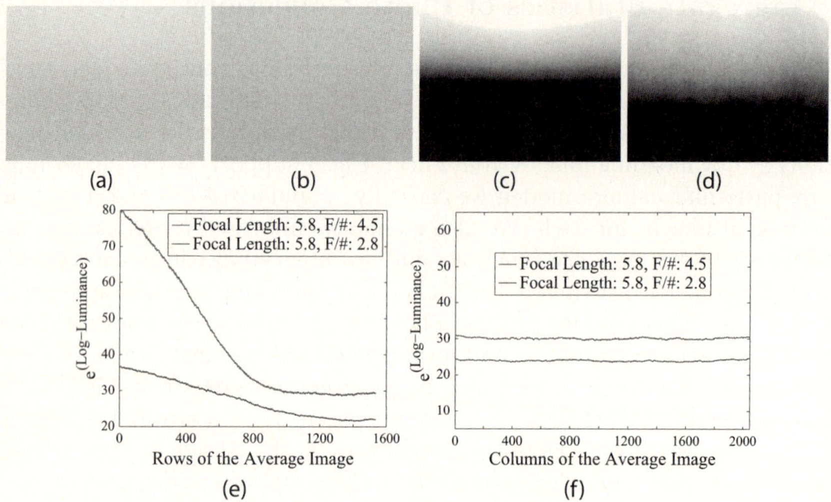

Fig. 1. (a) Average log-luminance of 15,550 photographs captured by Canon S1IS cameras with focal length 5.8 mm and f-number 4.5. The photographs, all 2048×1536, were linearized and vignette corrected before averaging. (b) Average log-luminance of 13,874 photographs captured by Canon S1IS cameras with focal length 5.8 mm and f-number 2.8. (c,d) Contrast-enhanced versions (for illustration only) of the images in (a) and (b), respectively. (e,f) Plots of the average log-luminances of respectively the 1000th column and 1000th row for the two settings in (a) and (b). Response functions were normalized so that luminance values were in the range (0,255) prior to averaging.

average images. This is possibly because illumination sources are typically above – outdoors, from the sun and sky, while indoors, from ceiling-mounted light fixtures. (ii) The average images do not have a horizontal gradient, illustrated by Figure 1 (f) which shows log-luminances along a row. We have found that these two observations are general and they hold true for all camera models and lens settings. In summary, in the absence of vignetting, average log-luminance images have a vertical gradient, but no horizontal gradient. This observation serves as the prior, which we exploit to recover vignetting in Section 4.2.

3.2 Joint Histogram of Irradiances at Neighboring Pixels

A prior on the distribution of gradients in a single image is commonly used in computer vision estimation tasks [16,17]. However, the larger data set of a photo collection allows us to measure how this gradient distribution varies as a function of irradiance values. Therefore, we compute the joint histogram of irradiances at neighboring pixels (where neighborhood is defined as 4-connected). Note that we characterize the joint histogram only for a small block of pixels, since we know from Section 3.1 that this statistic would also vary spatially.

We now describe how we compute the joint histogram of irradiances for a color channel of a camera model. We assume that we know the inverse response

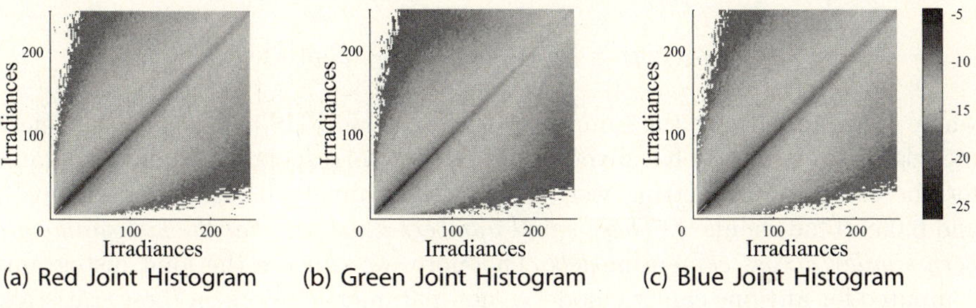

Fig. 2. Log of the joint histograms of (a) red, (b) green, and (c) blue irradiances computed from 15,550 photographs captured by Canon S1IS cameras with the extreme lens setting – smallest focal length (5.8 mm) and largest f-number (4.5). The inverse camera response functions used were normalized so that irradiance values were in the range (0,255). When computing the histograms we ignored irradiances less than 5 and greater than 250 to avoid the effects of under-exposure and saturation, respectively.

function, R, for that channel, where $R(i)$ is the irradiance value corresponding to intensity i. Using R we linearize that channel in photographs from that model and compute a joint histogram, JH, where $JH(i, j)$, gives the number of times irradiances $R(i)$ and $R(j)$ occur in neighboring pixels in a desired pixel block. We interpret the joint histogram as the joint probability distribution of irradiances by assuming that the distribution is piecewise uniform within each bin. However, since the values of R are typically non-uniformly spaced, the bins have different areas. Therefore, to convert the joint histogram to a probability distribution, we divide the value of each bin by its area. Note that the values of R determine the sampling lattice, so to enable comparisons between joint histograms for different response functions we resample the histogram on a regular grid in irradiance space. Finally, we normalize the resampled distribution so that it sums to one.

We computed joint histograms of red, green, and blue irradiances for several camera models using 31×31 pixel blocks at the center of photographs. Figure 3 shows the joint histograms for the Canon S1IS camera model computed from photographs with the smallest focal length and largest f-number. These histograms show that the probability of any two irradiances being incident on neighboring pixels varies depending on the values of the irradiances. Also, the probability of the same irradiance occuring at neighboring pixels is greater for low irradiance values and decreases slowly as the irradiance value increases. Finally, note that the histograms for different color channels differ slightly, illustrating that the visual world has different distributions for different colors. We have empirically observed that for any particular color channel, the joint histogram looks very similar across camera models, especially when computed for the extreme lens setting – smallest focal length and largest f-number. This is not surprising, because the extreme setting is chosen by different camera models for similar types of scenes. We quantified this similarity using the symmetric Kullback-Leibler (KL) divergence between corresponding histograms. The symmetric KL divergence between distributions p and q is defined as

$$KLDiv_{Sym}(p, q) = \Sigma_i q(i) \log(\frac{q(i)}{p(i)}) + \Sigma_i p(i) \log(\frac{p(i)}{q(i)}), \tag{1}$$

where $p(i)$ and $q(i)$ are the samples. For the Canon S1IS and Sony W1 camera models, the symmetric KL divergence between corresponding joint histograms for the extreme lens setting were 0.059 (red channel), 0.081 (green channel), and 0.068 (blue channel). *These small numbers illustrate that the histograms are very similar across camera models.* Therefore, we can use the joint histograms computed for any one camera model as non-parametric priors on these statistics.

4   Using the Priors for Radiometric Calibration

In this section we use these camera-independent statistical priors to recover the response function of a camera model, the vignetting of a camera model for different lens settings, and the bad pixels on the detector of a specific camera. We use the same basic approach for all three applications; given a photo collection, we estimate camera properties that minimize the difference between the statistics of the photo collection and the priors defined in the previous section.

4.1   Estimating Camera Response Function

We estimate a camera model response function by minimizing the difference between the joint histogram of irradiances (Section 3.2) for the camera model and the camera-independent prior joint histogram. To estimate the response for a color channel of a camera model, we first compute the joint histogram, JH, of intensities in a 31×31 pixel block at the center of photographs from a collection with the smallest focal length and largest f-number. Say R is an estimate of the inverse response function. Since R is a one-to-one mapping from image intensities to irradiances, JH can be used to compute the joint histogram of irradiances, as described in Section 3.2. We can then determine the 'goodness' of the estimate R by computing the symmetric KL Divergence (Equation 1) between this histogram and the prior histogram for that color channel. Therefore, we can estimate the response function using an optimization over R that minimizes this divergence. We use a simple polynomial [20] as the parametric representation of R, and optimize over its coefficients. We define $R(i) = 255 * \sum_{k=1}^{N} \alpha_k (\frac{i}{255})^k$, where $R(i)$ is the irradiance corresponding to intensity i, α_k are the coefficients, and N is the degree of the polynomial. We normalize $R(.)$ such that $R(255) = 255$. We have used $N = 5$ in our experiments, since we found it to be a good fit for all inverse response functions in our data set; the mean RMS fitting error was 0.41%. We use the Nelder-Mead Simplex method [33] for the optimization. Note that the joint histogram of image intensities has to be computed only once, though a resampling and normalization step must be performed at each iteration.

We used the priors obtained from the Canon S1IS model to estimate the inverse response functions of Sony W1, Canon G5, Casio Z120, and Minolta Z2

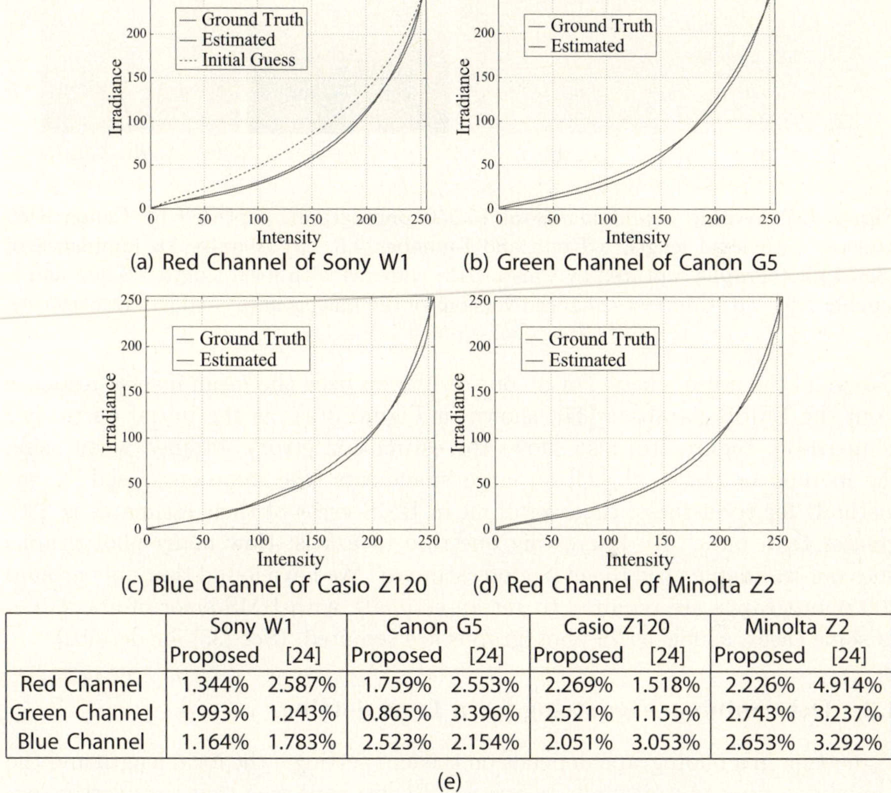

	Sony W1		Canon G5		Casio Z120		Minolta Z2	
	Proposed	[24]	Proposed	[24]	Proposed	[24]	Proposed	[24]
Red Channel	1.344%	2.587%	1.759%	2.553%	2.269%	1.518%	2.226%	4.914%
Green Channel	1.993%	1.243%	0.865%	3.396%	2.521%	1.155%	2.743%	3.237%
Blue Channel	1.164%	1.783%	2.523%	2.154%	2.051%	3.053%	2.653%	3.292%

(e)

Fig. 3. Estimated and ground truth inverse response functions of one channel for four camera models – (a) Sony W1, (b) Canon G5, (c) Casio Z120, and (d) Minolta Z2. For these estimates we used 17,819, 9,529, 1,315, and 3,600 photographs, respectively. (a) also shows the initial guess used by our optimization. (e) RMS percentage errors of the estimated inverse response functions for camera models from four different manufacturers obtained using our proposed method and the method of [24].

camera models. Due to space constraints, we only show the inverse responses of one of their channels in Figures 3(a-d). For comparison we also show the ground truth inverse response functions obtained using HDRShop [19][2]. As we can see, the estimated curves are very close to the ground truth curves. The difference between the two sets of curves is greater at higher image intensities, for which HDRShop typically provides very noisy estimates.

The RMS estimation errors are shown in Figure 3(e). Even though our estimation process uses a non-linear optimization, we have found it to be robust to

---

[2] Inverse response functions can only be estimated up to scale. To compare the inverse responses produced by our technique and HDRShop, we scaled the results from HDRShop by a factor that minimizes the RMS error between the two curves.

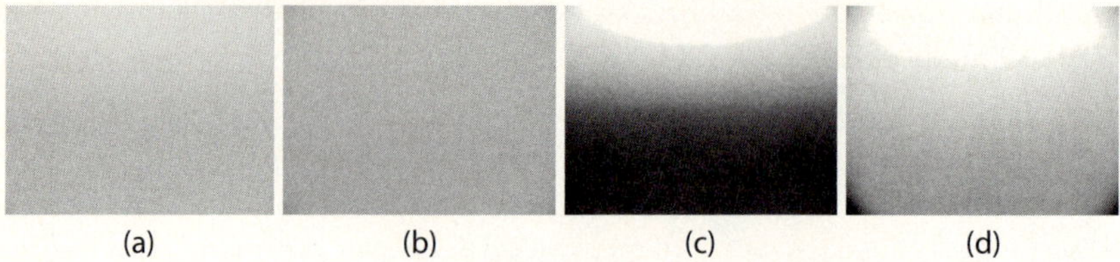

Fig. 4. (a) Average log-luminance of 15,550 photographs captured by Canon S1IS cameras with focal length 5.8 mm and f-number 4.5. (b) Average log-luminance of 13,874 photographs captured by Canon S1IS cameras with focal length 5.8 mm and f-number 2.8. (c,d) Contrast enhanced versions of the images in (a) and (b), respectively.

choices of the initial guess. For all our results we used the mean inverse response from the EMoR database [34], shown in Figure 3(a), as the initial guess. For comparison, Figure 3(e) also shows the estimation errors obtained when using the method of Lin et al. [24] on large image sets (the same ones used by our method) for robustness; the overall mean RMS error of their estimates is 28% greater than ours. An interesting question to ask is: How many photographs does our technique need to get a good estimate? We have found that only around 200 photographs are required to get an estimate with RMS error of about 2%. In some cases, as few as 25 photographs are required. (See [35] for details.)

4.2   Determining Vignetting for a Lens Setting

Vignetting in a photograph depends on the lens setting – the focal length and the f-number – used to capture it. In Section 3.1, we have seen that the average log-luminance of a group of linearized and vignette-corrected photographs captured with the same lens setting has a vertical gradient but no horizontal gradient. Using the technique in Section 4.1, we can recover response functions, linearize photographs and compute average log-luminance images. Figures 4 (a, b) show the average log-luminances for two groups of linearized photographs captured by Canon S1IS cameras with the same focal length, but different f-number. The photographs used were not vignette-corrected. The contrast-enhanced versions of these images are shown in Figures 4 (c) and (d), respectively. Note the darkening of the corners, which suggests that vignetting information is embedded in the average images. The average images now have a horizontal gradient in addition to a vertical gradient. This observation coupled with our prior model (Section 3.1) leads to a simple vignetting estimation algorithm: find a vignetting function that yields a corrected average log-luminance image with no horizontal gradient.

Since vignetting affects all color channels equally, we only need to analyze its effect on luminance. The measured luminance m at pixel (x, y) in photograph i can be written as:

$$m_i(x, y) = v(x, y) * l_i(x, y), \tag{2}$$

where $v(x, y)$ is the vignetting at that pixel and $l_i(x, y)$ is the luminance that would have been measured in the absence of vignetting. Taking the log on both sides of Equation 2 and computing the average log-luminance in N photographs with the same lens setting, we get

$$\frac{1}{N}\Sigma_i \log(m_i(x,y)) = \log(v(x,y)) + \frac{1}{N}\Sigma_i \log(l_i(x,y)). \tag{3}$$

Writing the measured average log-luminance, $\frac{1}{N}\Sigma_i \log(m_i(x,y))$, as $M(x,y)$, $\log(v(x,y))$ as $V(x,y)$, and the average log-luminance in the absence of vignetting, $\frac{1}{N}\Sigma_i \log(l_i(x,y))$, as $L(x,y)$, Equation 3 becomes

$$M(x,y) = V(x,y) + L(x,y). \tag{4}$$

According to our prior model, in the absence of vignetting an average log-luminance image does not have a horizontal gradient, i.e., all values in a row are equal. This implies that Equation 4 can be rewritten as

$$M(x,y) = V(x,y) + L(y). \tag{5}$$

Note that M is known, while V and L are unknown. We assume that vignetting is radially symmetric about the center of the image. Therefore, vignetting at pixel (x,y) can be expressed as a function of the distance, r, of the pixel from the image center. We model the log of the vignetting as a polynomial in r: $V(x,y) = \sum_{k=1}^{N} \beta_k r^k$, where β_k are the coefficients and N is the degree of the polynomial. In our experiments we have used $N = 9$. Note that the value of V is zero at the center of the image, modeling the fact that there is no vignetting there. This model reduces Equation 5 to a set of linear equations in the unknowns $L(y)$ and the vignetting coefficients β_k, which we can solve for efficiently.

The average log-luminance images in Figures 4 (a) and (b) can be used to estimate vignetting. However, we have observed that the top halves of photographs contain many saturated pixels, especially photographs taken with small focal lengths and large f-numbers (typically used for outdoor scenes with lots of light). For instance, photographs in our data set captured by Canon S1IS cameras with such a setting had pixels in the top half that were saturated approximately 30% of the time. This means that we significantly underestimate the average value for pixels in the top half. Since statistics of the top half of the average images are unreliable, we have used the bottom half to recover vignetting. Figures 5 (a-f) show the estimated vignetting curves obtained using this approach for two lens settings each of three camera models – Canon S1IS, Sony W1, and Canon G5. For comparison, ground truth vignetting curves obtained from photographs captured in an integrating sphere are also shown. As one can see, the estimated vignetting and ground truth curves are very close to each other. Figure 5(g) shows the RMS and mean estimation errors. We have found that our technique needs around 3000 photographs to get an estimate with RMS error of about 2%. (See [35] for details.)

We have observed that statistics at the center of photographs differ slightly from those of other portions of the image. We believe that this is due to a compositional bias – faces are usually captured in the center region. This deviation

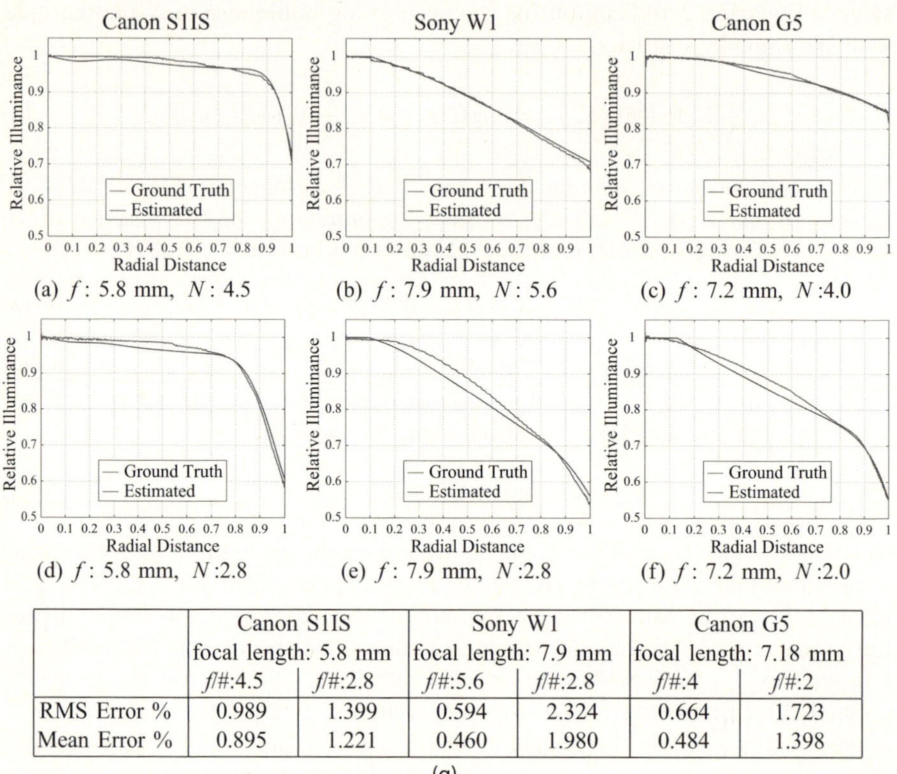

	Canon S1IS focal length: 5.8 mm		Sony W1 focal length: 7.9 mm		Canon G5 focal length: 7.18 mm	
	$f/\#$:4.5	$f/\#$:2.8	$f/\#$:5.6	$f/\#$:2.8	$f/\#$:4	$f/\#$:2
RMS Error %	0.989	1.399	0.594	2.324	0.664	1.723
Mean Error %	0.895	1.221	0.460	1.980	0.484	1.398

(g)

Fig. 5. (a-f) Vignetting estimated for two lens settings each of Canon S1IS, Sony W1, and Canon G5 cameras, using the bottom half of their respective average log-luminance images. 15,550, 13,874, 17,819, 15,434, 12,153, and 6,324 photographs, respectively were used for these estimates. (f and N stand for focal length and f-number respectively.) (g) RMS and mean percentage errors of the estimated vignetting for two lens settings each of three camera models; estimation errors are typically less than 2%.

in statistics sometimes causes relative illuminance near the image center to be incorrectly estimated as greater than one. We have handled this by clamping the curves to have a maximum value of one. Note that for lens settings with smaller f-numbers, the estimation is slightly poorer for a larger region near the image center. Such a setting is usually chosen for indoor scenes, where people are typically closer to the camera and their faces occupy a larger region near the image center, thus accentuating this compositional bias.

It is interesting to note from Figure 5 that for these camera models, at lens settings with small f-numbers (large apertures), the corners of the photograph get about 40% less light than the center! This large difference becomes very noticeable if overlapping photographs are stitched together without vignette correction. If photographs are corrected for vignetting, then the overlap seams become barely visible as was shown by [28,29].

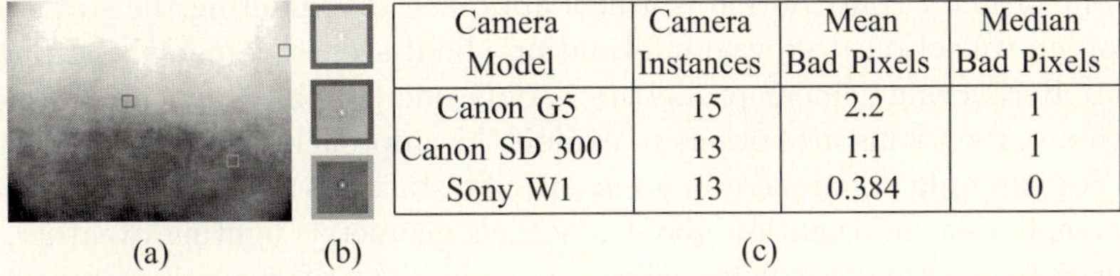

Camera Model	Camera Instances	Mean Bad Pixels	Median Bad Pixels
Canon G5	15	2.2	1
Canon SD 300	13	1.1	1
Sony W1	13	0.384	0

(a)   (b)   (c)

Fig. 6. (a) Contrast enhanced luminance of the average of 1,186 photographs from a particular Canon S1IS camera. (b) Zoomed in portions of the image in (a) in which we can clearly see bad pixels that have very different intensities from their neighbors. (c) A comparative study of the number of bad detector pixels in a particular camera instance for three different camera models.

4.3   Identifying Bad Pixels on a Camera Detector

During manufacturing, camera detectors are exposed to uniform illuminations so that bad pixels – pixels with abnormal sensitivities and biases – stand out and can be easily identified. However, some pixels develop defects later and it is difficult for consumers to create uniform environments to detect them. In Section 3.1 we saw that by averaging a large number of photographs, we average out particular scenes and noise to get a smoothly varying image. *Thus, a simple prior for bad pixel detection is that the average image should be smooth;* bad pixels should be identifiable as causing discontinuities in the average image.

We grouped photographs by the Flickr users who uploaded them, so that each group has pictures from the same camera instance. We then computed the average of each group. Figure 6 (a) shows the contrast enhanced luminance of the average of 1,186 photographs from a particular Canon S1IS camera. In this image, bad pixels clearly stand out, as can be seen in the zoomed-in portions shown in Figure 6 (b). We identify a pixel as bad if the difference between its average value and the median of the average values in a neighborhood around it is greater than a threshold (7 gray-levels). This technique can also be used to rank camera models by the number of bad pixels in each instance. The table in Figure 6(c) presents results from such a study, for which we picked camera instances which had at least 500 photographs in our collection.

5   Conclusion

In this paper, we have presented priors on two aggregate statistics of large photo collections, and exploited these statistics to recover the radiometric properties of camera models entirely from publicly available photographs, without physical access to the cameras themselves. In future work, we would like to develop statistics that reveal other camera properties such as radial distortion, chromatic aberration, spatially varying lens softness, etc.. There are, of course, a number of powerful and accurate approaches to camera calibration, and these existing techniques have both advantages and disadvantages relative to ours. In that light, our primary contribution is a new approach to exploiting the statistics of large photo collections to reveal information about scenes, cameras, and photographers. Recovering camera properties is only one possible application, and we hope that our work inspires others to exploit this approach in new and interesting ways. For example, differences in scene-specific statistics and scene-independent priors could yield information about a scene's geometry, lighting, weather, and motion. A photographer's photo collection could yield information on propensity for camera shake, typical field of view, and preferred camera orientation.

Statistical priors for single images have been useful for a number of computer vision tasks [16,17]. We argue that priors on the statistics of photo collections have the potential to be similarly powerful, since the deviation from these priors of a slice of the photo collection that holds one factor constant should reveal information about that factor. Computer vision problems that operate on a single image are often ill-posed because they must tease apart the influence of several confounding factors of the scene, the camera, and the photographer. For example, vignetting calibration is challenging because it is hard to know if darkening is caused by vignetting or changes in the scene. In effect, a photo collection allows us to *marginalize* over the factors that confound the task at hand. We believe that our work is only the first step in this exciting direction.

Acknowledgements. Thanks to William Freeman, Sylvain Paris, Anat Levin, Antonio Torralba, and Brian Curless for helpful discussions and comments.

References

1. Snavely, N., Seitz, S.M., Szeliski, R.: Photo tourism: Exploring photo collections in 3D. ACM Transactions on Graphics (SIGGRAPH), 835–846 (2006)
2. Goesele, M., Snavely, N., Curless, B., Hoppe, H., Seitz, S.M.: Multi-View Stereo for Community Photo Collections. In: ICCV (2007)
3. Hays, J., Efros, A.A.: Scene Completion Using Millions of Photographs. ACM Transactions on Graphics (SIGGRAPH) (2007)
4. Lalonde, J.-F., Hoiem, D., Efros, A.A., Rother, C., Winn, J., Criminisi, A.: Photo Clip Art. ACM Transactions on Graphics (SIGGRAPH) (2007)
5. Torralba, A., Fergus, R., Freeman, W.: Tiny Images. MIT Tech Report (2007)
6. Olshausen, B.A., Field, D.J.: Emergence of simple-cell receptive field properties by learning a sparse code for nature images. In: Nature, pp. 607–609 (1996)
7. Simoncelli, E.: Statistical Models for Images: Compression, Restoration and Synthesis. In: Asilomar Conference on Signals, Systems and Computers, pp. 673–678 (1997)
8. Switkes, E., Mayer, M.J., Sloan, J.A.: Spatial frequency analysis of the visual environment: anisotropy and the carpentered environment hypothesis. Vision Research, 1393–1399 (1978)
9. Baddeley, R.: The Correlational Structure of Natural Images and the Calibration of Spatial Representations. Cognitive Science, 351–372 (1997)
10. Burton, G.J., Moorhead, I.R.: Color and spatial structure in natural scenes. Applied Optics, 157–170 (1987)

11. Field, D.: Relations between the statistics of natural images and the response properties of cortical cells. J. of the Optical Society of America, 2379–2394 (1987)
12. Wackrow, R., Chandler, J.H., Bryan, P.: Geometric consistency and stability of consumer-grade digital cameras for accurate spatial measurement. The Photogrammetric Record, 121–134 (2007)
13. DxO Labs: www.dxo.com
14. PTLens: www.epaperpress.com/ptlens
15. Weiss, Y.: Deriving intrinsic images from image sequences. In: ICCV, pp. 68–75 (2001)
16. Tappen, M.F., Russell, B.C., Freeman, W.T.: Exploiting the sparse derivative prior for super-resolution and image demosaicing. In: Workshop on Statistical and Computational Theories of Vision (2003)
17. Fergus, R., Singh, B., Hertzmann, A., Roweis, S.T., Freeman, W.T.: Removing Camera Shake From A Single Photograph. SIGGRAPH, 787–794 (2006)
18. Torralba, A., Oliva, A.: Statistics of Natural Images Categories. Network: Computation in Neural Systems 14, 391–412 (2003)
19. Debevec, P.E., Malik, J.: Recovering high dynamic range radiance maps from photographs. SIGGRAPH, 369–378 (1997)
20. Mitsunaga, T., Nayar, S.K.: Radiometric self calibration. CVPR, 1374–1380 (1999)
21. Grossberg, M.D., Nayar, S.K.: Determining the Camera Response from Images: What is Knowable?. PAMI, 1455–1467 (2003)
22. Manders, C., Aimone, C., Mann, S.: Camera response function recovery from different illuminations of identical subject matter. ICIP, 2965–2968 (2004)
23. Farid, H.: Blind Inverse Gamma Correction. IEEE Transactions on Image Processing, 1428–1433 (2001)
24. Lin, S., Gu, J., Yamazaki, S., Shum, H.-Y.: Radiometric Calibration Using a Single Image. CVPR, 938–945 (2004)
25. Sawchuk, A.: Real-time correction of intensity nonlinearities in imaging systems. IEEE Transactions on Computers, 34–39 (1977)
26. Kang, S.B., Weiss, R.: Can we calibrate a camera using an image of a flat textureless lambertian surface? In: Vernon, D. (ed.) ECCV 2000. LNCS, vol. 1843, pp. 640–653. Springer, Heidelberg (2000)
27. Stumpfel, J., Jones, A., Wenger, A., Debevec, P.: Direct HDR capture of the sun and sky. Afrigraph, 145–149 (2004)
28. Goldman, D.B., Chen, J.H.: Vignette and exposure calibration and compensation. In: ICCV, pp. 899–906 (2005)
29. Litvinov, A., Schechner, Y.Y.: Addressing radiometric nonidealities: A unified framework. CVPR, 52–59 (2005)
30. Jia, J., Tang, C.K.: Tensor voting for image correction by global and local intensity alignment. IEEE Transactions PAMI 27(1), 36–50 (2005)
31. Zheng, Y., Lin, S., Kang, S.B.: Single-Image Vignetting Correction. CVPR (2006)
32. Dudas, J., Jung, C., Wu, L., Chapman, G.H., Koren, I., Koren, Z.: On-Line Mapping of In-Field Defects in Image Sensor Arrays. In: International Symposium on Defect and Fault-Tolerance in VLSI Systems, pp. 439–447 (2006)
33. Press, W., Teukolsky, S., Vetterling, W., Flannery, B.: Numerical Recipes in C: The Art of Scientific Computing (1992)
34. Grossberg, M.D., Nayar, S.K.: What is the Space of Camera Response Functions?. CVPR, 602–609 (2003)
35. http://www.cs.columbia.edu/CAVE/projects/photo_priors/

Understanding Camera Trade-Offs
through a Bayesian Analysis of Light Field Projections

Anat Levin[1], William T. Freeman[1,2], and Frédo Durand[1]

[1] MIT CSAIL
[2] Adobe Systems

Abstract. Computer vision has traditionally focused on extracting structure, such as depth, from images acquired using thin-lens or pinhole optics. The development of computational imaging is broadening this scope; a variety of unconventional cameras do not directly capture a traditional image anymore, but instead require the joint reconstruction of structure and image information. For example, recent coded aperture designs have been optimized to facilitate the joint reconstruction of depth and intensity. The breadth of imaging designs requires new tools to understand the tradeoffs implied by different strategies.

This paper introduces a unified framework for analyzing computational imaging approaches. Each sensor element is modeled as an inner product over the 4D light field. The imaging task is then posed as Bayesian inference: given the observed noisy light field projections and a prior on light field signals, estimate the original light field. Under common imaging conditions, we compare the performance of various camera designs using 2D light field simulations. This framework allows us to better understand the tradeoffs of each camera type and analyze their limitations.

1 Introduction

The flexibility of computational imaging has led to a range of unconventional camera designs. Cameras with coded apertures [1,2], plenoptic cameras [3,4], phase plates [5,6], and multi-view systems [7] record different combinations of light rays. Reconstruction algorithms then convert the data to viewable images, estimate depth and other quantites. These cameras involves tradeoffs among various quantites–spatial and depth resolution, depth of focus or noise. This paper describes a theoretical framework that will help to compare computational camera designs and understand their tradeoffs.

Computation is changing imaging in three ways. First, the information recorded at the sensor may not be the final image, and the need for a decoding algorithm must be taken into account to assess camera quality. Second, beyond 2D images, the new designs enable the extraction of 4D light fields and depth information. Finally, new *priors* can capture regularities of natural scenes to complement the sensor measurements and amplify decoding algorithms. The traditional evaluation tools based on the image point spread function (PSF) [8,9] are not able to fully model these effects. We seek tools for comparing camera designs, taking into account those three aspects. We want to evaluate the ability to recover a 2D image as well as depth or other information and we want to model the decoding step and use natural-scene priors.

D. Forsyth, P. Torr, and A. Zisserman (Eds.): ECCV 2008, Part IV, LNCS 5305, pp. 88–101, 2008.
© Springer-Verlag Berlin Heidelberg 2008

A useful common denominator, across camera designs and scene information, is the lightfield [7], which encodes the atomic entities (lightrays) reaching the camera. Light fields naturally capture some of the more common photography goals such as high spatial image resolution, and are tightly coupled with the targets of mid-level computer vision: surface depth, texture, and illumination information. Therefore, we cast the reconstruction performed in computational imaging as light field inference. We then need to extend prior models, traditionally studied for 2D images, to 4D light fields.

Camera sensors sum over sets of light rays, with the optics specifying the mapping between rays and sensor elements. Thus, a camera provides a linear projection of the 4D light field where each projected coordinate corresponds to the measurement of one pixel. The goal of decoding is to infer from such projections as much information as possible about the 4D light field. Since the number of sensor elements is significantly smaller than the dimensionality of the light field signal, prior knowledge about light fields is essential. We analyze the limitations of traditional signal processing assumptions [10,11,12] and suggest a new prior on light field signals which explicitly accounts for their structure. We then define a new metric of camera performance as follows: Given a light field prior, how well can the light field be reconstructed from the data measured by the camera? The number of sensor elements is of course a critical variable, and we chose to standardize our comparisons by imposing a fixed budget of N sensor elements to all cameras.

We focus on the information captured by each camera, and wish to avoid the confounding effect of camera-specific inference algorithms or the decoding complexity. For clarity and computational efficiency we focus on the 2D version of the problem (1D image/2D light field). We use simplified optical models and do not model lens aberrations or diffraction (these effects would still follow a linear projection model and can be accounted for with modifications to the light field projection function.)

Our framework captures the three major elements of the computational imaging pipeline – optical setup, decoding algorithm, and priors – and enables a systematic comparison on a common baseline.

1.1   Related Work

Approaches to lens characterization such as Fourier optics [8,9] analyze an optical element in terms of signal bandwidth and the sharpness of the PSF over the depth of field, but do not address depth information. The growing interest in 4D light field rendering has led to research on reconstruction filters and anti-aliasing in 4D [10,11,12], yet this research relies mostly on classical signal processing assumptions of band limited signals, and do not utilize the rich statistical correlations of light fields. Research on generalized camera families [13,14] mostly concentrates on geometric properties and 3D configurations, but with an assumption that approximately one light ray is mapped to each sensor element and thus decoding is not taken into account.

Reconstructing data from linear projections is a fundamental component in CT and tomography [15]. Fusing multiple image measurements is also used for super-resolution, and [16] studies uncertainties in this process.

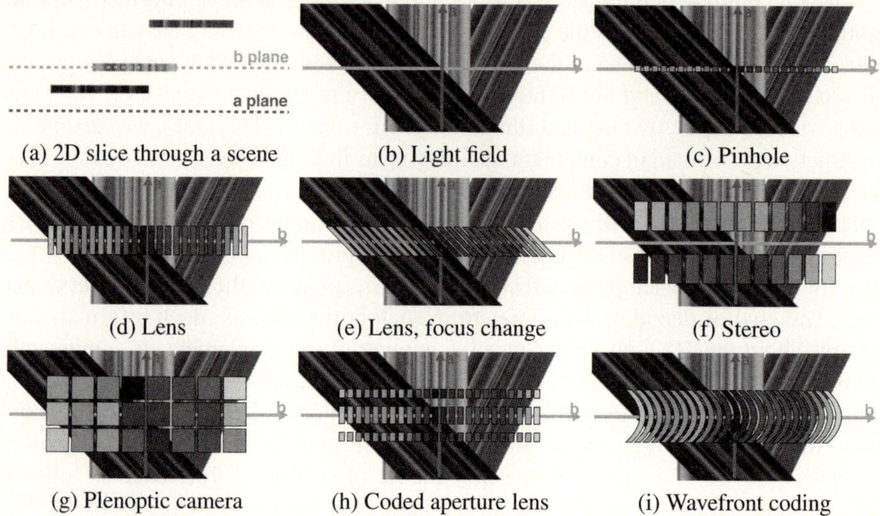

Fig. 1. (a) Flat-world scene with 3 objects. (b) The light field, and (c)-(i) cameras and the light rays integrated by each sensor element (distinguished by color).

2 Light Fields and Camera Configurations

Light fields are usually represented with a two-plane parameterization, where each ray is encoded by its intersections with two parallel planes. Figure 1(a,b) shows a 2D slice through a diffuse scene and the corresponding 2D slice of the 4D light field. The color at position (a_0, b_0) of the light field in fig. 1(b) is that of the reflected ray in fig. 1(a) which intersects the **a** and **b** lines at points a_0, b_0 respectively. Each row in this light field corresponds to a 1D view when the viewpoint shifts along **a**. Light fields typically have many elongated lines of nearly uniform intensity. For example the green object in fig. 1 is diffuse and the reflected color does not vary along the **a** dimension. The slope of those lines corresponds to the object depth [10,11].

Each sensor element integrates light from some set of light rays. For example, with a conventional lens, the sensor records an integral of rays over the lens aperture. We review existing cameras and how they project light rays to sensor elements. We assume that the camera aperture is positioned on the **a** line parameterizing the light field.

Pinhole Each sensor element collects light from a single ray, and the camera projection just slices a row in the light field (fig 1(c)). Since only a tiny fraction of light is let in, noise is an issue.

Lenses gather more light by focusing all light rays from a point at distance D to a sensor point. In the light field, $1/D$ is the slope of the integration (projection) stripe (fig 1(d,e)). An object is in focus when its slope matches this slope (e.g. green in fig 1(d)) [10,11,12]. Objects in front or behind the focus distance will be blurred. Larger apertures gather more light but can cause more defocus.

Stereo [17] facilitate depth inference by recording 2 views (fig 1(g), to keep a constant sensor budget, the resolution of each image is halved).

Plenoptic cameras capture multiple viewpoints using a microlens array [3,4]. If each microlens covers k sensor elements one achieves k different views of the scene, but the spatial resolution is reduced by a factor of k ($k = 3$ is shown in fig 1(g)).

Coded aperture [1,2] place a binary mask in the lens aperture (fig 1(h)). As with conventional lenses, objects deviating from the focus depth are blurred, but according to the aperture code. Since the blur scale is a function of depth, by searching for the code scale which best explains the local image window, depth can be inferred. The blur can also be inverted, increasing the depth of field.

Wavefront coding introduces an optical element with an unconventional shape so that rays from any world point do not converge. Thus, integrating over a curve in light field space (fig 1(i)), instead of the straight integration of lenses. This is designed to make defocus at different depths almost identical, enabling deconvolution without depth information, thereby extending depth of field. To achieve this, a cubic lens shape (or phase plate) is used. The light field integration curve, which is a function of the lens normal, can be shown to be a parabola (fig 1(i)), which is slope invariant (see [18] for a derivation, also independently shown by M. Levoy and Z. Zhu, personal communication).

3   Bayesian Estimation of Light Field

3.1   Problem Statement

We model an imaging process as an integration of light rays by camera sensors, or in an abstract way, as a linear projection of the light field

$$y = Tx + n \qquad (1)$$

where x is the light field, y is the captured image, n is an iid Gaussian noise $n \sim N(0, \eta^2 I)$ and T is the projection matrix, describing how light rays are mapped to sensor elements. Referring to figure 1, T includes one row for each sensor element, and this row has non-zero elements for the light field entries marked by the corresponding color (e.g. a pinhole T matrix has a single non-zero element per row).

The set of realizable T matrices is limited by physical constraints. In particular, the entries of T are all non-negative. To ensure equal noise conditions, we assume a maximal integration time, and the maximal value for each entry of T is 1. The amount of light reaching each sensor element is the sum of the entries in the corresponding T row. It is usually better to collect more light to increase the SNR (a pinhole is noisier because it has a single non-zero entry per row, while a lens has multiple ones).

To simplify notation, most of the following derivation will address a 2D slice in the 4D light field, but the 4D case is similar. While the light field is naturally continuous, for simplicity we use a discrete representation.

Our goal is to understand how well we can recover the light field x from the noisy projection y, and which T matrices (among the camera projections described in the previous section) allow better reconstructions. That is, if one is allowed to take N measurements (T can have N rows), which set of projections leads to better light field reconstruction? Our evaluation methodology can be adapted to a weight w which specifies how much we care about reconstructing different parts of the light field. For example, if the goal is an all-focused, high quality image from a single view point (as in wavefront coding), we can assign zero weight to all but one light field row.

The number of measurements taken by most optical systems is significantly smaller than the light field data, i.e. T contains many fewer rows than columns. As a result, it is impossible to recover the light field without prior knowledge on light fields. We therefore start by modeling a light field prior.

3.2   Classical Priors

State of the art light field sampling and reconstruction approaches [10,11,12] apply signal processing techniques, typically assuming band-limited signals. The number of non-zero frequencies in the signal has to be equal to the number of samples, and therefore before samples are taken, one has to apply a low-pass filter to meet the Nyquist limit. Light field reconstruction is then reduced to a convolution with a proper low-pass filter. When the depth range in the scene is bounded, these strategies can further bound the set of active frequencies within a sheared rectangle instead of a standard square of low frequencies and tune the orientation of the low pass filter. However, they do not address inference for a general projection such as the coded aperture.

One way to express the underlying band limited assumptions in a prior terminology is to think of an isotropic Gaussian prior (where by isotropic we mean that no direction in the light field is favored). In the frequency domain, the covariance of such a Gaussian is diagonal (with one variance per Fourier coefficient), allowing zero (or very narrow) variance at high frequencies above the Nyqusit limit, and a wider one at the lower frequencies. Similar priors can also be expressed in the spatial domain by penalizing the convolution with a set of high pass filters:

$$P(x) \propto exp(-\frac{1}{2\sigma_0} \sum_{k,i} |f_{k,i}x^T|^2) = exp(-\frac{1}{2}x^T \Psi_0^{-1} x) \tag{2}$$

where $f_{k,i}$ denotes the kth high pass filter centered at the ith light field entry. In sec 5, we will show that band limited assumptions and Gaussian priors indeed lead to equivalent sampling conclusions.

More sophisticated prior choices replace the Gaussian prior of eq 2 with a heavy-tailed prior [19]. However, as will be illustrated in section 3.4, such generic priors ignore the very strong elongated structure of light fields, or the fact that the variance along the disparity slope is significantly smaller than the spatial variance.

3.3   Mixture of Gaussians (MOG) Light Field Prior

To model the strong elongated structure of light fields, we propose using a mixture of oriented Gaussians. If the scene depth (and hence light field slope) is known we can define an anisotropic Gaussian prior that accounts for the oriented structure. For this, we define a slope field S that represents the slope (one over the depth of the visible point) at every light field entry (fig. 2(b) illustrates a sparse sample from a slope field).

Analytic camera evaluation tools may also permit the study of unexplored camera designs. One might develop new cameras by searching for linear projections that yield optimal light field inference, subject to physical implementation constraints. While the camera score is a very non-convex function of its physical characteristics, defining camera evaluation functions opens up these research directions.

Acknowledgments. We thank Royal Dutch/Shell Group, NGA NEGI-1582-04-0004, MURI Grant N00014-06-1-0734, NSF CAREER award 0447561. Fredo Durand acknowledges a Microsoft Research New Faculty Fellowship and a Sloan Fellowship.

References

1. Levin, A., Fergus, R., Durand, F., Freeman, W.: Image and depth from a conventional camera with a coded aperture. SIGGRAPH (2007)
2. Veeraraghavan, A., Raskar, R., Agrawal, A., Mohan, A., Tumblin, J.: Dappled photography: Mask-enhanced cameras for heterodyned light fields and coded aperture refocusing. SIGGRAPH (2007)
3. Adelson, E.H., Wang, J.Y.A.: Single lens stereo with a plenoptic camera. PAMI (1992)
4. Ng, R., Levoy, M., Bredif, M., Duval, G., Horowitz, M., Hanrahan, P.: Light field photography with a hand-held plenoptic camera. Stanford U. Tech. Rep. CSTR 2005-02 (2005)
5. Bradburn, S., Dowski, E., Cathey, W.: Realizations of focus invariance in optical-digital systems with wavefront coding. Applied optics 36, 9157–9166 (1997)
6. Dowski, E., Cathey, W.: Single-lens single-image incoherent passive-ranging systems. App. Opt. (1994)
7. Levoy, M., Hanrahan, P.M.: Light field rendering. SIGGRAPH (1996)
8. Goodman, J.W.: Introduction to Fourier Optics. McGraw-Hill Book Company, New York (1968)
9. Zemax: http://www.zemax.com
10. Chai, J., Tong, X., Chan, S., Shum, H.: Plenoptic sampling. SIGGRAPH (2000)
11. Isaksen, A., McMillan, L., Gortler, S.J.: Dynamically reparameterized light fields. SIGGRAPH (2000)
12. Ng, R.: Fourier slice photography. SIGGRAPH (2005)
13. Seitz, S., Kim, J.: The space of all stereo images. In: ICCV (2001)
14. Grossberg, M., Nayar, S.K.: The raxel imaging model and ray-based calibration. In: IJCV (2005)
15. Kak, A.C., Slaney, M.: Principles of Computerized Tomographic Imaging
16. Baker, S., Kanade, T.: Limits on super-resolution and how to break them. PAMI (2002)
17. Scharstein, D., Szeliski, R.: A taxonomy and evaluation of dense two-frame stereo correspondence algorithms. Intl. J. Computer Vision 47(1), 7–42 (2002)
18. Levin, A., Freeman, W., Durand, F.: Understanding camera trade-offs through a bayesian analysis of light field projections. MIT CSAIL TR 2008-049 (2008)
19. Roth, S., Black, M.J.: Fields of experts: A framework for learning image priors. In: CVPR (2005)
20. Levin, A., Sand, P., Cho, T.S., Durand, F., Freeman, W.T.: Motion invariant photography. SIGGRAPH (2008)
21. Schechner, Y., Kiryati, N.: Depth from defocus vs. stereo: How different really are they. IJCV (2000)

CenSurE: Center Surround Extremas for Realtime Feature Detection and Matching[*]

Motilal Agrawal[1], Kurt Konolige[2], and Morten Rufus Blas[3]

[1] SRI International, Menlo Park CA 94025, USA
agrawal@ai.sri.com
[2] Willow Garage, Menlo Park CA 94025, USA
konolige@willowgarage.com
[3] Elektro/DTU University, Lyngby, Denmark
mrb@elektro.dtu.dk

Abstract. We explore the suitability of different feature detectors for the task of image registration, and in particular for visual odometry, using two criteria: stability (persistence across viewpoint change) and accuracy (consistent localization across viewpoint change). In addition to the now-standard SIFT, SURF, FAST, and Harris detectors, we introduce a suite of scale-invariant center-surround detectors (CenSurE) that outperform the other detectors, yet have better computational characteristics than other scale-space detectors, and are capable of real-time implementation.

1 Introduction

Image matching is the task of establishing correspondences between two images of the same scene. This is an important problem in Computer Vision with applications in object recognition, image indexing, structure from motion and visual localization – to name a few. Many of these applications have real-time constraints and would benefit immensely from being able to match images in real time.

While the problem of image matching has been studied extensively for various applications, our interest in it has been to be able to reliably match two images in real time for camera motion estimation, especially in difficult off-road environments where there is large image motion between frames [1,2]. Vehicle dynamics and outdoor scenery can make the problem of matching images very challenging. The choice of a feature detector can have a large impact in the performance of such systems.

We have identified two criteria that affect performance.

- Stability: the persistence of features across viewpoint change
- Accuracy: the consistent localization of a feature across viewpoint change

[*] This material is based upon work supported by the United States Air Force under Contract No. FA8650-04-C-7136. Any opinions, findings and conclusions or recommendations expressed in this material are those of the author(s) and do not necessarily reflect the views of the United States Air Force.

D. Forsyth, P. Torr, and A. Zisserman (Eds.): ECCV 2008, Part IV, LNCS 5305, pp. 102–115, 2008.
© Springer-Verlag Berlin Heidelberg 2008

Stability is obviously useful in tracking features across frames. Accuracy of feature localization is crucial for visual odometry tasks, but keypoint operators such as SIFT typically subsample the image at higher scales, losing pixel-level precision.

Broadly speaking, we can divide feature classes into two types. *Corner detectors* such as Harris (based on the eigenvalues of the second moment matrix [3,4]) and FAST [5] (analysis of circular arcs [6]) find image points that are well localized, because the corners are relatively invariant to change of view. Both these detectors can be implemented very efficiently and have been used in structure-from-motion systems [2,7,8] because of their accuracy. However, they are not invariant to scale and therefore not very stable across scale changes, which happen constantly with a moving camera. The Harris-Laplace and the Hessian-Laplace features [9] combine scale-space techniques with the Harris approach. They use a scale-adapted Harris measure [10] or the determinant of the Hessian to select the features and the Laplacian to select the scale. Supposedly, visual odometry can benefit from scale-space features, since they can be tracked for longer periods of time, and should lead to improved motion estimates from incremental bundle adjustment of multiple frames.

While we expect scale-space features to be more stable than simple corner features, are they as accurate? The answer, at least for visual odometry, is "no". The reason is that, as typically implemented in an image pyramid, scale-space features are not well localized at higher levels in the pyramid. Obviously, features at high levels have less accuracy relative to the original image. The culprit in loss of accuracy is the image pyramid. If the larger features were computed at each pixel, instead of reducing the size of the image, accuracy could be maintained. However, computing features at all scales is computationally expensive, which is why SIFT features [11], one of the first scale-space proposals, uses the pyramid – each level incurs only 1/4 the cost of the previous one. SIFT attempts to recover some of the lost accuracy through subpixel interpolation.

Our proposal is to maintain accuracy by computing features at all scales *at every pixel* in the original image. The extrema of the Laplacian across scale have been shown to be very stable [12], so we consider this operator, or more generally, extrema of a center-surround response (CenSurE, or *Cen*ter *Sur*round *E*xtrema). We explore a class of simple center-surround filters that can be computed in time independent of their size, and show that, even when finding extrema across all scales, they are suitable for real-time tasks such as visual odometry. CenSurE filters outperform the best scale-space or corner features at this task in terms of track length and accuracy, while being much faster to compute; and they are also competitive in standard tests of repeatability for large-viewpoint changes.

While the main focus of this paper is on a novel feature detector, visual odometry (and other motion estimation tasks) can benefit from matching using a descriptor that is robust to viewpoint changes. In this paper, we develop a fast variant of the upright SURF descriptor, and show that it can be used in real-time tasks.

1.1　Related Work

The two scale-space detectors that are closest to our work, in technique and practicality, are SIFT [11] and SURF [13]. The main differences between approaches is summarized in the table below.

	CenSurE	SIFT	SURF
Spatial resolution at scale	full	subsampled	subsampled
Scale-space operator Approximation	Laplace (Center-surround)	Laplace (DOG)	Hessian (DOB)
Edge filter	Harris	Hessian	Hessian
Rotational invariance	approximate	yes	no

The key difference is the full spatial resolution achieved by CenSurE at every scale. Neither SIFT nor SURF computes responses at all pixels for larger scales, and consquently do not detect extrema across all scales. Instead, they consider each scale octave independently. Within an octave, they subsamples the responses, and find extrema only at the subsampled pixels. At each successive octave, the subsampling is increased, so that almost all computation is spent on the first octave. Consequently, the accuracy of features at larger scales is sacrificed, in the same way that it is for pyramid systems. While it would be possible for SIFT and SURF to forego subsampling, it would then be inefficient, with compute times growing much larger.

CenSurE also benefits from using an approximation to the Laplacian, which has been shown to be better for scale selection [12]. The center-surround approximation is fast to compute, while being insensitive to rotation (unlike the DOB Hessian approximation). Also, CenSurE uses a Harris edge filter, which gives better edge rejection than the Hessian.

Several simple center-surround filters exist in the literature. The bi-level Laplacian of Gaussian (BLoG) approximates the LoG filter using two levels. [14] describes circular BLoG filters and optimizes for the inner and outer radius to best approximate the LoG filter. The drawback is that the cost of BLoG depends on the size of the filter. Closer to our approach is that of Grabner et al. [15], who describe a difference-of-boxes (DOB) filter that approximates the SIFT detector, and is readily computed at all scales with integral images [16,17]. Contrary to the results presented in [15], we demonstrate that our DOB filters outperform SIFT in repeatability. This can be attributed to careful selection of filter sizes and using the second moment matrix instead of the Hessian to filter out responses along a line. In addition, the DOB filter is not invariant to rotation, and in this paper we propose filters that have better properties.

The rest of the paper is organized as follows. We describe our CenSurE features in detail in Section 2. We then discuss our modified upright SURF (MU-SURF) in Section 3. We compare the performance of CenSurE against several other feature detectors. Results of this comparison for image matching are presented in Section 4.1 followed by results for visual odometry in Section 4.2. Finally, Section 5 concludes this paper.

2   Center Surround Extrema (CenSurE) Features

Our approach to determining accurate large-scale features demands that we compute all features at all scales, and select the extrema across scale and location. Obviously, this strategy demands very fast computation, and we use simplified bi-level kernels as center-surround filters. The main concern is finding kernels that are rotationally invariant, yet easy to compute.

2.1   Finding Extrema

In developing affine-invariant features, Mikolajczyk and Schmid [18] report on two detectors that seem better than others in repeatability – the Harris-Laplace and Hessian-Laplace. Mikoljczyk and Schmid note that the Harris and Hessian detectors (essentially corner detectors) are good at selecting a location within a scale, but are not robust across scale. Instead, they show that the maximum of Laplacian operator across scales gives a robust characteristic scale - hence the hybrid operator, which they define as follows: first a peak in the Harris or Hessian operator is used to select a location, and then the Laplacian selects the scale at that location.

This strategy requires computing the Hessian/Harris measure at all locations and all scales, and additionally calculating the Laplacian at all scales where there are peaks in the corner detector. In our view, the Laplacian is easier to compute and to approximate than the Hessian, as was discovered by Lowe for SIFT features. So in our approach, we compute a simplified center-surround filter at all locations and all scales, and find the extrema in a local neighborhood. In a final step, these extrema are filtered by computing the Harris measure and eliminating those with a weak corner response.

2.2   Bi-level Filters

While Lowe approximated the Laplacian with the difference of Gaussians, we seek even simpler approximations, using center-surround filters that are bi-level, that is, they multiply the image value by either 1 or -1. Figure 1 shows a progression of bi-level filters with varying degrees of symmetry. The circular filter is the most faithful to the Laplacian, but hardest to compute. The other filters can be computed rapidly with integral images (Section 2.7), with decreasing cost from octagon to hexagon to box filter. We investigate the two endpoints: octagons for good performance, and boxes for good computation.

2.3   CenSurE Using Difference of Boxes

We replace the two circles in the circular BLoG with squares to form our CenSurE-DOB. This results in a basic center-surround Haar wavelet. Figure 1(d) shows our generic center-surround wavelet of block size n. The inner box is of size $(2n + 1) \times (2n + 1)$ and the outer box is of size $(4n + 1) \times (4n + 1)$. Convolution is done by multiplication and summing. If I_n is the inner weight

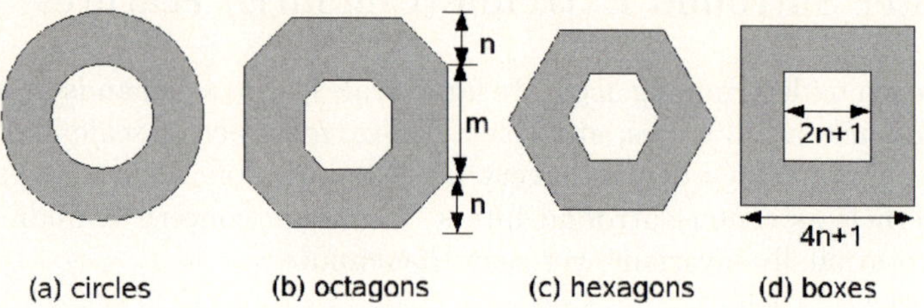

Fig. 1. Progression of Center-Surround bi-level filters. (a) circular symmetric BLoG (Bilevel LoG) filter. Successive filters (octagon, hexagon, box) have less symmetry.

and O_n is the weight in the outer box, then in order for the DC response of this filter to be zero, we must have

$$O_n(4n+1)^2 = I_n(2n+1)^2 \tag{1}$$

We must also normalize for the difference in area of each wavelet across scale.

$$I_n (2n+1)^2 = I_{n+1} (2(n+1)+1)^2 \tag{2}$$

We use a set of seven scales for the center-surround Haar wavelet, with block size $n = [1, 2, 3, 4, 5, 6, 7]$. Since the block sizes 1 and 7 are the boundary, the lowest scale at which a feature is detected corresponds to a block size of 2. This roughly corresponds to a LoG with a sigma of 1.885. These five scales cover $2\frac{1}{2}$ octaves, although the scales are linear. It is easy to add more filters with block sizes 8,9, and so on.

2.4   CenSurE Using Octagons

Difference of Boxes are obviously not rotationally invariant kernels. In particular, DOBs will perform poorly for 45 degrees in-plane rotation. Octagons, on the other hand are closer to circles and approximate LoG better than DOB.

In using octagons, the basic ideas of performing convolutions by inner and outer weighted additions remain the same. As in DOB, one has to find weights I_n and O_n such that the DC response is zero and all filters are normalized according to the area of the octagons.

An octagon can be represented by the height of the vertical side (m) and height of the slanted side (n) (Figure 1(b)). Table 1 shows the different octagon sizes corresponding to the seven scales. These octagons scale linearly and were experimentally chosen to correspond to the seven DOBs described in the previous section.

2.5   Non-maximal Suppression

We compute the seven filter responses at each pixel in the image. We then perform a non-maximal suppression over the scale space. Briefly, a response is

Table 1. CenSurE-OCT: inner and outer octagon sizes for various scales

scale	$n=1$	$n=2$	$n=3$	$n=4$	$n=5$	$n=6$	$n=7$
inner (m,n)	$(3,0)$	$(3,1)$	$(3,2)$	$(5,2)$	$(5,3)$	$(5,4)$	$(5,5)$
outer (m,n)	$(5,2)$	$(5,3)$	$(7,3)$	$(9,4)$	$(9,7)$	$(13,7)$	$(15,10)$

suppressed if there is a response greater (maxima case) or a response less than (minima case) its neighbors in a local neighborhood over the location and scales. Pixels that are either maxima or minima in this neighborhood are the feature point locations. We use a 3x3x3 neighborhood for our non-maximal suppression.

The magnitude of the filter response gives an indication of the strength of the feature. The greater the strength, the more likely it is to be repeatable. Weak responses are likely to be unstable. Therefore, we can apply a threshold to filter out the weak responses.

Since all our responses are computed on the original image without subsampling, all our feature locations are localized well and we do not need to perform subpixel interpolation.

2.6   Line Suppression

Features that lie along an edge or line are poorly localized along it and therefore are not very stable. Such poorly defined peaks will have large principal curvatures along the line but a small one in the perpendicular direction and therefore can be filtered out using the ratio of principal curvatures. We use the second moment matrix of the response function at the particular scale to filter out these responses.

$$H = \begin{bmatrix} \sum L_x^2 & \sum L_x L_y \\ \sum L_x L_y & \sum L_y^2 \end{bmatrix} \tag{3}$$

L_x and L_y are the derivatives of the response function L along x and y. The summation is over a window that is linearly dependent on the scale of the particular feature point: the higher the scale, the larger the window size. Note that this is the scale-adapted Harris measure [18,10] and is different from the Hessian matrix used by SIFT [11,15] to filter out line responses. Once the Harris measure is computed, its trace and determinant can be used to compute the ratio of principal curvatures. We use a threshold of 10 for this ratio and a 9×9 window at the smallest scale of block size 2.

The Harris measure is more expensive to compute than the Hessian matrix used by SIFT. However, this measure needs to be computed for only a small number of feature points that are scale-space maxima and whose response is above a threshold and hence does not present a computational bottleneck. In our experience it does a better job than Hessian at suppressing line responses.

2.7   Filter Computation

The key to CenSurE is to be able to compute the bi-level filters efficiently at all sizes. The box filter can be done using integral images [16,17]. An integral image

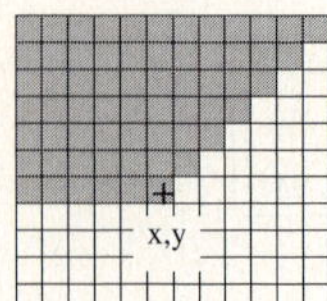
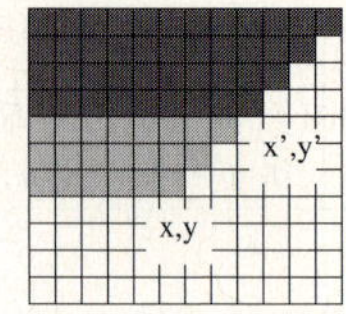
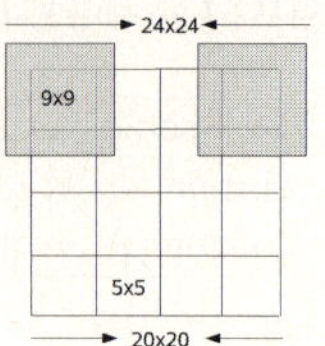

Fig. 2. Using slanted integral images to construct trapezoidal areas. Left is a slanted integral image, where the pixel x, y is the sum of the shaded areas; α is 1. Right is a half-trapezoid, from subtracting two slanted integral image pixels.

Fig. 3. Regions and subregions for MU-SURF descriptor. Each subregion (in blue) is 9x9 with an overlap of 2 pixels at each boundary. All sizes are relative to the scale of the feature s.

I is an intermediate representation for the image and contains the sum of gray scale pixel values of image N with height y and width x, i.e.,

$$I(x,y) = \sum_{x'=0}^{x} \sum_{y'=0}^{y} N(x',y') \tag{4}$$

The integral image is computed recursively, requiring only one scan over the image. Once the integral image is computed, it it takes only four additions to calculate the sum of the intensities over any upright, rectangular area, independent of its size.

Modified versions of integral images can be exploited to compute the other polygonal filters. The idea here is that any trapezoidal area can be computed in constant time using a combination of two different *slanted* integral images, where the sum at a pixel represents an angled area sum. The degree of slant is controlled by a parameter α:

$$I_\alpha(x,y) = \sum_{y'=0}^{y} \sum_{x'=0}^{x+\alpha(y-y')} N(x',y'). \tag{5}$$

When $\alpha = 0$, this is just the standard rectangular integral image. For $\alpha < 0$, the summed area slants to the left; for $\alpha > 0$, it slants to the right (Figure 2, left). Slanted integral images can be computed in the same time as rectangular ones, using incremental techniques.

Adding two areas together with the same slant determines one end of a trapezoid with parallel horizontal sides (Figure 2, right); the other end is done similarly, using a different slant. Each trapezoid requires three additions, just as in the rectangular case. Finally, the polygonal filters can be decomposed into 1 (box), 2 (hexagon), and 3 (octagon) trapezoids, which is the relative cost of computing these filters.

3   Modified Upright SURF (MU-SURF) Descriptor

Previously, we have demonstrated accurate visual odometry using ZNCC for feature matching [1] (using a 11×11 region). However, it is relatively sensitive to in-plane rotations (roll), larger changes in perspective, and inaccuracies in keypoint localization. The problems related to rolls and perspective changes become more significant as the region size increases. We have therefore decided to switch to an upright SURF type descriptor [13].

The SURF descriptor builds on from the SIFT descriptor by encoding local gradient information. It uses integral images to compute Haar wavelet responses, which are then summed in different ways in 4×4 subregions of the region to create a descriptor vector of length 64.

As pointed out by David Lowe [11], "it is important to avoid all boundary effects in which the descriptor abruptly changes as a sample shifts smoothly from being within one histogram to another or from one orientation to another." The SURF descriptor [13] weighs the Haar wavelet responses using a Gaussian centered at the interest point. This single weighting scheme gave poor results and we were unable to recreate the SURF descriptor results without accounting for these boundary effects.

To account for these boundary conditions, each boundary in our descriptor has a padding of $2s$, thereby increasing our region size from $20s$ to $24s$, s being the scale of the feature. The Haar wavelet responses in the horizontal (d_x) and vertical (d_y) directions are computed for each 24×24 point in the region with filter size $2s$ by first creating a summed image, where each pixel is the sum of a region of size s. The Haar wavelet output results in four fixed-size $d_x, d_y, |d_x|, |d_y|$ images that have the dimensions 24×24 pixels irrespective of the scale.

Each $d_x, d_y, |d_x|, |d_y|$ image is then split into 4×4 square overlapping subregions of size 9×9 pixels with an overlap of 2 pixels with each of the neighbors. Figure fig:descriptor shows these regions and subregions. For each subregion the values are then weighted with a precomputed Gaussian ($\sigma_1 = 2.5$) centered on the subregion center and summed into the usual SURF descriptor vector for each subregion: $v = (\sum d_x, \sum d_y, \sum |d_x|, \sum |d_y|)$. Each subregion vector is then weighted using another Gaussian ($\sigma_2 = 1.5$) defined on a mask of size 4×4 and centered on the feature point. Like the original SURF descriptor, this vector is then normalized.

The overlap allows each subregion to work on a larger area so samples that get shifted around are more likely to still leave a signature in the correct subregion vectors. Likewise, the subregion Gaussian weighting means that samples near borders that get shifted out of a subregion have less impact on the subregion descriptor vector.

From an implementation point of view the dynamic range of the vector was small enough that the end results could be scaled into C++ shorts. This allows for very fast matching using compiler vectorization.

CenSurE features themselves are signed based on their being dark or bright blobs. This is similar to SURF and can also be used to speed up the matching by only matching bright features to bright features and so forth.

We have compared the performance of MU-SURF with U-SURF for matching and found them to be similar. As will be pointed out in Section 4.3, our implementation of MU-SURF is significantly faster than U-SURF. It is unclear to us as to why MU-SURF is so much faster. We are currently looking into this.

4   Experimental Results

We compare CenSurE-DOB and CenSurE-OCT to Harris, FAST, SIFT, and SURF feature detectors for both image matching and visual odometry. Results for image matching are presented in Section 4.1 and VO in Section 4.2.

4.1   Image Matching

For image matching, we have used the framework of [12] to evaluate repeatability scores for each detector on the graffiti and boat sequences[1]. We have used the default parameters for each of these detectors. In addition, since each of these detectors has a single value that represents the strength of the feature, we have chosen a strength threshold such that each of these detectors results in the same number of features in the common overlapping regions. Figure 4 (a) & (b) shows a plot of the detector repeatability and number of correspondences for each detector using 800 features and an overlap threshold of 40% for the graffiti sequence. For Harris and FAST, the scale of all detected points was assumed to be the same and set at 2.0.

Both versions of CenSurE are better than SIFT or SURF, although for large viewpoint changes, the differences become only marginal. As can be expected, CenSurE-OCT does better than CenSurE-DOB.

The *boat* sequence is more challenging because of large changes in rotation and zoom. Figure 4 (c) & (d) shows the detector performance for this sequence for 800 features. On this challenging sequence, CenSurE performs slightly worse than either SIFT or SURF, especially for the larger zooms. This can be attributed to CenSurE's non-logarithmic scale sampling. Furthermore, CenSurE filters cover only $2\frac{1}{2}$ octaves and therefore has less degree of scale-invariance for large scale changes.

To evaluate the matching performance, we used our MU-SURF descriptor for each of those detectors and matched each detected point in one image to the one with the lowest error using Euclidean distance. A correspondence was deemed as matched if the true match was within a search radius r of its estimated correspondence. Note that this is a different criterion than considering overlap error and we have chosen this because this same criterion is used in visual odometry to perform image registration. Figure 5 shows the percentage of correct matches as a function of search radius when the number of features is fixed to 800.

---

[1] Available from `http://www.robots.ox.ac.uk/~vgg/research/affine/`

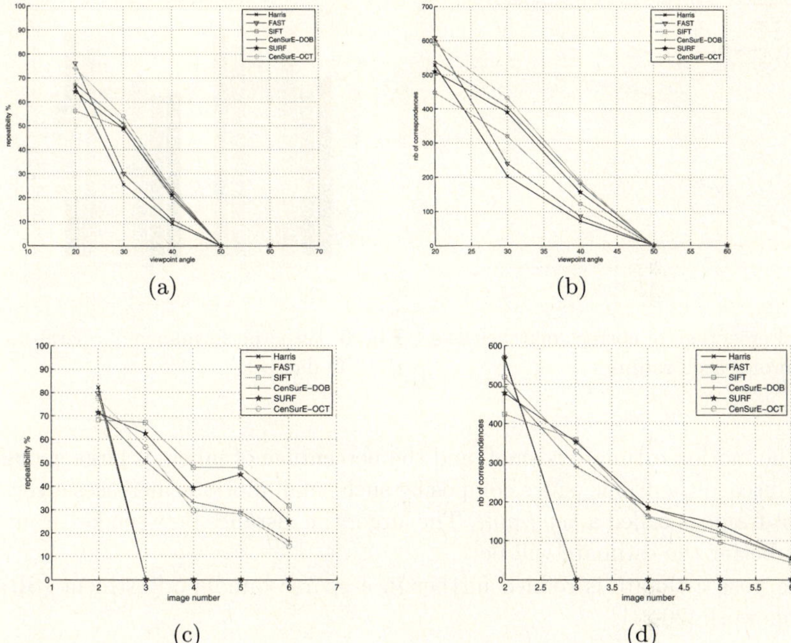

Fig. 4. Repeatability and number of correspondences for different detectors for the graffiti and boat sequences. The number of features is the same for each detector. (a) & (b) graffiti sequence. (c) & (d) boat sequence.

4.2   Visual Odometry

We evaluate the performance of CenSurE for performing visual odometry in challenging off-road environments. Because there can be large image motion between frames, including in-plane rotations, the tracking task is difficult: essentially, features must be re-detected at each frame. As usual, we compare our method against Harris, FAST, SIFT, and SURF features. Note that this is a test of the *detectors*; the same MU-SURF descriptor was used for each feature.

The Visual Odometry (VO) system derives from recent research by the authors and others on high-precision VO [1,2] using a pair of stereo cameras. For each new frame, we perform the following process.

1. Distinctive features are extracted from each new frame in the left image. Standard stereo methods are used to find the corresponding point in the right image.
2. Left-image features are matched to the features extracted in the previous frame using our descriptor. We use a large area, usually around 1/5 of the image, to search for matching features.
3. From these uncertain matches, we recover a consensus pose estimate using a RANSAC method [19]. Several thousand relative pose hypotheses are generated by randomly selecting three matched non-collinear features, and then scored using pixel reprojection errors.

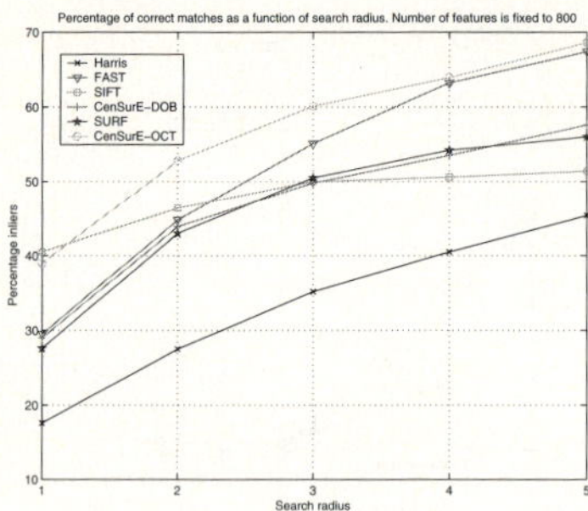

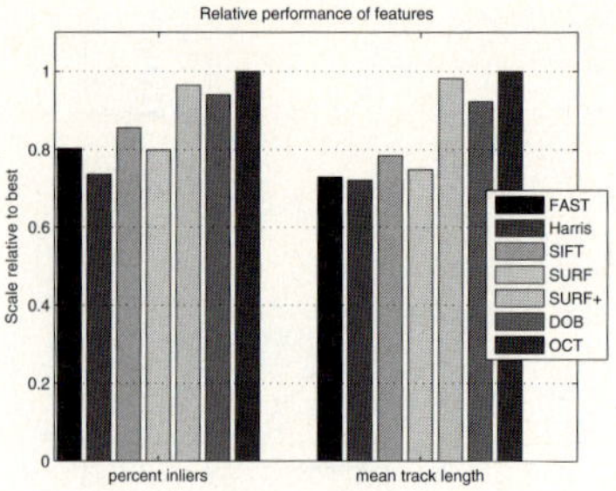

Fig. 5. Percentage of correct matches as a function of search radius

Fig. 6. Basic performance of operators in the VO dataset

4. If the motion estimate is small and the percentage of inliers is large enough, we discard the frame, since composing such small motions increases error. A kept frame is called a *key frame*. The larger the distance between key frames, the better the estimate will be.
5. The pose estimate is refined further in a sparse bundle adjustment (SBA) framework [20,21].

The dataset for this experiment consists of 19K frames taken over the course of a 3 km autonomous, rough-terrain run. The images have resolution 512x384, and were taken at a 10 Hz rate; the mean motion between frames was about 0.1m. The dataset also contains RTK GPS readings synchronized with the frames, so ground truth to within about 10 cm is available for gauging accuracy.

We ran each of the operators under the same conditions and parameters for visual odometry, and compared the results. Since the performance of an operator is strongly dependent on the number of features found, we set a threshold of 400 features per image, and considered the highest-ranking 400 features for each operator. We also tried hard to choose the best parameters for each operator. For example, for SURF we used doubled images and a subsampling factor of 1, since this gave the best performance (labeled "SURF+" in the figures).

The first set of statistics shows the raw performance of the detector on two of the most important performance measures for VO: the average percentage of inliers to the motion estimate, and the mean track length for a feature (Figure 6). In general, the scale-space operators performed much better than the simple corner detectors. CenSurE-OCT did the best, beating out SURF by a small margin. CenSurE-DOB is also a good performer, but suffers from lack of radial symmetry. Surprisingly, SIFT did not do very well, barely beating Harris corners.

Note that the performance of the scale-space operators is sensitive to the sampling density. For standard SURF settings (no doubled image, subsampling of 2) the performance is worse than the corner operators. Only when sampling densely for 2 octaves, by using doubled images and setting subsampling to 1, does performance approach that of CenSurE-OCT. Of course, this mode is much more expensive to compute for SURF (see Section 4.3).

5   Shape Matching

Our system produces an ordered list of species that are most likely to match the shape of a query leaf. It must be able to produce comparisons quickly for a dataset containing about 8,000 leaves from approximately 250 species. It is useful if we can show the user some initial results within a few seconds, and the top ten matches within a few seconds more. It is also important that we produce the correct species within the top ten matches as often as possible, since we are limited by screen size in displaying matches.

To perform matching, we make use of the *Inner Distance Shape Context* (IDSC, Ling and Jacobs [13]), which has produced close to the best published results for leaf recognition, and the best results among those methods quick enough to support real-time performance. IDSC samples points along the boundary of a shape, and builds a 2D histogram descriptor at each point. This histogram represents the distance and angle from each point to all other points, along a path restricted to lie entirely inside the leaf shape. Given n sample points, this produces n 2D descriptors, which can be computed in $O(n^3)$ time, using an all pairs shortest path algorithm. Note that this can be done off-line for all leaves in the dataset, and must be done on-line only for the query. Consequently, this run-time is not significant.

To compare two leaves, each sample point in each shape is compared to all points in the other shape, and matched to the most similar sample point. A shape distance is obtained by summing the χ^2 distance of this match over all sample points in both shapes, which requires $O(n^2)$ time.

Since IDSC comparison is quadratic in the number of sample points, we would like to use as few sample points as possible. However, IDSC performance decreases due to aliasing if the shape is under-sampled. We can reduce aliasing effects and boost performance by smoothing the IDSC histograms. To do this, we compute m histograms by beginning sampling at m different, uniformly spaced locations, and average the results. This increases the computation of IDSC for a single shape by a factor of m. However, it does not increase the size of the final IDSC, and so does not affect the time required to compare two shapes, which is our dominant cost.

We use a nearest neighbor classifier in which the species containing the most similar leaf is ranked first. Because the shape comparison algorithm does not imbed each shape into a vector space, we use a nearest neighbor algorithm designed for non-Euclidean metric spaces. Our distance does not actually obey the triangle inequality because it allows many-to-one matching, and so it is not really a metric (eg., all of shape A might match part of C, while B matches a different part of C, so A and B are both similar to C, but completely different from each other). However, in a set of 1161 leaves, we find that the triangle inequality is violated in only .025% of leaf triples, and these violations cause no errors in the nearest neighbor algorithm we use, the AESA algorithm (Ruiz [17]; Vidal [22]). In this method, we pre-compute and store the distance between all pairs of leaves in the dataset. This requires $O(N^2)$ space and time, for a dataset of N leaves, which is manageable for our datasets. At run time, a query is compared to one leaf, called a pivot. Based on the distance to the pivot, we can use the triangle inequality to place upper and lower bounds on the distance to all leaves and all species in the dataset. We select each pivot by choosing the leaf with the lowest current upper bound. When one species has an upper bound distance that is less than the lower bound to any other species, we can select this as the best match and show it to the user. Continuing this process provides an ordered list of matching species. In comparison to a brute force search, which takes nine

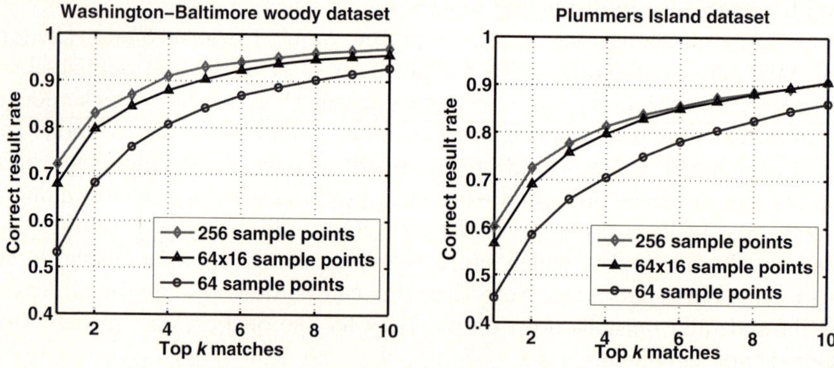

Fig. 4. Experimental results for two datasets

seconds with a dataset of 2004 leaves from 139 species, this nearest-neighbor algorithm reduces the time required to find the ten best matching species by a factor of 3, and reduces the time required to find the top three species by a factor of 4.4.

We have tested our algorithm using both the Plummers Island and Baltimore-Washington Woody Plants datasets. We perform a leave-one-out test, in which each leaf is removed from the dataset and used as a query. Figure 4 shows performance curves that indicate how often the correct species for a query is placed among the top k matches, as k varies. In this experiment, we achieve best performance using $n = 256$ sample points for IDSC. We reach nearly the same performance by computing the histograms using $n = 64$ sample points averaged over $m = 16$ starting points. The figure also shows that using $n = 64$ points without smoothing significantly degrades performance. Using 64 sample points is approximately 16 times faster than using 256 sample points. The correct answer appears in the top ten about 95%–97% of the time for woody plants of Baltimore-Washington and somewhat less (about 90% of the time) for the flora of Plummers Island. This is in part because shape matching is not very effective at discriminating between different species of grass (which are not woody plants). Overall, these results demonstrate effective performance. It seems that most errors occur for species in which the overall leaf shape is not sufficiently distinctive. We plan to address these issues by using additional cues, such as small scale features of the leaf margin (e.g., toothed or smooth) and the shape of the venation (vascular structure).

6   User Interfaces and Hardware

We have developed several prototype UIs to integrate the individual pieces of the matching system and investigate the performance of our interaction techniques and vision algorithms in real world situations. These prototypes are the result of collaboration with our botanist colleagues in an iterative process that has included ethnographic study of botanical species identification and collection in the field, user centered design, interaction technique development, and qualitative and quantitative feedback and user studies. We have pursued two primary research directions. The first focuses on existing mobile computing platforms for ongoing botanical field studies. The second develops mobile AR systems that are not appropriate for field use in their current form, but could provide significant advantages as hardware and software mature.

The conceptual model we use in our mobile computing platform is an extension of existing paper field guides. The system provides access to a library of knowledge about the physical world, and the physical leaf is the key to that information. In the AR prototype, virtual images representing matched species appear adjacent to the leaf in the physical world and can be manipulated directly through tangible interaction. In this case, the conceptual model is enhanced perception: the leaf anchors information embedded in the environment and accessed through augmented reality.

6.1   Mobile Computing

Our initial prototype, LeafView (Figure 1), provides four tabbed panes for interaction: browse, sample, search results, and history. The browse pane provides a zoomable UI (ZUI) (Bederson et al. [3]) with which the user can explore an entire flora dataset. When the user photographs a leaf with the system, the image is immediately displayed in the sample pane with contextual information including time, date, GPS location, and collector. The segmented image is displayed next to the captured leaf image to show the user what LeafView "sees" and provide feedback about image quality. As results are found, they are displayed with the

Fig. 5. AR user interface viewed through a video see-through display

original image in the search results pane. Each species result provides access to the matched leaf, type specimens, voucher images and information about the species in a ZUI to support detailed visual inspection and comparison, which is necessary when matching is imperfect. Selecting a match button associates a given species with the newly collected specimen in the collection database. The history pane displays a visual history of each collected leaf, along with access to previous search results, also in a ZUI. This represents the collection trip, which can be exported for botanical research, and provides a reference for previously collected specimens. Making this data available improves the long term use of the system by aiding botanists in their research.

LeafView was built with C#, MatLab, and Piccolo (Bederson, et al. [4]). Our first versions of the hardware used a Tablet PC with a separate Wi-Fi or Bluetooth camera and a Bluetooth WAAS GPS. However, feedback from botanists during field trials made it clear that it would be necessary to trade off the greater display area/processing power of the Tablet PC for the smaller size/weight of an Ultra-Mobile PC (UMPC) to make possible regular use in the field. We currently use a Sony VAIO VGN-UX390N, a UMPC with an integrated camera and small touch-sensitive screen, and an external GPS.

6.2 Augmented Reality

AR can provide affordances for interaction and display that are not available in conventional graphical UIs. This is especially true of Tangible AR (Kato et al. [12]), in which the user manipulates physical objects that are overlaid with additional information. Tangible AR is well matched to the hands-on environmental interaction typical of botanical field research. While current head-worn displays and tracking cannot meet the demands of daily fieldwork, we are developing experimental Tangible AR UIs to explore what might be practical in the future.

In one of our Tangible AR prototypes (Figure 5), a leaf is placed on a clipboard with optical tracking markers and a hand-held marker is placed next to the leaf to initiate a search. The results of matching are displayed alongside the physical leaf as a set of individual leaf images representing *virtual vouchers*, multifaceted representations of a leaf species that can be changed through tangible gestures. As the user passes the hand-held marker over a leaf image, the card visually transforms into that leaf's virtual voucher. The visual representation can be changed, through gestures such as a circular "reeling" motion, into images of the type specimen, entire tree, bark, or magnified view of the plant. Inspection and comparison is thus achieved through direct spatial manipulation of the virtual voucher—the virtual leaf in one hand and the physical leaf on the clipboard in the other hand. To accept a match, the virtual voucher is placed below the leaf and the system records the contextual data.

Different versions of our Tangible AR prototypes use a monoscopic Liteye-500 display, fixed to a baseball cap, and a stereoscopic Sony LDI-D100B display, mounted on a head-band, both of which support 800 × 600 resolution color imagery. The system runs on a UMPC, which fits with the display electronics into a fannypack. The markers are tracked in 6DOF using ARToolkit (Kato et al.

[12]) and ARTag (Fiala [7]), with a Creative Labs Notebook USB 2.0 camera attached to the head-worn display.

6.3   System Evaluation

Our prototypes have been evaluated in several ways during the course of the project. These include user studies of the AR system, field tests on Plummers Island, and expert feedback, building on previous work (White et al. [24]). In May 2007, both LeafView and a Tangible AR prototype were demonstrated and used to identify plants during the National Geographic BioBlitz in Rock Creek Park, Washington, DC, a 24-hour species inventory. Hundreds of people, from professional botanists to amateur naturalists, school children to congress-men, have tried both systems. While we have focused on supporting professional botanists, people from a diversity of backgrounds and interests have provided valuable feedback for the design of future versions.

7   Challenge Problem for Computer Vision

One goal of our project is to provide datasets that can serve as a challenge problem for computer vision. While the immediate application of such datasets is the identification of plant species, the datasets also provide a rich source of data for a number of general 2D and silhouette recognition algorithms.

In particular, our website includes three image datasets covering more than 500 plant species, with more than 30 leaves per species on average. Algorithms for recognition can be tested in a controlled fashion via leave-one-out tests, where the algorithms can train on all but one of the leaf images for each species and test on the one that has been removed. The web site also contains separate training and test datasets in order to make fair comparisons. Our IDSC code can also be obtained there, and other researchers can submit code and performance curves, which we will post. We hope this will pose a challenge for the community, to find the best algorithms for recognition in this domain.

Note that our system architecture for the electronic field guide is modular, so that we can (and will, if given permission) directly use the best performing methods for identification, broadening the impact of that work.

8   Future Plans

To date, we have focused on three regional floras. Yet, our goal is to expand the coverage of our system in temperate climates to include all vascular plants of the continental U.S. Other than the efforts involved in collecting the single leaf datasets, there is nothing that would prevent us from building a system for the U.S. flora. The visual search component of the system scales well: search can always be limited to consider only those species likely to be found in the current location, as directed by GPS.

In addition, we have begun to expand into the neotropics. The Smithsonian Center for Tropical Forest Science has set up twenty 50-hectare plots in tropical ecosystems around the world to monitor the changing demography of tropical forests. We aim to develop versions of the system for three neotropical floras: Barro Colorado Island, Panama; Yasuni National Park, Ecuador; and the Amazon River Basin in Brazil. This domain demands algorithms that not only consider leaf shape, but also venation (i.e., the leaf's vascular structure). Initial results are quite promising, but we have not yet developed a working system.

Finally, we have developed a prototype web-based, mobile phone version of our system, allowing anyone with a mobile phone equipped with a camera and browser to photograph leaves and submit them to a server version of our system for identification. We hope to develop a touch-based version on an iPhone or Android-based device in the near future. We feel that it should soon be possible to create a mobile phone-based system that covers the entire U.S., usable by the general population.

Acknowledgements

This work was funded in part by National Science Foundation Grant IIS-03-25867, An Electronic Field Guide: Plant Exploration and Discovery in the 21st Century, and a gift from Microsoft Research.

References

1. Abbasi, S., Mokhtarian, F., Kittler, J.: Reliable classification of chrysanthemum leaves through curvature scale space. In: ter Haar Romeny, B.M., Florack, L.M.J., Viergever, M.A. (eds.) Scale-Space 1997. LNCS, vol. 1252, pp. 284–295. Springer, Heidelberg (1997)
2. Agarwal, G., Belhumeur, P., Feiner, S., Jacobs, D., Kress, W.J., Ramamoorthi, R., Bourg, N., Dixit, N., Ling, H., Mahajan, D., Russell, R., Shirdhonkar, S., Sunkavalli, K., White, S.: First steps towards an electronic field guide for plants. Taxon 55, 597–610 (2006)
3. Bederson, B.: PhotoMesa: A zoomable image browser using quantum treemaps and bubblemaps. In: Proc. ACM UIST 2001, pp. 71–80 (2001)
4. Bederson, B., Grosjean, J., Meyer, J.: Toolkit design for interactive structured graphics. IEEE Trans. on Soft. Eng. 30(8), 535–546 (2004)
5. Belongie, S., Malik, J., Puzicha, J.: Shape Matching and Object Recognition Using Shape Context. IEEE Trans. on Patt. Anal. and Mach. Intell. 24(4), 509–522 (2002)
6. Edwards, M., Morse, D.R.: The potential for computer-aided identification in biodiversity research. Trends in Ecology and Evolution 10, 153–158 (1995)
7. Fiala, M.: ARTag, a fiducial marker system using digital techniques. In: Proc. CVPR 2005, pp. 590–596 (2005)
8. Felzenszwalb, P., Schwartz, J.: Hierarchical matching of deformable shapes. In: Proc. CVPR 2007, pp. 1–8 (2007)
9. Forsyth, D., Ponce, J.: Computer vision: A modern approach. Prentice Hall, Upper Saddle River (2003)

10. Galun, M., Sharon, E., Basri, R., Brandt, A.: Texture segmentation by multiscale aggregation of filter responses and shape elements. In: Proc. CVPR, pp. 716–723 (2003)
11. Heidorn, P.B.: A tool for multipurpose use of online flora and fauna: The Biological Information Browsing Environment (BIBE). First Monday 6(2) (2001), http://firstmonday.org/issues/issue6_2/heidorn/index.html
12. Kato, H., Billinghurst, M., Poupyrev, I., Imamoto, K., Tachibana, K.: Virtual object manipulation of a table-top AR environment. In: Proc. IEEE and ACM ISAR, pp. 111–119 (2000)
13. Ling, H., Jacobs, D.: Shape Classification Using the Inner-Distance. IEEE Trans. on Patt. Anal. and Mach. Intell. 29(2), 286–299 (2007)
14. Mokhtarian, F., Abbasi, S.: Matching shapes with self-intersections: Application to leaf classification. Proc. IEEE Trans. on Image 13(5), 653–661 (2004)
15. Nilsback, M., Zisserman, A.: A visual vocabulary for flower classification. In: Proc. CVPR, pp. 1447–1454 (2006)
16. Pankhurst, R.J.: Practical taxonomic computing. Cambridge University Press, Cambridge (1991)
17. Ruiz, E.: An algorithm for finding nearest neighbours in (approximately) constant average time. Patt. Rec. Lett. 4(3), 145–157 (1986)
18. Saitoh, T., Kaneko, T.: Automatic recognition of wild flowers. Proc. ICPR 2, 2507–2510 (2000)
19. Shi, J., Malik, J.: Normalized Cuts and Image Segmentation. IEEE Trans. on Patt. Anal. and Mach. Intell. 22(8), 888–905 (2000)
20. Söderkvist, O.: Computer vision classification of leaves from Swedish trees. Master Thesis, Linköping Univ. (2001)
21. Stevenson, R.D., Haber, W.A., Morris, R.A.: Electronic field guides and user communities in the eco-informatics revolution. Conservation Ecology 7(3) (2003), http://www.consecol.org/vol7/iss1/art3
22. Vidal, E.: New formulation and improvements of the nearest-neighbour approximating and eliminating search algorithm (AESA). Patt. Rec. Lett. 15(1), 1–7 (1994)
23. Wang, Z., Chi, W., Feng, D.: Shape based leaf image retrieval. IEE Proc. Vision, Image and Signal Processing 150(1), 34–43 (2003)
24. White, S., Feiner, S., Kopylec, J.: Virtual vouchers: Prototyping a mobile augmented reality user interface for botanical species identification. In: Proc. IEEE Symp. on 3DUI, pp. 119–126 (2006)
25. White, S., Marino, D., Feiner, S.: Designing a mobile user interface for automated species identification. In: Proc. CHI 2007, pp. 291–294 (2007)

A Column-Pivoting Based Strategy for Monomial Ordering in Numerical Gröbner Basis Calculations

Martin Byröd, Klas Josephson, and Kalle Åström

Centre For Mathematical Sciences,
Lund University, Lund, Sweden
{byrod,klasj,kalle}@maths.lth.se

Abstract. This paper presents a new fast approach to improving stability in polynomial equation solving. Gröbner basis techniques for equation solving have been applied successfully to several geometric computer vision problems. However, in many cases these methods are plagued by numerical problems. An interesting approach to stabilising the computations is to study basis selection for the quotient space $\mathbb{C}[\mathbf{x}]/I$. In this paper, the exact matrix computations involved in the solution procedure are clarified and using this knowledge we propose a new fast basis selection scheme based on QR-factorization with column pivoting. We also propose an adaptive scheme for truncation of the Gröbner basis to further improve stability. The new basis selection strategy is studied on some of the latest reported uses of Gröbner basis methods in computer vision and we demonstrate a fourfold increase in speed and nearly as good over-all precision as the previous SVD-based method. Moreover, we get typically get similar or better reduction of the largest errors[1].

1   Introduction

A large number of geometric computer vision problems can be formulated in terms of a system of polynomial equations in one or more variables. A typical example of this is minimal problems of structure from motion [1,2]. This refers to solving a specific problem with a minimal number of point correspondences. Further examples of minimal problems are relative motion for cameras with radial distortion [3] or for omnidirectional cameras [4]. Solvers for minimal problems are often used in the inner loop of a RANSAC engine to find inliers in noisy data, which means that they are run repeatedly a large number of times. There is thus a need for fast and stable algorithms to solve systems of polynomial equations.

Another promising, but difficult pursuit in computer vision (and other fields) is global optimization for *e.g.* optimal triangulation, resectioning and fundamental matrix estimation. See [5] and references therein. In some cases these

---

[1] This work has been funded by the Swedish Research Council through grant no. 2005-3230 'Geometry of multi-camera systems' and grant no. 2004-4579 'Image-Based Localization and Recognition of Scenes'.

D. Forsyth, P. Torr, and A. Zisserman (Eds.): ECCV 2008, Part IV, LNCS 5305, pp. 130–143, 2008.

optimization problems can be solved by finding the complete set of zeros of polynomial equations [6,7].

Solving systems of polynomial equations is known to be numerically very challenging and there exist no stable algorithm for the general case. Instead, specific solver algorithms are developed for each application. The state-of-the-art method for doing this is calculations with Gröbner bases. Typically, one obtains a floating point version of Buchberger's algorithm [8] by rewriting the various elimination steps using matrices and matrix operations [9]. These techniques have been studied and applied to vision problems in a number of cases [3,10,4]. However, for larger and more demanding problems Gröbner basis calculations are plagued by numerical problems [11,6].

A recently introduced, interesting approach to stabilisation of Gröbner basis computations is to study basis selection for the quotient space $\mathbb{C}[\mathbf{x}]/I$ [12], where I is the ideal generated by the set of equations. The choice of basis has been empirically shown to have a great impact on numerical performance and by adaptively selecting the basis for each instance of a problem one can obtain a dramatic increase in stability. In [12], a scheme based on singular value decomposition (SVD) was used to compute an orthogonal change of basis matrix. The SVD is a numerically very stable factorization method, but unfortunately also computationally rather expensive. Since the involved matrices tend to be large (around hundred rows and columns or more), the SVD computation easily dominates the running time of the algorithm.

In this paper, we propose a new fast strategy for selecting a basis for $\mathbb{C}[\mathbf{x}]/I$ based on QR-factorization with column pivoting. The Gröbner basis like computations employed to solve a system of polynomial equations can essentially be seen as matrix factorization of an under-determined linear system. Based on this insight, we combine the robust method of QR factorization from numerical linear algebra with the Gröbner basis theory needed to solve polynomial equations. More precisely, we employ QR-factorization with column pivoting in a crucial elimination step and obtain a simultaneous selection of basis and triangular factorization. With this method, we demonstrate an approximately fourfold increase in speed over the previous SVD based method while retaining good numerical stability.

Moreover, the technique of truncating the Gröbner basis to avoid large errors introduced in [13] fits nicely within the framework of column pivoting. Since the pivot elements are sorted in descending order, we get an adaptive criterion for where to truncate the Gröbner basis by setting a maximal threshold for the quotient between the largest and the smallest pivot element. When the quotient exceeds this threshold we abort the elimination and move the remaining columns into the basis. This way, we expand the basis only when necessary.

Factorization with column pivoting is a well studied technique and there exist highly optimized and reliable implementations of these algorithms in *e.g.* LAPACK [14], which makes this technique accessible and straight forward to implement. Matlab code for one of the applications, optimal triangulation from three views, is available at `http://www.maths.lth.se/vision/downloads`.

2 Review of Gröbner Basis Techniques for Polynomial Equation Solving

Solving systems of polynomial equations is a challenging problem in many respects and there exist no practical numerically stable algorithms for the general case. Instead special purpose algorithms need to be developed for specific applications. The state-of-the-art tool for doing this is calculations with Gröbner bases.

Our general goal is to find the complete set of solutions to a system

$$f_1(\mathbf{x}) = 0, \ldots, f_m(\mathbf{x}) = 0, \tag{1}$$

of m polynomial equations in s variables $\mathbf{x} = (x_1, \ldots, x_s)$. The polynomials $f_1, \ldots, f_m$ generate an *ideal* I in $\mathbb{C}[\mathbf{x}]$, the ring of multivariate polynomials in $\mathbf{x}$ over the field of complex numbers defined as the set

$$I = \{g : g(\mathbf{x}) = \Sigma_k h_k(\mathbf{x}) f_k(\mathbf{x})\}, \tag{2}$$

where the $h_k \in \mathbb{C}[\mathbf{x}]$ are any polynomials. The reason for studying the ideal I is that it has the same set of zeros as (1).

Consider now the space of equivalence classes modulo I. This space is denoted $\mathbb{C}[\mathbf{x}]/I$ and referred to as the *quotient space*. Two polynomials f and g are said to be equivalent modulo I if $f = g + h$, where $h \in I$. The logic behind this definition is that we get true equality, $f(\mathbf{x}) = g(\mathbf{x})$ on zeros of (1).

To do calculations in $\mathbb{C}[\mathbf{x}]/I$ it will be necessary to compute unique representatives of the equivalence classes in $\mathbb{C}[\mathbf{x}]/I$. Let $[\cdot] : \mathbb{C}[\mathbf{x}] \to \mathbb{C}[\mathbf{x}]/I$ denote the function that takes a polynomial f and returns the associated equivalence class $[f]$. We would now like to compose $[\cdot]$ with a mapping $\mathbb{C}[\mathbf{x}]/I \to \mathbb{C}[\mathbf{x}]$ that associates to each equivalence class a unique representative in $\mathbb{C}[\mathbf{x}]$. The composed map $\mathbb{C}[\mathbf{x}] \to \mathbb{C}[\mathbf{x}]$ should in other words take a polynomial f and return the unique representative $\overline{f}$ for the equivalence class $[f]$ associated with f. Assume for now that we can compute such a mapping. This operation will here be referred to as reduction modulo I.

A well known result from algebraic geometry now states that if the set of equations (1) has r zeros, then $\mathbb{C}[\mathbf{x}]/I$ will be a finite-dimensional linear space with dimension r [8]. Moreover, an elegant trick based on calculations in $\mathbb{C}[\mathbf{x}]/I$ yields the complete set of zeros of (1) in the following way: Consider multiplication by one of the variables x_k. This is a linear mapping from $\mathbb{C}[\mathbf{x}]/I$ to itself and since we are in a finite-dimensional space, by selecting an appropriate basis, this mapping can be represented as a matrix $\mathbf{m}_{x_k}$. This matrix is known as the *action matrix* and the eigenvalues of $\mathbf{m}_{x_k}$ correspond to x_k evaluated at the zeros of (1) [8]. Moreover, the eigenvectors of $\mathbf{m}_{x_k}$ correspond the vector of basis monomials/polynomials evaluated at the same zeros and thus the complete set of solutions can be directly read off from these eigenvectors. The action matrix can be seen as a generalization of the companion matrix to the multivariate case.

Given a linear basis $\mathcal{B} = \{[e_i]\}_{i=1}^r$ spanning $\mathbb{C}[\mathbf{x}]/I$, the action matrix $\mathbf{m}_{x_k}$ is computed by calculating $\overline{x_k e_i}$ for each of the basis elements $\overline{e_i}$. Performing this operation is the difficult part in the process. Traditionally, the reduction has been done by fixing a monomial ordering and then computing a Gröbner basis G for I, which is a canonical set of polynomials that generate I. Computing $\overline{f}$ is then done by polynomial division by G (usually written $\overline{f}^{G}$).

We now make two important observations: (i) We are not interested in finding the Gröbner basis *per se*; it is enough to get a well defined mapping $\overline{f}$ and (ii) it suffices to calculate reduction modulo I on the elements $x_k e_i$, *i.e.* we do not need to know what $\overline{f}$ is on all of $\mathbb{C}[\mathbf{x}]$. Note that if for some i, $x_k e_i \in \mathcal{B}$ then nothing needs to be done for that element. With this in mind, we denote by $\mathcal{R} = x_k \mathcal{B} \setminus \mathcal{B}$ the set of elements f for which we need to calculate representatives $\overline{f}$ of their corresponding equivalence classes $[f]$ in $\mathbb{C}[\mathbf{x}]/I$.

Calculating the Gröbner basis of I is typically accomplished by Buchberger's algorithm. This works well in exact arithmetic. However, in floating point arithmetic Buchberger's algorithm very easily becomes unstable. There exist some attempts to remedy this [15,16], but for more difficult cases it is necessary to study a particular class of equations (*e.g.* relative orientation for omnidirectional cameras [4], optimal three view triangulation [6], etc.) and use knowledge of what the structure of the Gröbner basis should be to design a special purpose Gröbner basis solver [9].

In this paper we move away from the goal of computing a Gröbner basis for I and focus on computing $\overline{f}$ for $f \in \mathcal{R}$ as mentioned above. However, it should be noted that the computations we do much resemble those necessary to get a Gröbner basis.

2.1   Computing Representatives for $\mathbb{C}[\mathbf{x}]/I$

In this section we show how representatives for $\mathbb{C}[\mathbf{x}]/I$ can be efficiently calculated in floating point arithmetic. The reason why Buchberger's algorithm breaks down in floating arithmetic is that eliminations of monomials are performed successively and this causes round-off errors to accumulate to the point where it is impossible to tell whether a certain coefficient should be zero or not. The trick introduced by Faugere [15] is to write the list of equations on matrix form

$$\mathbf{C}\mathbf{X} = 0, \tag{3}$$

where $\mathbf{X} = \left[\mathbf{x}^{\alpha_1} \ldots \mathbf{x}^{\alpha_n}\right]^{t}$ is a vector of monomials with the notation $\mathbf{x}^{\alpha_k} = x_1^{\alpha_{k1}} \cdots x_s^{\alpha_{ks}}$ and $\mathbf{C}$ is a matrix of coefficients. Elimination of leading terms now translates to matrix operations and we then have access to a whole battery of techniques from numerical linear algebra allowing us to perform many eliminations at the same time with control on pivoting etc.

By combining this approach with knowledge about a specific problem obtained in advance with a computer algebra system such as Macaulay2 [17] it is possible to write down a fixed number of expansion/elimination steps that will generate the necessary polynomials.

In this paper, we use a linear basis of monomials $\mathcal{B} = \{\mathbf{x}^{\alpha_1}, \ldots, \mathbf{x}^{\alpha_r}\}$ for $\mathbb{C}[\mathbf{x}]/I$. Recall now that we need to compute $\overline{x_k \mathbf{x}^{\alpha_i}}$ for $x_k \mathbf{x}^{\alpha_i} \notin \mathcal{B}$, *i.e.* for $\mathcal{R}$. This is the aim of the following calculations.

Begin by multiplying the equations (1) by a large enough set of monomials producing an equivalent (but larger) set of equations. We will come back to what *large enough* means. Thereafter, stack the coefficients of the new equations in an expanded coefficient matrix $\mathbf{C}_{\text{exp}}$, yielding

$$\mathbf{C}_{\text{exp}}\mathbf{X}_{\text{exp}} = 0. \tag{4}$$

Now partition the set of all monomials $\mathcal{M}$ occurring in the expanded set of equations as $\mathcal{M} = \mathcal{E} \bigcup \mathcal{R} \bigcup \mathcal{B}$ and order them so that $\mathcal{E} > \mathcal{R} > \mathcal{B}$ holds for all monomials in their respective sets. The monomials $\mathcal{E}$ ($\mathcal{E}$ for excessive) are simply the monomials which are neither in $\mathcal{R}$ nor in $\mathcal{B}$. This induces a corresponding partitioning and reordering of the columns of $\mathbf{C}_{\text{exp}}$:

$$\begin{bmatrix} \mathbf{C}_{\mathcal{E}} & \mathbf{C}_{\mathcal{R}} & \mathbf{C}_{\mathcal{B}} \end{bmatrix} \begin{bmatrix} \mathbf{X}_{\mathcal{E}} \\ \mathbf{X}_{\mathcal{R}} \\ \mathbf{X}_{\mathcal{B}} \end{bmatrix} = 0. \tag{5}$$

The $\mathcal{E}$-monomials are not in the basis and do not need to be reduced so we eliminate them by an LU decomposition on $\mathbf{C}_{\text{exp}}$ yielding

$$\begin{bmatrix} \mathbf{U}_{\mathcal{E}_1} & \mathbf{C}_{\mathcal{R}_1} & \mathbf{C}_{\mathcal{B}_1} \\ 0 & \mathbf{U}_{\mathcal{R}_2} & \mathbf{C}_{\mathcal{B}_2} \end{bmatrix} \begin{bmatrix} \mathbf{X}_{\mathcal{E}} \\ \mathbf{X}_{\mathcal{R}} \\ \mathbf{X}_{\mathcal{B}} \end{bmatrix} = 0, \tag{6}$$

where $\mathbf{U}_{\mathcal{E}_1}$ and $\mathbf{U}_{\mathcal{R}_2}$ are upper triangular. We can now discard the top rows of the coefficient matrix producing

$$\begin{bmatrix} \mathbf{U}_{\mathcal{R}_2} & \mathbf{C}_{\mathcal{B}_2} \end{bmatrix} \begin{bmatrix} \mathbf{X}_{\mathcal{R}} \\ \mathbf{X}_{\mathcal{B}} \end{bmatrix} = 0, \tag{7}$$

from which we get the elements of the ideal I we need since equivalently, if the submatrix $\mathbf{U}_{\mathcal{R}_2}$ is of full rank, we have

$$\mathbf{X}_{\mathcal{R}} = -\mathbf{U}_{\mathcal{R}_2}^{-1}\mathbf{C}_{\mathcal{B}_2}\mathbf{X}_{\mathcal{B}} \tag{8}$$

and then the $\mathcal{R}$-monomials can be expressed uniquely in terms of the $\mathcal{B}$-monomials. As previously mentioned, this is precisely what we need to compute the action matrix $\mathbf{m}_{x_k}$ in $\mathbb{C}[\mathbf{x}]/I$. In other words, the property of $\mathbf{U}_{\mathcal{R}_2}$ as being of full rank is sufficient to get the operation $\overline{f}$ on the relevant part of $\mathbb{C}[\mathbf{x}]$. Thus, in designing the set of monomials to multiply with (the first step in the procedure) we can use the rank of $\mathbf{U}_{\mathcal{R}_2}$ as a criterion for whether the set is large enough or not. However, the main problem in these computations is that even if $\mathbf{U}_{\mathcal{R}_2}$ is in principle invertible, it can be *very* ill conditioned.

A technique introduced in [12], which alleviates much of these problems uses basis selection for $\mathbb{C}[\mathbf{x}]/I$. The observation is that the right linear basis for $\mathbb{C}[\mathbf{x}]/I$ induces a reordering of the monomials, which has the potential to drastically improve the conditioning of $\mathbf{U}_{\mathcal{R}_2}$. Since $\mathbf{C}_{\text{exp}}$ depends on the data, the choice of linear basis cannot be made on beforehand, but has to be computed adaptively each time the algorithm is run. This leads to the difficult optimisation problem of selecting a linear basis so as to minimize the condition number of $\mathbf{U}_{\mathcal{R}_2}$. In [12] this problem was addressed by making use of SVD providing a numerically stable, but computationally expensive solution.

The advantage of the above exposition is that it makes explicit the dependence on the matrix $\mathbf{U}_{\mathcal{R}_2}$, both in terms of rank and conditioning. In particular, the above observations leads to the new fast strategy for basis selection which is the topic of the next section and a major contribution of this paper.

3   Column Pivoting as Basis Selection Strategy

In the one-variable case the monomials are given a natural ordering by their degree. In the multivariate case, there are several ways to order the monomials. To specify representatives for $\mathbb{C}[\mathbf{x}]/I$, one traditionally fixes one of these. The monomial order then automatically produces a linear basis for $\mathbb{C}[\mathbf{x}]/I$ in form of the set of monomials which are not divisible by the Gröbner basis in that monomial order.

For Buchberger's algorithm to make sense a monomial order is required to respect multiplication, *i.e.* $\mathbf{x}^\alpha > \mathbf{x}^\beta \Rightarrow x_k\mathbf{x}^\alpha > x_k\mathbf{x}^\beta$. Interestingly, when we relax the requirement of getting a strict Gröbner basis and compute $\overline{f}$ as outlined in the previous section, this property is unnecessarily strict. The crucial observation is that we can choose any linear basis for $\mathbb{C}[\mathbf{x}]/I$ we like, as long as we are able to compute well defined representatatives for the equivalence classes of $\mathbb{C}[\mathbf{x}]/I$. Thus, instead of letting the monomial order dictate the linear basis, we would like to do it the other way around and start by choosing a set of basis monomials $\mathcal{B}$.

After noting that we have some freedom in choosing $\mathcal{B}$, the first question is which monomials $\mathcal{P}$ (for permissible) in $\mathcal{M}$ are eligible for inclusion in the linear basis? Since we have to reduce the set $x_k\mathcal{B} \setminus \mathcal{B}$ to $\mathbb{C}[\mathbf{x}]/I$ we obviously have to require $x_k\mathcal{P} \subset \mathcal{M}$. Moreover, by making the construction leading up to (8), but replacing $\mathcal{B}$ by $\mathcal{P}$ we see that again the resulting $\mathbf{U}_{\mathcal{R}_2}$ needs to be of full rank to be able to guarantee reduction modulo I for all elements.

With these properties in place we aim at selecting $\mathcal{P}$ as large as possible and form $\begin{bmatrix} \mathbf{C}_\mathcal{E} & \mathbf{C}_\mathcal{R} & \mathbf{C}_\mathcal{P} \end{bmatrix}$. Any selection of basis monomials $\mathcal{B} \subset \mathcal{P}$ will then correspond to a matrix $\mathbf{C}_\mathcal{B}$ consisting of a subset of the columns of $\mathbf{C}_\mathcal{P}$.

By again performing an LU factorization and discarding the top rows to get rid of the $\mathcal{E}$-monomials, we get

$$\begin{bmatrix} \mathbf{U}_{\mathcal{R}_2} & \mathbf{C}_{\mathcal{P}_2} \\ \mathbf{0} & \mathbf{C}_{\mathcal{P}_3} \end{bmatrix} \begin{bmatrix} \mathbf{X}_\mathcal{R} \\ \mathbf{X}_\mathcal{P} \end{bmatrix} = 0, \tag{9}$$

in analogy with (7), where we now get zeros below $\mathbf{U}_{\mathcal{R}_2}$ since the larger $\mathbf{C}_\mathcal{P}$ means that we can still eliminate further. This is where the basis selection comes to play.

As noted above we can choose which monomials of the p monomials in $\mathcal{P}$ to put in the basis and which to reduce. This is equivalent to choosing a permutation Π of the columns of $\mathbf{C}_{\mathcal{P}_3}$ so that

$$\mathbf{C}_{\mathcal{P}_3}\varPi = \begin{bmatrix} c_{\pi(1)} & \cdots & c_{\pi(p)} \end{bmatrix}. \tag{10}$$

The goal must thus be to make this choice so as to minimize the condition number $\kappa(\begin{bmatrix} c_{\pi(1)} & \cdots & c_{\pi(p-r)} \end{bmatrix})$ of the first $p - r$ columns of the permuted matrix. In its generality, this is a difficult combinatorial optimization problem. However, the task can be approximately solved in an attractive way by QR factorization with column pivoting [18]. With this algorithm, $\mathbf{C}_{\mathcal{P}_3}$ is factorized as

$$\mathbf{C}_{\mathcal{P}_3}\varPi = \mathbf{QU}, \tag{11}$$

where $\mathbf{Q}$ is orthogonal and $\mathbf{U}$ is upper triangular. By solving for $\mathbf{C}_{\mathcal{P}_3}$ in (11) and substituting into (9) followed by multiplication from the left with $\begin{bmatrix} \mathbf{I} & \mathbf{0} \\ \mathbf{0} & \mathbf{Q}^t \end{bmatrix}$ and from the right with $\begin{bmatrix} \mathbf{I} & \mathbf{0} \\ \mathbf{0} & \varPi \end{bmatrix}$, we get

$$\begin{bmatrix} \mathbf{U}_{\mathcal{R}_2} & \mathbf{C}_{\mathcal{P}_2}\varPi \\ \mathbf{0} & \mathbf{U} \end{bmatrix} \begin{bmatrix} \mathbf{X}_{\mathcal{R}} \\ \varPi^t\mathbf{X}_{\mathcal{P}} \end{bmatrix} = 0, \tag{12}$$

We observe that $\mathbf{U}$ is not quadratic and emphasize this by writing $\mathbf{U} = \begin{bmatrix} \mathbf{U}_{\mathcal{P}_3} & \mathbf{C}_{\mathcal{B}_2} \end{bmatrix}$, where $\mathbf{U}_{\mathcal{P}_3}$ is quadratic upper triangular. We also write $\mathbf{C}_{\mathcal{P}_2}\varPi = \begin{bmatrix} \mathbf{C}_{\mathcal{P}_4} & \mathbf{C}_{\mathcal{B}_1} \end{bmatrix}$ and $\varPi^t\mathbf{X}_{\mathcal{P}_2} = \begin{bmatrix} \mathbf{X}_{\mathcal{P}'} & \mathbf{X}_{\mathcal{B}} \end{bmatrix}^t$ yielding

$$\begin{bmatrix} \mathbf{U}_{\mathcal{R}_2} & \mathbf{C}_{\mathcal{P}_4} & \mathbf{C}_{\mathcal{B}_1} \\ \mathbf{0} & \mathbf{U}_{\mathcal{P}_3} & \mathbf{C}_{\mathcal{B}_2} \end{bmatrix} \begin{bmatrix} \mathbf{X}_{\mathcal{R}} \\ \mathbf{X}_{\mathcal{P}'} \\ \mathbf{X}_{\mathcal{B}} \end{bmatrix} = 0 \tag{13}$$

and finally

$$\begin{bmatrix} \mathbf{X}_{\mathcal{R}} \\ \mathbf{X}_{\mathcal{P}'} \end{bmatrix} = - \begin{bmatrix} \mathbf{U}_{\mathcal{R}_2} & \mathbf{C}_{\mathcal{P}_4} \\ \mathbf{0} & \mathbf{U}_{\mathcal{P}_3} \end{bmatrix}^{-1} \begin{bmatrix} \mathbf{C}_{\mathcal{B}_1} \\ \mathbf{C}_{\mathcal{B}_2} \end{bmatrix} \mathbf{X}_{\mathcal{B}} \tag{14}$$

is the equivalent of (8) and amounts to solving r upper triangular equation systems which can be efficiently done by back substitution.

The reason why QR factorization fits so nicely within this framework is that it simultaneously solves the two tasks of reduction to upper triangular form and numerically sound column permutation and with comparable effort to normal Gaussian elimination.

Furthermore, QR factorization with column pivoting is a widely used and well studied algorithm and there exist free, highly optimized implementations, making this an accessible approach.

Standard QR factorization successively eliminates elements below the main diagonal by multiplying from the left with a sequence of orthogonal matrices (usually Householder transformations). For matrices with more columns than rows (under-determined systems) this algorithm can produce a rank-deficient $\mathbf{U}$ which would then cause the computations in this section to break down. QR with column pivoting solves this problem by, at iteration k, moving the column with greatest 2-norm on the last $m - k + 1$ elements to position k and then eliminating the last $m - k$ elements of this column by multiplication with an orthogonal matrix Q_k.

3.1   Adaptive Truncation

A further neat feature of QR factorization with column pivoting is that it provides a way of numerically estimating the conditioning of $\mathbf{C}_\mathcal{P}$ simultaneously with the elimination. In [13], it was shown that for reductions with a Gröbner basis, the Gröbner basis could be truncated yielding a larger representation of $\mathbb{C}[\mathbf{x}]/I$ (more than r basis elements), while retaining the original set of solutions. The advantage of this is that the last elements of the Gröbner basis often are responsible for a major part of the numerical instability and making use of the observation in [13], the last elements do not have to be computed.

As discussed earlier we do not calculate exactly a Gröbner basis, but the method of [13] is straightforward to adapt to the framework of this paper. However, both rank and conditioning of $\mathbf{C}_\mathcal{P}$ might depend on the data and we would therefore like to decide adaptively where to truncate, *i.e.* when to abort the QR factorization.

As a consequence of how the QR algorithm is formulated, the elements u_{kk} on the main diagonal of $\mathbf{U}$ will be sorted in decreasing absolute value. In exact arithmetic, if the rank is q, then $u_{kk} = 0$ for $k > q$. In floating point this will not be the case due to round-off errors. However, we can set a threshold τ and abort the elimination process once $|u_{kk}|/|u_{11}| < \tau$. The remaining columns (monomials) are then transfered to the basis which is correspondingly expanded.

Apart from being numerically sound, this strategy also spares some computational effort compared to setting a fixed larger basis. Truncating the set of polynomials means a higher dimensional representation of $\mathbb{C}[\mathbf{x}]/I$, which means we have to solve a larger eigenvalue problem. As will be shown in the experiments, the basis can usually be kept tight and only needs to be expanded in some cases.

4   Experiments

The purpose of this section is to verify the speed and accuracy of the QR-method. To this end, three different applications are studied. The first example is relative pose for generalised cameras, first solved by Stewénius *et al.* in 2005 [19]. The second one is the previously unsolved minimal problem of pose estimation with unknown focal length. The problem was formulated by Josephson *et al.* in [20], but not solved in floating point arithmetic. The last problem is optimal triangulation from three views [6].

Since the techniques described in this paper improve the numerical stability of the solver itself, but do not affect the conditioning of the actual problem, there is no point in considering the behavior under noise. Hence we will use synthetically generated examples without noise to compare the intrinsic numerical stability of the different methods.

In all three examples we compare with the "standard" method, by which we mean to fix a monomial order (typically grevlex) and use the basis dictated by that order together with straightforward gauss jordan elimination to express monomials in terms of the basis. Previous works have often used several expansion / elimination rounds. We have found this to have a negative effect on numerical stability so to make the comparison fair, we have implemented the standard method using a single elimination step in all cases.

For the adaptive truncation method, the threshold τ for the ratio between the k:th diagonal element and the first was set to 10^{-8}.

4.1 Relative Pose for Generalised Cameras

A generalised camera is a camera with no common focal point. This *e.g.* serves as a useful model for several ordinary cameras together with fixed relative locations [21]. For generalised cameras there is a minimal case for relative pose with two cameras and six points. This problem was solved in [19] and has 64 solutions. In [12] this problem was used to show how the SVD-method improved the numerics. We follow the methods of the later paper to get a single elimination step. This gives an expanded coefficient matrix of size 101×165 with the columns representing monomials up to degree eight in three variables. For details see [19] and [12].

The examples for this experiment were generated by picking six points from a normal distribution centered at the origin. Then six randomly chosen lines through these point were associated to each camera. This made up two generalised cameras with a relative orientation and translation.

Following this recipe, 10000 examples were generated and solved with the standard, QR- and SVD-method. The angular errors between true and estimated motion were measured. The results are shown in Figure 1.

The method with variable basis size was also implemented, but for this example the $\mathbf{U}_{\mathcal{R}_2}$ (see Equation 7) part of the coefficient matrix was always reasonably conditioned and hence the basis size was 64 in all 10000 test examples. There were no large errors for neither the SVD nor the QR method.

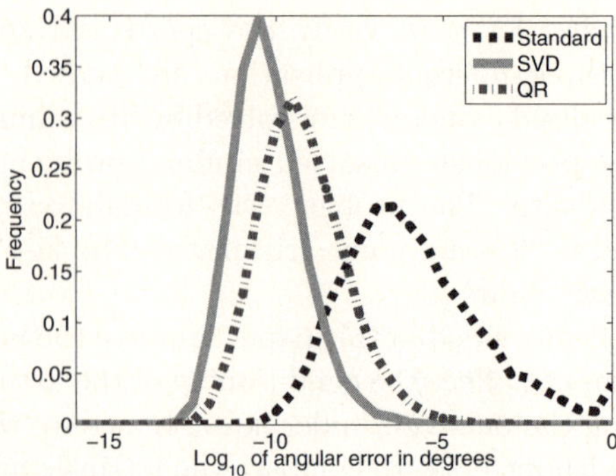

Fig. 1. Error distributions for the problem of relative pose with generalised cameras. The SVD-method yields the best results but the faster QR-method is not far behind and also eliminates all large errors.

4.2   Localisation with Hybrid Features

This problem was introduced in [20]. The problem is to find the pose of a calibrated camera with unknown focal length. One minimal setup for this problem is three point-correspondences with known world points and one correspondence to a world line. The last feature is equivalent to having a point correspondence with another camera. These types of mixed features are called hybrid features. In [20], the authors propose a parameterisation of the problem but no solution was given apart from showing that the problem has 36 solutions.

The parameterisation in [20] gives four equations in four unknowns. The unknowns are three quaternion parameters and the focal length. The equation derived from the line correspondence is of degree 6 and those obtained from the 3D points are of degree 3. The coefficient matrix $\mathbf{C}_{\mathrm{exp}}$ is then constructed by expanding all equations up to degree 10. This means that the equation derived from the line is multiplied with all monomials up to degree 4, but no single variable in the monomials is of higher degree than 2. In the same manner the point correspondence equations are multiplied with monomials up to degree 7 but no single variable of degree more than 5. The described expansion gives 980 equations in 873 monomials.

The next step is to reorder the monomials according to (5). In this problem $\mathbf{C}_{\mathcal{P}}$ corresponds to all monomials up to degree 4 except f^4 where f is the focal length, this gives 69 columns in $\mathbf{C}_{\mathcal{P}}$. The part $\mathbf{C}_{\mathcal{R}}$ corresponds to the 5:th degree monomials that appears when the monomials in $\mathcal{B}$ are multiplied with the first of the unknown quaternion parameters.

For this problem, we were not able to obtain a standard numerical solver. The reason for this was that even going to significantly higher degrees than mentioned above, we did not obtain an invertible $\mathbf{U}_{\mathcal{R}_2}$. In fact, with an exact linear basis (same number of basis elements as solutions), even the QR and SVD methods failed and truncation had to be used.

In this example we found that increasing the linear basis of $\mathbb{C}[\mathbf{x}]/I$ by a few elements over what was produced by the adaptive criterion was beneficial for the stability. In this experiment, we added three basis elements to the automatically produced basis. To get a working version of the SVD solver we had to adapt the truncation method to the SVD case as well. We did this by looking at the ratio of the singular values.

The synthetic experiments for this problem were generated by randomly drawing four points from a cube with side length 1000 centered at the origin and two cameras with a distance of approximately 1000 to the origin. One of these cameras was treated as unknown and one was used to get the camera to camera point correspondence. This gives one unknown camera with three point correspondences and one line correspondence. The experiment was run 10000 times.

In Figure 2 (right) the distribution of basis sizes is shown for the QR-method. For the SVD-method the basis size was identical to the QR-method in over 97% of the cases and never differed by more than one element.

Figure 2 (left) gives the distribution of relative errors in the estimated focal length. It can be seen that both the SVD-method and the faster QR-method

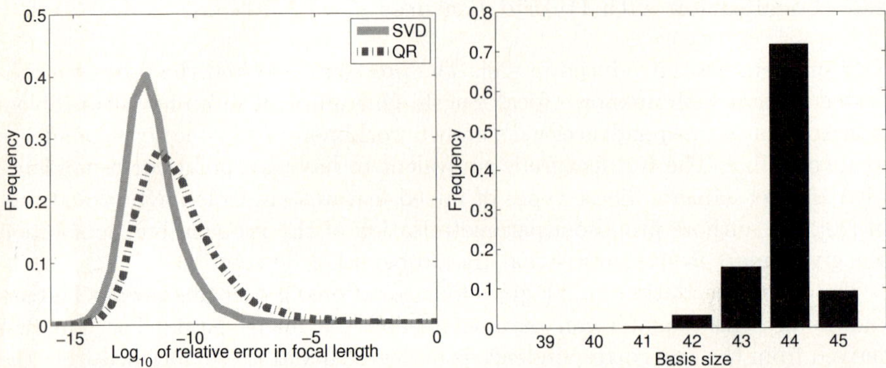

Fig. 2. Left: Relative error in focal length for pose estimation with unknown focal length. Both the SVD- and QR-methods uses adaptive truncatation. Right: The size of the adaptively chosen basis for the QR-method. For the SVD-method the size differs from this in less than 3% of the cases and by at most one element.

give useful results. We emphasize that we were not able to construct a solver with the standard method and hence no error distribution for that method is available.

4.3 Optimal Triangulation from Three Views

The last experiment does not concern a geometrical minimal case, but instead deals with an optimisation problem. Given noisy image measurements in three views, the problem is to find the world point that minimises the sum of squares of reprojection errors. This is the statistically optimal estimate under Gaussian noise.

We find the global minimum by calculating the complete set of stationary points of the reprojection error function. This was first done in [6], where the standard Gröbner basis method was used. However, because of numerical problems they were forced to use extremely slow, emulated 128 bit numerics to get accurate results. In [12] the SVD-method was later used to enable calculations in standard double precision. It should be mentioned that a more recently introduced and probably more practical global optimisation method for triangulation is given in [22]. Still though, this problem serves as an interesting test bed for equation solving.

For details on the construction of the coefficient matrix see [12,6]. The coefficient matrix constructed with this method is of size 225×209 and the number of solutions is 50. The QR-method was implemented as described earlier and the method with variable size basis was used. For reference, we implemented the method of [6] in standard double precision, with some small modifications to get a single elimination step (this made it slightly more stable).

The unknown point was randomly placed in a cubic box with side 1000 centered around the origin. The three cameras were placed approximately on a sphere with distance 1000 from origin and the focal lengths were also set to around 1000. The error in 3D placement over 10000 iterations is shown in Figure 3. It can be seen that the QR-method is almost as accurate as the SVD-method.

One important property of a solver is that the number of large errors is small. Thus in Table 1 the number of large errors are shown. The results show that the QR-method is better at suppressing large errors, probably due to the variable size of the basis.

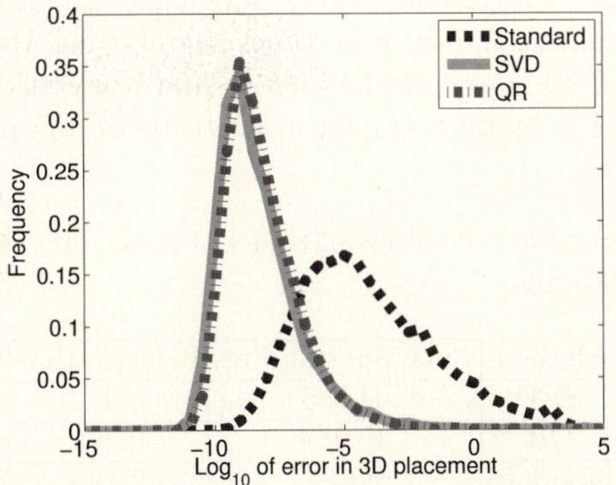

Fig. 3. The distribution of the error in 3D placement of the unknown point using optimal three view triangulation. The experiment was run 10000 times. The QR-method gives nearly identical results compared to the SVD-method.

Table 1. Number of errors larger than some levels. This shows that the QR-method gives fewer large errors probably due to the variable size of the basis.

Error	> 1	$> 10^{-1}$	$> 10^{-2}$	$> 10^{-3}$
QR	10	15	28	54
SVD	31	39	52	79

Table 2. Number of times a certain basis size appears in 10000 iterations. The largest basis size obtained in the experiment was 66.

Basis size	50	51	52	53	54	55	≥ 56
#	9471	327	62	34	26	17	58

4.4 Speed Comparison

In the problem of optimal three view triangulation the execution times for the three different algorithms were measured. Since the implementations were done in Matlab it was necessary to take care to eliminate the effect of Matlab being an interpreted language. To do this only the time after construction of the coefficient matrix was taken into account. This is because the construction of the coefficient matrix essentially amounts to copying coefficients to the right places which can be done extremely fast in *e.g.* a C language implementation.

In the routines that were measured no subroutines were called that were not built-in functions in Matlab. The measurements were done with Matlab's profiler.

The time measurements were done on an Intel Core 2 2.13 GHz machine with 2 GB memory. Each algorithm was executed with 1000 different coefficient matrices, these were constructed from the same type of scene setup as in the previous section. The same set of coefficient matrices was used for each method. The result is given in Table 3. Our results show that the QR-method with adaptive truncation is approximately four times faster than the SVD-method but 40% slower than the standard method. It should however be noted that here, the standard method is by far too inaccurate to be of any practical value.

Table 3. Time consumed in the solver part for the three different methods. The time is an average over 1000 calls.

Method	Time per call / ms	Relative time
SVD	41.685	1
QR	10.937	0.262
Standard	8.025	0.193

5   Conclusions

In this paper we have presented a new fast strategy for improving numerical stability of Gröbner basis polynomial equation solvers. The key contribution is a clarification of the exact matrix operations involved in computing an action matrix for $\mathbb{C}[\mathbf{x}]/I$ and the use of numerically sound QR factorization with column pivoting to obtain a simultaneous basis selection for $\mathbb{C}[\mathbf{x}]/I$ and reduction to upper triangular form. We demonstrate a nearly fourfold decrease in computation time compared to the previous SVD based method while retaining good numerical stability. Moreover, since the method is based on the well studied, freely available QR algorithm it is reasonably simple to implement and not much slower than using no basis selection at all.

The conclusion is thus that whenever polynomial systems arise and numerical stability is a concern, this method should be of interest.

References

1. Chasles, M.: Question 296. Nouv. Ann. Math. 14 (1855)
2. Kruppa, E.: Zur Ermittlung eines Objektes aus Zwei Perspektiven mit innerer Orientierung. Sitz-Ber. Akad. Wiss., Wien, math. naturw. Kl. Abt IIa, 1939–1948 (1913)
3. Kukelova, Z., Pajdla, T.: A minimal solution to the autocalibration of radial distortion. In: CVPR (2007)

4. Geyer, C., Stewénius, H.: A nine-point algorithm for estimating para-catadioptric fundamental matrices. In: CVPR, Minneapolis, USA (2007)
5. Hartley, R., Kahl, F.: Optimal algorithms in multiview geometry. In: Yagi, Y., Kang, S.B., Kweon, I.S., Zha, H. (eds.) ACCV 2007, Part I. LNCS, vol. 4843, pp. 13–34. Springer, Heidelberg (2007)
6. Stewénius, H., Schaffalitzky, F., Nistér, D.: How hard is three-view triangulation really? In: Proc. Int. Conf. on Computer Vision, Beijing, China, pp. 686–693 (2005)
7. Hartley, R., Sturm, P.: Triangulation. Computer Vision and Image Understanding 68, 146–157 (1997)
8. Cox, D., Little, J., O'Shea, D.: Ideals, Varieties, and Algorithms. Springer, Heidelberg (2007)
9. Stewénius, H.: Gröbner Basis Methods for Minimal Problems in Computer Vision. PhD thesis, Lund University (2005)
10. Stewénius, H., Kahl, F., Nistér, D., Schaffalitzky, F.: A minimal solution for relative pose with unknown focal length. In: Proc. Conf. Computer Vision and Pattern Recognition, San Diego, USA (2005)
11. Kukelova, Z., Pajdla, T.: Two minimal problems for cameras with radial distortion. In: Proceedings of The Seventh Workshop on Omnidirectional Vision, Camera Networks and Non-classical Cameras (OMNIVIS) (2007)
12. Byröd, M., Josephson, K., Åström, K.: Improving numerical accuracy of gröbner basis polynomial equation solvers. In: Proc.11th Int. Conf. on Computer Vision, Rio de Janeiro, Brazil (2007)
13. Byröd, M., Josephson, K., Åström, K.: Fast optimal three view triangulation. In: Asian Conference on Computer Vision (2007)
14. Anderson, E., et al.: LAPACK Users' Guide. Third edn. Society for Industrial and Applied Mathematics, Philadelphia, PA (1999)
15. Faugère, J.C.: A new efficient algorithm for computing gröbner bases (f_4). Journal of Pure and Applied Algebra 139, 61–88 (1999)
16. Faugère, J.C.: A new efficient algorithm for computing gröbner bases without reduction to zero (f5). In: ISSAC 2002, pp. 75–83. ACM Press, New York (2002)
17. Grayson, D., Stillman, M.: Macaulay 2 (1993)-2002),
http://www.math.uiuc.edu/Macaulay2
18. Golub, G.H., van Loan, C.F.: Matrix Computations, 3rd edn. The Johns Hopkins University Press (1996)
19. Stewénius, H., Nistér, D., Oskarsson, M., Åström, K.: Solutions to minimal generalized relative pose problems. In: OMNIVIS, Beijing, China (2005)
20. Josephson, K., Byröd, M., Kahl, F., Åström, K.: Image-based localization using hybrid feature correspondences. In: BenCOS 2007 (2007)
21. Pless, R.: Using many cameras as one. In: Proc. Conf. Computer Vision and Pattern Recognition, Madison, USA (2003)
22. Lu, F., Hartley, R.: A fast optimal algorithm for l_2 triangulation. In: Yagi, Y., Kang, S.B., Kweon, I.S., Zha, H. (eds.) ACCV 2007, Part II. LNCS, vol. 4844, pp. 279–288. Springer, Heidelberg (2007)

Co-recognition of Image Pairs
by Data-Driven Monte Carlo Image Exploration

Minsu Cho, Young Min Shin, and Kyoung Mu Lee

Department of EECS, ASRI, Seoul National University, 151-742, Seoul, Korea
`minsucho@diehard.snu.ac.kr`, `shinyoungmin@gmail.com`, `kyoungmu@snu.ac.kr`

Abstract. We introduce a new concept of 'co-recognition' for object-level image matching between an arbitrary image pair. Our method augments putative local region matches to reliable object-level correspondences without any supervision or prior knowledge on common objects. It provides the number of reliable common objects and the dense correspondences between the image pair. In this paper, generative model for co-recognition is presented. For inference, we propose data-driven Monte Carlo image exploration which clusters and propagates local region matches by Markov chain dynamics. The global optimum is achieved by a guiding force of our data-driven sampling and posterior probability model. In the experiments, we demonstrate the power and utility on image retrieval and unsupervised recognition and segmentation of multiple common objects.

1 Introduction

Establishing correspondences between image pairs is one of the fundamental and crucial issues for many vision problems. Although the development of various kinds of local invariant features [1,2,3] have brought about notable progress in this area, their local ambiguities remain hard to be solved. Thus, domain specific knowledge or human supervision has been generally required for accurate matching. Obviously, the best promising strategy to eliminate the ambiguities from local feature correspondences is to go beyond locality [4,5,6,7]. The larger image regions we exploit, the more reliable correspondences we can obtain. In this work we propose a novel data-driven Monte Carlo framework to augment naive local region correspondences to reliable object-level correspondences in an arbitrary image pair. Our method establishes multiple coherent clusters of dense correspondences to achieve recognition and segmentation of multiple common objects without any prior knowledge of specific objects.

For the purpose, we introduce a perceptually meaningful entity, which can be interpreted as a common object or visual pattern. We will refer to the entity in an image pair as a Maximal Common Saliency (MCS) and define it as follows: (1) An MCS is a semi-global region pair, composed of local region matches between the image pair. (2) The region pair should be mutually consistent in geometry and photometry. (3) Each region of the pair should be maximal in size. Now, the goal of our work is defined to obtain the set of MCSs from an image pair. According to the naming conventions of some related works [5,8], we term it *co-recognition*.

D. Forsyth, P. Torr, and A. Zisserman (Eds.): ECCV 2008, Part IV, LNCS 5305, pp. 144–157, 2008.
© Springer-Verlag Berlin Heidelberg 2008

Fig. 1. Result of co-recognition on our dataset *Mickey's*. Given an image pair, co-recognition detects all Maximal Common Saliencies without any supervision or prior knowledge. Each color represents each identity of the MCS, which means an object in this case. Note that the book (blue) is separated by occlusion but identified as one object. See the text for details.

As shown in Fig. 1, co-recognition is equivalent to recognizing and segmenting multiple common objects in a given image pair under two conditions: (1) All the common object appears mutually distinctive in geometry. (2) Each common object lies on different backgrounds in photometry.[1] Note that it can detect separated regions by occlusion as a single object without any prior knowledge or supervision. In this problem, local region correspondences can be established more easily if reliable poses of common objects are known in advance, and the converse is also true. We pose this chicken-and-egg problem in terms of data-driven Monte Carlo sampling with reversible jump dynamics [9,10] over the intra- and inter-image domain. A main advantage of our formulation is to combine bottom-up and top-down processes in a integrated and principled way. Thus, global MCS correspondences and their local region correspondences reinforce each other simultaneously so as to reach global optimum.

Among recent works related to ours are co-segmentation [8], co-saliency [5], and common visual pattern discovery [11]. Rother et al. [8] defined co-segmentation as segmenting common regions simultaneously in two images. They exploited a generative MRF-based graph model and the color histogram similarity measure. Toshev et al. [5] defined co-saliency matching as searching for regions which have strong intra-image coherency and high inter-image similarity. The method takes advantage of the segmentation cue to address the ambiguity of local feature matching. Yuan and Wu [11] used spatial random partition to discover common visual patterns from a collection of images. The common pattern is localized by aggregating the matched set of partitioned images. However, none of these methods recognize multiple common objects as distinct entities. Moreover, [8] and [11] do not consider geometrical consistency in the detected region.

---

[1] With first condition unsatisfied, several distinct common objects can be recognized as one. With second unsatisfied, objects can include a portion of similar background.

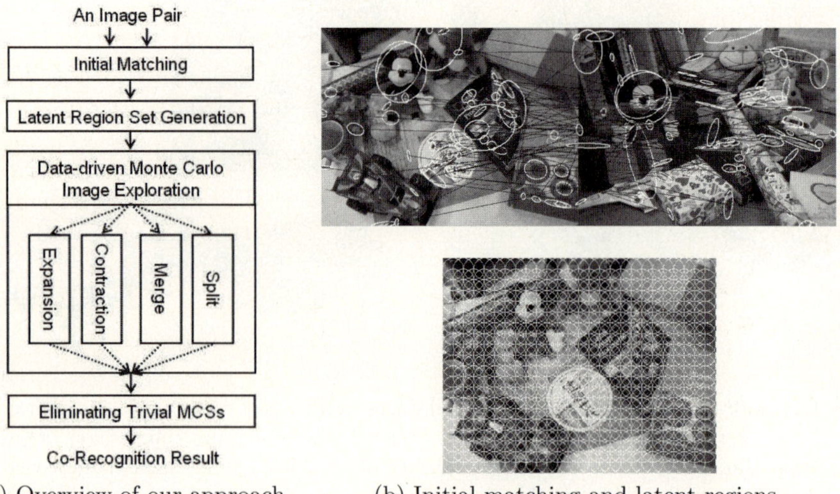

(a) Overview of our approach   (b) Initial matching and latent regions

Fig. 2. (a) Given two images, data-driven Monte Carlo image exploration solves co-recognition problem of the image pair. See the text for details. (b) Top: Several different types of local features can be used for initial matches. Bottom: Overlapping circular regions are generated covering the whole reference image for latent regions.

Our method has been inspired by the image exploration method for object recognition and segmentation, proposed by Ferrari et al [6]. The method is based on propagating initial local matches to neighboring regions by their affine homography. Even with few true initial matches, their iterative algorithm expands inliers and contracts outliers so that the recognition can be highly improved.[2] The similar correspondence growing approaches were proposed also in [4,7] for non-rigid image registration. Our new exploration scheme leads the image exploration strategy of [6] to unsupervised multi-object image matching by the Bayesian formulation and the DDMCMC framework [9]. Therefore, the co-recognition problem addressed by this paper can be viewed as a generalization of several other problems reported in the literature [5,6,8,11].

2   Overview of the Approach

Given an image pair, the goal of co-recognition is to recognize and segment all the MCSs and infer their dense correspondences in the pair simultaneously. Figure 2(a) illustrates the overview of our algorithm. First, we obtain initial affine region matches using several different types of local affine invariant features [3,2]. Then, each initial match forms an initial cluster by itself, which is a seed for an MCS. Second, one of the pair is set to be the reference image, and we generate

---

[2] Although recent object recognition and segmentation methods [12,13] based on local region features demonstrate more accurate results in segmentation, they require enough inliers to localize the object in their initialization step.

a grid of overlapping circular regions covering the whole reference image. All the overlapping regions are placed into a latent region set Λ, in which each element region waits to be included in one of the existing clusters (Fig. 2(b)). After these initialization steps, our data-driven Monte Carlo image exploration algorithm starts to search for the set of MCSs by two pairs of reversible moves; expansion/contraction and merge/split. In the expansion/contraction moves, a cluster obtains a new match or lose one. In merge/split moves, two clusters are combined into one cluster, or one cluster is divided into two clusters. Utilizing all these moves in a stochastic manner, our algorithm traverses the solution space efficiently to find the set of MCSs. The final solution is obtained by eliminating trivial MCSs from the result.

3   Generative Model of Co-recognition

We formulate co-recognition as follows. The set of MCSs is denoted by a vector of unknown variables θ which consists of clusters of matches:

$$\theta = (K, \{\Gamma_i \; ; \; i = 1, ..., K\}), \tag{1}$$

where Γ_i represents a cluster of matches, K means the number of clusters. Γ_i consists of local region matches across the image pair, expressed as follows:

$$\Gamma_i = \{(R_j, T_j); j = 1, ..., L_i\}, \tag{2}$$

where R_j denotes a small local region of the reference image, T_j indicates an affine transformation that maps the region R_j to the other image.[3] L_i denotes the number of local regions included in the cluster Γ_i.

In the Bayesian framework, we denote the posterior probability $p(\theta|I)$ as the probability of θ being the set of MCSs given an image pair I, which is proportional to the product of the prior $p(\theta)$ and the likelihood $p(I|\theta)$. Therefore, co-recognition is to find θ^* that maximizes this posterior as follows.

$$\theta^* = \arg\max_{\theta} p(\theta|I) = \arg\max_{\theta} p(I|\theta)p(\theta). \tag{3}$$

3.1   The Prior $p(\theta)$

The prior $p(\theta)$ models the geometric consistency and the maximality of MCSs.

Geometric Consistency of MCSs. To formulate the geometric constraint of a cluster Γ_i, we used the sidedness constraint of [6], and reinforced it with orientation consistency. Consider a triple (R_j, R_k, R_l) of local regions in the reference image and their corresponding regions $(R_j{}', R_k{}', R_l{}')$[4] in the other image. Let $c_j, c_j{}'$ be the centers of regions $R_j, R_j{}'$, respectively. Then, the sidedness constraint,

$$\mathrm{sign}((c_k \times c_l)c_j) = \mathrm{sign}((c_k{}' \times c_l{}')c_j{}') \tag{4}$$

---

[3] Registration of non-planar 3-d surfaces is approximated by a set of linear transformations of small local regions.

[4] That is, $R_i{}' = T_i R_i$.

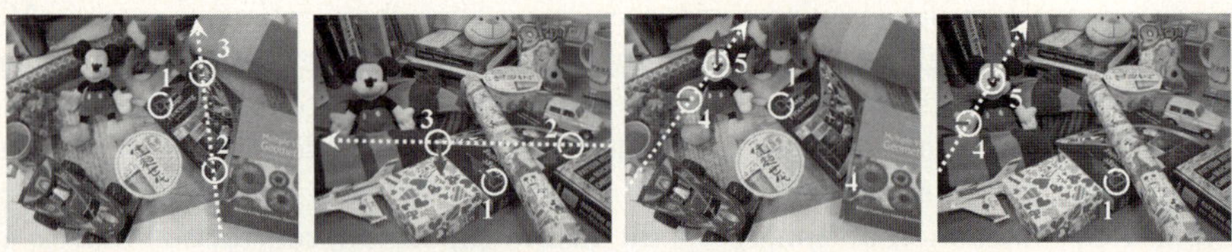

(a) Sidedness constraint   b) Reinforced with orientation consistency

Fig. 3. (a) 1 should be on the same side of the directed line from 2 to 3 in both images. (b) 4,5 and 1 satisfies sidedness constraint, while it does not lies on the same object. We can filter out this outlier triplet by checking if orientation(red arrow) changes in the triplet are mutually consistent.

means that the side of c_j w.r.t the directed line $(c_k \times c_l)$ should be just the same as the side of $c_j{}'$ w.r.t the directed line $(c_k{}' \times c_l{}')$ (Fig. 3(a)). This constraint holds for all correctly matching triplets of coplanar regions. Since the sidedness constraint is valid even for most non-planar regions, it is useful for sorting out triplets on a common surface. As illustrated in Fig. 3(b), we reinforce it with orientation consistency to deal with multiple common surfaces for our problem as follows:

$$\forall (m,n) \in \{(j,k),(j,l),(k,l)\}, \quad |\text{angle}(\text{angle}(o_m, o_m{}'), \text{angle}(o_n, o_n{}'))| < \delta_{\text{ori}} \tag{5}$$

where o_m means the dominant orientation of R_m in radian, while angle() denotes the function which calculates the clockwise angle diffrence in radian. Hence, the reinforced sidedness error with the orientation consistency is defined by

$$\text{err}_{\text{side}}(R_j, R_k, R_l) = \begin{cases} 0 & \text{if (4) and (5) hold} \\ 1 & \text{otherwise} \end{cases} \tag{6}$$

A triple violating the reinforced sidedness constraint has higher chances of having one or more mismatches in it. The geometric error of $R_j (\in \Gamma_i)$ is defined by the share of violations in its own cluster such that

$$\text{err}_{\text{geo}}(R_j) = \frac{1}{v} \sum_{R_k, R_l \in \Gamma_i \backslash R_j, k>l} \text{err}_{\text{side}}(R_j, R_k, R_l), \tag{7}$$

where $v = (L_i-1)(L_i-2)/2$ is the normalization factor that counts the maximum number of violations. When $L_i < 3$, $\text{err}_{\text{geo}}(R_j)$ is defined as 1 if the cluster $\Gamma_i(\ni R_j)$ violates the orientation consistency, otherwise 0.

The geometric error of a cluster is then defined by the sum of errors for all members in the cluster as follows:

$$\text{err}_{\text{geo}}(\Gamma_i) = \sum_{j=1}^{L_i} \text{err}_{\text{geo}}(R_j). \tag{8}$$

Maximality of MCSs. To encode the degree of maximality of θ, the relative area of each cluster should be examined. We approximate it by the number of matches in each cluster since all the latent regions have the same area and the number is constant after initialization. The maximality error is formulated as

$$\mathrm{err}_{\mathrm{maxi}}^{\cdot}(\theta) = \sum_{i=1}^{K} \left((\frac{L_i}{N})^{0.8} - \frac{L_i}{N} \right), \tag{9}$$

where N is the initial number of the latent region set Λ. The first term encourages the clusters of θ to merge, and the second term makes each cluster of θ to expand.

3.2   Likelihood $p(I|\theta)$

Photometric Consistency of MCSs. The likelihood encodes the photometric consistency of θ using the observation of the given image pair. Let us define the dissimilarity of two regions by

$$\mathrm{dissim}(R_1, R_2) = 1 - \mathrm{NCC}(R_1, R_2) + \frac{\mathrm{dRGB}(R_1, R_2)}{100}, \tag{10}$$

where NCC is the normalized cross-correlation between the gray patterns, while dRGB is the average pixel-wise Euclidean distance in RGB color-space after independent normalization of the 3 colorbands for photometric invariance [6]. R_1 and R_2 are normalized to unit circles with the same orientation before computation. Since a cluster of matches should have low dissimilarity in each match, the overall photometric error of a cluster is defined as follows.

$$\mathrm{err}_{\mathrm{photo}}(\Gamma_i) = \sum_{j=1}^{L_i} \mathrm{dissim}(R_j, R_j{}')^2. \tag{11}$$

Visual patterns in each MCS are assumed to be mutually independent in our model. Hence, the likelihood is defined as follows.

$$p(I|\theta) \propto \exp\left(-\lambda_{\mathrm{photo}} \sum_{i=1}^{K} \mathrm{err}_{\mathrm{photo}}(\Gamma_i) \right). \tag{12}$$

3.3   Integrated Posterior $p(\theta|I)$

From (8), (9), and (12), MCSs in a given image pair I can be obtained by maximizing the following posterior probability:

$$p(\theta|I) \propto \exp\left(-\lambda_{\mathrm{geo}} \sum_{i=1}^{K} \mathrm{err}_{\mathrm{geo}}(\Gamma_i) - \lambda_{\mathrm{maxi}} \mathrm{err}_{\mathrm{maxi}}(\theta) - \lambda_{\mathrm{photo}} \sum_{i=1}^{K} \mathrm{err}_{\mathrm{photo}}(\Gamma_i) \right). \tag{13}$$

This posterior probability reflects how well the solution generates the set of MCSs from the given image pair.

4   Data-Driven Monte Carlo Image Exploration

The posterior probability $p(\theta|I)$ in (13) has a high-dimensional and complicated landscape with a large number of local maxima. Moreover, maximizing the posterior is a trans-dimensional problem because neither the number of MCSs nor the number of matches in each MCS are known. To pursue the global optimum of this complex trans-dimensional posterior $p(\theta|I)$, we propose a new image exploration algorithm based on the reversible jump MCMC [10] with data-driven techniques [9].

The basic idea of MCMC is to design a Markov chain to sample from a probability distribution $p(\theta|I)$. At each sampling step, we propose a candidate state θ' from a proposal distribution $q(\theta'|\theta)$. Through the Metropolis-Hastings rule, the candidate state is accepted with the following acceptance probability.

$$\alpha = \min\left(1, \frac{q(\theta|\theta')p(\theta'|I)}{q(\theta'|\theta)p(\theta|I)}\right). \tag{14}$$

Theoretically, it is proven that the Markov chain constructed in this manner has its stationary distribution as $p(I|\theta)$ irrespective of the choice of the proposal $q(\theta'|\theta)$ and the initial state [10]. Nevertheless, in practice, the choice of the proposal significantly affects the efficiency of MCMC. Recently in computer vision area, data-driven MCMC [9] has been proposed and proven to improve the efficiency by incorporating domain knowledge in proposing new states of the Markov chain. In our algorithm, we adopt the data-driven techniques to guide our Markov chain using the current observation obtained by local region matches in the image pair. Our Markov chain kernel consists of two pairs of reversible jump dynamics which perform expansion/contraction and merge/split, respectively. At each sampling step, a move $m \in \{\text{expand, contract, split, merge}\}$ is selected with the constant probability $q(m)$.

4.1   Expansion/Contraction Moves

Expansion is to increase the size of an existing cluster by picking a region out of the latent region set Λ and propagating it with a support region in the cluster. Conversely, contraction functions to decrease the size by taking a region out of the members in the cluster and sending it back to Λ. Suppose, at a certain sampling step, that a cluster Γ_i is expanded to Γ_i', or conversely that Γ_i' is contracted to Γ_i, then this process can be expressed as the following form without loss of generality:

$$\theta = (K, \{\Gamma_i, ...\}) \leftrightarrow (K, \{\Gamma_i', ...\}) = \theta', \text{where } \Gamma_i \cup (R_k, T_k) = \Gamma_i'. \tag{15}$$

The Pathway to Propose Expansion. An expansion move is proposed by the following stochastic procedure with data-driven techniques. Firstly, a cluster is chosen among the current K clusters with the probability $q(\Gamma_i|\text{expand}) \propto \sqrt{L_i}$, which reflects a preference to larger clusters. Secondly, among the matches in the cluster, a support for propagation is selected with probability $q(R_j|\Gamma_i, \text{expand}) \propto$

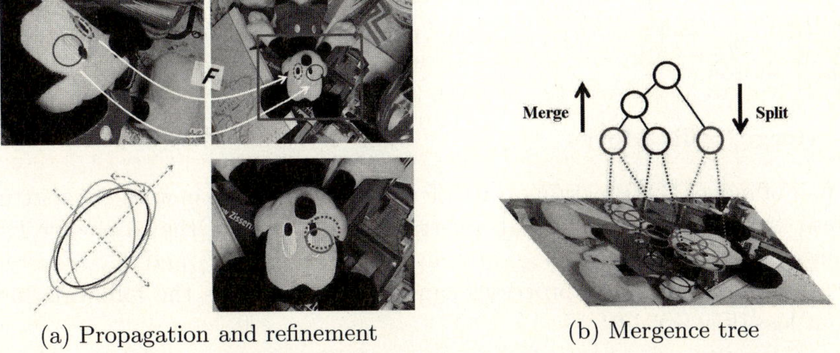

(a) Propagation and refinement   (b) Mergence tree

Fig. 4. (a) At the top, a support match (red dotted) propagates one of the latent regions (blue solid) by affine homography F. At the bottom, by adjusting the parameter of the ellipse, the initially propagated region (blue dotted) is refined into the more accurate region (green solid). (b) Each of the present clusters has its own mergence tree, which stores hierarchical information of the preceding clusters of itself. It helps to propose a simple and reversible merge/split moves at low cost.

$\sum_{R\in\Lambda}\exp\left(-\frac{\text{dist}(R_j,R)}{2\sigma^2_{\text{expand}}}\right)$, where dist() denotes the Euclidean distance between the region centers. In this stochastic selection, the supports that have more latent regions at nearer distance are favored. Finally, a latent region to propagate by the support is chosen with the probability $q(R_k|R_j,\Gamma_i,\text{expand}) \propto \exp\left(-\frac{\text{dist}(R_k,R_j)^2}{2\sigma^2_{\text{expand}}}\right)$, which means a preference to closer ones.

Propagation Attempt and Refinement. The building block of expansion is based on the propagation attempt and refinement in [6]. If an expansion move is proposed, we perform a propagation attempt followed by the refinement. As illustrated in Fig. 4(a), consider the case that a red dotted elliptical region R_1 in the reference image is already matched to R'_1 in the other image. Each R_1 and R'_1 has an affine transformation A and A', respectively, which transform the regions onto the orientation normalized unit circles. Thus, we can get the affine homography F between R_1 to R'_1 by $F = (A')^{-1}A$, satisfying $FR_1 = R'_1$. If a latent region R_2 is close enough to R_1 and lie on the same physical surface, we can approximate R'_2 in the other image by $R'_2 = FR_2$ as shown in Fig. 4(a). In that case, we state that the support match (R_1, R'_1) attempts to propagate the latent region R_2. Next, by locally searching the parameter space of the current affine homography F, the refiner adjusts it to find R'_2 with minimum dissimilarity such that $F_r = \arg\min_F \text{dissim}(R_2, FR_2)$ as shown at the bottom of Fig. 4(a).

The Pathway to Propose Contraction. An previously expanded region is proposed to contract by the following stochastic procedure with data-driven techniques. Firstly, a cluster is chosen among the current K clusters with the probability $q(\Gamma_i|\text{contract}) \propto \sqrt{L_i}$. Then, among matches supporting no other region in the cluster, one match is selected with the probability $q(R_k|\Gamma_i, \text{contract}) \propto$ $\exp\left(\frac{\text{err}_{\text{geo}}(R_k)^2 + \text{err}_{\text{photo}}(R_k)^2}{2\sigma^2_{\text{contract}}}\right)$, favoring the matches with higher error in geometry and photometry.

4.2 Merge/Split Moves

This pair of moves is for merging two different clusters into a new one or splitting one into two clusters. Suppose, at a certain sampling step, that a cluster Γ_i is split into two cluster Γ_l and Γ_m, or conversely that that Γ_l and Γ_m is merged into a cluster Γ_i, then the processes can be represented as the following form without loss of generality.

$$\theta = (K, \{\Gamma_i, ...\}) \leftrightarrow (K+1, \{\Gamma_l, \Gamma_m, ...\}) = \theta', \text{where } \Gamma_i = \Gamma_l \cup \Gamma_m. \qquad (16)$$

The Pathway to Propose Merge. We propose the merge of two clusters along the following stochastic procedure. Firstly, among the current K clusters, one cluster is chosen with the probability $q(\Gamma_l|\text{merge}) \propto 1/K$. Then, another cluster is selected with the probability $q(\Gamma_m|\Gamma_l, \text{merge}) \propto \exp\left(-\frac{\text{dist}(\Gamma_m, \Gamma_l)^2}{2\sigma^2_{\text{merge}}}\right)$, where dist() denotes the Euclidean distance between the cluster centroids. This represents the sampling from a Gaussian Parzen window centered at the centroid of the first cluster Γ_l.

Mergence Trees. Unlike merge, its reverse move, split, is complicated to propose since it involves classifying all the member regions of a cluster into two potential clusters. Moreover, to satisfy the detailed balance condition of MCMC [10], all the move sequences in dynamics should be reversible, which means that if a merge move can be proposed, then the exact reverse split move should be possible. To design efficient and reversible merge/split, we construct *mergence trees* for merge/split over all the process. Each cluster has its own mergence tree which stores the information of all the constituent clusters of itself in the tree structure (Fig. 4(b)). Utilizing the mergence trees, we can propose a simple but potential split move at low cost, that is the move to the state just before the latest merge move. Note that we always begin from the clusters with a single initial match, and the clusters are grown up gradually by the accepted moves among four types of proposals. Thus, one of the best split moves is simply tracing back to the past.

The Pathway to Propose Split. A previously merged cluster can be proposed to split into two as follows using the mergence tree. Firstly, a cluster among the current K clusters is chosen with the probability $q(\Gamma_i|\text{split}) \propto 1/K$. Then, the cluster is proposed to split into two clusters corresponding to child nodes in its mergence tree, with the probability $q(\Gamma_l, \Gamma_m|\Gamma_i, \text{split}, \text{mergence trees}) = 1$.

4.3 Overall Markov Chains Dynamics and Criterion of Reliable MCSs

Our DDMC image exploration algorithm simulates a Markov chain consisting of two pairs of sub-kernels, which continuously reconfigures θ according to $p(\theta|I)$.

At each sampling step, the algorithm chooses a move m with probability $q(m)$, then the sub-kernel of the move m is performed. The proposed move along its pathway is accepted with the acceptance probability (14). If the move is accepted, the current state jumps from θ to θ'. Otherwise, the current state is retained. In the early stage of sampling, we perform only expansion/contration moves without merge/split moves because the unexpanded clusters in the early stage are prone to unreliable merge/split moves. After enough iterations, merge/split moves incorporate with expansion/contraction moves, helping the Markov chains to have better chances of proposing reliable expansion/contraction moves and estimating correct MCSs.

To evaluate the reliability of MCSs in the best sample θ^*, we define the expansion ratio of an MCS as the expanded area of the MCS divided by the entire image area. Since a reliable MCS is likely to expand enough, we determine the reliable MCSs as those expanded more than the threshold ratio ϵ in both of two images. This criterion of our method eliminates the trivial or false correspondences effectively.

4.4   Implementation Details

For initialization, we used Harris-Affine [3] and MSER [2] detectors with SIFT as a feature descriptor. After nearest neighbor matching, potential outliers are filtered out through the ratio test with threshold 0.8 [1]. In our experiments, the grid for the latent region set is composed of regions of radius $h/25$, spaced $h/25$, where h denotes the height of the reference image. The radius trades correspondence density and segmentation quality for computational cost. It can be selected based on the specific purpose. The parameters in the posterior model were fixed as follows: $\delta_{\mathrm{ori}} = \pi/4, \lambda_{\mathrm{geo}} = 3, \lambda_{\mathrm{photo}} = 20, \lambda_{\mathrm{maxi}} = 6$. In the sampling stage, we set the probability of selecting each sub-kernel as $q(\mathrm{expand}) = q(\mathrm{contr}) = 0.4, q(\mathrm{split}) = q(\mathrm{merge}) = 0.1$, and the parameters of sub-kernels are set to $\sigma_{\mathrm{expand}} = l/100, \sigma_{\mathrm{contract}} = 0.5, \sigma_{\mathrm{merge}} = l/10$, where l means the diagonal length of the reference image. The results were obtained after 7000 iteration runs. Only the expansion/constration moves are performed in the first 1000 samplings. In most of our tests, the MAP θ^* was generated within about 5000 samplings. The expansion threshold ratio ϵ for reliable MCSs in all our experiments is set to 2% of each image.

5   Experiments

We have conducted two experiments: (i) unsupervised recognition and segmentation of multiple common objects and (ii) image retrieval for place recognition.

5.1   Unsupervised Recognition and Segmentation of Multiple Common Objects

Since there is no available public dataset for this problem yet, we built a new challenging dataset including multiple common objects with mutual occlusion

Fig. 5. Co-recognition results on *Minnie's*, *Jigsaws*, *Toys*, *Books*, and *Bulletins*. We built the datasets for evaluation of co-recognition except for *Bulletins*, which is borrowed from [11] for comparison.

Table 1. Performance evaluation of segmentation

Dataset	Mickey's	Minnie's	Jigsaws	Toys	Books	Bulletins	Average
Hit Ratio	80.7%	83.2%	80.0%	83.5%	94.6%	91.2%	85.5%
Bk Ratio	20.6%	37.4%	22.8%	25.2%	11.8%	16.8%	22.4%

and complex clutters. The ground truth segmentation of the common objects has been achieved manually[5]. Figure 5 and 1 show some of co-recognition results on them. Each color of the boundary represents identity of each MCS. The inferred MCSs, their segmentations (the 2nd column), and their dense correspondences (the 3rd column) are of good quality in all pairs of the dataset. On the average, the correct match ratio started from less than 5% in naive NN matches, growing to 42.2% after initial matching step, and finally reached to 92.8% in final reliable MCSs. The number of correct matches increased to 651%.

We evaluated segmentation accuracy by hit ratio h_r and background ratio b_r.[6] The results are summarized in Table 1. It also shows high accuracy in segmentation. For example, the dataset *Bulletins* is borrowed from [11], and our result of $h_r = 0.91, b_r = 0.17$ is much better than their result of $h_r = 0.76, b_r = 0.29$ in [11]. Moreover, note that our method provides object-level identities and

---

[5] The dataset with ground truth is available at http://cv.snu.ac.kr/~corecognition.

[6] $h_r = \frac{|\text{GroundTruth} \cap \text{Result}|}{|\text{GroundTruth}|}$, $b_r = \frac{|\text{Result}| - |\text{Result} \cap \text{GroundTruth}|}{|\text{Result}|}$.

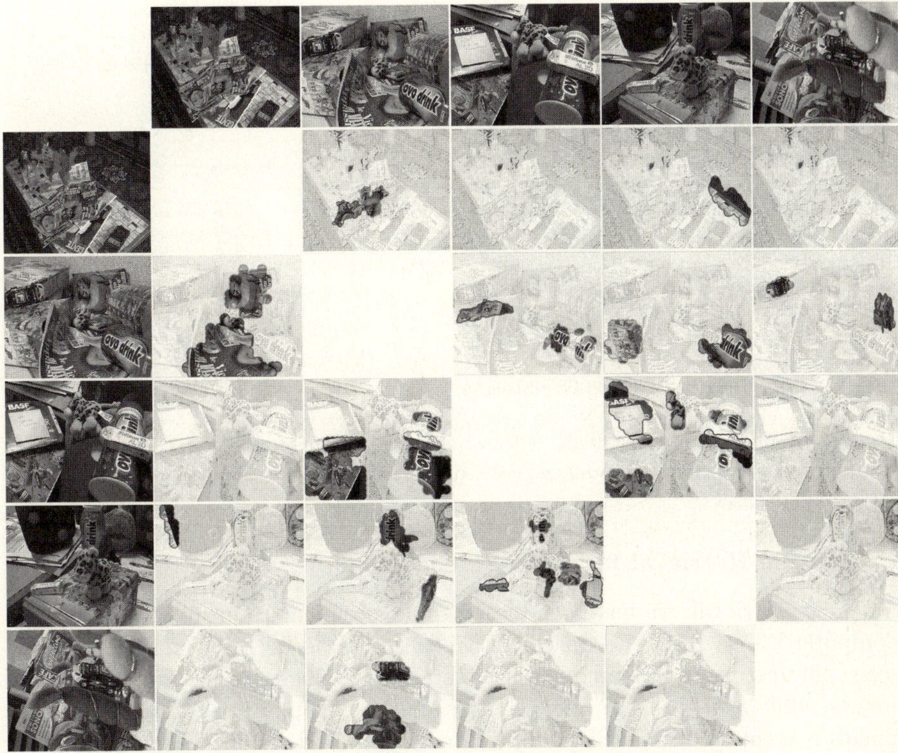

Fig. 6. Co-recognition on all combination pairs of 5 test images from the ETHZ Toys dataset. Both the detection rate and the precision are 93%.

dense correspondences, which are not provided by the method of [11]. Most of the over-expanded regions increasing the background ratio result from mutually similar background regions.

To demonstrate the unsupervised detection performance of co-recognition in view changes or deformation, we tested on all combination pairs of 5 complex images from the ETHZ toys dataset[7]. None of the model images in the dataset are included in this experiment. As shown in Fig. 6, although this task is very challenging even for human eyes, our method detected 13 true ones and 1 false one among 14 common object correspondences in the combination pairs. The detection rate and the precision are all 93%. Note that our method can recognize the separate regions as one MCS if mutual geometry of the regions is consistent according to the reinforced sidedness constraint (6). Thus, it can deal with complex partial occlusion which separates the objects into fragments. This allows us to estimate the correct number of identical entities of separate regions as in result of Fig. 1 and Fig. 6.

---

[7] http://www.robots.ox.ac.uk/~ferrari/datasets.html

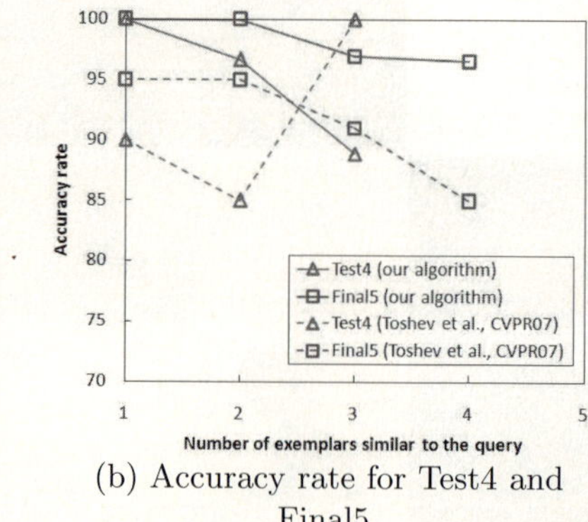

(a) Co-recognition on ICCV2005 datasets

(b) Accuracy rate for Test4 and Final5.

Fig. 7. (a) Co-recognition deals with object-level correspondence, which is higher than segment-level correspondence. (b) Comparison with co-saliency matching [5] on ICCV2005 datasets.

5.2 Image Retrieval for Place Recognition

For image retrieval, we have conducted the experiment as in [5] on ICCV 2005 Computer Vision Contest datasets[8]. Each of two datasets (*Test4* and *Final5*) has been split into exemplar and query set. *Test4* has 19 query images and 9 exemplar images, while *Final5* has 22 query images and 16 exemplar images. Each of query images is compared with all exemplar images, and all the matched image pairs are ranked according to the total area of reliable MCSs. For every query image having at least k similar examplars, the accuracy rate is evaluated with how many of them are included in top k ranks. The result in Fig. 7(b) reveals that our co-recognition outperforms co-saliency matching [5] largely in this experiment. The reason can be explained by comparing our result of the top in Fig. 7(a) with the result of the same pair in [5]. Co-recognition deals with object-level correspondences, which are higher than segment-level correspondences as [5], our method generates larger, denser, and more accurate correspondence without segmentation cue.

6 Conclusion

We have presented a novel notion of *co-recognition* and the algorithm, which recognizes and segments all the common salient region pairs with their maximal sizes in an arbitrary image pair. The problem is formulated as a Bayesian MAP problem and the solution is obtained by our stochastic image exploration algorithm using DDMCMC paradigm. Experiments on challenging datasets show promising results on the problem, some of which even humans cannot achieve easily. The proposed co-recognition has various applications for high-level image matching such as object-driven image retrieval.

---

[8] http://research.microsoft.com/iccv2005/Contest/

Acknowledgements

This research was supported in part by the Defense Acquisition Program Administration and Agency for Defense Development, Korea, through the Image Information Research Center under the contract UD070007AD, and in part by the MKE (Ministry of Knowledge Economy), Korea under the ITRC (Information Technolgy Research Center) Support program supervised by the IITA (Institute of Information Technology Advancement) (IITA-2008-C1090-0801-0018).

References

1. Lowe, D.G.: Object recognition from local scale-invariant features. In: ICCV, pp. 1150–1157 (1999)
2. Matas, J., Chum, O., Urban, M., Pajdla, T.: Robust wide baseline stereo from maximally stable extremal regions. In: BMVC (2002)
3. Mikolajczyk, K., Schmid, C.: An affine invariant interest point detector. In: Heyden, A., Sparr, G., Nielsen, M., Johansen, P. (eds.) ECCV 2002. LNCS, vol. 2350, pp. 128–142. Springer, Heidelberg (2002)
4. Vedaldi, A., Soatto, S.: Local features, all grown up. In: CVPR, pp. 1753–1760 (2006)
5. Toshev, A., Shi, J., Daniilidis, K.: Image matching via saliency region correspondences. In: CVPR (2007)
6. Ferrari, V., Tuytelaars, T., Gool, L.: Simultaneous object recognition and segmentation from single or multiple model views. IJCV 67(2), 159–188 (2006)
7. Yang, G., Stewart, C.V., Michal Sofka, C.L.T.: Registration of challenging image pairs:initialization, estimation, and decision. PAMI 29(11), 1973–1989 (2007)
8. Rother, C., Minka, T.P., Blake, A., Kolmogorov, V.: Cosegmentation of image pairs by histogram matching - incorporating a global constraint into MRFs. In: CVPR, pp. 993–1000 (2006)
9. Tu, Z., Chen, X., Yuille, A.L., Zhu, S.C.: Image parsing: unifying segmentation, detection, and recognition. In: ICCV, vol. 1, pp. 18–25 (2003)
10. Green, P.: Reversible jump markov chain monte carlo computation and bayesian model determination. Biometrica 82, 711–732 (1995)
11. Yuan, J., Wu, Y.: Spatial random partition for common visual pattern discovery. In: ICCV, pp. 1–8 (2007)
12. Simon, I., Seitz, S.M.: A probabilistic model for object recognition, segmentation, and non-rigid correspondence. In: CVPR (2007)
13. Cho, M., Lee, K.M.: Partially occluded object-specific segmentation in view-based recognition. In: CVPR (2007)

Movie/Script: Alignment and Parsing of Video and Text Transcription

Timothee Cour, Chris Jordan, Eleni Miltsakaki, and Ben Taskar

University of Pennsylvania, Philadelphia, PA 19104, USA
`{timothee,wjc,elenimi,taskar}@seas.upenn.edu`

Abstract. Movies and TV are a rich source of diverse and complex video of people, objects, actions and locales "in the wild". Harvesting automatically labeled sequences of actions from video would enable creation of large-scale and highly-varied datasets. To enable such collection, we focus on the task of recovering scene structure in movies and TV series for object tracking and action retrieval. We present a weakly supervised algorithm that uses the screenplay and closed captions to parse a movie into a hierarchy of shots and scenes. Scene boundaries in the movie are aligned with screenplay scene labels and shots are reordered into a sequence of long continuous tracks or *threads* which allow for more accurate tracking of people, actions and objects. Scene segmentation, alignment, and shot threading are formulated as inference in a unified generative model and a novel hierarchical dynamic programming algorithm that can handle alignment and jump-limited reorderings in linear time is presented. We present quantitative and qualitative results on movie alignment and parsing, and use the recovered structure to improve character naming and retrieval of common actions in several episodes of popular TV series.

1 Introduction

Hand-labeling images of people and objects is a laborious task that is difficult to scale up. Several recent papers [1,2] have successfully collected very large-scale, diverse datasets of faces "in the wild" using weakly supervised techniques. These datasets contain a wide variation in subject, pose, lighting, expression, and occlusions which is not matched by any previous hand-built dataset. Labeling and segmenting actions is perhaps an even more painstaking endeavor, where curated datasets are more limited. Automatically extracting large collections of actions is of paramount importance. In this paper, we argue that using movies and TV shows *precisely* aligned with easily obtainable screenplays can pave a way to building such large-scale collections. Figure 1 illustrates this goal, showing the top 6 retrieved video snippets for 2 actions (walk, turn) in TV series LOST using our system. The screenplay is parsed into a temporally aligned sequence of action frames (subject verb object), and matched to detected and named characters in the video sequence. Simultaneous work[3] explores similar goals in a more supervised fashion. In order to enable accurately localized action retrieval, we propose a much deeper analysis of the structure and syntax of both movies and transcriptions.

D. Forsyth, P. Torr, and A. Zisserman (Eds.): ECCV 2008, Part IV, LNCS 5305, pp. 158–171, 2008.
© Springer-Verlag Berlin Heidelberg 2008

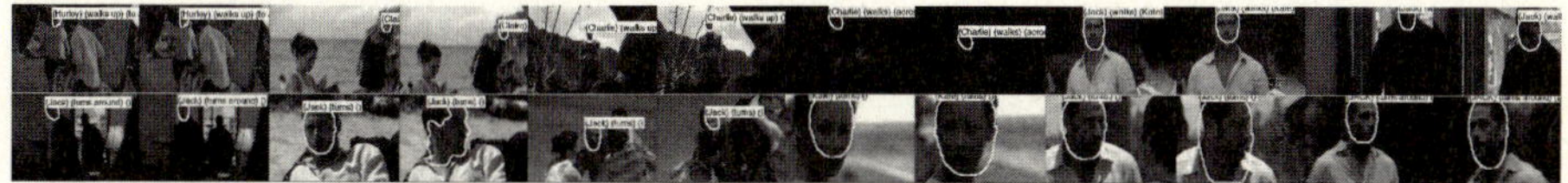

Fig. 1. Action retrieval using alignment between video and parsed screenplay. For each action verb (top: walk, bottom: turn), we display the top 6 retrieved video snippets in TV series LOST using our system. The screenplay and closed captions are parsed into a temporally aligned sequence of verb frames (subject-verb-object), and then matched to detected and named characters in the video sequence. The third retrieval, second row ("Jack turns") is counted as an error, since the face shows Boone instead of Jack. Additional results appear under www.seas.upenn.edu/~{}timothee.

Movies, TV series, news clips, and nowadays plentiful amateur videos, are designed to effectively communicate events and stories. A visual narrative is conveyed from multiple camera angles that are carefully composed and interleaved to create seamless action. Strong coherence cues and continuity editing rules are (typically) used to orient the viewer, guide attention and help follow the action and geometry of the scene. Video shots, much like words in sentences and paragraphs, must fit together to minimize perceptual discontinuity across cuts and produce a meaningful scene. We attempt to uncover elements of the inherent structure of scenes and shots in video narratives. This uncovered structure can be used to analyze the content of the video for tracking objects across cuts, action retrieval, as well as enriching browsing and editing interfaces.

We present a framework for automatic parsing of a movie or video into a hierarchy of shots and scenes and recovery of the shot interconnection structure. Our algorithm makes use of both the input image sequence, closed captions and the screenplay of the movie. We assume a hierarchical organization of movies into shots, threads and scenes, where each scene is composed of a set of interlaced threads of shots with smooth transitions of camera viewpoint inside each thread. To model the scene structure, we propose a unified generative model for joint scene segmentation and shot threading. We show that inference in the model to recover latent structure amounts to finding a Hamiltonian path in the sequence of shots that maximizes the "head to tail" shot similarity along the path, given the scene boundaries. Finding the maximum weight Hamiltonian path (reducible to the Traveling Salesman Problem or TSP) is intractable in general, but in our case, limited memory constraints on the paths make it tractable. In fact we show how to jointly optimize scene boundaries and shot threading in *linear time* in the number of shots using a novel hierarchical dynamic program.

We introduce textual features to inform the model with scene segmentation, via temporal alignment with screenplay and closed captions, see figure 2. Such text data has been used for character naming [4,5] and is widely available, which makes our approach applicable to a large number of movies and TV series. In order to retrieve temporally-aligned actions, we delve deeper into resolving textual ambiguities with pronoun resolution (determining whom or what 'he', 'she', 'it', etc. refer to in the screenplay) and extraction of verb frames. By detecting and naming characters, and resolving pronouns, we show promising results for more accurate action retrieval for several common verbs. We present quantitative and qualitative results for scene segmentation/alignment, shot segmentation/threading, tracking and character naming across shots and action retrieval in numerous episodes of popular TV series, and illustrate that shot reordering provides much improved character naming.

The main contributions of the paper are: 1) novel probabilistic model and inference procedure for shot threading and scene alignment driven by text, 2) extraction of verb frames and pronoun resolution from screenplay, and 3) retrieval of the corresponding actions informed by scene sctructure and character naming.

The paper is organized as follows. Section 2 proposes a hierarchical organization of movies into shots, threads and scenes. Sections 3 and 4 introduce a generative model for joint scene segmentation and shot threading, and a hierarchical dynamic program to solve it as a restricted TSP variant. Section 5 addresses the textual features used in our model. We report results in section 6 and conclude in section 7.

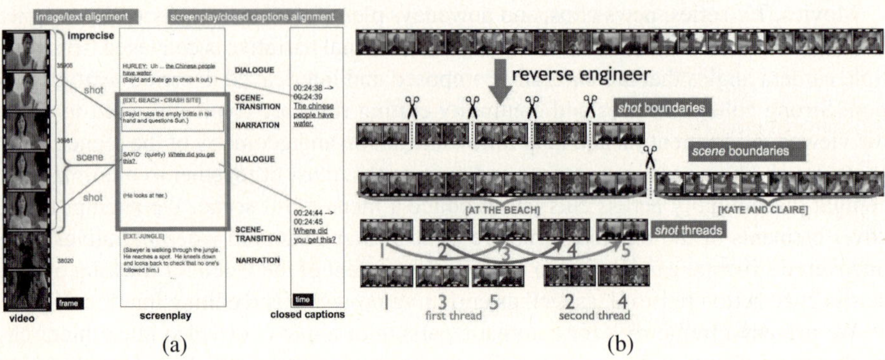

(a)   (b)

Fig. 2. (a) Alignment between video, screenplay and closed captions; (b) Deconstruction pipeline

2   Movie Elements: Shots, Threads, Scenes

Movies and TV series are organized in distinctive hierarchical and continuity structures consisting of elements such as scenes, threads and shots. Detecting and recovering these elements is needed for uninterrupted tracking of objects and people in a scene across multiple cameras, recovering geometric relationships of objects in a scene, intelligent video browsing, search and summarization.

Shot boundaries. The aim of shot segmentation is to segment the input frames into a sequence of shots (single unbroken video recordings) by detecting camera viewpoint discontinuities. A popular technique is to compute a set of localized color histograms for each image and use a histogram distance function to detect boundaries [6,7].

Shot threads. Scenes are often modeled as a sequence of shots represented as letters: **ABABAB** represents a typical dialogue scene alternating between two camera points of view A and B. More complex patterns are usually observed and in practice, the clustering of the shots into letters (camera angles/poses) is not always a very well defined problem, as smooth transitions between shots occur. Nevertheless we assume in our case that each shot in a scene is either a novel camera viewpoint or is generated from

Fig. 5. Shot reordering to recover continuity in 3 scenes of LOST

5.1   Text Data: Screenplay and Closed Captions

We use two sources of text for our segmentation-alignment problem: the screenplay, which narrates the actions and provides a transcript of the dialogues, and the closed captions, which provide time-stamped dialogues, as in figure 2(a). Both sources are essential since the screenplay reveals speaker identity, dialogues and scene transitions but no time-stamps, and closed captions reveal dialogues with time-stamps but nothing else. The screenplay and the closed captions are readily available for a majority of movies and TV series produced in the US. A similar approach was used in [5] to align faces with character names, with 2 differences: 1) they used the screenplay to reveal the speaker identity as opposed to scene transitions, and 2) subtitles were used instead of closed captions. Subtitles are encoded as bitmaps, thus require additional steps of OCR and spell-checking to convert them to text[5], whereas closed captions are encoded as ASCII text in DVDs, making our approach simpler and more reliable, requiring a simple modification of mplayer (`http://www.mplayerhq.hu/`).

5.2   Screenplay/Closed Captions Alignment

The alignment between the screenplay and the closed captions is non-trivial since the closed captions only contain the dialogues (without speaker) mentioned in the screenplay, often with wide discrepancies between both versions. We extend the dynamic time warping[12] approach in a straightforward way to time-stamp each element of the screenplay (as opposed to just the dialogues as in [5]). The screenplay is first parsed into a sequence of elements (either NARRATION, DIALOGUE, or SCENE-TRANSITION) using a simple grammar, and the dynamic programming alignment of the words in the screenplay and the closed captions provides a time interval $[T^{\text{start}}(i), T^{\text{end}}(i)]$ for each DIALOGUE element E_i. A NARRATION or SCENE-TRANSITION element E_j enclosed between two DIALOGUE elements E_{i_1}, E_{i_2} is assigned the following conservative time interval: $[T^{\text{start}}(i_1), T^{\text{end}}(i_2)]$.

5.3   Scene Segmentation Via Alignment

We determine the scene boundary term $P(b)$ from section 3 by aligning each SCENE-TRANSITION element mentioned in the screenplay to a scene start. $P(b)$ is uniform among the set of b satisfying the temporal alignment constraints:

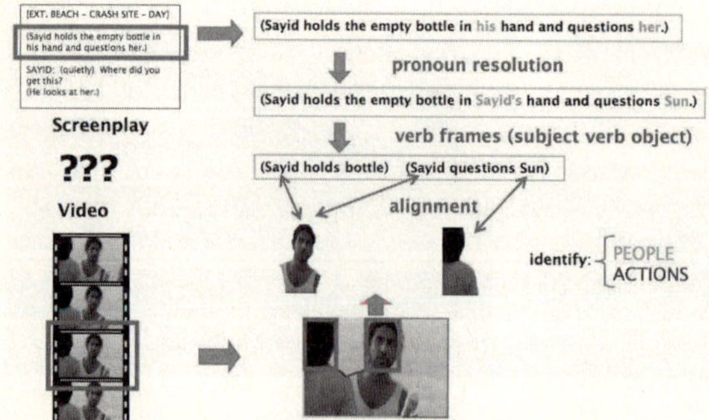

total verbs	25,000
distinct verbs	1,000
looks (most common)	2,000
turns	1,100
walks	800
takes	550
climbs	40
kisses	40
total dialogue lines	**16,000**
distinct speaker names	190
Jack (most common)	2,100

Fig. 6. Left: pronoun resolution and verb frames obtained from the parsed screenplay narrations. Right: statistics collected from 24 parsed screenplays (1 season of LOST).

$$1 \leq b_1 < \ldots < b_m = n \tag{9}$$
$$t^{\text{start}}(j) \leq b_{j-1} + 1 \leq t^{\text{end}}(j) \tag{10}$$

where $[t^{\text{start}}(j), [t^{\text{end}}(j)]$ is the time interval of the jth SCENE-TRANSITION element, converted into frame numbers, then to shot indexes.

Additional alignment constraints. Close inspection of a large number of screenplays collected for movies and TV series revealed a fairly regular vocabulary used to describe shots and scenes. One such example is FADE IN and FADE OUT corresponding to a transition between a black shot (where each frame is totally black) and a normal shot, and vice versa. Such black shots are easy to detect, leading to additional constraints in the alignment problem, and a performance boost.

5.4   Pronoun Resolution and Verb Frames

Alignment of the screenplay to dialog in closed captions and scene boundaries in the video helps to narrow down the scope of reference for other parts of the screenplay that are interspersed – the narration or scene descriptions, which contain mentions of actions and objects on the screen. In addition to temporal scope uncertainty for these descriptions, there is also ambiguity with respect to the subject of the verb, since personal pronouns (he, she) are commonly used. In fact, our analysis of common screenplays reveals there are more pronouns than occurences of character names in the narrations, and so resolving those pronouns is an important task. We employed a simple, deterministic scheme for pronoun resolution that uses a standard probabilistic context-free parser to analyze sentences and determine verb frames (subject-verb-object) and then scans the sentence for possible antecedents of each pronoun that agree in number and gender, see figure 6. The details of the algorithm are given in *supplemental materials*. Here is an example output of our implementation on a sentence extracted from screenplay narration (pronoun resolution shown in parenthesis): *On the side, Sun watches them. Jin reaches out and touches Sun 's chin, his (Jin's) thumb brushes her (Sun's) lips. She (Sun) looks at him (Jin) and pulls away a little. He (Jin) puts his (Jin's) hand down.*

Output verb frames: *(Sun - watches - something) (Jin - reaches out -) (Jin - touches - chin) (Sun - looks - at Jin) . (Sun - pulls away -) (Jin - puts down - hand).*

We report pronoun resolution accuracy on screenplay narrations of 3 different TV series (about half a screenplay for each), see table 1.

Table 1. Pronoun resolution accuracy on screenplay narrations of 3 different TV series

TV series screenplay	pronoun resolution accuracy	# pronouns	# sentences
LOST	75%	93	100
CSI	76 %	118	250
ALIAS	78%	178	250

6   Results

We experimented with our framework on a significant amount of data, composed of TV series (19 episodes from one season of LOST, several episodes of CSI), one feature length movie "The Fifth Element", and one animation movie "Aladdin", representing about 20 hours of video at DVD resolution. We report results on scene segmentation/alignment, character naming and tracking, as well as retrieval of query action verbs.

Shot segmentation. We obtain 97% F-score (harmonic mean of precision and recall) for shot segmentation, using standard color histogram based methods.

Scene segmentation and alignment. We hand labeled scene boundaries in one episode of LOST and one episode of CSI based on manual alignment of the frames with the screenplay. The accuracy for predicting the scene label of each shot was 97% for LOST and 91% for CSI. The F-score for scene boundary detection was 86% for LOST and 75% for CSI, see figure 7. We used $k = 9$ for the memory width, a value similar to the buffer size used in [10] for computing shot coherence. We also analyzed the effect on performance of the memory width k, and report results with and without alignment to screenplay in table 2. In comparison, we obtained an F-score of 43% for scene boundary detection using a model based on backward shot coherence [10] uninformed by screenplay, but optimized over buffer size and non-maximum suppression window size.

Scene content analysis. We manually labeled the scene layout in the same episodes of LOST and CSI, providing for each shot in a scene its generating shot (including

Table 2. % F-score (first number) for scene boundary detection and % accuracy (second number) for predicting scene label of shots (on 1 episode of LOST) as a function of the memory width k used in the TSP, and the prior $P(b)$. The case $k = 1$ corresponds to no reordering at all. Line 1: $P(b)$ informed by screenplay; line 2: $P(b)$ uniform; line 3: total computation time.

$P(b)$	$k = 1$	$k = 2$	$k = 3$	$k = 9$	$k = 12$
aligned	73/90	77/91	82/96	86/97	88/97
uniform	25/0	45/14	55/0	52/1	-/-
total time (s)	< 0.1	< 0.1	0.1	5	68

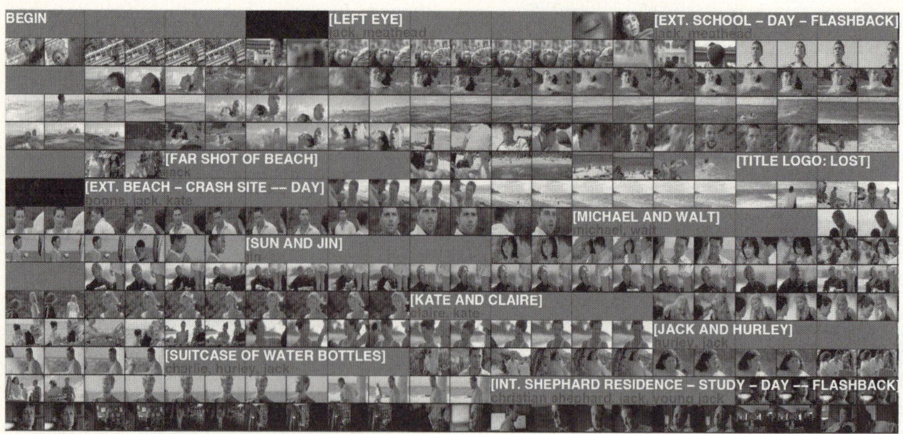

Fig. 7. Movie at a glance: scene segmentation-alignment and shot reordering for an episode of LOST (only a portion shown for readability). Scene boundaries are in red, together with the set of characters appearing in each scene, in blue.

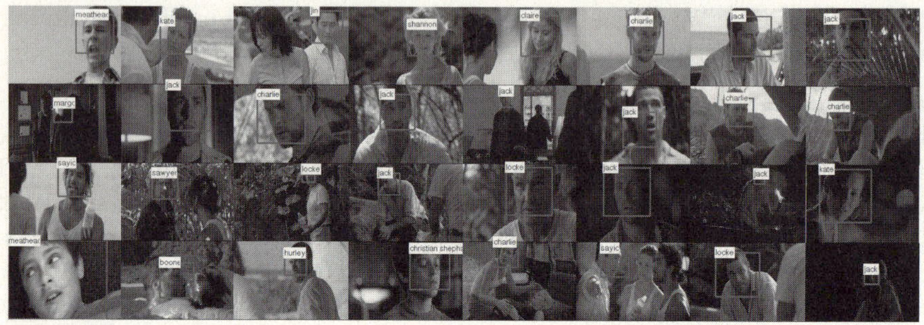

Fig. 8. Character naming using screenplay alignment and shot threading. Top 3 rows: correctly named faces; bottom row: incorrectly named faces. We detect face tracks in each shot and reorder them according to the shot threading permutation. Some face tracks are assigned a name prior based on the alignment between dialogues and mouth motion. We compute a joint assignment of names to face tracks using an HMM on the reordered face tracks.

the special case when this is a new viewpoint). We obtain a precision/recall of 75% for predicting the generating parent shot. See figure 5 for a sample of the results on 3 scenes. Note, to obtain longer tracks in figure 5, we recursively applied the memory limited TSP until convergence (typically a few iterations).

Character identification on reordered shots. We illustrate a simple speaker identification based on screenplay alignment and shot threading, see figure 8. We use a Viola-Jones[13] based face detector and tracking with normalized cross-correlation to obtain face tracks in each shot. We build a Hiddden Markov Model (HMM) with states

Fig. 9. Top 10 retrieved video snippets for 15 query action verbs: close eyes, grab, kiss, kneel, open, stand, cry, open door, phone, point, shout, sit, sleep, smile, take breath. Please zoom in to see screenplay annotation (and its parsing into verb frames for the first 6 verbs).

corresponding to assignments of face tracks to character names. The face tracks are ordered according to the shot threading permutation, and as a result there are much fewer changes of character name along this ordering. Following [14], we detect on-screen speakers as follows: 1) locate mouth for each face track using a mouth detector based on Viola-Jones, 2) compute a mouth motion score based on the normalized cross correlation between consecutive windows of the mouth track, averaged over temporal segments corresponding to speech portions of the screenplay. Finally we label the face tracks using Viterbi decoding for the Maximum a Posteriori (MAP) assignment (see website for more details). We computed groundtruth face names for one episode of LOST and compared our method against the following baseline that does not use shot reordering: each unlabeled face track (without a detected speaking character on screen) is labeled using the closest labeled face track in feature space (position of face track and color histogram). The accuracy over an episode of LOST is 76% for mainly dialogue scenes and 66% for the entire episode, as evaluated against groundtruth. The baseline model based using nearest neighbor performs at resp. 43% and 39%.

Retrieval of actions in videos. We consider a query-by-action verb retrieval task for 15 query verbs across 10 episodes of LOST, see figure 9. The screenplay is parsed into verb frames (subject-verb-object) with pronoun resolution, as discussed earlier. Each verb frame is assigned a temporal interval based on time-stamped intervening dialogues and tightened with nearby shot/scene boundaries. Queries are further refined to match the subject of the verb frame with a named character face. We report retrieval results as follows: for each of the following action verbs, we measure the number of times (out of 10) the retrieved video snippet correctly shows the actor on screen performing the action (we penalize for wrong naming): close eyes (9/10), grab (9/10), kiss (8/10), kneel (9/10), open (9/10), stand (9/10), cry (9/10), open door (10/10), phone (10/10), point (10/10), shout (7/10), sit (10/10), sleep (8/10), smile (9/10), take breath (9/10). The average is 90/100. Two additional queries are shown in figure 1 along with the detected and identified characters. We created a large dataset of retrieved action sequences combined with character naming for improved temporal and spatial localization, see `www.seas.upenn.edu/~{}timothee` for results and matlab code.

7   Conclusion

In this work we have addressed basic elements of movie structure: hierarchy of scenes and shots and continuity of shot threads. We believe that this structure can be useful for many intelligent movie manipulation tasks, such as semantic retrieval and indexing, browsing by character or object, re-editing and many more. We plan to extend our work to provide more fine-grained alignment of movies and screenplay, using coarse scene geometry, gaze and pose estimation.

References

1. Huang, G., Jain, V., Learned-Miller, E.: Unsupervised joint alignment of complex images. In: International Conference on Computer Vision, pp. 1–8 (2007)
2. Ramanan, D., Baker, S., Kakade, S.: Leveraging archival video for building face datasets. In: International Conference on Computer Vision, pp. 1–8 (2007)
3. Laptev, I., Marszałek, M., Schmid, C., Rozenfeld, B.: Learning realistic human actions from movies. In: IEEE Conference on Computer Vision and Pattern Recognition (2008), `http://lear.inrialpes.fr/pubs/2008/LMSR08`
4. Sivic, J., Everingham, M., Zisserman, A.: Person spotting: video shot retrieval for face sets. In: Leow, W.-K., Lew, M., Chua, T.-S., Ma, W.-Y., Chaisorn, L., Bakker, E.M. (eds.) CIVR 2005. LNCS, vol. 3568, Springer, Heidelberg (2005)
5. Everingham, M., Sivic, J., Zisserman, A.: Hello! my name is.. buffy – automatic naming of characters in tv video. In: Proceedings of the British Machine Vision Conference (2006)
6. Lienhart, R.: Reliable transition detection in videos: A survey and practitioner's guide. Int. Journal of Image and Graphics (2001)

7. Ngo, C.-W., Pong, T.C., Zhang, H.J.: Recent advances in content-based video analysis. International Journal of Image and Graphics 1, 445–468 (2001)
8. Zhai, Y., Shah, M.: Video scene segmentation using markov chain monte carlo. IEEE Transactions on Multimedia 8, 686–697 (2006)
9. Yeung, M., Yeo, B.L., Liu, B.: Segmentation of video by clustering and graph analysis. Comp. Vision Image Understanding (1998)
10. Kender, J., Yeo, B.: Video scene segmentation via continuous video coherence. In: IEEE Conference on Computer Vision and Pattern Recognition (1998)
11. Balas, E., Simonetti, N.: Linear time dynamic programming algorithms for new classes of restricted tsps: A computational study. INFORMS Journal on Computing 13, 56–75 (2001)
12. Myers, C.S., Rabiner, L.R.: A comparative study of several dynamic time-warping algorithms for connected word recognition. The Bell System Technical Journal (1981)
13. Viola, P.A., Jones, M.J.: Robust real-time face detection. International Journal of Computer Vision 57, 137–154 (2004)
14. Everingham, M.R., Sivic, J., Zisserman, A.: Hello! my name is buffy: Automatic naming of characters in tv video. In: BMVC, vol. III, p. 899 (2006)

Using 3D Line Segments for Robust and Efficient Change Detection from Multiple Noisy Images

Ibrahim Eden and David B. Cooper

Division of Engineering
Brown University
Providence, RI, USA
`{ieden,cooper}@lems.brown.edu`

Abstract. In this paper, we propose a new approach to change detection that is based on the appearance or disappearance of 3D lines, which may be short, as seen in a new image. These 3D lines are estimated automatically and quickly from a set of previously-taken learning-images from arbitrary view points and under arbitrary lighting conditions. 3D change detection traditionally involves unsupervised estimation of scene geometry and the associated BRDF at each observable voxel in the scene, and the comparison of a new image with its prediction. If a significant number of pixels differ in the two aligned images, a change in the 3D scene is assumed to have occurred. The importance of our approach is that by comparing images of lines rather than of gray levels, we avoid the computationally intensive, and some-times impossible, tasks of estimating 3D surfaces and their associated BRDFs in the model-building stage. We estimate 3D lines instead where the lines are due to 3D ridges or BRDF ridges which are computationally much less costly and are more reliably detected. Our method is widely applicable as man-made structures consisting of 3D line segments are the main focus of most applications. The contributions of this paper are: change detection based on appropriate interpretation of line appearance and disappearance in a new image; unsupervised estimation of "short" 3D lines from multiple images such that the required computation is manageable and the estimation accuracy is high.

1 Introduction

The change detection problem consists of building an appearance model of a 3D scene using n images, and then based on an $n+1^{st}$ image, determining whether a "significant" change has taken place. A fundamental approach to this problem is to estimate a 3D model for the scene and the associated BRDF; then based on the knowledge of the $n+1^{st}$ image viewing position and scene illumination, a decision is made as to whether there is a significant difference between the $n+1^{st}$ image and its prediction by the n-image based 3D geometry and BRDF (bidirectional reflectance distribution function) estimates.

In its general form, all learning is done in the unsupervised mode, and the n-image based learning is not done for a static 3D scene but rather for a functioning scene where changes are often taking place. A complicating factor in the change detection problem

D. Forsyth, P. Torr, and A. Zisserman (Eds.): ECCV 2008, Part IV, LNCS 5305, pp. 172–185, 2008.

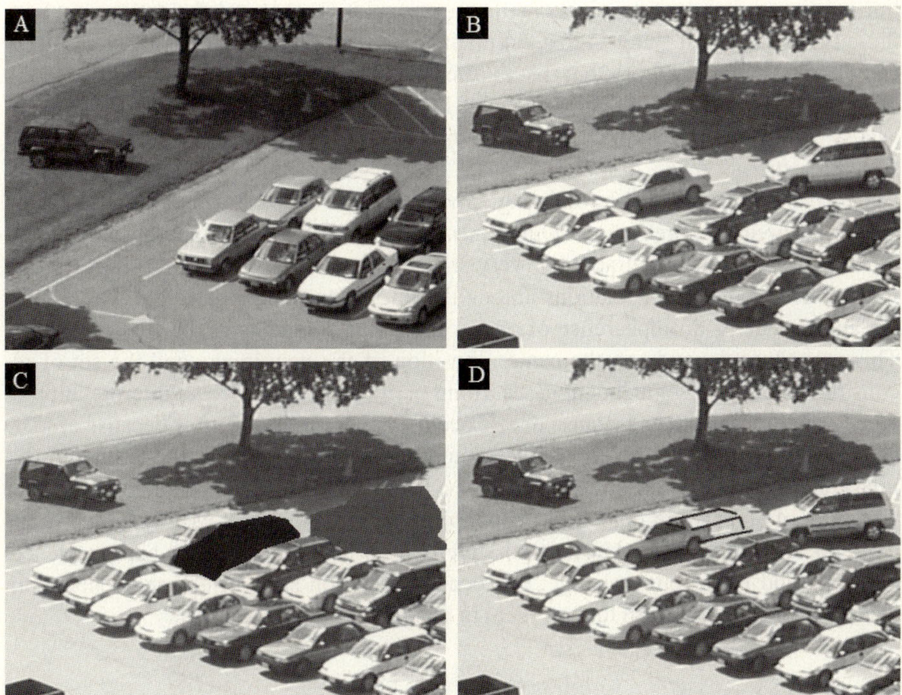

Fig. 1. Our line segment based change detection result after training on a sequence of 5 images. (A) A sample training image. (B) The test image. (C) Hand-marked ground truth for change where the new object is shown in "red" and the disappeared object is shown in "blue". (D) Result of our method. Lines associated with the new object are shown in "red" and lines associated with the disappeared object are shown in "blue". Two major change regions are detected with only a few false alarms due to specular highlights and object shadows. (This is a color image)

is that images can be taken at arbitrary time, under arbitrary lighting conditions and from arbitrary view points. Furthermore, they are usually single images and not video. For example, if they are taken from a flying aircraft, a 3D point in the scene is usually seen in one image and not in the immediately preceding or succeeding images, and is not seen again until the aircraft returns at some later time or until some other aircraft or satellite or moving camera on the ground sees the point at some later time.

In this paper, we assume n images are taken of a scene, and we then look for a change in the $n + 1^{st}$ image, and if one has occurred we try to explain its type (resulting from the arrival or from the departure of a 3D object). The learning is done in an unsupervised mode. We do not restrict ourselves to the case of buildings where the 3D lines are long, easy to detect, easy to estimate and are modest in number. Rather, we are interested in the case of many short lines where the lines can be portions of long curves or can be short straight line segments associated with complicated 3D objects, e.g., vehicles, scenes of damaged urban-scapes, natural structure, people, etc...

Why do we restrict this study to straight lines? We could deal with curves, but since curves can be decomposed into straight lines, and since straight lines – especially short line segments - appear extensively in 3D scenes and in images, we decided to start with those. The important thing is that estimating 3D structure and the associated BRDF can often be done in theory, but this is usually difficult to do computationally. On the other hand, estimating 3D line segments is much more tractable and can be considered as a system in its own right or as contributing to applications that require efficient 3D structure estimation.

Our paper consists of the following. Given n images, we estimate all 3D lines that appear in three or more images. Our approach to 3D line estimation emphasizes computational speed and accuracy. For very short lines, accuracy is greatly improved by making use of incidence relations among the lines. For change detection we look for the appearance or disappearance of one or more line segments in the $n + 1^{st}$ image. This procedure depends on the camera position of the new image and the set of reconstructed 3D line segments in the learning period, and therefore an interpretation of whether a line is not seen because of self occlusion within the scene or because of a 3D change. Usually, but not always, if an existing 3D line should be visible in the $n + 1^{st}$ image and is not, the reason is because of occlusion by the arrival of a new object or departure of an existing object. If a new object arrives, there will usually be new lines that appear because of it, but it is possible that no new straight lines appear. Hence, detecting and interpreting change, if it occurs, based on straight line segments is not clear cut, and we deal with that problem in this paper.

2   Related Work

Some of the earlier work on change detection focuses on image sequences taken from stationary cameras. The main drawback of these methods is their likelihood to create false alarms in cases where pixel values are affected by viewpoint, illumination, seasonal and atmospheric changes. This is the reason why pixel (intensity) and block (histogram) based change detection algorithms such as image differencing [1,2] and background modeling methods [3] fail in some applications.

Meanwhile, there exist change detection methods designed for non-stationary image sequences. There has been a lot of work in the literature on methods based on detecting moving objects [4,5], but these methods assume one or more moving objects in a continuous video sequence. On the other hand, 3D voxel based methods [6] where distributions of surface occupancy and associated BRDF are stored in each voxel can manage complex and changing surfaces, but these methods suffer from sudden illumination changes, perform poorly around specular highlights and object boundaries.

To our knowledge, line segment based change detection methods have rarely been studied in computer vision literature. Rowe and Grewe [7] make use of 2D line segments in their algorithm, but their method is specifically designed for aerial images where the images can be registered using an affine transformation. Li et al. [8] provided a method of detecting urban changes from a pair of satellite images by identifying changed line segments over time. Their method does not estimate the 3D geometry associated with the line segments and takes a pair of satellite (aerial) images as input where line matching can be done by estimating the homography between the two images. The change detection method we propose in this work is more generic, it can work on non-sequential image sequences where the viewpoint can change drastically between pairs of images and it is not based on any prior assumptions on the set of training images.

3 Multi-view Line Segment Matching and Reconstruction

Line segment matching over multiple images is known to be a difficult problem due to its exponential complexity requirement and challenging inputs. As a result of imperfections in edge detection and line fitting algorithms, lines are fragmented into small segments that diverge from the original line segments. When unreliable endpoints and topological relationships are given as inputs, exponential complexity search algorithms may fail to produce exact segment matching.

In this section, we present a generic, reliable and efficient method for multi-view line matching and reconstruction. Although our method is also suitable for small baseline problems (e.g. aerial images, continuous video sequences), such cases are not our primary focus as their line ordering along the epipolar direction does not change much and they can be solved efficiently by using planar homographies. In this paper, we focus on large baseline matching and reconstruction problems, where sudden illumination changes and specular highlights make it more difficult obtain consistent line segments in images of the same scene. These problems are more challenging as the line ordering in different images change due to differences in viewing angles. The following subsections describe three steps of our 3D line segment reconstruction method: an efficient line segment matching algorithm, reconstruction of single 3D lines segments and reconstruction of free form wire-frame structures.

3.1 An Efficient Line Segment Matching Algorithm

In general, the line matching problem is known to be exponential in the number of images. That is to say, given there are n images of the same scene and approximately m lines in each image, the total complexity of the line matching problem (the size of the search space) is $O(m^n)$. One way to reduce the combinatorial expansion of the matching problem is to use the epipolar beam [9,10]. Given the line $l = (x_1, x_2)$ in I, the corresponding line in I' should lie between $l'_1 = Fx_1$ and $l'_2 = Fx_2$ where F is the fundamental matrix between I and I' (see figure 2). While the epipolar beam reduces the combinatorial expansion of the matching algorithm, this reduction highly depends on the general alignment of line segments relative to epipolar lines. Plane sweep methods are also used to avoid the combinatorial expansion of the matching problem [11], but these methods do not perform well when the endpoints of 2D line segments are not consistent in different images of the same scene. Another way to increase the matching efficiency is to use color histogram based feature descriptors for 2D line segments [12], but these methods assume that colors only undergo slight changes and the data does not contain specular highlights. Our work focuses on more challenging real world problems where the above assumptions do not hold.

In this paper, we propose a new method that improves the multi-view line matching efficiency. Our method is based on the assumption that the 3D region of interest (ROI)

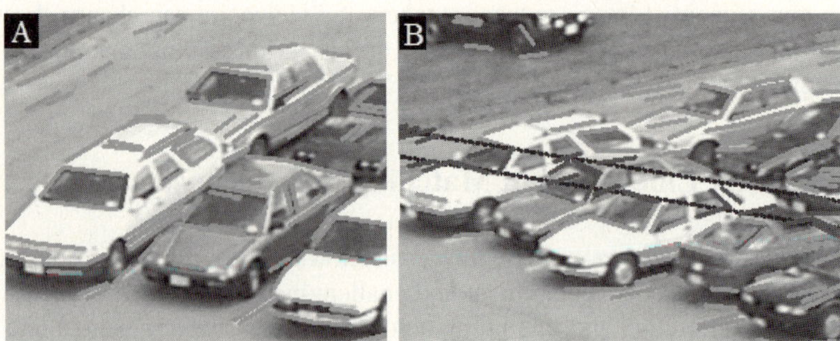

Fig. 2. An example of epipolar beam. (A) I_1: "selected line" for matching is shown in "red". (B) I_2: the epipolar beam associated with the selected line in I_1 is marked by "blue" and line segments that lie inside the epipolar beam (i.e., candidates for matching) are shown in "red". The epipolar beam in image (B) reduces the search space by 5.55. (This is a color image)

is approximately known, however this assumption is not a limitation for most multi-view applications, since the 3D ROI can be obtained by intersecting the viewing cones of the input images. The basic idea of our approach is to divide the 3D ROI into smaller cubes, and solve the matching problem for the line segments that lie inside each cube. The matching algorithm iteratively projects each cube into the set of training images, and extracts the set of 2D line segments in each image that lie (completely or partially) inside the convex polygon associated with the 3D cube.

Assuming that lines are distributed on the cubes homogeneously, the estimated number of lines inside each cube is $(\frac{m}{C})$, where C is the total number of cubes, and the algorithmic complexity of the introduced matching problem is $O(\frac{m^n}{C^n})$. It must be noted that under the assumption of homogenous distribution of lines segments in the 3D ROI the total matching complexity is reduced by a factor of $\frac{1}{C^n}$, where the efficiency gain is exponential. On the other hand, even in the existence of some dispersion over multiple cubes, our proposed algorithm substantially reduces the computational complexity of the matching algorithm. Figure 3 illustrates the quantitative comparison of different matching algorithms for 4 images of the same scene. The matching method we use in this work is a mixture of the algorithm described above and the epipolar beam method (EB+Cubes).

3.2   3D Line Segment Reconstruction

3D line segments are reconstructed by using sets of corresponding 2D line segments from different images obtained during the matching step. Each 3D line segment, $L = (X_1, X_2)$ is represented by 6 parameters in 3D space, where X_1 and X_2 are 3D points representing the end points. Here we assume that for each 3D line segment, corresponding set of 2D line segments are available in different images. We use the Nelder-Mead (Simplex) Method [13] to solve the minimization problem given in equation 1.

$$L^* = \underset{L \in \{R^3, R^3\}}{\arg\min} \sum_{i=1}^{n} d_l(l_i, l_i') + \beta \sum_{i=1}^{n} d_s(l_i, l_i'') \tag{1}$$

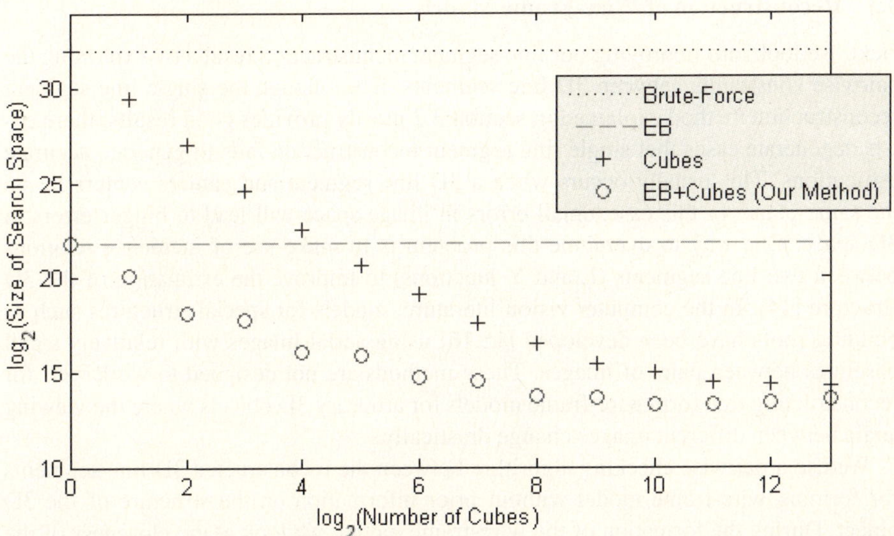

Fig. 3. Quantitative comparison of four different line segment matching algorithms using four images of the same scene. Brute-force: the simplest matching method, all combinations are checked over all images. EB: epipolar beam is used to reduce the search space. Cubes: the 3D space is splitted into smaller sections and matching is done for each section separately. EB+Cubes: (the method we use in this work) a combination of "Cubes" and "EB". It is shown that "Cubes+EB" method outperforms individual "EB" and "Cubes" methods and Brute-Force search. Notice that the size of the search space is given in logarithmic scale.

where $l'_i = (M_i X_1) \times (M_i X_2)$ is the projection of L to the ith image as an infinite line, $l''_i = (M_i X_1, M_i X_2)$ is the projection as a line segment, M_i is the projection matrix for the ith image, d_l is the distance metric between a line and a line segment and d_s is the distance metric between two line segments. The distance metrics d_l and d_s are defined as

$$d_l(l, l') = \sqrt{\frac{1}{|l|} \sum_{p \in l} d_p^2(p, l')}$$

$$d_s(l, l'') = \sqrt{\frac{1}{|l|} \sum_{p \in l} d_{ps}^2(p, l'')} + \sqrt{\frac{1}{|l''|} \sum_{p'' \in l''} d_{ps}^2(p'', l)}$$

where $d_p(p, l)$ is the perpendicular distance of a point (p) to an infinite 2D line and $d_{ps}(p, l)$ is the distance of a point to a line segment.

Note that, β in equation 1, is used to control the convergence of the local search algorithm. β is typically selected to be a number close to zero ($0 < \beta << 1$), so the first part of the objective function dominates the local search until the algorithm converges to the correct infinite line. Later, the second part (succeeded by β) of the objective function starts to dominate the local search algorithm in order to find the optimal end points for the 3D line segment.

3.3   Reconstruction of Wire-Frame Models

Next, we look into improving our line segment reconstruction results by exploiting the pairwise constraints between 3D line segments. Even though the single line segment reconstruction method explained in section 3.2 mostly provides good results, there exists degenerate cases that single line segment reconstruction fails to generate accurate estimations. This usually occurs when a 3D line segment and camera centers lie on the same plane. In this case, small errors in image space will lead to bigger errors in 3D space. One way to overcome this problem is to make use of incidence relations between two line segments (L and Y junctions) to improve the estimation of the 3D structure [14]. In the computer vision literature, models for special structures such as building roofs have been developed [15,16] using aerial images with relatively small baselines between pairs of images. These methods are not designed to work well for reconstructing free form wire-frame models for arbitrary 3D objects where the viewing angle between different images change drastically.

We use a pairwise checking algorithm between the reconstructed 3D line segments for forming wire-frame model without prior information on the structure of the 3D object. During the formation of the wire-frame model, we look at the closeness of the endpoints of reconstructed 3D line segments as well as the closeness of the endpoints of 2D line segments in images that are associated with the reconstructed 3D line segment. Two entities are joined if their endpoints are close enough both in the image space and the 3D space. The wireframe model keeps growing unless there are no line segments left that satisfy the closeness constraint.

Each wireframe model is formed as an undirected graph $G = (V, E)$ where the set of edges represent 3D line segments and vertices represent their endpoints. Instead of minimizing the objective function for each line segment separately, here we minimize an objective function for all lines (edges) in the wire-frame (graph) model. The criteria for wire-frame (graph) minimization problem is given in equation 2.

$$G^* = \operatorname*{arg\,min}_{V \in R^{3N_V}} \sum_{e \in E} \Big(\sum_{i=1}^{n} d_l(l_{i_e}, l'_{i_e}) + \beta \sum_{i=1}^{n} d_s(l_{i_e}, l''_{i_e}) \Big) \tag{2}$$

where N_V is the number of vertices in a wire-frame model and l_{i_e} is the line in i^{th} image associated with edge $e \in E$ in graph $G = (V, E)$.

Note that, the wireframe model introduces more constraints, therefore the sum of squares error of the individual line segment reconstruction is always smaller then the error of the wireframe model reconstruction. On the other hand, additional constraints provide better estimation of 3D line segments in terms of the actual 3D geometry. The result of 3D line segment reconstruction using wire-frame models is given in figure 4.

4   Change Detection

Our change detection method is based on the appearance and disappearance of line segments throughout an image sequence. We compare the geometry of lines rather than the gray levels of image pixels to avoid the computationally intensive, and some-times impossible, tasks of estimating 3D surfaces and their associated BRDFs in the model-building stage. Estimating 3D lines is computationally much less costly and is more

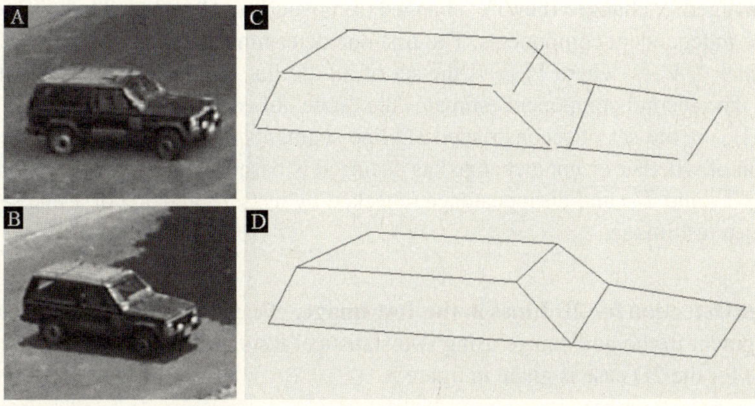

Fig. 4. Reconstruction results for 3D line segments of a Jeep. (A)-(B) The Jeep is shown from two different views. (C) Reconstruction results of single line segment reconstruction as explained in section 3.2. (D) Reconstruction results of the wire-frame model reconstruction as proposed in section 3.3.

reliable. Our method is widely applicable as man-made structures consisting of 3D line segments are the main focus of most applications.

4.1 Definition of the Problem

The definition of the general change detection problem is the following: A 3D surface model and BRDF are estimated for a region from a number of images taken by calibrated cameras in various positions at distinctly different times. This permits the prediction of the appearance of the entire region from a camera in an arbitrary new position. When a new image is taken by a calibrated camera from a new position, the computer must make a decision as whether a new object has appeared in the region or the image is of the same region [17]. In our change detection problem, which differs from the preceding, the algorithm must decide whether a new object has appeared in the region or whether an object in the region has left. We predict the 3D model which consists only of long and short straight lines, since estimating the complete 3D surface under varying illumination conditions and in the existence of specular highlights is often impractical. Moreover, image edges resulting from reflectance edges or from 3D surface ridges are less sensitive to image view direction and surface illumination, hence so are the 3D curves associated with these edges. For man-made objects and for general 3D curves, straight line approximations are usually appropriate and effective. Our method detects changes by interpreting reconstructed 3D line segments and 2D line segments detected in training and test images.

4.2 Decision Making Procedure

Our change detection method assigns a "state" to each 2D line segment in the test image and each reconstructed 3D line segment from the training images. These states are:

"not-changed", "changed (new)", "changed (removed)" and "occluded". The algorithm has two independent components. The first one determines the "state" of each 2D line segment $l \in V_{n+1}$ where V_{n+1} is the set of all 2D line segments detected in the test image. The second component estimates the "state" of each 3D line segment $L \in W_n$ where W_n is the set of reconstructed 3D line segments using the first n images. The decision of whether or not a change has occurred is based on the appearance (checked by tests T_1 and T_2) and disappearance (checked by tests T_3 and T_4) of a line segment in a given test image.

Change Detection for 2D Lines in the Test Image. We estimate the "state" of each 2D line segment in the new image using two statistical tests T_1 and T_2. The classification scheme for the 2D case is given in figure 5.

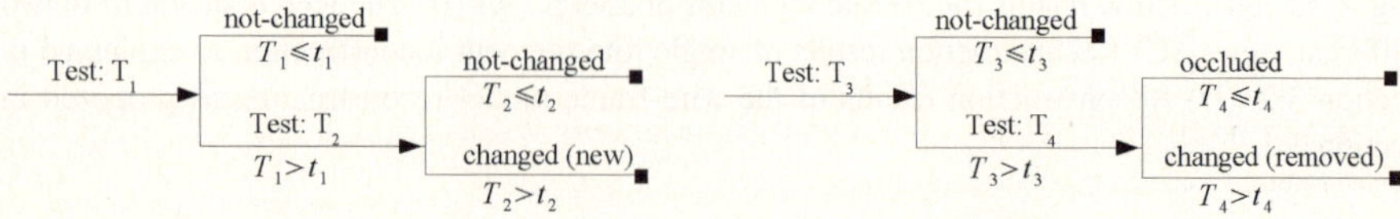

Fig. 5. General scheme for line segment based change detection. (Left) Change detection for 2D line segment in the test image. (Right) Change detection for reconstructed 3D line segments from training images. Threshold values t_1, t_2, t_3 and t_4 are selected to produce the desired change detection rate.

First we apply T_1 to test how well a 2D line segment in I_{n+1} fits to the 3D model W_n. Basically T_1 is the distance of the 2D line segment in the new image to the closest projection into I_{n+1} of the 3D line segment in the 3D model.

$$T_1 = \min_{L \in W_n} d_s(l, l''_{n+1})$$

The second step is, if necessary, to apply T_2 to test if there exists a 2D line segment in one of the past images that has taken from a similar viewing direction. Let's define the index of the image in the training set that has the closest camera projection matrix compared to the test image as c^*.

$$c^* = \arg\min_{i \in \{1,...,n\}} \|M_{n+1} - M_i\|_F$$

where $\| \cdot \|_F$ is the Frobenius norm. T_2 is defined as:

$$T_2 = \min_{l \in V_{c^*}} d_s(l, l''_{c^*})$$

Here we assume that for large training sets, there exists a camera close to the camera that took the test image assuming that camera matrices are normalized.

Change Detection for Reconstructed 3D Line Segments. We estimate the "state" of each reconstructed 3D line segment using two statistical tests T_3 and T_4. The classification scheme for the 3D case is given in figure 5.

First we apply T_3 to test how well a 3D line segment in W_n fits to the 2D lines in the new image. Basically T_3 is the distance of a reconstructed 3D line segment to the closest 2D line segment in the new image.

$$T_3 = \min_{l \in V_{n+1}} d_s(l, l''_{n+1})$$

here l''_{n+1} is the projection of L to the $(n+1)^{st}$ image.

The second step is, if necessary, to apply T_4 to test if there is an existing 3D line segment in W_n that occludes the current 3D line L.

$$T_4 = \min_{G \in W_n \backslash \{L\}} d_s(g''_{n+1}, l''_{n+1}) Z(M_{n+1}, L, G)$$

Here g''_{n+1} is the projection of G into I_{n+1} as a line segment and $Z(M_{n+1}, L, G)$ returns 1 if both endpoints of G are closer to the camera center of M_{n+1} than both endpoints of L, otherwise it returns ∞.

5   Experimental Results

In this section, we present the results of our change detection algorithm in three different image sequences (experiments). It is assumed that the scene geometry does not change during the training period. The reconstruction of the 3D model and the change detection is done using the methods explained in section 3 and section 4. The aim of each experiment is to show that our change detection method successfully detects the changes and type of changes in test images in which the scene geometry is significantly different than the training images.

The first sequence is a collection of 5 training images and 1 test image, all of which are taken in a two hour interval. The result of this experiment is shown in figure 1. The test image is taken at a different time of the day and from a different camera position; hence the illumination and viewpoint direction is significantly different compared to the training images. There are two important changes that have taken place in this test image. The first is the disappearance of a vehicle that was previously parked in the training images. The second change is the appearance of a vehicle close to the empty parking spot. Both major changes (and their types) are detected accurately with a low level of false alarm rate and main regions of change have been successfully predicted. Notice that small "new lines" (shown in "red") in the front row of the parking place are due to the specular highlights that did not exist in the set of training images. There is significant illumination difference between the test and training images, since the test image is taken a few hours after the training images. The red line on the ground of the empty parking spot is due to the shadow of another car. Geometrically, that line was occluded by the car that has left the scene, so it can be thought as an existing line which did not show up in training images due to self occlusion. The result of this experiment shows that our method is robust to severe illumination and viewpoint changes.

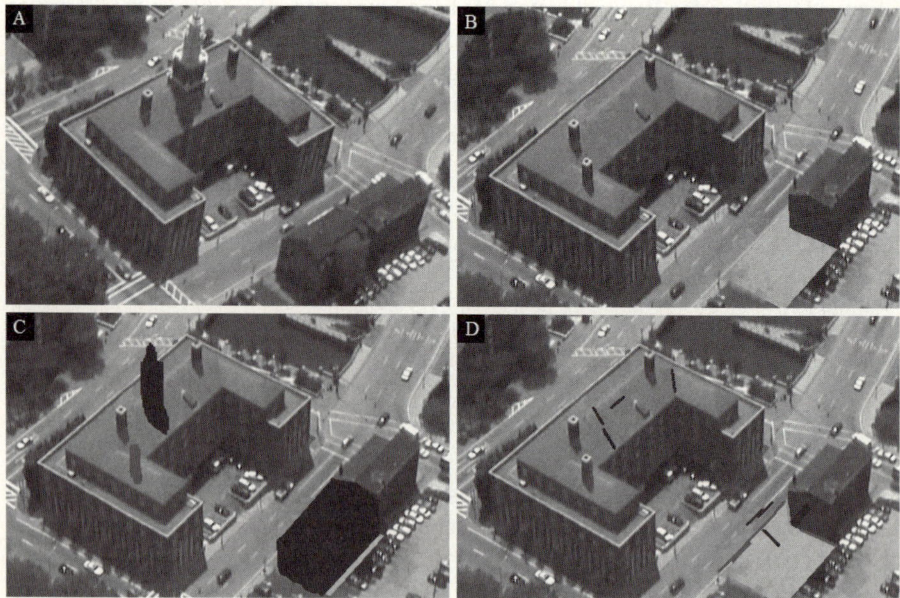

Fig. 6. Change detection results for an urban area after training on a sequence of 20 images. (A) A sample training image. (B) The test image. (C) Hand-marked ground truth for change where the new objects are labeled with "red" and the removed objects are labeled with "blue". (D) Line segment based change detection results in which "new" lines are shown in "red" and "removed" lines are shown in "blue". Our method detects the permanent change regions successfully and also recognizes a moving vehicle as an instance of temporal change. (This is a color image)

Unlike the first image sequence, the second and third image sequences do not have significant viewpoint and illumination changes between the training and test images. The result of the second experiment is shown in figure 6. To test our method, we manually created a few changes using image manipulation tools. We removed the town bell and added another chimney to a building. These change areas are marked with blue and red respectively in figure 6-C. Also the building at the bottom right corner of the test image is removed from the scene. This change is also marked with blue in figure 6-C. These major changes are successfully detected in this experiment. Also, despite the relatively small sizes of cars in this dataset, the change related to a moving vehicle (temporal change) is detected successfully.

The results of the third experiment is shown in figure 7. In this experiment, two new vehicles appear in the road and these changes are detected successfully. Similarly, we created manual changes in the scene geometry using image manipulation tools. We removed a few objects that existed on the terrace of the building in the bottom right part of the test image. These changes are also detected successfully by our change detection algorithm.

We also applied the Grimson change detection algorithm [3] to ground registered images for all sequences (see figure 8). Our 3D geometry based change detection method performs well for the first image sequence under significant viewpoint and illumination

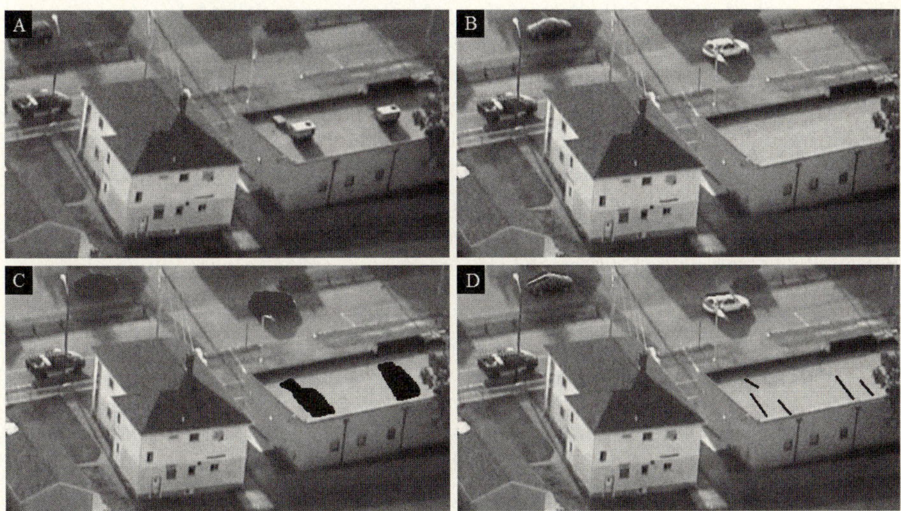

Fig. 7. Change detection results for an urban area after training on a sequence of 10 images. (A) A sample training image. (B) The test image. (C) Hand-marked ground truth for change where the new objects are labeled with "red" and the removed objects are labeled with "blue". (D) Line segment based change detection results in which "new" lines are shown in "red" and "removed" lines are shown in "blue". Our method detects the permanent change regions successfully and also recognizes two moving vehicles as instances of temporal change. (This is a color image)

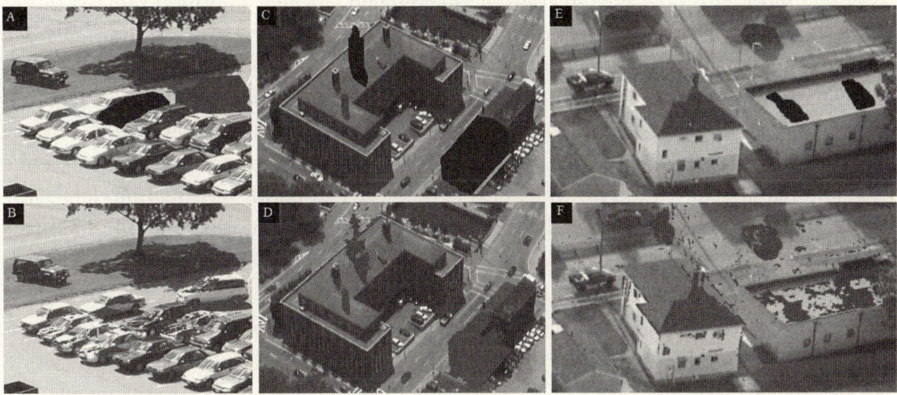

Fig. 8. Results of Grimson change detection algorithm applied to ground registered images. (A-B) Ground truth change and the result of the Grimson algorithm for the first sequence. (C-D) Ground truth change and the result of the Grimson algorithm for the second sequence. (E-F) Ground truth change and the result of the Grimson algorithm for the third sequence. (This is a color image)

differences between the training and test images. On the other hand, the Grimson method fails to detect changes reasonably due to viewpoint and illumination changes, and the existence of specular highlights. For second and third image sequences, the

Grimson method successfully detects changes except insignificant false alarms caused by the small viewpoint change between the training and test images.

To summarize our experimental results, we have shown that the significant change regions are successfully detected in all 3 experiments. The first experimental setup shows that our method can detect changes regardless of the changing viewpoint and illumination conditions. However, there are a few cases of insignificant false alarms possibly caused by shadows, specular highlights and lack of 3D geometry of newly exposed line segments in the test image. Also, it must be noted that unlike other change detection methods, our method detects the type of major changes successfully in almost all experiments.

6   Conclusion and Future Work

In this paper, we present the first generally applicable 3D line segment based change detection method for images taken from arbitrary viewing directions, at different times and under varying illumination conditions. Our method has been shown to detect significant changes with high accuracy on three different change detection experiments. Experiments indicate that our algorithm is capable of efficiently matching and accurately reconstructing small and large line segments, and successfully detecting changes (and their types) by interpreting 2D and 3D line segments. We show that our multi-view line segment matching algorithm works faster than other commonly used matching algorithms, and our 3D line segment reconstruction algorithm, exploring the connectivity of 2D and 3D line segments in order to constrain them in configurations, improve the accuracy of existing individual 3D line segment reconstruction techniques.

Future work will involve detection of shadow and specular highlight regions to improve the result of change detection by reducing the false alarm rate. Additionally, we are investigating ways of propagating the uncertainty of line segments during the 3D reconstruction process, and improving the change detection algorithm to work with uncertain projective geometry. Additionally, some initial experiments have already been performed to interpret the type of change associated with an existing pre-occluded line segment that appears in the test image. It is necessary to build statistical models using larger datasets to better interpret these changes related to newly appeared line segments in the test image.

Acknowledgments. This work was funded, in part, by the Lockheed Martin Corporation. We are grateful to Joseph L. Mundy for many insightful discussions on this work. We also thank Ece Kamar for her helpful comments on earlier drafts of this paper.

References

1. Bruzzone, L., Prieto, D.F.: Automatic analysis of the difference image for unsupervised change detection. IEEE Transactions on Geoscience and Remote Sensing 38(3), 1171–1182 (2000)
2. Bruzzone, L., Prieto, D.F.: An adaptive semiparametric and context-based approach to unsupervised change detection in multitemporal remote-sensing images. IEEE Transactions on Image Processing 11(4), 452–466 (2002)

3. Stauffer, C., Grimson, W.E.L.: Adaptive background mixture models for real-time tracking. In: CVPR, pp. 246–252 (1999)
4. Yalcin, H., Hebert, M., Collins, R.T., Black, M.J.: A flow-based approach to vehicle detection and background mosaicking in airborne video. In: CVPR, vol. II, p. 1202 (2005)
5. Broadhurst, A., Drummond, T., Cipolla, R.: A probabilistic framework for space carving. In: ICCV, pp. 388–393 (2001)
6. Pollard, T., Mundy, J.L.: Change detection in a 3-d world. In: CVPR, pp. 1–6 (2007)
7. Rowe, N.C., Grewe, L.L.: Change detection for linear features in aerial photographs using edge-finding. IEEE Transactions on Geoscience and Remote Sensing 39(7), 1608–1612 (2001)
8. Li, W., Li, X., Wu, Y., Hu, Z.: A novel framework for urban change detection using VHR satellite images. In: ICPR, pp. 312–315 (2006)
9. Schmid, C., Zisserman, A.: The geometry and matching of lines and curves over multiple views. International Journal of Computer Vision 40(3), 199–233 (2000)
10. Heuel, S., Förstner, W.: Matching, reconstructing and grouping 3D lines from multiple views using uncertain projective geometry. In: CVPR, pp. 517–524 (2001)
11. Taillandier, F., Deriche, R.: Reconstruction of 3D linear primitives from multiple views for urban areas modelisation. In: Photogrammetric Computer Vision, vol. B, p. 267 (2002)
12. Bay, H., Ferrari, V., Gool, L.J.V.: Wide-baseline stereo matching with line segments. In: CVPR, vol. I, pp. 329–336 (2005)
13. Nelder, J.A., Mead, R.: A simplex method for function minimization. Computer Journal 7, 308–313 (1965)
14. Taylor, C.J., Kriegman, D.J.: Structure and motion from line segments in multiple images. IEEE Transactions on Pattern Analysis and Machine Intelligence 17(11), 1021–1032 (1995)
15. Moons, T., Frère, D., Vandekerckhove, J., Gool, L.V.: Automatic modelling and 3d reconstruction of urban house roofs from high resolution aerial imagery. In: Burkhardt, H.-J., Neumann, B. (eds.) ECCV 1998. LNCS, vol. 1406, pp. 410–425. Springer, Heidelberg (1998)
16. Baillard, C., Schmid, C., Zisserman, A., Fitzgibbon, A.W.: Automatic line matching and 3D reconstruction of buildings from multiple views. In: ISPRS Congress, pp. 69–80 (1999)
17. Radke, R.J., Andra, S., Al-Kofahi, O., Roysam, B.: Image change detection algorithms: a systematic survey. IEEE Transactions on Image Processing 14(3), 294–307 (2005)

Action Recognition with a Bio–inspired Feedforward Motion Processing Model: The Richness of Center-Surround Interactions

Maria-Jose Escobar and Pierre Kornprobst

Odyssée project team, INRIA Sophia-Antipolis, France
{mjescoba,pkornp}@sophia.inria.fr

Abstract. Here we show that reproducing the functional properties of MT cells with various center–surround interactions enriches motion representation and improves the action recognition performance. To do so, we propose a simplified bio–inspired model of the motion pathway in primates: It is a feedforward model restricted to V1-MT cortical layers, cortical cells cover the visual space with a foveated structure and, more importantly, we reproduce some of the richness of center-surround interactions of MT cells. Interestingly, as observed in neurophysiology, our MT cells not only behave like simple velocity detectors, but also respond to several kinds of motion contrasts. Results show that this diversity of motion representation at the MT level is a major advantage for an action recognition task. Defining motion maps as our feature vectors, we used a standard classification method on the Weizmann database: We obtained an average recognition rate of 98.9%, which is superior to the recent results by Jhuang et al. (2007). These promising results encourage us to further develop bio–inspired models incorporating other brain mechanisms and cortical layers in order to deal with more complex videos.

1   Introduction

Action recognition in real scenes remains a challenging problem in computer vision. Until recently, most proposed approaches considered simplified sequence databases and relied on simplified assumptions or heuristics. Some examples of these kind of approaches are [1,2,3,4,5], where one could find therein other references and further information.

Motion is the key feature for a wide class of computer vision (CV) approaches: Existing methods consider different motion representations or characteristics, such as coarse motion estimation, global motion distribution, local motion feature detection or spatio-temporal structure learning [6,7,8,9,10,11,12]. Following this general idea which is to consider motion as an informative cue for action recognition (AR), we present a bio-inspired model for motion estimation and representation. Interestingly, it is confirmed that in the visual system the motion pathway is also very much involved in the AR task [10], but of course other

D. Forsyth, P. Torr, and A. Zisserman (Eds.): ECCV 2008, Part IV, LNCS 5305, pp. 186–199, 2008.

brain areas (e.g., the form pathway) and mechanisms (e.g., top-down attentional mechanisms) are also involved to analyze complex general scenes.

Among recent bio-inspired approaches for AR, [13] proposed a model for the visual processing in the dorsal (*motion*) and ventral (*form*) pathways. They validated their model in the AR task using stick figures constructed from real sequences. More recently, [14] proposed a feedforward architecture, which can be seen as an extension of [15]. In [14], the authors mapped their model to the cortical architecture, essentially V1 (with simple and complex cells). The only clear bio-inspired part is one of the models for S1 units and the pooling aspect. The use of spatio-temporal chunks seems to be supported also but the authors never claim any biological relevance for the corresponding subsequent processing stages (from S2 to C3). The max operator is also controversial and not supported in neurophysiology because it mainly does not allow feedbacks.

In this article, we follow the same objective as in [14], which is to propose a bio-inspired model of motion processing for AR in real sequences. Our model will be a connection-based network, in which a large number of neuron-like processing units operate in parallel. Each unit *neuron* will have an 'activation level' *membrane potential* that represents the strength of a particular feature in the environment. Here, our main contribution will be to better account for the visual system properties, and in particular, at MT layer level: We reproduce part of the variety of center-surround interactions [16,17]. Then, in order to prove the relevance of this extended motion description, we will show its benefits on the AR application, and compare our results with the ones obtained by [14].

This article presents the model described in Fig. 1 and it is organized as follows. Section 2 presents the core of the approach which is a biologically-inspired model of motion estimation, based on a feedforward architecture. As we previously mentioned, the aim of this article is to show how a bio-inspired model can be used in a real application such as AR. Note that we also studied some low-level properties of the model concerning motion processing [18] but those studies are out of the scope of this article. The first stage (Section 2.1) is the local motion extraction corresponding to the V1 layer, with a discrete foveated organization. The output of this layer is fed to the MT layer (Section 2.2), which is composed of a set of neurons whose dynamics are defined by a conductance-based neuron model. We define the connectivity between V1 and MT layers according to neurophysiology, which defines the center-surround interactions of a MT neuron. The output of the MT layer is a set of neuron membrane potentials, whose values indicate the presence of a certain velocity or contrasts of velocities. Then, in Section 3, we consider the problem of AR based on the MT layer activity. In this section we also present the experimental protocol, some validations and a comparison with the approach presented by [14]. Interestingly, we show how the variety of surround-interactions in MT cells found in physiology allows the improvement of the recognition performances. We conclude in Section 4.

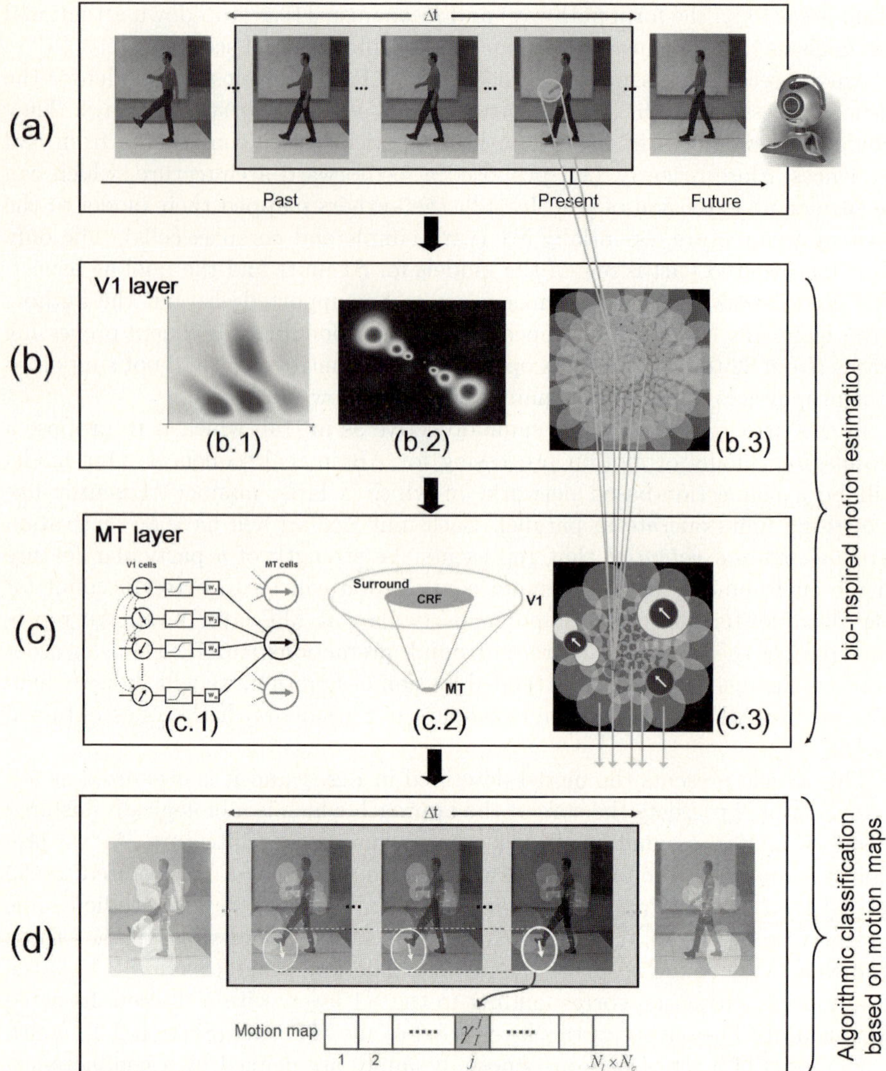

Fig. 1. Block diagram showing the different steps of our approach from the input image sequence as stimulus until the *motion map* encoding the motion pattern. (a) We use a real video sequence as input, the input sequences are preprocessed in order to have contrast normalization and centered moving stimuli. To compute the *motion map* representing the input image we consider a sliding temporal window of length Δt. (b) Directional-selectivity filters are applied over each frame of the input sequence in a log-polar distribution grid obtaining the activity of each V1 cell. (c) V1 outputs feed the MT cells which integrate the information in space and time. (d) The *motion map* is constructed calculating the mean activation of MT cells inside the sliding temporal window. The *motion map* has a length of $N_L \times N_c$ elements, where N_L is the number of MT layers of cells and N_c is the number of MT cells per layer. This *motion map* characterizes and codes the action stimulus.

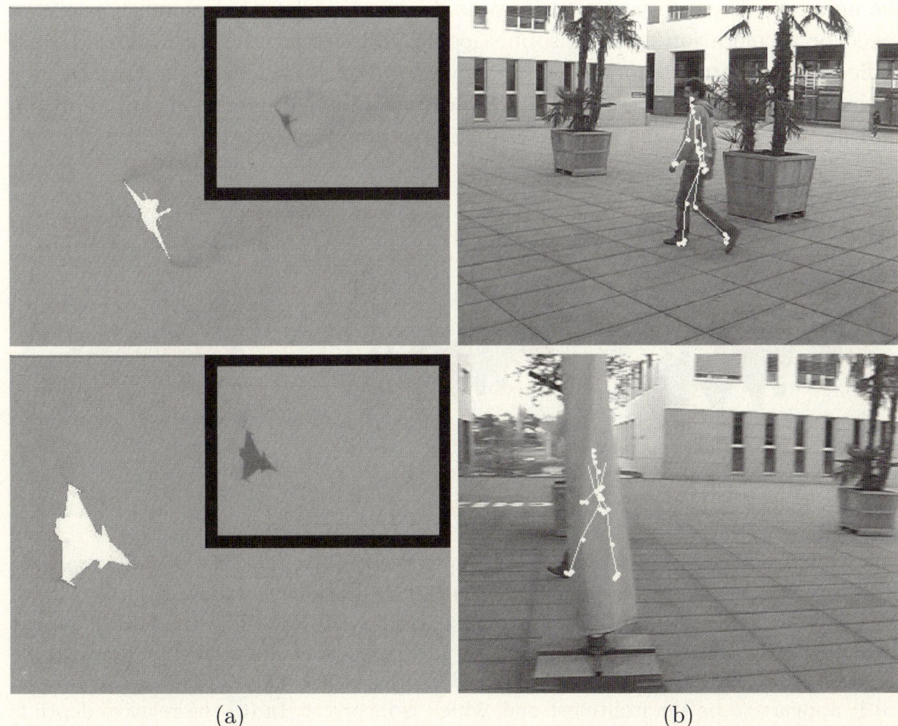

Fig. 1. Airplanes and people are examples of objects that exhibit a favored direction of motion. (a) We project the 3D aircraft model using the recovered pose to produce the white overlay. The original images are shown in the upper right corner. (b) We overlay the 3D skeleton in the recovered pose, which is correct even when the person is occluded.

that can zoom to keep the target object in the field of view, rendering the use of simple techniques such as background subtraction impractical.

2   Related Work and Approach

Non-holonomic constraints that link direction of travel and position have been widely used in fields such as radar-based tracking [4] or robot self-localization [5], often in conjunction with Kalman filtering. However, these approaches deal with points moving in space and do not concern themselves with the fact that they are extended 3D objects, whether rigid or deformable, that have an orientation, which conditions the direction in which they move. Such constraints have also been adopted for motion synthesis in the Computer Graphics community [6], but they are not directly applicable in a Computer Vision context since they make no attempt at fitting model to data.

Tracking rigid objects in 3D is now a well understood problem and can rely on many sources of image information, such as keypoints, texture, or edges [1]. If the image quality is high enough, simple dynamic models that penalize excessive speed or acceleration or more sophisticated Kalman filtering techniques [7] are sufficient to enforce temporal consistency. However, with lower quality data such as the plane videos of Fig. 1(a), the simple quadratic regularization constraints [8] that are used most often yield unrealistic results, as shown in Fig. 2.

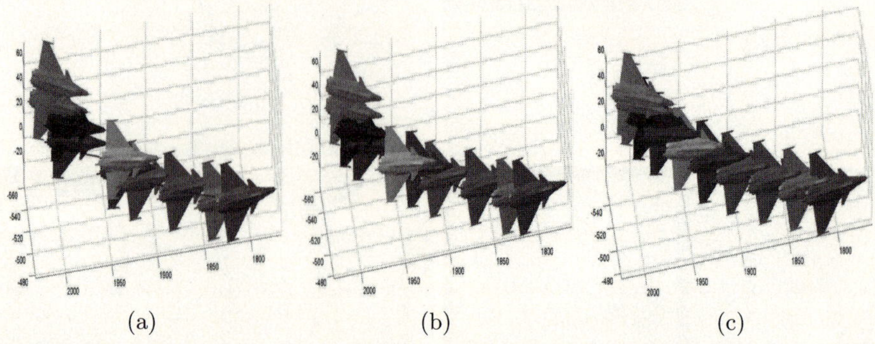

(a)   (b)   (c)

Fig. 2. The first 50 frames of the first airplane sequence. The 3D airplane model is magnified and plotted once every 5 frames in the orientation recovered by the algorithm: (a) Frame by Frame tracking without regularization. (b) Imposing standard quadratic regularization constraints. (c) Linking pose to motion produces a much more plausible set of poses. Note for example the recovered depth of the brightest airplane: In (a) and (b) it appears to be the frontmost one, which is incorrect. In (c) the relative depth is correctly retrieved.

Tracking a complex articulated 3D object such as a human body is much more complex and existing approaches remain brittle. Some of the problems are caused by joint reflection ambiguities, occlusion, cluttered backgrounds, non-rigidity of tissue and clothing, complex and rapid motions, and poor image resolution. The problem is particularly acute when using a single video to recover the 3D motion. In this case, incorporating motion models into the algorithms has been shown to be effective [2]. The models can be physics-based [9] or learned from training data [10,11,12,13]. However, all of these assume the joint angles, that define the body pose, and the global motion variables are independent. As is the case for rigid body tracking, they typically revert to second order Gauss-Markov modeling or Kalman filtering to smooth the global motion. Again, this can lead to unrealistic results as shown in Fig. 3. Some approaches implicitly take into account the relationship between pose and direction of travel by learning from training data a low-dimensional representation that includes both [3,14,15,16]. However, the set of motions that can be represented is heavily constrained by the contents of the training database, which limits their generality.

To remedy these problems, we explicitly link pose and motion as follows: Given an object moving along its trajectory as depicted by Fig. 4, the angle between $\dot{P}_t$, the derivative of its position, and its orientation Λ_t should in general be small. We can therefore write that

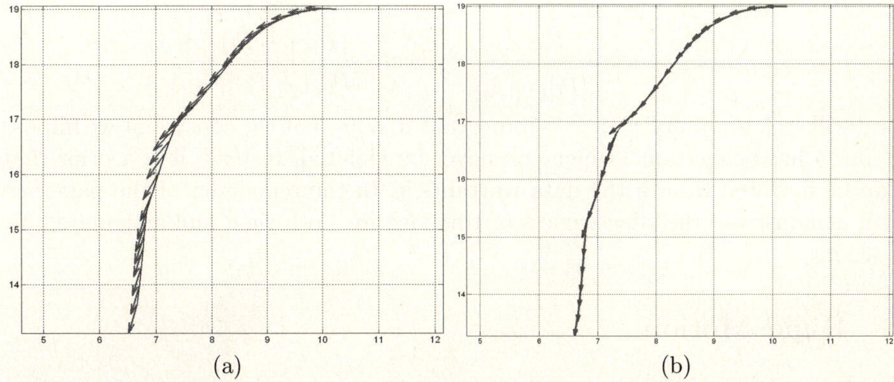

Fig. 3. Recovered 2D trajectory of the subject of Fig. 1(b). The arrows represent the direction he is facing. (a) When pose and motion are not linked, he appears to walk sideways. (b) When they are, he walks naturally. The underlying grid is made of 1 meter squares.

$$\frac{\dot{P}_t \cdot \Lambda_t}{||\dot{P}_t|| \cdot ||\Lambda_t||}$$

should be close to 1. To enforce this, we can approximate the derivative of the locations using finite differences between estimated locations $\hat{P}$ at different time instants. This approximation is appropriate when we can estimate the location at a sufficiently high frequency (e.g. 25 Hz).

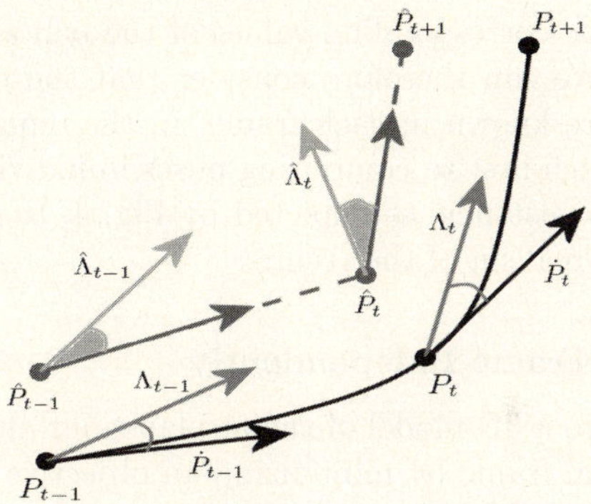

Fig. 4. The continuous curve represents the real trajectory of the object, while the dashed lines show its approximation by finite differences

Our constraint then reduces to minimizing the angle between the finite differences approximation of the derivative of the trajectory at time t, given by $\hat{P}_{t+1} - \hat{P}_t$, and the object's estimated orientation given by $\hat{\Lambda}_t$. We write this angle, which is depicted as filled both at time $t - 1$ and t in Fig. 4, as

$$\phi_{t \to t+1} = \text{acos} \frac{\hat{P}_t \cdot \hat{\Lambda}_t}{||\hat{P}_t|| \cdot ||\hat{\Lambda}_t||} = \text{acos} \frac{(\hat{P}_{t+1} - \hat{P}_t) \cdot \hat{\Lambda}_t}{||(\hat{P}_{t+1} - \hat{P}_t)|| \cdot ||\hat{\Lambda}_t||}$$

and will seek to minimize it. It is important to note that the constraint we impose is not a hard constraint, which can never be violated. Instead, it is a prior that can be deviated from if the data warrants it. In the remainder of the paper we will demonstrate the effectiveness of this idea for both rigid and articulated 3D tracking.

3   Rigid Motion

In the case of a rigid motion, we demonstrate our approach using video sequences of a fighter plane performing aerobatic maneuvers such as the one depicted by Fig. 5. In each frame of the sequences, we retrieve the pose which includes position expressed by cartesian coordinates and orientation defined by the roll, pitch and yaw angles. We show that these angles can be recovered from single viewpoint sequences with a precision down to a few degrees, and that linking pose and motion estimation contributes substantially to achieving this level of accuracy. This is extremely encouraging considering the fact that the videos we have been working with were acquired under rather unfavorable conditions: As can be seen in Fig. 5, the weather was poor, the sky gray, and the clouds many, all of which make the plane less visible and therefore harder to track. The airplane is largely occluded by smoke and clouds in some frames, which obviously has an adverse impact on accuracy but does not result in tracking failure.

The video sequences were acquired using a fully calibrated camera that could rotate around two axes and zoom on the airplane. Using a couple of encoders, it could keep track of the corresponding values of the pan and tilt angles, as well as the focal length. We can therefore consider that the intrinsic and extrinsic camera parameters are known in each frame. In the remainder of this section, we present our approach first to computing poses in individual frames and then imposing temporal consistency, as depicted by Fig. 4, to substantially improve the accuracy and the realism of the results.

3.1   Pose in Each Frame Independently

Since we have access to a 3D model of the airplane, our algorithm computes the pose in each individual frame by minimizing an objective function L_r that is a weighted sum of a color and an edge term:

- The color term is first computed as the Bhattacharyya distance [17] between the color histogram of the airplane that we use as a model, whose pose was captured manually in the first frame, and the color histogram of the image area corresponding to its projection in subsequent frames. To this we add a term that takes into account background information, also expressed as a difference of color histograms, which has proved important to guarantee robustness.

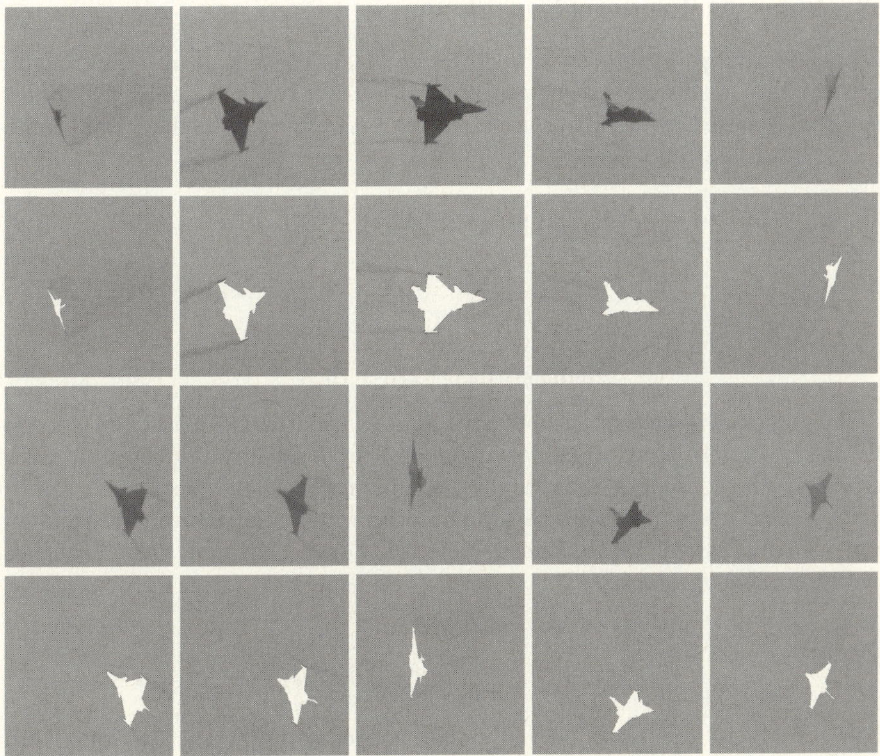

Fig. 5. Airplane video and reprojected model. **First and third rows:** Frames from the input video. Note that the plane is partially hidden by clouds in some frames, which makes the task more difficult. **Second and fourth rows:** The 3D model of the plane is reprojected into the images using the recovered pose parameters. The corresponding videos are submitted as supplemental material.

– The edge term is designed to favor poses such that projected model edges correspond to actual image edges and plays an important role in ensuring accuracy.

In each frame t, the objective function L_r is optimized using a particle-based stochastic optimization algorithm [18] that returns the pose corresponding to the best sample. The resulting estimated pose is a six-dimensional vector $\hat{S}_t = (\hat{P}_t, \hat{\Lambda}_t) = \text{argmin}_S L_r(S)$ where $\hat{P}_t = (\hat{X}_t, \hat{Y}_t, \hat{Z}_t)$ is the estimated position of the plane in an absolute world coordinate system and $\hat{\Lambda}_t = (\hat{\rho}_t, \hat{\theta}_t, \hat{\gamma}_t)$ is the estimated orientation expressed in terms of roll, pitch and yaw angles. The estimated pose $\hat{S}_t$ at time t is used to initialize the algorithm in the following frame $t + 1$, thus assuming that the motion of the airplane between two consecutive frames is relatively small, which is true in practice.

3.2 Imposing Temporal Consistency

Independently optimizing L_r in each frame yields poses that are only roughly correct. As a result, the reconstructed motion is extremely jerky. To enforce temporal consistency, we introduce a regularization term M defined over frames $t-1$, t, and $t+1$ as

$$M(S_t) = \alpha_1||A(P_t)||^2 + \alpha_2||A(\Lambda_t)||^2 + \beta(\phi^2_{t-1 \to t} + \phi^2_{t \to t+1}) \,, \tag{1}$$

$$A(P_t) = P_{t+1} - 2P_t + P_{t-1} \,, \tag{2}$$

$$A(\Lambda_t) = \Lambda_{t+1} - 2\Lambda_t + \Lambda_{t-1} \,. \tag{3}$$

The first two terms of (1) enforce motion smoothness. The third term is the one of Fig. 4, which links pose to motion by forcing the orientation of the airplane to be consistent with its direction of travel. In practice, α_1, α_2 and β are chosen to relate quantities that would otherwise be incommensurate and are kept constant for all the sequences we used. For an N-frame video sequence, ideally, we should minimize

$$f_r(S_1, \ldots, S_N) = \sum_{t=1}^{N} L_r(S_t) + \sum_{t=2}^{N-1} M(S_t) \tag{4}$$

with respect to the poses in individual images. In practice, for long video sequences, this represents a very large optimization problem. Therefore, in our current implementation, we perform this minimization in sliding temporal 3-frame windows using a standard simplex algorithm that does not require the computation of derivatives. We start with the first set of 3 frames, retain the resulting pose in the first frame, slide the window by one frame, and iterate the process using the previously refined poses to initialize each optimization step.

3.3 Tracking Results

The first sequence we use for the evaluation of our approach is shown in Fig. 5 and contains 1000 frames shot over 40 seconds, a time during which the plane performs rolls, spins and loops and undergoes large accelerations.

In Fig. 6(a) we plot the locations obtained in each frame independently. In Fig. 6(b) we imposed motion smoothness by using *only* the first two terms of (1). In Fig 6(c) we link pose to motion by using all three terms of (1). The trajectories are roughly similar in all cases. However, using the full set of constraints yields a trajectory that is both smoother and more plausible.

In Fig. 2, we zoom in on a portion of these 3 trajectories and project the 3D plane model in the orientation recovered every fifth frame. Note how much more consistent the poses are when we use our full regularization term.

The plane was equipped with sophisticated gyroscopes which gave us meaningful estimates of roll, pitch, and yaw angles, synchronized with the camera

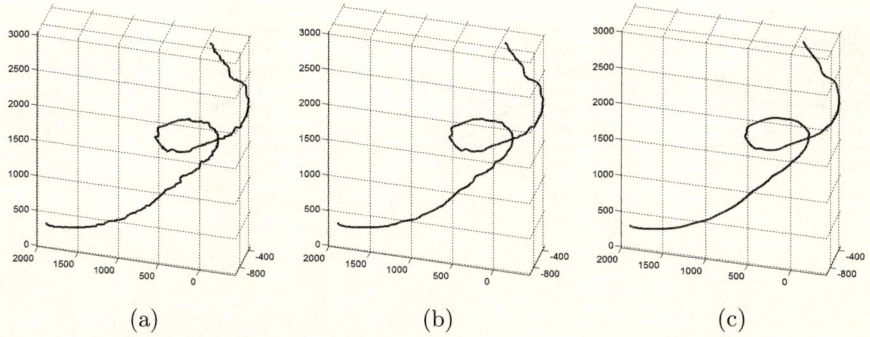

Fig. 6. Recovered 3D trajectory of the airplane for the 40s sequence of Fig. 5: (a) Frame by Frame tracking. (b) Imposing motion smoothness. (c) Linking pose to motion. The coordinates are expressed in meters.

and available every third frame. We therefore use them as ground truth. Table 1 summarizes the deviations between those angles and the ones our algorithm produces for the whole sequence. Our approach yields an accuracy improvement over frame by frame tracking as well as tracking with simple smoothness constraint. The latter improvement is in the order of 5 %, which is significant if one considers that the telemetry data itself is somewhat noisy and that we are therefore getting down to the same level of precision. Most importantly, the resulting sequence does not suffer from jitter, which plagues the other two approaches, as can be clearly seen in the videos given as supplemental material.

Table 1. Comparing the recovered pose angles against gyroscopic data for the sequence of Fig. 5. Mean and standard deviation of the absolute error in the 3 angles, in degrees.

	Roll Angle Error		Pitch Angle Error		Yaw Angle Error	
	Mean	Std. Dev.	Mean	Std. Dev.	Mean	Std. Dev.
Frame by Frame	2.291	2.040	1.315	1.198	3.291	2.245
Smoothness Constraint only	2.092	1.957	1.031	1.061	3.104	2.181
Linking Pose to Motion	1.974	1.878	0.975	1.000	3.003	2.046

In Fig. 7 we show the retrieved trajectory for a second sequence, which lasts 20 seconds. As before, in Table 2, we compare the angles we recover against gyroscopic data. Again, linking pose to motion yields a substantial improvement.

4   Articulated Motion

To demonstrate the effectiveness of the constraint we propose in the case of articulated motion, we start from the body tracking framework proposed in [19]. In this work, it was shown that human motion could be reconstructed in 3D

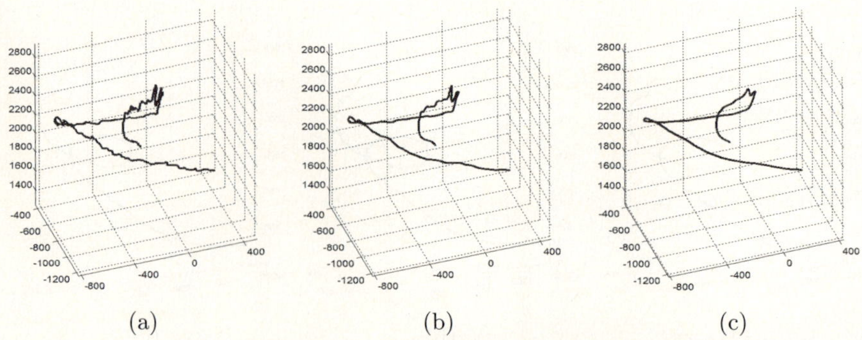

Fig. 7. Recovered 3D trajectory of the airplane for a 20s second sequence: (a) Frame by Frame tracking. (b) Imposing motion smoothness. (c) Linking pose to motion. The coordinates are expressed in meters.

Table 2. Second sequence: Mean and standard deviation of the absolute error in the 3 angles, in degrees

	Roll Angle Error		Pitch Angle Error		Yaw Angle Error	
	Mean	Std. Dev.	Mean	Std. Dev.	Mean	Std. Dev.
Frame by Frame	3.450	2.511	1.607	1.188	3.760	2.494
Smoothness Constraint only	3.188	2.445	1.459	1.052	3.662	2.237
Linking Pose to Motion	3.013	2.422	1.390	0.822	3.410	2.094

by detecting canonical poses, using a motion model to infer the intermediate poses, and then refining the latter by maximizing an image-based likelihood in each frame independently. In this section, we show that, as was the case for rigid motion recovery, relating the pose to the direction of motion leads to more accurate and smoother 3D reconstructions.

In the remainder of the section, we first introduce a slightly improved version of the original approach on which our work is based. We then demonstrate the improvement that the temporal consistency constraint we advocate brings about.

4.1   Refining the Pose in Each Frame Independently

We rely on a coarse body model in which individual limbs are modeled as cylinders. Let $S_t = (P_t, \Theta_t)$ be the state vector that defines its pose at time t, where Θ_t is a set of joint angles and P_t a 3D vector that defines the position and orientation of the root of the body in a 2D reference system attached to the ground plane.

In the original approach [19], a specific color was associated to each limb by averaging pixel intensities in the projected area of the limb in the frames where a canonical pose was detected. Then S_t was recovered as follows: A rough initial state was predicted by the motion model. Then the sum-of-squared-differences between the synthetic image, obtained by reprojecting the model, and the actual one was minimized using a simple stochastic optimization algorithm.

Here, we replace the single color value associated to each limb by a histogram, hereby increasing generality. As in Sect. 3.1, we define an objective function L_a that measures the quality of the pose using the Bhattacharyya distance to express the similarity between the histogram associated to a limb and that of the image portion that corresponds to its reprojection. Optimizing L_a in each frame independently leads, as could be expected, to a jittery reconstruction as can be seen in the video given as supplemental material.

4.2   Imposing Temporal Consistency

In order to improve the quality of our reconstruction, we perform a global optimization on all N frames between two key-pose detections, instead of minimizing L_a independently in each frame. To model the relationship between poses we learn a PCA model from a walking database and consider a full walking cycle as a single data point in a low-dimensional space [20,11]. This lets us parameterize all the poses S_i between consecutive key-pose detections by n PCA coefficients $(\alpha_1 \ldots \alpha_n)$, plus a term, η, that represents possible variations of the walking speed during the walking cycle ($n = 5$ in our experiments). These coefficients do not take into account the global position and orientation of the body, which needs to be parameterized separately. Since the walking trajectory can be obtained by a 2D spline curve lying on the ground plane, defined by the position and orientation of the root at the two endpoints of the sequence, modifying these endpoints P_{start} and P_{end} will yield different trajectories. The root position and orientation corresponding to the different frames will then be picked along the spline curve according to the value of η. It in fact defines where in the walking cycle the subject is at halftime between the two detections. For a constant speed during a walking cycle the value of η is 0.5, but it can go from 0.3 to 0.7 depending on change in speed between the first and the second half-cycle.

We can now formulate an objective function that includes both the image likelihood and a motion term, which, in this case, constrains the person to move in the direction he is facing. This objective function is then minimized with respect to the parameters introduced above $(\alpha_1, \ldots, \alpha_n, P_{\text{start}}, P_{\text{end}}, \eta)$ on the full sequence between two consecutive key-pose detections. In other words, we seek to minimize

$$f_a(S_1, \ldots, S_N) = \sum_{t=1}^{N} L_a(S_t) + \sum_{t=2}^{N} \beta(\phi_{t-1 \to t}^2) \tag{5}$$

with respect to $(\alpha_1, \ldots, \alpha_n, P_{\text{start}}, P_{\text{end}}, \eta)$, where the second term is defined the same way as in the airplane case and β is as before a constant weight that relates incommensurate quantities. The only difference is that in this case both the estimated orientation and the expected motion, that define the angle ϕ, are 2-dimensional vectors lying on the ground plane. This term is the one that links pose to motion. Note that we do not need quadratic regularization terms such as the first two of (1) because our parameters control the entire trajectory, which is guaranteed to be smooth.

4.3   Tracking Results

We demonstrate our approach on a couple of very challenging sequences. In the sequence of Fig. 8, the subject walks along a circular trajectory and the camera is following him from its center. At a certain point the subject undergoes a total occlusion but the algorithm nevertheless recovers his pose and position thanks to its global motion model. Since the tracking is fully 3D, we can also recover the trajectory of the subject on the ground plane and his instantaneous speed at each frame.

In Fig. 3 we examine the effect of linking or not pose to motion on the recovered trajectory: That is, setting β to zero or not in (5). The arrows represent the orientation of the subject on the ground plane. They are drawn every fifth frame.

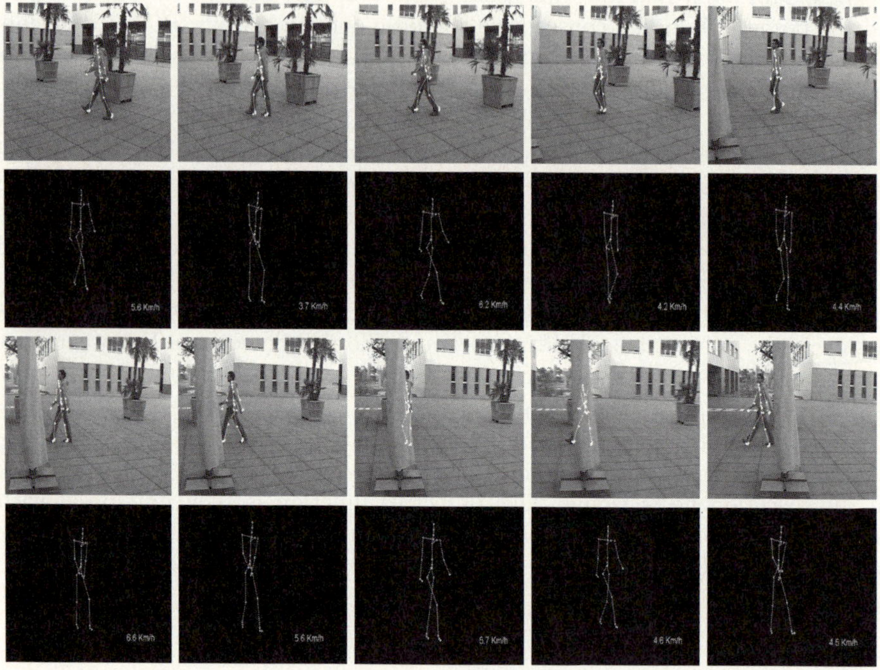

Fig. 8. Pedestrian tracking and reprojected 3D model for the sequence of Fig. 1 **First and third rows:** Frames from the input video. The recovered body pose has been reprojected on the input image. **Second and fourth rows:** The 3D skeleton of the person is seen from a different viewpoint, to highlight the 3D nature of the results. The numbers in the bottom right corner are the instantaneous speeds derived from the recovered motion parameters. The corresponding videos are submitted as supplementary material.

The images clearly show that, without temporal consistency constraints, the subject appears to slide sideways while when the constraints are enforced the motion is perfectly consistent with the pose. This can best be evaluated from the videos given as supplemental material.

Fig. 9. Pedestrian tracking and reprojected 3D model in a second sequence. **First and third rows:** Frames from the input video. The recovered body pose has been reprojected on the input image. **Second and fourth rows:** The 3D skeleton of the person is seen from a different viewpoint, to highlight the 3D nature of the results. The numbers in the bottom right corner are the instantaneous speeds derived from the recovered motion parameters.

To validate our results, we manually marked the subject's feet every 10 frames in the sequence of Fig. 8 and used their position with respect to the tiles on the ground plane to estimate their 3D coordinates. We then treated the vector joining the feet as an estimate of the body orientation and the midpoint as an estimated of its location. As can be seen in Table 3, linking pose to motion produces a small improvement in the position estimate and a much more substantial one in the orientation estimate, which is consistent with what can be observed in Fig. 3.

In the sequence of Fig. 9 the subject is walking along a curvilinear path and the camera follows him, so that the viewpoint undergoes large variations. We are nevertheless able to recover pose and motion in a consistent way, as shown in Fig. 10 which represents the corresponding recovered trajectory.

Table 3. Comparing the recovered pose angles against manually recovered ground truth data for the sequence of Fig. 8. It provides the mean and standard deviation of the absolute error in the X and Y coordinates, in centimeters, and the mean and standard deviation of the recovered orientation, in degrees.

	X Error		Y Error		Orientation Error	
	Mean	Std. Dev.	Mean	Std. Dev.	Mean	Std. Dev.
Not Linking Pose to Motion	12.0	7.1	16.8	11.9	11.7	7.6
Linking Pose to Motion	11.8	7.3	14.9	9.3	6.2	4.9

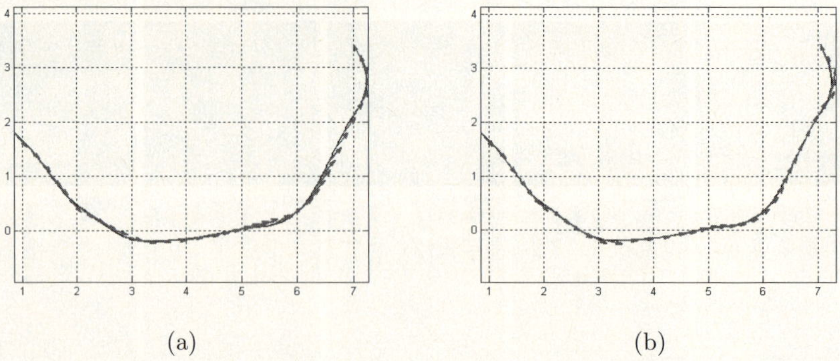

(a)   (b)

Fig. 10. Recovered 2D trajectory of the subject of Fig. 9. As in Fig. 3, when orientation and motion are not linked, he appears to walk sideway (a) but not when they are (b).

5   Conclusion

In this paper, we have used two very different applications to demonstrate that jointly optimizing pose and direction of travel substantially improves the quality of the 3D reconstructions that can be obtained from video sequences. We have also shown that we can obtain accurate and realistic results using a single moving camera.

This can be done very simply by imposing an explicit constraint that forces the angular pose of the object or person being tracked to be consistent with their direction of travel. This could be naturally extended to more complex interactions between pose and motion. For example, when a person changes orientation, the motion of his limbs is not independent of the turn radius. Similarly, the direction of travel of a ball will be affected by its spin. Explicitly modeling these subtle but important dependencies will therefore be a topic for future research.

References

1. Lepetit, V., Fua, P.: Monocular model-based 3d tracking of rigid objects: A survey. Foundations and Trends in Computer Graphics and Vision (2005)
2. Moeslund, T.B., Hilton, A., Krüger, V.: A survey of advances in vision-based human motion capture and analysis. CVIU 104(2), 90–126 (2006)

3. Sidenbladh, H., Black, M.J., Sigal, L.: Implicit Probabilistic Models of Human Motion for Synthesis and Tracking. In: Heyden, A., Sparr, G., Nielsen, M., Johansen, P. (eds.) ECCV 2002. LNCS, vol. 2350, pp. 784–800. Springer, Heidelberg (2002)

4. Bar-Shalom, Y., Kirubarajan, T., Li, X.R.: Estimation with Applications to Tracking and Navigation. John Wiley & Sons, Inc., Chichester (2002)

5. Zexiang, L., Canny, J.: Nonholonomic Motion Planning. Springer, Heidelberg (1993)

6. Ren, L., Patrick, A., Efros, A.A., Hodgins, J.K., Rehg, J.M.: A data-driven approach to quantifying natural human motion. ACM Trans. Graph. 24(3) (2005)

7. Koller, D., Daniilidis, K., Nagel, H.H.: Model-Based Object Tracking in Monocular Image Sequences of Road Traffic Scenes. IJCV 10(3), 257–281 (1993)

8. Poggio, T., Torre, V., Koch, C.: Computational Vision and Regularization Theory. Nature 317 (1985)

9. Brubaker, M., Fleet, D., Hertzmann, A.: Physics-based person tracking using simplified lower-body dynamics. In: CVPR (2007)

10. Urtasun, R., Fleet, D., Fua, P.: 3D People Tracking with Gaussian Process Dynamical Models. In: CVPR (2006)

11. Ormoneit, D., Sidenbladh, H., Black, M.J., Hastie, T.: Learning and tracking cyclic human motion. In: NIPS (2001)

12. Agarwal, A., Triggs, B.: Tracking articulated motion with piecewise learned dynamical models. In: Pajdla, T., Matas, J(G.) (eds.) ECCV 2004. LNCS, vol. 3023, pp. 54–65. Springer, Heidelberg (2004)

13. Taycher, L., Shakhnarovich, G., Demirdjian, D., Darrell, T.: Conditional Random People: Tracking Humans with CRFs and Grid Filters. In: CVPR (2006)

14. Rosenhahn, B., Brox, T., Seidel, H.: Scaled motion dynamics for markerless motion capture. In: CVPR (2007)

15. Brox, T., Rosenhahn, B., Cremers, D., Seidel, H.: Nonparametric density estimation with adaptive, anisotropic kernels for human motion tracking. In: Workshop on HUMAN MOTION Understanding, Modeling, Capture and Animation (2007)

16. Howe, N.R., Leventon, M.E., Freeman, W.T.: Bayesian reconstructions of 3D human motion from single-camera video. In: NIPS (1999)

17. Djouadi, A., Snorrason, O., Garber, F.: The quality of training sample estimates of the bhattacharyya coefficient. PAMI 12(1), 92–97 (1990)

18. Isard, M., Blake, A.: CONDENSATION - conditional density propagation for visual tracking. IJCV 29(1), 5–28 (1998)

19. Fossati, A., Dimitrijevic, M., Lepetit, V., Fua, P.: Bridging the Gap between Detection and Tracking for 3D Monocular Video-Based Motion Capture. In: CVPR (2007)

20. Urtasun, R., Fleet, D., Fua, P.: Temporal Motion Models for Monocular and Multiview 3–D Human Body Tracking. CVIU 104(2-3), 157–177 (2006)

Automated Delineation of Dendritic Networks in Noisy Image Stacks

Germán González[1], François Fleuret[2,*], and Pascal Fua[1]

[1] Ecole Polytechnique Fédérale de Lausanne, Computer Vision Laboratory,
Bâtiment BC, CH-1015 Lausanne, Switzerland
`{german.gonzalez,pascal.fua}@epfl.ch`
[2] IDIAP Research Institute, P.O. Box 592, CH-1920, Martigny, Switzerland
`francois.fleuret@idiap.ch`

Abstract. We present a novel approach to 3D delineation of dendritic networks in noisy image stacks. We achieve a level of automation beyond that of state-of-the-art systems, which model dendrites as continuous tubular structures and postulate simple appearance models. Instead, we learn models from the data itself, which make them better suited to handle noise and deviations from expected appearance.

From very little expert-labeled ground truth, we train both a classifier to recognize individual dendrite voxels and a density model to classify segments connecting pairs of points as dendrite-like or not. Given these models, we can then trace the dendritic trees of neurons automatically by enforcing the tree structure of the resulting graph. We will show that our approach performs better than traditional techniques on brighfield image stacks.

1 Introduction

Full reconstruction of neuron morphology is essential for the analysis and understanding of their functioning. In its most basic form, the problem involves processing stacks of images produced by a microscope, each one showing a slice of the same piece of tissue at a different depth.

Currently available commercial products such as Neurolucida[1], Imaris[2], or Metamorph[3] provide sophisticated interfaces to reconstruct dendritic trees and rely heavily on manual operations for initialization and re-initialization of the delineation procedures. As a result, tracing dendritic trees in noisy images remains a tedious process. It can take an expert up to 10 hours for each one. This limits the amount of data that can be processed and represents a significant bottleneck in neuroscience research on neuron morphology.

Automated techniques have been proposed but are designed to work on very high quality images in which the dendrites can be modeled as tubular structures [1,2]. In

---

* Supported by the Swiss National Science Foundation under the National Centre of Competence in Research (NCCR) on Interactive Multimodal Information Management (IM2).

[1] http://www.microbrightfield.com/prod-nl.htm

[2] http://www.bitplane.com/go/products/imaris

[3] http://www.moleculardevices.com/pages/software/metamorph.html

D. Forsyth, P. Torr, and A. Zisserman (Eds.): ECCV 2008, Part IV, LNCS 5305, pp. 214–227, 2008.

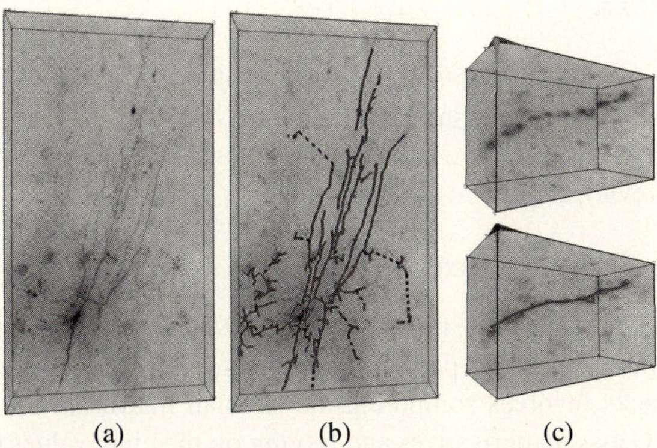

(a)   (b)   (c)

Fig. 1. (a) Minimum intensity projection of an image stack. Each pixel value is the minimum intensity value of the voxels that are touched by the ray cast from the camera through the pixel. (b) 3D tree reconstructed by our algorithm, which is best viewed in color. (c) Detail of the data volume showing the non-tubular aspect of a dendrite with the corresponding automatically generated delineation.

practice, however, due to the underlying neuron structure, irregularities in the dyeing process, and other sources of noise, the filaments often appear as an irregular series of blobs surrounded by other non-neuron structures, as is the case of the brightfield image stacks depicted by Fig. 1. Yet, such images are particularly useful for analyzing large samples. More generally, very high resolution images take a long time to acquire and require extremely expensive equipment, such as confocal microscopes. The ability to automatically handle lower resolution and noisier ones is therefore required to make these techniques more accessible. Ideally, the painstaking and data-specific tuning that many existing methods require should also be eliminated.

In this paper, we therefore propose an approach to handling the difficulties that are inherent to this imaging process. We do not assume an *a priori* dendrite model but rely instead on supervised and unsupervised statistical learning techniques to construct models as we go, which is more robust to unpredictable appearance changes. More specifically, we first train a classifier that can distinguish dendrite voxels from others using a very limited amount of expert-labeled ground truth. At run-time, it lets us detect such voxels, some of which should be connected by edges to represent the dendritic tree. To this end, we first find the minimum spanning tree connecting dendrite-like voxels. We then use an Expectation-Maximization approach to learn an appearance model for the edges that correspond to dendrites and those that do not. Finally, given these appearance models, we re-build and prune the tree to obtain the final delineation, such as the one depicted by Fig. 1(b), which is beyond what state-of-the-art techniques can produce automatically.

To demonstrate the versatility of our approach, we also ran our algorithm on retinal images, which we were able to do by simply training our classifier to recognize 2D blood vessel pixels instead of 3D dendrite voxels.

2   Related Work

Reconstructing networks of 3D filaments, be they blood vessels or dendrites, is an important topic in Biomedical Imaging and Computer Vision [3,4]. This typically involves measuring how filament-like voxels are and an algorithm connecting those that appear to be. We briefly review these two aspects below.

2.1   Finding Dendrite-Like Voxels

Most automated methods assume the filaments to be locally tubular and model them as generalized cylinders. The most popular approach to detecting such cylindrical structures in image stacks involves computing the Hessian matrix at individual voxels by convolution with Gaussian derivatives and relying on the eigenvalues of the Hessian to classify voxels as filament-like or not [5,6,7]. The Hessians can be modified to create an oriented filter in the direction of minimum variance, which should correspond to the direction of any existing filament [8,9]. To find filaments of various widths, these methods perform the computation using a range of variances for the Gaussian masks and select the most discriminant one. The fact that intensity changes inside and outside the filaments has also been explicitly exploited by locally convolving the image with differential kernels [1], finding parallel edges [10], and fitting *superellipsoids* or cylinders to the vessel based on its surface integral [2,11].

All these methods, however, assume image regularities that are present in high-quality images but not necessarily in noisier ones. Furthermore, they often require careful parameter tuning, which may change from one data-set to the next. As a result, probabilistic approaches able to learn whether a voxel belongs to a filament or not have begun to be employed. Instead of assuming the filaments to be cylinders, they aim at learning their appearance from the data. In [12], the eigenvalues of the structure tensor, are represented by a mixture model whose parameters are estimated via E-M. Support Vector Machines that operates on the Hessian's eigenvalues have also been used to discriminate between filament and non-filament voxels [13].

The latter approach [13] is closest to our dendrite detection algorithm. We however go several steps further to increase robustness: First, we drop the Hessian and train our classifier directly on the intensity data, thereby making fewer assumptions and being able to handle structures that are less visibly tubular. Second, we also learn an appearance model for the filament itself as opposed to individual voxels.

2.2   Reconstructing Filaments

Existing approaches to building the dendritic tree all rely on a *dendritness* measure of how dendrite-like filaments look, usually based on the voxel-based measures discussed above. They belong to one of two main classes.

The first class involves growing filaments from seed points [2,14,15,16]. This has been successfully demonstrated for confocal fluorescent microscopy images. It is computationally effective because the dendritness of filaments need only be evaluated in a small subset of the voxels. However, it may easily fail in noisy data because of its sequential nature. If the growing process diverges at one voxel, the rest of the dendritic tree will be lost.

The second class requires optimizing the path between seed points, often provided by the operator, to maximize the overall dendritness [8,11,17]. In these examples, the authors use active contour models, geometrical constraints and the live-wire algorithm between to connect the seeds.

By contrast to these methods that postulate an *a priori* cost function for connecting voxels, our approach learns a model at run-time, which lets it deal with the potentially changing appearance of the filaments depending on experimental conditions. Furthermore, we do this fully automatically, which is not the case for any of the methods discussed above.

3 Methodology

Our goal is to devise an algorithm that is fully automatic and can adapt to noisy data in which the appearance of the dendrites is not entirely predictable. Ideally we would like to find the tree maximizing the probability of the image under a consistent generative model. Because such an optimization is intractable, we propose an approximation that involves the three following steps:

1. We use a hand-labeled training image stack to train once and for all a classifier that computes a voxel's probability to belong to a dendrite from its neighbors intensities.
2. We run this classifier on our stacks of test images, use a very permissive threshold to select potential dendrite voxels, apply non-maximum suppression, and connect all the surviving voxels with a minimum spanning tree. Some of its edges will correspond to actual dendritic filaments and other will be spurious. We use both the correct and spurious edges to learn filament appearance models in an EM framework.
3. Under a Markovian assumption, we combine these edge appearance models to jointly model the image appearance and the true presence of filaments. We then optimize the probability of the latter given the former and prune spurious branches.

As far as detecting dendrite voxels is concerned, our approach is related to the Hessian-based approach of [13]. However, dropping the Hessian and training our classifier directly on the intensity data lets us relax the cylindrical assumption and allows us to handle structures that are less visibly tubular. As shown in Fig. 2, this yields a marked improvement over competing approaches.

In terms of linking, our approach can be compared to those that attempt to find optimal paths between seeds [11,8] using a dendrite appearance model, but with two major improvements: First our seed points are detected automatically instead of being manually supplied, which means that some of them may be spurious and that the connectivity has to be inferred from the data. Second we do not assume an *a priori* filament model but learn one from the data as we go. This is much more robust to unpredictable appearance changes. Furthermore, unlike techniques that model filaments as tubular structures [1,2], we do not have to postulate regularities that may not be present in our images.

3.1 Notations

Given the three step algorithm outlined above, we now introduce the notations we will use to describe it in more details.

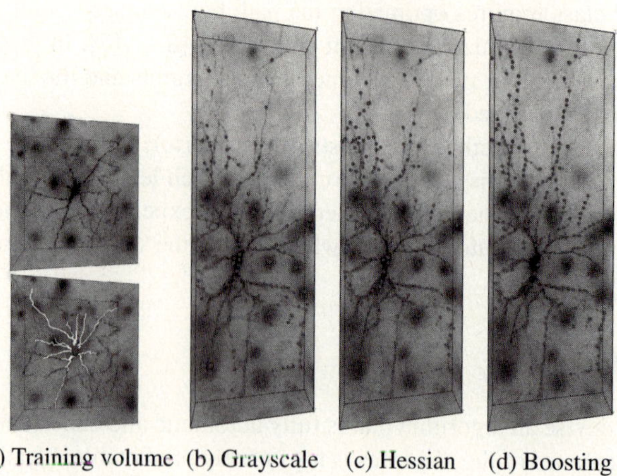

(a) Training volume   (b) Grayscale   (c) Hessian   (d) Boosting

Fig. 2. (a) Training data. On top: image stack representing one neuron. Bellow: Manually delineated filaments overlaid in white. (b,c,d) Voxels labeled as potentially belonging to a dendrite. (b) By thresholding the grayscale images. (c) By using the Hessian. (d) By using our classifier. Note that the seed points obtained with our method describe better the underlying neuron structure.

Let $Z_1, \ldots, Z_N$ be the voxels corresponding to the local maxima of the classifier response and will serve as vertices for the dendritic tree we will build. For $1 \leq n \leq N$, let X_n be a Boolean random variable standing for whether or not there truly is a filament at location Z_n. Finally, Let $\boldsymbol{x} = (x_1, \ldots, x_N)$ and $\boldsymbol{x}_{\backslash i} = (x_1, \ldots, x_{i-1}, x_{i+1}, x_N)$.

For $1 \leq i \leq N$ and $1 \leq j \leq N$, let $J_{i,j}$ denote a random variable standing for the appearance of the edge going from Z_i to Z_j and let $L_{i,j} = ||Z_i - Z_j||$ be its length. $J_{i,j}$ is obtained by sampling the voxel response of the classifier in a regular lattice between (Z_i, Z_j). Let $A_{i,j}$ be a vector composed by the projection of $J_{i,j}$ in a latent space and $L_{i,j}$.

Let T denote the true dendritic tree we are trying to infer. It is a graph whose vertices are a subset of $Z_1, \ldots, Z_N$ and whose edges are defined by $\mathcal{G}$, a set of pairs of indexes in $\{1, \ldots, N\} \times \{1, \ldots, N\}$.

3.2   Local Dendrite Model

As discussed in Section 2, the standard approach to deciding whether voxels are inside a dendrite or not is to compute the Hessian of the intensities and look at its eigenvalues. This however implicitly makes strong assumptions on the expected intensity patterns. Instead of using such a hand-designed model, we train a classifier from a small quantity of hand-labeled neuron data with AdaBoost [18], which yields superior classification performance as shown in Fig. 2.

More specifically, the resulting classifier f is a linear combination of *weak learners* h_i:

$$f(x, y, z) = \sum_{i=1}^{N} \alpha_i h_i(x, y, z) \, , \tag{1}$$

where the h_i represent differences of the integrals of the image intensity over two cubes in the vicinity of (x, y, z) and T_i is the weak classifier threshold. We write

$$h_i(x, y, z) = \sigma\left(\sum_{V_i^1} I(x', y', z') - \sum_{V_i^2} I(x'', y'', z'') - T_i\right) \tag{2}$$

where σ is the sign function, V_i^1, V_i^2 are respectively the two volumes defining h_i, translated according to (x, y, z). These weak classifiers can be calculated with just sixteen memory accesses by using precomputed integral cubes, which are natural extensions of integral images.

During training, we build at each iteration 10^3 h_i weak learners by randomly picking volume pairs and finding an optimal T_i threshold for each. After running Adaboost, $N = 1000$ weak learners are retained in the f classifier of 1. The training samples are taken from the manual reconstruction of Fig. 2. They consist of filaments at different orientations and of a certain width. The final classifier responds to filaments of the pre-defined width, independently of the orientation.

At run time, we apply f on the whole data volume and perform non-maximum suppression by retaining only voxels that maximize it within a $8 \times 8 \times 20$ neighborhood, such as those shown in Fig. 2. The anisotropy on the neighborhood is due to the low resolution of the images in the z axis, produced by the point spread function of the microscope.

3.3   Learning an Edge Appearance Model

The process described above yields $Z_1, \ldots, Z_N$, a set of voxels likely, but not guaranteed to belong to dendrites. To build an edge appearance model, we compute their minimum spanning tree. Some of its edges will correspond to filaments and some not. We therefore create a low dimensional descriptor for the edges, and use it to learn a gaussian mixture model that we can use to distinguish the two classes of edges.

To obtain an edge descriptor, we first sample the voxel response on a regular lattice centered around each edge and perform PCA on the resulting set of vectors. For each edge, we retain the first N PCA components. We construct a $N + 1$-D edge feature vector, $A_{i,j}$ by appending the edge length $L_{i,j}$ to this N-D vector.

This population of $N + 1$-D vectors is a mixture of edges truly located on filaments, and of edges located elsewhere. We therefore apply an E-M procedure to derive both a prior and a Gaussian model for both. The only specificity of this unsupervised training is to force the covariance between the length and the other N components to be zero, since the length of an edge is only weakly correlated with its length-normalized appearance.

Hence, given a subgraph $\mathcal{G}$ with a population of edges that are both in the dendrite and elsewhere, this E-M procedure produces two Gaussian models μ_0 and μ_1 on R^{N+1} that represent respectively the edges truly on filaments and those elsewhere.

3.4   Building and Pruning the Tree

We can now use the edge appearance model to reconstruct the dendritic tree. To this end we first compute the maximum spanning tree using as weight for the edges their

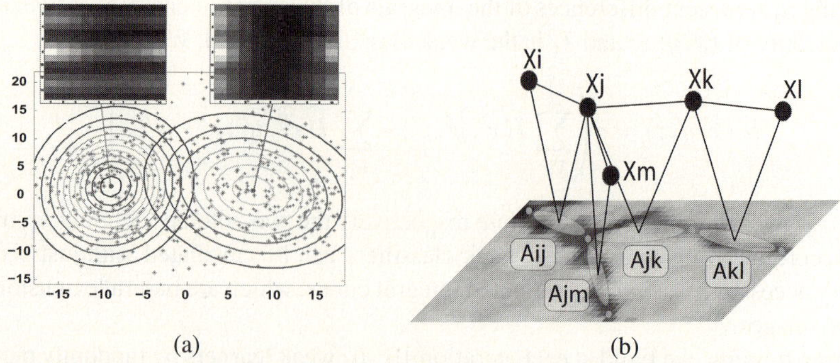

(a)   (b)

Fig. 3. (a) First two dimensions of the PCA space of the edge appearance models. The Gaussian models are shown as contour lines. The two small figures at the top represent the projection of the means in the original lattice. The top-left one represents the model μ_1 for filaments, which appear as a continuous structure. The top-right one represents the non-filament model μ_0. Since, by construction the endpoints of the edges are local maxima, the intensity there is higher than elsewhere. (b) Hidden Markov Model used to estimate the probability of a vertex to belong to the dendritic tree.

likelihood to be part of a dendrite. Nevertheless, the tree obtained with this procedure is over-complete, spanning vertices that are not part of the dendrites, Fig. 4(b). In order to eliminate the spurious branches, we use the tree to evaluate the probability that individual vertices belong to a dendrite, removing those with low probability. We iterate between the tree reconstruction and vertex elimination until convergence, Fig. 4(c).

We assume that the relationship between the hidden state of the vertices and the edge appearance vectors can be represented in terms of a hidden Markov model such

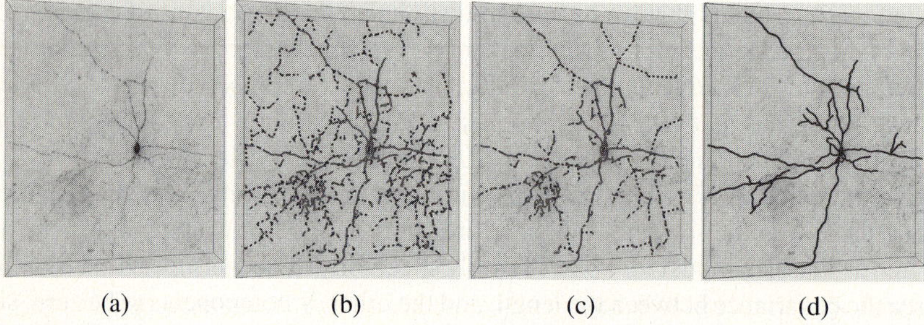

(a)   (b)   (c)   (d)

Fig. 4. Building and pruning the tree. (a) Image stack (b) Initial maximum spanning tree. (c) After convergence of the iterative process. (d) Manually delineated ground truth. Red solid lines denote edges that are likely to be dendrites due to their appearance. Blue dashed lines represent edges retained by the minimum spanning tree algorithm to guarantee connectivity.The main filaments are correctly recovered. Note that our filament detector is sensitive to filaments thinner than the ones in the ground truth data. This produces the structures in the right part of the images that are not part of the ground truth data.

as the one depicted by Fig. 3(b). More precisely, we take $\mathcal{N}(\mathcal{G}, i)$ to be the neighboring vertices of i in $\mathcal{G}$ and assume that

$$P(X_i \mid \boldsymbol{X}_{\setminus i}, (A_{k,l})_{(k,l)\in\mathcal{G}}) = P(X_i \mid (X_k)_{k\in\mathcal{N}(\mathcal{G},i)}, (A_{i,k})_{k\in\mathcal{N}(\mathcal{G},i)}) , \qquad (3)$$

$$P(A_{i,j} \mid \boldsymbol{X}, (A_{k,l})_{(k,l)\in\mathcal{G}\setminus(i,j)}) = P(A_{i,j} \mid X_i, X_j) . \qquad (4)$$

Under these assumptions, we are looking for a tree consistent with the edge appearance model of section 3.3. This means that the labels of its vector of maximum posterior probabilities $\boldsymbol{x}$ are all 1s. To do so we alternate the building of a tree spanning the vertices currently labeled 1 and the re-labeling of the vertices to maximize the posterior probability. The tree we are looking for is a fixed point of this procedure.

Building the Tree. We are looking for maximum likelihood tree that spans all vertices. Formally:

$$\operatorname*{argmax}_{\mathcal{T}} \{\log P(T = \mathcal{T} \mid (A_{i,j})_{1\leq i,j\leq N})\}$$

$$= \operatorname*{argmax}_{\mathcal{T}} \{\log P((A_{i,j})_{1\leq i,j\leq N} \mid T = \mathcal{T})\} = \operatorname*{argmax}_{\mathcal{T}} \sum_{i,j\in\mathcal{T}} \log \frac{\mu_1(A_{i,j})}{\mu_0(A_{i,j})} .$$

To this end, we use a slightly modified version of the minimum spanning tree algorithm. Starting with an empty graph, we add to it at every iteration the edge (i, j) that does not create a cycle and maximizes

$$\log(\mu_1(A_{i,j})/\mu_0(A_{i,j})) .$$

While this procedure is not guaranteed to find a global optimum, it gives good results in practice. The main weakness we have to deal with is the over-completeness of the resulting tree. While it is very rare to miss an important vertex or part of filament, we have to discard many spurious branches spanned on non-filaments.

Eliminating Unlikely Vertices. From the appearance models μ_0 and μ_1 learned in section 3.3, and the Markovian assumption of Section 3.3, we can estimate for any graph $\mathcal{G}$ the most probable subset of nodes truly on filaments. More specifically, we are looking for the labeling $\boldsymbol{x}$ of maximum posterior probability given the appearance, defined as follow

$$\operatorname*{argmax}_{\boldsymbol{x}} P(\boldsymbol{X} = \boldsymbol{x} \mid (A_{i,j})_{i,j\in\mathcal{G}})$$

Since full optimization is intractable we propose an iterative greedy search. We loop through each point i, flipping the value of x_i if it increases the posterior probability. This can be seen as a component-wise optimization where the updating rule consists of fixing all $x_j, j \neq i$ and applying the following update to x_i

$$x_i \leftarrow \operatorname*{argmax}_{x} P(X_i = x, \boldsymbol{X}_{\setminus i} = \boldsymbol{x}_{\setminus i} \mid (A_{i,j})_{i,j\in\mathcal{G}})$$

$$= \operatorname*{argmax}_{x} P(X_i = x \mid \boldsymbol{X}_{\setminus i} = \boldsymbol{x}_{\setminus i}, (A_{i,j})_{i,j\in\mathcal{G}}),$$

and under assumptions (3) and (4), we have

$$P(X_i = x \mid \boldsymbol{X}_{\backslash i} = \boldsymbol{x}_{\backslash i}, (A_{i,j})_{i,j \in \mathcal{G}})$$
$$= \prod_{j \in \mathcal{N}(\mathcal{G},i)} P(X_j = x_j \mid X_i = x)P(A_{i,j} \mid X_i = x, X_j = x_j),$$

where $P(X_j = 0 \mid X_i = 0) = P(X_j = 1 \mid X_i = 1) = 1 - \epsilon$ and $P(X_j = 1 \mid X_i = 0) = P(X_j = 0 \mid X_i = 1) = \epsilon$. ϵ is a parameter chosen to be 0.2. $P(A_{i,j} \mid X_i = x, X_j = x_j)$ comes from our appearance model, with the assumption that the only true filaments correspond to $X_i = X_j = 1$.

The initialization of each x_i is done according to the posterior probability of the edges going through it. If there is an edge with $\mu_1(a_{i,j}) > \mu_0(a_{i,j})$, then $x_i = 1$. The termination condition for the loop is that all points are visited without any flip, or that the number of flips excess ten times the number of points. In practice the second condition is never met, and only $10 - 20\%$ of the points flip their hidden variable.

4   Results

In this section we first describe the images we are using. We then compare the discriminative power of our dendrite model against simple grayscale thresholding and the baseline Hessian based method [6]. Finally, we validate our automated tree reconstruction results by comparing them against a manual delineation.

4.1   Image Data

Our image database consists of six neuron image stacks, in two of which the dendritic tree has been manually delineated. We use one of those trees for training and the other for validation purposes.

The neurons are taken from the somatosensory cortex of Wistar-han rats. The image stacks are obtained with a standard brightfield microscope. Each image of the stack shows a slice of the same piece of tissue at a different depth. The tissue is transparent enough so that these pictures can be acquired by simply changing the focal plane.

Each image stack has an approximate size of $5 * 10^9$ voxels, and is downsampled to a size of 10^8 voxels to make the evaluation of the image functional in every voxel computationally tractable. After down-sampling, each voxel has the same width, height and depth, of $0.8\,\mu$m.

4.2   Image Functional Evaluation

The f classifier of 1 is trained using the manual delineation of Fig. 2. As positive samples, we retain 500 voxels belonging to filaments of width ranging from two to six voxels and different orientations. As negative samples, we randomly pick 1000 voxels that are no closer to a neuron than three times the neuron width and are representative of the image noise. Since the training set contains filaments of many different orientations, Adaboost produces a classifier that is orientation independent.

Fig. 2 depicts the candidate dendrite voxels obtained by performing non maxima suppression of images calculated by simply thresholding the original images, computing

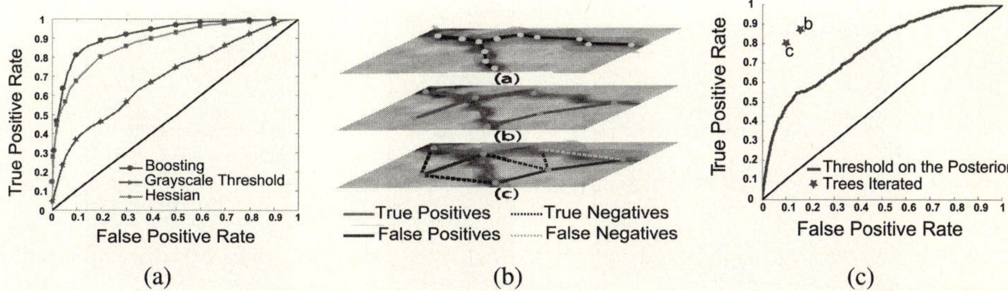

Fig. 5. (a) ROC curve for all three measures using the validation data of figure 4(d). The boosting classifier outperforms the baseline hessian method of [6] in noisy brightfield images. (b) Defining a metric to compare our results against a manual delineation. Top: portion of a manual delineation in which the vertices are close to each other and the tolerance width painted in red. Middle: Portion of the tree found by our algorithm at the same location. Bottom: The fully-connected graph we use to evaluate our edge appearance model and plot the corresponding ROC curves. (c) ROC curve for the detection of edges on filament obtained by thresholding the individual estimated likelihood of the edges of the graph of (b). The individual points represent the iterations of the tree reconstruction algorithm. Two of them are depicted by Fig. 4(b,c). After five iterations we reach a fixed point, which is our final result.

a Hessian-based measure [6], or computing the output of our classifier at each voxel. The same procedure is applied in the validation data of Fig 4(d). Considering correct the vertices that are within $5\,\mu$m (6 voxels) of the neuron, we can plot the three ROC curves of Fig. 5(a) that show that our classifier outperforms the other two.

4.3   Tree Reconstruction

To evaluate the quality of the tree, we compare it against the validation data of Fig. 4(d), which is represented as a set of connected points. As shown in Fig. 5(a,b), performing this comparison is non-trivial because in the manual delineation the vertices are close to each other whereas our algorithm allows for distant points to be connected.

To overcome this difficulty, we introduce a measure of whether an edge linking X_i to X_j is present in the manual delineation. First, we use the manually labeled points to construct a volume in which every voxel closer than $5\,\mu$m to one such point is assigned the value 1, and 0 otherwise. We then compute the average value in the straight line linking X_i and X_j in that volume. If it is greater than a threshold, we consider that the edge is described by the graph. Here, we take the threshold to be 0.8.

Given this measure, labeling the edges of the tree returned by our algorithm as true or false positives is straightforward. However, since we also need to compute rates of true and false negatives to build ROC curves such as the one of Fig. 5, we create graphs such as the one depicted by Fig. 5(c) in which each vertex is connected to all its nearest neighbors.

In Fig. 5, we plot a ROC curve obtained by thresholding the likelihood that the edges of the graph of Fig. 5(c) belong to a neuron based on the edge appearance model of Section 3.3. Note that this model is not very discriminative by itself. The individual points in Fig. 5 represent true and false positive rates for the successive trees built by the procedure of Section 3.4 and depicted by Fig. 4(b,c,d). As the iterations proceed, the false

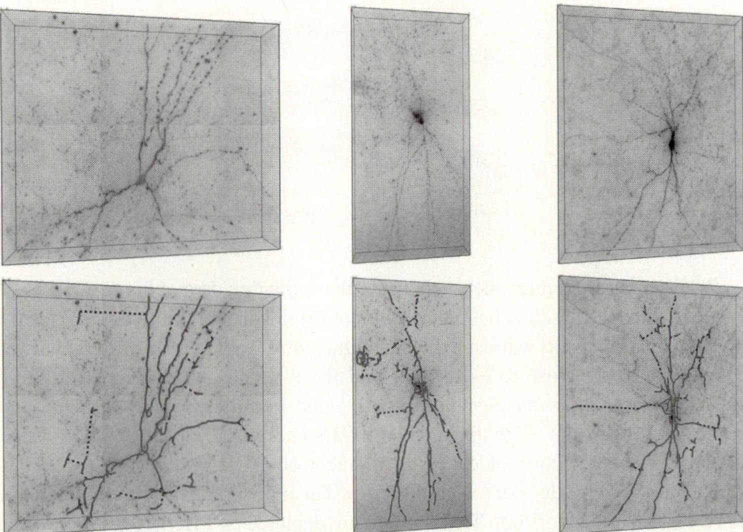

Fig. 6. Three additional reconstructions without annotations. Top row: Image stacks. Bottom row: 3D Dendritic tree built by our algorithm. As in Fig. 4, the edges drawn with a solid red lines are those likely to belong to a dendrite given their appearance. The edges depicted with dashed blue lines are kept to enforce the tree structure through all the vertices. This figure is best viewed in color.

positive rate is progressively reduced. Unfortunately, so is the true positive rate as we loose some of the real dendrite edges. However, the main structure remains and cleaning up this result by hand is much faster than manually delineating the tree of Fig. 4(e).

In Fig. 6, we show reconstruction results in four more image stacks. Our algorithm recovers the main dendrites despite their irregularities and the high noise level and, again, cleaning up this tree is much easier than producing one from scratch. Some of incorrect edges are also retained because the minimum spanning algorithm enforces connectivity of all the vertices, even when it is not warranted.

4.4   From Dendrites to Blood Vessels

Since we learn filament models as we go, one of the strengths of our approach is its generality. To demonstrate it, we ran our algorithm on the retina images of Fig. 7 and 8 without any changes, except for the fact that we replaced the 3D weak classifiers of Section 3.2 by 2D ones, also based on Haar wavelets. The algorithm learned both a local blood-vessel model and 2D filament model.

In Fig. 7(b), we evaluate the performance of our boosted classifier against that of other approaches discussed in [19]. It performs similarly to most of them, but a bit worse than the best. This can be attributed to the fact that it operates at a single scale and is optimized to detect large vessels, whereas the others are multiscale. As a consequence, when we run the full algorithm we obtain the results of Fig. 8 in which the large vessels are correctly delineated, but some of the small ones are missed. This would be fixed by training our initial classifier to handle different widths.

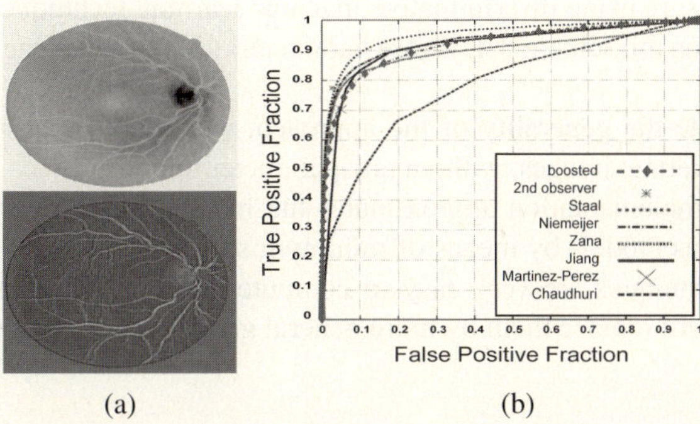

(a)   (b)

Fig. 7. (a) Top: image of the retina. Bottom: response of our boosting classifier in this image. (b) Comparison of our classifier against other algorithms evaluated in the DRIVE database [19]. It performs similarly to most of them, but worse than algorithms designed specifically to trace blood vessels in images of the retina. This can be attributed to the fact that our boosted classifier operates at a single scale and is optimized to detect large vessels, whereas the others are multiscale.

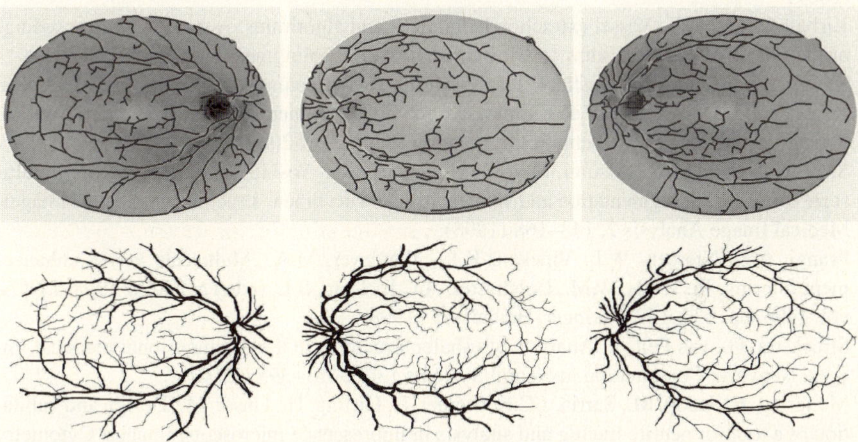

Fig. 8. Retinal trees reconstructed with our method. Top row: original image with the reconstructed tree overlay. As in Fig. 6, edges likely to belong to filaments are drawn in red, while edges kept to enforce the tree structure are colored in blue. Bottom row: manually obtained ground truth. Note that thick filaments are correctly delineated, whereas thin filaments are prone to errors because our classifier is trained only for the thick ones.

5   Conclusion

We have proposed a novel approach to fully-automated 3D delineation of dendritic networks in noisy brightfield images based on statistical machine learning techniques and tree-optimization methods.

By contrast to state-of-the-art methods, we do not postulate *a priori* models for either the dendrite or the edge model between dendrite-like voxels. Instead, we generate the dendrite measure using discriminative machine learning techniques. We model the edges as a gaussian mixture model, whose parameters are learned using E-M on neuron-specific samples.

To demonstrate the generality of the approach, we showed that it also works for blood vessels in retinal images, without any parameter tuning.

Our current implementation approximates the maximum likelihood dendritic tree under the previous models by means of minimum spanning trees and markov random fields. Those techniques are very easy to compute, but tend to produce artifacts. In future work we will replace them by more general graph optimization techniques.

References

1. Al-Kofahi, K., Lasek, S., Szarowski, D., Pace, C., Nagy, G., Turner, J., Roysam, B.: Rapid automated three-dimensional tracing of neurons from confocal image stacks. IEEE Transactions on Information Technology in Biomedicine (2002)
2. Tyrrell, J., di Tomaso, E., Fuja, D., Tong, R., Kozak, K., Jain, R., Roysam, B.: Robust 3-d modeling of vasculature imagery using superellipsoids. Medical Imaging 26(2), 223–237 (2007)
3. Kirbas, C., Quek, F.: Vessel extraction techniques and algorithms: A survey. In: Proceedings of the Third IEEE Symposium on BioInformatics and BioEngineering, p. 238 (2003)
4. Krissian, K., Kikinis, R., Westin, C.F.: Algorithms for extracting vessel centerlines. Technical Report 0003, Department of Radiology, Brigham and Women's Hospital, Harvard Medical School, Laboratory of Mathematics in Imaging (September 2004)
5. Sato, Y., Nakajima, S., Atsumi, H., Koller, T., Gerig, G., Yoshida, S., Kikinis, R.: 3d multi-scale line filter for segmentation and visualization of curvilinear structures in medical images. Medical Image Analysis 2, 143–168 (1998)
6. Frangi, A.F., Niessen, W.J., Vincken, K.L., Viergever, M.A.: Multiscale vessel enhancement filtering. In: Wells, W.M., Colchester, A.C.F., Delp, S.L. (eds.) MICCAI 1998. LNCS, vol. 1496, pp. 130–137. Springer, Heidelberg (1998)
7. Streekstra, G., van Pelt, J.: Analysis of tubular structures in three-dimensional confocal images. Network: Computation in Neural Systems 13(3), 381–395 (2002)
8. Meijering, E., Jacob, M., Sarria, J.C.F., Steiner, P., Hirling, H., Unser, M.: Design and validation of a tool for neurite tracing and analysis in fluorescence microscopy images. Cytometry Part A 58A(2), 167–176 (2004)
9. Aguet, F., Jacob, M., Unser, M.: Three-dimensional feature detection using optimal steerable filters. In: Proceedings of the 2005 IEEE International Conference on Image Processing (ICIP 2005), Genova, Italy, September 11-14, 2005, vol. II, pp. 1158–1161 (2005)
10. Dima, A., Scholz, M., Obermayer, K.: Automatic segmentation and skeletonization of neurons from confocal microscopy images based on the 3-d wavelet transform. IEEE Transaction on Image Processing 7, 790–801 (2002)
11. Schmitt, S., Evers, J.F., Duch, C., Scholz, M., Obermayer, K.: New methods for the computer-assisted 3d reconstruction of neurons from confocal image stacks. NeuroImage 23, 1283–1298 (2004)
12. Agam, G., Wu, C.: Probabilistic modeling-based vessel enhancement in thoracic ct scans. In: CVPR 2005: Proceedings of the 2005 IEEE Computer Society Conference on Computer Vision and Pattern Recognition (CVPR 2005), vol. 2, pp. 684–689. IEEE Computer Society, Washington (2005)

13. Santamaría-Pang, A., Colbert, C.M., Saggau, P., Kakadiaris, I.A.: Automatic centerline extraction of irregular tubular structures using probability volumes from multiphoton imaging. In: Ayache, N., Ourselin, S., Maeder, A. (eds.) MICCAI 2007, Part II. LNCS, vol. 4792, pp. 486–494. Springer, Heidelberg (2007)
14. Al-Kofahi, K.A., Can, A., Lasek, S., Szarowski, D.H., Dowell-Mesfin, N., Shain, W., Turner, J.N., et al.: Median-based robust algorithms for tracing neurons from noisy confocal microscope images (December 2003)
15. Flasque, N., Desvignes, M., Constans, J., Revenu, M.: Acquisition, segmentation and tracking of the cerebral vascular tree on 3d magnetic resonance angiography images. Medical Image Analysis 5(3), 173–183 (2001)
16. McIntosh, C., Hamarneh, G.: Vessel crawlers: 3d physically-based deformable organisms for vasculature segmentation and analysis. In: CVPR 2006: Proceedings of the 2006 IEEE Computer Society Conference on Computer Vision and Pattern Recognition, pp. 1084–1091. IEEE Computer Society Press, Washington (2006)
17. Szymczak, A., Stillman, A., Tannenbaum, A., Mischaikow, K.: Coronary vessel trees from 3d imagery: a topological approach. Medical Image Analisys (08 2006)
18. Freund, Y., Schapire, R.: Experiments with a New Boosting Algorithm. In: International Conference on Machine Learning, pp. 148–156. Morgan Kaufmann, San Francisco (1996)
19. Staal, J., Abramoff, M., Niemeijer, M., Viergever, M., van Ginneken, B.: Ridge based vessel segmentation in color images of the retina. IEEE Transactions on Medical Imaging 23, 501–509 (2004)

Calibration from Statistical Properties
of the Visual World

Etienne Grossmann[1,*], José António Gaspar[2], and Francesco Orabona[3]

[1] Tyzx, Inc., Menlo Park, USA
[2] ISR, Instituto Superior Técnico, Lisbon, Portugal
[3] Idiap Research Institute, Martigny, Switzerland

Abstract. What does a blind entity need in order to determine the geometry of
the set of photocells that it carries through a changing lightfield? In this paper, we
show that very crude knowledge of some statistical properties of the environment
is sufficient for this task.

We show that some dissimilarity measures between pairs of signals produced
by photocells are strongly related to the angular separation between the photocells. Based on real-world data, we model this relation quantitatively, using dissimilarity measures based on the correlation and conditional entropy. We show
that this model allows to estimate the angular separation from the dissimilarity.
Although the resulting estimators are not very accurate, they maintain their performance throughout different visual environments, suggesting that the model
encodes a very general property of our visual world.

Finally, leveraging this method to estimate angles from signal pairs, we show
how distance geometry techniques allow to recover the complete sensor geometry.

1   Introduction

This paper departs from traditional computer vision by not considering images or image
features as input. Instead, we take signals generated by photocells with unknown orientation and a common center of projection, and explore the information these signals
can shed on the sensor and its surrounding world.

We are particularly interested in determining whether the signals allow to determine
the geometry of the sensor, that is, to calibrate a sensor like the one shown in Figure 1.
Psychological experiments [1] showed that a person wearing distorting glasses for a
few days, after a very confusing and disturbing period, could learn the necessary image
correction to restart interacting effectively with the environment. Can a computer do the
same when, rather than distorted images, it is given the signals produced by individual
photocells? In this situation, it is clear that traditional calibration techniques [2,3] are
out of the question.

Less traditional non-parametric methods that assume a smooth image mapping and
smooth motion [4] can obviously not be applied either. Using controlled-light stimuli

---

* This work was partially supported by TYZX, Inc, by the Portuguese FCT POS_C program that
includes FEDER funds, and by the EU-project URUS FP6-EU-IST-045 062.

D. Forsyth, P. Torr, and A. Zisserman (Eds.): ECCV 2008, Part IV, LNCS 5305, pp. 228–241, 2008.

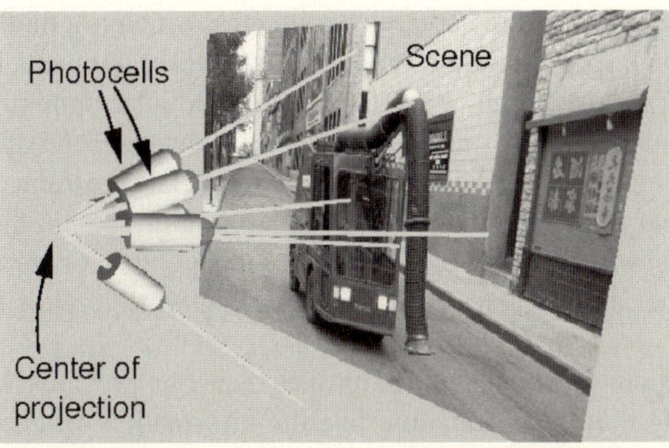

Fig. 1. A discrete camera consists of a number of photocells (pixels) that measure the light traveling along pencil of lines

or known calibration, matches could be obtained, allowing to use match-based non-parametric techniques [5]. In this study however, we wish to exclude known calibration objects and other controlled stimuli.

Our approach is inspired from the work of Pierce and Kuipers [6], who measure the dissimilarity, or distance, between sensor elements that are not necessarily light sensors. The elements are then embedded in a metric space using metric scaling [7], which also determines the dimension of the space. A relaxation method then improves this embedding, so that the Euclidean distance between sensor elements better matches the dissimilarity between the sensor inputs. Getting close to the problem addressed in the present paper, the authors use this method to reconstitute the geometry of a rectangular array of visual sensors that scans a fronto-parallel image.

Going further, Olsson et al. [8] use the information distance of [9] as a more appropriate method to measure the distance between visual or other sensor elements. They also show how visual sensors -the pixels of the camera of a mobile robot- can be mapped to a plane, either using the method of [6], or their own, that embeds sensor elements specifically in a square grid.

The works of Olsson et al. and of Pierce and Kuipers are very interesting to computer vision researchers, but they cannot calibrate an arbitrary discrete camera, since the embedding space is either abstract or fixed to a grid. In both cases, it lacks an explicit connection to the geometry of the sensor.

Grossmann et al [10] partially fill this gap by showing that the information distance can be used to estimate the angular separation between pairs of photocells, and from there, estimate the geometry of a sensor of limited angular radius.

Because the present work exploits statistical properties of the light-field of the world surrounding a light sensor, it is also related to research on the statistical properties of real-world images. In that area, a model of image formation is used, but images, rather than sequences, are studied. That research has put in evidence fundamental properties, in terms of local, global and spectral statistics, of real-world images, and found ways to exploit these properties for computer vision tasks, such as classification [11], image restoration [12] and 3D inference [13]. Although these results are of great interest, they are not directly applicable in our case, mainly because we lack images.

Moreover, these statistics are about planar images, which is a hindrance in our case: first, we do not want to exclude the case of visual sensor elements that are separated by more than 180 degrees, such as the increasingly popular omnidirectional cameras. Also, the local statistical properties of perspective images depend of the orientation of the image plane with respect to the scene, except in special constrained cases such as the fronto-parallel "leaf world" of Wu et al. [14]. Defining images on the unit sphere thus appears as a natural way to render image statistics independent of the sensor orientation, at least with proper assumptions on the surrounding world and/or the motion of the sensor.

The present article elaborates and improves over our previous work [10]. We innovate by showing that the correlation, like the information distance, can be used to provide geometric information about a sensor. Also, we use a simpler method to model to relation between angles and signal statistics.

More important, we go much beyond [15] in showing that this model generalizes well to diverse visual environments, and can thus be considered to be a reliable characteristic of our visual world. In addition, we show that the presented calibration method performs much better, for example by allowing to calibrate sensors that cover more than one hemisphere.

1.1   Proposed Approach

The present work relies on statistical properties of the data streams produced by pairs of sensor elements that depend only on the *angular separation* between the photocells. For example, if the sampled lightfield is a homogeneous random field defined on the sphere [16], then the covariance between observations depends only on the angular separation between the sampled points.

This assumption does not hold in general in our anisotropic world, but it does hold, e.g. if the orientation of the sensor is uniformly distributed amongst all unitary transformations of the sphere, that is, if the sensor is randomly oriented, so that each photocell is just as likely to sample the light-field in any direction.

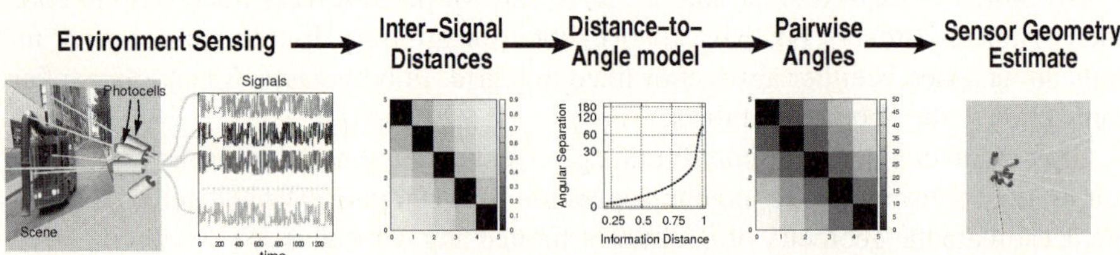

Fig. 2. The process of estimating the geometry of an unknown discrete camera

This assumption of homogeneity -or isotropy- of the sampled lightfield is of great practical utility, in conjunction with a few other assumptions of good behavior: in this work, we only use statistics that converge properly (e.g. in probability or more strongly) when signal lengths tend to infinity.

Perhaps more importantly we are only interested in statistics that have an expectancy that is a strictly monotonous function of the angular separation of the pair of photocells. That is, if x, y are two signals (random variables) generated by two photocells separated by an angle θ, and $d(x, y)$ is the considered statistic, then the expectancy of $d(x, y)$ is a strictly monotonous function of θ, for $0 \leq \theta \leq \pi$. The importance of this last point is that this function can be inverted, resulting in a functional model that links the value of the statistic to the angle.

The statistic-to-angle graph of such statistics is the a-priori knowledge about the world that we leverage to estimate the geometry of discrete cameras. In the present work, we use discrepancy measures based on the correlation or conditional entropy, defined in Section 3. In Section 4, we show how to build the considered graph.

Having obtained angle estimates, we recover the sensor geometry, in Section 5.1, by embedding the angles in a sphere. This is done using simple techniques from distance geometry [17]. Experimental results are presented in Section 5.2. Finally, Section 6 presents some conclusions and possible directions for future research. The calibration process considered in the present work is outlined in Figure 2. The statistic-to-angle modeling produces the crucial functional relation used in the third-from right element of Figure 2.

2 Discrete Camera Model and Simulation

Before entering into the details of our methodology for estimating the sensor geometry, we define the discrete camera and explain how to simulate it using an omnidirectional image sensor.

We define a discrete camera [10] as a set of N photocells indexed by $i \in \{1, \ldots, N\}$, pointing in directions $X_i \in \mathbb{R}^3$ and having a unique center of projection. These photocells acquire along the time t, brightness measurements $x(i, t)$ in the range $\{0, \ldots, 255\}$. The directions of the light rays, contrarily to conventional cameras, are not necessarily organized in a regular grid. Many examples of cameras can be found under these definitions. One example is the linear camera, where all the X_i are co-planar. Another example is the conventional perspective camera which comprises a rectangular grid of photocells that are enumerated in our model by a single index i,

$$\left\{ X_i \mid X_i \sim K^{-1} \begin{bmatrix} i\%W \\ \lfloor i/W \rfloor \\ 1 \end{bmatrix}, \ 0 \leq i < HW \right\}$$

where W, H are the image width and height, K is the intrinsic parameters matrix, $\%$ represents the integer modulo operation and $\lfloor . \rfloor$ is the lower-rounding operation. Cameras equipped with fisheye lenses, or having log- polar sensors, can also be modeled again by setting X_i to represent the directions of the light-rays associated to the image pixels. In the same vein, omnidirectional cameras having a single projection center, as the ones represented by the unified projection model [18], also fit in the proposed model. In this paper we use a calibrated omnidirectional camera to simulate various discrete cameras.

2.1 Image Sensor

We simulate a discrete camera with known Euclidean geometry by sampling a calibrated panoramic image with unique projection center at fixed locations. Since the camera is calibrated, it is straightforward to locate the position (u, v) in the panoramic image corresponding to the 3D direction X of a photocell that is part of the simulated discrete camera. In the present work, we use bilinear interpolation to measure the graylevel value at non-integer coordinates (u, v).

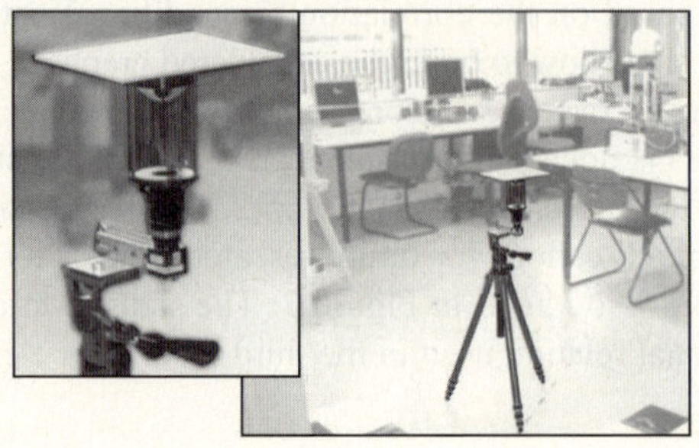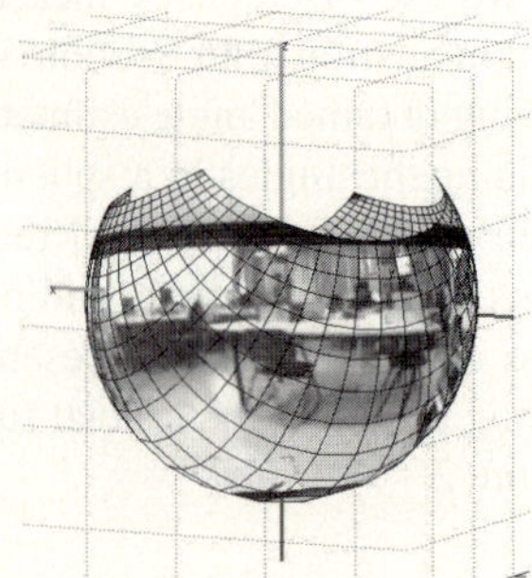

Fig. 3. Left: The camera used to sample omnidirectional images (image mirrored). **Right:** A calibrated omnidirectional image mapped to a sphere.

Images are acquired by a VStone catadiopric camera consisting of a perspective camera fitted to a hyperbolic mirror, shown in Figure 3, left. This system is modeled as single projection center camera [18] with a $360° \times 210°$ field of view and a $\sim 45°$ blind spot at the south pole (Fig. 3, right). The mirror occupies a 453×453 pixel region of the image. The angular separation between neighboring pixels in the panoramic image is usually slightly smaller than $0.5°$. Also, some mild vignetting occurs, that could be corrected. Apart for these minor inconveniences, simulating a discrete camera by an omnidirectional camera presents many advantages: no other specialized hardware is needed and each omnidirectional image can be used to simulate many discrete camera "images", as in Fig. 4, right. With respect to perspective cameras, the available field of view allows to study very-wide-angle discrete cameras.

3   Distances between Pairs of Signals

In this section, we define the measures of distance between signals, correlation and information distance, that will later be used to estimate angles.

3.1   Correlation Distance

We call correlation distance between signals $x(t)$ and $y(t)$, $1 \leq t \leq T$, the quantity

$$d_c(x, y) = \frac{1}{2}(1 - C(x, y)),$$

where $C\left(x,y\right)$ is the correlation between the signals. It is easy to verify that $d_c\left(.,.\right)$ is a distance.

For the task considered in this paper, it is natural to prefer the correlation distance over the variance or the (squared) Euclidean distance $\|x-y\|^2$, because both vary with signal amplitude (and offset, for the latter), whereas $d_c\left(.,.\right)$ is offset- and scale-invariant.

3.2   Information Distance

Given two random variables x and y (in our case, the values produced by individual pixels of a discrete camera) taking values in a discrete set $\{1,\ldots,Q\}$, the *information distance* between x and y is [9]:

$$d\left(x,y\right) = H\left(x|y\right) + H\left(y|x\right) \;=\; 2H\left(x,y\right) - H\left(y\right) - H\left(x\right), \qquad (1)$$

where $H\left(x,y\right)$ is the Shannon entropy of the paired random variable $\left(x,y\right)$, and $H\left(x\right)$ and $H\left(y\right)$ are the entropies of x and y, respectively. It is easy to show that Eq. (1) defines a distance over random variables. This distance is bounded by $H\left(x,y\right) \leq \log_2 Q$, and is conveniently replaced thereafter by the *normalized information distance* :

$$d_I\left(x,y\right) = d\left(x,y\right)/H\left(x,y\right), \qquad (2)$$

which is bounded by 1, independently of Q [9].

It should be noted that estimating the information distance is non-trivial: naively replacing unknown probabilities $p_x\left(q\right)$ by sample frequencies $\hat{p}_x\left(q\right) = |\{t|x\left(t\right) = q\}|/T$, where T is the signal length and $|.|$ denotes the set cardinal, yields a biased estimator $\hat{H}\left(x\right)$. This estimator has expectancy

$$E\left\{\hat{H}\right\} = H - \frac{Q-1}{2T} + \frac{1 - \sum_q \frac{1}{p_x\left(q\right)}}{12T^2} + O\left(\frac{1}{T^3}\right). \qquad (3)$$

This expression shows the slow convergence rate and strong bias of $\hat{H}\left(x\right)$. We somewhat alleviate these problems by first, correcting for the first bias term $\left(Q-1\right)/2T$, i.e. applying the Miller-Madow correction; and by re-quantizing the signal to a much smaller number of bins, $Q = 4$. Extensive benchmarking in [15] has shown these choices to be beneficial.

4   Estimating Angular Separation from Inter-signal Distance

As explained earlier, our a-priori knowledge of the world will be encoded in a graph mapping a measure of discrepancy between two signals, to the angular separation between the photocells that generated the signals. We now show how to build this graph, and assess its effectiveness at estimating angles.

For this purpose, we use the 31-pixel planar discrete camera (or "probe") shown in Fig. 4, left. This probe design allows to study the effect of angular separations ranging from 0.5 to 180 degrees and each sample provides 465=31(31-1)/2 pixel pairs. In

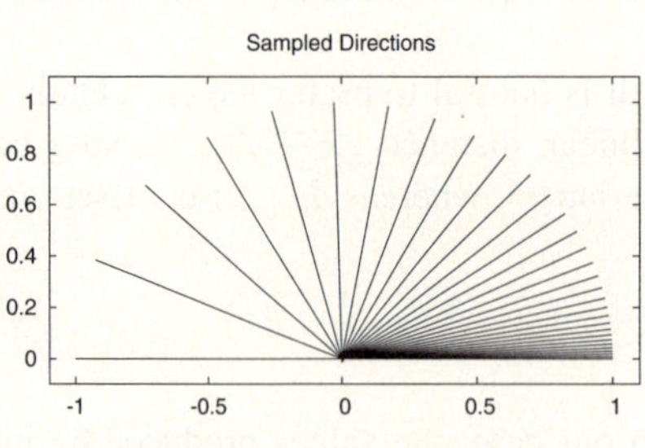

Fig. 4. Left: Geometry of a discrete camera consisting of a planar array of thirty one (31) pixels, spanning $180°$ in the plane. The first two pixels are separated by $0.5°$, the separation between consecutive photocells increases geometrically (ratio $\simeq 1.14$), so that the 31st photocell is antipodal with respect to the first. **Right:** Two instances of the linear discrete camera, inserted in an omnidirectional image. Pixels locations are indicated by small crosses connected by white lines.

the "tighter" part of the discrete camera layout, there exists a slight linear dependence between the values of consecutive pixels due to aliasing.

The camera is hand-held and undergoes "random" general rotation and translation, according to the author´s whim, while remaining near the middle of the room, at 1.0 to 1.8 meters from the ground. We acquired three sequences consecutively, in very similar conditions and joined them in a single sequence totaling 1359 images, i.e. approximately 5 minutes of video at ˜4.5 frames per second.

To simulate the discrete camera, we randomly choose an orientation (i.e. half a great circle) such that all pixels of the discrete camera fall in the field of view of the panoramic camera. Figure 4 shows two such choices of orientations. For each choice of orientation, we produce a sequence of 31 samples $x(i, t)$, $1 \leq i \leq 31$, $1 \leq t \leq 1359$, where each $x(i, t) \in \{0, \ldots, 255\}$. Choosing 100 different orientations, we obtain 100 discrete sensors and 100 arrays of data $x_n(i, t)$, $1 \leq n \leq 100$. Appending these arrays we obtain 31 signals $x(i, t)$ of length to 135900.

We then compute, for each pair of pixels (indices) $1 \leq i, j \leq 31$, the correlation and information distances, $d_c(i, j)$ and $d_I(i, j)$. Joining to these the known angular separations $\theta_{i,j}$, we obtain a set of pairs $(\theta_{i,j}, d(i, j))$, $1 \leq i, j \leq 31$.

From this dataset, we build a constant by parts model of the expectancy of the distance, knowing the angle. For the correlation distance, we limit the abscissa to values in $[0, 1/2]$. After verifying and, if needed enforcing, the monotonicity of this model, we invert it, obtaining a graph of angles as a function of (correlation or information) distances. Strict monotonicity has to be enforced for the correlation-based data, owing to the relatively small number of data points used for each quantized angle.

Figure 5 shows the resulting graphs. This figure shows one of the major issues that appear when estimating the angular separation between pixels from the correlation or information distance: the graphs become very steep for large values of the distance, indicating that small changes of the distance result in large changes in the estimated angle. On the other hand, for small distance values, the curves are much flatter, suggesting

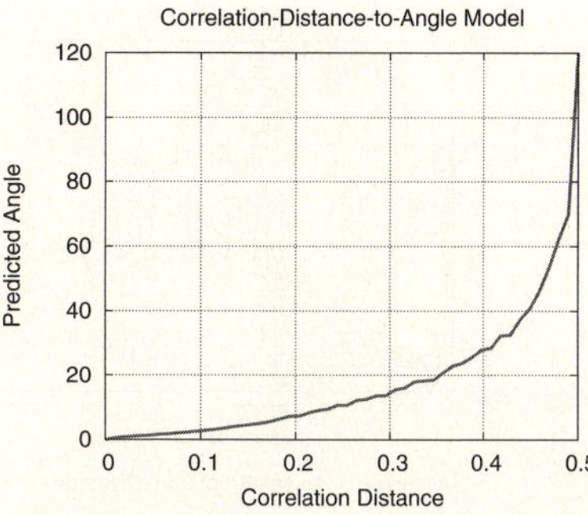

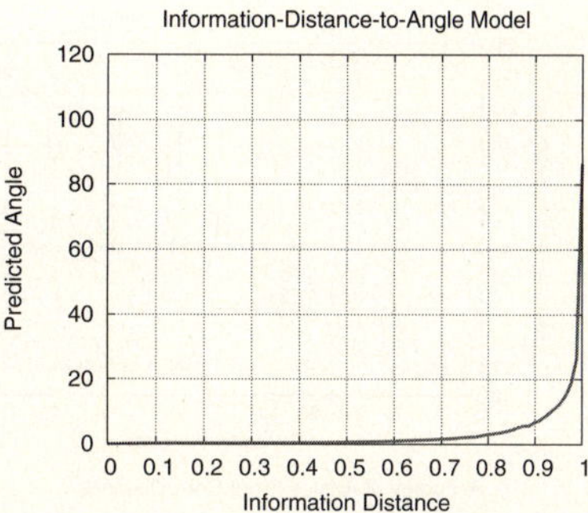

Fig. 5. Models relating correlation (left) or information distance (right) to angular separation between photocells. These models were build from simulated signals produced by the linear probe of Fig. 4, left. Signals of length $T = 135900$, acquired indoors were used.

that small angles can be determined with greater accuracy. Both trends are particularly true for the information distance.

4.1 Experimental Validation

We now assess how well angles can be estimated from the graphs obtained in the previous section. For this purpose, we use 100 sets of 31 signals $x_n(i,t)$, $1 \leq n \leq 100$, $1 \leq i \leq 31$, $1 \leq t \leq 1359$ acquired in the same conditions as above. We compute the correlation and information distances of pairs of signals $d_c(n,i,j)$ and $d_I(n,i,j)$ and, using the models in Fig. 5, angular estimates $\hat{\theta}_c(n,i,j)$ and $\hat{\theta}_I(n,i,j)$.

Figure 6 shows the precision and accuracy of the estimated angles. This figure shows that the estimated angles are fairly accurate for angular separations smaller than $5°$, but degrades sharply for greater values. As could be expected from our comments at the beginning of the section, the curves confirm that the information distance yields better estimates of small angles, while correlation distance does best (but still not very well) for larger angles.

We now turn to the generalization ability of the models in Fig. 5. For this purpose, we use 100 31-uplets of signals of length 2349, taken from an out- and indoor sequence, four images of which are shown in Fig. 7. In this sequence, and contrarily to the previous sequence, the camera remains mostly horizontal. Also, the scene is usually farther away and more textured. A lot of saturation is also apparent.

Following the previous procedure, we estimate angles from these new signals and show the precision and accuracy statistics in Figure 8.

The striking resemblance between Figures 8 and 6 indicates that the models in Fig. 5 generalize pretty well to outdoors scenes. We surmise that the fact that the correlation distance yields more accurate estimates outdoors than indoors is due to the extra texture, which increases the correlation distance for small angles, and corrects the bias in angular estimates observed near the origin of the top left curve of Fig. 6.

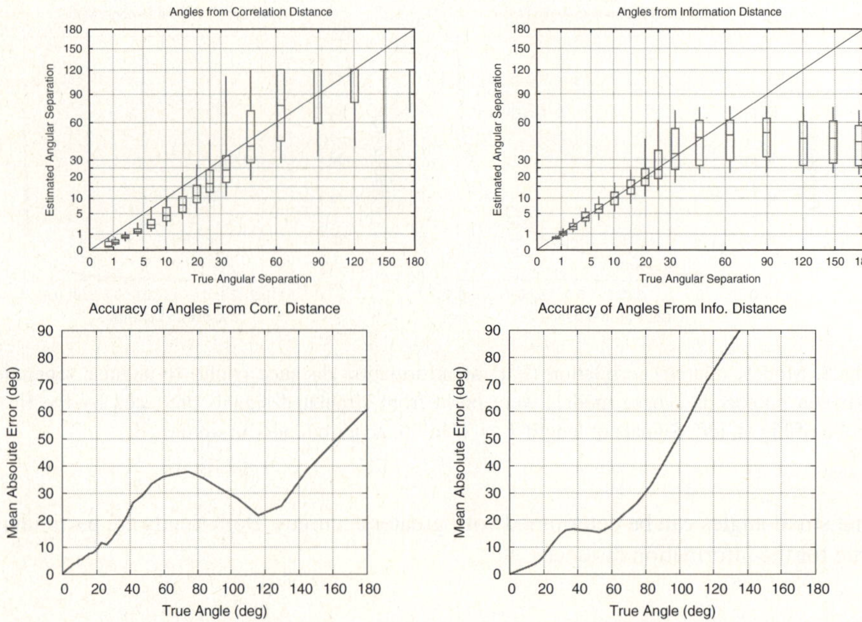

Fig. 6. Precision and accuracy of angles estimated from correlation (left) or information distance (right). The boxplots at the top show the 5th percentile, first quartile, median, third quartile and 95th percentile of the estimated angles, plotted against the true angles. The bottom curves show the mean absolute error in the estimated angles. These statistics were generated from 100 planar probes (Fig. 4, left) and signals of length $T = 1359$. The angles were estimated using the models of Fig. 5. The signals were acquired in the same conditions as those used to build the models.

Fig. 7. Four images from a sequence of 2349 images acquired indoors and outdoors at approximately 4.5FPS

5 Calibrating a Discrete Camera

Having seen the qualities and shortcomings of the proposed angle estimators, we now show how to use them to calibrate a discrete camera.

To stress the generalization ability of the angle estimators, all the reconstructions produced by the above method are obtained from the in- and outdoors sequence of Fig. 7, rather than from the indoors sequence used to build the distance-to-angle models.

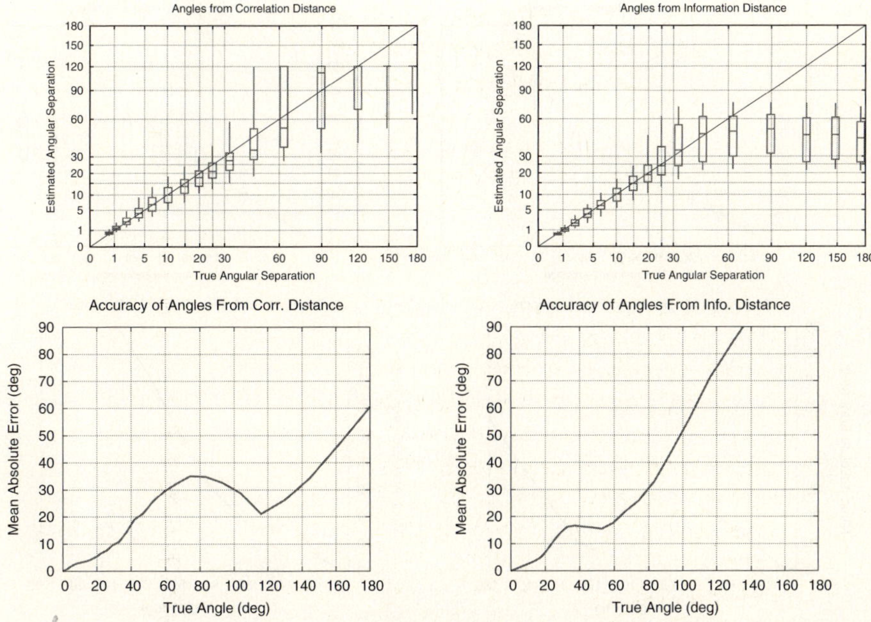

Fig. 8. Precision and accuracy of angles estimated in the same conditions as in Fig. 6, except that signals extracted from an indoors-and-outdoors sequence (Fig. 7) were used. These figures show that the models in Fig. 5 generalize fairly well to signals produced in conditions different from that in which the models were produced. In particular, the angles estimated from the correlation distance are improved w.r.t. those of Fig. 6 (see text).

5.1   Embedding Points in the Sphere

The last step we take to calibrate a discrete camera requires solving the problem:

Problem 1) Spherical embedding problem: Given angle estimates θ_{ij}, $1 \leq i, j \leq N$, find points X_i on the unit sphere, separated by angles approximately equal to θ_{ij}, i.e. $X_i^\top X_j \simeq \cos \theta_{ij}$, for all i, j.

This problem can be reduced to the classical problem of distance geometry [17]:

Problem 2) Euclidean embedding problem: Given distance estimates D_{ij}, $1 \leq i, j \leq N$, find points Y_i in a metric vector space, such that, for all i, j, $\|Y_i - Y_j\| \simeq D_{ij}$

Indeed, by defining an extra point $Y_0 = (0, 0, 0)$, and distances $D_{ij} = \sqrt{2 - 2 \cos \theta_{ij}}$ for $i, j \neq 0$ and $D_{oi} = 1$, the mapping of the first problem to the second is immediate. Solutions to both problems (with exact equality, rather than approximate) were published in 1935 [19][1]. Schoenberg´s Theorem 2 [19] states that if the matrix C with terms $C_{ij} = \cos \theta_{ij}$ is positive semidefinite with rank $r \geq 1$, then there exist points on

---

[1] Schoenberg cites previous work by Klanfer and by Menger, to which we did have access.

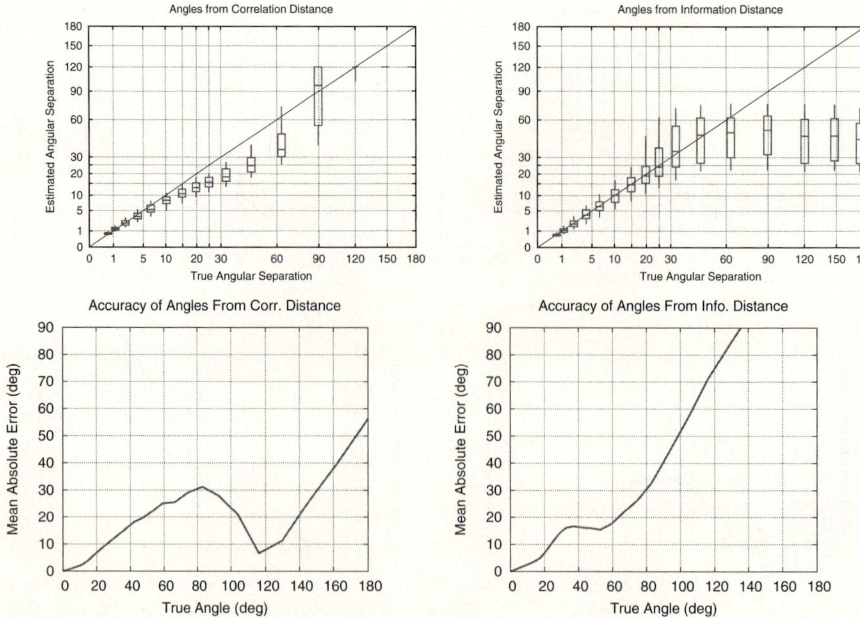

Fig. 9. Precision and accuracy of angles estimated in the same conditions as in Fig. 8, except that the planar probes are constrained to remain approximately horizontal. These figures show that the models in Fig. 5 are usable even if the isotropy assumption of the moving entity is not valid.

the unit $(r-1)$−dimensional sphere that verify $X_i^\top X_j = C_{ij}$ for all i, j. This result directly suggests the following method for embedding points in the 2-sphere:

1. Build the matrix C with terms $C_{ij} = \cos\theta_{ij}, 1 \le i, j \le N$.
2. Compute, using the SVD decomposition, the rank-3 approximation $\tilde{C} = UU^\top$ of C, where U is $N \times 3$.
3. Define $X_i = (U_{i1}, U_{i2}, U_{i3}) / \|(U_{i1}, U_{i2}, U_{i3})\|$.

One should note that this very simple algorithm is not optimal in many ways. In particular, it does not take into account that the error in the angles θ_{ij} is greater in some cases than in others. It is easy to verify that the the problem is not directly tractable by variable-error factorization methods used in computer vision.

Noting that the error in the estimated angles is approximately proportional to the actual angle suggests an embedding method that weighs less heavily large angular estimates. One such method is Sammon´s algorithm [20], which we adapt and modify for the purpose of spherical embedding from our noisy data. In this paper, we minimize the sum

$$\sum_{i,j} w_{i,j} \left(X_i^\top X_j - C_{ij}\right)^2, \text{ where } w_{ij} = \begin{cases} \max\left\{0, \frac{1}{1-C_{ij}} - \frac{1}{1-C_o}\right\} & \text{if } C_{ij} \ne 1 \\ \frac{1}{\eta} & \text{otherwise.} \end{cases}$$

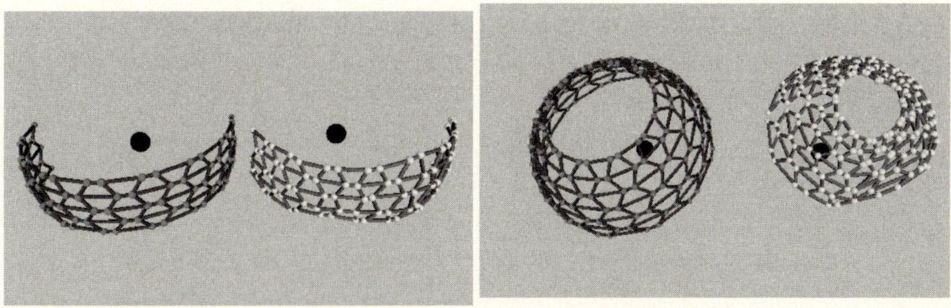

Fig. 10. Calibrations of two different sensors covering more than one hemisphere. On the left, a band-like sensor consisting of 85 photocells, calibrated from correlations (estimated: smaller, true: bigger). On the right, a discrete camera covering more than $180 \times 360°$, of 168 photocells, calibrated from the information distance (estimated: smaller, true: bigger). Each ball represents a photocell except the big black balls, representing the optical center.

To reflect the fact that big angles are less well estimated, we set $C_0 = 0.9$, so that estimates greater than $\mathrm{acos}\,(0.9) \simeq 25°$ be ignored. The other parameter, η is set to 1, allowing the points X_i to stray a little bit away from the unit sphere. Our implementation is inspired by the second-order iterative method of Cawley and Talbot (`http://theoval.sys.uea.ac.uk/~gcc/matlab/default.html`). For initialization, we use an adaptation of [21] to the spherical metric embedding problem, which will be described in detail elsewhere.

5.2   Sensor Calibration

We now evaluate the results of this embedding algorithm on data produced by the angle-estimating method of Sec. 4. For this purpose, we produce sequences of pixel signals in the same conditions as previously, using the outdoors and indoors sequence shown in Figure 7, except that the sensor shape is different. The information and correlation distances between pixels is then estimated from these signals, the angular separation between the pixels is estimated using Sec. 4, and the embedding method of Sec. 5.1 is applied to these angle estimates.

Figure 10 shows the results of our calibration method on sensors covering more than a hemisphere, which thus cannot be embedded in a plane without significant distortion. It should be noted that, although the true sensor is each time more than hemispheric, the estimated calibration is in both cases smaller. This shrinkage is a known effect of some embedding algorithms, which we could attempt to correct.

Figure 11 shows how our method applies to signals produced by a different sensor from the one used to build the distance-to-angle models, namely an Olympus Stylus 300 camera. An 8-by-8 square grid pixels spanning 34 degrees was sampled along a 22822 image sequence taken indoors and outdoors. From this sequence, the estimated angles were generally greater than the true angles, which explains the absence of shrinkage. The higher angle estimates were possibly due to higher texture contents of the sequence. The estimated angles were also fairly noisy, possibly due to the sequence length, and we surmise that longer sequences would yield better results.

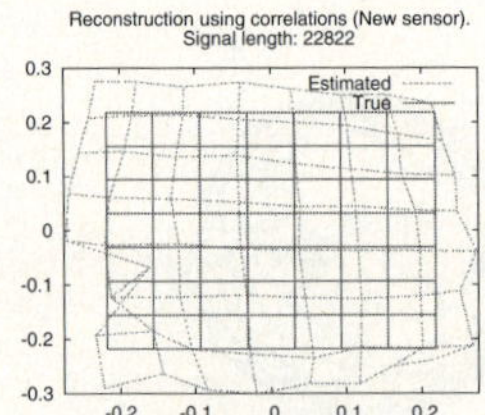

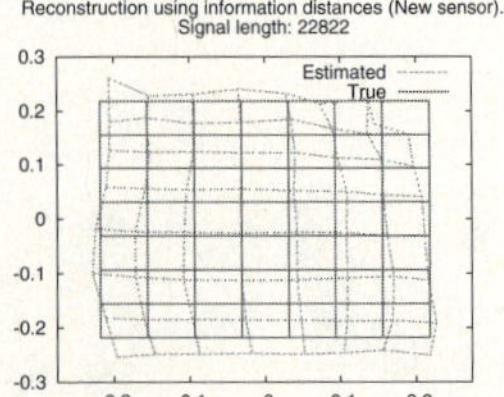

Correlation-Based   Information Distance-Based

Fig. 11. Reconstructed and true pixel layouts of a discrete camera consisting of photocells lying on a rectangular grid. The sensor used differs from that with which the models of Fig 5 were built. The reconstructions are obtained by first estimating the pairwise angular distances, then embedding the angles in the sphere (see text). For visualization, the reconstructions are aligned by the usual procrustes method, mapped to the plane by projective mapping with unit focal length. Added line segments show the true pixel neighborhood relations. The left plot is obtained from the correlation distance, and the right from the information distance.

These results represent typical results that researchers reproducing our method may encounter. Results from other experiments will be presented elsewhere.

6   Discussion

In this paper, we have shown that simple models exist that relate signal discrepancy to angular separation, and are valid in indoors and outdoors scenes. This suggests the existence of near-universal properties of our visual world, in line with other work showing statistical properties of natural images. Contrarily to previous works, we consider statistics of the lightfield taken as a function defined on the sphere, rather than the plane, a choice that allows us to consider fields of view greater than 180 degrees.

We addressed the problem of determining the geometry of a set of photocells in a very general setting. We have confirmed that a discrete camera can be calibrated to a large extent, using just two pieces of data: a table relating signal distances to angles; and a long enough signal produced by the camera.

The presented results are both superior and of a much wider scope than that of [15]: we have shown that it is necessary neither to strictly enforce the assumptions that the camera directs each pixel uniformly in all directions, nor that statistically similar environments be used to build the statistic-to-angle table and to calibrate the discrete camera. This flexibility reinforces the impression that models such as those shown in Figure 5 have a more general validity than the context of calibration.

We showed also that angle estimators based on correlation and information distance (entropy) have different performance characteristics. It would be very interesting to apply machine learning techniques to leverage the power of many such weak estimators.

Finally a more curious question is worth asking in the future: can the problem of angle estimation be altogether bypassed in a geometrically meaningful calibration procedure? Embedding methods based on rank or connectivity [17,22], e.g. correlation or information distance, suggest that this is possible.

References

1. Kohler, I.: Experiments with goggles. Scientific American 206, 62–72 (1962)
2. Tsai, R.: An efficient and accurate camera calibration technique for 3D machine vision. In: IEEE Conf. on Computer Vision and Pattern Recognition (1986)
3. Hartley, R., Zisserman, A.: Multiple View Geometry in Computer Vision. Cambridge University Press, Cambridge (2000)
4. Nistér, D., Stewenius, H., Grossmann, E.: Non-parametric self-calibration. In: Proc. ICCV (2005)
5. Ramalingam, S., Sturm, P., Lodha, S.: Towards complete generic camera calibration. In: Proc. CVPR, vol. 1, pp. 1093–1098 (2005)
6. Pierce, D., Kuipers, B.: Map learning with uninterpreted sensors and effectors. Artificial Intelligence Journal 92(169–229) (1997)
7. Krzanowski, W.J.: Principles of Multivariate Analysis: A User's Perspective. Statistical Science Series. Clarendon Press (1988)
8. Olsson, L., Nehaniv, C.L., Polani, D.: Sensory channel grouping and structure from uninterpreted sensor data. In: NASA/NoD Conference on Evolvable Hardware (2004)
9. Crutchfield, J.P.: Information and its metric. In: Lam, L., Morris, H.C. (eds.) Nonlinear Structures in Physical Systems–Pattern Formation, Chaos and Waves, pp. 119–130. Springer, Heidelberg (1990)
10. Grossmann, E., Orabona, F., Gaspar, J.A.: Discrete camera calibration from the information distance between pixel streams. In: Proc. Workshop on Omnidirectional Vision, Camera Networks and Non-classical Cameras, OMNIVIS (2007)
11. Torralba, A., Oliva, A.: Statistics of natural image categories. Network: Computation in Neural Systems 14, 391–412 (2003)
12. Freeman, W.T., Pasztor, E.C., Carmichael, O.T.: Learning low-level vision. International Journal of Computer Vision 40(1), 25–47 (2000)
13. Potetz, B., Lee, T.S.: Scaling laws in natural scenes and the inference of 3d shape. In: NIPS – Advances in Neural Information Processing Systems, pp. 1089–1096. MIT Press, Cambridge (2006)
14. Wu, Y.N., Zhu, S.C., Guo, C.E.: From information scaling of natural images to regimes of statistical models. Technical Report 2004010111, Department of Statistics, UCLA (2004)
15. Grossmann, E., Gaspar, J.A., Orabona, F.: Discrete camera calibration from pixel streams. In: Computer Vision and Image Understanding (submitted, 2008)
16. Roy, R.: Spectral analysis for a random process on the sphere. Annals of the institute of statistical mathematics 28(1) (1976)
17. Dattorro, J.: Convex Optimization & Euclidean Distance Geometry. Meboo Publishing (2005)
18. Geyer, C., Daniilidis, K.: A unifying theory for central panoramic systems and practical applications. In: Vernon, D. (ed.) ECCV 2000. LNCS, vol. 1843, pp. 445–461. Springer, Heidelberg (2000)
19. Schoenberg, I.J.: Remarks to Maurice Fréchet's article "Sur la définition axiomatique d'une classe d'espaces distanciés vectoriellement applicable sur l'espace de Hilbert". Annals of Mathematics 36(3), 724–732 (1935)
20. Sammon, J.W.J.: A nonlinear mapping for data structure analysis. IEEE Transactions on Computers C-18, 401–409 (1969)
21. Lee, R.C.T., Slagle, J.R., Blum, H.: A triangulation method for the sequential mapping of points from n-space to two-space. IEEE Trans. Computers 26(3), 288–292 (1977)
22. Shang, Y., Ruml, W., Zhang, Y., Fromherz, M.P.J.: Localization from mere connectivity. In: MobiHoc 2003: Proc. ACM Intl. Symp. on Mobile Ad Hoc Networking & Computing, pp. 201–212. ACM Press, New York (2003)

Regular Texture Analysis as Statistical Model Selection

Junwei Han, Stephen J. McKenna, and Ruixuan Wang

School of Computing, University of Dundee, Dundee DD1 4HN, UK
{jeffhan,stephen,ruixuanwang}@computing.dundee.ac.uk
http://www.computing.dundee.ac.uk

Abstract. An approach to the analysis of images of regular texture is proposed in which lattice hypotheses are used to define statistical models. These models are then compared in terms of their ability to explain the image. A method based on this approach is described in which lattice hypotheses are generated using analysis of peaks in the image autocorrelation function, statistical models are based on Gaussian or Gaussian mixture clusters, and model comparison is performed using the marginal likelihood as approximated by the Bayes Information Criterion (BIC). Experiments on public domain regular texture images and a commercial textile image archive demonstrate substantially improved accuracy compared to two competing methods. The method is also used for classification of texture images as regular or irregular. An application to thumbnail image extraction is discussed.

1 Introduction

Regular texture can be modelled as consisting of repeated texture elements, or *texels*. The texels tesselate (or tile) the image (or more generally a surface). Here we consider so-called wallpaper patterns. Wallpaper patterns can be classified into 17 groups depending on their symmetry [1]. Translationally symmetric regular textures can always be generated by a pair of shortest vectors (two linearly independent directions), $\mathbf{t}_1$ and $\mathbf{t}_2$, that define the size, shape and orientation (but not the position) of the texel and the lattice which the texel generates. The lattice topology is always then quadrilateral. Geometric deformations, varying illumination, varying physical characteristics of the textured surface, and sensor noise all result in images of textured patterns exhibiting approximately regular, as opposed to exactly regular, texture. This paper considers the problem of automatically inferring texels and lattice structures from images of planar, approximately regular textures viewed under orthographic projection. While this might at first seem restrictive, this problem is, as will become apparent, far from solved. There exists no fully automatic and robust algorithm to the best of the authors' knowledge. Furthermore, solutions will find application, for example in analysis, retrieval and restoration of images of printed textiles, wallpaper and tile designs.

D. Forsyth, P. Torr, and A. Zisserman (Eds.): ECCV 2008, Part IV, LNCS 5305, pp. 242–255, 2008.

1.1 Related Work

Extraction of periodicity plays an important role in understanding texture and serves as a key component in texture recognition [2], synthesis [3] and segmentation [4]. Previous work proposed for texel and lattice extraction can be grouped broadly into two categories: the local feature-based approach [5,6,7,8,9,10,11] and the global structure-based approach [1,12,13,14,15,16]. All texture analysis is necessarily both local and global. The categorisation is in terms of the computational approach: whether it starts by identifying local features and proceeds to analyse global structure, or starts with a global analysis and proceeds by refining estimates of local structure.

The local feature-based approach starts by identifying a number of texel candidates. Matching based on visual similarity between these potential texels and their neighbours is then performed. Successful matching leads to the connection of texels into a lattice structure. The approach iterates until no more new texels are found. Methods vary in the way they initialise texel candidates and in the parametric models used to cope with geometric and photometric variation. Lin et al. [6] asked users to provide an initial texel. Interest points and edges have been used to generate texel candidates automatically [7,8,9]. However, Hays et al. [5] pointed out that interest points often fail to find texel locations and instead initialized by combining interest points and normalized cross correlation patches. Affine models have been adopted to deal with local variation among texels [7,10,11]. Global projective transformation models have also been used, taking advantage of the spatial arrangement of texels [8,9]. Hays et al. [5] formulated lattice detection as a texel correspondence problem and performed texel matching based on visual similarity and geometric consistency. Lin et al. [6] proposed a Markov random field model with a lattice structure to model global topological relationships among texels and an image observation model able to handle local variations.

The global structure-based approach [1,12,13,14,15,16] tries to extract texels using methods that emphasise the idea of periodic patterns as global processes. Starovoitov et al. [16] used features derived from cooccurrence matrices to extract texels. Charalampidis et al. [15] used a Fourier transform and made use of peaks corresponding to fundamental frequencies to identify texels. The autocorrelation (AC) function is generally more robust than the Fourier transform for the task of texel extraction especially in cases in which a regular texture image contains only a few texel repetitions [1,12]. Peaks in the AC function of a regular texture image can identify the shape and arrangement of texels. Chetverikov [13] developed a regularity measure by means of finding the maximum over all directions on the AC function. Leu [14] used the several highest peaks in the AC function computed on the gradient field of the image to capture translation vectors. A promising approach was presented by Lin et al. [12] in which salient peaks were identified using Gaussian filters to iteratively smooth the AC function. The generalized Hough transform was then applied to find translation vectors, t_1 and t_2. Liu et al. [1] highlighted the fact that spurious peaks often result in incorrect lattice vectors. Therefore, they proposed a "region of dominance" operator to select a list of dominant peaks. The translation vectors were estimated based on these dominant peaks. However, the important problem of how to determine the number of dominant peaks was not addressed. Whilst it is usually relatively easy for a human to select an appropriate subset of peaks, automating this process is difficult. Fig. 1 shows three different texels obtained similarly to Lin et al. [12] from the same image by using different numbers of peaks. The peaks were obtained using the region of dominance method [1]. Whilst using only the first ten peaks can result in success, the method is rather sensitive to this choice.

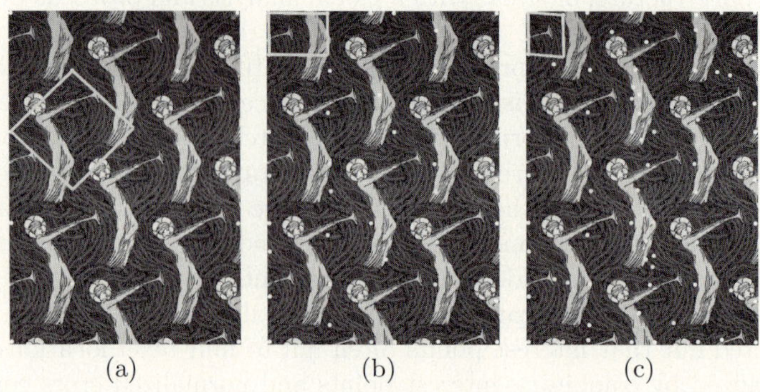

(a)   (b)   (c)

Fig. 1. Texels obtained using (a) ten, (b) forty, and (c) seventy dominant peaks in the autocorrelation function. The peak locations are marked with white dots.

Available local feature-based methods can be effective under significant texture surface deformation and are more suited to such situations. However, they require texels that can be identified based on local features (such as corners) and perform matching between individual texels. Therefore they often fail to detect larger, non-homogeneous texels. Fig. 2 shows examples of such failures.

Global structure-based methods are suitable for textures that do not exhibit large geometric deformation and often successfully identify larger texels with more complicated appearances. However, existing methods have free parameters

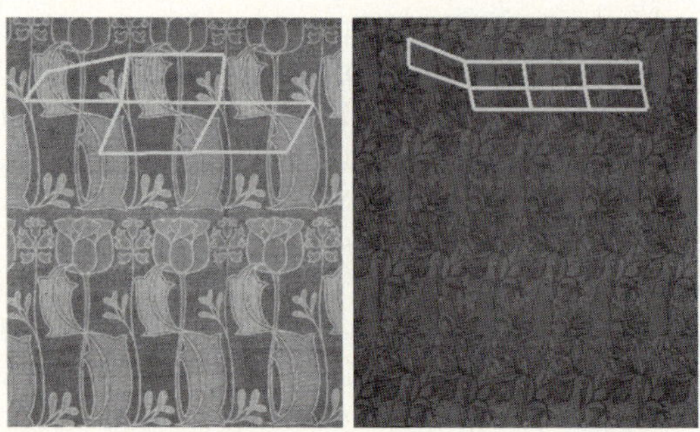

Fig. 2. Two examples of a local feature-based method [5] extracting incorrect lattices

for which a fixed value that works on a wide range of images can often not be found. Methods based on finding peaks in an AC function often yield many unreliable peaks and the number which are reliable can vary dramatically between images. This serious drawback currently makes these methods difficult to apply to large image collections.

1.2   Contributions

We propose a novel model comparison framework to test texel hypotheses and find the optimal one. Hypotheses can be constructed using existing methods according to different subsets of AC peaks by varying the number of peaks used. A statistical model is defined for each lattice hypothesis. The most probable hypothesis given the image observation will be selected. The design of the statistical model takes account of photometric and (to a lesser extent) geometric variations between texels. Hence, our method is robust and completely automatic.

The contributions of this paper can be summarized as follows. (i) A Bayesian model comparison framework is proposed to extract texels from regular texture images based on statistical models defined to handle variations between texels. (ii) Lattice comparison is also used to classify texture images as regular or irregular. (iii) Empirical comparison of the proposed method with two existing methods is performed on a challenging regular texture image database. (iv) The method is applied to generate smart thumbnails for an image browsing and retrieval system.

The rest of this paper is organized as follows. Section 2 presents the Bayesian model comparison framework. Section 3 describes details of lattice model comparison. Section 4 describes the method used in our experiments for generating lattice hypotheses. Experimental results are given in Section 5. An application in which the proposed method is used to generate smart thumbnails for regular texture images is reported in Section 6. Finally, conclusions are drawn in Section 7.

2   Bayesian Model Comparison Framework

Our approach is to formulate texel hypotheses as statistical models and then compare these models given the image data. It is not sufficient for a model to be able to fit the data well. The best texel hypothesis under this criterion would be the image itself whereas our purpose is to extract the smallest texture element. Therefore, overfitting must be guarded against by penalising model complexity. Texel hypothesis comparison can be regarded as a typical model comparison problem for unsupervised statistical modelling of data. Such a problem can be formulated as Bayesian model comparison which naturally penalises complexity (Occam's razor).

Let $I = \{x_1, x_2, \ldots, x_N\}$ be an image with N pixels. Here, $x_n, 1 \leq n \leq N$ is the intensity of the nth pixel. Let $H \equiv (\mathbf{t}_1, \mathbf{t}_2)$ denote a texel hypothesis for I, H_k the kth in a set of hypotheses, and M_k a statistical model defined based on H_k with parameters θ_k. Texel extraction can be formulated as choosing the most probable texel hypothesis given the image. According to Bayes' theorem, the posterior probability is proportional to the likelihood of the hypothesis times a prior:

$$p(H_k|I) = \frac{p(I|H_k)p(H_k)}{p(I)} \propto p(I|H_k)p(H_k) \tag{1}$$

In the absence of prior knowledge favouring any of the texel hypotheses, the (improper) prior is taken to be uniform. For each H_k, we define a unique M_k deterministically so $p(M_k|H_k)$ is a delta function. Hence,

$$p(H_k|I) \propto p(I|M_k) = \int p(I|\theta_k, M_k)p(\theta_k|M_k)d\theta_k \tag{2}$$

Texel hypotheses can be compared by comparing the marginal likelihoods, $p(I|M_k)$, for their models. Here $p(I|\theta_k, M_k)$ is the probability density function of the image data given the model M_k and its parameters θ_k, and $p(\theta_k|M_k)$ is the prior probability density function of parameters θ_k given the model M_k.

The integral in Equation (2) can only be computed analytically in certain cases such as exponential likelihoods with conjugate priors. Otherwise, approximations can be obtained using sampling methods, for example. While it would be interesting to explore these alternatives in future work, this paper uses the Bayes Information Criterion (BIC) as a readily computable approximation. BIC approximates the marginal likelihood integral via Laplace's method and the reader is referred to the papers by Schwarz [17] and Raftery [18] for full details of its derivation. Given a maximum likelihood parameter estimate, $\hat{\theta}$, we have

$$\log p(I|M) \approx \log p(I|\hat{\theta}, M) + \log p(\hat{\theta}) + \frac{d}{2}\log 2\pi - \frac{d}{2}\log N - \frac{1}{2}\log |\mathbf{i}| + O(N^{-1/2}) \tag{3}$$

where d is the number of parameters and $\mathbf{i}$ is the expected Fisher information matrix for one observation. The subscript k has been dropped here for clarity. The term $\log p(I|\hat{\theta}, M)$ is of order $O(N)$, $(d/2)\log N$ is of order $O(\log N)$, and the remaining terms are of order $O(1)$ or less. The log marginal likelihood can be approximated by removing all terms of order $O(1)$ or less. The BIC for the model is then

$$BIC(M) = -\log p(I|\hat{\theta}, M) + (d/2)\log N \approx -\log p(I|M_k) \tag{4}$$

The first term can be interpreted as an error of fit to the data while the second term penalises model complexity.

The proposed approach to regular texture analysis involves (i) generation of multiple texel hypotheses, and (ii) comparison of hypotheses based on statistical models. The hypothesis with the model that has the largest marginal likelihood is selected. Using the BIC approximation, hypothesis $H_{\hat{k}}$ is selected where,

$$\hat{k} = \arg\max_k\{p(H_k|I)\} = \arg\min_k\{BIC(M_k)\} \tag{5}$$

This method can also be used to classify textures as regular or irregular. If a 'good' lattice can be detected in an image then it should be classified as regular. The proposed lattice comparison framework can be adopted for this purpose by comparing the most probable lattice found with a reference hypothesis in which the entire image is a single 'texel'. If the reference hypothesis has a higher BIC value then the image is classified as regular. Otherwise, it is classified as irregular, i.e.

$$BIC(M_R) \leq BIC(M_{\hat{k}}) \qquad Irregular\ texture$$
$$BIC(M_R) > BIC(M_{\hat{k}}) \qquad Regular\ texture \qquad (6)$$

where M_R refers to the model corresponding to the reference lattice and $M_{\hat{k}}$ is the best lattice hypothesis selected by Equation (5).

3   Lattice Models

The lattice model should be able to account for both regularity from periodic arrangement and statistical photometric and geometric variability. Let us first suppose a regular texture image I with N pixels $x_1, x_2, \ldots, x_N$, and a hypothesis H with Q pixels per texel. Based on H, each pixel of the image is assigned to one of Q positions on the texel according to the lattice structure. Thus, the N pixels are partitioned into Q disjoint sets, or clusters. If we choose to assume that the N pixels are independent given the model, we have,

$$p(I|M) = \prod_{n=1}^{N} p(x_n|M) = \prod_{q=1}^{Q} \prod_{n:f(n,H)=q} p(x_n|M) \qquad (7)$$

where $f(n, H) \in \{1, \ldots, Q\}$ maps n to its corresponding index in the texel. Fig. 3 illustrates this assigment of pixels to clusters.

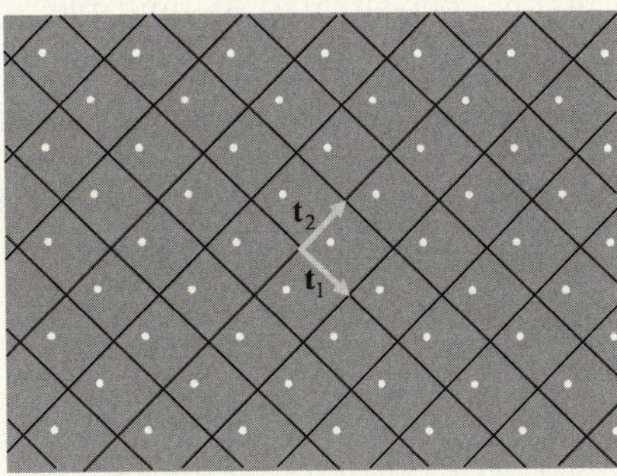

Fig. 3. An example of cluster allocation according to a texel hypothesis, $H \equiv (\mathbf{t}_1, \mathbf{t}_2)$. The value of $f(n, H)$ is the same for each of the highlighted pixels. There are Q pixels in each parallelogram.

Modelling each of the Q clusters as Gaussian with fixed variance gives:

$$BIC(M) = (Q/2)\log N - \sum_{q=1}^{Q} \sum_{n:f(n,H)=q} \log p(x_n|\hat{\mu}_q, \sigma^2) \qquad (8)$$

$$= (Q/2)\log N + C_1 + \frac{1}{2\sigma^2} \sum_{q=1}^{Q} \sum_{n:f(n,H)=q} (x_n - \hat{\mu}_q)^2 \qquad (9)$$

where C_1 is a constant that depends on σ^2, and $\hat{\mu}_q$ is a maximum likelihood estimate of the mean of the qth cluster.

Alternatively, a more heavy-tailed distribution can be used for each cluster. This might better model outliers due to physical imperfections in the texture surface and variations due to small geometric deformations. For example, a cluster can be modelled as a mixture of two Gaussians with the same mean but different variances, (σ_1^2, σ_2^2), and a mixing weight, π_1 that places greater weight on the low variance Gaussian. In that case,

$$BIC(M) = -\sum_{q=1}^{Q} \sum_{n:f(n,H)=q} \log p(x_n|\hat{\mu}_q, \sigma_1^2, \sigma_2^2, \pi_1) + (Q/2)\log N \qquad (10)$$

$$= (Q/2)\log N + C_2 \qquad (11)$$

$$-\sum_{q=1}^{Q} \sum_{n:f(n,H)=q} \log\left(\frac{\pi_1}{\sigma_1} \exp \frac{-(x_n - \hat{\mu}_q)^2}{2\sigma_1^2} + \frac{1-\pi_1}{\sigma_2} \exp \frac{-(x_n - \hat{\mu}_q)^2}{2\sigma_2^2}\right)$$

where C_2 is a constant.

4   Lattice Hypothesis Generation

In principle, there is an unlimited number of lattice hypotheses. However, probability density will be highly concentrated at multiple peaks in the hypothesis space. The posterior distribution can therefore be well represented by only considering a, typically small, number of hypotheses at these peaks. In the maximum a posteriori setting adopted here, the approach taken is to identify multiple hypotheses in a data-driven manner and then compare these hypotheses using BIC. The approach is general in that any algorithms that generate a variety of reasonable hypotheses can be used.

In the experiments reported here, aspects of the methods of Lin et al. [12] and Liu et al. [1] were combined to generate hypotheses. Peaks in AC functions are associated with texture periodicity but automatically deciding which peaks can characterize the arrangement of texels is problematic and has not been properly addressed in the literature [1,12,13,14]. In particular, changing the number of peaks considered can result in different lattice hypotheses. Since the total number of peaks is limited, we can only obtain a limited number of hypotheses.

Given a grey-scale image $I(x, y), 1 \leq x \leq L, 1 \leq y \leq W$ where L and W are image height and width, its AC function can be computed as follows:

$$AC(x, y) = \frac{\sum_{i=1}^{L} \sum_{j=1}^{W} I(i, j) I(i + x, j + y)}{\sum_{i=1}^{L} \sum_{j=1}^{W} I^2(i, j)} \tag{12}$$

Applying the fast Fourier transform (FFT) to calculate the AC function is a more efficient alternative.

$$AC(x, y) = F^{-1}[F[I(x, y)]^* F[I(x, y)]] \tag{13}$$

where F and F^{-1} denote FFT and inverse FFT, respectively.

Lin et al. [12] used iterative smoothing with Gaussian filters to obtain salient peaks. However, Liu et al. [1] advised to take into account the spatial relationships among peaks and used a "region of dominance" operator. The basic idea behind this operator is that peaks that dominate large regions of the AC function are more perceptually important. In this paper, we combine these two algorithms. First, we apply Gaussian filters to iteratively smooth the AC function. Then, salient peaks obtained from the first stage are ranked according to their dominance. The most highly ranked peaks are selected as input for lattice hypothesis construction using a Hough transform [12]. The number of peaks in the rank-ordered list to use was varied in order to generate multiple hypotheses. Typically a few tens of the generated hypotheses will be distinct.

5   Experiments

A dataset of 103 regular texture images was used for evaluation, comprising 68 images of printed textiles from a commercial archive and 35 images taken from three public domain databases (the Wikipedia Wallpaper Groups page, a Corel database, and the CMU near regular texture database). These images ranged in size from 352×302 pixels to 2648×1372 pixels. The number of texel repeats per image ranged from 5 to a few hundreds. This data set includes images that are challenging because of (i) appearance variations among texels, (ii) small geometric deformations, (iii) texels that are not distinctive from the background and are large non-homogeneous regions, (iv) occluding labels, and (v) stains, wear and tear in some of the textile images.

Systematic evaluations of lattice extraction are lacking in the literature. We compared the proposed method with two previously published algorithms. Two volunteers (one male and one female) qualitatively scored and rank ordered the algorithms. In cases of disagreement, they were forced to reach agreement through discussion. (Disagreement happened in very few cases).

When the proposed method used Gaussians to model clusters, the only free parameter was the variance, σ^2. A suitable value for σ^2 was estimated from a set of 20 images as follows. Many texel hypotheses were automatically generated using different numbers of AC peaks and a user then selected from them the best translation vectors, $\mathbf{t}_1, \mathbf{t}_2$. Pixels were allocated to clusters according to the resulting lattice and a maximum likelihood estimation of σ^2 was computed. The result was $\sigma^2 = 264$. Since this semi-automatic method might not be using precise texel estimates, it might overestimate the variance compared to that which would be obtained using optimal lattices. Therefore, further values for σ^2 (100, 144 and 196) were also used for evaluation in order to test the sensitivity of the method. In any particular experiment, σ^2 was fixed for all 103 test images. The method was also evaluated using a Gaussian mixture to model each cluster, with free parameters set to $\sigma_1^2 = 60$, $\sigma_2^2 = 800$, and $\pi_1 = 0.9$.

The observers were shown lattices overlaid on images and were asked to label each lattice as obviously correct (OC), obviously incorrect (OI), or neutral. They were to assign OC if the lattice was exactly the same or very close to what they expected, OI if the result was far from their expectations, and *neutral* otherwise. The presentation of results to the observers was randomised so as to hide from them which algorithms produced which results. The proposed method was compared with two related algorithms [12,1]. Liu et al. [1] did not specify how to determine the number of peaks in the autocorrelation function. Results are reported here using three different values for the number of peaks, namely 10, 40, and 70. Table 1 summarises the results. It seems clear that the method proposed in this paper has superior accuracy to the two other methods. The value of σ^2 had little effect on the results. Fig. 4 shows some examples of lattices obtained. The two images displayed in the first row have clear intensity variations between texels. The two examples in the second row have labels in the image and appearance varies among texels. Examples shown in rows 3 to 5 contain large non-homogenous texels. The left example in the last row is a *neutral* result. This example has a significant geometric deformation among texels. The right example in the last row is an *OI* result since it did not find the smallest texel.

Table 1. Comparison of proposed algorithm with related algorithms. Accuracy is defined as the number of OC results divided by the total number of test images.

Algorithm variant	# OC results	# OI results	# Neutral results	Accuracy
Gaussian ($\sigma^2 = 100$)	83	9	11	0.81
Gaussian ($\sigma^2 = 144$)	83	14	6	0.81
Gaussian ($\sigma^2 = 196$)	82	14	7	0.80
Gaussian ($\sigma^2 = 264$)	79	18	6	0.77
Gaussian mixture	81	17	5	0.79
Liu et al. [1] (10 peaks)	45	54	4	0.44
Liu et al. [1] (40 peaks)	50	47	6	0.49
Liu et al. [1] (70 peaks)	28	70	5	0.27
Lin et al. [12]	22	70	11	0.21

A further experiment was performed to compare the proposed method to the two other methods. For each image, lattice results from our algorithm using Gaussians, our algorithm using Gaussian mixtures, the algorithm of Liu et al. [1], and the algorithm of Lin et al. [12], respectively, were shown on the screen simultaneously. The two subjects rank ordered those four results. Algorithms

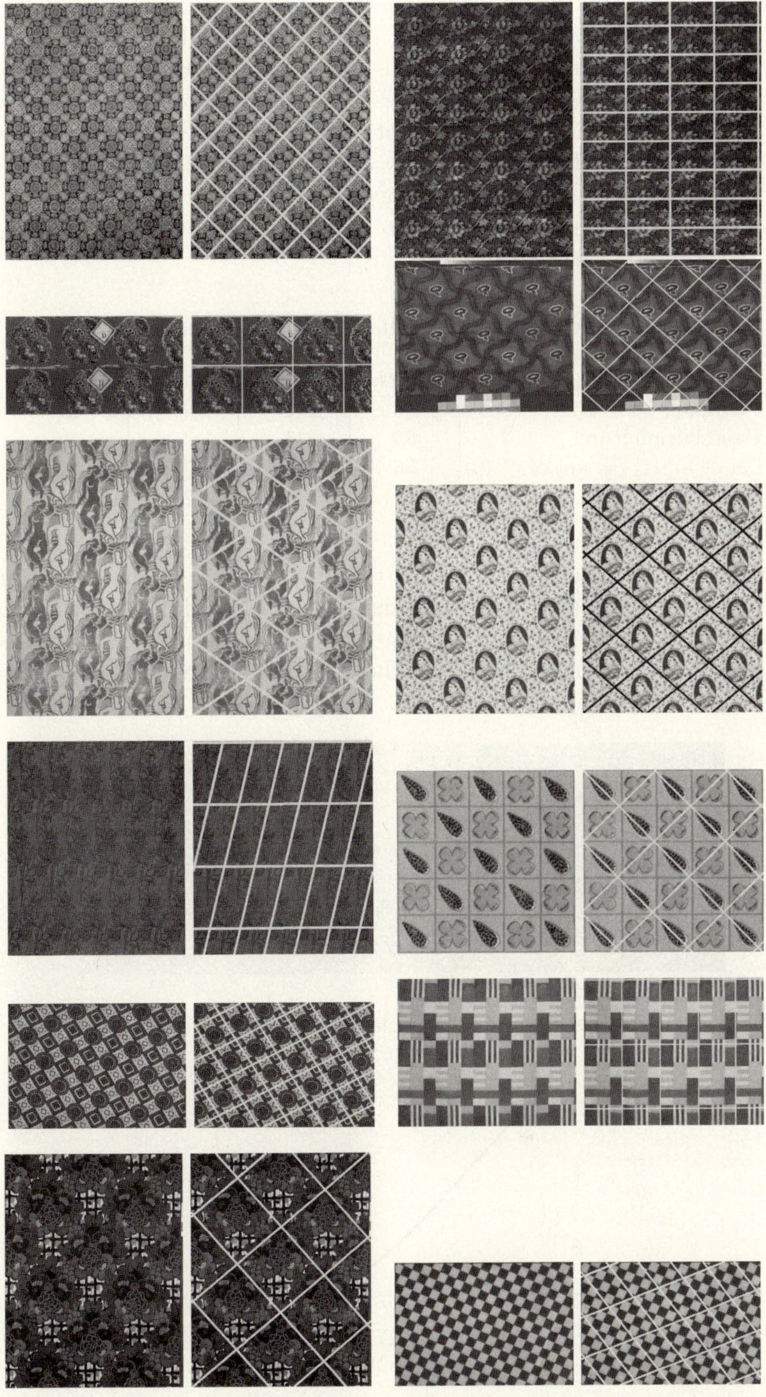

Fig. 4. Results from the proposed algorithm using Gaussian models

shared the same rank if they yielded equally good results. For example, if three of the algorithms gave good lattices of equal quality and the fourth algorithm gave a poor lattice then three algorithms shared rank 1 and the other algorithm was assigned rank 4. Table 2 summarizes the rankings. For the Gaussian model, we set $\sigma^2 = 264$ which yields the worst accuracy of the variance values tried. For the algorithm of Liu et al. [1], we set the number of dominant peaks to 40, which achieved the best performance of the values tried. Even with these parameter settings which disadvantage the proposed method, Table 2 shows that it is superior to the other algorithms.

Table 2. Comparisons by ranking results of different algorithms

Algorithm	# Rank 1	# Rank 2	# Rank 3	# Rank 4
Gaussian, $\sigma^2 = 264$	83	12	6	2
Gaussian mixture	86	11	5	1
Liu et al. [1] (# peaks = 40)	56	5	23	19
Lin et al. [12]	18	2	24	59

The method was also used to classify texture images as regular or irregular as described in Equation (6). A set of 62 images was selected randomly from a museum fine art database and from the same commercial textile archive as used earlier. Figure 5 shows some examples of these images. A classification experiment

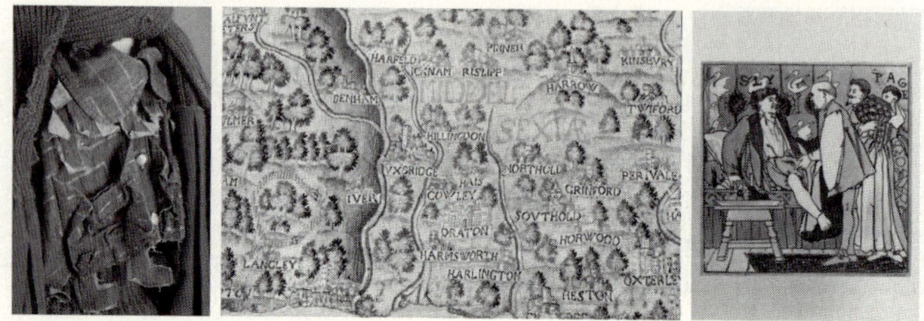

Fig. 5. Examples of images to be classified as having *irregular* texture

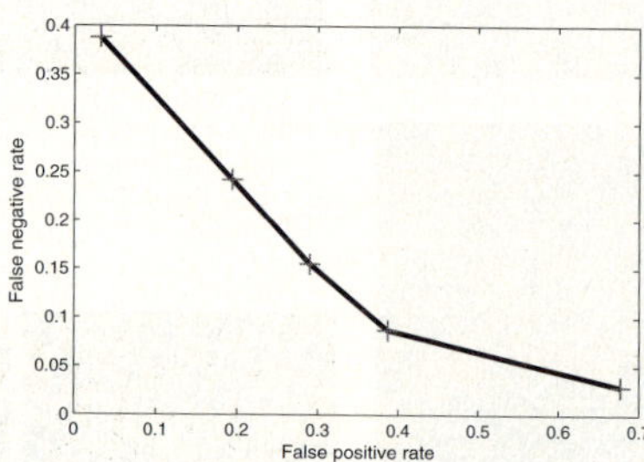

Fig. 6. Classification of texture as regular or irregular. The curve was plotted by varying the value of σ^2 and characterises the trade-off between the two types of error.

was performed using these images as negative examples and the 103 regular texture images as positive examples. Figure 6 shows the ROC curve obtained by varying the value of σ^2 in the Gaussian model ($\sigma^2 \in \{49, 64, 81, 100, 144\}$). The equal error rate was approximately 0.22.

The computational speed depends on the number of lattice hypotheses (and many different subsets of peaks lead to the same lattice hypothesis). A Matlab implementation typically takes a few minutes per image on a $2.4GHz$, $3.5GB$ PC which is adequate for off-line processing.

6 Smart Thumbnail Generation for Regular Texture Images

Thumbnail images are widely used when showing lots of images on a display device of limited size. Most traditional approaches generate thumbnails by directly sub-sampling the original image which often reduces the recognisability of meaningful objects and patterns in the image. Suh et al. [19] developed a novel thumbnail generation method by taking into account human visual attention. A saliency map and a face detector were used to identify regions expected to attract visual attention. Although this method is effective for many images, it is not appropriate for images with regular texture that often comprise abstract patterns.

In an informal experiment, 9 human observers of varied age were asked to draw a rectangle on each of 14 regular texture images to delineate the region they would like to see as a thumbnail on a limited display. Most users tended to select regions a little larger than a single texel, or containing a few texels. This suggests that thumbnails might usefully be generated from regular texture images automatically by cropping based on texel extraction. Currently, we are exploring the use of such thumbnails for content-based image browsing and retrieval. Thumbnails are generated by cropping a rectangular sub-image that bounds a region a little larger than a texel, $(1.5\mathbf{t}_1, 1.5\mathbf{t}_2)$. Fig. 7 compares two thumbnails generated in this way with the standard method of directly reducing

Fig. 7. Comparisons of two thumbnail generation methods. In each set, the first image is the original image, the second image is the thumbnail generated by our method, and the third image is the thumbnail generated by the standard method.

the resolution. Thumbnails extracted using knowledge of the texels can convey more detailed information about the pattern design.

7 Conclusions

A fully automatic lattice extraction method for regular texture images has been proposed using a framework of statistical model selection. Texel hypotheses were generated based on finding peaks in the AC function of the image. BIC was adopted to compare various hypotheses and to select a 'best' lattice. The experiments and comparisons with previous work have demonstrated the promise of the approach. Various extensions to this work would be interesting to investigate in future work. Alternative methods for generating hypotheses could be explored in the context of this approach. Further work is needed to explore the relative merits of non-Gaussian models. This should enable better performance on images of damaged textiles, for example. BIC can give poor approximations to the marginal likelihood and it would be worth exploring alternative approximations based on sampling methods, for example. Finally, it should be possible in principle to extend the approach to analysis of near-regular textures on deformed 3D surfaces by allowing relative deformation between texels. This could be formulated as a Markov random field over texels, for example. Indeed, Markov random field models have recently been applied to regular texture tracking [6].

Acknowledgments. The authors thank J. Hays for providing his source code, and Chengjin Du and Wei Jia for helping to evaluate the algorithm. This research was supported by the UK Technology Strategy Board grant "FABRIC: Fashion and Apparel Browsing for Inspirational Content" in collaboration with Liberty Fabrics Ltd., System Simulation Ltd. and Calico Jack Ltd. The Technology Strategy Board is a business-led executive non-departmental public body, established by the government. Its mission is to promote and support research into, and development and exploitation of, technology and innovation for the benefit of UK business, in order to increase economic growth and improve the quality of life. It is sponsored by the Department for Innovation, Universities and Skills (DIUS). Please visit www.innovateuk.org for further information.

References

1. Liu, Y., Collins, R.T., Tsin, Y.: A computational model for periodic pattern perception based on frieze and wallpaper groups. IEEE Transactions on Pattern Analysis and Machine Intelligence 26, 354–371 (2004)
2. Leung, T., Malik, J.: Recognizing surfaces using three-dimensional textons. In: IEEE International Conference on Computer Vision, Corfu, Greece, pp. 1010–1017 (1999)
3. Liu, Y., Tsing, Y., Lin, W.: The promise and perils of near-regular texture. International Journal of Computer Vision 62, 145–159 (2005)
4. Malik, J., Belongie, S., Shi, J., Leung, T.: Textons, contours and regions: cue integration in image segmentation. In: IEEE International Conference of Computer Vision, Corfu, Greece, pp. 918–925 (1999)

5. Hays, J., Leordeanu, M., Efros, A., Liu, Y.: Discovering texture regularity as a higher-order correspondance problem. In: European Conference on Computer Vision, Graz, Austria, pp. 533–535 (2006)
6. Lin, W., Liu, Y.: A lattice-based MRF model for dynamic near-regular texture tracking. IEEE Transactions on Pattern Analysis and Machine Intelligence 29, 777–792 (2007)
7. Leung, T., Malik, J.: Detecting, localizing and grouping repeated scene elements from an image. In: European Conference on Computer Vision, Cambridge, UK, pp. 546–555 (1996)
8. Tuytelaars, T., Turina, A., Gool, L.: Noncombinational detection of regular repetitions under perspective skew. IEEE Transactions on Pattern Analysis and Machine Intelligence 25, 418–432 (2003)
9. Schaffalitzky, F., Zisserman, A.: Geometric grouping of repeated elements within images. In: Shape, Contour and Grouping in Computer Vision. Lecture Notes In Computer Science, pp. 165–181. Springer, Heidelberg (1999)
10. Forsyth, D.A.: Shape from texture without boundries. In: European Conference in Computer Vision, Copenhagen, Denmark, pp. 225–239 (2002)
11. Lobay, A., Forsyth, D.A.: Recovering shape and irradiance maps from rich dense texton fields. In: Computer Vision and Pattern Recognition, Washington, USA, pp. 400–406 (2004)
12. Lin, H., Wang, L., Yang, S.: Extracting periodicity of a regular texture based on autocorrelation functions. Pattern Recognition Letters 18, 433–443 (1997)
13. Chetverikov, D.: Pattern regularity as a visual key. Image and Vision Computing 18, 975–985 (2000)
14. Leu, J.: On indexing the periodicity of image textures. Image and Vision Computing 19, 987–1000 (2001)
15. Charalampidis, D.: Texture synthesis: Textons revisited. IEEE Transactions on Image Processing 15, 777–787 (2006)
16. Starovoitov, V., Jeong, S.Y., Park, R.: Texture periodicity detection: features, properties, and comparisons. IEEE Transactions on Systems, Man, and Cybernetics-A 28, 839–849 (1998)
17. Schwarz, G.: Estimating the dimensions of a model. Annals and Statistics 6, 461–464 (1978)
18. Raftery, A.E.: Bayesian model selection in social research. Sociological Methodology 25, 111–163 (1995)
19. Suh, B., Ling, H., Benderson, B.B., Jacobs, D.W.: Automatic thumbnail cropping and its effectiveness. In: ACM Symposium on User Interface Software and Technology, pp. 95–104 (2003)

Higher Dimensional Affine Registration and Vision Applications

Yu-Tseh Chi[1], S.M. Nejhum Shahed[1], Jeffrey Ho[1], and Ming-Hsuan Yang[2]

[1] CISE Department, University of Florida, Gainesville, 32607
{ychi,smshahed,jho}@csie.ufl.edu
[2] EECS, University of California, Merced, CA 95344
mhyang@ucmerced.edu

Abstract. Affine registration has a long and venerable history in computer vision literature, and extensive work have been done for affine registrations in $\mathbb{R}^2$ and $\mathbb{R}^3$. In this paper, we study affine registrations in $\mathbb{R}^m$ for $m > 3$, and to justify breaking this dimension barrier, we show two interesting types of matching problems that can be formulated and solved as affine registration problems in dimensions higher than three: stereo correspondence under motion and image set matching. More specifically, for an object undergoing non-rigid motion that can be linearly modelled using a small number of shape basis vectors, the stereo correspondence problem can be solved by affine registering points in $\mathbb{R}^{3n}$. And given two collections of images related by an unknown linear transformation of the image space, the correspondences between images in the two collections can be recovered by solving an affine registration problem in $\mathbb{R}^m$, where m is the dimension of a PCA subspace. The algorithm proposed in this paper estimates the affine transformation between two point sets in $\mathbb{R}^m$. It does not require continuous optimization, and our analysis shows that, in the absence of data noise, the algorithm will recover the exact affine transformation for almost all point sets with the worst-case time complexity of $O(mk^2)$, k the size of the point set. We validate the proposed algorithm on a variety of synthetic point sets in different dimensions with varying degrees of deformation and noise, and we also show experimentally that the two types of matching problems can indeed be solved satisfactorily using the proposed affine registration algorithm.

1 Introduction

Matching points, particularly in low-dimensional settings such as 2D and 3D, has been a classical problem in computer vision. The problem can be formulated in a variety of ways depending on the allowable and desired deformations. For instance, the orthogonal and affine cases have been studied already awhile ago, e.g., [1][2], and recent research activities have been focused on non-rigid deformations, particularly those that can be locally modelled by a family of well-known basis functions such as splines, e.g., [3]. In this paper, we study the more classical problem of matching point sets[1] related by affine transformations. The novel viewpoint taken here is the emphasis on affine registrations

---

[1] In this paper, the two point sets are assumed to have the same size.

D. Forsyth, P. Torr, and A. Zisserman (Eds.): ECCV 2008, Part IV, LNCS 5305, pp. 256–269, 2008.
© Springer-Verlag Berlin Heidelberg 2008

in $\mathbb{R}^m$ for $m > 3$, and it differs substantially from the past literature on this subject, which has been overwhelmingly devoted to registration problems in $\mathbb{R}^2$ and $\mathbb{R}^3$.

To justify breaking this dimension barrier, we will demonstrate that two important and interesting types of matching problems can be formulated and solved as affine registration problems in $\mathbb{R}^m$ with $m > 3$: stereo correspondence under motion and image set matching (See Figure 1). In the stereo correspondence problem, two video cameras are observing an object undergoing some motion (rigid or non-rigid), and a set of k points on the object are tracked consistently in each view. The problem is to match the tracking results across two views so that the k feature points can be located and identified correctly. In the image set matching problem, two collections of images are given such that the unknown transformation between corresponding pairs of images can be approximated by some linear transformation $\mathcal{F} : \mathbb{R}^m \to \mathbb{R}^{m'}$ between two (high-dimensional) image spaces. The task is to compute the correspondences directly from the images. Both problems admit quick solutions. For example, for stereo correspondence under motion, one quick solution would be to select a pair of corresponding frames and compute the correspondences directly between these two frames. This approach is clearly since there is no way to know *a priori* which pair of frames is optimal for computing the correspondences. Furthermore, if the baseline between cameras is large, direct stereo matching using image features does not always produce good results, even when very precise tracking result are available. Therefore, there is a need for a principled algorithm that can compute the correspondences directly using all the tracking results simultaneously instead of just a pair of frames.

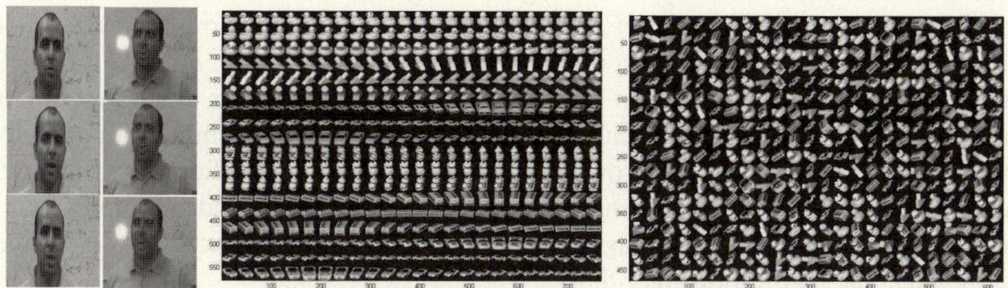

Fig. 1. Left: Stereo Correspondence under Motion. A talking head is observed by two (affine) cameras. Feature points are tracked separately on each camera and the problem is to compute the correspondences between observed feature points across views. **Center and Right:** Image Set Matching. Two collections (432 images each) of images are given. Each image on the right is obtained by rotating and down-sizing an image on the left. The problem is to recover the correspondences. These two problems can be formulated as affine registration problems in $\mathbb{R}^m$ with $m > 3$.

An important point to realize is that in each problem there are two linear subspaces that parameterize the input data. For nonrigid motions that can be modelled using linear shape basis vectors, this follows immediately from the work of [4][5]. For image set matching, each set of images can usually be approximated by a linear subspace with dimension that is considerably smaller than that of the ambient image space. We will show that the correspondences can be computed (or be approximated) by affine registering point sets in these two linear subspaces. Therefore, instead of using quantities derived from image intensities, our solution to these two matching problems is to first formulate them as affine point set matching problems in $\mathbb{R}^m$, with $m > 3$, and solve the resulting affine registration problems.

Let $\mathcal{P} = \{p_1, \cdots, p_k\}$ and $\mathcal{Q} = \{q_1, \cdots, q_k\}$ denote two point sets in $\mathbb{R}^m$ with equal number of points. The affine registration problem is typically formulated as an optimization problem of finding an affine transformation $\mathbf{A}$ and a correspondence map π between points in $\mathcal{P}, \mathcal{Q}$ such that the following registration error function is minimized

$$\mathcal{E}(\mathbf{A}, \pi) = \sum_{i=1}^{k} \mathbf{d}^2(\mathbf{A}p_i, q_{\pi(i)}), \tag{1}$$

where $\mathbf{d}(\mathbf{A}p_i, q_{\pi(i)})$ denotes the usual L^2-distance between $\mathbf{A}p_i$ and $q_{\pi(i)}$. The venerable iterative closest point (ICP) algorithm [6][7] can be easily generalized to handle high-dimensional point sets, and it gives an algorithm that iteratively solves for correspondences and affine transformation. However, the main challenge is to produce good initial correspondences and affine transformation that will guarantee the algorithm's convergence and the quality of the solution. For dimensions two and three, this is already a major problem and the difficulty increases exponentially with dimension. In this paper, we propose an algorithm that can estimate the affine transformation (and hence the correspondences π) directly from the point sets $\mathcal{P}, \mathcal{Q}$. The algorithm is algebraic in nature and does not require any optimization, which is its main strength. Furthermore, it allows for a very precise analysis showing that for generic point sets and in the absence of noise, it will recover the exact affine transformation and the correspondences. For noisy data, the algorithm's output can serve as a good initialization for the affine-ICP algorithm. While the algorithm is indeed quite straightforward, it is to the best of our knowledge that there has not been published algorithm which is similar to ours in its entirety. In this paper, we will provide experimental results that validate the proposed affine registration algorithm and show that both the stereo correspondence problem under motion and image set matching problem can be solved quite satisfactorily using the proposed affine registration algorithm.

2   Affine Registrations and Vision Applications

In this section, we provide the details for formulating the stereo correspondence under motion and image set matching problems as affine registration problems.

2.1   Stereo Correspondences under Motion

For clarity of presentation, we will first work out the simpler case of rigid motions. We assume two stationary affine cameras C_1, C_2 observing an object $\mathcal{O}$ undergoing some (rigid or nonrigid) motion. On each camera, we assume that some robust tracking algorithm is running so that a set $\{X_1, \cdots, X_k\}$ of k points on $\mathcal{O}$ are tracked over T frames separately on both cameras. Let (x_{ij}^t, y_{ij}^t) $1 \leq i \leq 2, 1 \leq j \leq k, 1 \leq t \leq T$ denote the image coordinates of $X_j \in \mathcal{O}$ in the tth frame from camera i. For each camera, the tracker provides the correspondences $(x_{ij}^t, y_{ij}^t) \leftrightarrow (x_{ij}^{t'}, y_{ij}^{t'})$ across different frames t and t'. Our problem is to compute correspondences across two views so that the corresponding points $(x_{1j}^t, y_{1j}^t) \leftrightarrow (x_{2j}^t, y_{2j}^t)$ are the projections of the scene point X_j in the images. We show next that it is possible to compute the correspondences directly using only the high-dimensional geometry of the point sets (x_{ij}^t, y_{ij}^t) without referencing to image features such as intensities.

For each view, we can stack the image coordinates of one tracked point over T frames vertically into a $2T$-dimensional vector:

$$\mathbf{p}_j = (\, x_{1j}^1 \ y_{1j}^1 \ \cdots, x_{1j}^T \ y_{1j}^T \,)^t, \quad \mathbf{q}_j = (\, x_{2j}^1 \ y_{2j}^1 \ \cdots, x_{2j}^T \ y_{2j}^T \,)^t \tag{2}$$

In motion segmentation (e.g., [8]), the main objects of interest are the 4-dimensional subspaces L_p, L_q spanned by these $2T$-dimensional vectors

$$\mathcal{P} = \{\mathbf{p}_1, \cdots, \mathbf{p}_k\}, \qquad \mathcal{Q} = \{\mathbf{q}_1, \cdots, \mathbf{q}_k\},$$

and the goal is to cluster motions by determining the subspaces L_p, L_q given the set of vectors $\mathcal{P} \cup \mathcal{Q}$. Our problem, on the hand, is to determine the correspondences between points in $\mathcal{P}$ and $\mathcal{Q}$. It is straightforward to show that there exists an affine transformation $\mathbf{L} : L_p \rightarrow L_q$ that produces the correct correspondences, i.e., $\mathbf{L}(\mathbf{p}_i) = \mathbf{q}_i$ for all i. To see this, we fix an arbitrary world frame with respect to which we can write down the camera matrices for C_1 and C_2. In addition, we also fix an object coordinates system with orthonormal basis $\{\mathbf{i}, \mathbf{j}, \mathbf{k}\}$ centered at some point $o \in \mathcal{O}$. Since $\mathcal{O}$ is undergoing a rigid motion, we denote by $o_t, \mathbf{i}_t, \mathbf{j}_t, \mathbf{k}_t$, the world coordinates of $o, \mathbf{i}, \mathbf{j}, \mathbf{k}$ at frame t. The point X_j, at frame t, with respect to the fixed world frame is given by

$$X_j^t = o_t + \alpha_j \mathbf{i}_t + \beta_j \mathbf{j}_t + \gamma_j \mathbf{k}_t, \tag{3}$$

for some real coefficients $\alpha_j, \beta_j, \gamma_j$ that are independent of time t. The corresponding image point is then given as

$$(\, x_{ij}^t, y_{ij}^t \,)^t = \tilde{o}_{it} + \alpha_j \tilde{\mathbf{i}}_{it} + \beta_j \tilde{\mathbf{j}}_{it} + \beta_j \tilde{\mathbf{k}}_{it},$$

where $\tilde{o}_{it}, \tilde{\mathbf{i}}_{it}, \tilde{\mathbf{j}}_{it}, \tilde{\mathbf{k}}_{it}$ are the projections of the vectors $o_t, \mathbf{i}_t, \mathbf{j}_t, \mathbf{k}_t$ onto camera i. In particular, if we define the $2T$-dimensional vectors $\mathbf{O}_i, \mathbf{I}_i, \mathbf{J}_i, \mathbf{K}_i$ by stacking the vectors $\tilde{o}_{it}, \tilde{\mathbf{i}}_{it}, \tilde{\mathbf{j}}_{it}, \tilde{\mathbf{k}}_{it}$ vertically as before, we have immediately,

$$\mathbf{p}_j = \mathbf{O}_1 + \alpha_j \mathbf{I}_1 + \beta_j \mathbf{J}_1 + \gamma_j \mathbf{K}_1, \quad \mathbf{q}_j = \mathbf{O}_2 + \alpha_j \mathbf{I}_2 + \beta_j \mathbf{J}_2 + \gamma_j \mathbf{K}_2. \tag{4}$$

The two linear subspaces L_p, L_q are spanned by the basis vectors $\{\mathbf{O}_1, \mathbf{I}_1, \mathbf{J}_1, \mathbf{K}_1\}$, $\{\mathbf{O}_2, \mathbf{I}_2, \mathbf{J}_2, \mathbf{K}_2\}$, respectively. The linear map that produces the correct correspondences is given by the linear map $\mathbf{L}$ such that $\mathbf{L}(\mathbf{O}_1) = \mathbf{O}_2, \mathbf{L}(\mathbf{I}_1) = \mathbf{I}_2, \mathbf{L}(\mathbf{J}_1) = \mathbf{J}_2$ and $\mathbf{L}(\mathbf{K}_1) = \mathbf{K}_2$. A further reduction is possible by noticing that the vectors $\mathbf{p}_j, \mathbf{q}_j$ belong to two three-dimensional affine linear subspaces L_p', L_q' in $\mathbb{R}^{2T}$, affine subspaces that pass through the points $\mathbf{O}_1, \mathbf{O}_2$ with bases $\{\mathbf{I}_1, \mathbf{J}_1, \mathbf{K}_1\}$ and $\{\mathbf{I}_2, \mathbf{J}_2, \mathbf{K}_2\}$, respectively. These two subspaces can be obtained by computing the principle components for the collections of vectors $\mathcal{P}, \mathcal{Q}$. By projecting points in $\mathcal{P}, \mathcal{Q}$ onto L_p', L_q', respectively, it is clear that the two sets of projected points are now related by an affine map $\mathbf{A} : L_p' \rightarrow L_q'$. In other words, the correspondence problem can now be solved by solving the equivalent affine registration problem for these two sets of projected points (in $\mathbb{R}^3$).

Non-Rigid Motions. The above discussion generalizes immediately to the types of non-rigid motions that can be modelled (or approximated) using linear shape basis [2,5,9]. In this model, for k feature points, a shape basis element $\mathbf{B}_l$ is a $3 \times k$ matrix. For a model that employs m linear shape basis elements, the 3D world coordinates of the k feature points at t^{th} frame can be written as a linear combination of these shape basis elements:

$$\left[X_1^t \cdots X_k^t\right] = \sum_{l=1}^{m} a_l^t \mathbf{B}_l, \tag{5}$$

for some real numbers a_l^t. Using affine camera model, the imaged points (disregarding the global translation) are given by the following equation [9]

$$\left[\mathbf{x}_1^t \cdots \mathbf{x}_k^t\right] = (\mathbf{a} \otimes P)\mathbf{B}, \tag{6}$$

where $\mathbf{a}^t = (a_1^t, \cdots, a_m^t)$, P is the first 2×3 block of the affine camera matrix and $\mathbf{B}$ is the $3m \times k$ matrix formed by vertically stacking the shape basis matrices $\mathbf{B}_l$. The right factor in the above factorization is independent of the camera (and the images), and we have the following equations similar to Equations 4:

$$\mathbf{p}_j = \mathbf{O}_1 + \sum_{l=1}^{m}(\alpha_{jl}\mathbf{I}_{1l} + \beta_{jl}\mathbf{J}_{1l} + \gamma_{jl}\mathbf{K}_{1l}), \quad \mathbf{q}_j = \mathbf{O}_2 + \sum_{l=1}^{m}(\alpha_{jl}\mathbf{I}_{2l} + \beta_{jl}\mathbf{J}_{2l} + \gamma_{jl}\mathbf{K}_{2l}), \tag{7}$$

where $\mathbf{I}_{il}, \mathbf{J}_{il}\mathbf{K}_{il}$ are the projections of the three basis vectors in the lth shape basis element $\mathbf{B}_l$ onto camera i. The numbers α_{jl}, β_{jl} and γ_{jl} are in fact entries in the matrix $\mathbf{B}_l$. These two equations then imply, using the same argument as before, that we can recover the correspondences directly using a $3m$-dimensional affine registration provided that the vectors $\mathbf{O}_i, \mathbf{I}_{il}, \mathbf{J}_{il}, \mathbf{K}_{il}$ are linearly independent for each i, which is typically the case when the number of frames is sufficiently large.

2.2 Image Set Matching

In the image set matching problem, we are given two sets of images $\mathcal{P} = \{I_1, \cdots, I_k\} \subset \mathbb{R}^m$, $\mathcal{Q} = \{I_1', \cdots, I_k'\} \subset \mathbb{R}^{m'}$ and the corresponding pairs of images I_i, I_i' are related by a linear transformation $\mathcal{F} : \mathbb{R}^m \rightarrow \mathbb{R}^{m'}$ between two high-dimensional image spaces:

$$I_k' \approx \mathcal{F}(I_k).$$

Examples of such sets of images are quite easy to come by, and Figure 1 gives an example in which I_i' is obtained by rotating and downsizing I_i. It is easy to see that many standard image processing operations such as image rotation and down-sampling can be modelled as (or approximated by) a linear map $\mathcal{F}$ between two image spaces. The problem here is to recover the correspondences $I_i \leftrightarrow I_i'$ without actually computing the linear transformation $\mathcal{F}$, which will be prohibitively expensive since the dimensions of the image spaces are usually very high.

Many interesting sets of images can in fact be approximated well by low-dimensional linear subspaces in the image space. Typically, such linear subspaces can be computed readily using principal component analysis (PCA). Let L_p, L_q denote two such low-dimensional linear subspaces approximating $\mathcal{P}, \mathcal{Q}$, respectively and we will use the same notations $\mathcal{P}, \mathcal{Q}$ to denote their projections onto the subspace L_p, L_q. A natural question to ask is how are the (projected) point sets $\mathcal{P}, \mathcal{Q}$ related? Suppose that $\mathcal{F}$ is orthogonal and L_p, L_q are the principle subspaces of the same dimension. If the data is "noiseless", i.e., $I'_k = \mathcal{F}(I_k)$, it is easy to show that $\mathcal{P}, \mathcal{Q}$ are then related by an orthogonal transformation. In general, $\mathcal{F}$ may not be orthogonal and data points are noisy, the point sets $\mathcal{P}, \mathcal{Q}$ are related by a transformation $\mathbf{T} = \mathbf{A} + \mathbf{r}$, which is a sum of an affine transformation $\mathbf{A}$ and a nonrigid transformation $\mathbf{r}$. If the nonrigid part is small, we can recover the correspondences by affine registering the two point sets $\mathcal{P}, \mathcal{Q}$. Note that this gives an algorithm for computing the correspondences without explicitly using the image contents, i.e., there is no feature extraction. Instead, it works directly with the geometry of the point sets.

3   Affine Registrations in $\mathbb{R}^m$

The above discussion provides the motivation for studying affine registration in $\mathbb{R}^m$ for $m > 3$. Let $\mathcal{P} = \{p_1, \cdots, p_k\}$ and $\mathcal{Q} = \{q_1, \cdots, q_k\}$ be two point sets in $\mathbb{R}^m$ related by an unknown affine transformation

$$q_{\pi(i)} = \mathbf{A}p_i + \mathbf{t}, \tag{8}$$

where $\mathbf{A} \in \mathbf{GL}(m), \mathbf{t} \in \mathbb{R}^m$ the translational component of the affine transformation and $\pi : \mathcal{P} \to \mathcal{Q}$, the unknown correspondence to be recovered. We assume that the point sets $\mathcal{P}, \mathcal{Q}$ have same number of points and π is a bijective correspondence.

Iterative closest point (ICP) algorithm is a very general point registration algorithm that generalizes easily to higher dimensions. Several papers have been published recently [10,11,12,13,14] on ICP-related point registration algorithms in $\mathbb{R}^2$ and $\mathbb{R}^3$. While these works concern exclusively with rigid transformations, it is straightforward to incorporate affine transformation into ICP algorithm, which iterative solves for correspondences and affine transformation[2]. Given an assignment (correspondences) $\pi : \{1, \cdots, k\} \to \{1, \cdots, k\}$ the optimal affine transformation $\mathbf{A}$ in the least squares sense can be solved by minimizing

$$\mathcal{E}(\mathbf{A}, \mathbf{t}, \pi) = \sum_{i=1}^{k} \mathbf{d}^2(\mathbf{A}p_i + \mathbf{t}, q_{\pi(i)}). \tag{9}$$

Solving $\mathbf{A}, \mathbf{t}$ separately while holding π fixed, the above registration error function gives a quadratic programming problem in the entries of $\mathbf{A}$, and the optimal solution can be computed readily by solving a linear system. With a fixed $\mathbf{A}$, $\mathbf{t}$ can be solved immediately. On the hand, given an affine transformation, a new assignment π can be defined using closest points:

$$\pi(i) = \arg \min_{1 \leq j \leq k} \mathbf{d}^2(\mathbf{A}p_i + \mathbf{t}, q_j).$$

---

[2] We will call this algorithm affine-ICP.

Once an initial affine transformation and assignment is given, affine-ICP is easy to implement and very efficient. However, the main difficulty is the initialization, which can significantly affect the algorithm's performance. With a poor initialization, the algorithm almost always converges to an undesirable local minimum and as the group of affine transformations is noncompact, it is also possible that it diverges to infinity, i.e., the linear part of the affine transformation converges to a singular matrix. One way to generate an initial affine transformation (disregarding $\mathbf{t}$) is to randomly pick m pairs of points from $\mathcal{P}, \mathcal{Q}, \{(x_1, y_1), \cdots, (x_m, y_m)\}$, $x_i \in \mathcal{P}, y_i \in \mathcal{Q}$ and define $\mathbf{A}$ as $y_i = \mathbf{A}(x_i)$. It is easy to see that the probability of picking a good set of pairs that will yield good initialization is roughly in the order of $1/C(k, m)$. For small dimensions $m = 2, 3$ and medium-size point sets (k in the order of hundreds), it is possible to exhaustively sample all these initial affine transformations. However, as $C(k, m)$ depends exponentially on the dimension m, this approach becomes impractical once $m > 3$. Therefore, for affine-ICP approach to work, we need a novel way to generate good initial affine transformation and correspondences.

Our solution starts with a novel affine registration algorithm. The outline of the algorithm is straightforward: we first reduce the problem to orthogonal case and spectral information is then used to narrow down the correct orthogonal transformation. This algorithm does not require continuous optimization (e.g., solving linear systems) and we can show that for generic point sets without noise, it will recover the exact affine transformation. This latter property suggests that for noisy point sets, the affine transformation estimated by the proposed algorithm should not be far from the optimal one. Therefore, the output of our proposed algorithm can be used as the initial affine transformation for the affine-ICP algorithm.

3.1   Affine Registration Algorithm

Let $\mathcal{P}, \mathcal{Q}$ be two point sets as above related by an unknown affine transformation as in Equation 8. By centering the point sets with respect to their respective centers of mass $\mathbf{m}_p, \mathbf{m}_q$,

$$\mathbf{m}_p = \frac{1}{k} \sum_{i=1}^{k} p_i, \qquad \mathbf{m}_q = \frac{1}{k} \sum_{i=1}^{k} q_i,$$

the centered point sets $\mathcal{P}^c = \{p_1 - \mathbf{m}_p, \cdots, p_k - \mathbf{m}_p\}$ and $\mathcal{Q}^c = \{q_1 - \mathbf{m}_q, \cdots, q_k - \mathbf{m}_q\}$ are related by the same $\mathbf{A}$: $q_{\pi(i)} - \mathbf{m}_q = \mathbf{A}(p_i - \mathbf{m}_p)$. That is, we can work with centered point sets $\mathcal{P}^c$ and $\mathcal{Q}^c$. Once $\mathbf{A}$ and π have been recovered from the point sets $\mathcal{P}^c$ and $\mathcal{Q}^c$, the translational component $\mathbf{t}$ can be estimated easily. In the absence of noise, determining the matrix $\mathbf{A}$ is in fact a combinatorial search problem. We can select m linearly independent points $\{p_{i_1}, \cdots, p_{i_m}\}$ from $\mathcal{P}$. For every ordered m points $\omega = \{q_{i_1}, \cdots, q_{i_m}\}$ in $\mathcal{Q}$, there is a (nonsingular) matrix $\mathbf{B}_\omega$ sending p_{i_j} to q_{i_j} for $1 \leq j \leq m$. The desired matrix $\mathbf{A}$ is among the set of such matrices, which numbers roughly k^m (k is the number of points). For generic point sets, this exponential dependence on dimension can be avoided if $\mathbf{A}$ is assumed to be orthogonal. Therefor, we will first use the covariance matrices computed from $\mathcal{P}$ and $\mathcal{Q}$ to reduce the problem to the 'orthogonal case'. Once the problem has been so reduced, there are various ways to finish off the problem by exploiting invariants of the orthogonal matrices, namely, distances. Let $\mathbf{S_P}$ and $\mathbf{S_Q}$ denote the covariance matrices for $\mathcal{P}$ and $\mathcal{Q}$, respectively:

$$\mathbf{S_P} = \sum_{i=1}^{k} p_i p_i^t, \qquad \mathbf{S_Q} = \sum_{i=1}^{k} q_i q_i^t.$$

We make simple coordinates changes using their inverse square-roots:

$$p_i \rightarrow \mathbf{S}_{\mathcal{P}}^{-\frac{1}{2}} p_i, \qquad q_i \rightarrow \mathbf{S}_{\mathcal{Q}}^{-\frac{1}{2}} q_i. \tag{10}$$

We will use the same notations to denote the transformed points and point sets. If the original point sets are related by $\mathbf{A}$, the transformed point sets are then related by $\bar{\mathbf{A}} = \mathbf{S}_{\mathcal{Q}}^{-\frac{1}{2}} \mathbf{A} \mathbf{S}_{\mathcal{P}}^{\frac{1}{2}}$. The matrix $\bar{\mathbf{A}}$ can be easily shown to be orthogonal:

Proposition 1. *Let $\mathcal{P}$ and $\mathcal{Q}$ denote two point sets (of size k) in $\mathbb{R}^m$, and they are related by an unknown linear transformation $\mathbf{A}$. Then, the transformed point sets (using Equation 10) are related by a matrix $\bar{\mathbf{A}}$, whose rows are orthonormal vectors in $\mathbb{R}^m$.*

The proof follows easily from the facts that 1) the covariance matrices $\mathbf{S_P}$ and $\mathbf{S_Q}$ are now identity matrices for the transformed point sets, and 2) $\mathbf{S_Q} = \bar{\mathbf{A}} \mathbf{S}_{\mathcal{P}} \bar{\mathbf{A}}^t$. They together imply that the rows of $\bar{\mathbf{A}}$ must be orthonormal.

3.2   Determining the Orthogonal Transformation $\bar{\mathbf{A}}$

Since the point sets $\mathcal{P}, \mathcal{Q}$ have unit covariance matrices, the invariant approach in [1] cannot be applied to solve for the orthogonal transformation $\bar{\mathbf{A}}$. Nevertheless, there are other invariants that can be useful. For example, if the magnitudes of points in $\mathcal{P}$ are all different, registration becomes particularly easy: each point p_{i_j} is matched to the point q_{i_j} with the same magnitude. Of course, one does not expect to encounter such nice point sets very often. However, for orthogonal matrices, there is a very general way to produce a large number of useful invariants.

Let p_1, p_2 be any two points in $\mathcal{P}$ and q_1, q_2 their corresponding points in $\mathcal{Q}$. Since $\bar{\mathbf{A}}$ is orthogonal, the distance $d(p_1, p_2)$ between p_1 and p_2 equals the distance $d(q_1, q_2)$ between q_1 and q_2. Although we do not know the correspondences between points in $\mathcal{P}$ and $\mathcal{Q}$, the above observation naturally suggests the idea of canonically constructing two symmetric matrices, $L_{\mathcal{P}}$ and $L_{\mathcal{Q}}$, using pairwise distances between points in $\mathcal{P}$ and $\mathcal{Q}$, respectively. The idea is that the matrices so constructed differ only by an unknown permutation of their columns and rows. Their eigenvalues, however, are not effected by such permutations, and indeed, the two matrices $L_{\mathcal{P}}$ and $L_{\mathcal{Q}}$ have the same eigenvalues. Furthermore, there are also correspondences between respective eigenspaces $E_\lambda^{\mathcal{P}}$ and $E_\lambda^{\mathcal{Q}}$ associated with eigenvalue λ. If λ is a non-repeating eigenvalue, we have two associated (unit) eigenvectors $v_{\mathcal{P}}^\lambda$ and $v_{\mathcal{Q}}^\lambda$ of $L_{\mathcal{P}}$ and $L_{\mathcal{Q}}$, respectively. The vector $v_{\mathcal{P}}^\lambda$ differs from $v_{\mathcal{Q}}^\lambda$ by a permutation of its components and a possible multiplicative factor of -1.

There are many ways to construct the matrices $L_{\mathcal{P}}$ and $L_{\mathcal{Q}}$. Let $f(x)$ be any function. We can construct a $k \times k$ symmetric matrix $L_{\mathcal{P}}(f)$ from pairwise distances using the formula

$$L_{\mathcal{P}}(f) = \mathbf{I}_k - \mu \begin{pmatrix} f(d(p_1, p_1)) & \cdots & f(d(p_1, p_k)) \\ \vdots & \cdots & \vdots \\ f(d(p_k, p_1)) & \cdots & f(d(p_k, p_k)) \end{pmatrix}, \qquad (11)$$

where $\mathbf{I}_k$ is the identity matrix and μ some real constant. One common choice of f that we will use here is the Gaussian exponential $f(x) = \exp(-x^2/\sigma^2)$, and the resulting symmetric matrix $L_{\mathcal{P}}$ is related to the well-known (unnormalized) discrete Laplacian associated with the point set $\mathcal{P}$ [15]. Denote $U_p D_p U_p^t = L_{\mathcal{P}}, U_q D_q U_q^t = L_{\mathcal{Q}}$ the eigen-decompositions of $L_{\mathcal{P}}$ and $L_{\mathcal{Q}}$. When the eigenvalues are all distinct, up to sign differences, U_p and U_q differ only by some unknown row permutation if we order the columns according to the eigenvalues. This unknown row permutation is exactly the desired correspondence π. In particular, we can determine m correspondences by matching m rows of U_p and U_q, and from these m correspondences, we can recover the orthogonal transformation $\bar{\mathbf{A}}$. The complexity of this operation is $O(mk^2)$ and we have the following result

Proposition 2. *For a generic pair of point sets* $\mathcal{P}, \mathcal{Q}$ *with equal number of points in* $\mathbb{R}^m$ *related by some orthogonal transformation* $\mathbf{L}$ *and correspondences* π *such that* $q_{\pi(i)} = \mathbf{L}p_i$, *the above method will recover* $\mathbf{L}$ *and* π *exactly for some choice of* σ.

The proof (omitted here) is an application of Sard's theorem and transversality in differential topology [16]. The main idea is to show that for almost all point sets $\mathcal{P}$, the symmetric matrix $L_{\mathcal{P}}$ will not have repeating eigenvalues for some σ. This will guarantee that the row-matching procedure described above will find the m needed correspondences after examining all rows of U_p m times. Since the time complexity for matching one row is $O(k)$, the total time complexity is no worse than $O(mk^2)$.

3.3 Dealing with Noises

The above method breaks down when noise is present. In this case, the sets of eigenvalues for $L_{\mathcal{P}}, L_{\mathcal{Q}}$ are in general different, and the matrices U_p, U_q are no longer expected to differ only by a row permutation. Nevertheless, for small amount of noise, one can expect that the matrices $L_{\mathcal{P}}, L_{\mathcal{Q}}$ are small perturbations of two corresponding matrices for noiseless data. For example, up to a row permutation, U_q is a small perturbation of U_p. For each eigenvalue λ^p of $L_{\mathcal{P}}$, there should be an eigenvalue λ^q of $L_{\mathcal{Q}}$ such that the difference $|\lambda_p - \lambda_q|$ is small, and this will allow us to establish correspondences between eigenvalues of $L_{\mathcal{P}}, L_{\mathcal{Q}}$. The key idea is to define a reliable matching measure $\mathbf{M}$ using eigenvectors of $L_{\mathcal{P}}, L_{\mathcal{Q}}$, e.g., if p, q are two corresponding points, $\mathbf{M}(p, q)$ will tend to be small. Otherwise, it is expected to be large. Once a matching measure $\mathbf{M}$ is defined, it will allow us to establish tentative correspondences $p_i \longleftrightarrow q_j$: $q_j = \arg\min_{i'} \mathbf{M}(p_i, q_{i'})$. Similar to the homography estimation in structure from motion [2], some of the tentative correspondences so established are incorrect while a good portion of them are expected to be correct. This will allow us to apply RANSAC [17] to determine the orthogonal transformation: generate a small number of hypotheses (orthogonal matrices from sets of randomly generated m correspondences) and pick the one that gives the smallest registration error. We remark that in our approach, the tentative correspondences are computed from the geometry of the point sets $\mathcal{P}, \mathcal{Q}$ embedded in $\mathbb{R}^m$. In stereo matching and homography estimation [2], they are computed using image features such as image gradients and intensity values.

More precisely, let $\lambda_1^p < \lambda_2^p < \cdots < \lambda_l^p$ ($l \leq k$) be l non-repeating eigenvalues of $L_{\mathcal{P}}$ and likewise, $\lambda_1^q < \lambda_2^q < \cdots < \lambda_l^q$ the l eigenvalues of $L_{\mathcal{Q}}$ such that $|\lambda_i^q - \lambda_i^p| < \epsilon$ for some threshold value ϵ. Let $v_{\lambda_1}^{\mathcal{P}}, v_{\lambda_2}^{\mathcal{P}}, \cdots, v_{\lambda_l}^{\mathcal{P}}$, and $v_{\lambda_1}^{\mathcal{Q}}, v_{\lambda_2}^{\mathcal{Q}}, \cdots, v_{\lambda_l}^{\mathcal{Q}}$ denote the corresponding eigenvectors. We stack these eigenvectors horizontally to form two $k \times l$ matrices VP and VQ:

$$\text{VP} = [v_{\lambda_1}^{\mathcal{P}} \; v_{\lambda_2}^{\mathcal{P}} \; \cdots \; v_{\lambda_l}^{\mathcal{P}}], \qquad \text{VQ} = [v_{\lambda_1}^{\mathcal{Q}} \; v_{\lambda_2}^{\mathcal{Q}} \; \cdots \; v_{\lambda_l}^{\mathcal{Q}}]. \tag{12}$$

Denote the i, j-entry of VP (and also VQ) by $\text{VP}(i, j)$. We define the matching measure **M** as

$$\mathbf{M}(p_i, q_j) = \sum_{h=1}^{l} \min\{ (\text{VP}(i, h) - \text{VQ}(j, h))^2, (\text{VP}(i, h) + \text{VQ}(j, h))^2 \}.$$

Note that if $l = k$, **M** is comparing the ith row of $L_{\mathcal{P}}$ with j^{th} row of $L_{\mathcal{Q}}$. For efficiency, one does not want to compare the entire row; instead, only a small fragment of it. This would require us to use those eigenvectors that are most discriminating for picking the right correspondences. For discrete Laplacian, eigenvectors associated with smaller eigenvalues can be considered as smooth functions on the point sets, while those associated with larger eigenvalues are the non-smooth ones since they usually exhibit greater oscillations. Typically, the latter eigenvectors provide more reliable matching measures than the former ones and in many cases, using one or two such eigenvectors ($l = 2$) is already sufficient to produce good results.

Table 1. Experimental Results I. For each dimension and each noise setting, one hundred trials, each with different point sets and matrix **A**, were performed. The averaged relative error and percentage of mismatched points as well as standard deviations (in parenthesis) are shown.

Dim →	3	3	5	5	10	10
Noise ↓	Matrix Error	Matching Error	Matrix Error	Matching Error	Matrix Error	Matching Error
0%	0 (0)	0 (0)	0 (0)	0 (0)	0 (0)	0 (0)
1%	0.001 (0.0005)	0 (0)	0.002 (0.0006)	0 (0)	0.004 (0.0008)	0 (0)
2%	0.003 (0.001)	0 (0)	0.004 (0.001)	0 (0)	0.008 (0.001)	0 (0)
5%	0.008 (0.003)	0 (0)	0.01 (0.003)	0 (0)	0.02 (0.003)	0 (0)
10%	0.017 (0.01)	0.008 (0.009)	0.05 (0.05)	0.009 (0.04)	0.04 (0.009)	0 (0)

4 Experiments

In this section, we report four sets of experimental results. First, with synthetic point sets, we show that the proposed affine registration algorithm does indeed recover exact affine transformations and correspondences for noiseless data. Second, we show that the proposed algorithm also works well for 2D point sets. Third, we provide two sequences of nonrigid motions and show that the feature point correspondences can be

Table 2. Experimental Results II. Experiments with point sets of different sizes with 5% noise added. All trials match point sets in $\mathbb{R}^{10}$ with settings similar to Table 1. Average errors for one hundred trials are reported with standard deviations in parenthesis.

# of Pts $\rightarrow$ Errors $\downarrow$	100 Points	150 Points	200 Points	250 Points	300 Points	400 Points
Matrix Error	0.02 (0.003)	0.05(0.008)	0.05 (0.009)	0.05 (0.01)	0.05 (0.01)	0.04 (0.009)
Matching Error	0 (0)	0 (0)	0 (0)	0 (0)	0 (0)	0 (0)

satisfactorily solved using affine registration in $\mathbb{R}^9$. And finally, we use images from COIL database to show that the image set matching problem can also be solved using affine registration in $\mathbb{R}^8$. We have implemented the algorithm using MATLAB without any optimization. The sizes of the point sets range from 20 to 432, and on a DELL desktop with single 3.1GHz processor, each experiment does not run longer than one minute.

4.1   Affine Registration in $\mathbb{R}^m$

In this set of experiments, our aim is to give a qualitative as well as quantitative analysis on the accuracy and robustness of the proposed method. We report our experimental results on synthetic data in several different dimensions and using various different noise settings. Tables 1 and 2 summarize the experimental results. In Table 1, the algorithm is tested in three dimensions, 3, 5 and 10, and five different noise settings, $0\%, 1\%, 2\%, 5\%, 10\%$. For each pair of dimension and noise setting, we ran 100 trials, each with a randomly generated non-singular matrix $\mathbf{A}$ and a point set containing 100 points. In trials with $x\%$ noise setting, we add a uniform random noise ($\pm x\%$) to each coordinate of every point independently. Let $\mathbf{A}'$ denote the estimated matrix. A point $p \in \mathcal{P}$ is matched to the point $q \in \mathcal{Q}$ if $q = \min_{q_i \in \mathcal{Q}} \mathbf{dist}(\mathbf{A}'p, q_i)$. For each trial, we report the percentage of mismatched points and the relative error of the estimated matrix $\mathbf{A}'$: $\frac{\|\mathbf{A}'-\mathbf{A}\|}{\|\mathbf{A}\|}$, using the Frobenius norm.

The number of (RANSAC) samples drawn in each trial has been fixed at 800 for the results reported in Table 1. This is the number of samples needed to produce zero mismatch for dimension 10 with 10% noise setting. In general, for lower dimensions, a much smaller number of samples (around 200) would also have produced similar results. In Table 2, we vary the sizes of the point sets and work in $\mathbb{R}^{10}$. The setting is similar to that of Table 1 except with fixed 5% noise setting for all trials. The results clearly show that the proposed algorithm consistently performs well with respect to the sizes of the point sets. Note also that for noiseless point sets the exact affine transformations are always recovered.

4.2   2D Point Sets

In the second set of experiments, we apply the proposed algorithm to 2D image registration. It is known that the effect of a small view change on an image can be approximated by a 2D affine transformation of the image [2]. Using images from COIL database, we manually click feature points on pairs of images with $15°$ to $30°$ difference in view

Fig. 2. 2D Image Registration. **1st column**: Source images (taken from COIL database) with feature points marked in red. **2nd and 4th column**: Target images with feature points marked in blue. **3rd and 5th column**: Target images with corresponding feature points marked in blue. The affine transformed points from the source images are marked in red. Images are taken with $15°$ and $30°$ differences in viewpoint. The RMS errors for these four experiments (from left to right) 2.6646, 3.0260, 2.0632, 0.7060, respectively.

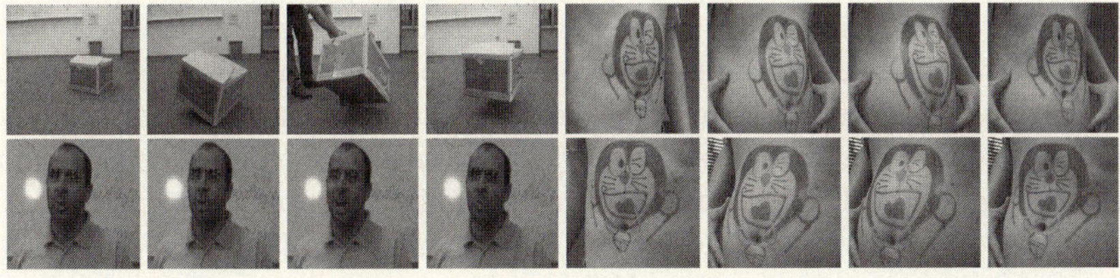

Fig. 3. Top: Sample frames from two video sequences of two objects undergoing nonrigid motions. **Bottom:** Sample frames from another camera observing the same motions.

point. The registration results for four pairs of images are shown in Figure 2. Notice the small RMS registration errors for all these results given that the image size is 128×128.

4.3 Stereo Correspondences under Nonrigid Motions

In this experiment, we apply affine registration algorithm to compute correspondences between tracked feature points in two image sequences. We gathered four video sequences from two cameras observing two objects undergoing nonrigid motions (Figure 3). One is a talking head and the other is a patterned tatoo on a man's belly. A simple correlation-based feature point tracker is used to track twenty and sixty points for these two sequences, respectively. Seventy frames were tracked in both sequences and manual intervention was required several times in both sequences to correct and adjust tracking results. We use three shape basis for both sequences [5], and to compute the correspondences, we affine register two point sets in $\mathbb{R}^9$ as discussed before. For the two point sets $\mathcal{P}, \mathcal{Q} \subset \mathbb{R}^9$, we applied the proposed algorithm to obtain initial correspondences and affine transformation. This is followed by running an affine-ICP algorithm with fifty iterations. For comparison, the affine-ICP algorithm initialized using closest points[3] is run for one hundred iterations. For the talking sequence, the

---

[3] Given two point sets in $\mathbb{R}^9$, the initial correspondence $p_i \leftrightarrow q_j$ is computed by taking q_j to be the point in $\mathcal{Q}$ closest to p_i.

proposed algorithm recovers all the correspondences correctly, while for the tatoo sequence, among the recovered sixty feature point correspondences, nine are incorrect. This can be explained by the fact that in several frames, some of the tracked feature points are occluded and missing and the subsequent factorizations produce relatively noisy point sets in $\mathbb{R}^9$. On the other hand, affine-ICP with closest point initialization fails poorly for both sequences. In particular, more than three quarters of the estimated correspondences are incorrect.

4.4   Image Set Matching

In this experiment, images from the first six objects in the COIL database are used. They define the image set $\mathcal{A}$ with 432 images. Two new sets $\mathcal{B}, \mathcal{C}$ of images are generated from $\mathcal{A}$: the images are 80% down-sampled and followed by 45° and 90° rotations, respectively. The original images have size 128×128 and the images in the two new sets have size 100×100. An eight-dimensional PCA subspace is used to fit each set of images with relative residue smaller than 1%. Images in each set are projected down to their respective PCA subspaces and the correspondences are automatically computed by affine registering the projected point sets. The two experiments shown in Figure 4 match point sets $\mathcal{A}, \mathcal{B}$ and $\mathcal{A}, \mathcal{C}$. We apply the proposed affine registration algorithm to obtain an initial estimate on correspondences and affine transformation. Since the data is noisy, we follow this with the affine-ICP algorithm running fifty iterations as above. For comparison, we apply the affine-ICP algorithm using closest points as initialization. In both experiments, the affine-ICP algorithm, not surprisingly, performs poorly with substantial L^2-registration errors (Equation 9) and large number of incorrect correspondences. The proposed algorithm recovers all correspondences correctly and it yields small L^2-registration errors.

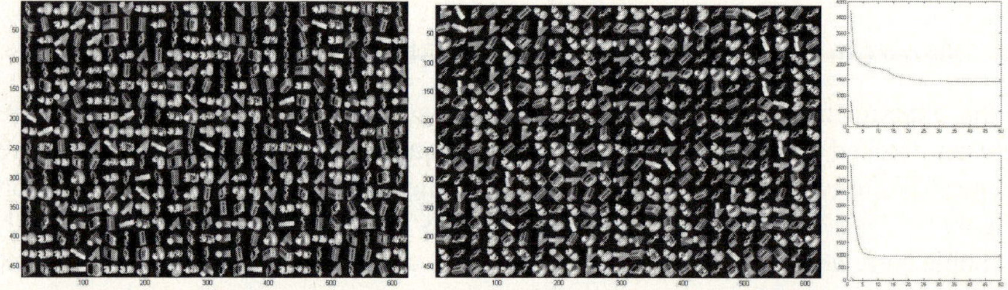

Fig. 4. Image Set Matching. The original image set $\mathcal{A}$ is shown in Figure 1. Image sets $\mathcal{B}, \mathcal{C}$ are shown above. The plots on the right show the L^2-registration error for each of the fifty iterations of running affine-ICP algorithm using different initializations. Using the output of the proposed affine registration as the initial guess, the affine-ICP algorithm converges quickly to the desired transformation (blue curves) and yields correct correspondences. Using closest points for initial correspondences, the affine-ICP algorithm converges (red curves) to incorrect solutions in both experiments.

5   Conclusion and Future Work

In this paper, we have shown that the stereo correspondence problem under motion and image set matching problem can be solved using affine registration in $\mathbb{R}^m$ with $m > 3$. We have also proposed an algorithm for estimating an affine transformation directly from two point sets without using continuous optimization. In the absence of noise, it will recover the exact affine transformation for generic pairs of point sets in $\mathbb{R}^m$. For noisy data, the output of the proposed algorithm often provides good initializations for the affine-ICP algorithm. Together, they provide us with an efficient and effective algorithm for affine registering point sets in $\mathbb{R}^m$ with $m > 3$. We have applied the proposed algorithm to the two aforementioned problems. Preliminary experimental results are encouraging and they show that these two problems can indeed be solved satisfactorily using the proposed affine registration algorithm.

References

1. Scott, G., Lonquiet-Higgins, C.: An algorithm for associating the features of two images. Proc. of Royal Society of London B244, 21–26 (1991)
2. Hartley, R., Zisserman, A.: Multiple View Geometry in Computer Vision. Cambridge University Press, Cambridge (2003)
3. Chui, H., Rangarajan, A.: A new algorithm for non-rigid point matching. In: Proc. IEEE Conf. on Comp. Vision and Patt. Recog, vol. 2, pp. 44–51 (2000)
4. Toamsi, C., Kanade, T.: Shape and motion from image streams under orthography—a factorization method. Int. J. Computer Vision 9(2), 137–154 (1992)
5. Bregler, C., Hertzmann, A., Biermann, H.: Recovering non-rigid 3d shape from image streams. In: Proc. IEEE Conf. on Comp. Vision and Patt. Recog., pp. 2690–2696 (2000)
6. Besel, P.J., Mckay, H.D.: A method for registration of 3-d shapes. PAMI 14, 239–256 (1992)
7. Zhang, Z.: Iterative point matching for registration of free-form curves and surfaces. Int. J. Computer Vision 13, 119–152 (1994)
8. Kanatani, K.: Motion segmentation by subspace separation and model selection. In: Proc. Int. Conf. on Computer Vision, vol. 2, pp. 586–591 (2001)
9. Brand, M.: Morphable 3d models from video. In: Proc. IEEE Conf. on Comp. Vision and Patt. Recog., vol. 2, pp. 456–463 (2001)
10. Fitzgibbon, A.W.: Robust registration of 2d and 3d point sets. Computer Vision and Image Understanding 2, 1145–1153 (2003)
11. Sharp, G.C., Lee, S.W., Wehe, D.K.: Icp registration using invariant features. IEEE Transactions on Pattern Analysis and Machine Intelligence 24, 90–102 (2002)
12. Rusinkiewicz, S., Levoy, M.: Efficient variants of the icp algorithm. In: Proc. Third International Conference on 3D Digital Imaging and Modeling (3DIM), pp. 145–152 (2001)
13. Granger, S., Pennec, X.: Multi-scale em-icp: A fast and robust approach for surface registration. In: Proc. European Conf. on Computer Vision, vol. 3, pp. 418–432 (2002)
14. Makadia, A., Patterson, A.I., Daniilidis, K.: Fully automatic registration of 3d point clouds. In: Proc. IEEE Conf. on Comp. Vision and Patt. Recog., vol. 1, pp. 1297–1304 (2006)
15. Chung, F.R.K.: Spectral Graph Theory. American Mathematical Society (1997)
16. Hirsch, M.: Differential Topology. Springer, Heidelberg (1976)
17. Fischler, M., Bolles, R.: Random sample consensus: A paradigm for model fitting with applications to image analysis and automated cartography. Communications of the ACM 24, 381–395 (1981)

Semantic Concept Classification by Joint Semi-supervised Learning of Feature Subspaces and Support Vector Machines

Wei Jiang[1], Shih-Fu Chang[1], Tony Jebara[1], and Alexander C. Loui[2]

[1] Columbia University, New York, NY 10027, USA
[2] Eastman Kodak Company, Rochester, NY 14650, USA

Abstract. The scarcity of labeled training data relative to the high-dimensionality multi-modal features is one of the major obstacles for semantic concept classification of images and videos. Semi-supervised learning leverages the large amount of unlabeled data in developing effective classifiers. Feature subspace learning finds optimal feature subspaces for representing data and helping classification. In this paper, we present a novel algorithm, Locality Preserving Semi-supervised Support Vector Machines (LPSSVM), to jointly learn an optimal feature subspace as well as a large margin SVM classifier. Over both labeled and unlabeled data, an optimal feature subspace is learned that can maintain the smoothness of local neighborhoods as well as being discriminative for classification. Simultaneously, an SVM classifier is optimized in the learned feature subspace to have large margin. The resulting classifier can be readily used to handle unseen test data. Additionally, we show that the LPSSVM algorithm can be used in a Reproducing Kernel Hilbert Space for nonlinear classification. We extensively evaluate the proposed algorithm over four types of data sets: a toy problem, two UCI data sets, the Caltech 101 data set for image classification, and the challenging Kodak's consumer video data set for semantic concept detection. Promising results are obtained which clearly confirm the effectiveness of the proposed method.

1   Introduction

Consider one of the central issues in semantic concept classification of images and videos: the amount of available unlabeled test data is large and growing, but the amount of labeled training data remains relatively small. Furthermore, the dimensionality of the low-level feature space is generally very high, the desired classifiers are complex and, thus, small sample learning problems emerge.

There are two primary techniques for tackling above issues. *Semi-supervised learning* is a method to incorporate knowledge about unlabeled test data into the training process so that a better classifier can be designed for classifying test data [1], [2], [3], [4], [5]. *Feature subspace learning*, on the other hand, tries to learn a suitable feature subspace for capturing the underlying data manifold over which distinct classes become more separable [6], [7], [8], [9].

D. Forsyth, P. Torr, and A. Zisserman (Eds.): ECCV 2008, Part IV, LNCS 5305, pp. 270–283, 2008.
© Springer-Verlag Berlin Heidelberg 2008

One emerging branch of semi-supervised learning methods is graph-based techniques [2], [4]. Within a graph, the nodes are labeled and unlabeled samples, and weighted edges reflect the feature similarity of sample pairs. Under the assumption of label smoothness on the graph, a discriminative function f is often estimated to satisfy two conditions: the loss condition – it should be close to given labels y_L on the labeled nodes; and the regularization condition – it should be smooth on the whole graph, i.e., close points in the feature space should have similar discriminative functions. Among these graph-based methods, *Laplacian Support Vector Machines (LapSVM)* and *Laplacian Regularized Least Squares (LapRLS)* are considered state-of-the-art for many tasks [10]. They enjoy both high classification accuracy and extensibility to unseen out-of-sample data.

Feature subspace learning has been shown effective for reducing data noise and improving classification accuracy [6], [7], [8], [9]. Finding a good feature subspace can also improve semi-supervised learning performance. As in classification, feature subspaces can be found by supervised methods (e.g., LDA [8]), unsupervised methods (e.g., graph-based manifold embedding algorithms [6], [9]), or semi-supervised methods (e.g., generalizations of graph-based embedding by using the ground-truth labels to help the graph construction process [7]).

In this paper, we address both issues of feature subspace learning and semi-supervised classification. We pursue a new way of feature subspace and classifier learning in the semi-supervised setting. A novel algorithm, *Locality Preserving Semi-supervised SVM (LPSSVM)*, is proposed to jointly learn an optimal feature subspace as well as a large margin SVM classifier in a semi-supervised manner. A joint cost function is optimized to find a smooth and discriminative feature subspace as well as an SVM classifier in the learned feature subspace. Thus, the local neighborhoods relationships of both labeled and unlabeled data can be maintained while the discriminative property of labeled data is exploited. The following highlight some aspects of the proposed algorithm:

1. The target of LPSSVM is both feature subspace learning and semi-supervised classification. A feature subspace is jointly optimized with an SVM classifier so that in the learned feature subspace the labeled data can be better classified with the optimal margin, and the locality property revealed by both labeled and unlabeled data can be preserved.

2. LPSSVM can be readily extended to classify novel unseen test examples. Similar to LapSVM and LapRLS and other out-of-sample extension methods [5], [10], this extends the algorithm's flexibility in real applications, in contrast with many traditional graph-based semi-supervised approaches [4].

3. LPSSVM can be learned in the original feature space or in a Reproducing Kernel Hilbert Space (RKHS). In other words, a kernel-based LPSSVM is formulated which permits the method to handle real applications where nonlinear classification is often needed.

To evaluate the proposed LPSSVM algorithm, extensive experiments are carried out over four different types of data sets: a toy data set, two UCI data sets [11], the Caltech 101 image data set for image classification [12], and the large scale Kodak's consumer video data set [13] from real users for video concept detection. We compare our algorithm with several state of the arts, including the standard SVM [3], semi-supervised LapSVM and LapRLS [10], and the naive approach of first learning a feature subspace (unsupervised) and then solving an SVM (supervised) in the learned feature subspace. Experimental results demonstrate the effectiveness of our LPSSVM algorithm.

2   Related Work

Assume we have a set of data points $X = [\mathbf{x}_1, \ldots, \mathbf{x}_n]$, where $\mathbf{x}_i$ is represented by a d-dimensional feature vector, i.e., $\mathbf{x}_i \in \mathbb{R}^d$. X is partitioned into labeled subset X_L (with n_L data points) and unlabeled subset X_U (with n_U data points), $X = [X_L, X_U]$. y_i is the class label of $\mathbf{x}_i$, e.g., $y_i \in \{-1, +1\}$ for binary classification.

2.1   Supervised SVM Classifier

The SVM classifier [3] has been a popular approach to learn a classifier based on the labeled subset X_L for classifying the unlabeled set X_U and new unseen test samples. The primary goal of an SVM is to find an optimal separating hyperplane that gives a low generalization error while separating the positive and negative training samples. Given a data vector $\mathbf{x}$, SVMs determine the corresponding label by the sign of a linear decision function $f(\mathbf{x}) = \mathbf{w}^T \mathbf{x} + b$. For learning non-linear classification boundaries, a kernel mapping ϕ is introduced to project data vector $\mathbf{x}$ into a high dimensional feature space as $\phi(\mathbf{x})$, and the corresponding class label is given by the sign of $f(\mathbf{x}) = \mathbf{w}^T \phi(\mathbf{x}) + b$. In SVMs, this optimal hyperplane is determined by giving the largest margin of separation between different classes, i.e. by solving the following problem:

$$\min_{\mathbf{w}, b, \epsilon} Q_d = \min_{\mathbf{w}, b, \epsilon} \left\{ \frac{1}{2} \|\mathbf{w}\|_2^2 + C \sum_{i=1}^{n_L} \epsilon_i \right\}, s.t.\ y_i(\mathbf{w}^T \phi(\mathbf{x}_i) + b) \geq 1 - \epsilon_i, \epsilon_i \geq 0, \forall\ \mathbf{x}_i \in X_L\ . \quad (1)$$

where $\epsilon = \epsilon_1, \ldots, \epsilon_{n_L}$ are the slack variables assigned to training samples, and C controls the scale of the empirical error loss the classifier can tolerate.

2.2   Graph Regularization

To exploit the unlabeled data, the idea of graph Laplacian [6] has been shown promising for both subspace learning and classification. We briefly review the ideas and formulations in the next two subsections. Given the set of data points X, a weighted undirected graph $G = (V, E, W)$ can be used to characterize the pairwise similarities among data points, where V is the vertices set and each node v_i corresponds to a data point $\mathbf{x}_i$; E is the set of edges; W is the set of weights measuring the strength of the pairwise similarity.

Regularization for feature subspace learning. In feature subspace learning, the objective of graph Laplacian [6] is to embed original data graph into an m-dimensional Euclidean subspace which preserves the locality property of original data. After embedding, connected points in original G should stay close. Let $\hat{X}$ be the $m{\times}n$ dimensional embedding, $\hat{X}=[\hat{\mathbf{x}}_1,\ldots,\hat{\mathbf{x}}_n]$, the cost function is:

$$\min_{\hat{X}}\left\{\sum_{i,j=1}^{n}||\hat{\mathbf{x}}_i-\hat{\mathbf{x}}_j||_2^2 W_{ij}\right\}, s.t.\hat{X}D\hat{X}^T=I \Rightarrow \min_{\hat{X}}\left\{\mathrm{tr}(\hat{X}L\hat{X}^T)\right\}, s.t.\hat{X}D\hat{X}^T=I. \quad (2)$$

where L is the Laplacian matrix and $L=D-W$, D is the diagonal weight matrix whose entries are defined as $D_{ii}=\sum_j W_{ij}$. The condition $\hat{X}D\hat{X}^T=I$ removes an arbitrary scaling factor in the embedding [6]. The optimal embedding can be obtained as the matrix of eigenvectors corresponding to the lowest eigenvalues of the generalized eigenvalue problem: $L\hat{\mathbf{x}}=\lambda D\hat{\mathbf{x}}$. One major issue of this graph embedding approach is that when a novel unseen sample is added, it is hard to locate the new sample in the embedding graph. To solve this problem, the *Locality Preserving Projection (LPP)* is proposed [9] which tries to find a linear projection matrix $\mathbf{a}$ that maps data points $\mathbf{x}_i$ to $\mathbf{a}^T\mathbf{x}_i$, so that $\mathbf{a}^T\mathbf{x}_i$ can best approximate graph embedding $\hat{\mathbf{x}}_i$. Similar to Eq(2), the cost function of LPP is:

$$\min_{\mathbf{a}} Q_s = \min_{\mathbf{a}}\left\{\mathrm{tr}(\mathbf{a}^T XLX^T\mathbf{a})\right\}, \quad s.t.\ \mathbf{a}^T XDX^T\mathbf{a}=I\ . \quad (3)$$

We can get the optimal projection as the matrix of eigenvectors corresponding to the lowest eigenvalues of generalized eigenvalue problem: $XLX^T\mathbf{a}=\lambda XDX^T\mathbf{a}$.

Regularization for classification. The idea of graph Laplacian has been used in semi-supervised classification, leading to the development of Laplacian SVM and Laplacian RLS [10]. The assumption is that if two points $\mathbf{x}_i,\mathbf{x}_j\in X$ are close to each other in the feature space, then they should have similar discriminative functions $f(\mathbf{x}_i)$ and $f(\mathbf{x}_j)$. Specifically the following cost function is optimized:

$$\min_{f}\frac{1}{n_L}\sum_{i=1}^{n_L} V(\mathbf{x}_i,y_i,f) + \gamma_A||f||_2^2 + \gamma_I\mathbf{f}^T L\mathbf{f}\ . \quad (4)$$

where $V(\mathbf{x}_i,y_i,f)$ is the loss function, e.g., the square loss $V(\mathbf{x}_i,y_i,f)=(y_i-f(\mathbf{x}_i))^2$ for LapRLS and the hinge loss $V(\mathbf{x}_i,y_i,f) = \max(0,1-y_if(\mathbf{x}_i))$ for LapSVM; $\mathbf{f}$ is the vector of discriminative functions over the entire data set X, i.e., $\mathbf{f} = [f(\mathbf{x}_1),\ldots,f(\mathbf{x}_{n_U+n_L})]^T$. Parameters γ_A and γ_I control the relative importance of the complexity of f in the ambient space and the smoothness of f according to the feature manifold, respectively.

2.3   Motivation

In this paper, we pursue a new semi-supervised approach for feature subspace discovery as well as classifier learning. We propose a novel algorithm, *Locality Preserving Semi-supervised SVM (LPSSVM)*, aiming at joint learning of both an optimal feature subspace and a large margin SVM classifier in a semi-supervised manner. Specifically, the graph Laplacian regularization condition in Eq(3) is adopted to maintain the smoothness of the neighborhoods over both labeled and unlabeled data. At the same time, the discriminative constraint in Eq(1) is used to maximize the discriminative property of the learned feature subspace over the labeled data. Finally, through optimizing a joint cost function, the semi-supervised feature subspace learning and semi-supervised classifier learning can work together to generate a smooth and discriminative feature subspace as well as a large-margin SVM classifier.

In comparison, standard SVM does not consider the manifold structure presented in the unlabeled data and thus usually suffers from small sample learning problems. The subspace learning methods (e.g. LPP) lack the benefits of large margin discriminant models. Semi-supervised graph Laplacian approaches, though incorporating information from unlabeled data, do not exploit the advantage of feature subspace discovery. Therefore, the overarching motivation of our approach is to jointly explore the merit of feature subspace discovery and large-margin discrimination. We will show through four sets of experiments such approach indeed outperforms the alternative methods in many classification tasks, such as semantic concept detection in challenging image/video sets.

3   Locality Preserving Semi-supervised SVM

In this section we first introduce the linear version of the proposed LPSSVM technique then show it can be readily extended to a nonlinear kernel version.

3.1   LPSSVM

The smooth regularization term Q_s in Eq(3) and discriminative cost function Q_d in Eq(1) can be combined synergistically to generate the following cost function:

$$\min_{\mathbf{a},\mathbf{w},b,\epsilon} Q = \min_{\mathbf{a},\mathbf{w},b,\epsilon} \{Q_s + \gamma Q_d\} = \min_{\mathbf{a},\mathbf{w},b,\epsilon} \left\{ \mathrm{tr}(\mathbf{a}^T X L X^T \mathbf{a}) + \gamma [\frac{1}{2}||\mathbf{w}||_2^2 + C\sum_{i=1}^{n_L} \epsilon_i] \right\} \tag{5}$$

$$s.t.\ \mathbf{a}^T X D X^T \mathbf{a} = I, \quad y_i(\mathbf{w}^T \mathbf{a}^T \mathbf{x}_i + b) \geq 1 - \epsilon_i, \ \epsilon_i \geq 0, \ \forall\ \mathbf{x}_i \in X_L\ .$$

Through optimizing Eq(5) we can obtain the optimal linear projection $\mathbf{a}$ and classifier $\mathbf{w}, b$ simultaneously. In the following, we develop an iterative algorithm to minimize over $\mathbf{a}$ and $\mathbf{w}, b, \epsilon$ which will monotonically reduce the cost Q by coordinate ascent towards a local minimum. First, using the method of Lagrange multipliers, Eq(5) can be rewritten as the following:

$$\min_{\mathbf{a},\mathbf{w},b,\epsilon} Q = \min_{\mathbf{a},\mathbf{w},b,\epsilon} \max_{\alpha,\mu} \left\{ \mathrm{tr}(\mathbf{a}^T X L X^T \mathbf{a}) + \gamma [\frac{1}{2}||\mathbf{w}||_2^2 - F^T(X_L^T \mathbf{a}\mathbf{w} - B) + M] \right\}, s.t. \mathbf{a}^T X D X^T \mathbf{a} = I.$$

where we have defined quantities: $F = [\alpha_1 y_1, \ldots, \alpha_{n_L} y_{n_L}]^T$, $B = [b, \ldots, b]^T$, $M = C\sum_{i=1}^{n_L} \epsilon_i + \sum_{i=1}^{n_L} \alpha_i(1 - \epsilon_i) - \sum_{i=1}^{n_L} \mu_i \epsilon_i$, and non-negative Lagrange multipliers $\alpha = \alpha_1, \ldots, \alpha_{n_L}$, $\mu = \mu_i, \ldots, \mu_{n_L}$. By differentiating Q with respect to $\mathbf{w}, b, \epsilon_i$ we get:

$$\frac{\partial Q}{\partial \mathbf{w}} = 0 \Rightarrow \mathbf{w} = \sum_{i=1}^{n_L} \alpha_i y_i \mathbf{a}^T \mathbf{x}_i = \mathbf{a}^T X_L F\ . \tag{6}$$

$$\frac{\partial Q}{\partial b} = 0 \Rightarrow \sum_{i=1}^{n_L} \alpha_i y_i = 0, \quad \frac{\partial Q}{\partial \epsilon_i} = 0 \Rightarrow C - \alpha_i - \mu_i = 0\ . \tag{7}$$

Note Eq(6) and Eq(7) are the same as those seen in SVM optimization [3], with the only difference that the data points are now transformed by $\mathbf{a}$ as $\tilde{\mathbf{x}}_i = \mathbf{a}^T \mathbf{x}_i$. That is, given a known $\mathbf{a}$, the optimal $\mathbf{w}$ can be obtained through the standard SVM optimization process. Secondly, by substituting Eq(6) into Eq(5), we get:

$$\min_{\mathbf{a}} Q = \min_{\mathbf{a}} \left\{ \mathrm{tr}(\mathbf{a}^T X L X^T \mathbf{a}) + \frac{\gamma}{2} F^T X_L^T \mathbf{a} \mathbf{a}^T X_L F \right\}, \quad s.t.\ \mathbf{a}^T X D X^T \mathbf{a} = I \ . \quad (8)$$

$$\frac{\partial Q}{\partial \mathbf{a}} = 0 \ \Rightarrow \ (X L X^T + \frac{\gamma}{2} X_L F F^T X_L^T)\mathbf{a} = \lambda X D X^T \mathbf{a} \ . \quad (9)$$

It is easy to see that $X L X^T + \frac{\gamma}{2} X_L F F^T X_L^T$ is positive semi-definite and we can update $\mathbf{a}$ by solving the generalized eigenvalue problem described in Eq(9).

Combining the above two components, we have a two-step interative process to optimize the combined cost function:

Step-1. With the current projection matrix $\mathbf{a}_t$ at the t-th iteration, train an SVM classifier to get $\mathbf{w}_t$ and $\alpha_{1,t}, \ldots, \alpha_{n_L,t}$.

Step-2. With the current $\mathbf{w}_t$ and $\alpha_{1,t}, \ldots, \alpha_{n_L,t}$, update the projection matrix $\mathbf{a}_{t+1}$ by solving the generalized eigenvalue problem in Eq(9).

3.2   Kernel LPSSVM

In this section, we show that the LPSSVM method proposed above can be extended to a nonlinear kernel version. Assume that $\phi(\mathbf{x}_i)$ is the projection function which maps the original data point $\mathbf{x}_i$ into a high-dimension feature space. Similar to the approach used in Kernel PCA [14] or Kernel LPP [9], we pursue the projection matrix $\mathbf{a}$ in the span of existing data points, i.e.,

$$\mathbf{a} = \sum_{i=1}^{n} \phi(\mathbf{x}_i) v_i = \phi(X) \mathbf{v} \ . \quad (10)$$

where $\mathbf{v} = [v_1, \ldots, v_n]^T$. Let K denote the kernel matrix over the entire data set $X = [X_L, X_U]$, where $K_{ij} = \phi(\mathbf{x}_i) \cdot \phi(\mathbf{x}_j)$. K can be written as: $K = \begin{bmatrix} K_L & K_{LU} \\ K_{UL} & K_U \end{bmatrix}$, where K_L and K_U are the kernel matrices over the labeled subset X_L and the unlabeled subset X_U respectively; K_{LU} is the kernel matrix between the labeled data set and the unlabeled data set and K_{UL} is the kernel matrix between the unlabeled data and the labeled data ($K_{LU} = K_{UL}^T$).

In the kernel space, the projection updating equation (i.e., Eq(8)) turns to:

$$\min_{\mathbf{a}} Q = \min_{\mathbf{a}} \left\{ \mathrm{tr}(\mathbf{a}^T \phi(X) L \phi^T(X) \mathbf{a}) + \frac{\gamma}{2} F^T \phi^T(X_L) \mathbf{a} \mathbf{a}^T \phi(X_L) F \right\}, s.t. \mathbf{a}^T \phi(X) D \phi^T(X) \mathbf{a} = I \ .$$

By differentiating Q with respect to $\mathbf{a}$, we can get:

$$\phi(X) L \phi^T(X) \mathbf{a} + \frac{\gamma}{2} \phi(X_L) F F^T \phi^T(X_L) \mathbf{a} = \lambda \phi(X) D \phi^T(X) \mathbf{a}$$

$$\Rightarrow \left(K L K + \frac{\gamma}{2} K^{LU|L} F F^T (K^{LU|L})^T \right) \mathbf{v} = \lambda K D K \mathbf{v} \ . \quad (11)$$

where $K^{LU|L} = [K_L^T, K_{UL}^T]^T$. Eq(11) plays a role similar to Eq(9) that it can be used to update the projection matrix.

Likewise, similar to Eq(6) and Eq(7) for the linear case, we can find the maximum margin solution in the kernel space by solving the dual problem:

$$\tilde{Q}_{svm}^{dual} = \sum_{i=1}^{n_L} \alpha_i - \frac{1}{2} \sum_{i=1}^{n_L} \sum_{j=1}^{n_L} \alpha_i \alpha_j y_i y_j \phi^T(\mathbf{x}_i) \mathbf{a}\mathbf{a}^T \phi(\mathbf{x}_j)$$

$$= \sum_{i=1}^{n_L} \alpha_i - \frac{1}{2} \sum_{i=1}^{n_L} \sum_{j=1}^{n_L} \alpha_i \alpha_j y_i y_j \left[\sum_{g=1}^{n} K_{ig}^{L|LU} v_g \right] \left[\sum_{g=1}^{n} K_{gj}^{LU|L} v_g \right] .$$

where $K^{L|LU} = [K_L, K_{LU}]$. This is the same with the original SVM dual problem [3], except that the kernel matrix is changed from original K to:

$$\hat{K} = \left[K^{L|LU} \mathbf{v} \right] \left[\mathbf{v}^T K^{LU|L} \right] . \tag{12}$$

Combining the above two components, we can obtain the kernel-based two-step optimization process as follows:

Step-1: With the current projection matrix $\mathbf{v}_t$ at iteration t, train an SVM to get $\mathbf{w}_t$ and $\alpha_{1,t}, \ldots, \alpha_{n_L,t}$ with the new kernel described in Eq(12).

Step-2: With the current $\mathbf{w}_t$, $\alpha_{1,t}, \ldots, \alpha_{n_L,t}$, update $\mathbf{v}_{t+1}$ by solving Eq(11).

In the testing stage, given a test example $\mathbf{x}_j$ ($\mathbf{x}_j$ can be an unlabeled training sample, i.e., $\mathbf{x}_j \in X_U$ or $\mathbf{x}_j$ can be an unseen test sample), the SVM classifier gives classification prediction based on the discriminative function:

$$f(\mathbf{x}_j) = \mathbf{w}^T \mathbf{a}^T \phi(\mathbf{x}_j) = \sum_{i=1}^{n_L} \alpha_i y_i \phi(\mathbf{x}_i) \mathbf{a}\mathbf{a}^T \phi(\mathbf{x}_j) = \sum_{i=1}^{n_L} \alpha_i y_i \left[\sum_{g=1}^{n} K_{ig}^{L|LU} v_g \right] \left[\sum_{g=1}^{n} K(\mathbf{x}_g, \mathbf{x}_j) v_g \right]^T .$$

Thus the SVM classification process is also similar to that of standard SVM [3], with the difference that the kernel function between labeled training data and test data is changed from $K^{L|test}$ to: $\hat{K}^{L|test} = \left[K^{L|LU} \mathbf{v} \right] \left[\mathbf{v}^T K^{LU|test} \right]$. $\mathbf{v}$ plays the role of modeling kernel-based projection $\mathbf{a}$ before computing SVM.

3.3   The Algorithm

The LPSSVM algorithm is summarized in Fig. 1. Experiments show usually LPSSVM converges within 2 or 3 iterations. Thus in practice we may set $T = 3$. γ controls the importance of SVM discriminative cost function in feature subspace learning. If $\gamma = 0$, Eq(11) becomes traditional LPP. In experiments we set $\gamma = 1$ to balance two cost components. The dimensionality of the learned feature subspace is determined by controlling the energy ratio of eigenvalues kept in solving the eigenvalue problem of Eq(11). Note that in LPSSVM, the same Gram matrix is used for both graph construction and SVM classification, and later (Sec.4) we will see without extensive tuning of parameters LPSSVM can get good performance. For example, the default parameter setting in LibSVM [15] may be used. This is very important in real applications, especially for large-scale image/video sets. Repeating experiments to tune parameters can be time and resource consuming.

Input: n_L labeled data X_L, and n_U unlabeled data X_U.

1 Choose a kernel function $K(x, y)$, and compute Gram matrix $K_{ij} = K(\mathbf{x}_i, \mathbf{x}_j)$, e.g. RBF kernel $K(\mathbf{x}_i, \mathbf{x}_j) = \exp\{-\theta||\mathbf{x}_i - \mathbf{x}_j||_2^2\}$ or Spatial Pyramid Match Kernel [16].

2 Construct data adjacency graph over entire $X_L \cup X_U$ using kn nearest neighbors. Set edge weights W_{ij} based on the kernel matrix described in step 1.

3 Compute graph Laplacian matrix: $L = D - W$ where D is diagonal, $D_{ii} = \sum_j W_{ij}$.

4 Initialization: train SVM over Gram matrix of labeled X_L, get $\mathbf{w}_0$ and $\alpha_{1,0}, \ldots, \alpha_{n_L,0}$.

5 Iteration: for $t = 1, \ldots, T$
  - Update $\mathbf{v}_t$ by solving problem in Eq(11) with $\mathbf{w}_{t-1}$ and $\alpha_{1,t-1}, \ldots, \alpha_{n_L,t-1}$.
  - Calculate new kernel by Eq(12) using $\mathbf{v}_t$. Train SVM to get $\mathbf{w}_t$, $\alpha_{1,t}, \ldots, \alpha_{n_L,t}$.
  - Stop iteration if $\sum_{i=1}^{n_L}(\alpha_{i,t-1} - \alpha_{i,t})^2 < \tau$.

Fig. 1. The LPSSVM algorithm

In terms of speed, LPSSVM is very fast in the testing stage, with complexity similar to that of standard SVM classification. In training stage, both steps of LPSSVM are fast. The generalized eigenvalue problem in Eq(11) has a time complexity of $O(n^3)$ ($n = n_L + n_U$). It can be further reduced by exploiting the sparse implementation of [17]. For step 1, the standard quadratic programming optimization for SVM is $O(n_L^3)$, which can be further reduced to linear complexity (about $O(n_L)$) by using efficient solvers like [18].

4   Experiments

We conduct experiments over 4 data sets: a toy set, two UCI sets [11], Caltech 101 for image classification [12], and Kodak's consumer video set for concept detection [13]. We compare with some state-of-the-arts, including supervised SVM [3], semi-supervised LapSVM and LapRLS [10]. We also compare with a naive LPP+SVM: first apply kernel-based LPP to get projection and then learn SVM in projected space. For fair comparison, all SVMs in different algorithms use RBF kernels for classifying UCI data, Kodak's consumer videos, and toy data, and use the Spatial Pyramid Match (SPM) kernel [16] for classifying Caltech 101 (see Sec.4.3 for details). This is motivated by the promising performance in classifying Caltech 101 in [16] by using SPM kernels. In LPSSVM, $\gamma = 1$ in Eq(5) to balance the consideration on discrimination and smoothness, and $\theta = 1/d$ in RBF kernel where d is feature dimension. This follows the suggestion of the popular toolkit LibSVM [15]. For all algorithms, the error control parameter $C = 1$ for SVM. This parameter setting is found robust for many real applications [15]. Other parameters: γ_A, γ_I in LapSVM, LapRLS [10] and kn for graph construction, are determined through cross validation. LibSVM [15] is used for SVM, and source codes from [17] is used for LPP.

4.1   Performance over Toy Data

We construct a "three suns" toy problem in Fig. 2. The data points with each same color (red, blue or cyan) come from one category, and we want to separate

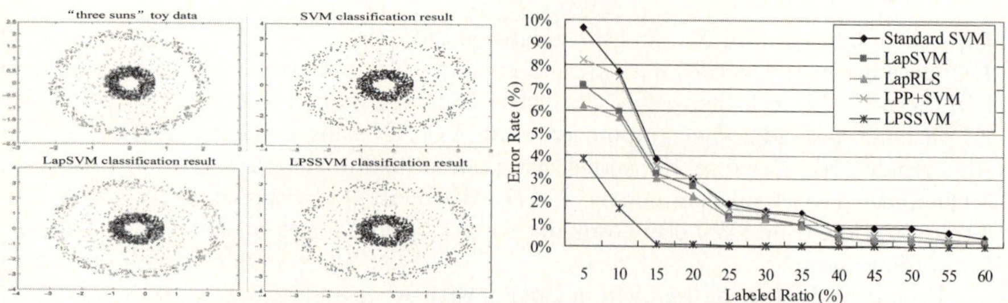

Fig. 2. Performance over toy data. Compared with others, LPSSVM effectively discriminates 3 categories. Above results are generated by using the SVM Gram matrix directly for constructing Laplacian graph. With more deliberate tuning of the Laplacian graph, LapSVM, LapRLS, and LPSSVM can give better results. Note that the ability of LPSSVM to maintain good performance without graph tuning is important.

the three categories. This data set is hard since data points around the class boundaries from different categories (red and cyan, and blue and cyan) are close to each other. This adds great difficulty to manifold learning. The one-vs.-all classifier is used to classify each category from others, and each test data is assigned the label of the classifier with the highest classification score. Fig. 2 gives an example of the classification results using different methods with 10% samples from each category as labeled data (17 labeled samples in total). The averaged classification error rates (over 20 randomization runs) when varying the number of labeled data are also shown. The results clearly show the advantage of our LPSSVM in discriminative manifold learning and classifier learning.

4.2   Performance over UCI Data

This experiment is performed on two UCI data sets [11]: Johns Hopkins Ionosphere (351 samples with 34-dimension features), and Sonar (208 samples with 60-dimension features). Both data sets are binary classification problems. In Fig. 3 we randomly sample N points from each category ($2N$ points in total) as labeled data and treat the rest data as unlabeled data as well as test data for evaluation. The experiments are repeated for 20 randomization runs, and the averaged classification rates (1 - error rates) are reported. From the result, our LPSSVM consistently outperforms all other competing methods over different numbers of labeled data in both data sets.

4.3   Performance over Caltech 101

The Caltech 101 set [12] consists of images from 101 object categories and an additional background class. This set contains some variations in color, pose and lighting. The bag-of-features representation [19] with local SIFT descriptors [20] has been proven effective for classifying this data set by previous works [16]. In this paper we adopt the SPM approach proposed in [16] to measure the image similarity and compute the kernel matrix. In a straightforward implementation

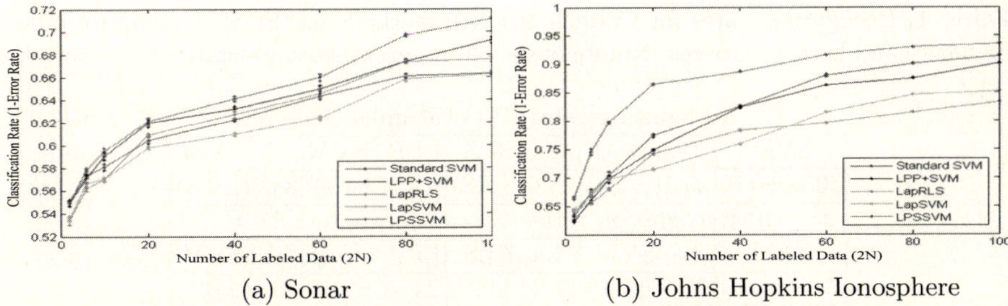

(a) Sonar (b) Johns Hopkins Ionosphere

Fig. 3. Classification rates over UCI data sets. The *vertical dotted line over each point* shows the standard deviation over 20 randomization runs.

of SPM, only the labeled data is fed to the kernel matrix for standard SVM. For other methods, the SPM-based measure is used to construct kernel matrices for both labeled and unlabeled data (i.e., K_L, K_U, K_{LU}) before various semi-supervised learning methods are applied. Specifically, for each image category, 5 images are randomly sampled as labeled data and 25 images are randomly sampled as unlabeled data for training. The remaining images are used as novel test data for evaluation (we limit the maximum number of novel test images in each category to be 30). Following the procedure of [16], a set of local SIFT features of 16×16 pixel patches are uniformly sampled from these images over a grid with spacing of 8 pixels. Then for each image category, a visual codebook is constructed by clustering all SIFT features from 5 labeled training images into 50 clusters (codewords). Local features in each image block are mapped to the codewords to compute codeword histograms. Histogram intersections are calculated at various locations and resolutions (2 levels), and are combined to estimate similarity between image pairs. One-vs.-all classifiers are built for classifying each image category from the other categories, and a test image is assigned the label of the classifier with the highest classification score.

Table 1 (a) and (b) give the average recognition rates of different algorithms over 101 image categories for the unlabeled data and the novel test data, respectively. From the table, over the unlabeled training data LPSSVM can improve baseline SVM by about 11.5% (on a relative basis). Over the novel test data, LPSSVM performs quite similarly to baseline SVM[1].

It is interesting to notice that all other competing semi-supervised methods, i.e., LapSVM, LapRLS, and naive LPP+SVM, get worse performance than LPSSVM and SVM. Please note that extensive research has been conducted for supervised classification of Caltech 101, among which SVM with SPM kernels gives one of the top performances. To the best of our knowledge, there is no report showing that the previous semi-supervised approaches can compete this state-of-the-art SPM-based SVM in classifying Caltech 101. The fact that our LPSSVM can outperform this SVM, to us, is very encouraging.

---

[1] Note the performance of SPM-based SVM here is lower than that reported in [16]. This is due to the much smaller training set than that in [16]. We focus on scenarios of scarce training data to access the power of different semi-supervised approaches.

Table 1. Recognition rates for Caltech 101. All methods use SPM to compute image similarity and kernel matrices. Numbers shown in parentheses are standard deviations.

(a) Recognition rates (%) over unlabeled data

SVM	LapSVM	LapRLS	LPP+SVM	LPSSVM
30.2($\pm$0.9)	25.1($\pm$1.1)	28.6($\pm$0.8)	14.3($\pm$4.7)	33.7($\pm$0.8)

(b) Recognition rates (%) over novel test data

SVM	LapSVM	LapRLS	LPP+SVM	LPSSVM
29.8($\pm$0.8)	24.5($\pm$0.9)	26.1($\pm$0.8)	11.7($\pm$3.9)	30.1($\pm$0.7)

The reason other competing semi-supervised algorithms have a difficult time in classifying Caltech 101 is because of the difficulty in handling small sample size in high dimensional space. With only 5 labeled and 25 unlabeled high dimensional training data from each image category, curse of dimensionality usually hurts other semi-supervised learning methods as the sparse data manifold is difficult to learn. By simultaneously discovering lower-dimension subspace and balancing class discrimination, our LPSSVM can alleviate this small sample learning difficulty and achieve good performance for this challenging condition.

4.4 Performance over Consumer Videos

We also use the challenging Kodak's consumer video data set provided in [13], [21] for evaluation. Unlike the Caltech images, content in this raw video source involves more variations in imaging conditions (view, scale, lighting) and scene complexity (background and number of objects). The data set contains 1358 video clips, with lengths ranging from a few seconds to a few minutes. To avoid shot segmentation errors, keyframes are sampled from video sequences at a 10-second interval. These keyframes are manually labeled to 21 semantic concepts. Each clip may be assigned to multiple concepts; thus it represents a multi-label corpus. The concepts are selected based on actual user studies, and cover several categories like activity, occasion, scene, and object.

To explore complementary features from both audio and visual channels, we extract similar features as [21]: visual features, e.g., grid color moments, Gabor texture, edge direction histogram, from keyframes, resulting in 346-dimension visual feature vectors; Mel-Frequency Cepstral Coefficients (MFCCs) from each audio frame (10ms) and delta MFCCs from neighboring frames. Over the video interval associated with each keyframe, the mean and covariance of the audio frame features are computed to generate a 2550-dimension audio feature vector . Then the visual and audio feature vectors are concatenated to form a 2896-dimension multi-modal feature vector. 136 videos (10%) are randomly sampled as training data, and the rest are used as unlabeled data (also for evaluation). No videos are reserved as novel unseen data due to the scarcity of positive samples for some concepts. One-vs.-all classifiers are used to detect each concept, and average precision (AP) and mean of APs (MAP) are used as performance metrics, which are official metrics for video concept detection [22].

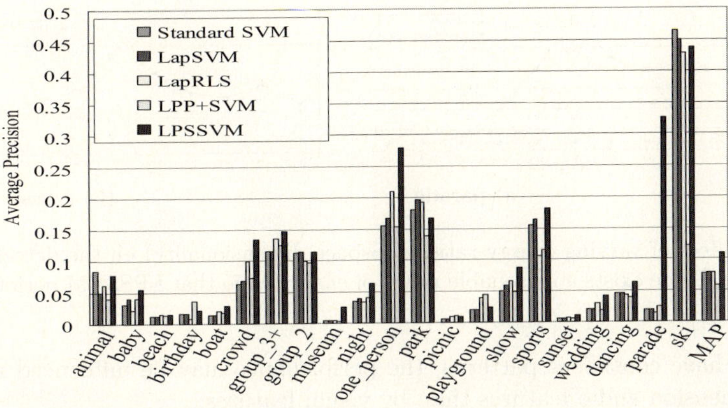

Fig. 4. Performance over consumer videos: per-concept AP and MAP. LPSSVM gets good performance over most concepts with strong cues from both visual and audio channels, where LPSSVM can find discriminative feature subspaces from multi-modalities.

Fig. 4 gives the per-concept AP and the overall MAP performance of different algorithms[2]. On average, the MAP of LPSSVM significantly outperforms other methods - 45% better than the standard SVM (on a relative basis), 42%, 41% and 92% better than LapSVM, LapRLS and LPP+SVM, respectively. From Fig. 4, we notice that our LPSSVM performs very well for the "parade" concept, with a 17-fold performance gain over the 2nd best result. Nonetheless, even if we exclude "parade" and calculate MAP over the other 20 concepts, our LPSSVM still does much better than standard SVM, LapSVM, LapRLS, and LPP+SVM by 22%, 15%, 18%, and 68%, respectively.

Unlike results for Caltech 101, here semi-supervised LapSVM and LapRLS also slightly outperform standard SVM. However, the naive LPP+SVM still performs poorly - confirming the importance of considering subspace learning and discriminative learning simultaneously, especially in real image/video classification. Examining individual concepts, LPSSVM achieves the best performance for a large number of concepts (14 out of 21), with a huge gain (more than 100% over the 2nd best result) for several concepts like "boat", "wedding", and "parade". All these concepts generally have strong cues from both visual and the audio channels, and in such cases LPSSVM takes good advantage of finding a discriminative feature subspace from multiple modalities, while successfully harnessing the challenge of the high dimensionality associated with the multi-modal feature space. As for the remaining concepts, LPSSVM is 2nd best for 4 additional concepts. LPSSVM does not perform as well as LapSVM or LapRLS for the rest 3 concepts (i.e., "ski", "park", and "playground"), since there are no consistent audio cues associated with videos in these classes, and thus it is difficult to learn an effective feature subspace. Note although for "ski" visual

---

[2] Note the SVM performance reported here is lower than that in [21]. Again, this is due to the much smaller training set than that used in [21].

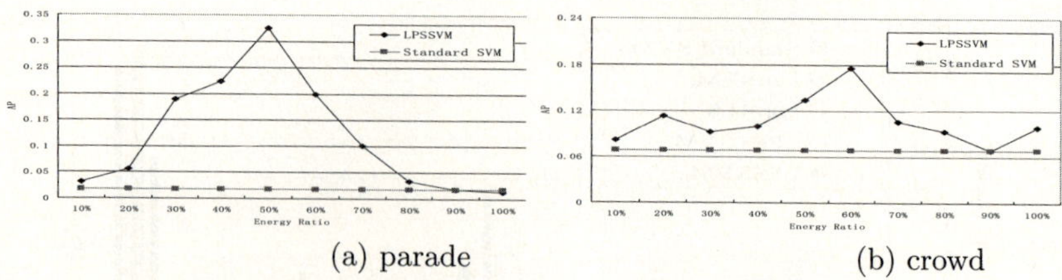

(a) parade   (b) crowd

Fig. 5. Effect of varying energy ratio (subspace dimensionality) on the detection performance. There exists a reasonable range of energy ratio that LPSSVM performs well.

features have consistent patterns, the performance may be influenced more by high-dimension audio features than by visual features.

Intriguing by the large performance gain for several concepts like "parade", "crowd", and "wedding", we analyze the effect of varying dimensionality of the subspace on the final detection accuracy. The subspace dimensionality is determined by the energy ratio of eigenvalues kept in solving the generalized eigenvalue problem. As shown in Fig. 5, even if we keep only 10% energy, LPSSVM still gets good performance compared to standard SVM - 73% gain for "parade" and 20% gain for "crowd". On the other hand, when we increase the subspace dimensionality by setting a high energy ratio exceeding 0.7 or 0.8, the performances start to decrease quickly. This further indicates that there exist effective low-dimension manifolds in high-dimension multi-modal feature space, and LPSSVM is able to take advantage of such structures. In addition, there exists a reasonable range of energy ratio (subspace dimension) that LPSSVM will outperform competing methods. How to automatically determine subspace dimension is an open issue and will be our future work.

5   Conclusion

We propose a novel learning framework, LPSSVM, and optimization methods for tackling one of the major barriers in large-scale image/video concept classification - combination of small training size and high feature dimensionality. We develop an effective semi-supervised learning method for exploring the large amount of unlabeled data, and discovering subspace structures that are not only suitable for preserving local neighborhood smoothness, but also for discriminative classification. Our method can be readily used to evaluate unseen test data, and extended to incorporate nonlinear kernel formulation. Extensive experiments are conducted over four different types of data: a toy set, two UCI sets, the Caltech 101 set and the challenging Kodak's consumer videos. Promising results with clear performance improvements are achieved, especially under adverse conditions of very high dimensional features with very few training samples where the state-of-the-art semi-supervised methods generally tend to suffer.

Future work involves investigation of automatic determination of the optimal subspace dimensionality (as shown in Fig. 5). In addition, there is another way to optimize the proposed joint cost function in Eq(5). With relaxation $\mathbf{a}^T X D X^T \mathbf{a} - I \succeq 0$ instead of $\mathbf{a}^T X D X^T \mathbf{a} - I = 0$, the problem can be solved via SDP (Semidefinite Programming), where all parameters can be recovered without resorting to iterative processes. In such a case, we can avoid the local minima, although the solution may be different from that of the original problem.

References

1. Joachims, T.: Transductive inference for text classification using support vector machines. In: ICML, pp. 200–209 (1999)
2. Chapelle, O., et al.: Semi-supervised learning. MIT Press, Cambridge (2006)
3. Vapnik, V.: Statistical learning theory. Wiley-Interscience, New York (1998)
4. Zhu, X.: Semi-supervised learning literature survey. Computer Sciences Technique Report 1530. University of Wisconsin-Madison (2005)
5. Bengio, Y., Delalleau, O., Roux, N.: Efficient non-parametric function induction in semi-supervised learning. Technique Report 1247, DIRO. Univ. of Montreal (2004)
6. Belkin, M., Niyogi, P.: Laplacian eigenmaps for dimensionality reduction and data representation. Neural Computation 15, 1373–1396 (2003)
7. Cai, D., et al.: Spectral regression: a unified subspace learning framework for content-based image retrieval. ACM Multimedia (2007)
8. Duda, R.O., et al.: Pattern classification, 2nd edn. John Wiley and Sons, Chichester (2001)
9. He, X., Niyogi, P.: Locality preserving projections. Advances in NIPS (2003)
10. Belkin, M., Niyogi, P., Sindhwani, V.: Manifold regularization: A geometric framework for learning from labeled and unlabeled examples. Journal of Machine Learning Research 7, 2399–2434 (2006)
11. Blake, C., Merz, C.: Uci repository of machine learning databases (1998), http://www.ics.uci.edu/~mlearn/MLRepository.html
12. Li, F., Fergus, R., Perona, P.: Learning generative visual models from few training examples: An incremental bayesian approach tested on 101 object categories. In: CVPR Workshop on Generative-Model Based Vision (2004)
13. Loui, A., et al.: Kodak's consumer video benchmark data set: concept definition and annotation. In: ACM Int'l Workshop on Multimedia Information Retrieval (2007)
14. Schölkopf, B., Smola, A., Müller, K.: Nonlinear component analysis as a kernel eigenvalue problem. Neural Computation 10, 1299–1319 (1998)
15. Hsu, C., Chang, C., Lin, C.: A practical guide to support vector classification, http://www.csie.ntu.edu.tw/~cjlin/libsvm/
16. Lazebnik, S., Schmid, C., Ponce, J.: Beyond bags of features: Spatial pyramid matching for recognizing natural scene categories. In: CVPR, vol, 2, pp. 2169–2178
17. Cai, D., et al.: http://www.cs.uiuc.edu/homes/dengcai2/Data/data.html
18. Joachims, T.: Training linear svms in linear time. ACM KDD, 217–226 (2006)
19. Fergus, R., et al.: Object class recognition by unsupervised scale-invariant learning. In: CVPR, pp. 264–271 (2003)
20. Lowe, D.: Distinctive image features from scale-invariant keypoints. International Journal of Computer Vision 60, 91–110 (2004)
21. Chang, S., et al.: Large-scale multimodal semantic concept detection for consumer video. In: ACM Int'l Workshop on Multimedia Information Retrieval (2007)
22. NIST TRECVID (2001 – 2007), http://www-nlpir.nist.gov/projects/trecvid/

Learning from Real Images to Model Lighting Variations for Face Images

Xiaoyue Jiang[1,2], Yuk On Kong[3], Jianguo Huang[1],
Rongchun Zhao[1], and Yanning Zhang[1]

[1] School of Computer Science, Northwestern Polytechnical University,
Xi'an, 710072, China
[2] School of Psychology, University of Birmingham, Birmingham B15 2TT, UK
[3] Department of Electronics and Informatics, Vrije Universiteit Brussel,
Brussels 1050, Belgium

Abstract. For robust face recognition, the problem of lighting variation is considered as one of the greatest challenges. Since the nine points
of light (9PL) subspace is an appropriate low-dimensional approximation to the illumination cone, it yielded good face recognition results
under a wide range of difficult lighting conditions. However building the
9PL subspace for a subject requires 9 gallery images under specific lighting conditions, which are not always possible in practice. Instead, we
propose a statistical model for performing face recognition under variable illumination. Through this model, the nine basis images of a face
can be recovered via maximum-a-posteriori (MAP) estimation with only
one gallery image of that face. Furthermore, the training procedure requires only some real images and avoids tedious processing like SVD
decomposition or the use of geometric (3D) or albedo information of a
surface. With the recovered nine dimensional lighting subspace, recognition experiments were performed extensively on three publicly available
databases which include images under single and multiple distant point
light sources. Our approach yields better results than current ones. Even
under extreme lighting conditions, the estimated subspace can still represent lighting variation well. The recovered subspace retains the main
characteristics of 9PL subspace. Thus, the proposed algorithm can be
applied to recognition under variable lighting conditions.

1   Introduction

Face recognition is difficult due to variations caused by pose, expression, occlusion and lighting (or illumination), which make the distribution of face object
highly nonlinear. Lighting is regarded as one of the most critical factors for robust
face recognition. Current attempt to handle lighting variation by either finding
the invariant features or modeling the variation. The edge based algorithm [1]
and the algorithm based on quotient image [2,3,4]belong to the first type. But
these methods cannot extract sufficient features for accurate recognition.

Early work on modeling lighting variation [5,6] showed that a 3D linear
subspace can represent the variation of a Lambertian object under a fixed

D. Forsyth, P. Torr, and A. Zisserman (Eds.): ECCV 2008, Part IV, LNCS 5305, pp. 284–297, 2008.

pose when there is no shadow. With the same Lambertian assumption, Belhumeur and Kriegman [7] showed that images illuminated by an arbitrary number of point light sources formed a convex polyhedral cone, i.e. the illumination cone. In theory, the dimensionality of the cone is finite. They also pointed out that the illumination cone can be approximated by a few properly chosen images. Good recognition results of the illumination cone in [8] demonstrated its representation for lighting variation. [9] indicated that lighting subspace of Lambertian object can be approximated by a linear subspace with dimension between three and seven. Recent research is mainly focused on the application of low-dimensional subspace to lighting variation modeling. With the assumption of Lambertian surface and non-concavity, Ramamoorith and Hanrahan [10] and Basri and Jacobs[11] independently introduced the spherical harmonic (SH) subspace to approximate the illumination cone. However, the harmonic images (basis images of SH subspace) are computed from the geometric and albedo information of the subject's surface. In order to use the SH subspace theory, a lot of algorithms applied the 3D model of faces to handling lighting variations [12,13,14,15,16]. However, recovering the 3D shape from images is still an open problem in computer vision.

Lee et al.[19] built up a subspace that is nearest to the SH subspace and has the largest intersection with the illumination cone, called the nine points of light (9PL) subspace. It has a universal configuration for different subjects, i.e. the subspace is spanned by images under the same lighting conditions for different subjects. In addition, the basis images of 9PL subspace can be duplicated in real environments, while those of the SH subspace cannot because its the basis images contain negative values. Therefore the 9PL subspace can overcome the inherent limitation of SH subspace. Since the human face is neither completely Lambertain nor entirely convex, SH subspace can hardly represent the specularities or cast shadows (not to mention inter-reflection). The basis images of 9PL subspace are taken from real environment, they already contain all the complicated reflections of the objects. Therefore the 9PL subspace can give a more detailed and accurate description of lighting variation.

In practice, the requirement of these nine real images cannot always be fulfilled. Usually there are fewer gallery images (e.g. one gallery image) per subject, which can be taken under arbitrary lighting conditions. In this paper, we propose a statistical model for recovering the 9 basis images of the 9PL subspace from only one gallery image. Zhang and Samaras [12] presented a statistical method for recovering the basis images of SH subspace instead. In their training procedure, geometric and albedo information is still required for synthesizing the harmonic images. In contrast, the proposed method requires only some real images that can be easily obtained in real environment. Since the recovered basis images of the 9PL subspace contain all the reflections caused by the shape of faces, such as cast shadows, specularities, and inter-reflections, better recognition results are obtained, even under extreme lighting conditions. Compared with other algorithms based on 3D model [12,15,16], the proposed algorithm is entirely a 2D algorithm, which has much lower computational complexity. The proposed algorithm also has comparable recognition results. Note that we do not consider pose variation in this paper and assume that all subjects are in the frontal pose.

This paper is organized as follows. In Section 2, we briefly summarize the methods of low-dimensional linear approximation of the illumination cone, including the SH subspace and the 9PL subspace. The training of our statistical model and the application of the model for recovering basis images from only one gallery image are described in Sections 3 and 4 respectively. Section 5 is dedicated to the experimental results. The conclusion is given in Section 6.

2   Approximation of the Illumination Cone

Belhumeur and Kriegman [7] proved that the set of n-pixel images of a convex object that had a Lambertian surface illuminated by an arbitrary number of point light sources at infinity formed a convex polyhedral cone, called the illumination cone $\mathcal{C}$ in $\mathcal{R}^n$. Each point in the cone is an image of the object under a particular lighting condition, and the entire cone is the set of images of the object under all possible lighting conditions. Any images in the illumination cone $\mathcal{C}$ (including the boundary) can be determined as a convex combination of extreme rays (images) given by

$$I_{ij} = max(\tilde{B}\tilde{s}_{ij}, 0) \tag{1}$$

where $\tilde{s}_{ij} = \tilde{b}_i \times \tilde{b}_j$ and $\tilde{B} \in \Re^{n \times 3}$. Every row $\tilde{b}_i$ of $\tilde{B}$ is a three element row vector determined by the product of the albedo with the inward pointing unit normal vector of a point on the surface. There are at most $q(q-1)$ extreme rays for $q \leq n$ distinct surface normal vectors. Therefore the cone can be constructed with finite extreme rays and the dimensionality of the lighting subspace is finite. However, building the full illumination cone is tedious, and the low dimensional approximation of the illumination cone is applied in practice.

From the view of signal processing, the reflection equation can be considered as the rotational convolution of incident lighting with the albedo of the surface [10]. The spherical harmonic functions $Y_{lm}(\theta, \phi)$ are a set of orthogonal basis functions defined in the unit sphere, given as follows,

$$Y_{lm}(\theta, \phi) = N_{lm} P_l^m(\cos \theta) \exp^{im\phi} \tag{2}$$

where $N_{lm} = \sqrt{\frac{2l+1}{4\pi} \frac{(l-m)!}{(l+m)!}}$, $(\theta, \phi)$is the spherical coordinate (θ is the elevation angle, which is the angle between the polar axis and the z-axis with range $0 \leq \theta \leq 180^o$, and ϕ is the azimuth angle with the range $-180^o \leq \phi \leq 180^o$). P_l^m is the associated Legendre function, and the two indices meet the conditions $l \geq 0$ and $l \geq m \geq -l$. Then functions in the sphere, such as the reflection equation, can be expanded by the spherical harmonic functions, which are basis functions on the sphere. Images can be represented as a linear combination of spherical harmonic functions. The first three order ($l \leq 3$) basis can account for 99% energy of the function. Therefore the first three order basis functions (altogether

9) can span a subspace for representing the variability of lighting. This subspace is called the spherical harmonic (SH) subspace .

Good recognition results reported in [11] indicates that the SH subspace $\mathcal{H}$ is a good approximation to the illumination cone $\mathcal{C}$. Given the geometric information of a face, its spherical harmonic functions can be calculated with Eq.(2). These spherical harmonic functions are synthesized images, also called harmonic images. Except the first harmonic image, all the others have negative values, which cannot be obtained in reality. To avoid the requirement of geometric information, Lee et al.[19] found a set of real images which can also serve as a low dimensional approximation to illumination cone based on linear algebra theory.

Since the SH subspace $\mathcal{H}$ is good for face recognition, it is reasonable to assume that a subspace $\mathcal{R}$ close to $\mathcal{H}$ would be likewise good for recognition. $\mathcal{R}$ should also intersect with the illumination cone $\mathcal{C}$ as much as possible. Hence a linear subspace $\mathcal{R}$ which is meant to provide a basis for good face recognition will also be a low dimensional linear approximation to the illumination cone $\mathcal{C}$. Thus subspace should satisfy the following two conditions [19]:

1. The distance between $\mathcal{R}$ and $\mathcal{H}$ should be minimized.
2. The unit volume ($vol(\mathcal{C} \cap \mathcal{R})$) of $\mathcal{C} \cap \mathcal{R}$ should be maximized (the unit volume is defined as the volume of the intersection of $\mathcal{C} \cap \mathcal{R}$ with the unit ball)

Note that $\mathcal{C} \cap \mathcal{R}$ is always a subcone of $\mathcal{C}$; therefore maximizing its unit volume is equivalent to maximize the solid angle subtended by the subcone $\mathcal{C} \cap \mathcal{R}$. If $\{\tilde{I}_1, \tilde{I}_2, \cdots, \tilde{I}_k\}$ are the basis images of $\mathcal{R}$. The cone $\mathcal{R}_c \subset \mathcal{R}$ is defined by $\tilde{I}_k$,

$$\mathcal{R}_c = \{I | I \in \mathcal{R}, I = \sum_{k=1}^{M} \alpha_k \tilde{I}_k, \alpha_k \geq 0\} \tag{3}$$

is always a subset of $\mathcal{C} \cap \mathcal{R}$. In practice the subcone $\mathcal{C} \cap \mathcal{R}$ is taken as $\mathcal{R}_c$ and the subtended angle of $\mathcal{R}_c$ is maximized. $\mathcal{R}$ is computed as a sequence of nested linear subspace $\mathcal{R}_0 \subseteq \mathcal{R}_1 \subseteq \cdots \subseteq \mathcal{R}_i \subseteq \cdots \subseteq \mathcal{R}_9 = \mathcal{R}$, with $\mathcal{R}_k(k > 0)$ a linear subspace of dimension i and $\mathcal{R}_0 = \emptyset$. First, EC denotes the set of (normalized) extreme rays in the illumination cone $\mathcal{C}$; and EC_k denotes the set obtained by deleting k extreme rays from EC, where $EC_0 = EC$. With $\mathcal{R}_{k-1}$ and EC_{k-1}, the sets EC_k and R_k can be defined iteratively as follows:

$$\tilde{I}_k = \arg \max_{I \in EC_{k-1}} \frac{dist(I, \mathcal{R}_{k-1})}{dist(I, \mathcal{H})} \tag{4}$$

where $\tilde{I}_k$ denotes the element in EC_{k-1}. $\mathcal{R}_k$ is defined as the space spanned by $\mathcal{R}_{k-1}$ and $\tilde{I}_k$. $EC_k = EC_{k-1} \backslash \tilde{I}_k$. The algorithm stops when $\mathcal{R}_9 \equiv \mathcal{R}$ is reached. The result of Eq.(4) is a set of nine extreme rays that span $\mathcal{R}$ and there are nine directions corresponding to these nine extreme rays. For different subjects, the nine lighting directions are qualitatively very similar. By averaging Eq.(4) of different subjects and maximizing this function as follows:

$$\tilde{I}_k = \arg \max_{I \in EC_{k-1}} \sum_{p=1}^{N} \frac{dist(I^p, \mathcal{R}_{k-1}^p)}{dist(I^p, \mathcal{H}^p)} \tag{5}$$

where I^p denotes the image of subject p taken under a single light source. H^p is the SH subspace of subject p. $\mathcal{R}_{k-1}^p$ denotes the linear subspace spanned by images $\{\tilde{I}_1^p, \cdots, \tilde{I}_k^p\}$ of subject p. The universal configuration of nine light source direction is obtained. They are $(0,0)$, $(68,-90)$, $(74,108)$, $(80,52)$, $(85,-42)$, $(85,-137)$, $(85,146)$, $(85,-4)$, $(51,67)$[14]. The directions are expressed in spherical coordinates as pairs of (ϕ, θ), Figure 1(a) illustrates the nine basis images of a person from the Yale Face Database B [8].

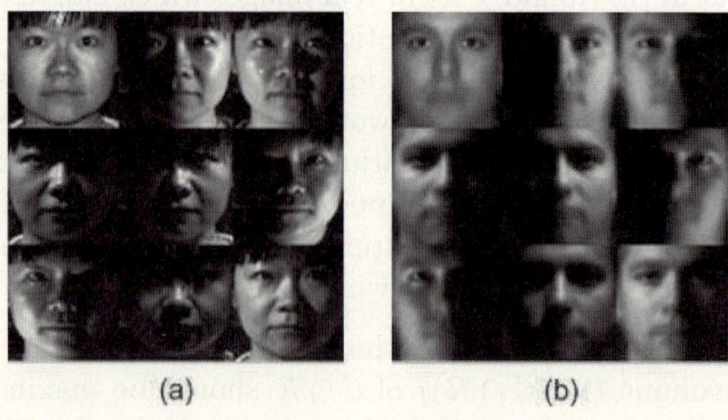

(a)   (b)

Fig. 1. the basis images of 9PL subspace. (a) images taken under certain lighting conditions can serve as the basis images of the object. (b) the mean images of the basis images estimated from the bootstrap data set.

3   Statistical Model of Basis Images

According to the universal configuration of lighting directions, we can apply nine images taken under controlled environment to spanning the 9PL linear subspace. However, even these nine images may not be available in some situations. Thus, we propose a statistical method for estimating the basis images from one gallery image. To build the statistical model, we must find the probability density function (pdf) of basis images and the pdf of the error term. Due to the limited amount of the training data, we use the bootstrap method to estimate the statistics of basis images. The recovering step is to estimate the corresponding basis images from one single image of a novel subject under arbitrary lighting conditions. For a given image, we first estimate its lighting coefficient. Then according to the maximum a posteriori (MAP) estimation, we obtain an estimation of the basis images. Finally, we apply the recovered subspace to face recognition. The probe image is identified as the face whose lighting subspace is closest in distance to the image.

Given nine basis images, we can reconstruct images under arbitrary lighting conditions as follows,

$$I = B\mathbf{s} + e(\mathbf{s}) \tag{6}$$

where $I \subset \Re^{d \times 1}$ is the image vector. $B \subset \Re^{d \times 9}$ is the matrix of nine basis images, every column of which is the vector of the basis image. $\mathbf{s} \subset \Re^{d \times 1}$ is the vector of lighting coefficients which denotes the lighting conditions of the image. Error term $e(\mathbf{s}) \subset \Re^{d \times 1}$ is related to the pixels' position and lighting conditions.

For a novel image, we estimate its basis images through the maximum a posterior (MAP) estimation. That is

$$B_{MAP} = \arg \max_B P(B|I) \tag{7}$$

According to the Bayes rule

$$P(B|I) = \frac{P(I|B)P(B)}{P(I)} \tag{8}$$

where $P(I)$ is the evidence factor which guarantees that posterior probabilities would sum to one. Then Eq.(7) can become

$$B_{MAP} = \arg \max_B (P(I|B)P(B)) \tag{9}$$

In order to recover basis images from an image with Eq.(9), one should know the pdf of the basis images, i.e. $P(B)$, and the pdf of the likelihood, i.e. $P(I|B)$. Assuming the error term of Eq.(6) is normally distributed with mean $\mu_e(\mathbf{s})$ and variance $\sigma_e^2(\mathbf{s})$, we can deduce that the pdf of the likelihood $P(I|B)$ is also Gaussian with mean $B\mathbf{s} + \mu_e(\mathbf{s})$ and variance $\sigma_e^2(\mathbf{s})$ according to Eq.(6).

We assume that the pdf of the basis images B are Gaussians of means μ_B and covariances C_B as in [12,20]. The probability $P(B)$ can be estimated from the basis images in the training set. In our experiments, the basis images of 20 different subjects from the extented Yale face database B [8] are introduced to the bootstrap set. Note that, the basis images of every subject are real images which were taken under certain lighting conditions. The lighting conditions are determined by the universal configurations of the 9PL subspace. The sample mean μ_B and sample covariance matrix C_B are computed. Figure 1(b) shows the mean basis images, i.e. μ_B.

The error term $e(\mathbf{s}) = I - B\mathbf{s}$ models the divergence between the real image and the estimated image which is reconstructed by the low dimensional subspace. The error term is related to the lighting coefficients. Hence, we need to know the lighting coefficients of different lighting conditions. In the training set, there are 64 different images that taken under different lighting condition for every subject. Under a certain lighting condition, we calculate the lighting coefficients of every subject's image, i.e. $\mathbf{s}_k^p$ (the lighting coefficients of the pth subject's image under the lighting condition $\mathbf{s}_k$). For a training image, its lighting coefficients can be estimated by solving the linear equation $I = B\mathbf{s}$. The mean value of different subjects' lighting coefficients can be the estimated coefficients $(\bar{\mathbf{s}}_k)$ for that lighting condition, i.e. $\bar{\mathbf{s}}_k = \sum_{p=1}^{N} \mathbf{s}_k^p / N$. Then, under a certain lighting condition, the error term the of the pth subject's image is

$$e_p(\bar{\mathbf{s}}_k) = I_k^p - B_p \bar{\mathbf{s}}_k \tag{10}$$

where I_k^p is the training image of the pth subject under lighting condition $\mathbf{s}_k$ and B_p is the basis images of the pth subject. Following the above assumption, we estimate the mean $\mu_e(\bar{\mathbf{s}}_k)$ and variance $\sigma_e^2(\bar{\mathbf{s}}_k)$ of the error term.

4   Estimating the Basis Images

As described in the previous section, the basis images of a novel image can be recovered by using the MAP estimation. Since the error term is related to lighting condition, we need to estimate the lighting condition, i.e. the lighting coefficients, of every image before calculating its basis images.

4.1   Estimating Lighting Coefficients

Lighting influences greatly the appearance of an image. Under similar illumination, images of different subjects will appear almost the same. The difference between the images of the same subject under different illuminations is always larger than that between the images of different subjects under the same illumination [21]. Therefore we can estimate the lighting coefficients of a novel image with an interpolation method. The kernel regression is a smooth interpolation method [22]. It is applied to estimating the lighting coefficients. For every training image, we have their corresponding lighting coefficients. For a novel image I_n, its lighting coefficient is given by

$$\mathbf{s} = \frac{\sum_{k=1}^{M} w_k \mathbf{s}_k^p}{\sum_{k=1}^{M} w_k} \tag{11}$$

$$w_k = \exp(-\frac{[D(I_n, I_k^p)]^2}{2(\sigma_{I_k^p})^2}) \tag{12}$$

where $D(I_n, I_k^p) = \|I_n - I_k^p\|_2$ is the L_2 norm of the image distance. $\sigma_{I_k^p}$ determines the weight of test image I_k^p in the interpolation. In the training set, every subject has 64 different images and there are altogether 20 different subjects. Thus, for a novel image, there are 20 images with similar illumination. In our experiment, we assign the farthest distance of these 20 images from the probe image to $\sigma_{I_k^p}$. $\mathbf{s}_k^p$ is the lighting coefficient of image I_k^p.

4.2   Estimating the Error Term

The error term denotes the difference between the reconstructed image and the real image. This divergence is caused by the fact that the 9PL subspace is the low-dimensional approximation to the lighting subspace, and it only accounts for the low frequency parts of the lighting variance. The statistics of the error under a new lighting condition can be estimated from those of the error under known illumination, i.e. $\mu_e(\bar{\mathbf{s}}_k)$, $\sigma_e^2(\bar{\mathbf{s}}_k)$, also via the kernel regression method [20].

$$\mu_e(\mathbf{s}) = \frac{\sum_{k=1}^{M} w_k \mu_e(\bar{\mathbf{s}}_k)}{\sum_{k=1}^{M} w_k} \tag{13}$$

$$\sigma_e^2(\mathbf{s}) = \frac{\sum_{k=1}^{M} w_k \sigma_e^2(\bar{\mathbf{s}}_k)}{\sum_{k=1}^{M} w_k} \tag{14}$$

$$w_k = \exp(-\frac{[D(\mathbf{s}, \bar{\mathbf{s}}_k)]^2}{2[\sigma_{\bar{\mathbf{s}}_k}]^2}) \tag{15}$$

where $D(\mathbf{s}, \bar{\mathbf{s}}_k) = \|\mathbf{s} - \bar{\mathbf{s}}_2\|_2$ is the L_2 norm of the lighting coefficient distance. Like $\sigma_{I_k^p}$, $\sigma_{\bar{\mathbf{s}}_k}$ determines the weight of the error term related to the lighting coefficients $\bar{\mathbf{s}}_k$. Also, we assign the farthest lighting coefficient distance of these 20 images from the probe image to $\sigma_{\bar{\mathbf{s}}_k}$.

4.3   Recovering the Basis Images

Given the estimated lighting coefficients $\mathbf{s}$ and the corresponding error term $\mu_e(\mathbf{s})$, $\sigma_e^2(\mathbf{s})$, we can recover the basis images via the MAP estimation. If we apply the log probability, omit the constant term, and drop $\mathbf{s}$ for compactness, Eq.(9) can become

$$\arg\max_B \left(-\frac{1}{2}(\frac{I - B\mathbf{s} - \mu_e}{\sigma_e})^2 - \frac{1}{2}(B - \mu_B)C_B^{-1}(B - \mu_B)^T \right) \tag{16}$$

To solve Eq.(16), we estimate the derivatives,

$$-\frac{2}{\sigma_e^2}(I - B\mathbf{s} - \mu_e)\mathbf{s}^T + 2(B - \mu_B)C_B^{-1} = 0 \tag{17}$$

Then we rewrite Eq.(17) as a linear equation,

$$AB = b \tag{18}$$

where $A = \frac{\mathbf{s}\mathbf{s}^T}{\sigma_e^2} + C_B^{-1}$ and $b = \frac{I - \mu_e}{\sigma_e^2}\mathbf{s} + C_B^{-1}\mu_B$. The solution of the linear equation is $B = A^{-1}b$. Using the Woodbury's identity [25], we can obtain an explicit solution

$$B_{MAP} = A^{-1}b \tag{19}$$
$$= \left(C_B - \frac{C_B\mathbf{s}\mathbf{s}^T C_B}{\sigma_e^2 + \mathbf{s}^T C_B\mathbf{s}} \right) \left(\frac{I - \mu_e}{\sigma_e^2}\mathbf{s} + C_B^{-1}\mu_B \right)$$
$$= \left(\frac{I - \mu_B\mathbf{s} - \mu_e}{\sigma_e^2 + \mathbf{s}^T C_B\mathbf{s}} \right) C_B\mathbf{s} + \mu_B$$

From Eq.(19), the estimated basis image is composed of the term of characteristics, $\left(\frac{I - \mu_B\mathbf{s} - \mu_e}{\sigma_e^2 + \mathbf{s}^T C_B\mathbf{s}} C_B\mathbf{s} \right)$, and the term of mean, μ_B. In the term of characteristics, $(I - \mu_B\mathbf{s} - \mu_e)$ is the difference between the probe image and the image reconstructed by the mean basis images.

4.4   Recognition

The most direct way to perform recognition is to measure the distance between probe images and the subspace spanned by the recovered basis images. Every column of B is one basis image. However, the basis images are not orthonormal vectors. Thus we perform the QR decomposition on B to obtain a set of orthonormal basis, i.e. the matrix Q. Then the projection of probe image I to the subspace spanned by B is $QQ^T I$, and the distance between the probe image I and the subspace spanned by B can be computed as $\|QQ^T I - I\|_2$. In the recognition procedure, the probe image is identified as the subspace with minimum distance from it.

5   Experiments

The statistical model is trained by images from the extended Yale Face Database B. With the trained statistical model, we can reconstruct the lighting subspace from only one gallery image. This estimation is insensitive to lighting variation. Thus, recognition can be achieved across illumination conditions.

5.1   Recovered Basis Images

To recover the basis images from a single image, the lighting coefficients of the image are estimated first. Then we estimate the error terms of the image Finally, the basis images of the image can be obtained with Eq.(19).

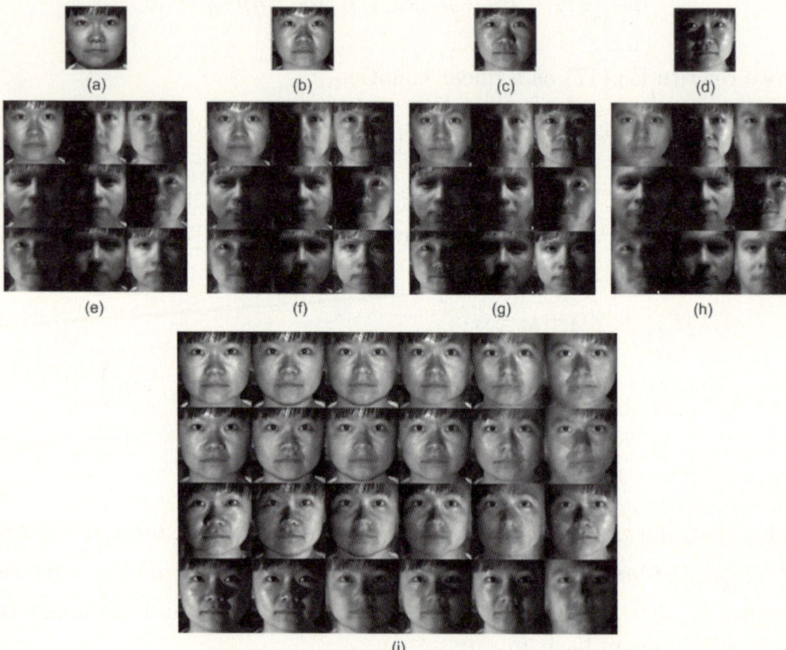

Fig. 2. Recovered basis images. (a)∼(d) are images in subset 1∼4 of Yale Face Database B respectively. (e)∼(h) are recovered basis images from image (a)∼(d) respectively. (i) are the reconstruction results: from left to right, the columns are the original images, the reconstruction results from the real basis images and the estimated basis images(e)∼(h), respectively.

Although the images of the same object are under different lighting conditions, the recovered basis images should be similar. The probe images are from the Yale face database B. There are 10 subjects and 45 probe images per subject. According to the lighting conditions of the probe images, they can be grouped into 4 subsets as in [8]. The details can be found in Table 1. From subset1 to subset4, the lighting conditions become extreme. For every subject, we recover its basis images from only one of its probe images each time. Then we can obtain 45 sets of basis images for every subject. Fig.2(e)$\sim$(h) are the basis images recovered from an image of each subset. $\bar{\sigma}_{basis}$(the mean standard deviation of the 45 sets of basis images of 10 subjects) is 7.76 intensity levels per pixel, while $\bar{\sigma}_{image}$(the mean standard deviation of the original 45 probe images of 10 subjects) is 44.12 intensity levels per pixel. From the results, we can see that the recovered basis images are insensitive to the variability of lighting. Thus we can recover the basis images of a subject from its images under arbitrary lighting conditions. Fig.2(i) are the reconstruction results from different basis images.The reconstructed images also contain shadows and inter-reflections because the recovered basis images contain detailed reflection information. As a result, good recognition results can be obtained.

Table 1. The subsets of Yale Face Database B

	subset1	subset2	subset3	subset4
illumination	0$\sim$12	13$\sim$25	26$\sim$50	50$\sim$77
Number of images	70	120	120	140

5.2   Recognition

Recognition is performed on the Yale Face Database B [8] first. We take the frontal images (pose 0) as the probe set, which is composed of 450 images (10 subjects, 45 images per subject). For every subject, one image is used for recovering its lighting subspace and the 44 remaining images are used for recognition. The comparison of our algorithm with the reported results is shown in Table 2.

Our algorithm reconstructed the 9PL subspace for every subject. The recovered basis images also contained complicated reflections on faces, such as cast shadows, specularities, and inter-reflection. Therefore the recovered 9PL subspace can give a more detailed and accurate description for images under different lighting conditions. As a result, we can get good recognition results on images with different lighting conditions. Also, the reported results of 'cone-cast', 'harmonic images-cast' and '9PL-real' showed that better results can be obtained when cast shadows were considered. Although [15,16] also use only one image to adjust lighting conditions, they need to recover the 3D model of the face first. The performance of our algorithm is comparable to that of these algorithms, which are based on high-resolution rendering [15,16] and better than that of those algorithms based on normal rendering [14]. Our algorithm is a completely 2D-based approach. Computationally, it is much less expensive compared with

Table 2. The Recognition Error Rate of Different Recognition Algorithms on Yale Face Database B

Algorithms	subset1&2	subset3	subset4
Correlation[8]	0.0	23.3	73.6
Eigenfaces[8]	0.0	25.8	75.7
Linear Subspace[8]	0.0	0.0	15.0
Cones-attached[8]	0.0	0.0	8.6
Cones-cast[8]	0.0	0.0	0.0
harmonic images-cast[8]	0.0	0.0	2.7
3D based SH model [12]	0.0	0.3	3.1
BIM(30 Bases)[15]	0.0	0.0	0.7
Wang *et al.*[16]	0.0	0.0	0.1
Chen *et al.*[17]	0.0	0.0	1.4
9PL-real[19]	0.0	0.0	0.0
our algorithm	0.0	0.0	0.72

those 3D based methods. The basis images of a subject can be directly computed with Eq.19 while the recognition results are comparable to those from the 3D-based methods.

5.3   Multiple Lighting Sources

An image taken under multiple lighting sources can be considered as images taken under a single lighting source being superimposed. Through interpolation, the lighting coefficients of images taken under single lighting are linearly combined to approximate those of the image taken under multiple-lighting. Here we also apply the statistical model trained on the extended Yale Database B to basis images estimation. Similarly the lighting coefficients of images are estimated through interpolation. Then the error term can be estimated according to the lighting coefficients. Finally, the basis images are recovered.

In the PIE face database [23], there are 23 images per subject taken under multiple lighting sources, and altogether 69 subjects. We recover 23 sets of the basis images from the 23 images of every subject respectively. With these estimated basis images, we perform recognition on the 1587 images (23 images per person) 23 times. We also estimate basis for images in the AR database [24]. We select randomly 4 images under different illumination per subject (image 1, 5, 6, 7) and recover the respective basis images from those images. Recognition is performed on 504 images (126 subjects and 4 images per subject) 4 times. Samples of the recovered basis images from images in the PIE and AR databases are shown in Fig.3. The average recognition rates, the mean standard deviation of the recovered basis images ($\bar{\sigma}_{basis}$) and the mean standard deviation of the gallery images ($\bar{\sigma}_{images}$) are presented in Table 3. Also [12] reported a recognition rate of 97.88% on part of PIE and [18] reported his recognition rate as 97% on PIE database. Our recognition results are better. The results show that the statistical model trained by images taken under a single lighting source can also be generalized to images taken under multiple lighting sources.

Table 3. Recognition Rate on Different Databases

Face Database	PIE	AR
$\bar{\sigma}_{basis}$	11.01	11.34
$\bar{\sigma}_{image}$	285	38.59
Recognition rate	98.21%	97.75%

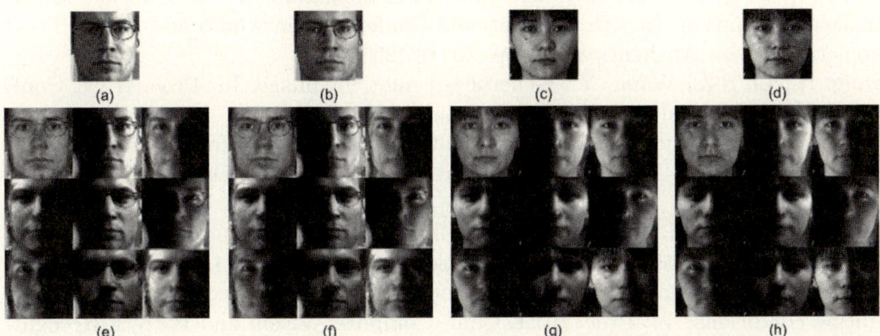

Fig. 3. Recovered basis images. (a) and (b) are images in PIE database, (e) and (f) are estimated basis images from image (a) and (b), respectively. (c) and (d)are images in AR database, (g) and (h) are estimated basis images from image (c) and (d), respectively.

6   Conclusion

The 9PL provides a subspace which is useful for recognition and is spanned by real images. Based on this framework, we built a statistical model for these basis images. With the MAP estimation, we can recover the basis images from one gallery image under arbitrary lighting conditions, which could be single lighting source or multiple lighting sources. The experimental results based on the recovered subspace are comparable to those from other algorithms that require lots of gallery images or the geometric information of the subjects. Even in extreme lighting conditions, the recovered subspace can still appropriately represent lighting variation. The recovered subspace retains the main characteristics of the 9PL subspace.

Based on our statistical model, we can build the lighting subspace of a subject from only one gallery image. It avoids the limitation of requiring tedious training or complex training data, such as many gallery images or the geometric information of the subject. After the model has been trained well, the computation for recovering the basis images is quite simple and without the need of 3D models. The proposed framework can also potentially be used to deal with pose and lighting variations together, with training images in different poses taken under different lighting for building the statistical model.

Acknowledgement

This work is funded by China Postdoctoral Science Foundation(No.20070421129).

References

1. Guo, X., Leung, M.: Face recognition using line edge map. IEEE Trans. Pattern Recognition and Machine Intelligence 24(6), 764–799 (2002)
2. Shashua, A., Tammy, R.: The quotient images: class-based rendering and recognition with varying illuminations. IEEE Trans. Pattern Recognition and Machine Intelligence 23(2), 129–139 (2001)
3. Gross, R., Brajovic, V.: An image processing algorithm for illumination invariant face recognition. In: 4th International Conference on Audio and Video Based Biometric Person Authentication, pp. 10–18 (2003)
4. Wang, H., Li, S.Z., Wang, Y.: Generalized quotient image. In: Proc. IEEE Conf. Computer Vision and Pattern Recognition (2004)
5. Hallinan, P.: A low-dimensional representation of human faces for arbitrary lighting conditions. In: Proc. IEEE Conf. Computer Vision and Pattern Recognition, pp. 995–999 (1994)
6. Nayar, S., Murase, H.: Dimensionality of illumination in appearance matching. In: Proc. IEEE Conf. Robotics and Automation, pp. 1326–1332 (1996)
7. Belhumeur, P., Kriegman, D.J.: What is set of images of an object under all possible lighting conditions? In: Proc. IEEE Conf. Computer Vision and Pattern Recognition, pp. 270–277 (1996)
8. Georghiads, A., Belhumeur, P., Kriegman, D.: From few to many: illumination cone models for face recognition under variable lighting and pose. IEEE Trans. Pattern Recognition and Machine Intelligence 23(6), 643–660 (2001)
9. Yuille, A., Snow, D., Epstein, R., Belhumeur, P.: Determing generative models of objects under varying illumination: shape and albedo from multiple images using SVD and integrability. International Journal of Computer Vision 35(3), 203–222 (1999)
10. Ramamoorthi, R., Hanrahan, P.: On the relationship between radiance and irradiance: determine the illumination from images of a convex Lambertian object. J. Optical. Soc. Am. A 18(10), 2448–2459 (2001)
11. Basri, R., Jacobs, D.: Lambertian reflectance and linear subspaces. IEEE Trans. Pattern Recognition and Machine Intelligence 25(2), 218–233 (2003)
12. Zhang, L., Samaras, D.: Face recognition under variable lighting using harmonic image exemplars. In: Proc. IEEE Conf. Computer Vision and Pattern Recognition (2003)
13. Wen, Z., Liu, Z., Huang, T.: Face relighting with radiance environment map. In: Proc. IEEE Conf. Computer Vision and Pattern Recognition (2003)
14. Zhang, L., Wang, S., Samaras, D.: Face synthesis and recognition from a single image under arbitrary unknown lighting using a spherical harmonic basis morphable model. In: Proc. IEEE Conf. Computer Vision and Pattern Recognition (2005)
15. Lee, J., Moghaddam, B., Pfister, H., Machiraju, R.: A bilinear illumination model for robust face recognition. In: Proc. IEEE International Conference on Computer Vision (2005)
16. Wang, Y., Liu, Z., Hua, G., et al.: Face re-lighting from a single image under harsh lighting conditions. In: Proc. IEEE Conf. Computer Vision and Pattern Recognition (2007)
17. Chen, H.F., Belhumeur, P.N., Jacobs, D.W.: In search of illumination invariants. In: Proc. IEEE Conf. Computer Vision and Pattern Recognition (2000)
18. Zhou, S., Chellappa, R.: Illuminating light field: image-based face recognition across illuminations and poses. In: Proc. IEEE Intl. Conf. on Automatic Face and Gesture Recognition (May 2004)

19. Lee, K., Ho, J., Kriegman, D.: Acquiring linear subspaces for face recognition under variable lighting. IEEE Trans. Pattern Recognition and Machine Intelligence 27(5), 684–698 (2005)
20. Sim, T., Kanade, T.: Combining models and exemplars for face recognition: an illumination example. In: Proc. Of Workshop on Models versus Exemplars in Computer Vision, CVPR 2001 (2001)
21. Adini, Y., Moses, Y., Ullman, S.: Face recognition: the problem of compensating for changes in illumination directions. IEEE Trans. Pattern Analysis and Machine Intelligence 19(7), 721–733 (1997)
22. Atkenson, C., Moore, A., Schaal, S.: Locally weighted learning. Artificial Intelligence Review (1996)
23. Sim, T., Baker, S., Bsat, M.: The CMU pose, illumination, and expression (pie) database. In: Proc. IEEE International Conference on Automatic Face and Gesture Recognition (May 2002)
24. Martinez, A.M., Benavente, R.: The AR face database. CVC Tech. Report No.24 (1998)
25. Scharf, L.: Statistical signal processing: detection, estimation and time series analysis, p. 54. Addison-Wesley, Reading (1991)

Toward Global Minimum through Combined Local Minima

Ho Yub Jung, Kyoung Mu Lee, and Sang Uk Lee

Department of EECS, ASRI, Seoul National University, 151-742, Seoul, Korea
hoyub@diehard.snu.ac.kr, kyoungmu@snu.ac.kr, sanguk@ipl.snu.ac.kr

Abstract. There are many local and greedy algorithms for energy minimization over Markov Random Field (MRF) such as iterated condition mode (ICM) and various gradient descent methods. Local minima solutions can be obtained with simple implementations and usually require smaller computational time than global algorithms. Also, methods such as ICM can be readily implemented in a various difficult problems that may involve larger than pairwise clique MRFs. However, their short comings are evident in comparison to newer methods such as graph cut and belief propagation. The local minimum depends largely on the initial state, which is the fundamental problem of its kind. In this paper, disadvantages of local minima techniques are addressed by proposing ways to combine multiple local solutions. First, multiple ICM solutions are obtained using different initial states. The solutions are combined with random partitioning based greedy algorithm called Combined Local Minima (CLM). There are numerous MRF problems that cannot be efficiently implemented with graph cut and belief propagation, and so by introducing ways to effectively combine local solutions, we present a method to dramatically improve many of the pre-existing local minima algorithms. The proposed approach is shown to be effective on pairwise stereo MRF compared with graph cut and sequential tree re-weighted belief propagation (TRW-S). Additionally, we tested our algorithm against belief propagation (BP) over randomly generated 30×30 MRF with 2×2 clique potentials, and we experimentally illustrate CLM's advantage over message passing algorithms in computation complexity and performance.

1   Introduction

Recently, there are great interests in energy minimization methods over MRF. The pairwise MRF is currently the most prominent MRF which became most frequent subject of study in computer vision. Also, in the forefront, there is a movement toward 2×2 and higher clique potentials for de-noising and segmentation problems [1,2,3,4,5,6]. They claim better performance through larger clique potentials that can give more specified constraints.

However, the conventional belief propagation which has been so effective in the pairwise MRF, is shown to have severe computational burden over large cliques. In a factor graph belief propagation, the computational load increases exponentially as the size of clique increases, although for the linear constraint

D. Forsyth, P. Torr, and A. Zisserman (Eds.): ECCV 2008, Part IV, LNCS 5305, pp. 298–311, 2008.

MRFs, the calculation can be reduced to time linear [6,3]. Graph cut based methods are also introduced for energy functions with global constraints and larger clique potentials with pair-wise elements [5,4]. However, these methods are targeted toward a specific category of energy functions and the applicability limitations are high.

A practical and proven method for minimizing even the higher order MRFs is simulated annealing. Gibbs sampler, generalized Gibbs sampler, data-driven Markov chain Monte Carlo and Swendsen-Wang cut were respectively applied to de-noising, texture synthesizing and segmentation problems that involved large clique potentials [7,8,9,10]. However, simulated annealing is considered impractically slow compared to belief propagation and graph cut even in pairwise MRFs [10,11]. More recently, simulated annealing has been modified by localized temperature scheduling and additional window scheduling to increase its effectiveness [12,13].

Another approach that is often being ignored is the greedy local minimum algorithms. With the introductions of theoretically sound graph cut and belief propagation over pairwise MRF, older methods such as ICM [14] and various gradient descent methods are often disregarded as an under-performing alternatives [11]. However, methods like ICM and other local minimum algorithms do not have any constraints over the size of cliques in MRF. Gradient descent method was readily implemented over 5×5 and 3×2 clique potential in the denoising problem [2,1]. Texture synthesis and segmentation problems were model by high order MRF and the energy was minimized using ICM [15]. Thus, when considering both the computational time and performance, local greedy methods that depend largely on the initial states are still viable in many of the high order MRFs.

In this paper we propose a new algorithm to effectively combine these local minima to obtain a solution that is closer to the global minimum state. First, local solutions are calculated from various initial states. Then, they are combined by random partitioning process such that the energy is minimized. The proposed Combined Local Minima (CLM) approach is very simple but it can effectively find lower energy state than graph cut and belief propagation. CLM is tested on the pairwise stereo MRFs provided by [16,17,18,19,20], and it is shown that the performance can be better than graph cut [21] and sequential tree reweighted belief propagation (TRWS) [22]. We also performed tests over randomly generated 2×2 clique MRFs, and showed that the proposed method converges not only faster but finds lower energy state than belief propagation. However, the biggest advantage of the proposed algorithm is that it can bring further improvement over various local minima algorithms that are applicable to general energy functions.

Section 2 will review ICM algorithm. Section 3 presents proposed CLM. In the experiment section, CLM is shown to be competative over pairwise MRF and superior over 2×2 clique MRF. The paper will close with conclusion and possible future work.

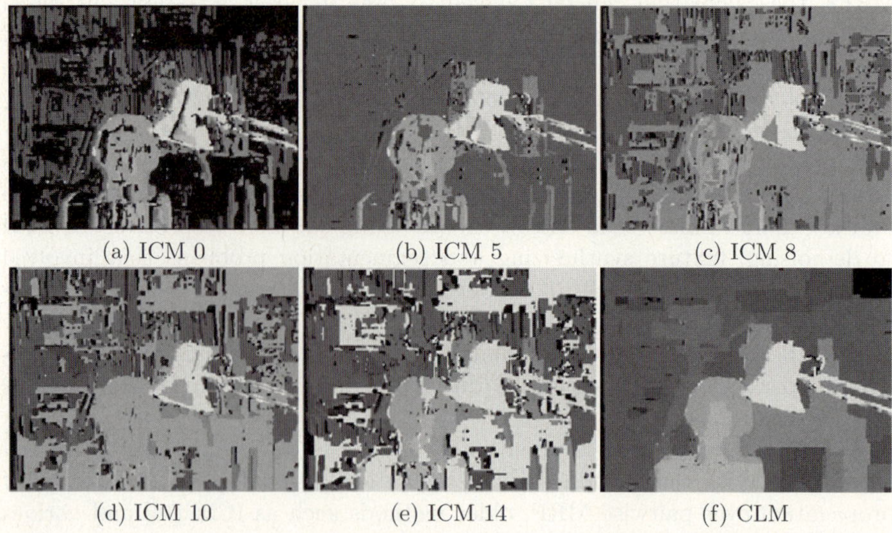

<table>
<tr><td>(a) ICM 0</td><td>(b) ICM 5</td><td>(c) ICM 8</td></tr>
<tr><td>(d) ICM 10</td><td>(e) ICM 14</td><td>(f) CLM</td></tr>
</table>

Fig. 1. (a) to (e) show ICM solutions from different initial states. Homogeneous states of disparity 0, 5, 8, 10, and 14 are respectively used as the initial states of (a), (b), (c), (d), and (e). Combined local minima algorithm effectively combines these ICM solutions into an lower energy state (f).

2   Iterated Conditional Mode (ICM)

For obtaining the local minima states, there are various different methods to choose from. However, in this section, iterated conditional mode will be reviewed for discrete MRF.

MRF consists of a set of nodes $V = \{v_1, v_2, ...v_N\}$. For each nodes $v \in V$, a label l can be assigned from a set L, producing a state x. The number of nodes in V is denoted as N, and the number of labels in L is Q. In a discrete labelling problem, the number of possible states will be Q^N. The energy function $\varphi(x)$ is a function of N dimension vector $x = (x_1, x_2, x_3, ..., x_N)$.

ICM is a simple algorithm that determines the minimum energy at each vertex $v \in V$. For a non-convex energies (such as pairwise energy function), ICM produces a local minima solutions that depends upon the starting state. Following pseudo code minimizes the energy function $\varphi(x)$ in a labelling problem with nodes $v_i \in \{v_1, v_2, ...v_N\}$ and labels $l_j \in L = \{l_1, l_2, ...l_Q\}$.

Iterated Conditional Modes: ICM
1. Determine the initial state x
2. Repeat until no further energy minimization.
3.   For $i = 1$ to $i = N$
4.   For $j = 1$ to $j = Q$
5.   Assign l_j to v_i
                if $\varphi(x_1, ..., x_i = l_j, .., x_N) < \varphi(x_1, ..., x_N)$.

The problem of choosing the right initial state is the big disadvantage of ICM. Figure 1 (a) to (e) show the ICM solutions for Tsukuba stereo MRF. The solutions in Figure 1 are obtained with different initial homogeneous states. Even though the energy minimization cannot be low as graph cut or belief propagation, the computational time is very small because the comparative inequality of step 5 can be evaluated in $O(1)$ for most of the energy functions, including pairwise functions. ICM guarantees to converge but the performance is very poor as shown in figure 1. Also, because of its simplicity, ICM can be applied to high order MRF with larger cliques where graph cut and BP are having problems with.

3   Combined Local Minima

The simplest way to overcome the initial state dilemma of greedy algorithm is to take multiple initial states. Among the multiple local minima obtained from ICM, the lowest energy state can be chosen as the final solution. However, this approach is problematic for MRF with very large dimensions, and obtaining comparable solutions to graph cut and belief propagation is near impossible. Thus, greedy algorithms are not often used for the MRF problems. In this section, however, we assume that each local minima solution has a subset that is a match to a subset of global minima state. We believe that a random partition combination of local minima solution can be used to obtain energy level closer to global minima.

3.1   Combined Space

In a typical labelling problems such as segmentation and stereo, the nodes are presented by the pixel positions. The number of all possible states for such set up will be Q^N. However, the combination of local minima will produce a smaller space. In this section, the general notations will be defined for the proposed algorithm.

The solution space for N number of nodes and set of labels $L = \{l_1, l_2, ..., l_Q\}$ is $\Omega = \{L \times L \times, ..., \times L\}$, where Ω is N dimension space. However, we are proposing to minimize energy over reduced solution space that is obtained from the combinations of local solutions. First, k number of local minima set $\{s_1, s_2, ..., s_k\}$ are found using ICM such that each s_i is N dimension vector having following labels.

$$s_i = \left(l_{s_i}^1, l_{s_i}^2, l_{s_i}^3, ..., l_{s_i}^N\right). \tag{1}$$

$l_{s_i}^j$ is the label value for $v_j \in V$ node of s_i local minima state. $\Omega_S \subseteq \Omega$ is the new solution space composed of the new sets of labels $L_j \subseteq L$.

$$\Omega_S = (L_1 \times L_2 \times L_3 \times, ..., \times L_N). \tag{2}$$

L_j is obtained from the set of the local solutions such that $L_j = \{l_{s_1}^j, l_{s_2}^j, l_{s_3}^j, ..., l_{s_k}^j\}$.

The search for the minimum energy state will be over Ω_S, although there is no guarantee that the global minima is in the reduced space. Choosing the right combinations of local minima for CLM will admittedly be heuristic for each problem. More on the choices of local minima will be discussed in the later sections. However, when the sufficient number and variety of local minima are present in the proposed CLM, the solution space will be the original Ω.

3.2   Combined Local Minima

The proposed combinatorial algorithm for local minima is very simple and intuitive, however, it is shown to be very effective over traditional pairwise MRF and randomly generated 2×2 clique MRF. For the pairwise MRF, QPBO algorithm can effectively combine two minima solutions together [23]. However, QPBO algorithm is viable only for pairwise MRF, thus we rely on random partitioning technique which is simpler and can be applicable to higher order MRF.

We propose following algorithm to minimize energy from a set of local minima. CLM partitions both current state and local minima states and replaces a part of current state to one of the local minima states' such that energy is reduced for current state. It is a basic greedy algorithm over partitioned states.

Combined Local Minima: CLM

1. Given k number of local minima states from k different initial states,
   $$s_1 = \left(l_{s_1}^1, l_{s_1}^2, l_{s_1}^3, ..., l_{s_1}^N\right), \ s_2 = \left(l_{s_2}^1, l_{s_2}^2, l_{s_2}^3, ..., l_{s_2}^N\right),$$
   $$...,s_k = \left(l_{s_k}^1, l_{s_k}^2, l_{s_k}^3, ..., l_{s_k}^N\right).$$
   and the current state $x = \left(l_x^1, l_x^2, l_x^3, ..., l_x^N\right),$
   repeat for specified number of iterations.

2. Randomly partition both the current state x and local minima states $s_1, s_2, ..., s_k$ into same m number of partitions such that
   $$x = \left(V_x^1, V_x^2, V_x^3, ..., V_x^m\right), \ s_1 = \left(V_{s_1}^1, V_{s_1}^2, V_{s_1}^3, ..., V_{s_1}^m\right),$$
   $$s_2 = \left(V_{s_2}^1, V_{s_2}^2, V_{s_2}^3, ..., V_{s_2}^m\right), ...,s_k = \left(V_{s_k}^1, V_{s_k}^2, V_{s_k}^3, ..., V_{s_k}^m\right).$$

3. Repeat for $i = 1$ to $i = m$.

4. Make $k + 1$ proposal states $\{x_0, x_1, x_2, ..., x_k\}$ in combinations of current state x and $s_1, ..., s_k$ such that V_x^i vector partition of x is replaced by the V_s^i of local minima states. See below.
   $$x_0 = x = \left(V_x^1, V_x^2, V_x^3, ..., V_x^m\right), \ x_1 = \left(V_x^1, V_x^2, ..., V_{s_1}^i, ..., V_x^m\right),$$
   $$x_2 = \left(V_x^1, V_x^2, ..., V_{s_2}^i, ..., V_x^m\right), ..., x_k = \left(V_x^1, V_x^2, ..., V_{s_k}^i, ..., V_x^m\right).$$
   Among set $S = \{x_0, x_1, ..., x_k\}$, take the lowest energy state as the current state.

The computational complexity of CLM depends largely on the complexity of evaluating $\varphi(x_i)$. If $\varphi(x_i)$ is needed to be calculated in $O(N)$, ICM's complexity will be $O(kmN)$. If m is randomly chosen, the worst case would be for $m = N$, and the time complexity will be $O(kN^2)$ per iteration. However, if the maximum clique size is small compared to MRF size, both the worst and best complexity will be $O(kN)$ because only V^i and areas around V^i are needed to be evaluated to find the lowest energy among $S = \{x_0, x_1, ..., x_k\}$. Also, the complexity can still be lowered using various computation techniques such as integral image method [24].

The proposed algorithm is greedy and guarantees that the energy does not increase for each iteration. Figure 2 shows the iterative results of the proposed

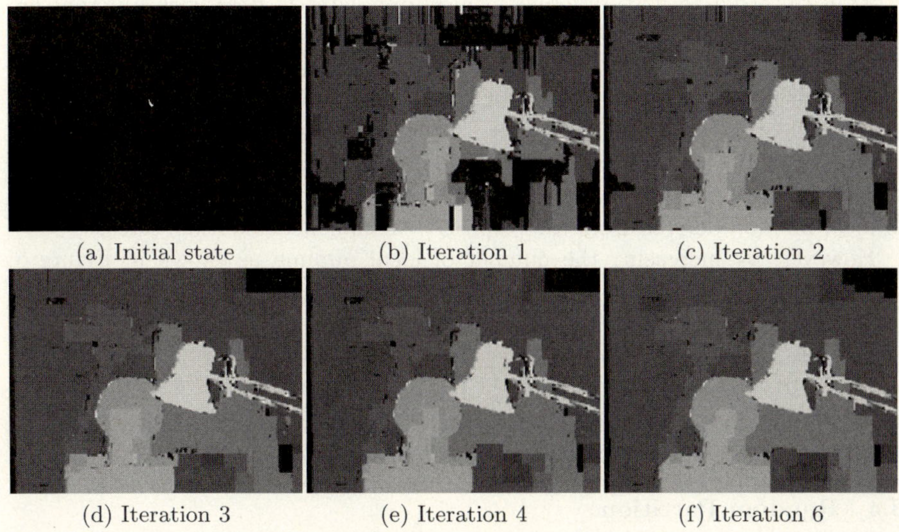

(a) Initial state   (b) Iteration 1   (c) Iteration 2

(d) Iteration 3   (e) Iteration 4   (f) Iteration 6

Fig. 2. (a) shows the initial state of CLM. (b), (c), (d), (e), and (f) show respectively the first, second, third, fourth, and sixth iterations of combined local minima algorithm.

CLM over Tsukuba stereo pair MRF. $k = 16$ number of local minima were used. Few of local minima are shown in Figure 1. With only a small number of iterations, CLM can output energy minimization result far superior to ICM method, and with enough iterations it can be effective as the message passing and graph cut algorithms.

However, there are two heuristics that must be resolved for CLM. First, it is unclear how current state x and $\{s_1, s_2, ..., s_k\}$ should be randomly partitioned in step 2 of the algorithm. Second, the choice of local minima and the value of k are subject to question. These two issues are important to the performance of the proposed algorithm and the basic guidelines are provided in next subsections.

3.3   Obtaining k Local Minima

It is intuitive to assert that if large number of local minima is used for CLM, the obtained energy will be lower. However, for the price of lower energy, more computational time and memory are required. The right tradeoff between computation resources and desired energy level is essential to CLM. This is both advantage and disadvantage of proposed method because by using CLM, you can control the level of performance and computing resources.

Another factor that contributes to the performance of CLM is the variety of local minima. For example, if all the local minima solutions are same, the energy will not be lowered no matter how many times they are combined. Usually, variety of initial states for ICM result in the variety of minima solutions. However, some heuristics may be needed for obtaining different local minima. We have empirically developed few precept for both of these issues.

Thus, in order to have different local minima states, ICM with different homogeneous initial states were used. See experimental section and Figure 4 and 5. In both of the comparison tests, the number of local minima are set to Q, the number of labels. $\{s_1, ..., s_Q\}$ are obtained from ICM with homogeneous initial state, respectively having labels $l_1, l_2, ..., l_Q$. In both stereo MRF and randomly generated MRF, such initial states resulted in the energy minimization comparable to message passing algorithms. Thus, the rule of thumb is to use Q number of local minima derived from the respective homogeneous initial states.

However, by increasing the number of local minima as shown in Figure 5, much lower energy can be achieved with incremental addition to computation time. In Figure 5, CLM200 minimizes energy using total of 200 local minima composed of Q homogeneous initial states and $200 - Q$ number of ICM solutions obtained from random initial states. CLM200 achieves much lower energy than belief propagation. Although, random initial states are used here, more adaptive initial states can also be applied for different problems.

3.4   Random Partition

In this paper, we use rectangular partition method for step 2 of CLM algorithm, much like window annealing of [13]. See Figure 3 (a). 4 integers are randomly chosen, and MRF can be partitioned accordingly. Such method is used because of simplicity of computation and the fact that it can commodate the square lattice structure of digital images. Furthermore, by having rectangular partitions, the energy value of state can be obtained very fast using integral image technique [24], which was used for the stereo pair experiment. However, the integral image technique is not essential to the CLM. In Figure 5, integral image technique is not used during the operations of CLM, and it has superior performance over belief propagation.

For MRFs with random structure, rectangular partition can not be applied. A possible random partitioning algorithm that can be used is the one that was applied in Swendsen-Wang cut algorithm [10]. In Swendsen-Wang cut, the edges between the nodes are probabilistically cut, and the connected nodes after the random cut would make a single cluster. This method was not used in the experimentation section because of needless complexity over square lattice MRF. Again, the partition method can be specified to each problem at hand. However, V^i should be no larger than N obviously, and there should be a positive probability that size of V^i could be 1, so that the optimization can be occur over single nodes.

4   Experiments

In order to show the effectiveness of the proposed CLM, we compared it's performance with graph cut and TRW-S over pairwise stereo MRF. Additionally, window annealing (WA) [13] results are included in the test. Pairwise stereo MRF is known to be effectively optimized by alpha expansion graph cut (GC) and TRW-S [21,22], but very ill posed for greedy algorithms such as ICM. The experiments were performed over stereo pairs provided by [16,17,18,20,19].

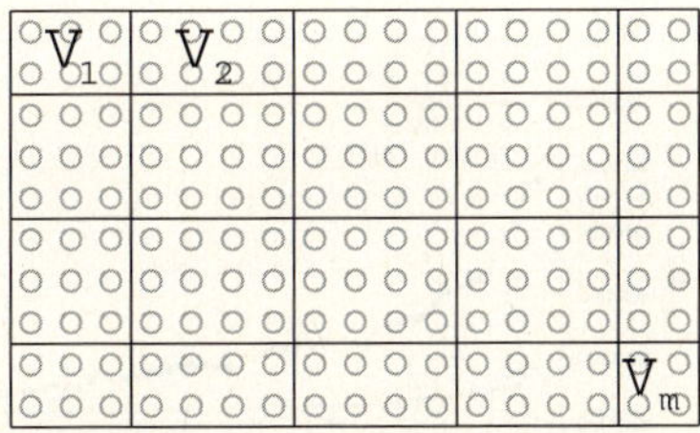

(a) Rectangular Partition

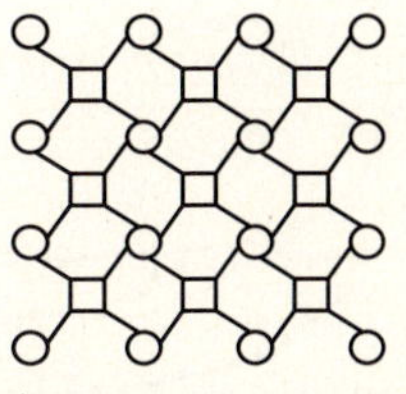

(b) 2 × 2 clique MRF

Fig. 3. (a) shows an example of rectangular partitioning of a square lattice MRF. A state can be partitioned into rectangular clusters $V = \{V^1, V^2, ..., V^m\}$. Such partition method allows simple calculation of energy function by integral image technique which was used for the pairwise MRF test. In this Figure (b), 4 × 4 MRF with 2 × 2 clique potentials is depicted. The circle nodes are $v \in V$. The square factor nodes define the cliques of MRF by connected the neighbors, N_g. MRF is built by assigning random clique potentials from an uniform distribution. In the randomly generated 2 × 2 clique MRF, integral image technique is not used for computational speed up.

Also recently, larger than pairwise clique models are often proposed for vision problems. Gradient descent and belief propagation are used over 2 × 2 and larger clique MRF to attack such problems as de-noising, and shape from shading [6,1,2,3]. Thus, we tested our algorithm over randomly generated MRF with 2 × 2 clique potentials, see Figure 5 (a). Alpha expansion algorithms cannot deal with randomly generated larger than pairwise MRF, and it was excluded from the test. CLM reaches a lower energy faster than belief propagation (BP) and WA methods. The computational complexity of proposed method is $O(kN)$, allowing CLM to be a practical minimization scheme over large clique MRFs. All computations are done over a 3.4GHz desktop.

4.1   Pair-Wise Stereo MRF

Pairwise and sumodular MRF is most common MRF used in computer vision. Also, it has been the subject to many comparative tests. Particularly, the stereo MRF has been an frequent in comparison tests of energy minimization methods [11,25,22]. However, the performance differences between two state of art methods, graph cut and message passing algorithms, are still not clear when the computational time is an issue. Although, TRW-S may eventually find lower energy than graph cut, it can take many more iterations to do so. In some cases, TRW-S is faster and finds lower energy than graph cut. In this test, we tried to present energy functions that are fair to both graph cut and TRW-S. As shown in Figure 4, for Cones and Bowling2 MRF, TRW-S clearly outperforms the graph cut. Otherhand, for Teddy and Art MRF, the graph cut finds lower energy much faster. The performance of each methods seem to depend largely upon the strength of discontinuity costs. Simulated annealing, otherhand, depends large on the temperature scheduling. Awhile WA is competitive with previous methods in speed, usually it could not find lower energy. Although it is possible to tweak

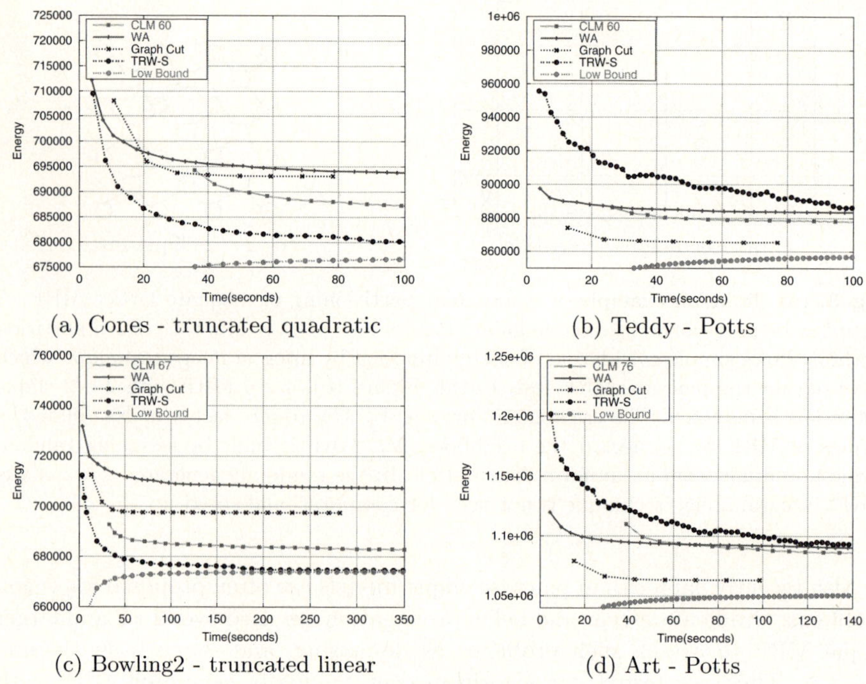

(a) Cones - truncated quadratic   (b) Teddy - Potts

(c) Bowling2 - truncated linear   (d) Art - Potts

Fig. 4. (a)Cones uses truncated quadratic discontinuity cost. (c) Bowling2 is the result for truncated linear discontinuity cost. (b) Teddy and (d) Art use Potts discontinuity cost. CLM 60 means 60 local minima are used in CLM algorithm. The CLM's performance is shown to be in-between TRW-S and GC. The performance difference to state-of-art methods are very small, however, CLM performance does not seem to strongly vary according to the discontinuity model apposed to TRW-S and graph cut.

the annealing scheduling for lower minimization, we kept the same temperature and window scheduling of [13].

For the energy function, gray image Birchfield and Tomasi matching costs [26] and Potts, truncated linear and truncated quadratic discontinuity cost are used.

$$\varphi(x) = \sum_{p \in V} D(p) + \sum_{(p,q) \in N_g} V(p,q). \tag{3}$$

$D(p)$ is a pixel-wise matching cost between left and right image. $V(p,q)$ is pair-wise discontinuity costs. The implementations of graph cut and TRW-S by [11,21,27,28,29,22] are used in this experiment.

For the implementation of CLM, Q number of local minima ICM solutions are obtained from following set of initial states $\{(0,0,...,0),(1,1,...,1),...,(Q-1,Q-1,...,Q-1)\}$. As mentioned before, a rule of thumb seems to be Q number of local minima with homogeneous initial states, especially if the MRF is known to have smoothness constraint. For the state partition technique of step 2 of CLM, a simple rectangular partitioning method is used, see Figure 3.

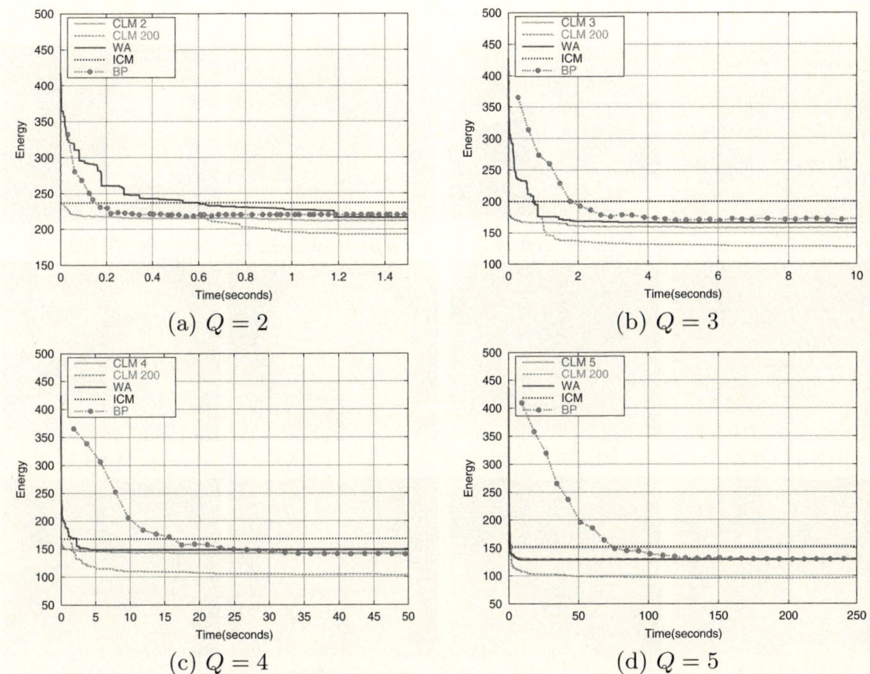

Fig. 5. Energy versus time results of max product BP, ICM, WA, and CLM over 30×30 randomly generated MRF. Figure (a), (b), (c), and (d) respectively have label size $Q = 2$, $Q = 3$, $Q = 4$, and $Q = 5$. CLM using $k = Q$ and $k = 200$ number of local minima are performed for each random MRF. The increase in the local minima allows lower energy state to be achieved in exchange for computation time and memory. However, such price is very small compared to the computation time of BP.

Figure 4 shows energy vs time graph results using Potts, truncated linear, truncated quadratic discontinuity model. Qualitatively, there is a very small difference between TRW-S, graph cut, WA, and CLM, see Figure 6. However, the energy versus time graphs show more edifying comparison. The first iteration of the CLM takes much longer time than the other iterations because all the local solutions are needed to be computed. Overall performance of the proposed CLM stands in the middle of graph cut and TRW-S. However, compared with window annealing, CLM outperforms it everywhere except for the initial calculations.

4.2   Randomly Generated 2 × 2 Clique MRF

However, the biggest advantage of proposed CLM is that the computational complexity does not increase exponentially. Belief propagation based methods, however, the time complexity of message calculation goes up exponentially as the number of clique size increases [6]. In this section, the proposed CLM is tested over randomly generated 2×2 clique MRF. Below equation describes the energy function as sum of clique potentials.

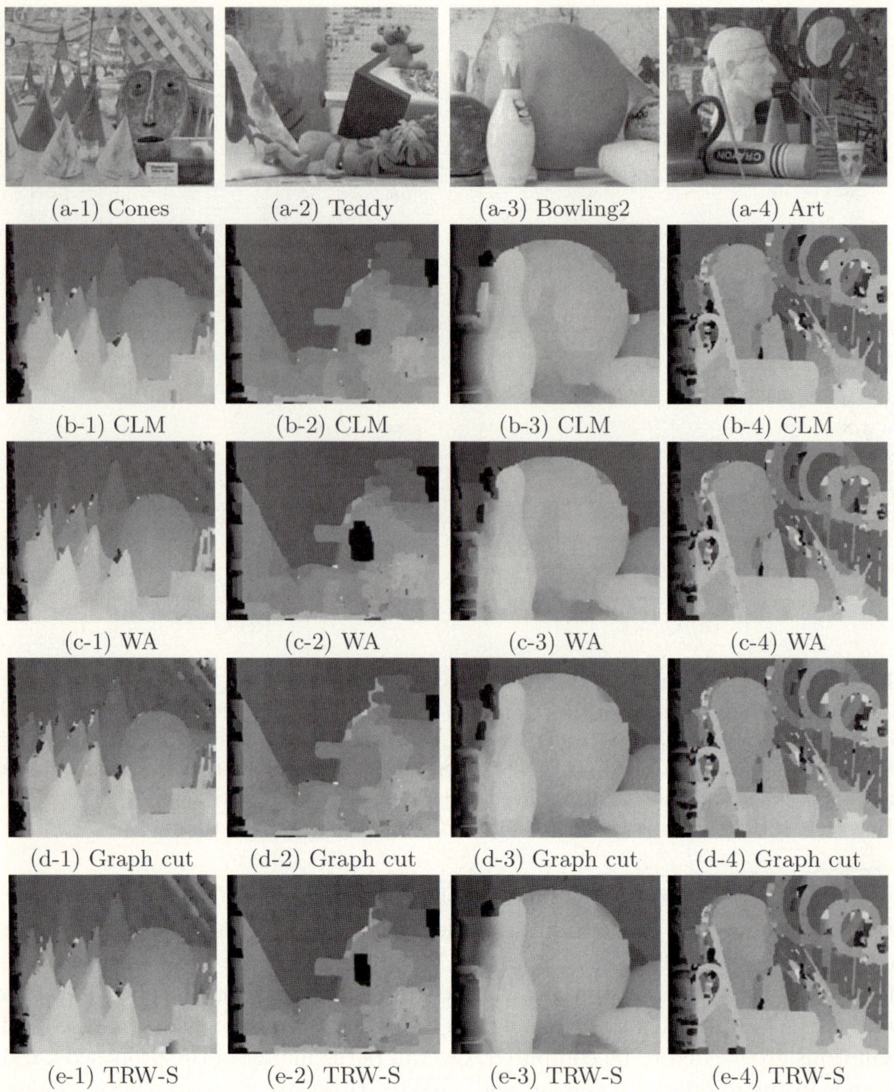

Fig. 6. This figure shows the qualitative stereo energy minimization results at roughly at same computation time. (a-1) to (a-4) are left reference stereo images. (b-1) to (b-4) are the results of proposed CLM. (c), (d), and (e) respectively show the results of window annealing, graph cut and TRW-S. For each stereo pair, the same energy function is used. The qualitative differences between 4 methods are very small, except for Teddy image where graph cut's lower energy makes a difference over the roof area of image. Otherwise, the energy difference between 4 methods are small enough to make no visible differences.

$$\varphi(x) = \sum_{(p,q,r,s)\in N_g} V(p,q,r,s) \tag{4}$$

The clique potential $V(p,q,r,s)$ is a function of 4 dimensioned vector. The value of each $V(p,q,r,s)$ is randomly assigned from an uniform distribution $[0,1]$. In Figure 3 (b), the square factor nodes are connected to 4 variable nodes p,q,r,s. 30×30 variable nodes with 2×2 clique potentials are generated for the comparison test. The energy minimization results of CLM, ICM, and BP are shown in Figure 5; (a) to (d) are the results obtained for MRFs with label size $Q = 2$ to $Q = 5$, respectively.

For the implementation of belief propagation, the factor nodes are transformed into variable nodes with Q^4 number of labels having corresponding $V(p,q,r,s)$ as the unary costs. The pairwise potentials are assigned either 0 or ∞ based on the consistency requirement. CLM is implemented using $k = Q$ local minima and also for $k = 200$ local minima. For $k = Q$, local minima are found by ICM over Q homogeneous states like the stereo problem, even though the smoothness assumption is no longer viable in this problem. For CLM200, additional $200 - Q$ ICM minima obtained from random initial states are used. Same rectangular partitioning is used but the integral image technique is not used.

In these 4 tests, it is clear that the proposed CLM converges faster than BP and WA. The difference from BP is more evident for Figure 5 (c) and (d), because even though BP is fast as CLM200 for $Q = 2$ Figure 5 (a), but as the label size increases, BP could not keep up with speed of CLM. Thus, as the number of labels and clique size become larger, message passing algorithms will become practically ineffective awhile proposed CLM can maintain reasonable computational time. Furthermore, with larger number of local minima, the CLM can reach much lower energy than BP and WA with comparably insignificant addition to computation resources.

5   Conclusion and Future Work

In this paper, we propose a new a method to combine local minima solutions toward more global minimum by random partition method. CLM's performance is compared with state-of-art energy minimization methods over most well known pairwise stereo MRF. Combined local minima is shown to be effective as graph cut and TRW-S. Furthermore, tests over randomly generated 2×2 clique MRFs show that the computation complexity of CLM is much smaller than traditional message passing algorithms as the clique and label size become larger.

Additionally, we included window annealing method in the experiment. However, due to heuristics of simulated annealing and the proposed method, it is hard to say which method is better. Nevertheless, both algorithms show clear advantages over the high ordered MRF compared to existing methods awhile maintaining competitiveness in the pairwise MRFs. We hope that such conclusion will encourage other computer vision researchers to explore more complex MRFs involving larger clique potentials. In the future, MRF with random structure (non square lattice) will be studied using Swendsen Wang cut like partition method.

Acknowledgement

This research was supported in part by the Defense Acquisition Program Administration and Agency for Defense Development, Korea, through the Image Information Research Center under the contract UD070007AD, and in part by the MKE (Ministry of Knowledge Economy), Korea under the ITRC (Information Technolgy Research Center) Support program supervised by the IITA (Institute of Information Technology Advancement) (IITA-2008-C1090-0801-0018).

References

1. Roth, S., Black, M.J.: Steerable random fields. In: ICCV (2007)
2. Roth, S., Black, M.J.: Field of experts: A framework for learning image priors. In: CVPR (2005)
3. Potetz, B.: Efficient belief propagation for vision using linear constraint nodes. In: CVPR (2007)
4. Kohli, P., Mudigonda, P., Torr, P.: p^3 and beyond: Solving energies with higher order cliques. In: CVPR (2007)
5. Rother, C., Kolmogorov, V., Minka, T., Blake, A.: Cosegmenation of image pairs by histogram matching- incorporating a global constraint into mrfs. In: CVPR (2006)
6. Lan, X., Roth, S., Huttenlocher, D., Black, M.J.: Efficient belief propagation with learned higher-order markov random fields. In: Leonardis, A., Bischof, H., Pinz, A. (eds.) ECCV 2006. LNCS, vol. 3952, pp. 269–282. Springer, Heidelberg (2006)
7. Geman, S., Geman, D.: Stochastic relaxation, gibbs distributions, and the bayesian restoration of images. PAMI 6 (1984)
8. Zhu, S.C., Liu, X.W., Wu, Y.N.: Exploring texture ensembles by efficent markov chain monte carlo: Toward a trichromacy theory of texture. PAMI 22(6) (2000)
9. Tu, Z., Zhu, S.C.: Image segmentation by data-driven markov chain monte carlo. PAMI 24 (2002)
10. Barbu, A., Zhu, S.C.: Generalizing swendsen-wang cut to sampling arbitrary posterior probabilities. PAMI 27 (2005)
11. Szeliski, R., Zabih, R., Scharstein, D., Veksler, O., Kolmogorov, V., Agarwala, A., Tappen, M., Rother, C.: A comparative study of energy minimization methods for markov random fields. In: Leonardis, A., Bischof, H., Pinz, A. (eds.) ECCV 2006. LNCS, vol. 3952, pp. 16–29. Springer, Heidelberg (2006)
12. Woodford, O.J., Reid, I.D., Torr, P.H.S., Fitzgibbon, A.W.: Field of experts for image-based rendering. BMVC (2006)
13. Jung, H.Y., Lee, K.M., Lee, S.U.: Window annealing over square lattice markov random field. ECCV (2008)
14. Besag, J.: On the statistical analysis of dirty pictures (with discussion). Journal of the Royal Statistical Society Series B 48 (1986)
15. Mignotte, M.: Nonparametric multiscale energy-based model and its application in some imagery problems. PAMI 26 (2004)
16. http://vision.middlebury.edu/stereo/
17. Scharstein, D., Szeliski, R.: A taxonomy and evaluation of dense two-frame stereo correspondence algorithms. In: IJCV (2002)
18. Scharstein, D., Szeliski, R.: High-accuracy stereo depth maps using structured light. In: CVPR (2003)

19. Hirshmuller, H., Szeliski, R.: Evaluation of cost functions for stereo matching. In: CVPR (2007)
20. Scharstein, D., Pal, C.: Learning conditional random fields for stereo. In: CVPR (2007)
21. Boykov, Y., Veksler, O., Zabih, R.: Fast approximate energy minimization via graph cuts. PAMI 23 (2001)
22. Kolmogorov, V.: Convergent tree-reweighted message passing for energy minimization. PAMI 28 (2006)
23. Lempitsky, V., Rother, C., Blake, A.: Logcut - efficient graph cut optimization for markov random fields. In: ICCV (2007)
24. Crow, F.: Summed-area tables for texture mapping. SIGGRAPH (1984)
25. Tappen, M.F., Freeman, W.T.: Comparison of graph cuts with belief propagation for stereo, using identical mrf parameters. In: ICCV (2003)
26. Birchfield, S., Tomasi, C.: A pixel dissimilarity measure that is insensiitive to image samplin. PAMI 20 (1998)
27. Kolmogorov, V., Zabih, R.: What energy functions can be minimized via graph cuts? PAMI 26 (2004)
28. Boykov, Y., Kolmogorov, V.: An experimental comparison of min-cut/max-flow algorithms for energy minimization in vision. PAMI 26 (2004)
29. Wainwright, M.J., Jaakkola, T.S., Willsky, A.S.: Map estimation via agreement on trees: Message-passing and linear-programming approaches. IEEE Trans. Information Theory 51(11) (2005)

Differential Spatial Resection - Pose Estimation Using a Single Local Image Feature

Kevin Köser and Reinhard Koch

Institute of Computer Science
Christian-Albrechts-University of Kiel
24098 Kiel, Germany
`{koeser,rk}@mip.informatik.uni-kiel.de`

Abstract. Robust local image features have been used successfully in robot localization and camera pose estimation; region tracking using affine warps is considered state of the art also for many years. Although such correspondences provide a warp of the local image region and are quite powerful, in direct pose estimation they are so far only considered as points and therefore three of them are required to construct a camera pose. In this contribution we show how it is possible to directly compute a pose based upon one such feature, given the plane in space where it lies. This *differential correspondence concept* exploits the texture warp and has recently gained attention in estimation of conjugate rotations. The approach can also be considered as the limiting case of the well-known spatial resection problem when the three 3D points approach each other infinitesimally close. We show that the differential correspondence is more powerful than conic correspondences while its exploitation requires nothing more complicated than the roots of a third order polynomial. We give a detailed sensitivity analysis, a comparison against state-of-the-art pose estimators and demonstrate real-world applicability of the algorithm based on automatic region recognition.

1 Introduction

Since the first description of spatial resection from 3 points by Grunert[7] in 1841, many people have worked on pose estimation or the so called P3P problem [5,31,6,8]. PnP stands for pose estimation from n points and is underconstrained for $n < 3$ unless further information is incorporated. In this work we derive how a variation of the problem may be solved, namely when only a single affine image feature (cf. to [24] for a discussion) can be identified with a known 3D space surface with orthophoto texture. Additionally to the traditionally used 2D-3D point correspondence, such an image-model relation provides a local linear texture warp between the image and an orthophoto of the surface. This warp can be interpreted as the Jacobian of the perspectivity between the image and the 3D surface's tangent plane and we show that it determines the open degrees of freedom. The novel approach allows to estimate a perspective camera's pose based upon only one image-model correspondence, which is particularly interesting in

D. Forsyth, P. Torr, and A. Zisserman (Eds.): ECCV 2008, Part IV, LNCS 5305, pp. 312–325, 2008.
© Springer-Verlag Berlin Heidelberg 2008

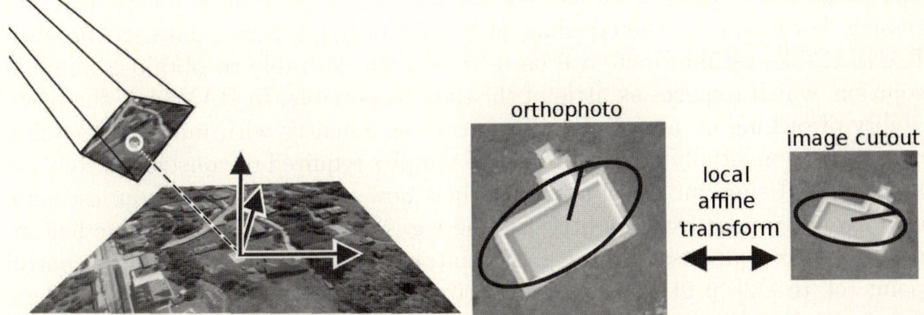

Fig. 1. Differential Spatial Resection exploiting Perspectivity. This figure shows an aerial camera observing a ground plane (left image). If the internal camera calibration is removed, the two images are related by a perspectivity. The projection of some point on the plane and the linear transform of the surrounding region provide 6 constraints for the 6 degrees of freedom for pose estimation. In the right part we see an MSER feature correspondence between an orthophoto and the unknown camera image providing a locally affine texture transform. The primitive for correspondence can be imagined as an infinitesimally small ellipse with orientation.

robot localization [29], initialization or recovery in camera tracking[3,32] or determining the pose of a detected object[30]. In these applications, often SIFT[18] or MSER[21] features are used nowadays, which cover some image *region* ideally corresponding to a surface in the scene. In [3] even the normal of such local surface regions is estimated and also [29] performs stereo from three cameras on a robot. However, in all of the above cited approaches, the correspondences are geometrically handled as points when it comes to initialization or direct pose estimation, although they carry much more information. Therefore, by now at least three of such robust feature correspondences were required to directly estimate a camera or object pose. In contrast, in this contribution we demonstrate how one affine image-model correspondence is already sufficient to estimate the pose.

The exploited primitive can also be seen as the limiting case where the three 3D points of Grunert's solution come infinitesimally close, allowing for what we call *differential spatial resection*. The concept of such correspondences has lately been proposed in [15] for estimation of the infinite homography and is displayed in fig.1. The question we answer is: Given a local affine transform between a region in some view and an orthophoto, how can we compute a homography with this transform as its local linearization and what camera pose belongs to it, given that the homography maps from a known world plane to the camera plane ? Furthermore, we show in section 4.1 the relation to pose estimation from conics, which is essentially a squared formulation of our approach although providing one degree of freedom less.

The proposed algorithm belongs to the set of minimal solvers, which exploit n DOF (degrees of freedom) in some observation to estimate a model with also n DOF. Such solutions are not targeted to produce ultimate optimal estimates but initial start values from as little data as possible, suitable for further processing. For instance, when dealing with small or hand-clicked data sets or when RANSAC-like estimators[6] are used, it is often desirable to obtain a minimal solution, which requires as little of the data as possible. In RANSAC, the probability of picking an all-inlier-set from correspondences with many mismatches depends exponentially on the number of samples required to construct a solution hypothesis. Using our novel approach, it is now possible to obtain an estimate of a camera's pose from as little as one e.g. MSER[21] or comparable feature (cf. to [24] for a discussion) or e.g. one suitable photogrammetric ground control point (cf. to [22], p.1111) in an image, given the local plane in 3D space where it is located and its texture.

For instance, when a feature descriptor is recognized in an unknown image, the 6 DOF camera or object pose can be obtained by the methods given here. To improve the pose estimation result, gradient based optimization techniques[19,16] can be applied between the current view and a reference texture. The reference texture can either be an orthophoto (cf. to [22], p.758) or any other view with sufficient resolution for which the warp to an orthophoto is known. When several such feature correspondences and the camera poses are optimized at once, this is similar to the approach of Jin et al.[11]. However, their approach is formulated in a nonlinear fashion only and requires an initialization, comparable to the requirements for bundle adjustment. Since we exploit the perspectivity concept, a plane-to-plane mapping in euclidian space, in section 3 we also present the related work in homography estimation [33,13,10] and projective reconstruction[27], which did not inspect the differential constraints on the perspectivity, because often the calibrated camera case is not considered in projective approaches. The exploitation of the Jacobian of the texture warp has been proposed though for the estimation of a conjugate rotation in [15].

Notation. To improve the readability of the equations we use the following notation: Boldface italic serif letters $\boldsymbol{x}$ denote Euclidean vectors while boldface upright serif letters $\mathbf{x}$ denote homogeneous vectors. For matrices we do not use serifs, so that Euclidean matrices are denoted as A and homogeneous matrices are denoted as A, while functions $\mathtt{H}\,[\boldsymbol{x}]$ appear in typewriter font.

2   Perspectivity

The contribution is based on estimating a transformation between two theoretical planes: The first plane is tangent to a textured surface in 3D and the second plane is orthogonal to the optical axis of a camera. The estimation of the pose is then formulated as the problem of obtaining a perspectivity between these two planes (see figure 1). A 2D perspectivity is a special kind of homography (cf. also to [9], pp. 34), which has only 6 degrees of freedom and which is particularly important for mappings between planes in Euclidian space. We assume a locally planar geometry at the origin of 3D space facing into z-direction and attach x, y-coordinates onto it, which coincide with the x, y coordinates in 3D space. If we now move a perspective pinhole camera to position C with orientation R

(which has rows r_i^T) and with internal camera calibration K, a point $\mathbf{p}_s$ in space is mapped to an image point $\mathbf{p}_i$ by the camera as follows(cf. to [9], p. 157 for details):

$$\mathbf{p}_i = \mathsf{K}(R^\mathsf{T}| - R^\mathsf{T}C)\mathbf{p}_s \tag{1}$$

We assume the internal parameters of our camera to be known and without loss of generality set K to the identity in the following. The method is not restricted to straight-line preserving ideal cameras, but can also be applied with real lenses with distortion, fish-eye lenses or even omni-directional cameras, as long as they have a single center of projection and the equivalent function of the matrix K, which maps rays in the camera coordinate system to positions in the image, is differentiable and invertible.

Now we have a look at the points on our $z = 0$ plane to derive the perspectivity:

$$\mathbf{p}_i = (R^\mathsf{T}| - R^\mathsf{T}C)\mathbf{p}_{s,z=0} = (r_1\ r_2\ r_3\ - R^\mathsf{T}C)(x\ y\ 0\ 1)^\mathsf{T} \tag{2}$$

$$= (r_1\ r_2\ - R^\mathsf{T}C)(x\ y\ 1)^\mathsf{T} \simeq (\tilde{r}_1\ \tilde{r}_2\ t)(x\ y\ 1)^\mathsf{T} = \mathsf{H}\ \mathbf{p}_p \tag{3}$$

$\tilde{r}_i$ are scaled versions of r_i such that $t_z = 1$ and $\simeq$ means equality up to scale. Obviously, the homography H maps points $\mathbf{p}_p$ of the plane coordinate system to points $\mathbf{p}_i$ in the image coordinate system. H is a perspectivity and depends only on 6 parameters, the pose of the camera. Since H is an object of projective space, it can be scaled without changing the actual transformation. While the perspectivity H acts linearly in projective space $\mathbb{P}^2$, in Euclidian 2D space H is a nonlinear mapping from $\mathbb{R}^2 \to \mathbb{R}^2$ because of the nonlinear homogenization:

$$\mathsf{H}\left[p_p\right] = p_i = \frac{(\mathsf{H}\mathbf{p}_p)|_{1..2}}{(\mathsf{H}\mathbf{p}_p)|_3} \tag{4}$$

In the next section we describe the differential correspondence and how it can be exploited to obtain constraints on H.

3   Differential Correspondence

Progress in robust local features (cf. to [24,23] for a thorough discussion) allows automatic matching of images in which appearance of local regions undergoes approximately affine changes of brightness and/or of shape, e.g. for automated panorama generation[1], scene reconstruction[30] or wide-baseline matching[18,21]. The idea is that interesting features are detected in each image and that the surrounding region of each feature is normalized with respect to the local image structure in this region, leading to about the same normalized regions for correspondences in different images, which can be exploited for matching. The concatenation of the normalizations provides affine correspondences between different views, i.e. not only a point-to-point relation but also a relative transformation of the local region (e.g. scale, shear or rotation). Although such correspondences carry more information than the traditional point correspondence used in estimation of multiple view geometry [9], this additional information is rarely used. Approaches not using point correspondences deal with conic correspondences [14,12], which typically lead to systems of quadratic equations or require lots of matrix factorizations. Schmid and Zisserman[28] investigated the behavior of local curvature under homography mapping. Chum et al. noted in [2] that an affine correspondence is somehow equivalent to three point correspondences: in addition to the center point two further points can be detected in the feature coordinate system (the *local affine frame*). This allowed the estimation of a fundamental matrix from 3 affine feature correspondences (from which 9 point correspondence were generated). A similar idea was also exploited recently in projective reconstruction, where the projection matrix was locally linearized[27] leading to additional constraints in non-linear optimization. The "local sampling" of the affine feature concept on the other hand was also adopted for other epipolar geometry problems, e.g. in [26]. In contrast to the latter we do not sample but use a compact analytic expression for the whole correspondence: We observe that the concatenation of the normalization transformations provides a good approximation to the first order Taylor expansion of the perspectivity, i.e. that the resulting affine transform is the local linearization of the perspectivity, as it has been recently proposed for estimation of the infinite homography[15]:

$$\mathbf{H}\left[\boldsymbol{x}\right] = \mathbf{H}\left[\boldsymbol{x}_0\right] + \left.\frac{\partial \mathbf{H}}{\partial \boldsymbol{x}}\right|_{\boldsymbol{x}_0} (\boldsymbol{x} - \boldsymbol{x}_0) + \ldots \tag{5}$$

$$A \approx \left.\frac{\partial \mathbf{H}}{\partial \boldsymbol{x}}\right|_{\boldsymbol{x}_0} \qquad A \in \mathbb{R}^{2 \times 2} \tag{6}$$

Here $\mathbf{H} : \mathbb{R}^2 \rightarrow \mathbb{R}^2$ is the homography mapping between the image and the orthophoto in Euclidean coordinates and A represents local shear, scale and rotation between the two corresponding features. This fact has been exploited in matching for quite some time but has not been used for pose estimation before.

The considerations so far apply to affine features (e.g. MSER[21]). However, if matches result from weaker features (e.g. DoG/SIFT[18]), the proposed method can also be applied. The main insight is that if a correct match has been established such that the local regions are approximately aligned the affine transform based upon the relative parameters is already nearly correct.

However, since we need an accurate estimate of the Jacobian of the image transformation, it is reasonable even for already affine features to apply a gradient-based optimization of A using the Lucas-Kanade approach [19,16]. When using affine trackers, e.g. such as [3], the optimized information is readily available. We will call the point correspondence plus the local linear warp a *differential correspondence* in the remainder.

4   Pose Estimation from a Differential Correspondence

Having obtained a differential correspondence between a camera image and the textured plane in the origin, the local warp equals the derivative of the perspectivity. This derivative $\partial\mathsf{H}/\partial\boldsymbol{p}_p$ tells us something about the relative scaling of coordinates between the plane in the origin and the image, e.g. if $\boldsymbol{C}$ is large and the camera is far away from the origin $\partial\mathsf{H}/\partial\boldsymbol{p}_p$ will be small, because a large step on the origin plane will result in a small step in the image far away. Actually, $\partial\mathsf{H}/\partial\boldsymbol{p}_p$ carries information about rotation, scale and shear through perspective effects. Since H can be scaled arbitrarily without changing H, we set $\mathsf{H}_{3,3} = 1$ without loss of generality[1] and compute the derivative at the origin:

$$\frac{\partial\mathsf{H}}{\partial\boldsymbol{p}_p}\bigg|_0 = \begin{pmatrix} \tilde{r}_{11} - \tilde{r}_{13}t_1 & \tilde{r}_{12} - \tilde{r}_{13}t_1 \\ \tilde{r}_{21} - \tilde{r}_{23}t_2 & \tilde{r}_{22} - \tilde{r}_{23}t_2 \end{pmatrix} = \begin{pmatrix} a_{11} & a_{12} \\ a_{21} & a_{22} \end{pmatrix} \tag{7}$$

Also, we compute where the origin is projected in our image:

$$\boldsymbol{p}_{origin} = \mathsf{H}(0\ 0\ 1)^{\mathsf{T}} = -R^{\mathsf{T}}C \simeq \boldsymbol{t} \tag{8}$$

Given a differential correspondence, the derivative as well as the projection of the origin are given by the relative parameters of the detected features. This can determine all degrees of freedom of the camera pose, however the overparameterization of the rotation must be resolved: Since $\tilde{R}$ is a scaled rotation matrix, $\tilde{\boldsymbol{r}}_1$ and $\tilde{\boldsymbol{r}}_2$ must be of same length and orthogonal:

$$\tilde{r}_{11}^2 + \tilde{r}_{12}^2 + \tilde{r}_{13}^2 = \tilde{r}_{21}^2 + \tilde{r}_{22}^2 + \tilde{r}_{23}^2 \qquad \wedge \qquad \tilde{\boldsymbol{r}}_1^{\mathsf{T}}\tilde{\boldsymbol{r}}_2 = 0 \tag{9}$$

We can now compute H by first substituting $\boldsymbol{t}$ into eq. (7), then solving for $\tilde{r}_{11}$, $\tilde{r}_{21}$, $\tilde{r}_{12}$ and $\tilde{r}_{22}$ and substituting into eq.(9), leaving us with two quadratic equations in the two unknowns $\tilde{r}_{13}$ and $\tilde{r}_{23}$:

$$(\tilde{r}_{13}t_1 + a_{11})^2 + (\tilde{r}_{13}t_1 + a_{12})^2 + \tilde{r}_{13}^2 = (\tilde{r}_{23}t_2 + a_{21})^2 + (\tilde{r}_{23}t_2 + a_{22})^2 + \tilde{r}_{23}^2 \tag{10}$$

$$(\tilde{r}_{13}t_1 + a_{11})(\tilde{r}_{23}t_2 + a_{21}) + (\tilde{r}_{13}t_1 + a_{12})(\tilde{r}_{23}t_2 + a_{22}) + \tilde{r}_{13}\tilde{r}_{23} = 0 \tag{11}$$

The first equation is about the length and the second about the orthogonality of the $\tilde{r}$-vectors as typical for constraints on rotation matrices. We find it instructive to interpret them as the intersection problem of two planar conics, the *length conic* C_l and the *orthogonality conic* C_o:

$$(\tilde{r}_{13}\ \tilde{r}_{23}\ 1)\mathsf{C}_l(\tilde{r}_{13}\ \tilde{r}_{23}\ 1)^{\mathsf{T}} = 0 \tag{12}$$

$$(\tilde{r}_{13}\ \tilde{r}_{23}\ 1)\mathsf{C}_o(\tilde{r}_{13}\ \tilde{r}_{23}\ 1)^{\mathsf{T}} = 0 \tag{13}$$

$$\mathsf{C}_l = \begin{pmatrix} 2t_1^2 + 1 & 0 & t_1(a_{11} + a_{12}) \\ 0 & -2t_2^2 - 1 & -t_2(a_{21} + a_{22}) \\ t_1(a_{11} + a_{12}) & -t_2(a_{21} + a_{22}) & a_{11}^2 + a_{12}^2 - a_{21}^2 - a_{22}^2 \end{pmatrix} \tag{14}$$

$$\mathsf{C}_o = \begin{pmatrix} 0 & t_1 t_2 + \frac{1}{2} & (a_{21} + a_{22})t_1 \\ t_1 t_2 + \frac{1}{2} & 0 & (a_{11} + a_{12})t_2 \\ (a_{21} + a_{22})t_1 & (a_{11} + a_{12})t_2 & a_{11}a_{21} + a_{12}a_{22} \end{pmatrix} \tag{15}$$

---

[1] This is not a restriction because the only unrepresented value $\mathsf{H}_{3,3} = 0$ maps the origin to the line at infinity and therefore such a feature would not be visible.

Solving for the Pose Parameters. Two conics cannot have more than four intersection points, therefore, we can obtain at most four solutions for our camera pose. To solve the intersection of the two conics we use the elegant method of Finsterwalder and Scheufele[5], which proved also to be the numerically most stable method of the six different 3-point algorithms for spatial resection [8]: Since a common solution of equations (12) and (13) must also fulfill any linear combination of both, we construct a linear combination of both conics, which does not have full rank (zero determinant), but which still holds all solutions. This creates a third order polynomial, which has at least one real root and which can be solved easily:

$$det(\lambda \mathsf{C}_o + (1 - \lambda)\mathsf{C}_l) = 0 \tag{16}$$

The resulting degenerate conic will in general consist of two lines. The intersection of these lines with the original conics is only a quadratic equation and determines the solutions. The resulting R and C have to be selected and normalized in such a way that we obtain an orthonormal rotation matrix (determinant $+1$) and the camera looks towards the plane. We have now obtained up to four hypotheses for the pose of the camera in the object coordinate system (relative to the feature). If there is a world coordinate system, in which the plane is not at the origin, the rigid world transformation has to be appended to the computed pose of the camera. Computing the relative pose in the object coordinate system in general also improves conditioning since the absolute numbers of the object's pose in the world become irrelevant.

Optimization and Tracking. Once initial parameters are obtained it is straightforward to use a 6-parametric gradient-based minimization technique [19,16] to further optimize the camera pose. Note that if we are using a pinhole camera and the feature in 3D is locally planar, instead of optimizing an approximate affine transform we might as well use a 6-parametric homography. Thus measurements may be incorporated from a larger region without making a mistake or an approximation. Even better, since it is possible to use global camera pose parameters, it is easy to optimize even multiple rigidly coupled features (e.g. in a rigid scene). Or, if robustness against outliers is a concern, each of the features provides an individual pose estimate and robust estimation techniques such as RANSAC[6] can be used to obtain a fused solution. If video data is available, the parameters can directly be used for tracking the regions, objects or camera pose over time similar to what is proposed in [11]. However, in this contribution we focus on the geometric aspects of the minimal solution, i.e. where we see a single feature in a single image, without prior knowledge.

4.1   Relation to Conic Correspondence

In this section the differential feature concept is shown to be a simplified version of correspondences of conics, providing more constraints in a linear (instead of quadratic) fashion: In [20] Ma derived a way to determine the pose of a camera from two conics. He noted that a conic has only 5 DOF and thus a single conic is not sufficient to determine the 6 DOF of the camera pose uniquely. A conic

C_S on the space plane of the previous section maps to a conic C_I in the image with the equation

$$C_I = H^T C_S H, \tag{17}$$

where H is the perspectivity of the previous sections. First, we show how the two primitives used in our differential correspondence can be related to conic representations: For each affine feature, e.g. MSER, there exists a local image coordinate system, the *local affine frame*[2], such that coordinates can be specified relative to the size, shear, position and orientation of a feature. Imagine that L takes (projective) points from local feature coordinates to image coordinates:

$$x_I = L x_{LAF} \tag{18}$$

If the same feature is seen in two images, points with identical feature (LAF) coordinates will have the same grey value. The local affine frames of the features in the different images are then called L_1 and L_2 and their concatenation is the first order Taylor approximation H_{Taylor} of the texture warp (e.g. a homography) between the two images at the feature positions:

$$H_{Taylor} = L_1 L_2^{-1} \tag{19}$$

If we now just think of a single image and imagine a small ellipse through the points $(0; \lambda)^T, (\lambda; 0)^T, (0; -\lambda)^T$ and $(-\lambda; 0)^T$ of the local feature coordinate system, this ellipse can be represented by a conic equation in homogeneous coordinates such that points at the ellipse contour fulfill the quadratic constraint:

$$0 = x_{LAF}^T \begin{pmatrix} 1 & & \\ & 1 & \\ & & -\lambda^2 \end{pmatrix} x_{LAF} \tag{20}$$

The LAF described as a conic matrix in image coordinates therefore is

$$C_\lambda = L^T \begin{pmatrix} 1 & & \\ & 1 & \\ & & -\lambda^2 \end{pmatrix} L = L^T R^T \begin{pmatrix} 1 & & \\ & 1 & \\ & & -\lambda^2 \end{pmatrix} RL \tag{21}$$

where R is an arbitrary (homogeneous 2D) rotation matrix, which cancels out. Therefore the first thing to observe is that 2D orientation of the feature is lost in conic representation. A conic has only five degrees of freedom and a conic correspondence therefore imposes at most five constraints on any H. Furthermore, these constraints are quadratic in the entries of H as can be seen from eq. (17). This equation is also essentially a squared version of equation (19). On the other hand, the differential correspondence is only valid locally and introduces inacurracies for larger regions, but it is available when sufficient texture is in the image, while a conic may have any size. However, conics traditionally exploit a special geometric shape (typically an ellipse contour) and ideal perspective cameras and ideal planes, because conic curve estimation in distorted cameras is more involved. In contrast, the differential feature concept can also directly be applied in fish-eye or omnidirectional cameras.

5   Evaluation

In this section the differential correspondence-based pose estimation is evaluated first using synthetic sensitivity experiments. Next, rendered images with known ground truth information are used to evaluate the real-world applicability, where everything has to be computed from image data. In the final experiments, object pose estimation from one feature is shown qualitatively using non-ideal cameras.

Sensitivity to Noise and Internal Calibration Errors. Our evaluation starts with an analysis of the sensitivity to different disturbances. Since the algorithm provides a minimal solution, which translates a 6 DOF differential correspondence into a 6 DOF pose, the pose will adapt to noise in the correspondence. In figure (5) it is shown that for localization accuracies better than 1 pixel in a camera with focal length 500 pixel the camera orientation is on average better than 1 degree and also the direction of the camera center is better than 1 degree. The orientation error is computed from the axis-angle representation of the rotation which transforms the ground truth orientation into the estimated orientation and therefore incorporates all directions. The center error is the angle between the ground truth camera center and the estimated camera center as seen from the 3D feature's position.

To obtain a reasonable noise magnitude for the differential correspondence parameters, we assume that the center of a patch can be localized with a Gaussian uncertainty of zero mean and variance σ_p^2 and that the corners of a square patch of size $(2w + 1) \times (2w + 1)$ pixels can be localized with about the same uncertainty, which can then be propagated to uncertainty for the affine parameters. When creating noisy 6D affine features, we therefore sample the noise from a Gaussian distribution with diagonal covariance depending on one parameter σ_p, which is printed on the x-axis of figure 5. It is remarkable that the errors in orientation and position are highly correlated. This can be explained from the fact that a slightly different differential correspondence results in a slightly different camera orientation. However, since the feature must be projected to about the same position, the camera center has to adapt accordingly. As figure 5 shows, the pose estimation is stable even when the camera is not calibrated correctly, although it can be seen that the resulting pose is disturbed as inherent in minimal solutions. In particular it is clear that an error in principal point results in an error in the pose when the reference feature in 3D is correct. Keep in mind that at focal length 500 a principal point error of ten pixel means that the optical axis is more than $1°$ mis-calibrated.

Solid Angle, Approximation by 3 Points and Comparison with Spatial Resection/POSIT. Using the proposed approach, the affine warp must be measured between the orthophoto and the image under inspection and this requires a region upon which this is done. If the alignment is done using an affine warp, the region should be chosen as small as possible, particularly when the feature is seen from an oblique angle, because in the affine warp model it is assumed that the warp (the Jacobian of the homography) does not change between the

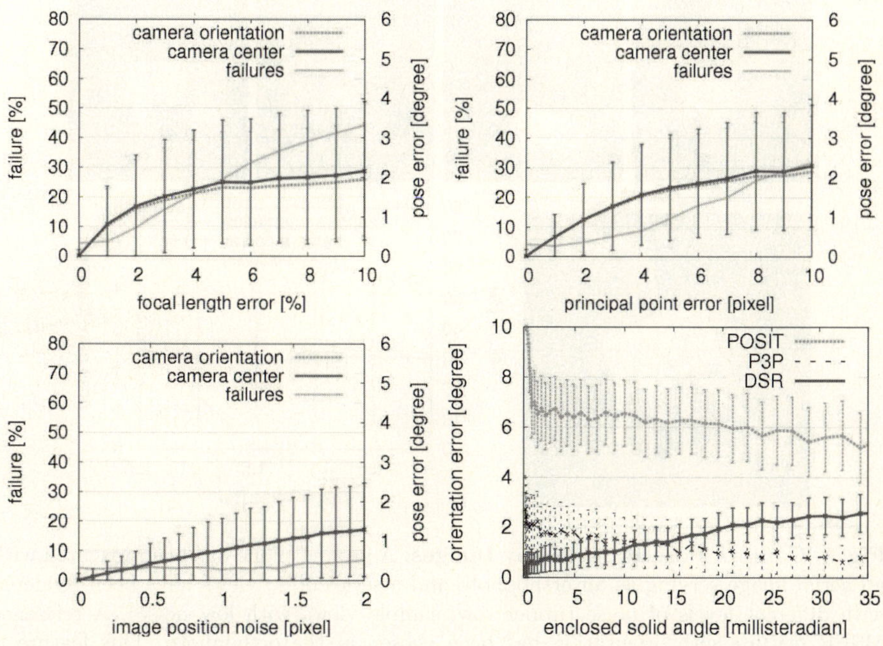

Fig. 2. Sensitivity with Respect to Noise, Calibration and Feature Area. In these experiments, 105.000 random camera poses in front of the $z = 0$ plane have been synthesized (providing the ground truth differential correspondences). In the two top graphs, focal length (500) and principal point (100;100) have been disturbed up to 10% and the resulting error in the best pose is displayed as well as the number of cases, where no solution was possible (or the best pose was more than 5° off). In the lower left graph, Gaussian noise has been added to the 6 parameters of the differential correspondence, where we assume that the position accuracy σ_p of the center of the patch is the same as for the corners of a patch of half window size w and therefore disturb the 4 affine parameters with $\sigma_p/(\sqrt{2}w)$, where we assume a 21×21 window. The error bars indicate the size of the standard deviation. In the bottom right figure, we compare the 3-point solution proposed in the Manual of Photogrammetry[22, pp.786] (P3P), the planar POSIT algorithm [25] based on the 4 patch corners (which already includes the parallel projection approximation by [17] in the POS step) and our novel solution applied to the case that we use 3 or more close points: The differential correspondence is approximated using the four patch corners only, while we vary the size (given as the solid angle) of the patch and fix σ_p for the corners at 0.5. The error bars show 1/3 standard deviation. As expected, it can be seen that for large solid angle spatial resection performs best while for decreasing solid angles the novel solution gets better and better, outperforming the other approaches for very narrow constellations.

corners of the local patch. On the other hand, when the 3 individual 3D points of Grunert's solution approach each other, the standard spatial resection can become unstable, because it is based on the difference of the distances to the 3 points. To overcome this issue, Kyle [17] proposed an approximate initial guess

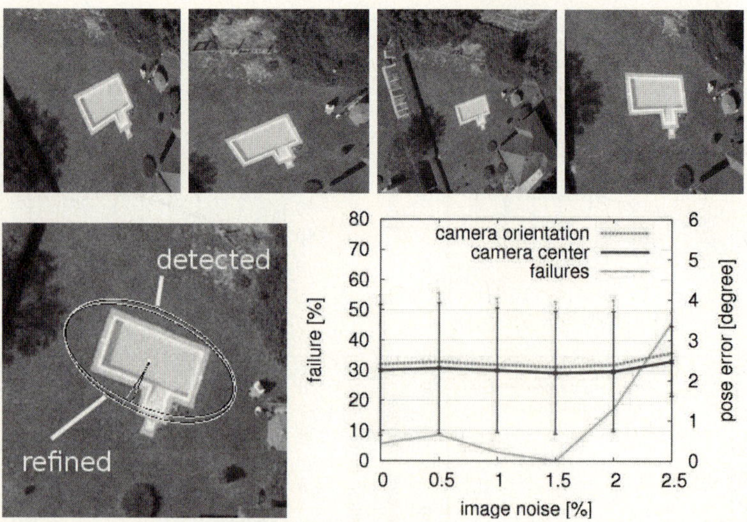

Fig. 3. Camera Pose From Noisy Images. A ground plane has been textured with an aerial image serving as an orthophoto and a series of 40 views have been rendered with different levels of noise (upper row: sample views with low noise). A reference MSER feature with orientation has been chosen in the orthophoto. This feature is then detected in the other views and refined using a simple 6-parametric affine warp (see ellipses in bottom left image) according to [16] based upon a half window size of 10 pixels. From such differential correspondences, the camera pose is estimated and compared against the known ground truth value as explained earlier. Whenever the error was above 20° or the algorithm did not come up with a solution a failure was recorded. The bottom right graph shows the average pose errors in dependence of the added image noise. When adding much more image noise, the MSER detector is no longer able to find the feature. This experiment is particularly interesting because it shows that the concept does still work when the ellipse is not infinitely small.

for narrow angle images, which is the same as the POS (Pose from Orthography and Scaling) in the POSIT[4] algorithm: Both require 4 non-coplanar points. For the POSIT algorithm however, there exists also a planar variant[25], which copes with planar 3D points.

Therefore we compare our novel algorithm (well-suited for small solid angles) to the spatial resection[7,8] implemented as proposed in the Manual of Photogrametry[22, pp. 786] and the planar POSIT[25] algorithm kindly provided on the author's homepage, which are both desiged for larger solid angles. We vary the size of a local square image patch from ten to several hundred pixels and use the corners as individual 2D-3D correspondences in the existing algorithms. For our new method the patch corner points are used to compute a virtual local affine transform which approximates the required Jacobian. An evaluation of the quality of the approximation can be seen in the bottom right of fig. 5, which shows that for small solid angles the novel solution outperforms spatial resection, while for large solid angles - as expected - the affine

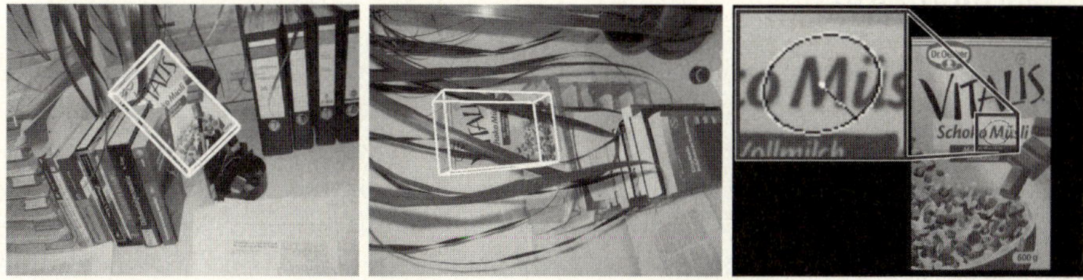

Fig. 4. Object Pose Estimation from a Single Feature. This figure shows that in a real camera with radial distortion object pose estimation is possible from a single feature. The orthophoto of the object is displayed in the right image with the local feature region enlarged. The two left images show cluttered views with the object partially occluded. The "M" has been detected using MSER and refined, the resulting object poses from this single differential correspondence are then displayed by augmenting a contour model (white).

approximation is not suitable. It is however still better in average than the orthographic approximation in the planar POSIT algorithm. Particularly, when the solid angle approaches zero, the error in the novel solution tends to zero, while for the other algorithms no solution can be obtained or the best solution is worse than the robust error threshold of $10°$.

Normal or Pose Error of the Local Plane. An error of the normal of the 3D reference plane, for which the orthophoto exists or an error of the pose of this plane cannot be detected within the algorithm. The pose is computed relative to this plane and an error of the plane in global coordinates will consequently result in a relative error of the camera pose in global coordinates.

Warp Measured From Real Texture. In the next experiment, we rendered views from a ground plane and apply automatic matching and pose estimation based upon a single prominent image feature. Since we do have the ground truth data the estimated pose can be analyzed in this case. The experiments are explained in fig.3 and show that even in presence of image noise using automatic matching and refinement approaches the pose can be estimated quite reliably, given the minimal local texture data which is used.

Images with Radial Distortion. In the final experiment we took photographs of an office scene, where we detect a cereal box, which is partially occluded. As in the previous experiment, an MSER feature is obtained from an orthophoto of the cereal box. Next this feature is automatically found in the test image and refined using gradient-based (affine) optimization. Again, from this differential correspondence the object pose is estimated, which might be interesting in applications where only small portions of an object are visible. The augmentation in fig.4 shows quite good results.

6   Conclusion

A method for estimating a camera pose based upon a single local image feature
has been proposed which exploits the often readily available local affine warp
between two images. This differential correspondence provides more constraints
than a point or a conic and can be used easily in calibrated cameras even if they
deviate from the linear projection model. The algorithm proved to be stable
under several kinds of disturbance and can also be applied when the 3 individual
3D points of a general spatial resection problem come very close because the
novel formulation avoids directly computing the 3 distances, which can lead to
numerical difficulties in practise. Another benefit of the novel minimal solution
is that it allows now for computing the pose from a single image-model match
of common robust features which could reduce RANSAC complexity compared
to the previously required set of 3 correspondences.

References

1. Brown, M., Lowe, D.G.: Automatic panoramic image stitching using invariant features. International Journal of Computer Vision 74(1), 59–73 (2007)
2. Chum, O., Matas, J., Obdrzalek, S.: Epipolar geometry from three correspondences. In: Computer Vision Winter Workshop, Prague, pp. 83–88 (2003)
3. Davison, A.J., Reid, I.D., Molton, N.D., Stasse, O.: Monoslam: Real-time single camera slam. IEEE Transactions on Pattern Analysis and Machine Intelligence 29(6), 1052–1067 (2007)
4. DeMenthon, D., Davis, L.S.: Model-based object pose in 25 lines of code. International Journal of Computer Vision 15, 123–141 (1995)
5. Finsterwalder, S., Scheufele, W.: Das Rückwärtseinschneiden im Raum. In: Bayerische, K., der Wissenschaften, A. (eds.) Sitzungsberichte der mathematisch-physikalischen Klasse, vol. 23/4, pp. 591–614 (1903)
6. Fischler, M., Bolles, R.: RANdom SAmpling Consensus: a paradigm for model fitting with application to image analysis and automated cartography. Communications of the ACM 24(6), 381–395 (1981)
7. Grunert, J.A.: Das Pothenot'sche Problem, in erweiterter Gestalt; nebst Bemerkungen über seine Anwendung in der Geodäsie. In: Archiv der Mathematik und Physik, vol. 1, pp. 238–248, Greifswald. Verlag C.A. Koch (1841)
8. Haralick, B., Lee, C., Ottenberg, K., Nölle, M.: Review and analysis of solutions of the three point perspective pose estimation problem. International Journal of Computer Vision 13(3), 331–356 (1994)
9. Hartley, R., Zisserman, A.: Multiple View Geometry in Computer Vision, 2nd edn. Cambridge University Press, Cambridge (2004)
10. Irani, M., Rousso, B., Peleg, S.: Recovery of ego-motion using region alignment. Transact. on Pattern Analysis and Machine Intelligence 19(3), 268–272 (1997)
11. Jin, H., Favaro, P., Soatto, S.: A semi-direct approach to structure from motion. The Visual Computer 19(6), 377–394 (2003)
12. Kahl, F., Heyden, A.: Using conic correspondence in two images to estimate the epipolar geometry. In: Proceedings of ICCV, pp. 761–766 (1998)
13. Kähler, O., Denzler, J.: Rigid motion constraints for tracking planar objects. In: Hamprecht, F.A., Schnörr, C., Jähne, B. (eds.) DAGM 2007. LNCS, vol. 4713, pp. 102–111. Springer, Heidelberg (2007)

14. Kannala, J., Salo, M., Heikkila, J.: Algorithms for computing a planar homography from conics in correspondence. In: Proceedings of BMVC 2006 (2006)
15. Koeser, K., Beder, C., Koch, R.: Conjugate rotation: Parameterization and estimation from an affine feature corespondence. In: Proceedings of IEEE Conference on Computer Vision and Pattern Recognition (CVPR) (2008)
16. Koeser, K., Koch, R.: Exploiting uncertainty propagation in gradient-based image registration. In: Proc. of BMVC 2008 (to appear, 2008)
17. Kyle, S.: Using parallel projection mathematics to orient an object relative to a single image. The Photogrammetric Record 19, 38–50 (2004)
18. Lowe, D.G.: Distinctive image features from scale-invariant keypoints. International Journal of Computer Vision 60(2), 91–110 (2004)
19. Lucas, B.D., Kanade, T.: An iterative image registration technique with an application to stereo vision. In: IJCAI 1981, pp. 674–679 (1981)
20. De Ma, S.: Conics-based stereo, motion estimation, and pose determination. International Journal of Computer Vision 10(1), 7–25 (1993)
21. Matas, J., Chum, O., Urban, M., Pajdla, T.: Robust wide baseline stereo from maximally stable extremal regions. In: Proceedings of BMVC 2002 (2002)
22. McGlone, J.C. (ed.): Manual of Photogrammetry, 5th edn. ASPRS (2004)
23. Mikolajczyk, K., Schmid, C.: A performance evaluation of local descriptors. Transact. on Pattern Analysis and Machine Intell. 27(10), 1615–1630 (2005)
24. Mikolajczyk, K., Tuytelaars, T., Schmid, C., Zisserman, A., Matas, J., Schaffalitzky, F., Kadir, T., van Gool, L.: A comparison of affine region detectors. International Journal of Computer Vision 65(1-2), 43–72 (2005)
25. Oberkampf, D., DeMenthon, D., Davis, L.S.: Iterative pose estimation using coplanar feature points. CVGIP 63(3) (1996)
26. Riggi, F., Toews, M., Arbel, T.: Fundamental matrix estimation via TIP - transfer of invariant parameters. In: Proceedings of the 18th International Conference on Pattern Recognition, Hong Kong, August 2006, pp. 21–24 (2006)
27. Rothganger, F., Lazebnik, S., Schmid, C., Ponce, J.: Segmenting, modeling, and matching video clips containing multiple moving objects. IEEE Transactions on Pattern Analysis and Machine Intelligence 29(3), 477–491 (2007)
28. Schmid, C., Zisserman, A.: The geometry and matching of lines and curves over multiple views. International Journal of Computer Vision 40(3), 199–234 (2000)
29. Se, S., Lowe, D.G., Little, J.: Vision-based global localization and mapping for mobile robots. IEEE Transactions on Robotics 21(3), 364–375 (2005)
30. Skrypnyk, I., Lowe, D.G.: Scene modelling, recognition and tracking with invariant image features. In: IEEE and ACM International Symposium on Mixed and Augmented Reality, pp. 110–119 (2004)
31. Thompson, E.H.: Space resection: Failure cases. The Photogrammetric Record 5(27), 201–207 (1966)
32. Williams, B., Klein, G., Reid, I.: Real-time slam relocalisation. In: Proceedings of ICCV, Rio de Janeiro, Brazil, pp. 1–8 (2007)
33. Zelnik-Manor, L., Irani, M.: Multiview constraints on homographies. IEEE Transactions on Pattern Analysis and Machine Intelligence 24(2), 214–223 (2002)

Riemannian Anisotropic Diffusion for Tensor Valued Images

Kai Krajsek[1], Marion I. Menzel[1], Michael Zwanger[2], and Hanno Scharr[1]

[1] Forschungszentrum Jülich, ICG-3, 52425 Jülich, Germany
`{k.krajsek,m.i.menzel,h.scharr}@fz-juelich.de`
[2] Siemens AG, Healthcare Sector
MR Application Development, 91052 Erlangen, Germany
`Michael.Zwanger@siemens.com`

Abstract. Tensor valued images, for instance originating from diffusion tensor magnetic resonance imaging (DT-MRI), have become more and more important over the last couple of years. Due to the nonlinear structure of such data it is nontrivial to adapt well-established image processing techniques to them. In this contribution we derive anisotropic diffusion equations for tensor-valued images based on the intrinsic Riemannian geometric structure of the space of symmetric positive tensors. In contrast to anisotropic diffusion approaches proposed so far, which are based on the Euclidian metric, our approach considers the nonlinear structure of positive definite tensors by means of the intrinsic Riemannian metric. Together with an intrinsic numerical scheme our approach overcomes a main drawback of former proposed anisotropic diffusion approaches, the so-called *eigenvalue swelling effect*. Experiments on synthetic data as well as real DT-MRI data demonstrate the value of a sound differential geometric formulation of diffusion processes for tensor valued data.

1 Introduction

In this paper anisotropic diffusion driven by a diffusion tensor is adapted to tensor-valued data in a way respecting the Riemannian geometry of the data structure. *Nonlinear diffusion* has become a widely used technique with a well understood theory (see e.g. [1,2] for overviews). It was introduced in [3] and has been frequently applied to scalar-, color- or vector-valued data. *Anisotropic diffusion*[1] driven by a diffusion tensor [2] is the most general form of diffusion processes. *Tensor-valued data* frequently occur in image processing, e.g. covariance matrices or structure tensors in optical flow estimation (see e.g. [4]). Due to rapid technological developments in magnetic resonance imaging (MRI) also interest in tensor-valued measurement data increases. Due to the increasing need of processing tensor valued data, the development of appropriate regularization techniques become more and more important (e.g. see [5,6,7,8] and [9] as well as references therein). *Riemannian geometry* refers to the fact that the set of positive definite tensors $P(n)$ of size n does not form a vector space but a nonlinear manifold embedded in the vector space of all symmetric matrices. The nonlinear

---

[1] Please note that the term 'anisotropic diffusion' is not uniquely defined in literature. In this contribution we use the term in accordance with the definition given in [2].

D. Forsyth, P. Torr, and A. Zisserman (Eds.): ECCV 2008, Part IV, LNCS 5305, pp. 326–339, 2008.
© Springer-Verlag Berlin Heidelberg 2008

structure of $P(n)$ is studied from a differential geometric point of view for a long time [10]. Due to the nonlinear structure of $P(n)$, well established image processing techniques for scalar and vector valued data might destroy the positive definiteness of the tensors. Approaches for processing tensor valued images can be classified into two groups: using extrinsic [5,11,12,13,14] or intrinsic view [15,16,17,18,19,20,21,7,22]. Methods using the extrinsic point of view consider the space of positive definite symmetric tensors as an embedding in the space of all symmetric tensors which constitute a vector space. Distances, as e.g. required for derivatives, are computed with respect to the flat Euclidian metric of the space of symmetric matrices. To keep tensors on the manifold of positive definite tensors, solutions are projected back onto the manifold [5], selected only on the manifold in a stochastic sampling approach [11], or processing is restricted to operations not leading out of $P(n)$, e.g. convex filters [12,13,14]. Although then tensors stay positive definite the use of a flat metric is not appropriate to deal with $P(n)$. For instance in regularization, the processed tensors become deformed when using the flat Euclidian metric [7] known as *eigenvalue swelling effect* [5,6,7,8]. Tschumperlé and Deriche [5] avoid the eigenvalue swelling effect by applying a spectral decomposition and regularizing eigenvalues and eigenvectors separatively. Chefd'hotel et al. [6] proposed to take the metric of the underlying manifold for deriving evolution equations from energy functionals that intrinsically fulfill the constraints upon them (e.g. rank or eigenvalue preserving) as well as for the numerical solution scheme. However, they consider the Euclidian metric for measuring distances between tensors such that their methods suffer from the eigenvalue swelling effect for some of the proposed evolution equations. Methods using the intrinsic point of view consider $P(n)$ as a Riemannian symmetric space (see [23] and Sect. 3 for an introduction in symmetric Riemannian spaces) equipped with an affine invariant metric on the tangent space at each point. Consequently, using this metric the eigenvalue swelling effect is avoided. The symmetry property of the Riemannian manifold easily allows to define evolution equations on the tangent spaces, approximate derivatives by tangent vectors as well as construct intrinsic gradient descent schemes as we will show for anisotropic diffusion in the following.

Related work. Differential geometric approaches have been introduced to different fields in image processing and computer vision [24,25,26]. Only quite recently, methods based on the Riemannian geometry of $P(n)$ have been introduced independently by different authors [16,17,18,19,20,21,7,22]. For instance, in [20,7] a 'Riemannian framework for tensor computing', has been proposed in which several well established image processing approaches including interpolation, restoration and isotropic nonlinear diffusion filtering have been generalized to $P(n)$ in an intrinsic way. Furthermore, an anisotropic regularization approach has been proposed by adapting the isotropic Laplace-Beltrami operator that can be identified with a second order Markov Random field approach. A quite similar approach has been proposed in [27] by formulating diffusion filtering directly on a discrete graph structure. In [8], a weighted mean has been proposed that allows to smooth the image in an anisotropic way. However, all these approaches [7,27,8], do not allow to construct diffusion tensors from model based structure estimation, as common in literature for scalar data [2]. To do so in an intrinsic way, one cannot do without a numerical scheme for mixed second order derivatives, first introduced in the current paper. A computational more efficient approach than the framework of Pennec et al. [7] based on the so called *log-Euclidean* metric has been introduced in [28]. There, the positive definite tensors are mapped onto the space of symmetric matrices by means of the matrix logarithmic map. In this new space common vector valued approaches can be applied. The final result is obtained by mapping the transformed symmetric matrices back onto the space of positive definite matrices using the matrix exponential map. However, the log-Euclidean metric is not affine invariant. As a consequence the approach might suffer from a change of coordinates. However, the formulation of anisotropic diffusion for tensor valued data based on the log-Euclidean metric might be a computational efficient alternative not proposed in literature so far. In [22,29] a Riemannian framework based on local coordinates has been proposed (see also in [30] for a variational framework for general manifolds). Although, the authors in [22,29] consider the affine invariant metric their approach may only be classified as intrinsic in a continuous formulation. For computing discrete data, a simple finite difference approximation is applied. Inferring from a continuous formulation without a proof to a discrete approximation can be misleading as constraints holding in the continuous case may be relaxed by discretization. As a consequence, the proposed approaches not necessarily preserve positive definiteness of tensors (for a detailed discussion of this topic for scalar valued signals we refer to [2]). Furthermore, the approach of [29] shows no significant difference with the log Euclidean framework whereas our approach clearly outperforms it. We refer to our approach as the full intrinsic scheme in order to distinguish it from schemes that are only intrinsic in the continuous setting. Anisotropic diffusion based on an extrinsic view [12,31] and by means of the exponential map [6] has been proposed. In both cases the Euclidian metric is used to measure distances between tensors. As a consequence, both approaches suffer from the eigenvalue swelling effect.

Our contribution. We derive an intrinsic anisotropic diffusion equation for the manifold of positive definite tensors. To this end, second order derivatives in the continuous as well as discrete approximations are derived as they occur in the anisotropic diffusion equation. The derived numerical scheme could also be used to generalize other PDEs involving mixed derivatives from scalar valued images to the manifold $P(n)$ without the need of local coordinates. In the experimental part, we provide a study in which we compare different state of the art regularization approaches with our approach.

2   Diffusion for Scalar Valued Images

We review diffusion filtering which is a well established image processing technique for scalar valued images [3,32,2]. We formulate the diffusion equation by means of a gradient descent of some energy functional that later allows us to generalize this concept to tensor valued data. Let f be a scalar valued image defined on a N-dimensional domain. Diffusion filtering image processing creates a family of images $\{u(x,t)|t \geq 0\}$ from the solution of the physical diffusion equation

$$\partial_t u = \mathrm{div}\left(\mathbf{D}\nabla u\right) \tag{1}$$

with initial condition $f = u(x,0)$ and diffusion tensor $\mathbf{D}$ with components d_{ij}. Note that we could also formulate the image restoration task as a solution of a diffusion reaction equation by adding a data depending term to (1). We will discuss the pure diffusion process only. All following results keep valid also for a formulation with data depending reaction terms. The diffusion equation can be reformulated applying the chain rule in the form $\partial_t u = \sum_{i,j}(\partial_i d_{ij})(\partial_j u) + d_{ij}\partial_i\partial_j u$ which will be more convenient for the formulation on tensor valued data. The diffusion process can be classified according to the diffusion tensor $\mathbf{D}$. If the diffusion tensor does not depend upon the evolving image, the diffusion process is denoted as *linear* due to the linearity of (1) otherwise it is termed *nonlinear*. The diffusion process can furthermore be classified into *isotropic* when the diffusion tensor is proportional to the identity matrix otherwise it is denoted as *anisotropic*. Except for the nonlinear anisotropic diffusion scheme, the diffusion equation can be derived from a corresponding energy functional $E(u)$ via calculus of variation, i.e. the gradient descent scheme of these energy functionals can be identified with a diffusion equation. Let $L(u)$ denote the energy density such that $E(u) = \int L(u)\,dx$, $w : \mathbb{R}^N \rightarrow \mathbb{R}$ a test function and ε a real valued variable. The functional derivative $\delta E := \left.\frac{\delta E(u+\varepsilon w)}{\delta\varepsilon}\right|_{\varepsilon=0}$ of an energy functional $E(u)$ can be written as

$$\delta E = \int \langle \nabla L(u), w\rangle_u \, d\mathbf{x} \; , \tag{2}$$

where $\nabla L(u)$ defines the gradient of the energy density and $\langle \nabla L(u), w\rangle_u$ denotes the scalar product of the energy density gradient $\nabla L(u)$ and the test function evaluated at $\mathbf{x}$. Note that w as well as $\nabla L(u)$ are elements of the tangent space at u which is the Euclidian space itself for scalar valued images. As we will see in Sect. 4, this formulation allows a direct generalization to the space of symmetric positive definite tensors. The gradient descent scheme of the energy functional leads to the diffusion equation in terms of the energy density

$$\partial_t u = -\nabla L(u) \; . \tag{3}$$

Let us now consider the linear anisotropic diffusion equation (1), i.e. $\mathbf{D}$ not depending on the evolving signal. The corresponding energy function is known to be

$$E(u) = \frac{1}{2}\int \nabla u^T \mathbf{D}\nabla u\,d\mathbf{x} \; . \tag{4}$$

The functional derivative of (4) can be brought into the form

$$\delta E(u) = \int \langle -\mathrm{div}\,(\mathbf{D}\nabla u)\,, w\rangle_u \,d\mathbf{x} \tag{5}$$

assuming homogenous Neumann boundary conditions and applying Green's formula. Comparing (5) with (2) gives together with (3) the diffusion equation (1). Our objective is now to generalize the linear anisotropic diffusion process to the space of positive definite tensors by means of the energy functional formulation. The nonlinear anisotropic diffusion equation on $P(n)$, can then be deduced from the linear one.

3   The Space of Positive Definite Tensors

In the following we review the structure of the space of positive definite tensors $P(n)$ and introduce the differential geometric tools necessary for deriving anisotropic diffusion equations for $P(n)$. By introducing a basis, any tensor can be identified with its corresponding matrix representation $A \in \mathbb{R}^{n \times n}$. The space of $n \times n$ matrices constitutes a vector space embodied with a scalar product $\langle A, B \rangle = \mathrm{Tr}\left(A^T B\right)$, inducing the norm $||A|| = \sqrt{\langle A, A \rangle}$. However, tensors Σ frequently occurring in computer vision and image processing applications, e.g. covariance matrices and DT-MRI tensors, embody further structure on the space of tensors: they are symmetric $\Sigma^T = \Sigma$ and positive definite, i.e. it holds $x^T \Sigma x > 0$ for all nonzero $x \in \mathbb{R}^n$. The approach to anisotropic diffusion presented here, measures distances between tensors by the length of the shortest path, the geodesic, with respect to $GL(n)$ (affine) invariant Riemannian metric on $P(n)$. This metric takes the nonlinear structure of $P(n)$ into account and it has demonstrated in several other application its superiority over the flat Euclidean matric [17,18,20,21,7,22]. Such an *intrinsic* treatment requires the formulation of $P(n)$ as a Riemannian manifold, i.e. each tangent space is equipped with an inner product that smoothly varies from point to point. A geodesic $\Gamma_{\mathbf{X}}(t)$ parameterized by the 'time' t and going through the tensor $\Gamma(0) = \Sigma$ at time $t = 0$ is uniquely defined by its tangent vector $\mathbf{X}$ at Σ. This allows one to describe each geodesic by a mapping from the subspace $\mathcal{A} = (t\mathbf{X})$, $t \in \mathbb{R}$ spanned by the tangent vector onto the manifold $P(n)$. The $GL(n)$ invariant metric is induced by the scalar product

$$\langle W_1, W_2 \rangle_{\Sigma} = \mathrm{Tr}\left(\Sigma^{-\frac{1}{2}} W_1 \Sigma^{-1} W_2 \Sigma^{-\frac{1}{2}}\right) , \qquad (6)$$

as one can easily verify. The $GL(n)$ invariant metric allows to derive an expression of the geodesic equation going through Σ by tangent vectors $\mathbf{X}$ [7]

$$\Gamma_{\Sigma}(t) = \Sigma^{\frac{1}{2}} \exp(t\Sigma^{-\frac{1}{2}} \mathbf{X} \Sigma^{-\frac{1}{2}}) \Sigma^{\frac{1}{2}} . \qquad (7)$$

For $t = 1$ this map is denoted as the exponential map which is one to one in case of the space of positive definite tensors. Its inverse, denoted as the logarithmic map, reads

$$\mathbf{X} = \Sigma^{\frac{1}{2}} \log\left(\Sigma^{-\frac{1}{2}} \Gamma_{\Sigma}(1) \Sigma^{-\frac{1}{2}}\right) \Sigma^{\frac{1}{2}} . \qquad (8)$$

As the gradient of any energy density ∇L is element of the tangent space [33], we can formulate a diffusion process as $\partial_t \Sigma = -\nabla L$ on the tangent space. The evolution of the tensor Σ is obtained by going a small step in the negative direction of the gradient $-dt\nabla L$ and mapping this point back on the manifold using the geodesic equation (7). The energy density is then computed for the tangent vector at $\Gamma_{\Sigma}(dt)$ which in turn can then be used for finding the next tensor in the evolving scheme as described above. This is a gradient descent approach, denoted as the geodesic marching scheme, for energy densities defined on $P(n)$ and which per construction assures that we cannot leave the manifold.

4   Riemannian Anisotropic Diffusion

After reviewing the necessary differential geometric tools, we will derive anisotropic diffusion equations for a tensor field $P(n)$ over $\mathbb{R}^N$. As done for the diffusion equation for the scalar valued signals (Sect. 2), we derive the linear diffusion equation by variation of the corresponding energy functional and infer from the linear equation to the nonlinear counterpart. Let $\partial_i \Sigma(\mathbf{x})$, $i = 1, ..., N$ denote partial derivative of the tensor field in direction i, elements of the tangent space at Σ. We define the energy functional

$$E(\Sigma) = \int \sum_{i,j} d_{ij} \langle \partial_i \Sigma, \partial_j \Sigma \rangle_\Sigma \, d\mathbf{x} \tag{9}$$

$$\text{with} \quad \langle \partial_i \Sigma, \partial_j \Sigma \rangle_\Sigma = \mathrm{Tr}\left((\partial_i \Sigma)\Sigma^{-1}(\partial_j \Sigma)\Sigma^{-1}\right) \; . \tag{10}$$

The components of the diffusion tensor d_{ij} (please do not confuse d_{ij} with the elements of the tensor field) locally controls the direction of smoothing and for the moment being does not depend on the evolving tensor field. The gradient of the energy functional is then derived by defining a 'test function' W that is actually a tangent vector in the tangent space at Σ and computing the functional derivative

$$\delta E = 2 \int \sum_{ij} d_{ij} \mathrm{Tr}((\partial_i W)\, \Sigma^{-1}\, (\partial_j \Sigma)\, \Sigma^{-1} \tag{11}$$

$$- (\partial_i \Sigma)\, \Sigma^{-1}\, (\partial_j \Sigma)\, \Sigma^{-1} W \Sigma^{-1})\, d\mathbf{x} \tag{12}$$

In order to get rid of the derivatives on the 'test function' W we integrate by parts with respect to x_j. Assuming homogenous Neumann boundary conditions the functional derivative can be brought in the form

$$\delta E = -2 \sum_{i,j} \int \langle W, \Sigma \partial_i (d_{ij} \Sigma^{-1}(\partial_j \Sigma)\Sigma^{-1})\Sigma \tag{13}$$

$$+ (\partial_i \Sigma)\Sigma^{-1}(\partial_j \Sigma) \rangle_\Sigma \, d\mathbf{x} \tag{14}$$

Comparing the inner product with the general form in (2) identifies the gradient of the energy density

$$\nabla L = -2 \sum_{i,j} \Sigma \partial_i (d_{ij} \Sigma^{-1}(\partial_j \Sigma)\Sigma^{-1})\Sigma + (\partial_i \Sigma)\Sigma^{-1}(\partial_j \Sigma) \; . \tag{15}$$

Inserting this energy density in (3) results in the desired diffusion equation. Using the identity $\partial_i \Sigma^{-1} = -\Sigma^{-1}(\partial_i \Sigma)\Sigma^{-1}$ the energy density gradient can be simplified to

$$\nabla L = -2 \sum_{i,j} \left(\partial_i \partial_j \Sigma - (\partial_i \Sigma)\Sigma^{-1}(\partial_j \Sigma)\right) - 2 \sum_{i,j} (\partial_i d_{ij})(\partial_j \Sigma) \tag{16}$$

The terms on the right side of (16) for which $i = j$ hold $\Delta_i \Sigma = \partial_i^2 \Sigma - (\partial_i \Sigma)\Sigma^{-1}$ $(\partial_i \Sigma)$ are components of the Laplace Beltrami operator $\Delta = \sum_i \Delta_i$ derived in [7]. In addition to the work in [20,7], we also derived mixed components

$$\Delta_{ij}\Sigma = \partial_i \partial_j \Sigma - (\partial_i \Sigma)\Sigma^{-1}(\partial_j \Sigma), \quad i \neq j \tag{17}$$

needed for the linear anisotropic diffusion equation. The nonlinear anisotropic diffusion equation is defined exchanging the diffusion tensor components in (4) with components depending on the evolved tensor field. So we have all components to define an anisotropic diffusion equation on the space of positive definite matrices in an intrinsic way. To this end, only the second order derivatives ∂_i^2 and $\partial_i \partial_j$ occurring in (1) need to be exchanged by their counterparts Δ_i and Δ_{ij}. So far we have not specified the explicit form of the diffusion tensor which should be made up here. We generalize the structure tensor to the nonlinear space and afterwards, as in the case of scalar valued images, construct the diffusion tensor from the spectral decomposition of the structure tensor. Let $\nabla \Sigma = (\partial_1 \Sigma, ..., \partial_N \Sigma)^T$ denote the gradient and $\mathbf{a}$ a unite vector in $\mathbb{R}^N$ such that we can express the derivative in direction $\mathbf{a}$ as $\partial_a = \mathbf{a}^T \nabla$. The direction of less variation in the tensor space can then analogous to the structure tensor in linear spaces, be estimated by minimizing the local energy

$$E(\mathbf{a}) = \int_V \langle \partial_a \Sigma, \partial_a \Sigma \rangle_\Sigma d\mathbf{x} = \mathbf{a}^T J \mathbf{a} \ , \tag{18}$$

where we defined the components of the structure tensors J on $P(n)$ by $J_{ij} = \int_V \langle \partial_i \Sigma, \partial_j \Sigma \rangle_\Sigma d\mathbf{x}$. The diffusion tensor $\mathbf{D}$ is then designed as usual by exchanging the eigenvalues λ_j of the structure tensor by a decreasing diffusivity function $g(\lambda_j)$. For our numerical experiments (in 2D) we choose $g(\lambda_l) = 1/\sqrt{1 + \lambda_l/\beta^2}$ for the larger eigenvalue and $g(\lambda_s) = 1$ for the smaller eigenvalue with the heuristically chosen contrast parameter $\beta = 0.05$.

5   Numerical Issues

So far we have assumed the tensor to be defined on a continuous domain. In the experiential setting we are confronted with tensor fields defined on a discrete grid. The application of Riemanian anisotropic diffusion requires a discrete approximation for the derivatives derived in Sect. 4. In principle, we could use matrix differences to approximate the derivatives but this would contradict our effort to derive an intrinsic expression of the anisotropic diffusion equation. The finite differences are extrinsic since they are based on Euclidian differences between tensors, i.e. they use the difference in the space of symmetric matrices and not the Riemannian metric of the space $P(n)$. In order to approximate the gradient ∇L in (16) on a discrete grid, we need discrete approximations of derivatives of first and second order. Intrinsic approximations to first order derivatives have already proposed in [20] and is reviewed here with the following preposition. Let us denote with $T\Sigma_x^{e_j} := \overrightarrow{\Sigma(x)\Sigma(x + \varepsilon e_j)}$ the tangent vector defined by the logarithmic map as

$$T\Sigma_x^{e_j} = \Sigma^{\frac{1}{2}} \log \left(\Sigma^{-\frac{1}{2}} \Sigma(x + \varepsilon e_j) \Sigma^{-\frac{1}{2}} \right) \Sigma^{\frac{1}{2}} \tag{19}$$

Preposition 1. The first order discrete approximation of the first order derivative of Σ in direction j reads

$$\partial_j \Sigma = \frac{1}{2\varepsilon} \left(\overrightarrow{\Sigma(x)\Sigma(x + \varepsilon e_j)} - \overrightarrow{\Sigma(x)\Sigma(x - \varepsilon e_j)} \right) + O(\varepsilon) \tag{20}$$

A second order discrete approximation scheme to the second order derivative in direction e_j has been derived in [7]. We state it here as a second preposition, for the proof see [7].

Preposition 2. The second order discrete approximation of the second order derivative in direction e_j is

$$\Delta_j \Sigma = \frac{1}{\varepsilon^2}(\overrightarrow{\Sigma(x)\Sigma(x+\varepsilon e_j)} + \overrightarrow{\Sigma(x)\Sigma(x-\varepsilon e_j)}) + O(\varepsilon^2) \ . \tag{21}$$

For the anisotropic diffusion equation we also need mixed derivatives $\Delta_{ij}\Sigma$ that can be approximated according to preposition 3.

Preposition 3. The second order discrete approximation of the second order mixed derivative in direction i and j is given by

$$\frac{\Delta_{ij}\Sigma + \Delta_{ji}\Sigma}{2} = \frac{1}{\varepsilon^2}(\overrightarrow{\Sigma(x)\Sigma(x+\varepsilon e_n)} + \overrightarrow{\Sigma(x)\Sigma(x-\varepsilon e_n)} \tag{22}$$

$$-\overrightarrow{\Sigma(x)\Sigma(x+\varepsilon e_p)} - \overrightarrow{\Sigma(x)\Sigma(x-\varepsilon e_p)}) + O(\varepsilon^2) \ ,$$

with the abbreviation $e_n = \frac{1}{\sqrt{2}}(e_i + e_j)$, $e_p = \frac{1}{\sqrt{2}}(e_i - e_j)$.

Proof. We expend the tangent vector as

$$T\Sigma_x^{e_n} = \varepsilon \partial_n \Sigma + \frac{\varepsilon^2}{2}\partial_n^2 \Sigma - \frac{\varepsilon^2}{2}(\partial_n \Sigma)\Sigma^{-\frac{1}{2}}(\partial_n \Sigma) + O(\varepsilon^3) \ . \tag{23}$$

Now, we express the derivative in direction n by derivatives along the coordinate axes in i and j direction , $\partial_n = \frac{1}{\sqrt{2}}\partial_i + \frac{1}{\sqrt{2}}\partial_j$, yielding

$$T\Sigma_x^{e_n} = \frac{\varepsilon}{\sqrt{2}}(\partial_i \Sigma + \partial_j \Sigma) + \frac{\varepsilon^2}{4}((\partial_i^2 \Sigma + \partial_j^2 \Sigma + 2\partial_i\partial_j \Sigma)$$

$$-(\partial_i \Sigma)\Sigma^{-\frac{1}{2}}(\partial_i \Sigma) - (\partial_j \Sigma)\Sigma^{-\frac{1}{2}}(\partial_j \Sigma)$$

$$-(\partial_i \Sigma)\Sigma^{-\frac{1}{2}}(\partial_j \Sigma) - (\partial_j \Sigma)\Sigma^{-\frac{1}{2}}(\partial_i \Sigma)) + O(\varepsilon^3) \ .$$

Computing the sum $T\Sigma_x^{\Delta e_n} := T\Sigma_x^{e_n} + T\Sigma_x^{-e_n}$ becomes a fourth order approximation as all uneven terms cancel out

$$T\Sigma_x^{\Delta e_n} = \frac{\varepsilon^2}{4}((\partial_i^2 \Sigma + \partial_j^2 \Sigma + 2\partial_i\partial_j \Sigma) - (\partial_i \Sigma)\Sigma^{-\frac{1}{2}}(\partial_i \Sigma) - \tag{24}$$

$$(\partial_j \Sigma)\Sigma^{-\frac{1}{2}}(\partial_j \Sigma) - (\partial_i \Sigma)\Sigma^{-\frac{1}{2}}(\partial_j \Sigma) - (\partial_j \Sigma)\Sigma^{-\frac{1}{2}}(\partial_i \Sigma)) + O(\varepsilon^4)$$

Expanding $T\Sigma_x^{\Delta e_p} := T\Sigma_x^{e_p} + T\Sigma_x^{-e_p}$ in the same way yields

$$T\Sigma_x^{\Delta e_p} = \frac{\varepsilon^2}{4}((\partial_i^2 \Sigma + \partial_j^2 \Sigma - 2\partial_i\partial_j \Sigma) - (\partial_i \Sigma)\Sigma^{-\frac{1}{2}}(\partial_i \Sigma) - \tag{25}$$

$$(\partial_j \Sigma)\Sigma^{-\frac{1}{2}}(\partial_j \Sigma) + (\partial_i \Sigma)\Sigma^{-\frac{1}{2}}(\partial_j \Sigma) + (\partial_j \Sigma)\Sigma^{-\frac{1}{2}}(\partial_i \Sigma)) + O(\varepsilon^4)$$

By subtracting (25) from (24) and dividing the square of the grid size ε^2 we obtained the claimed second order approximation for the mixed derivatives which concludes the proof.

6   Experiments

Performance of our Riemannian anisotropic diffusion (RAD) approach is demonstrated on synthetic tensor fields and real DT-MRI data. We compare our Riemannian anisotropic diffusion scheme with three state of the art tensor valued regularization schemes: the anisotropic diffusion (EAD) scheme based on the flat Euclidean metric [12,31], the intrinsic nonlinear isotropic diffusion (RID) scheme [20] and the nonlinear isotropic diffusion (LEID) scheme based on the log-Euclidean metric [34]. As a computational effective alternative to our Riemannian anisotropic diffusion scheme, we propose to combine the diffusion scheme proposed in [12,31] with the log-Euclidean metric [34] which is considered as a fourth reference method (LEAD). As a performance measure for the regularized tensor field, we choose the fractional anisotropy (FA) [35]. Measures derived from DT-MRI such as the FA are used to generate additional image contrast required for detection of brain lesions, or to delineate white matter (highly directional structures) from non-white matter tissue, which is important for surgery. FA takes on values between 0 (corresponding to perfect isotropy) and 1 indicating maximal anisotropy. For solving the diffusion equations, we used the same time step of $dt = 0.01$ for all experiments and computed the evolving tensor field for 1000 time steps. As shown in [21], the linear gradient descent scheme realizes a first order approximation to the intrinsic marching scheme, such that for small time steps diffusion processes based on different metrics should be comparable for distinct times.

Fig. 1. Line reconstruction experiment; upper row (from left to right): original tensor field, EAD scheme, LEAD scheme; lower row (from left to right): LEID scheme, RID scheme, RAD scheme

Fig. 2. Denoising experiment; upper row (from left to right): noise corrupted tensor field, EAD scheme, our LEAD scheme; lower row (from left to right): LEID scheme, RID scheme, RAD scheme

6.1 Synthetic Data

Experiment 1. In the first experiment on synthetic data we examine the ability of the different diffusion processes to complete interrupted line structures. To this end, we generate a 32×32 large tensor field of 3×3 tensors (see Fig. 2 upper left; in order to visualize details more precise only a cutout of the tensor field is shown). Each tensor is represented by an ellipsoid and the orientation of its main axis is additional color coded whereas the FA is encoded in the saturation of the depicted tensors. The line structure is interrupted by isotropic tensors with small eigenvalues ($\lambda_j = 0.05$) that are hardly visible due to the saturation encoding of the FA. The results for all diffusion processes are shown in Fig. 1. The nonlinear isotropic processes LEID and RID stops at the line interruption and is not able to complete the line. This results from the fact that, although the smoothing process is also anisotropic for nonlinear isotropic diffusion processes [20], the diffusivity function depends only on its direct neighbors and therefore does not 'see' the line behind the gap. The anisotropic diffusion schemes are steered by the diffusion tensor which encodes the directional information of a neighborhood depending on the average region for the structure tensor. The anisotropic diffusion approaches fill the gap and reconstruct the line. However, again the EAD-process suffers from the eigenvalue swelling effect and only one tensor connects both interrupted line structures. However increasing the average region of the structure tensor might fill the gap more clearly. Our RAD and LEAD schemes reconstruct the line structure. However, we observe a small decreasing of the anisotropy for the log-Euclidean metric, whereas the anisotropy for the affine invariant metric increases in the vicinity of image borders.

Experiment 2. In this experiment we examine the ability of the different diffusion schemes to reconstruct the tensor field from noisy data. To this end, we corrupt the tangent vector of each tensor by Gaussian noise (with standard deviations $\sigma = 0.6$). Fig. 2 shows the noise corrupted field (the noise free tensor field is the same as in experiment 1) and the evolved tensor fields for the different diffusion schemes. The anisotropic schemes manage (more or less) to close the gap in the line structure despite the noise whereas the isotropic schemes does not. The schemes based on the log Euclidean metric lead to a slight decrease of the anisotropy whereas the RAD schemes leads to an increase of the anisotropy in the tensor field. How this effect influences further processing steps, e.g. fiber tracking algorithm, is left to be examined for future research.

6.2 Real Data

Experiment 3. In our last experiment, the different algorithms were applied to DT-MRI data measured from a *human* brain *in-vivo*. DT-MRI of the brain of a healthy volunteer (written informed consent was obtained) was performed on a 1.5 T Magnetom Avanto scanner (Siemens Medical Solutions). A single-shot diffusion-weighted twice-refocused spin-echo planar imaging sequence was used. Measurement parameters were

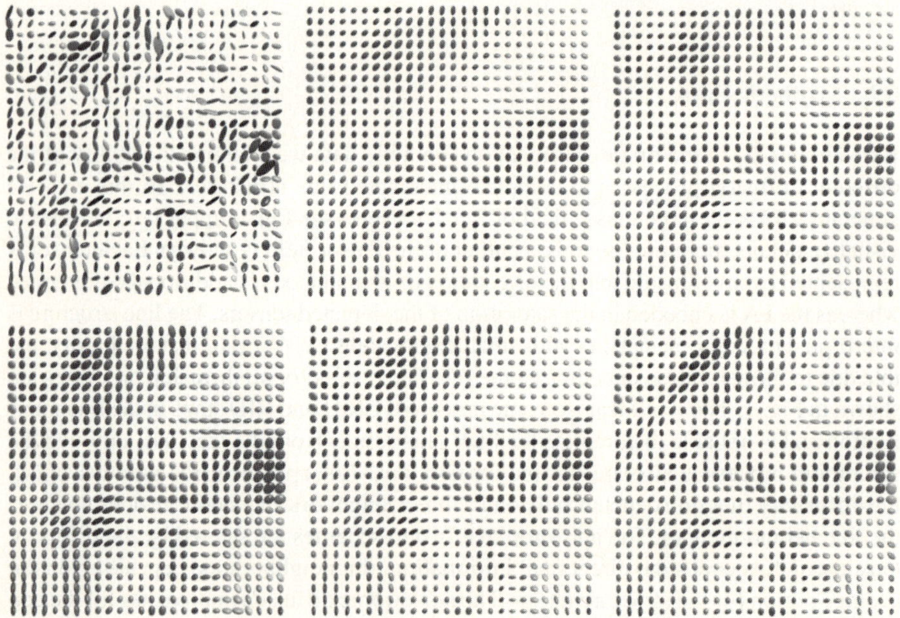

Fig. 3. Denoising experiment 3; (upper row, from left to right): noisy DT-MRI image, LEID scheme, RID scheme; (lower row, from left to right) EAD scheme, LEAD scheme, RAD scheme

as follows : TR = 6925 ms / TE=104ms / 192 matrix with 6/8 phase partial fourier, 23 cm field of view (FOV), and 36 2.4-mm-thick contiguous axial slices. The in-plane resolution was 1.2 mm/pixel. We estimate a volumetric tensor field of size $192 \times 192 \times 36$ and take one slice for further processing. For evaluation purposes we recorded tensor fields of the brain with 6 different signal-to-noise ratios (SNR), denoted as DTI1-6 in the following. Thus, we can use the DT-MRI-images (DTI6) from the long measurement (i.e. good SNR) as a reference data set, where we compare the FA of the tensor with the results obtained from the lower SNR data set (DTI1-5), which can be obtained in a clinical feasible measurement time. We compute, starting from the five different noisy tensor fields, the evolved tensor fields for all considered diffusion schemes (Fig. 3 shows cutouts of the noisy field and evolved fields) and compare its FA with the reference field. All schemes lead to rather smooth tensor fields. However, the anisotropic diffusion schemes (EAD, LEAD and RAD) lead to an enhancement of orientated structures within in the tensor fields which is most distinct for our RAD scheme. As in the previous experiments, the eigenvalue swelling effect in case of the EAD scheme can be observed. Our RAD/LEAD schemes yield the best results among anisotropic regularization schemes with respect to the FA measure as shown in Tab. 1.

Table 1. Results of experiment 3: The average and standard deviation of the the fractional anisotropy error $|FA - \underline{FA}|$ ($\underline{FA}$ belongs to the reference tensor field) over 1000 time steps for each diffusion scheme as well as for five different noise levels are computed

Method	DTI1	DTI2	DTI3	DTI4	DTI5
EAD	0.098 ± 0.007	0.100 ± 0.008	0.103 ± 0.008	0.109 ± 0.009	0.112 ± 0.010
RID	0.112 ± 0.016	0.119 ± 0.015	0.116 ± 0.013	0.114 ± 0.012	0.113 ± 0.013
LEID	0.099 ± 0.017	0.108 ± 0.017	0.107 ± 0.014	0.106 ± 0.012	0.105 ± 0.012
LEAD	$\mathbf{0.078 \pm 0.005}$	$\mathbf{0.079 \pm 0.006}$	$\mathbf{0.081 \pm 0.006}$	$\mathbf{0.084 \pm 0.007}$	$\mathbf{0.086 \pm 0.007}$
RAD	$\mathbf{0.089 \pm 0.004}$	$\mathbf{0.089 \pm 0.005}$	$\mathbf{0.093 \pm 0.007}$	$\mathbf{0.096 \pm 0.007}$	$\mathbf{0.098 \pm 0.009}$

7   Conclusion

We generalized the concept of anisotropic diffusion to tensor valued data with respect to the affine invariant Riemannian metric. We derived the intrinsic mixed second order derivatives as they are required for the anisotropic diffusion process. Furthermore, we derived a discrete intrinsic approximation scheme for the mixed second order derivatives. Since mixed second order derivatives appear also in other methods based on partial differential equation, this contribution could also serve as a basis for generalizing these methods in an intrinsic way in a discrete formulation. Experiments on synthetic as well as real world data demonstrate the value of our full intrinsic differential geometrical formulation of the anisotropic diffusion concept. As a computational effective alternative, we proposed an anisotropic diffusion scheme based on the log-Euclidean metric. Summing up, our proposed anisotropic diffusion schemes show promising results on the given test images. Further work might examine the reconstruction properties of other tensor characteristics as well as the influence on so far heuristically chosen parameters, e.g. the diffusivity function.

References

1. Berger, M.-O., Deriche, R., Herlin, I., Jaffré, J., Morel, J.-M. (eds.): Icaos 1996: Images and wavelets and PDEs. Lecture Notes in Control and Information Sciences, vol. 219 (1996)
2. Weickert, J.: Anisotropic diffusion in image processing. Teubner, Stuttgart (1998)
3. Perona, P., Malik, J.: Scale-space and edge detection using anisotropic diffusion. IEEE Transactions on Pattern Analysis and Machine Intelligence 12 (1990)
4. Bigün, J., Granlund, G.H.: Optimal orientation detection of linear symmetry. In: ICCV, London, UK, pp. 433–438 (1987)
5. Tschumperlé, D., Deriche, R.: Diffusion tensor regularization with constraints preservation. In: Proc. IEEE Computer Society Conference on Computer Vision and Pattern Recognition (CVPR 2001), pp. 948–953 (2001)
6. Chefd'hotel, C., Tschumperlé, D., Deriche, R., Faugeras, O.: Regularizing flows for constrained matrix-valued images. J. Math. Imaging Vis. 20(1-2), 147–162 (2004)
7. Pennec, X., Fillard, P., Ayache, N.: A Riemannian framework for tensor computing. International Journal of Computer Vision 66(1), 41–66 (2006)
8. Castano-Moraga, C.A., Lenglet, C., Deriche, R., Ruiz-Alzola, J.: A Riemannian approach to anisotropic filtering of tensor fields. Signal Processing 87(2), 263–276 (2007)
9. Weickert, J., Hagen, H.: Visualization and Processing of Tensor Fields (Mathematics and Visualization). Springer, New York (2005)
10. Rao, C.: Information and accuracy attainable in estimation of statistical parameters. Bull. Calcutta Math. Soc. 37, 81–91 (1945)
11. Martin-Fernandez, M., San-Jose, R., Westin, C.F., Alberola-Lopez, C.: A novel Gauss-Markov random field approach for regularization of diffusion tensor maps. In: Moreno-Díaz Jr., R., Pichler, F. (eds.) EUROCAST 2003. LNCS, vol. 2809, pp. 506–517. Springer, Heidelberg (2003)
12. Weickert, J., Brox, T.: Diffusion and regularization of vector- and matrix-valued images. In: Inverse Problems, Image Analysis, and Medical Imaging. Contemporary Mathematics, pp. 251–268 (2002)
13. Westin, C.-F., Knutsson, H.: Tensor field regularization using normalized convolution. In: Moreno-Díaz Jr., R., Pichler, F. (eds.) EUROCAST 2003. LNCS, vol. 2809, pp. 564–572. Springer, Heidelberg (2003)
14. Burgeth, B., Didas, S., Florack, L., Weickert, J.: A generic approach to the filtering of matrix fields with singular PDEs. In: Sgallari, F., Murli, A., Paragios, N. (eds.) SSVM 2007. LNCS, vol. 4485, pp. 556–567. Springer, Heidelberg (2007)
15. Gur, Y., Sochen, N.A.: Denoising tensors via Lie group flows. In: Paragios, N., Faugeras, O., Chan, T., Schnörr, C. (eds.) VLSM 2005. LNCS, vol. 3752, pp. 13–24. Springer, Heidelberg (2005)
16. Moakher, M.: A differential geometric approach to the geometric mean of symmetric positive-definite matrices. SIAM J. Matrix Anal. Appl (2003)
17. Fletcher, P., Joshi, S.: Principle geodesic analysis on symmetric spaces: Statistics of diffusion tensors. In: Computer Vision and Mathematical Methods in Medical and Biomedical Image Analysis, ECCV 2004 Workshops CVAMIA and MMBIA, pp. 87–98 (2004)
18. Lenglet, C., Rousson, M., Deriche, R., Faugeras, O.D., Lehericy, S., Ugurbil, K.: A Riemannian approach to diffusion tensor images segmentation. In: Christensen, G.E., Sonka, M. (eds.) IPMI 2005. LNCS, vol. 3565, pp. 591–602. Springer, Heidelberg (2005)
19. Batchelor, P.G., Moakher, M., Atkinson, D., Calamante, F., Connelly, A.: A rigorous framework for diffusion tensor calculus. Magn. Reson. Med. 53(1), 221–225 (2005)
20. Fillard, P., Arsigny, V., Ayache, N., Pennec, X.: A Riemannian framework for the processing of tensor-valued images. In: Fogh Olsen, O., Florack, L.M.J., Kuijper, A. (eds.) DSSCV 2005. LNCS, vol. 3753, pp. 112–123. Springer, Heidelberg (2005)

21. Lenglet, C., Rousson, M., Deriche, R., Faugeras, O.: Statistics on the manifold of multivariate normal distributions: Theory and application to diffusion tensor MRI processing. J. Math. Imaging Vis. 25(3), 423–444 (2006)
22. Zéraï, M., Moakher, M.: Riemannian curvature-driven flows for tensor-valued data. In: Sgallari, F., Murli, A., Paragios, N. (eds.) SSVM 2007. LNCS, vol. 4485, pp. 592–602. Springer, Heidelberg (2007)
23. Helgason, S.: Differential Geometry, Lie groups and symmetric spaces. Academic Press, London (1978)
24. El-Fallah, A., Ford, G.: On mean curvature diffusion in nonlinear image filtering. Pattern Recognition Letters 19, 433–437 (1998)
25. Sochen, N., Kimmel, R., Malladi, R.: A geometrical framework for low level vision. IEEE Transaction on Image Processing, Special Issue on PDE based Image Processing 7(3), 310–318 (1998)
26. Begelfor, E., Werman, M.: Affine invariance revisited. In: CVPR '06: Proceedings of the 2006 IEEE Computer Society Conference on Computer Vision and Pattern Recognition, pp. 2087–2094. IEEE Computer Society Press, Washington (2006)
27. Zhang, F., Hancock, E.: Tensor MRI regularization via graph diffusion. In: BMVC 2006, pp. 578–589 (2006)
28. Arsigny, V., Fillard, P., Pennec, X., Ayache, N.: Log-Euclidean metrics for fast and simple calculus on diffusion tensors. Magnetic Resonance in Medicine 56(2), 411–421 (2006)
29. Gur, Y., Sochen, N.A.: Fast invariant Riemannian DT-MRI regularization. In: Proc. of IEEE Computer Society Workshop on Mathematical Methods in Biomedical Image Analysis (MM-BIA), Rio de Janeiro, Brazil, pp. 1–7 (2007)
30. Mémoli, F., Sapiro, G., Osher, S.: Solving variational problems and partial differential equations mapping into general target manifolds. Journal of Computational Physics 195(1), 263–292 (2004)
31. Brox, T., Weickert, J., Burgeth, B., Mrázek, P.: Nonlinear structure tensors. Revised version of technical report no. 113. Saarland University, Saarbrücken, Germany (2004)
32. Nielsen, M., Johansen, P., Olsen, O., Weickert, J. (eds.): Scale-Space 1999. LNCS, vol. 1682. Springer, Heidelberg (1999)
33. Maaß, H.: Siegel's Modular Forms and Dirichlet Series. Lecture notes in mathematics, vol. 216. Springer, Heidelberg (1971)
34. Arsigny, V., Fillard, P., Pennec, X., Ayache, N.: Fast and simple calculus on tensors in the log-Euclidean framework. In: Duncan, J.S., Gerig, G. (eds.) MICCAI 2005. LNCS, vol. 3749, pp. 115–122. Springer, Heidelberg (2005)
35. Bihan, D.L., Mangin, J.F., Poupon, C., Clark, C.A., Pappata, S., Molko, N., Chabriat, H.: Diffusion tensor imaging: Concepts and applications. Journal of Magnetic Resonance Imaging 13(4), 534–546 (2001)

FaceTracer: A Search Engine for Large Collections of Images with Faces

Neeraj Kumar[1,*], Peter Belhumeur[1], and Shree Nayar[1]

Columbia University

Abstract. We have created the first image search engine based entirely on faces. Using simple text queries such as "smiling men with blond hair and mustaches," users can search through over 3.1 million faces which have been automatically labeled on the basis of several facial attributes. Faces in our database have been extracted and aligned from images downloaded from the internet using a commercial face detector, and the number of images and attributes continues to grow daily. Our classification approach uses a novel combination of Support Vector Machines and Adaboost which exploits the strong structure of faces to select and train on the optimal set of features for each attribute. We show state-of-the-art classification results compared to previous works, and demonstrate the power of our architecture through a functional, large-scale face search engine. Our framework is fully automatic, easy to scale, and computes all labels off-line, leading to fast on-line search performance. In addition, we describe how our system can be used for a number of applications, including law enforcement, social networks, and personal photo management. Our search engine will soon be made publicly available.

1 Introduction

We have created the first *face* search engine, allowing users to search through large collections of images which have been automatically labeled based on the appearance of the faces within them. Our system lets users search on the basis of a variety of facial attributes using natural language queries such as, "men with mustaches," or "young blonde women," or even, "indoor photos of smiling children." This face search engine can be directed at all images on the internet, tailored toward specific image collections such as those used by law enforcement or online social networks, or even focused on personal photo libraries.

The ability of current search engines to find images based on facial appearance is limited to images with text annotations. Yet, there are many problems with annotation-based search of images: the manual labeling of images is time-consuming; the annotations are often incorrect or misleading, as they may refer to other content on a webpage; and finally, the vast majority of images are

---

[*] Supported by the National Defense Science & Engineering Graduate Fellowship.

D. Forsyth, P. Torr, and A. Zisserman (Eds.): ECCV 2008, Part IV, LNCS 5305, pp. 340–353, 2008.

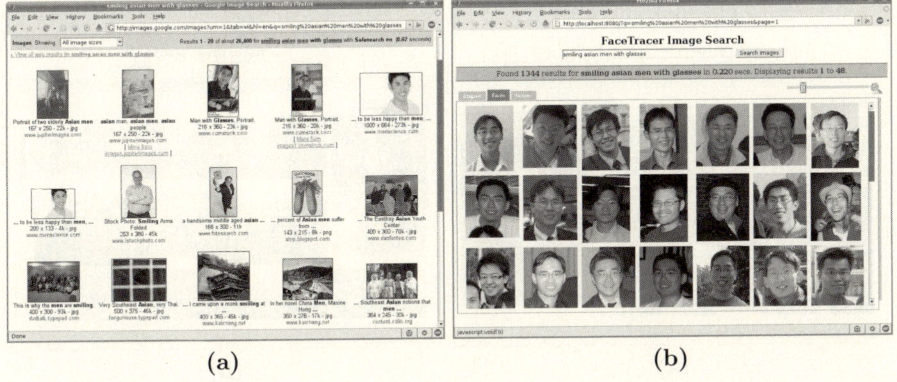

(a)   (b)

Fig. 1. Results for the query "smiling asian men with glasses," using (a) the Google image search engine and (b) our face search engine. Our system currently has over 3.1 million faces, automatically detected and extracted from images downloaded from the internet, using a commercial face detector [1]. Rather than use text annotations to find images, our system has automatically labeled a large number of different facial attributes on each face (off-line), and searches are performed using only these labels. Thus, search results are returned almost instantaneously. The results also contain links pointing back to the original source image and associated webpage.

simply not annotated. Figures 1a and 1b show the results of the query, "smiling asian men with glasses," using a conventional image search engine (Google Image Search) and our search engine, respectively. The difference in quality of search results is clearly visible. Google's reliance on text annotations results in it finding images that have no relevance to the query, while our system returns only the images that match the query.

Like much of the work in content-based image retrieval, the power of our approach comes from automatically labeling images off-line on the basis of a large number of attributes. At search time, only these labels need to be queried, resulting in almost instantaneous searches. Furthermore, it is easy to add new images and face attributes to our search engine, allowing for future scalability. Defining new attributes and manually labeling faces to match those attributes can also be done collaboratively by a community of users.

Figures 2a and 2b show search results of the queries, "young blonde women" and "children outdoors," respectively. The first shows a view of our extended interface, which displays a preview of the original image in the right pane when the user holds the mouse over a face thumbnail. The latter shows an example of a query run on a personalized set of images. Incorporating our search engine into photo management tools would enable users to quickly locate sets of images and then perform bulk operations on them (e.g., edit, email, or delete). (Since current tools depend on manual annotation of images, they are significantly more time-consuming to use.) Another advantage of our attribute-based search on personal collections is that with a limited number of people, simple queries can often find images of a particular person, without requiring any form of face recognition.

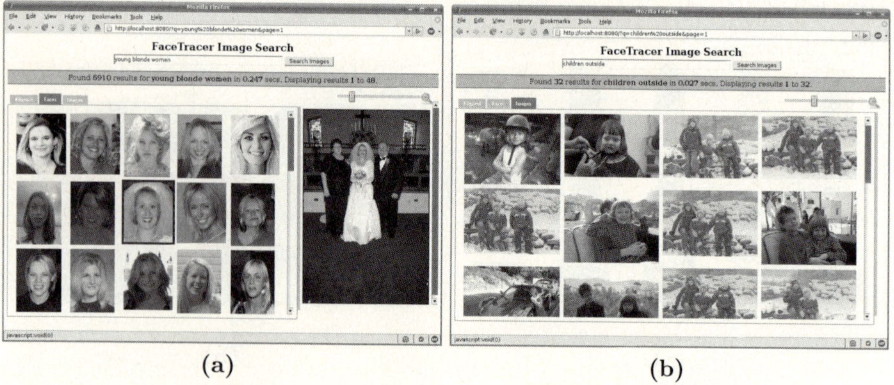

(a)   (b)

Fig. 2. Results of queries (a) "young blonde women" and (b) "children outside," using our face search engine. In (a), search results are shown in the left panel, while the right panel shows a preview of the original image for the selected face. (b) shows search results on a personalized dataset, displaying the results as thumbnails of the original images. Note that these results were correctly classified as being "outside" using only the cropped face images, showing that face images often contain enough information to describe properties of the image which are not directly related to faces.

Our search engine owes its superior performance to the following factors:

- **A large and diverse dataset of face images with a significant subset containing attribute labels.** We currently have over 3.1 million aligned faces in our database – the largest such collection in the world. In addition to its size, our database is also noteworthy for being a completely "real-world" dataset. The images are downloaded from the internet and encompass a wide range of pose, illumination, imaging conditions, and were taken using a large variety of cameras. The faces have been automatically extracted and aligned using a commercial face and fiducial point detector [1]. In addition, 10 attributes have been manually labeled on more than 17,000 of the face images, creating a large dataset for training and testing classification algorithms.

- **A scalable and fully automatic architecture for attribute classification.** We present a novel approach tailored toward face classification problems, which uses a boosted set of Support Vector Machines (SVMs) [2] to form a strong classifier with high accuracy. We describe the results of this algorithm on a variety of different attributes, including demographic information such as gender, age, and race; facial characteristics such as eye wear and facial hair; image properties such as blurriness and lighting conditions; and many others as well. A key aspect of this work is that classifiers for new attributes can be trained automatically, requiring only a set of labeled examples. Yet, the flexibility of our framework does not come at the cost of reduced accuracy – we compare against several state-of-the-art classification methods and show the superior classification rates produced by our system.

We will soon be releasing our search engine for public use.

2   Related Work

Our work lies at the intersection of several fields, including computer vision, machine learning, and content-based image retrieval. We present an overview of the relevant work, organized by topic.

Attribute Classification. Prior works on attribute classification have focused mostly on gender and ethnicity classification. Early works such as [3]used neural networks to perform gender classification on small datasets. The Fisherfaces work of [4] showed that linear discriminant analysis could be used for simple attribute classification such as glasses/no glasses. More recently, Moghaddam and Yang [5] used Support Vector Machines (SVMs) [2] trained on small "face-prints" to classify the gender of a face, showing good results on the FERET face database [6]. The works of Shakhnarovich et al. [7] and Baluja & Rowley [8] used Adaboost [9] to select a linear combination of weak classifiers, allowing for almost real-time classification of faces, with results in the latter case again demonstrated on the FERET database. These methods differ in their choice of weak classifiers: the former uses the Haar-like features of the Viola-Jones face detector [10], while the latter uses simple pixel comparison operators.

In contrast, we develop a method that combines the advantages of SVMs and Adaboost (described in Sect. 4). We also present results of an extensive comparison against all three of these prior methods in Sect. 5. Finally, we note that this is an active area of research, and there are many other works on attribute classification which use different combinations of learning techniques, features, and problem formulations [11,12]. An exploration of the advantages and disadvantages of each is beyond the scope of this paper.

Content-Based Image Retrieval (CBIR). Our work can also be viewed as a form of CBIR, where our content is limited to images with faces. Interested readers can refer to the work of Datta et al. [13] for a recent survey of this field. Most relevant to our work is the "Photobook" system [14], which allows for similarity-based searches of faces and objects using parametric eigenspaces. However, their goal is different from ours. Whereas they try to find objects similar to a chosen one, we locate a set of images starting only with simple text queries. Although we use vastly different classifiers and methods for feature selection, their division of the face into functional parts such as the eyes, nose, etc., is echoed in our approach of training classifiers on functional face regions.

3   Creating the Face Database

To date, we have built a large database of over 3.1 million face images extracted from over 6.2 million images collected from the internet. This database continues to grow as we automatically collect, align, and assign attributes to face images daily. An overview of the database creation process is illustrated in Fig. 3. We download images using two different methods – keyword searches and random downloads. The first allows us to build datasets related to particular terms

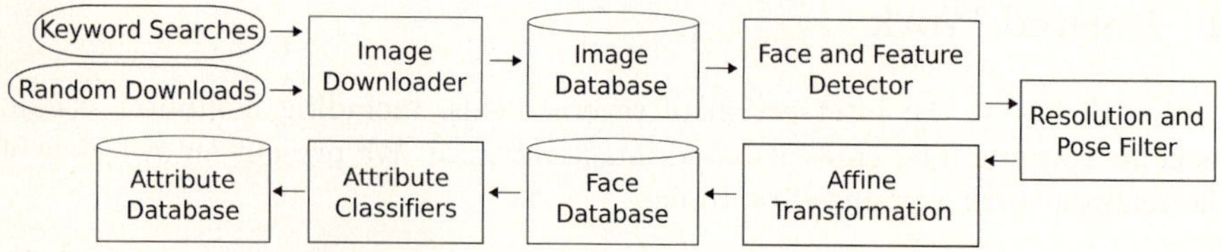

Fig. 3. Overview of database creation. See text for details.

(e.g., celebrity names and professions). The latter allows us to sample from the more general distribution of images on the internet. In particular, it lets us include images that have no corresponding textual information, i.e., that are effectively invisible to current image search engines. Our images are downloaded from a wide variety of online sources, such as Google Images, Microsoft Live Image Search, and Flickr, to name a few. Relevant metadata such as image and page URLs are stored in the EXIF tags of the downloaded images.

Next, we apply the OKAO face detector [1] to the downloaded images to extract faces. This detector also gives us the pose angles of each face, as well as the locations of six fiducial points (the corners of both eyes and the corners of the mouth). We filter the set of faces by resolution and face pose ($\pm 10°$ from front-center). Finally, the remaining faces are aligned to a canonical pose by applying an affine transformation. This transform is computed using linear least squares on the detected fiducial points and corresponding points defined on a template face. (In future work, we intend to go beyond near frontal poses.)

We present various statistics of our current face database in Table 1, divided by image source. We would like to draw attention to three observations about our data. First, from the statistics of randomly downloaded images, it appears that a significant fraction of them contain faces (25.7%), and on average, each image contains 0.5 faces. Second, our collection of aligned faces is the largest such collection of which we are aware. It is truly a "real-world" dataset, with completely uncontrolled lighting and environments, taken using unknown cameras and in unknown imaging conditions, with a wide range of image resolutions. In this respect, our database is similar to the LFW dataset [15], although ours is larger by 2 orders of magnitude and not targeted specifically for face recognition. In contrast, existing face datasets such as Yale Face A&B [16], CMU PIE [17], and FERET [6] are either much smaller in size and/or taken in highly controlled settings. Even the more expansive FRGC version 2.0 dataset [18] has a limited number of subjects, image acquisition locations, and all images were taken with the same camera type. Finally, we have labeled a significant number of these images for our 10 attributes, enumerated in Table 2. In total, we have over 17,000 attribute labels.

4   Automatic Attribute Classification for Face Images

Our approach to image search relies on labeling each image with a variety of attributes. For a dataset as large as ours, it is infeasible to manually label every

Table 1. Image database statistics. We have collected what we believe to be the largest set of aligned real-world face images (over 3.1 million so far). These faces have been extracted using a commercial face detector [1]. Notice that more than 45% of the downloaded images contain faces, and on average, there is one face per two images.

Image Source	# Images Downloaded	# Images With Faces	% Images With Faces	Total # Faces Found	Average # Faces Found Per Image
Randomly Downloaded	4,289,184	1,102,964	25.715	2,156,287	0.503
Celebrities	428,312	411,349	96.040	285,627	0.667
Person Names	17,748	7,086	39.926	10,086	0.568
Face-Related Words	13,028	5,837	44.804	14,424	1.107
Event-Related Words	1,658	997	60.133	1,335	0.805
Professions	148,782	75,105	50.480	79,992	0.538
Series	7,472	3,950	52.864	8,585	1.149
Camera Defaults	895,454	893,822	99.818	380,682	0.425
Miscellanous	417,823	403,233	96.508	194,057	0.464
Total	**6,219,461**	**2,904,343**	**46.698**	**3,131,075**	**0.503**

Table 2. List of labeled attributes. The labeled face images are used for training our classifiers, allowing for automatic classification of the remaining faces in our database. Note that these were labeled by a large set of people, and thus the labels reflect a group consensus about each attribute rather than a single user's strict definition.

Attribute/ Options	Number Labeled	Attribute/ Options	Number Labeled	Attribute/ Options	Number Labeled
Gender	**1,954**	**Smiling**	**1,571**	**Race**	**1,309**
Male	867	True	832	White	433
Female	1,087	False	739	Black	399
Age	**3,301**	**Mustache**	**1,947**	Asian	477
Baby	577	True	618	**Eye Wear**	**2,360**
Child	636	False	1,329	None	1,256
Youth	784	**Blurry**	**1,763**	Eyeglasses	665
Middle Aged	815	True	763	Sunglasses	439
Senior	489	False	1,000	**Environment**	**1,583**
Hair Color	**1,033**	**Lighting**	**633**	Outdoor	780
Black	717	Flash	421	Indoor	803
Blond	316	Harsh	212	**Total**	**17,454**

image. Instead, we use our large sets of manually-labeled images to build accurate classifiers for each of the desired attributes.

In creating a classifier for a particular attribute, we could simply choose all pixels on the face, and let our classifier figure out which are important for the task and which are not. This, however, puts too great a burden on the classifier, confusing it with non-discriminative features. Instead, we create a rich set of local feature options from which our classifier can automatically select the best ones. Each option consists of four choices: the region of the face to extract features from, the type of pixel data to use, the kind of normalization to apply to the data, and finally, the level of aggregation to use.

Face Regions. We break up the face into a number of functional regions, such as the nose, mouth, etc., much like those defined in the work on modular eigenspaces

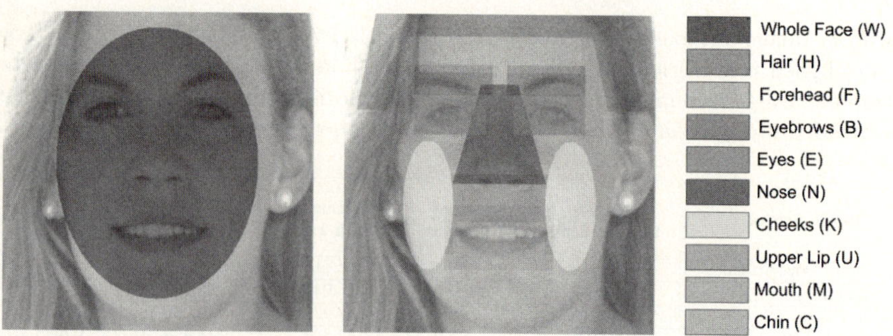

Fig. 4. The face regions used for automatic feature selection. On the left is one region corresponding to the whole face, and on the right are the remaining regions, each corresponding to functional parts of the face. The regions are large enough to be robust against small differences between individual faces and overlap slightly so that small errors in alignment do not cause a feature to go outside of its region. The letters in parentheses denote the code letter for the region, used later in the paper.

[19]. The complete set of 10 regions we use are shown in Fig. 4. Our coarse division of the face allows us to take advantage of the common geometry shared by faces, while allowing for differences between individual faces, as well as robustness to small errors in alignment.

Types of Pixel Data. We include different color spaces and image derivatives as possible feature types. These can often be more discriminative than standard RGB values for certain attributes. Table 3 lists the various options.

Normalizations. Normalizations are important for removing lighting effects, allowing for better generalization across images. We can remove illumination gains by using mean normalization, $\hat{x} = \frac{x}{\mu}$, or both gains and offsets by using energy normalization, $\hat{x} = \frac{x-\mu}{\sigma}$. In these equations, x refers to the input value, μ and σ are the mean and standard deviation of all the x values within the region, and $\hat{x}$ refers to the normalized output value.

Aggregations. For some attributes, aggregate information over the entire region might be more useful than individual values at each pixel. This includes histograms of values over the region, or simply the mean and variance.

To concisely refer to a complete feature option, we define a shorthand notation using the format, "Region:pixel type.normalization.aggregation." The region notation is shown in Fig. 4; the notation for the pixel type, normalization, and aggregation is shown in Table 3.

4.1 Classifier Architecture

In recent years, Support Vector Machines (SVMs) [2] have been used successfully for many classification tasks [20,21]. SVMs aim to find the linear hyperplane which best separates feature vectors of two different classes, so as to

Table 3. Feature type options. A complete feature type is constructed by first converting the pixels in a given region to one of the pixel value types from the first column, then applying one of the normalizations from the second column, and finally aggregating these values into the output feature vector using one of the options from the last column. The letters in parentheses are used as code letters in a shorthand notation for concisely designating feature types.

Pixel Value Types	Normalizations	Aggregation
RGB (r)	None (n)	None (n)
HSV (h)	Mean-Normalization (m)	Histogram (h)
Image Intensity (i)	Energy-Normalization (e)	Statistics (s)
Edge Magnitude (m)		
Edge Orientation (o)		

simultaneously minimize the number of misclassified examples (training error) and maximize the distance between the classes (the *margin*).

As with many classification algorithms, SVMs perform best when given only the relevant data – too many extraneous inputs can confuse or overtrain the classifier, resulting in poor accuracy on real data. In particular, if we would like to train a classifier for an attribute that is only dependent on a certain part of the face (e.g., "is smiling?"), giving the SVM a feature vector constructed from all the pixels of the face is unlikely to yield optimal results. Given the large number of regions and feature types described in the previous section, an efficient and automatic selection algorithm is needed to find the optimal combination of features for each attribute. Following the successes of [10,7,8,11], we use Adaboost [9] for this purpose.

Adaboost is a principled, iterative approach for building strong classifiers out of a collection of "weak" classifiers. In each iteration of Adaboost, the weak classifier that best classifies a set of weighted examples is greedily picked to form part of the final classifier. The weights on the examples are then adjusted to make misclassified examples more important in future iterations, and the process is repeated until a given number of weak classifiers has been picked. A major advantage of Adaboost is that it is resistant to overtraining [22,23].

We combine the strengths of these two methods by constructing a number of "local" SVMs and letting Adaboost create an optimal classifier using a linear combination of them. We create one SVM for each region, feature type, and SVM parameter combination, using the LibSVM library [24]. Normally, Adaboost is performed using weak classifiers, which need to be retrained at the beginning of each round. However, we rely on the fact that our local SVMs will either be quite powerful (if created using the relevant features for the current attribute), or virtually useless (if created from irrelevant features). Retraining will not significantly improve the classifiers in either case.

Accordingly, we precompute the results of each SVM on all examples, one SVM at a time. Thus, our classifiers remain fixed throughout the Adaboost process, and we do not need to keep a large number of SVMs in memory. Once all SVM outputs have been computed, we run our Adaboost rounds to obtain the

Table 4. Error rates and top feature combinations for each attribute, computed by training on 80% of the labeled data and testing on the remaining 20%, averaging over 5 runs (5-fold cross-validation). Note that the attribute-tuned global SVM performs as well as, or better than, the local SVMs in all cases, and requires much less memory and computation than the latter. The top feature combinations selected by our algorithm are shown in ranked order from more important to less as "Region:feature type" pairs, where the region and feature types are listed using the code letters from Fig. 4 and Table 3. For example, the first combination for the hair color classifier, "H:r.n.s," takes from the hair region (H) the RGB values (r) with no normalization (n) and using only the statistics (s) of these values.

Attribute	Error Rates for Attribute-Tuned Local SVMs	Error Rates for Attribute-Tuned Global SVM	Top Feature Combinations in Ranked Order Each combination is represented as Region:pixtype.norm.aggreg
Gender	9.42%	8.62%	W:i.m.n | W:o.n.n | W:i.n.n | W:i.e.n
Age	17.34%	16.65%	W:i.m.n | W:i.n.n | H:r.e.n | E:r.m.n | H:r.e.s | W:o.n.n
Race	7.75%	6.49%	W:i.m.n | E:r.e.n | C:o.n.n | M:r.m.n | W:o.n.n
Hair Color	7.85%	5.54%	H:r.n.s | W:i.m.n | E:r.m.n | H:r.n.n | W:i.n.n | H:r.m.n
Eye Wear	6.22%	5.14%	W:m.n.n | W:i.n.n | K:o.n.h | W:m.m.n | N:r.n.n
Mustache	6.42%	4.61%	U:r.e.n | M:r.m.n
Smiling	4.60%	4.60%	M:r.m.n | M:r.n.n | M:r.e.n | W:i.n.n | W:i.e.n | M:i.n.n
Blurry	3.94%	3.41%	W:m.m.n | H:m.n.n | W:m.n.n | H:m.m.n | M:m.m.n
Lighting	2.82%	1.61%	W:i.n.n | W:i.e.n | K:r.n.n | C:o.n.n | E:o.n.n
Environment	12.25%	12.15%	N:r.m.n | K:r.e.n | K:r.m.n | W:r.m.n | E:r.m.n

weights on each SVM classifier. We use the formulation of Adaboost described in [8], with the modification that errors are computed in a continuous manner (using the confidence values obtained from the SVM classifier), rather than discretely as is done in [8]. We found this change improves the stability of the results, without adversely affecting the error rates.

The error rates of these "attribute-tuned local SVMs" are shown in the second column of Table 4. The rates were computed by dividing the labeled examples for each attribute into 5 parts, using 4 parts to train and the remaining one to test, and then rotating through all 5 sets (5-fold cross-validation). Note that in most cases, our error rates are below 10%, and for many attributes, the error rate is under 5%. (The higher error rates for age are due to the fact that different people's labels for each of the age categories did not match up completely.)

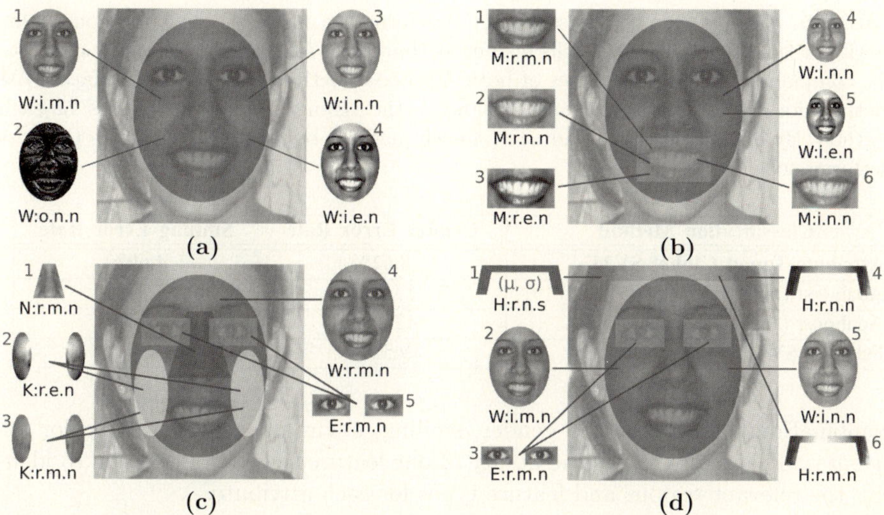

Fig. 5. Illustrations of automatically-selected region and feature types for (a) gender, (b) smiling, (c) environment, and (d) hair color. Each face image is surrounded by depictions of the top-ranked feature combinations for the given attribute, along with their corresponding shorthand label (as used in Table 4). Notice how each classifier uses different regions and feature types of the face.

We emphasize the fact that these numbers are computed using our real-world dataset, and therefore reflect performance on real images.

A limitation of this architecture is that classification will require keeping a possibly large number of SVMs in memory, and each one will need to be evaluated for every input image. Furthermore, one of the drawbacks of the Adaboost formulation is that different classifiers can only be combined linearly. Attributes which might depend on non-linear combinations of different regions or feature types would be difficult to classify using this architecture.

We solve both of these issues simultaneously by training one "global" SVM on the union of the features from the top classifiers selected by Adaboost. We do this by concatenating the features from the N highest-weighted SVMs (from the output of Adaboost), and then training a single SVM classifier over these features (optimizing over N). In practice, the number of features chosen is between 2 (for "mustache") and 6 (e.g., for "hair color"). Error rates for this algorithm, denoted as "Attribute-Tuned Global SVM," are shown in the third column of Table 4. Notice that for each attribute, these rates are equal to, or less than, the rates obtained using the combination of local SVMs, despite the fact that these classifiers run significantly faster and require only a fraction of the memory (often less by an order of magnitude).

The automatically-selected region and feature type combinations for each attribute are shown in the last column of Table 4. Listed in order of decreasing importance, the combinations are displayed in a shorthand notation using the codes given in Fig. 4 and Table 3. In Fig. 5, we visually illustrate the top feature

Table 5. Comparison of classification performance against prior methods. Our attribute-tuned global SVM performs better than prior state-of-the-art methods. Note the complementary performances of both Adaboost methods versus the full-face SVM method for the different attributes, showing the strengths and weaknesses of each method. By exploiting the advantages of each method, our approach achieves the best performance.

Classification Method	Gender Error Rate	Smiling Error Rate
Attribute-Tuned Global SVM	**8.62%**	**4.60%**
Adaboost (pixel comparison feats.) [9]	13.13%	7.41%
Adaboost (Haar-like feats.) [8]	12.88%	6.40%
Full-face SVM [6]	9.52%	13.54%

combinations chosen for the gender, smiling, environment, and hair color attributes. This figure shows the ability of our feature selection approach to identify the relevant regions and feature types for each attribute.

5   Comparison to Prior Work

While we have designed our classifier architecture to be flexible enough to handle a large variety of attributes, it is important to ensure that we have not sacrificed accuracy in the process. We therefore compare our approach to three state-of-the-art methods for attribute classification: full-face SVMs using brightness normalized pixel values [5], Adaboost using Haar-like features [7], and Adaboost using pixel comparison features [8]. Since these works have mostly focused on gender classification, we use that attribute as our first testing criteria.

The error rates for gender classification using our training and testing data on all methods are shown in the second column of Table 5. We note that our method performs slightly better than the prior SVM method and significantly better than both Adaboost methods. The difference between the Adaboost and SVM methods may reflect one limitation of using linear combinations of weak classifiers – the classifiers might be too weak to capture all the nuances of gender differences.

To see how these methods do on a localized attribute, we also applied each of them to the "smiling" attribute. Here, while once again our method has the lowest error rate, we see that the Adaboost methods perform significantly better than the prior SVM method. This result highlights the power of Adaboost to correctly find the important features from a large set of possibilities, as well as the degradation in accuracy of SVMs when given too much irrelevant data.

6   The FaceTracer Engine

We have trained attribute-tuned global SVM classifiers for each attribute listed in Table 4. In an offline process, all images in our database are sent through the classifiers for each attribute, and the resulting attribute labels are stored for fast online searches using the FaceTracer engine.

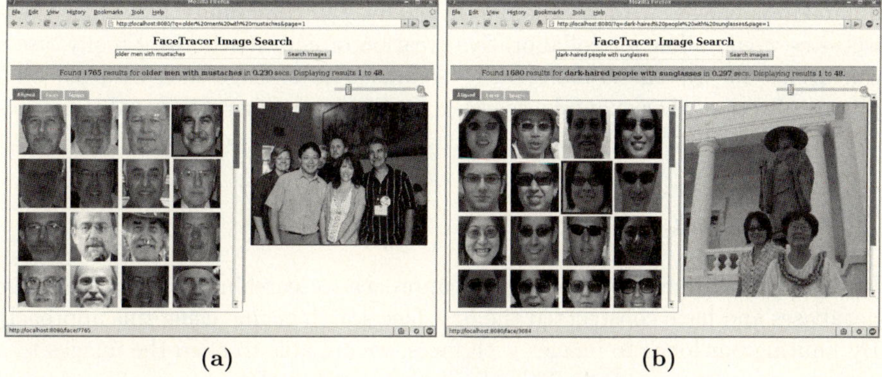

(a)   (b)

Fig. 6. Results of queries (a) "older men with mustaches" and (b) "dark-haired people with sunglasses" on our face search engine. The results are shown with aligned face images on the left, and a preview of the original image for the currently selected face on the right. Notice the high quality of results in both cases.

For a search engine, the design of the user interface is important for enabling users to easily find what they are looking for. We use simple text-based queries, since these are both familiar and accessible to most internet users. Search queries are mapped onto attribute labels using a dictionary of terms. Users can see the current list of attributes supported by the system on the search page, allowing them to construct their searches without having to guess what kinds of queries are allowed. This approach is simple, flexible, and yields excellent results in practice. Furthermore, it is easy to add new phrases and attributes to the dictionary, or maintain separate dictionaries for different languages.

Results are ranked in order of decreasing confidence, so that the most relevant images are shown first. (Our classifier gives us confidence values for each labeled attribute.) For searches with multiple query terms, we combine the confidences of different labels such that the final ranking shows images in decreasing order of relevance to all search terms. To prevent high confidences for one attribute from dominating the search results, we convert the confidences into probabilities, and then use the product of the probabilities as the sort criteria. This ensures that the images with high confidences for *all* attributes are shown first.

Example queries on our search engine are shown in Figs. 1b, 2, and 6. The returned results are all highly relevant, and the user can view the results in a variety of ways, as shown in the different examples. Figure 2b shows that we can learn useful things about an image using just the appearance of the faces within it – in this case determining whether the image was taken indoors or outdoors.

Our search engine can be used in many other applications, replacing or augmenting existing tools. In law enforcement, eyewitnesses to crimes could use our system to quickly narrow a list of possible suspects and then identify the actual criminal from this reduced list, saving time and increasing the chances of finding the right person. On the internet, our face search engine is a perfect match for social networking websites such as Facebook and Myspace, which contain large numbers of images with people. Additionally, the community aspect of these websites would allow for collaborative creation of new attributes. Finally, users can utilize our system to more easily organize and manage their own personal photo collections. For example, searches for blurry or other poor-quality images can be used to find and remove all such images from the collection.

7   Discussion

In this work, we have described a new approach to searching for images in large databases and have constructed the first face search engine using this approach. By limiting our focus to images with faces, we are able to align the images to a common coordinate system. This allows us to exploit the commonality of facial structures across people to train accurate classifiers for real-world face images. Our approach shows the power of combining the strengths of different algorithms to create a flexible architecture without sacrificing classification accuracy.

As we continue to grow and improve our system, we would also like to address some of our current limitations. For example, to handle more than just frontal faces would require that we define the face regions for each pose bin. Rather than specifying the regions manually, however, we can define them once on a 3D model, and then project the regions to 2D for each pose bin. The other manual portion of our architecture is the labeling of example images for training classifiers. Here, we can take advantage of communities on the internet by offering a simple interface for both defining new attributes and labeling example images. Finally, while our dictionary-based search interface is adequate for most simple queries, taking advantage of methods in statistical natural language processing (NLP) could allow our system to map more complex queries to the list of attributes.

Acknowledgements. We are grateful to Omron Technologies for providing us the OKAO face detection system. This work was supported by NSF grants IIS-03-08185 and ITR-03-25867.

References

1. Omron: OKAO vision (2008), http://www.omron.com/rd/vision/01.html
2. Cortes, C., Vapnik, V.: Support-vector networks. Machine Learning 20(3) (1995)
3. Golomb, B.A., Lawrence, D.T., Sejnowski, T.J.: Sexnet: A neural network identifies sex from human faces. NIPS, 572–577 (1990)
4. Belhumeur, P.N., Hespanha, J., Kriegman, D.J.: Eigenfaces vs. fisherfaces: Recognition using class specific linear projection. In: Buxton, B.F., Cipolla, R. (eds.) ECCV 1996. LNCS, vol. 1065, pp. 45–58. Springer, Heidelberg (1996)
5. Moghaddam, B., Yang, M.-H.: Learning gender with support faces. TPAMI 24(5), 707–711 (2002)
6. Phillips, P., Moon, H., Rizvi, S., Rauss, P.: The FERET evaluation methodology for face-recognition algorithms. TPAMI 22(10), 1090–1104 (2000)

7. Shakhnarovich, G., Viola, P.A., Moghaddam, B.: A unified learning framework for real time face detection and classification. ICAFGR, 14–21 (2002)
8. Baluja, S., Rowley, H.: Boosting sex identification performance. IJCV (2007)
9. Freund, Y., Shapire, R.E.: Experiments with a new boosting algorithm. In: ICML (1996)
10. Viola, P., Jones, M.: Rapid object detection using a boosted cascade of simple features. In: CVPR (2001)
11. Bartlett, M.S., Littlewort, G., Fasel, I., Movellan, J.R.: Real time face detection and facial expression recognition: Development and applications to human computer interaction. CVPRW 05 (2003)
12. Wang, Y., Ai, H., Wu, B., Huang, C.: Real time facial expression recognition with adaboost. In: ICPR, pp. 926–929 (2004)
13. Datta, R., Li, J., Wang, J.Z.: Content-based image retrieval: Approaches and trends of the new age. Multimedia Information Retrieval, 253–262 (2005)
14. Pentland, A., Picard, R., Sclaroff, S.: Photobook: Content-based manipulation of image databases. IJCV, 233–254 (1996)
15. Huang, G.B., Ramesh, M., Berg, T., Learned-Miller, E.: Labeled faces in the wild: A database for studying face recognition in unconstrained environments. Technical Report 07-49 (2007)
16. Georghiades, A.S., Belhumeur, P.N., Kriegman, D.J.: From few to many: Illumination cone models for face recognition under variable lighting and pose. TPAMI 23(6), 643–660 (2001)
17. Sim, T., Baker, S., Bsat, M.: The CMU pose, illumination, and expression (PIE) database. In: ICAFGR, pp. 46–51 (2002)
18. Phillips, P.J., Flynn, P.J., Scruggs, T., Bowyer, K.W., Chang, J., Hoffman, K., Marques, J., Min, J., Worek, W.: Overview of the face recognition grand challenge. CVPR, 947–954 (2005)
19. Pentland, A., Moghaddam, B., Starner, T.: View-based and modular eigenspaces for face recognition. CVPR, 84–91 (1994)
20. Huang, J., Shao, X., Wechsler, H.: Face pose discrimination using support vector machines (SVM). In: ICPR, pp. 154–156 (1998)
21. Osuna, E., Freund, R., Girosi, F.: Training support vector machines: An application to face detection. CVPR (1997)
22. Schapire, R., Freund, Y., Bartlett, P., Lee, W.S.: Boosting the margin: A new explanation for the effectiveness of voting methods. The Annals of Statistics 26(5), 1651–1686 (1998)
23. Drucker, H., Cortes, C.: Boosting decision trees. NIPS, 479–485 (1995)
24. Chang, C.C., Lin, C.J.: LIBSVM: a library for support vector machines (2001), http://www.csie.ntu.edu.tw/cjlin/libsvm/

What Does the Sky Tell Us about the Camera?

Jean-François Lalonde, Srinivasa G. Narasimhan, and Alexei A. Efros

School of Computer Science, Carnegie Mellon University
http://graphics.cs.cmu.edu/projects/sky

Abstract. As the main observed illuminant outdoors, the sky is a rich source of information about the scene. However, it is yet to be fully explored in computer vision because its appearance depends on the sun position, weather conditions, photometric and geometric parameters of the camera, and the location of capture. In this paper, we propose the use of a physically-based sky model to analyze the information available within the visible portion of the sky, observed over time. By fitting this model to an image sequence, we show how to extract camera parameters such as the focal length, and the zenith and azimuth angles. In short, the sky serves as a geometric calibration target. Once the camera parameters are recovered, we show how to use the same model in two applications: 1) segmentation of the sky and cloud layers, and 2) data-driven sky matching across different image sequences based on a novel similarity measure defined on sky parameters. This measure, combined with a rich appearance database, allows us to model a wide range of sky conditions.

1   Introduction

When presented with an outdoor photograph (such as images on Fig. 1), an average person is able to infer a good deal of information just by looking at the sky. Is it morning or afternoon? Do I need to wear a sunhat? Is it likely to rain? A professional, such as a sailor or a pilot, might be able to tell even more: time of day, temperature, wind conditions, likelihood of a storm developing, etc. As the main observed illuminant in an outdoor image, the sky is a rich source of information about the scene. However it is yet to be fully explored in computer vision. The main obstacle is that the problem is woefully under-constrained. The appearance of the sky depends on a host of factors such as the position of the sun, weather conditions, photometric and geometric parameters of the camera, and location and direction of observation. Unfortunately, most of these factors remain unobserved in a single photograph; the sun is rarely visible in the picture, the camera parameters and location are usually unknown, and worse yet, only a small fraction of the full hemisphere of sky is actually seen.

However, if we were to observe the same small portion of the sky *over time*, we would see the changes in sky appearance due to the sun and weather that are not present within a single image. In short, this is exactly the type of problem that might benefit from observing a time-lapse image sequence. Such a sequence is typically acquired by a static camera looking at the same scene over a period of

D. Forsyth, P. Torr, and A. Zisserman (Eds.): ECCV 2008, Part IV, LNCS 5305, pp. 354–367, 2008.

Fig. 1. The sky appearance is a rich source of information about the scene illumination

time. When the scene is mostly static, the resulting sequence of images contains a wealth of information that has been exploited in several different ways, the most commonly known being background subtraction, but also shadow detection and removal [1], video factorization and compression [2], radiometric calibration [3], camera geo-location [4], temporal variation analysis [5] and color constancy [6]. The main contribution of this paper is to show what information about the camera is available in the visible portion of the sky in a time-lapse image sequence, and how to extract this information to calibrate the camera.

The sky appearance has long been studied by physicists. One of the most popular physically-based sky model was introduced by Perez et al [7]. This model has been used in graphics for relighting [8] and rendering [9]. Surprisingly however, very little work has been done on extracting information from the visible sky. One notable exception is the work of Jacobs et al [10] where they use the sky to infer the camera azimuth by using a correlation-based approach. In our work, we address a broader question: what does the sky tell us about the camera? We show how we can recover the viewing geometry using an optimization-based approach. Specifically, we estimate the camera focal length, its zenith angle (with respect to vertical), and its azimuth angle (with respect to North). We will assume that a static camera is observing the same scene over time, with no roll angle (i.e. the horizon line is parallel to the image horizontal axis). Its location (GPS coordinates) and the times of image acquisition are also known. We also assume that the sky region has been segmented, either manually or automatically [5].

Once the camera parameters are recovered, we then show how we can use our sky model in two applications. First, we present a novel sky-cloud segmentation algorithm that identifies cloud regions within an image. Second, we show how we can use the resulting sky-cloud segmentation in order to find matching skies across different cameras. To do so, we introduce a novel bi-layered sky model which captures both the physically-based sky parameters and cloud appearance, and determine a similarity measure between two images. This distance can then be used for finding images with similar skies, even if they are captured by different cameras at different locations. We show qualitative cloud segmentation and sky matching results that demonstrate the usefulness of our approach.

In order to thoroughly test our algorithms, we require a set of time-lapse image sequences which exhibit a wide range of skies and cameras. For this, we use the AMOS (Archive of Many Outdoor Scenes) database [5], which contains image sequences taken by static webcams over more than a year.

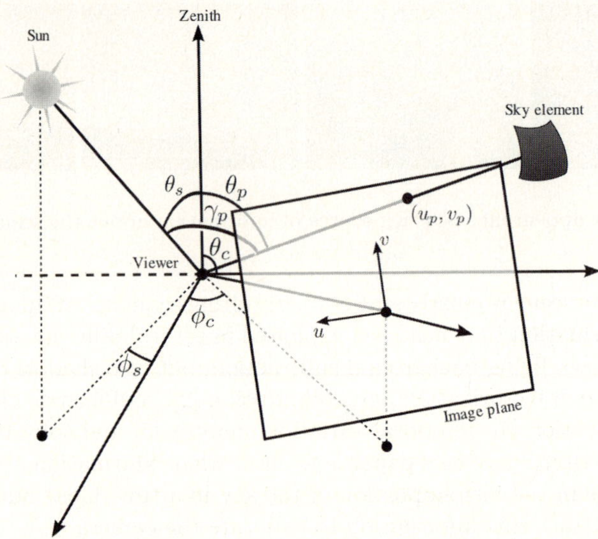

Fig. 2. Geometry of the problem, when a camera is viewing a sky element (blue patch in the upper-right). The sky element is imaged at pixel (u_p, v_p) in the image, and the camera is rotated by angles (θ_c, ϕ_c). The camera focal length f_c, not shown here, is the distance between the origin (center of projection), and the image center. The sun direction is given by (θ_s, ϕ_s), and the angle between the sun and the sky element is γ_p. Here (u_p, v_p) are known because the sky is segmented.

2   Physically-Based Model of the Sky

First, we introduce the physically-based model of the sky that lies at the foundation of our approach. We will first present the model in its general form, then in a useful simplified form, and finally demonstrate how it can be written as a function of camera parameters. We will consider clear skies only, and address the more complicated case of clouds at a later point in the paper.

2.1   All-Weather Perez Sky Model

The Perez sky model [7] describes the luminance of any arbitrary sky element as a function of its elevation, and its relative orientation with respect to the sun. It is a generalization of the CIE standard clear sky formula [11], and it has been found to be more accurate for a wider range of atmospheric conditions [12]. Consider the illustration in Fig. 2. The *relative* luminance l_p of a sky element is a function of its zenith angle θ_p and the angle γ_p with the sun:

$$l_p = f(\theta_p, \gamma_p) = [1 + a \exp(b/\cos\theta_p)] \times [1 + c \exp(d\gamma_p) + e \cos^2 \gamma_p] \ , \qquad (1)$$

where the 5 constants (a, b, c, d, e) specify the current atmospheric conditions. As suggested in [9], those constants can also be expressed as a linear function of a single parameter, the turbidity t. Intuitively, the turbidity encodes the amount of scattering in the atmosphere, so the lower t, the clearer the sky. For clear skies, the constants take on the following values: $a = -1$, $b = -0.32$, $c = 10$, $d = -3$, $e = 0.45$, which corresponds approximately to $t = 2.17$.

The model expresses the *absolute* luminance L_p of a sky element as a function of another arbitrary reference sky element. For instance, if the zenith luminance L_z is known, then

$$L_p = L_z \frac{f(\theta_p, \gamma_p)}{f(0, \theta_s)} \ , \tag{2}$$

where θ_s is the zenith angle of the sun.

2.2 Clear-Weather Azimuth-Independent Sky Model

By running synthetic experiments, we were able to determine that the influence of the second factor in (1) becomes negligible when the sun is more than $100°$ away from a particular sky element. In this case, the sky appearance can be modeled by using only the first term from (1):

$$l'_p = f'(\theta_p) = 1 + a \exp(b/\cos\theta_p) \ . \tag{3}$$

This equation effectively models the sky *gradient*, which varies from light to dark from horizon to zenith on a clear day. L'_p is obtained in a similar fashion as in (2):

$$L'_p = L_z \frac{f'(\theta_p)}{f'(0)} \ . \tag{4}$$

2.3 Expressing the Sky Model as a Function of Camera Parameters

Now suppose a camera is looking at the sky, as in Fig. 2. We can express the general (1) and azimuth-independent (3) models as functions of camera parameters. Let us start with the simpler azimuth-independent model.

If we assume that the camera zenith angle θ_c is independent of its azimuth angle ϕ_c, then $\theta_p \approx \theta_c - \arctan\left(\frac{v_p}{f}\right)$. This can be substituted into (3):

$$l'_p = g'(v_p, \theta_c, f_c) = 1 + a \exp\left(\frac{b}{\cos(\theta_c - \arctan(v_p/f_c))}\right) \ , \tag{5}$$

where, v_p is the v-coordinate of the sky element in the image, and f_c is the camera focal length.

In the general sky model case, deriving the equation involves expressing γ_p as a function of camera parameters:

$$\gamma_p = \arccos\left(\cos\theta_s \cos\theta_p + \sin\theta_s \sin\theta_p \cos\Delta\phi_p\right) \ , \tag{6}$$

where $\Delta\phi_p \approx \phi_c - \phi_s - \arctan\left(\frac{u_p}{f}\right)$, and u_p is the sky element u-coordinate in the image. We substitute (6) into (1) to obtain the final equation. For succinctness, we omit writing it in its entirety, but do present its general form:

$$l_p = g(u_p, v_p, \theta_c, \phi_c, f_c, \theta_s, \phi_s) \ , \tag{7}$$

where θ_c, ϕ_c (θ_s, ϕ_s) are the camera (sun) zenith and azimuth angles.

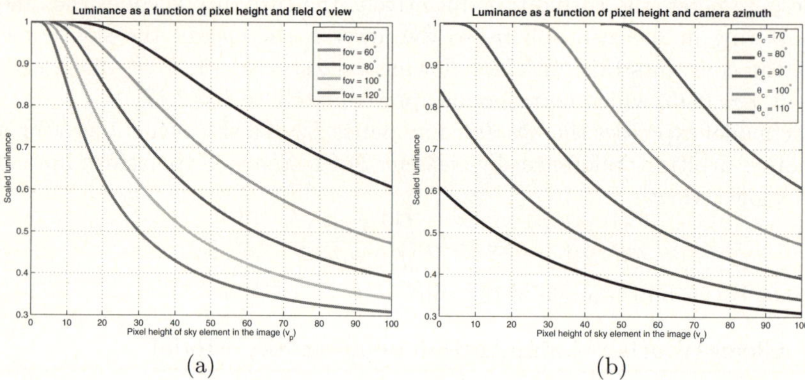

Fig. 3. Luminance profiles predicted by the azimuth-independent model (5). For clear skies, intensity diminishes as pixel height (x-axis) increases. (a) The camera zenith angle is kept constant at $\theta_c = 90°$, while the field of view is varied. (b) The field of view is kept constant at $80°$, while the camera zenith angle is varied. Both parameters have a strong influence on the shape and offset of the predicted sky gradient.

Before we present how we use the models presented above, recall that we are dealing with *ratios* of sky luminances, and that a reference element is needed. Earlier, we used the zenith luminance L_z as a reference in (2) and (4), which unfortunately is not always visible in images. Instead, we can treat this as an additional unknown in the equations. Since the denominators in (2) and (4) do not depend on camera parameters, we can combine them with L_z into a single unknown scale factor k.

3   Using the Clear Sky as a Calibration Target

In the previous section, we presented a physically-based model of the clear sky that can be expressed as a function of camera parameters. Now if we are given a set of images taken from a static camera, can we use the clear sky as a calibration target and *recover* the camera parameters, from the sky appearance only?

3.1   Recovering Focal Length and Zenith Angle

Let us first consider the simple azimuth-independent model (5). If we plot the predicted luminance profile for different focal lengths as in Fig. 3-(a) (or, equivalently, for different fields of view), we can see that there is a strong dependence between the focal length f_c and the shape of the luminance gradient. Similarly, the camera azimuth θ_c dictates the vertical offset, as in Fig. 3-(b). From this intuition, we devise a method of recovering the focal length and zenith angle of a camera from a set of images where the sun is far away from its field of view (i.e. at least $100°$ away). Suppose we are given a set $\mathcal{I}$ of such images, in which the sky is visible at pixels in set $\mathcal{P}$, also given. We seek to find the camera parameters (θ_c, f_c) that minimize

$$\min_{\theta_c, f_c, k^{(i)}} \sum_{i \in \mathcal{I}} \sum_{p \in \mathcal{P}} \left(y_p^{(i)} - k^{(i)} g'(v_p, \theta_c, f_c) \right)^2 , \tag{8}$$

where $y_p^{(i)}$ is the observed intensity of pixel p in image i, and $k^{(i)}$ are unknown scale factors (Sect. 2.3), one per image. This non-linear least-squares minimization can be solved iteratively using standard optimization techniques such as Levenberg-Marquadt, or `fminsearch` in MATLAB. f_c is initialized to a value corresponding to a $35°$ field of view, and θ_c is set such that the horizon line is aligned with the lowest visible sky pixel. All $k^{(i)}$'s are initialized to 1.

3.2 Recovering Azimuth Angle

From the azimuth-independent model (5) and images where the sun is far from the camera field of view, we were able to estimate the camera focal length f_c and its zenith angle θ_c. Now if we consider the general model (7) that depends on the sun position, we can also estimate the camera azimuth angle using the same framework as before.

Suppose we are given a set of images $\mathcal{J}$ where the sky is clear, but where the sun is now closer to the camera field of view. Similarly to (8), we seek to find the camera azimuth angle which minimizes

$$\min_{\phi_c, k^{(j)}} \sum_{j \in \mathcal{J}} \sum_{p \in \mathcal{P}} \left(y_p^{(j)} - k^{(j)} g(u_p, v_p, \theta_c, \phi_c, f_c, \theta_s, \phi_s) \right)^2 . \tag{9}$$

We already know the values of f_c and θ_c, so we do not need to optimize over them. Additionally, if the GPS coordinates of the camera and the time of capture of each image are known, the sun zenith and azimuth (θ_s, ϕ_s) can be computed using [13]. Therefore, the only unknowns are $k^{(j)}$ (one per image), and ϕ_c. Since this equation is highly non-linear, we have found that initializing ϕ_c to several values over the $[-\pi, \pi]$ interval and keeping the result that minimizes (9) works the best.

4 Evaluation of Camera Parameters Estimation

In order to thoroughly evaluate our model, we have performed extensive tests on synthetic data generated under a very wide range of operating conditions. We also evaluated our model on real image sequences to demonstrate its usefulness in practice.

4.1 Synthetic Data

We tested our model and fitting technique on a very diverse set of scenarios using data synthetically generated by using the original Perez sky model in (1). During these experiments, the following parameters were varied: the camera focal length f_c, the camera zenith and azimuth angles (θ_c, ϕ_c), the number of

Table 1. Camera calibration from the sky on 3 real image sequences taken from the AMOS database [5]. Error in focal length, zenith and azimuth angle estimation is shown for each sequence. The error is computed with respect to values obtained by using the sun position to estimate the same parameters [14].

Sequence name	Focal length error (%)	Zenith angle error (°)	Azimuth angle error (°)
257	1.1	< 0.1	2.6
414	3.1	< 0.1	2
466	2.5	< 0.1	4.5

input images used in the optimization, the number of visible sky pixels, and the camera latitude (which effects the maximum sun height). In all our experiments, 1000 pixels are randomly selected from each input image, and each experiment is repeated for 15 random selections.

The focal length can be recovered with at most 4% error even in challenging conditions: 30% visibility, over a wide range of field of view ($[13°, 93°]$ interval), zenith angles ($[45°, 135°]$), azimuth angles ($[-180°, 180°]$), and sun positions (entire hemisphere). We note a degradation in performance at wider fields of view ($> 100°$), because the assumption of independent zenith and azimuth angles starts to break down (Sect. 2.3). Less than $0.1°$ error for both zenith and azimuth angles is obtained in similar operating conditions.

4.2   Real Data

Although experiments on synthetic data are important, real image sequences present additional challenges, such as non-linear camera response functions, non-gaussian noise, slight variations in atmospheric conditions, etc. We now evaluate our method on real image sequences and show that our approach is robust to these noise sources and can be used in practice.

First, the camera response function may be non-linear, so we need to radio-metrically calibrate the camera. Although newer techniques [3] might be more suitable for image sequences, we rely on [15] which estimates the inverse response function by using color edges gathered from a single image. For additional ro-bustness, we detect edges across several frames. Recall that the optimization procedures in (8) and (9) requires clear sky image sets $\mathcal{I}$ and $\mathcal{J}$, where the sun is far or close to the camera respectively. We approximate (5) by a vertical quadratic in image space, and automatically build set $\mathcal{I}$ by keeping images with low residual fitting error. Similarly, set $\mathcal{J}$ is populated by finding images with a good fit to horizontal quadratic. It is important that the effect of the moving sun be visible in the selected images $\mathcal{J}$.

We present results from applying our algorithm on three image sequences taken from the AMOS database [5]. Since ground truth is not available on those sequences, we compare our results with those obtained with the method de-scribed in [14], which uses hand-labelled sun positions to obtain high-accuracy es-timates. Numerical results are presented in Table 1, and Fig. 4 shows a

Fig. 5. Sky-cloud separation example results. *First row*: input images (radiometrically corrected). *Second row*: sky layer. *Third row*: cloud segmentation. The clouds are color-coded by weight: 0 (blue) to 1 (red). Our fitting algorithm is able to faithfully extract the two layers in all these cases.

is reached. The process typically converges in 3 iterations, and the final value for w_p is used as the cloud segmentation. Cloud coverage is then computed as $\frac{1}{|\mathcal{P}|} \sum_{p \in \mathcal{P}} w_p$.

5.3   Segmentation Results

Figure 5 shows typical results of cloud layers extracted using our approach. Note that unweighted least-squares (10) fails on all these examples because the clouds occupy a large portion of the sky, and the optimization tries to fit them as much as possible, since the quadratic loss function is not robust to outliers. A robust loss function behaves poorly because it treats the sky pixels as outliers in the case of highly-covered skies, such as the examples shown in the first two columns of Fig. 6. Our approach injects domain knowledge into the optimization by using a data-driven sky prior, forcing it to fit the visible sky. Unfortunately, since we do not model sunlight, the estimation does not converge to a correct segmentation when the sun is very close to the camera, as illustrated in the last two columns of Fig. 6.

6   Application: Matching Skies across Image Sequences

After obtaining a sky-cloud segmentation, we consider the problem of finding matching skies between images taken by different cameras. Clearly, appearance-based matching algorithms such as cross-correlation would not work if the cameras have different parameters. Instead, we use our sky model along with cloud

Fig. 6. More challenging cases for the sky-cloud separation, and failure cases. *First row*: input images (radiometrically corrected). *Second row*: sky layer. *Third row*: cloud layer. The clouds are color-coded by weight: 0 (blue) to 1 (red). Even though the sky is more than 50% occluded in the input images, our algorithm is able to recover a good estimate of both layers. The last two columns illustrate a failure case: the sun (either when very close or in the camera field of view) significantly alters the appearance of the pixels such that they are labeled as clouds.

statistics in order to find skies that have similar properties. We first present our novel bi-layered representation for sky and clouds, which we then use to define a similarity measure between two images. We then present qualitative matching results on real image sequences.

6.1   Bi-layered Representation for Sky and Clouds

Because clouds can appear so differently due to weather conditions, a generative model such as the one we are using for the sky is likely to have a large number of parameters, and thus be difficult to fit to image data. Instead, we propose a hybrid model: our physically-based sky model parameterized by the turbidity t for the sky appearance, and a non-parametric representation for the clouds.

Taking inspiration from Lalonde et al [16], we represent the cloud layer by a joint color histogram in the xyY space over all pixels which belong to the cloud regions. While they have had success with color histograms only, we have found this to be insufficient on our richer dataset, so we also augment the representation with a texton histogram computed over the same regions. A 1000-word texton dictionary is built from a set of skies taken from training images different than the ones used for testing. In our implementation, we choose 21^3 bins for the color histograms.

Fig. 7. Sky matching results across different cameras. The left-most column shows several images taken from different days of sequence 466 in the AMOS database. The three other columns are the nearest-neighbor matches in sequences 257, 407 and 414 respectively, obtained using our distance measure. Sky conditions are well-matched, even though cameras have different parameters.

Once this layered sky representation is computed, similar images can be retrieved by comparing their turbidities and cloud statistics (we use χ^2 distance for histogram comparison). A combined distance is obtained by taking the sum of cloud and turbidity distance, with the relative importance between the two determined by the cloud coverage.

6.2 Qualitative Evaluation

The above algorithm was tested on four sequences from the AMOS database. Since we do not have ground truth to evaluate sky matching performance, we provide qualitative results in Fig. 7. Observe that sky conditions are matched correctly, even though cameras have different horizons, focal lengths, and camera response functions. A wide range of sky conditions can be matched successfully, including clear, various amounts of clouds, and overcast conditions. We provide additional segmentation and matching results on our project website.

7 Summary

In this paper, we explore the following question: what information about the camera is available in the visible sky? We show that, even if a very small portion of the hemisphere is visible, we can reliably estimate three important camera parameters by observing the sky *over time*. We do so by expressing a well-known physically-based sky model in terms of the camera parameters, and by fitting it to clear sky images using standard minimization techniques. We then demonstrate the accuracy of our approach on synthetic and real data. Once the camera parameters are estimated, we show how we can use the same model to segment out clouds from sky and build a novel bi-layered representation, which can then be used to find similar skies across different cameras.

We plan to use the proposed sky illumination model to see how it can help us predict the illumination of the scene. We expect that no parametric model will be able to capture this information well enough, so data-driven methods will become even more important.

Acknowledgements

This research is supported in parts by an ONR grant N00014-08-1-0330 and NSF grants IIS-0643628, CCF-0541307 and CCF-0541230. A. Efros is grateful to the WILLOW team at ENS Paris for their hospitality.

References

1. Weiss, Y.: Deriving intrinsic images from image sequences. In: IEEE International Conference on Computer Vision (2001)
2. Sunkavalli, K., Matusik, W., Pfister, H., Rusinkiewicz, S.: Factored time-lapse video. ACM Transactions on Graphics (SIGGRAPH 2007) 26(3) (August 2007)
3. Kim, S.J., Frahm, J.M., Polleyfeys, M.: Radiometric calibration with illumination change for outdoor scene analysis. In: IEEE Conference on Computer Vision and Pattern Recognition (2008)
4. Jacobs, N., Satkin, S., Roman, N., Speyer, R., Pless, R.: Geolocating static cameras. In: IEEE International Conference on Computer Vision (2007)
5. Jacobs, N., Roman, N., Pless, R.: Consistent temporal variations in many outdoor scenes. In: IEEE Conference on Computer Vision and Pattern Recognition (2007)

6. Sunkavalli, K., Romeiro, F., Matusik, W., Zickler, T., Pfister, H.: What do color changes reveal about an outdoor scene? In: IEEE Conference on Computer Vision and Pattern Recognition (2008)

7. Perez, R., Seals, R., Michalsky, J.: All-weather model for sky luminance distribution – preliminary configuration and validation. Solar Energy 50(3), 235–245 (1993)

8. Yu, Y., Malik, J.: Recovering photometric properties of architectural scenes from photographs. Proceedings of ACM SIGGRAPH 1998 (July 1998)

9. Preetham, A.J., Shirley, P., Smits, B.: A practical analytic model for daylight. Proceedings of ACM SIGGRAPH 1999 (August 1999)

10. Jacobs, N., Roman, N., Pless, R.: Toward fully automatic geo-location and geo-orientation of static outdoor cameras. In: Workshop on applications of computer vision (2008)

11. Committee, C.T.: Spatial distribution of daylight – luminance distributions of various reference skies. Technical Report CIE-110-1994, International Commission on Illumination (1994)

12. Ineichen, P., Molineaux, B., Perez, R.: Sky luminance data validation: comparison of seven models with four data banks. Solar Energy 52(4), 337–346 (1994)

13. Reda, I., Andreas, A.: Solar position algorithm for solar radiation applications. Technical Report NREL/TP-560-34302, National Renewable Energy Laboratory (November 2005)

14. Lalonde, J.F., Narasimhan, S.G., Efros, A.A.: Camera parameters estimation from hand-labelled sun positions in image sequences. Technical Report CMU-RI-TR-08-32, Robotics Institute. Carnegie Mellon University (July 2008)

15. Lin, S., Gu, J., Yamazaki, S., Shum, H.Y.: Radiometric calibration from a single image. In: IEEE Conference on Computer Vision and Pattern Recognition (2004)

16. Lalonde, J.F., Hoiem, D., Efros, A.A., Rother, C., Winn, J., Criminisi, A.: Photo clip art. ACM Transactions on Graphics (SIGGRAPH 2007) 26(3) (August 2007)

Three Dimensional Curvilinear Structure Detection Using Optimally Oriented Flux

Max W.K. Law and Albert C.S. Chung

Lo Kwee-Seong Medical Image Analysis Laboratory,
Department of Computer Science and Engineering,
The Hong Kong University of Science and Technology, Hong Kong
{maxlawwk,achung}@cse.ust.hk

Abstract. This paper proposes a novel curvilinear structure detector, called Optimally Oriented Flux (OOF). OOF finds an optimal axis on which image gradients are projected in order to compute the image gradient flux. The computation of OOF is localized at the boundaries of local spherical regions. It avoids considering closely located adjacent structures. The main advantage of OOF is its robustness against the disturbance induced by closely located adjacent objects. Moreover, the analytical formulation of OOF introduces no additional computation load as compared to the calculation of the Hessian matrix which is widely used for curvilinear structure detection. It is experimentally demonstrated that OOF delivers accurate and stable curvilinear structure detection responses under the interference of closely located adjacent structures as well as image noise.

1 Introduction

Analysis of curvilinear structures in volumetric images has a wide range of applications, for instance centerline extraction [1,3], detection and segmentation [7,15,9], vascular image enhancement [12,8,11] or visualization [2]. In particular, low-level detectors which are sensitive to curvilinear structures are the foundations of the aforementioned applications. One classic low-level detector is the multiscale based image intensity second-order statistics. Lindeberg [10] conducted in depth research regarding the use of the Gaussian smoothing function with various scale factors for extracting multiscale second-order statistics. Koller *et al.* [7] exploited the image intensity second-order statistics to form Hessian matrices for the analysis of curvilinear structures in three dimensional image volumes. Frangi *et al.* [6] introduced the vesselness measure based on eigenvalues extracted from the Hessian matrix in a multiscale fashion. Krissian *et al.* [9] studied the relation between the Hessian matrix and the image gradient computed in multiple scales for the detection of tubular structures. Manniesing *et al.* [11] made use of the multiscale Hessian matrix based features to devise a nonlinear scale space representation of curvilinear structures for vessel image enhancement.

Another recently developed low-level detector for the curvilinear structure analysis is the image gradient flux. It is a scalar measure which quantifies the amount of image gradient flowing in or out of a local spherical region. A large magnitude of the image gradient flux is an indication of the presence of a curvilinear structure disregarding the

D. Forsyth, P. Torr, and A. Zisserman (Eds.): ECCV 2008, Part IV, LNCS 5305, pp. 368–382, 2008.

structure direction. Bouix *et al.* proposed to compute the image gradient flux for extracting centerlines of curvilinear structures [3]. Siddqi *et al.* [15] showed promising vascular segmentation results by evolving an image gradient flux driven active surface model. But the major disadvantage of the image gradient flux is its regardless of directional information.

Grounded on the multiscale based Hessian matrix, Sato *et al.* [12] presented a thorough study on the properties of the eigenvalues extracted from the Hessian matrix in different scales, and their performance in curvilinear structure segmentation and visualization. The study showed that the eigenvalues extracted from the Hessian matrix can be regarded as the results of convolving the image with the second derivative of a Gaussian function. This function offers differential effects which compute the difference between the intensity inside an object and in the vicinity of the object. However, if the intensity around the objects is not homogeneous due to the presence of closely located adjacent structures, the differential effect given by the second derivatives of Gaussian is adversely affected.

In this paper, we propose a novel detector of curvilinear structures, called *optimally oriented flux* (OOF). Specifically, the oriented flux encodes directional information by projecting the image gradient along some axes, prior to measuring the amount of the projected gradient that flows in or out of a local spherical region. Meanwhile, OOF discovers the structure direction by finding an optimal projection axis which minimizes the oriented flux. OOF is evaluated for each voxel in the entire image. The evaluation of OOF is based on the projected image gradient at the boundary of a spherical region centered at a local voxel. When the local spherical region boundary touches the object boundary of a curvilinear structure, the image gradient at the curvilinear object boundary produces an OOF detection response. Depending on whether the voxels inside the local spherical region have stronger intensity, the sign of the OOF detection response varies. It can be utilized to distinguish between regions inside and outside curvilinear structures.

The major advantage of the proposed method is that the OOF based detection is localized at the boundary of the local spherical region. Distinct from the Hessian matrix, OOF does not consider the region in the vicinity of the structure where a nearby object is possibly present. As such, OOF detection result is robust against the disturbance introduced by closely located objects. With this advantage, utilizing OOF for curvilinear structure analysis is highly beneficial when closely located structures are present. Moreover, the computation of OOF does not introduce additional computation load compared to the Hessian matrix. Validated by a set of experiments, OOF is capable of providing more accurate and stable detection responses than the Hessian matrix, with the presence of closely located adjacent structures.

2   Methodology

2.1   Optimally Oriented Flux (OOF)

The notion of oriented flux along a particular direction refers to the amount of image gradient projected along that direction at the surface of an enclosed local region. The image gradient can flow either in or out of the enclosed local region. Without loss of generality, our elaboration focuses on the situation where the structures have stronger intensity than background regions. As such, *optimally oriented flux* (OOF) aims at finding an optimal projection direction that minimizes the inward oriented flux for the detection of curvilinear structure.

The outward oriented flux along a direction $\hat{\rho}$ is calculated by projecting the image gradient $v(\cdot)$ along the direction of $\hat{\rho}$ prior to the computation of flux in a local spherical region S_r with radius r. Based on the definition of flux [13], the computation of the outward oriented flux along the direction of $\hat{\rho}$ is,

$$f(\boldsymbol{x}; r, \hat{\rho}) = \int_{\partial S_r} \Big((\boldsymbol{v}(\boldsymbol{x} + \boldsymbol{h}) \cdot \hat{\rho})\hat{\rho} \Big) \cdot \hat{n} dA, \tag{1}$$

where dA is the infinitesimal area on ∂S_r, $\hat{n}$ is the outward unit normal of ∂S_r at the position $\hat{h}$. As ∂S_r is a sphere surface, $\boldsymbol{h} = r\hat{n}$, thus

$$f(\boldsymbol{x}; r, \hat{\rho}) = \int_{\partial S_r} \left\{ \sum_{k=1}^{3} \sum_{l=1}^{3} \Big(v_k(\boldsymbol{x} + r\hat{n})\rho_k \rho_l n_l \Big) \right\} dA = \hat{\rho}^T \mathbf{Q}_{r,\boldsymbol{x}} \hat{\rho}, \tag{2}$$

where $\hat{\rho} = (\rho_1, \rho_2, \rho_3)^T$, $\boldsymbol{v}(\boldsymbol{x}) = (v_1(\boldsymbol{x}), v_2(\boldsymbol{x}), v_3(\boldsymbol{x}))^T$, $\hat{n} = (n_1, n_2, n_3)^T$, $\mathbf{Q}_{r,\boldsymbol{x}}$ is a matrix that the entry at the ith row and jth column ($i, j \in \{1, 2, 3\}$) is,

$$q_{r,\boldsymbol{x}}^{i,j} = \int_{\partial S_r} v_i(\boldsymbol{x} + r\hat{n})n_j dA. \tag{3}$$

2.2   Analytical Computation of OOF

The idea of OOF is to identify the direction $\hat{p}$ that the inward oriented flux attains the minimum. It is not easy to discretize any one of the surface integrals of Equations 1 and 3 to estimate oriented flux and find the optimal axis which minimizes the inward oriented flux. Nevertheless, computation of OOF can be achieved analytically by acquiring the values of the entries of $\mathbf{Q}_{r,\boldsymbol{x}}$, that only involves convolving an image with a set of filters $\psi_{r,i,j}$,

$$q_{r,\boldsymbol{x}}^{i,j} = \psi_{r,i,j}(\boldsymbol{x}) * I(\boldsymbol{x}). \tag{4}$$

The above formulation avoids discretization and reduces computation complexity as compared with the discretization of either Equation 1 or Equation 3. By using fast Fourier transform, the complexity of evaluating Equation 4 and thus $\mathbf{Q}_{r,\boldsymbol{x}}$ is $O(N \log N)$, where $\forall \boldsymbol{x} \in \Omega$ and Ω is the image domain having N voxels. The proposed method introduces no additional computation load compared to some traditional approaches, such as Hessian matrix based methods [12,9,6].

We begin the elaboration of the filters $\psi_{r,i,j}(\boldsymbol{x})$ from Equation 3,

$$q_{r,\boldsymbol{x}}^{i,j} = \int_{\partial S_r} v_i(\boldsymbol{x} + r\hat{n})n_j dA = \int_{\partial S_r} [v_i(\boldsymbol{x} + r\hat{n})\hat{a}_j] \cdot \hat{n} dA, \tag{5}$$

where $\hat{a}_1$, $\hat{a}_2$ and $\hat{a}_3$ are the unit vectors along the x-, y- and z-directions respectively. Assuming that $\boldsymbol{v}$ is continuous, by the divergence theorem,

$$q_{r,\boldsymbol{x}}^{i,j} = \int_{S_r} \nabla \cdot [v_i(\boldsymbol{x} + \boldsymbol{y})\hat{a}_j] dV = \int_{S_r} \frac{\partial}{\partial \hat{a}_j} v_i(\boldsymbol{x} + \boldsymbol{y}) dV, \tag{6}$$

where $\boldsymbol{y}$ is the position vector inside the sphere S_r and dV is the infinitesimal volume in S_r. The continuous image gradient $\boldsymbol{v}(\boldsymbol{x})$ is acquired by convolving the discrete image with the first derivatives of Gaussian with a small scale factor, i.e. $v_i(\boldsymbol{x}) = (g_{\hat{a}_i,\sigma} * I)(\boldsymbol{x})$, where $*$ is the convolution operator, $g_{\hat{a}_i,\sigma}$ is the first derivative of Gaussian along the direction of $\hat{a}_i$ and $\sigma = 1$ in all our implementations. Furthermore, the volume integral of Equation 6 is extended to the entire image domain Ω by employing a step function, $b_r(\boldsymbol{x}) = \begin{cases} 1, & \|\boldsymbol{x}\| \leq r, \\ 0, & \text{otherwise}, \end{cases}$ hence

$$q_{r,\boldsymbol{x}}^{i,j} = \int_{\Omega} b_r(\boldsymbol{y})((g_{\hat{a}_i\hat{a}_j,\sigma} * I)(\boldsymbol{x} + \boldsymbol{y}))dV = \left((b_r * g_{\hat{a}_i\hat{a}_j,\sigma})(\boldsymbol{x})\right) * I(\boldsymbol{x}), \quad (7)$$

where $g_{\hat{a}_i\hat{a}_j,\sigma}$ is the second derivative of Gaussian along the axes $\hat{a}_i$ and $\hat{a}_j$. Therefore, the set of linear filters of Equation 4 is $\psi_{r,i,j}(\boldsymbol{x}) = (b_r * g_{\hat{a}_i,\hat{a}_j,\sigma})(\boldsymbol{x})$. The next step is to obtain the analytical Fourier expression of $\psi_{r,i,j}(\boldsymbol{x})$ in order to compute the convolution in Equation 4 by Fourier coefficient multiplication. Denote $\Psi_{r,i,j}(\boldsymbol{u})$ be the Fourier expression of $\psi_{r,i,j}(\boldsymbol{x})$, where $\boldsymbol{u} = (u_1, u_2, u_3)^T$ is the position vector in the frequency domain. The values of u_1, u_2 and u_3 are in "cycle per unit voxel" and in a range of $[-0.5, 0.5)$. By employing Fourier transforms on $g_{\hat{a}_i\hat{a}_j,\sigma}$ and Hankel transforms [4] on $b_r(\boldsymbol{x})$,

$$\Psi_{r,i,j}(\boldsymbol{u}) = 4\pi r u_i u_j e^{-2(\pi\|\boldsymbol{u}\|\sigma)^2} \frac{1}{\|\boldsymbol{u}\|^2} \left(\cos(2\pi r\|\boldsymbol{u}\|) - \frac{\sin(2\pi r\|\boldsymbol{u}\|)}{2\pi r\|\boldsymbol{u}\|} \right). \quad (8)$$

Based on the above formulation, the optimal projection axis which minimizes inward oriented flux can be computed analytically. Denote the optimal direction is $\boldsymbol{w}_{r,\boldsymbol{x}}$, minimizing inward oriented flux is equivalent to maximizing $f_r(\boldsymbol{x}; \boldsymbol{w}_{r,\boldsymbol{x}})$ subject to the constraint $\|\boldsymbol{w}_{r,\boldsymbol{x}}\| = [\boldsymbol{w}_{r,\boldsymbol{x}}]^T \boldsymbol{w}_{r,\boldsymbol{x}} = 1$. The solution is found by taking the first derivative on the Lagrange equation,

$$\mathcal{L}(\boldsymbol{w}_{r,\boldsymbol{x}}) = [\boldsymbol{w}_{r,\boldsymbol{x}}]^T \mathbf{Q}_{r,\boldsymbol{x}} \boldsymbol{w}_{r,\boldsymbol{x}} + \lambda_{r,\boldsymbol{x}}(1 - [\boldsymbol{w}_{r,\boldsymbol{x}}]^T \boldsymbol{w}_{r,\boldsymbol{x}}), \quad (9)$$

for $\nabla\mathcal{L}(\boldsymbol{w}_{r,\boldsymbol{x}}) = 0$, and since $q_{r,\boldsymbol{x}}^{i,j} = q_{r,\boldsymbol{x}}^{j,i}$ (see Equation 7), and thus $\mathbf{Q} = \mathbf{Q}^T$,

$$\mathbf{Q}_{r,\boldsymbol{x}} \boldsymbol{w}_{r,\boldsymbol{x}} = \lambda_{r,\boldsymbol{x}} \boldsymbol{w}_{r,\boldsymbol{x}}. \quad (10)$$

Equation 10 is in turn solved as a generalized eigenvalue problem. For volumetric images, there are at most three distinct pairs of $\lambda_{r,\boldsymbol{x}}$ and $\boldsymbol{w}_{r,\boldsymbol{x}}$. The eigenvalues can be positive, zero or negative. These eigenvalues are denoted as $\lambda_i(\boldsymbol{x}; r)$, for $\lambda_1(\cdot) \leq \lambda_2(\cdot) \leq \lambda_3(\cdot)$, and the corresponding eigenvectors are $\boldsymbol{w}_i(\boldsymbol{x}; r)$. Inside a curvilinear structure having stronger intensity than the background, the first two eigenvalues would be much smaller than the third one, $\lambda_1(\cdot) \leq \lambda_2(\cdot) << \lambda_3(\cdot)$ and $\lambda_3(\cdot) \approx 0$. The first two eigenvectors span the normal plan of the structure and the third eigenvector is the structure direction.

2.3   Eigenvalues and Eigenvectors

The major difference between the eigenvalues and eigenvectors extracted from OOF and those from the Hessian matrix is that the computation of OOF and thus, its eigenvalues and eigenvectors are grounded on the analysis of image gradient on the local sphere surface (∂S_r in Equation 3). In contrast, as pointed out by Sato *et al.* [12], the computation of the Hessian matrix is closely related to the results of applying the second derivative of Gaussian function on the image. This function computes the weighted intensity average difference between the regions inside the structure and in the vicinity of the structure. As such, the coverage of this function extends beyond the boundary of target structures and possibly includes structures nearby. As a result, the weighted intensity average difference computed by the second derivative of Gaussian function can be affected by the adjacent objects. It can be harmful to the detection accuracy of the Hessian matrix when closely located adjacent structures are present.

On the contrary, the evaluation of OOF is performed on the boundary of a local spherical region ∂S_r. Detection response of OOF is induced from the intensity discontinuities at the object boundary when the local sphere surface touches the object boundary of the structure. The detection of OOF is localized at the boundary of the local spherical region. The localized detection avoids the inclusion of objects nearby. Therefore, the OOF based detection is robust against the disturbance introduced by closely located adjacent structures.

The eigenvalues extracted from OOF are the values of oriented flux along the corresponding eigenvectors,

$$\lambda_i(\boldsymbol{x};r) = [\boldsymbol{\omega}_i(\boldsymbol{x};r)]^T \mathbf{Q}_{r,\boldsymbol{x}} \boldsymbol{\omega}_i(\boldsymbol{x};r) = f(\boldsymbol{x};r,\boldsymbol{\omega}_i(\boldsymbol{x};r)). \tag{11}$$

The image gradient at the object boundary of a strong intensity curvilinear structure points to the centerline of the structure. Inside the structure, when the local spherical region boundary ∂S_r (see Equation 1) touches the object boundary, at the contacting position of these two boundaries, the image gradient $\boldsymbol{v}(\cdot)$ is aligned in the opposite direction of the outward normal $\hat{n}$, hence $\lambda_1(\cdot) \leq \lambda_2(\cdot) << 0$. On the other hand, the image gradient is perpendicular to the structure direction, the projected image gradient along $\boldsymbol{\omega}_3(\cdot)$ has zero or very small magnitude, thus $\lambda_3(\cdot) \approx 0$. In contrast, if OOF is computed for a voxel which is just outside the curvilinear structure, at the position where ∂S_r touches the curvilinear structure boundary, the image gradient $\boldsymbol{v}(\cdot)$ is in the same direction as the outward normal $\hat{n}$. It results in a large positive eigenvalue, that is $\lambda_3(\cdot) >> 0$.

Combining multiple eigenvalues to tailor a measure for identifying structures in a specific shape is now possible. For instance $\Lambda_{12}(\boldsymbol{x};r) = \lambda_1(\boldsymbol{x};r) + \lambda_2(\boldsymbol{x};r)$ can provide responses at the curvilinear object centerline with circular cross section. According to Equations 1 and 11,

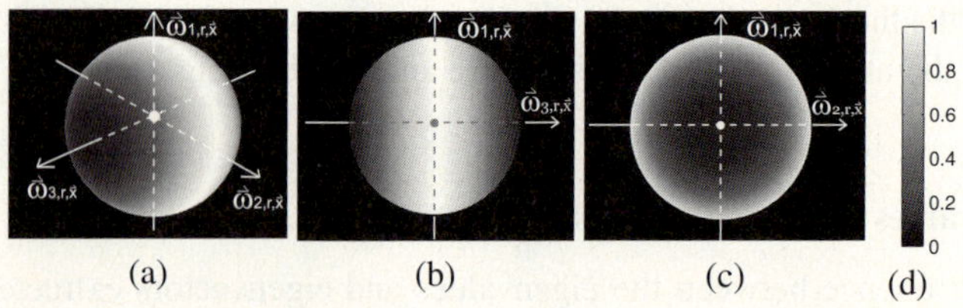

(a)   (b)   (c)   (d)

Fig. 1. (a, b, c) The values of $||[\mathbf{W}_{12}(\cdot)]^T \hat{n}||$. (d) The intensity scale of the images in (a-c).

$$\Lambda_{12}(\boldsymbol{x};r) = \int_{\partial S_r} \left([\mathbf{W}_{12}(\boldsymbol{x};r)]^T \boldsymbol{v}(\boldsymbol{x}+\boldsymbol{h})\right) \cdot \left([\mathbf{W}_{12}(\boldsymbol{x};r)]^T \hat{n}\right) dA.$$

where $\mathbf{W}_{12}(\boldsymbol{x};r) = [\boldsymbol{\omega}_1(\boldsymbol{x};r)\ \ \boldsymbol{\omega}_2(\boldsymbol{x};r)]$. The term involving the projection of $\hat{n}$ in the second half of the surface integral of the above equation is independent to the image gradient. This term varies along the boundary of the spherical region ∂S_r. It is a weighting function that makes the projected image gradients at various positions on the sphere surface contribute differently to the resultant values of $\Lambda_{12}(\boldsymbol{x};r)$. The values of $\|[\mathbf{W}_{12}(\boldsymbol{x};r)]^T \hat{n}\|$ on the local spherical region surface are shown in Figures 1a-c. A large value of $\|[\mathbf{W}_{12}(\boldsymbol{x};r)]^T \hat{n}\|$ represents the region where $\Lambda_{12}(\boldsymbol{x};r)$ is sensitive, as the projected image gradient at that region receives a higher weight for the computation of $\Lambda_{12}(\boldsymbol{x};r)$. The large valued regions of $\|[\mathbf{W}_{12}(\boldsymbol{x};r)]^T \hat{n}\|$ are distributed in a ring shape around the axis $\boldsymbol{\omega}_3(\boldsymbol{x};r)$. In a curvilinear structure having circular cross section, the image gradient at the object boundary points to the centerline of the structure. Therefore, at the centerline of the structure, $\Lambda_{12}(\boldsymbol{x};r)$ delivers the strongest response if r and the radius of the structure are matched.

Finally, it is worth mentioning that the elaboration of $\Lambda_{12}(\cdot)$ merely demonstrates a possibility to integrate different eigenvalues to facilitate the analysis of curvilinear structures. It is possible to devise other combinations of eigenvalues of the proposed method analogous to those presented in [12] and [6].

2.4   Regarding Multiscale Detection

Multiscale detection is an essential technique for handling structures with various sizes. The multiscale detection of OOF involves repetitive computations of OOF using a set of radii (r in Equation 1). The radius set should cover both the narrowest and the widest curvilinear structures in an image volume. Since the evaluation of OOF is localized at the spherical region boundary, the spherical region has to touch the target structure boundary to obtain detection responses of OOF. As such, linear radius samples should be taken for OOF with the consideration of the voxel length in order to properly detect vessels in a given range of radii. It also ensures that a structure with non-circular cross section can induce detection responses of OOF obtained in at least one radius sample. We suggest that radius samples are taken in every 0.5 voxel length according to the Nyquist sampling rate.

For different values of r, the area covered by the surface integral of Equation 1 varies. Dividing the computation result of Equation 1 by $4\pi r^2$ (the surface area of the spherical region) is an appropriate mean to normalize the detection response over radii and hence, the computation of Equation 1 is scale-invariant. Such normalization is essential to aggregating OOF responses in a multiple scale setting. For the same reason, the eigenvalues of $\mathbf{Q}_{r,\boldsymbol{x}}$, $\lambda_i(r, \boldsymbol{x})$ are divided by $4\pi r^2$ prior to being utilized in any multiscale framework. This OOF normalization scheme is distinct to the average-outward-flux (AOF) measure [5], which divides the outward flux by the surface area of the spherical region to attain the AOF-limiting-behavior. The AOF measure works only on the gradient of a distance function of a shape with its boundary clearly delineated. OOF, in contrast, is applied to a gradient of a gray-scale image, where no explicit shape boundary is embedded and noise is possibly present.

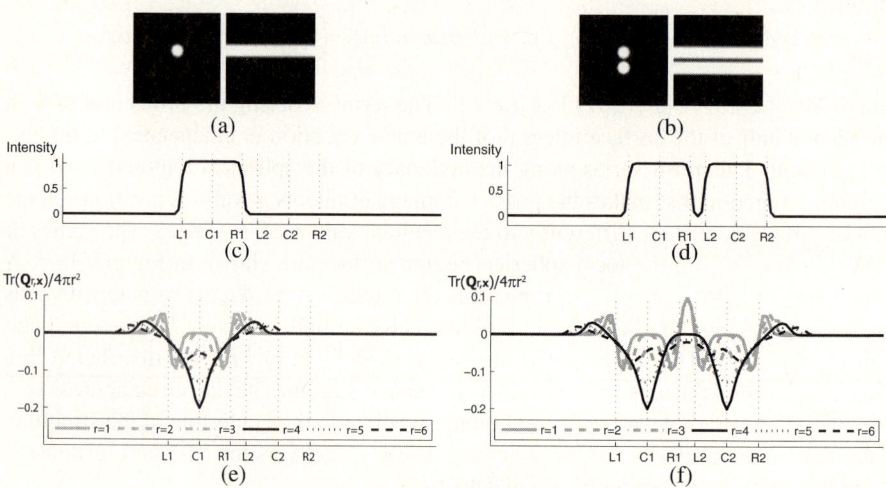

Fig. 2. Examples of evaluating OOF using multiple radii. (a, b) The slices of $z = 0$ (left) and $x = 0$ (right) of two synthetic image volumes consisting of synthetic tubes with a radius of 4 voxel length. C1 and C2 are the positions of the centers of the tubes. L1, R1 and L2, R2 are the positions of the boundaries of the tubes centered at C1 and C2 respectively. (b) The width of the separation between the closely located tubes is 2 voxel length. (c, d) The intensity profiles along the line $x = 0, z = 0$ of the synthetic image volumes shown in (a) and (b) respectively. (e, f) The normalized trace of $\mathbf{Q}_{r,x}$ along the line $x = 0, z = 0$ of the image volumes shown in (a) and (b) respectively.

In Figures 2a-f, we show two examples of evaluating OOF on image volumes consisting of one synthetic tube (Figures 2a and c) and two closely located synthetic tubes (Figures 2b and d) using multiple radii. The normalized trace of the matrix $\mathbf{Q}_{r,x}$ (Equations 9), which is equal to the sum of the normalized eigenvalues of $\mathbf{Q}_{r,x}$, is utilized to quantify the detection response strength of OOF. The normalized trace of the matrix $\mathbf{Q}_{r,x}$ is computed using multiple radii in both of the synthetic image volumes. In Figures 2e and f, it is observed that the normalized trace of $\mathbf{Q}_{r,x}$ is negative for all radii inside the tubes. It attains its maximal negative values at the tube centers and with the radius r matching the tube radius, i.e. $r = 4$. The magnitudes of the normalized trace of $\mathbf{Q}_{r,x}$ with $r = 4$ decline at positions away from the tube centers. In these positions, it attains its maximal magnitudes with smaller values of r when approaching the tube boundaries. Therefore, making use of the normalized trace of $\mathbf{Q}_{r,x}$ as well as the normalized eigenvalues of $\mathbf{Q}_{r,x}$, (the trace of $\mathbf{Q}_{r,x}$ is equal to the sum of its eigenvalues), with maximal negative values or maximal magnitudes over radii is capable of delivering a strong detection responses inside curvilinear structures.

When OOF is computed using multiple radii, the spherical regions of OOF with large radii possibly overshoot the narrow structure boundaries. The computation of OOF with overshot radii can include the objects nearby and adversely affects the detection responses of OOF (see Figure 2e, $r = 5$ and 6 versus Figure 2f, $r = 5$ and 6). In which, utilizing the normalized eigenvalues or the normalized trace of the matrix $\mathbf{Q}_{r,x}$ with the maximal negative values or maximal magnitudes over radii as mentioned above can eliminate the responses obtained by using overshot radii. Furthermore, it excludes the OOF responses associated with undersized radii at the center of curvilinear structures (see Figures 2e and f, $r = 1, 2$ and 3). In the case that the radius of the spherical region r matches the target structures, OOF avoids the inclusion of objects nearby. It therefore reports the same response at the centerlines of the tubes with $r = 4$ despite the presence of closely located structures (see Figure 2e, $r = 4$ versus Figure 2f, $r = 4$).

3   Experimental Results

In this section, we compare the performance of OOF and the Hessian matrix by using both synthetic data and real clinical cases. The differential terms of the Hessian matrix are obtained by employing the central mean difference scheme on the image smoothed by a Gaussian kernel with scale factor ϱ. The eigenvalues and eigenvectors extracted from the Hessian matrix and $\mathbf{Q}$ for OOF (Equation 10) are represented as $\lambda_i^{\mathbf{H}}(\boldsymbol{x}; r)$, $\omega_i^{\mathbf{H}}(\boldsymbol{x}; r)$ and $\lambda_i^{\mathbf{Q}}(\boldsymbol{x}; r)$, $\omega_i^{\mathbf{Q}}(\boldsymbol{x}; r)$, respectively. The order of the eigenvalues and the notation of sums of the first two eigenvalues ($\Lambda_{12}^{\mathbf{H}}(\boldsymbol{x}; r)$ and $\Lambda_{12}^{\mathbf{Q}}(\boldsymbol{x}; r)$) are analogous to those described in Section 2.2.

3.1   Synthetic Data

The proposed method, OOF, is examined in this section using synthetic images containing tori with various sizes. There are 10 synthetic volumetric images in the size of $100 \times 100 \times 100$ voxels being generated for the synthetic experiments. The main purpose is to verify the performance of OOF and compare OOF with the Hessian matrix when closely located structures are present.

The configurations of the tori in the synthetic images are shown in Figure 3. The number of tori in different synthetic images varies and depends on the values of d and R. The tori are placed in a layer fashion along the z-direction. The strategy to generate the first layer of tori is to place a torus with $D = 10$ at the altitude $z = 8$. The center

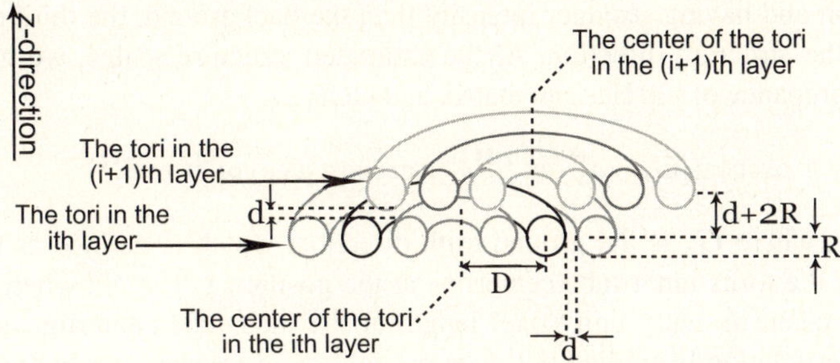

Fig. 3. The description of the tori. These tori have been used in the synthetic data experiments. The center of the tori in each layer is randomly selected from the positions of ($x = 35, y = 35$), ($x = 45, y = 35$), ($x = 35, y = 45$) and ($x = 45, y = 45$). The values of d and R are fixed to generate a torus image. In the experiments, there are 10 torus images generated by using 10 pairs of $\{d, R\}$, $\{2, 1\}$, $\{2, 2\}$, $\{2, 3\}$, $\{2, 4\}$, $\{2, 5\}$, $\{5, 1\}$, $\{5, 2\}$, $\{5, 3\}$, $\{5, 4\}$ and $\{5, 5\}$.

of that torus is randomly selected among the positions $(x = 45, y = 45, z = 8)$, $(x = 35, y = 45, z = 8)$, $(x = 45, y = 35, z = 8)$ and $(x = 35, y = 35, z = 8)$. We keep deploying adjacent tori centered at the same position of the first torus but having larger values of D in an interval of $2R + d$ until $D \leq 42$. Each successive layer of tori is generated in a $2R + d$ interval of altitude z for $z \leq 90$. The center of each layer of tori is randomly selected among the positions of $(x = 35, y = 35)$, $(x = 45, y = 35)$, $(x = 35, y = 45)$ and $(x = 45, y = 45)$. The background intensity of these images is 0 and the intensity inside the tori is assigned to 1. The torus images are smoothed by a Gaussian kernel with scale factor 1 to mimic the smooth intensity transition from structures to background. Each synthetic image is corrupted by two levels of additive Gaussian noise, with standard deviations of $\sigma_{\text{noise}} = \{0.75, 1\}$. Finally, 20 testing cases are generated for this experiment.

The experiment results are based on the measures obtained in the estimated object scales of the both methods. For the testing objects with circular cross section such as the tori used in this experiment, computing the sums of the first two eigenvalues $\Lambda_{12}^{\mathbf{H}}(\cdot)$ and $\Lambda_{12}^{\mathbf{Q}}(\cdot)$ at structure centerlines is useful to determine the structure scales. The reason is that $\Lambda_{12}^{\mathbf{H}}(\cdot)$ of the Hessian matrix quantifies the second order intensity change occurred along the radial direction of a circle on the normal plane of the structure. Meanwhile, for OOF, $\Lambda_{12}^{\mathbf{Q}}(\cdot)$ evaluates the amount of gradient pointing to the centerlines of tubes with circular cross section. Based on the above observation, the object scale is obtained as $S_{\boldsymbol{x}}^{\mathbf{H}} = \arg \max_{s \in E}(-\frac{s^2}{3} \Lambda_{12}^{\mathbf{H}}(\boldsymbol{x}; \frac{s}{\sqrt{3}}))$ for the Hessian matrix (see [7,14] for details regarding the structure scale detection and [10] for Hessian matrix based feature normalization over scales) and $S_{\boldsymbol{x}}^{\mathbf{Q}} = \arg \max_{s \in F}(-\frac{1}{4\pi s^2} \Lambda_{12}^{\mathbf{Q}}(\boldsymbol{x}; s))$ for OOF. The set of discrete detection scales of OOF and detection scales of the Hessian matrix are represented as F and E respectively. These scales cover the structure radii ranged from 1 to 6 voxel length. The radii of OOF are taken for each 0.5 voxel length and there are in total 11 different radii in F. Meanwhile, the same number of scales are logarithmically sampled for the Hessian matrix scale set E so as to minimize the detection error of the Hessian matrix [12].

There are two measures being studied for the comparison of OOF and the Hessian matrix, "Angular discrepancy" and "Response fluctuation". For objects with circular cross section and having stronger intensity than the background, the third eigenvector represents the structure direction. At the estimated structure scales, we measure the angular discrepancy of the Hessian matrix and OOF by

$$\arccos(|\boldsymbol{G}_t \cdot \boldsymbol{\omega}_3^{\mathbf{H}}(\boldsymbol{x}; S_t^{\mathbf{H}})|), \quad \arccos(|\boldsymbol{G}_t \cdot \boldsymbol{\omega}_3^{\mathbf{Q}}(\boldsymbol{x}; S_t^{\mathbf{Q}})|), \tag{12}$$

respectively, where $\boldsymbol{G}_t$ is the ground truth direction, which is defined as the tangent direction of the torus inner-tube centerline at the position t, $t \in \mathcal{T}$, where $\mathcal{T}$ is a set of samples taken in every unit voxel length at the inter-tube centerlines of the tori. Bilinear interpolation is applied if t does not fall on an integer coordinate. The value of the angular discrepancy is in a range of $[0, \pi/2]$ and a small value of the angular discrepancy represents an accurate estimation of structure direction.

The second measure, "Response fluctuation" for the tori having circular cross section is defined as the ratio between the variance and the mean absolute value of $\Lambda_{12}(\cdot)$. The "Response fluctuation" of the Hessian matrix and OOF are defined as

Table 1. The performance of *optimally oriented flux* and the Hessian matrix obtained in the synthetic data experiments. The entries in the columns of "Angular discrepancy" include two values, the mean and the standard deviation (the bracketed values) of the resultant values of Equation 12. The values in the columns of "Response fluctuation" are the results based on Equation 13.

$d = 5, \sigma_{\text{noise}} = 0.75$				
	Angular discrepancy		Response fluctuation	
R	Hessian matrix	**OOF**	Hessian matrix	**OOF**
1	0.406 (0.250)	0.309 (0.176)	0.270	0.246
2	0.232 (0.197)	0.180 (0.093)	0.166	0.160
3	0.109 (0.111)	0.110 (0.065)	0.092	0.095
4	0.063 (0.068)	0.062 (0.054)	0.059	0.054
5	0.054 (0.075)	0.059 (0.027)	0.052	0.056

$d = 2, \sigma_{\text{noise}} = 0.75$				
	Angular discrepancy		Response fluctuation	
R	Hessian matrix	**OOF**	Hessian matrix	**OOF**
1	0.408 (0.260)	0.304 (0.178)	0.283	0.252
2	0.305 (0.215)	0.227 (0.129)	0.218	0.195
3	0.162 (0.155)	0.135 (0.072)	0.133	0.117
4	0.098 (0.127)	0.087 (0.055)	0.092	0.085
5	0.079 (0.125)	0.065 (0.033)	0.086	0.069

$d = 5, \sigma_{\text{noise}} = 1$				
	Angular discrepancy		Response fluctuation	
R	Hessian matrix	**OOF**	Hessian matrix	**OOF**
1	0.518 (0.288)	0.409 (0.239)	0.321	0.291
2	0.331 (0.252)	0.246 (0.148)	0.210	0.200
3	0.204 (0.218)	0.169 (0.109)	0.129	0.105
4	0.112 (0.158)	0.110 (0.080)	0.089	0.080
5	0.107 (0.159)	0.082 (0.044)	0.073	0.061

$d = 2, \sigma_{\text{noise}} = 1$				
	Angular discrepancy		Response fluctuation	
R	Hessian matrix	**OOF**	Hessian matrix	**OOF**
1	0.532 (0.305)	0.414 (0.243)	0.338	0.298
2	0.435 (0.278)	0.319 (0.192)	0.272	0.239
3	0.279 (0.243)	0.200 (0.132)	0.177	0.134
4	0.181 (0.220)	0.125 (0.095)	0.127	0.108
5	0.157 (0.217)	0.097 (0.088)	0.107	0.085

$$\frac{\underset{t\in\mathcal{T}}{\text{Var}}\left(\Lambda_{12}^{\mathbf{H}}(\boldsymbol{x};S_t^{\mathbf{H}})\right)}{\underset{t\in\mathcal{T}}{\text{Mean}}\left(\left|\Lambda_{12}^{\mathbf{H}}(\boldsymbol{x};S_t^{\mathbf{H}})\right|\right)}, \quad \frac{\underset{t\in\mathcal{T}}{\text{Var}}\left(\Lambda_{12}^{\mathbf{Q}}(\boldsymbol{x};S_t^{\mathbf{Q}})\right)}{\underset{t\in\mathcal{T}}{\text{Mean}}\left(\left|\Lambda_{12}^{\mathbf{Q}}(\boldsymbol{x};S_t^{\mathbf{Q}})\right|\right)}, \tag{13}$$

respectively. A small value of fluctuation implies a stable response, which is robust against the adverse effects introduced by the interference of closely located structures as well as image noise.

The results based on the above measurements for different combinations of noise levels and torus separations are presented and listed in Table 1. In Table 1, it is observed that both the Hessian matrix and OOF perform better when the inner-tube radii of tori rise. It is because structures having low curvature surfaces such as large inner-tube radius tori are easier to be detected than the tori having small inner-tube radii. To evaluate the performance drops of OOF and the Hessian matrix in handling images having closely located structures, the changes of the mean angular discrepancy and response fluctuation in various cases are investigated in Table 2. In the entries of Table 2, a small value represents high robustness against the reduction of torus separation (Table 2a); the increment of noise level (Table 2b); and both of them (Table 2c).

As previously mentioned, the detection of OOF is localized at the boundary of local spherical regions. The OOF detection responses are merely induced from the intensity discontinuities taken place at the structure boundary, when the local sphere surface of OOF touches the structure boundary. In contrast to OOF, the Hessian matrix based detection relies on the computation of the weighted intensity average difference between the regions inside the structure and in the vicinity of the structure, where a nearby object is possibly present. As the correct detection scale of the Hessian matrix increases for recognizing large scale structures, the detection coverage of the correct scale of the Hessian matrix expands. It increases the chances to include adjacent structures. Hence, the increments of mean angular discrepancies and response fluctuations of the Hessian

Table 2. The changes of mean angular discrepancy and response fluctuation from the case of "$d = 5, \sigma_{\text{noise}} = 0.75$" to other three cases presented in Table 1

(a)

From "$d = 5, \sigma_{\text{noise}} = 0.75$" to "$d = 2, \sigma_{\text{noise}} = 0.75$"				
	Changes of mean angular discrepancy		Changes of response fluctuation	
R	Hessian matrix	OOF	Hessian matrix	OOF
1	+0.002	-0.005	+0.013	+0.006
2	+0.073	+0.048	+0.052	+0.035
3	+0.053	+0.025	+0.041	+0.023
4	+0.035	+0.024	+0.033	+0.031
5	+0.025	+0.005	+0.034	+0.012

(b)

From "$d = 5, \sigma_{\text{noise}} = 0.75$" to "$d = 5, \sigma_{\text{noise}} = 1$"				
	Changes of mean angular discrepancy		Changes of response fluctuation	
R	Hessian matrix	OOF	Hessian matrix	OOF
1	+0.112	+0.100	+0.050	+0.045
2	+0.099	+0.067	+0.044	+0.040
3	+0.095	+0.059	+0.036	+0.010
4	+0.049	+0.047	+0.030	+0.026
5	+0.053	+0.023	+0.021	+0.004

(c)

From "$d = 5, \sigma_{\text{noise}} = 0.75$" to "$d = 2, \sigma_{\text{noise}} = 1$"				
	Changes of mean angular discrepancy		Changes of response fluctuation	
R	Hessian matrix	OOF	Hessian matrix	OOF
1	+0.126	+0.104	+0.068	+0.052
2	+0.203	+0.139	+0.106	+0.079
3	+0.170	+0.090	+0.085	+0.039
4	+0.118	+0.062	+0.068	+0.054
5	+0.103	+0.037	+0.054	+0.029

matrix are larger than those of OOF, especially when R increases, in the cases that the torus separation is reduced from 5 voxel length to 2 voxel length (the second and the forth columns versus the first and the third columns of Table 2a).

Moreover, in the situation where noise is increased (Table 2b), it is observed that OOF (the second and the forth columns) has less increment of the mean angular discrepancies than the Hessian matrix (the first and the third columns), particularly when R increases. Although the Gaussian smoothing taken by the Hessian matrix partially eliminates noise from the image volume, the smoothing process also reduces the edge sharpness of the structure boundaries. In particular, the scale factor of the Gaussian smoothing process of the Hessian matrix has to rise to deal with large scale structures. Consequently, the Hessian matrix performs detection based on the smoothed object boundaries which are easier to be corrupted by image noise. For OOF, the detection does not require Gaussian smoothing using a large scale factor ($\sigma = 1$ for OOF). It retains the edge sharpness of the structure boundaries. Therefore, the OOF detection has higher robustness against image noise than the Hessian matrix. As expected, when the torus separation is reduced to 2 voxel length and the noise level is raised to $\sigma_{\text{noise}} = 1$, OOF has higher robustness than the Hessian matrix, against the presence of both closely located adjacent structures and high level noise than the Hessian matrix (Table 2c).

To summarize the results of the synthetic data experiments (Tables 1 and 2), OOF is validated in several aspects, the structure direction estimation accuracy, the stability of responses, the robustness against the disturbance introduced by closely located structures and the increment of noise levels. In some applications, an accurate structure direction estimation is vital. For instance, vascular image enhancement, the estimated structure direction is to avoid smoothing along the directions across object boundaries. Furthermore, for tracking curvilinear structure centerlines (a centerline tracking example is in [1]), estimated structure direction is to guide the centerline tracking process. Also, small response fluctuation facilitates the process to extract curvilinear structures or locate object centerlines by discovering the local maxima or ridges of the response.

On the other hand, the structure direction estimation accuracy and the stability of structure responses of OOF are robust against the reduction of structure separation and the increment of noise levels. As such, employing OOF to provide information of curvilinear structures is highly beneficial for curvilinear structure analysis.

3.2 Application Example - Blood Vessel Extraction

In this section, we demonstrate an example on utilizing OOF to supply information of curvilinear structures for extracting vessels in a vascular image. The vascular image utilized in this example is a phase contrast magnetic resonance angiographic (PCMRA) image volume (Figure 4a) and the image intensity represents the blood flow speed inside the vasculature. The challenges to extraction algorithms are the presence of closely located vessels due to the complicated geometry of vascular structures, and the small and low intensity vessels in images with relatively high background noise level.

To perform comparison between OOF and the Hessian matrix, we replace the Hessian matrix based information used by a vessel extraction measure with the similar information extracted from OOF. It is reminded that the main goal of this paper is to propose OOF as a general curvilinear structure detector. Therefore, measures having heuristic parameters which involve different values for certain kinds of structures are not preferred in this example, such as the vesselness measure [6] or majority of techniques in [12] for integrating multiple eigenvalues which involve heuristic parameters. On the other hand, the sum of the first two eigenvalues employed in the synthetic experiments is designed to provide responses at centerlines of curvilinear structures. It is not suitable for vessel extraction, which requires a measure to give vessel detection responses in the entire image region. We make use of the geometric mean of the first two eigenvalues, which was proposed for the detection of vessels in [12,7],

$$\mathcal{M}(\boldsymbol{x}; s) = \begin{cases} \sqrt{|\lambda_1(\boldsymbol{x}; s)\lambda_2(\boldsymbol{x}; s)|}, & \lambda_1(\boldsymbol{x}; s) \leq \lambda_2(\boldsymbol{x}; s) < 0, \\ 0, & \text{otherwise}, \end{cases} \tag{14}$$

This measure is computed in a set of discrete scales to obtain the object scales, $S_{\boldsymbol{x}}^{\mathbf{H}} = \arg\max_{s \in E'} \left(\frac{s^2}{3} \mathcal{M}^{\mathbf{H}}(\boldsymbol{x}; \frac{s}{\sqrt{3}}) \right)$ for Hessian matrix and $S_{\boldsymbol{x}}^{\mathbf{Q}} = \arg\max_{s \in F'} \left(\frac{1}{4\pi s^2} \mathcal{M}^{\mathbf{Q}}(\boldsymbol{x}; s) \right)$ for OOF. There are 15 radii and scales being employed for F' and E' respectively to cover the vessel radii ranged from 1 to 8 voxel length. Linear radius samples for F' and logarithmic scale samples for E' are utilized analogous to those described in the synthetic experiments. The vessel measure response is retrieved as the resultant values of Equation 14 obtained in the estimated object scales. The binary extraction results are obtained by thresholding the vessel measure responses. The thresholding value is found empirically so that neither over-segmentation nor under-segmentation of major vessels is observed and the same amount of voxels for both the methods are selected. Finally, 4% of voxels having the highest vessel measure responses among all voxels are thresholded as the extraction results.

The vessel extraction results are shown in Figures 4b and c. The interesting positions in the results are highlighted by the numbered arrows in Figures 4b and c. In the regions pointed at by the fifth and sixth arrows in Figures 4b and c, the Hessian based method misidentifies closely located vessels as merged structures. On the contrary, the OOF based method is capable of discovering the small separation between the closely located vessels. This result is consistent with the findings in the synthetic experiments, where OOF is more robust than the Hessian matrix when handling closely located structures (Table 2a).

In Figure 4c, it is found that several vessels with weak intensity (arrows 1, 2, 3, 4 and 7) are missed by the Hessian based method where the OOF based method has no problem to extract them (Figure 4b). The reason is that the noise level relative to the weak intensity structures is higher than those relative to strong intensity structures.

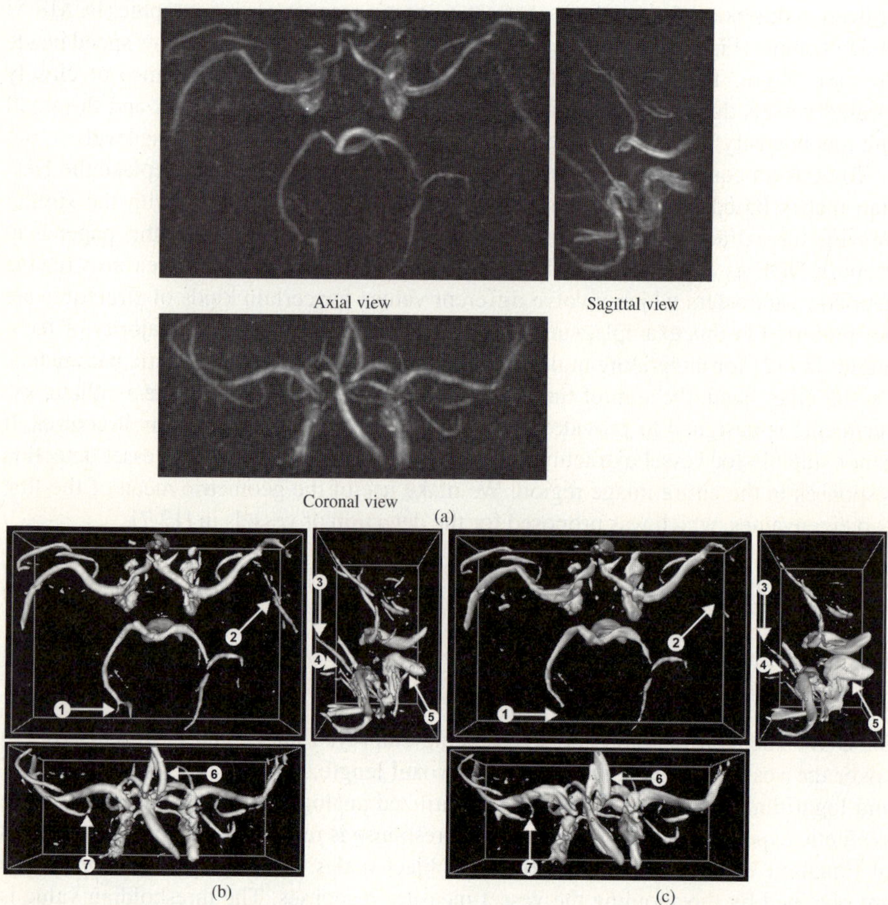

Fig. 4. (a) A phase contrast magnetic resonance angiographic image volume with the size of $213 \times 143 \times 88$ voxels. (b) The vessel extraction results obtained by using the *optimally oriented flux* based method. (c) The vessel extraction results obtained by using the Hessian matrix based method.

Coherent to the synthetic experiments, in which OOF shows higher robustness against image noise as compared to Hessian matrix (see Table 2b). The vessel extraction results in this real case experiment reflects that robustness against image noise is important on extracting vessels with weak intensity.

4 Future Developments and Conclusion

In this paper, we have presented the use of *optimally oriented flux* (OOF) for detecting curvilinear structures. With the aid of the analytical Fourier expression of OOF, no discretization and orientation sampling are needed. It therefore leads to a highly efficient computation of OOF. Computation-wise, it has the same complexity as in the computation of the commonly used approach, Hessian matrix. Furthermore, computation of OOF is based on the image gradient at the boundary of local spheres. It focuses on the detection of intensity discontinuities occurred at the object boundaries of curvilinear structures.

The OOF based detection avoids including the adjacent objects. Thus, it exhibits the robustness against the interference introduced by closely located adjacent structures. This advantage is validated and demonstrated by a set of experiments on synthetic and real image volumes. In addition, in the experiments, it is observed that OOF has higher structure direction estimation accuracy and stable detection responses under the disturbance of high level image noise. With the aforementioned high detection accuracy and robustness, OOF as opposed to the Hessian matrix, to supply information of curvilinear structures, is more beneficial for curvilinear structure analysis.

In this paper, our current focus is on formulating OOF as a general detector for extracting reliable information of curvilinear structures. Identifying branches, high curvature curvilinear structures or distinguishing between blob-like, sheet-like and tubular structures would involve post-processing steps of the information extracted by the curvilinear structure detector, such as those presented in [12]. Considering the robustness of OOF against image noise and interference of closely located adjacent structures, tailoring appropriate post-processing steps of OOF for various kinds of structures will be an interesting direction for the future developments of this work.

References

1. Aylward, S., Bullitt, E.: Initialization, noise, singularities, and scale in height ridge traversal for tubular object centerline extraction. TMI 21(2), 61–75 (2002)
2. Bouix, S., Siddiqi, K., Tannenbaum, A.: Flux driven fly throughs. CVPR 1, 449–454 (2003)
3. Bouix, S., Siddiqi, K., Tannenbaum, A.: Flux driven automatic centerline extraction. MedIA 9(3), 209–221 (2005)
4. Bracewell, R.: The Fourier Transform and Its Application. McGraw-Hill, New York (1986)
5. Dimitrov, P., Damon, J.N., Siddiqi, K.: Flux invariants for shape. CVPR 1, I–835–I–841 (2003)
6. Frangi, A., Niessen, W., Viergever, M.: Multiscale vessel enhancement filtering. In: Wells, W.M., Colchester, A.C.F., Delp, S.L. (eds.) MICCAI 1998. LNCS, vol. 1496, pp. 130–137. Springer, Heidelberg (1998)

7. Koller, T., Gerig, G., Szekely, G., Dettwiler, D.: Multiscale detection of curvilinear structures in 2-d and 3-d image data. In: IEEE International Conference on Computer Vision, pp. 864–869 (1995)
8. Krissian, K.: Flux-based anisotropic diffusion applied to enhancement of 3-d angiogram. TMI 21(11), 1440–1442 (2002)
9. Krissian, K., Malandain, G., Ayache, N., Vaillant, R., Trousset, Y.: Model-based multiscale detection of 3d vessels. CVPR 3, 722–727 (1998)
10. Lindeberg, T.: Edge detection and ridge detection with automatic scale selection. IJCV 30(2), 117–156 (1998)
11. Manniesing, W.N.R., Viergever, M.A.: Vessel enhancing diffusion a scale space representation of vessel structures. MedIA 10(6), 815–825 (2006)
12. Sato, Y., Nakajima, S., Shiraga, N., Atsumi1, H., Yoshida, S., Koller, T., Gerig, G., Kikinis, R.: Three-dimensional multi-scale line filter for segmentation and visualization of curvilinear structures in medical images. MedIA 2(2), 143–168 (1998)
13. Schey, H.M.: div, grad, curl, and all that, 3rd edn. W.W.Norton & Company (1997)
14. Steger, C.: An unbiased detector of curvilinear structures. PAMI 20(2), 113–125 (1998)
15. Vasilevskiy, A., Siddiqi, K.: Flux maximizing geometric flows. PAMI 24(12), 1565–1578 (2002)

Scene Segmentation for Behaviour Correlation

Jian Li, Shaogang Gong, and Tao Xiang

Department of Computer Science
Queen Mary College, University of London, London, E1 4NS, UK
{jianli,sgg,txiang}@dcs.qmul.ac.uk

Abstract. This paper presents a novel framework for detecting abnormal pedestrian and vehicle behaviour by modelling cross-correlation among different co-occurring objects both locally and globally in a given scene. We address this problem by first segmenting a scene into semantic regions according to how object events occur globally in the scene, and second modelling concurrent correlations among regional object events both locally (within the same region) and globally (across different regions). Instead of tracking objects, the model represents behaviour based on classification of atomic video events, designed to be more suitable for analysing crowded scenes. The proposed system works in an unsupervised manner throughout using automatic model order selection to estimate its parameters given video data of a scene for a brief training period. We demonstrate the effectiveness of this system with experiments on public road traffic data.

1   Introduction

Automatic abnormal behaviour detection has been a challenging task for visual surveillance. Traditionally, anomaly is defined according to how individuals behave in isolation over space and time. For example, objects can be tracked across a scene and if a trajectory cannot be matched by a set of known trajectory model templates, it is considered to be abnormal [1,2]. However, due to scene complexity, many types of abnormal behaviour are not well defined by only analysing how individuals behave alone. In other words, many types of anomaly definition are only meaningful when behavioural interactions/correlations among different objects are taken into consideration. In this paper, we present a framework for detecting abnormal behaviour by examining correlations of behaviours from multiple objects. Specifically, we are interested in subtle multiple object abnormality detection that is only possible when behaviours of multiple objects are interpreted in correlation as the behaviour of each object is normal when viewed in isolation. To that end, we formulate a novel approach to representing visual behaviours and modelling behaviour correlations among multiple objects.

In this paper, a type of behaviour is represented as a class of visual events bearing similar features in position, shape and motion information [3]. However, instead of using per frame image events, atomic video events as groups of image events with shared attributes over a temporal window are extracted and utilised

D. Forsyth, P. Torr, and A. Zisserman (Eds.): ECCV 2008, Part IV, LNCS 5305, pp. 383–395, 2008.
© Springer-Verlag Berlin Heidelberg 2008

as the basic units of representation in our approach. This reduces the sensitivity of events to image noise in crowded scenes. The proposed system relies on both globally and locally classifying atomic video events. Behaviours are inherently context-aware, exhibited through constraints imposed by scene layout and the temporal nature of activities in a given scene. In order to constrain the number of meaningful behavioural correlations from potentially a very large number of all possible correlations of all the objects appearing everywhere in the scene, we first decompose semantically the scene into different spatial regions according to the spatial distribution of atomic video events. In each region, events are re-clustered into different groups with ranking on both types of events and their dominating features to represent how objects behave locally within each region. As shown in Section 5, by avoiding any attempt to track individual objects over a prolonged period in space, our representation provides an object-independent representation that aims to capture categories of behaviour regardless contributing objects that are associated with scene location. We demonstrate in our experiments that such an approach is more suitable and effective for discovering unknown and detecting subtle abnormal behaviours attributed by unusual presence of and correlation among multiple objects.

Behavioural correlation has been studied before, although it is relatively new compared to the more established traditional trajectory matching based techniques. Xiang and Gong [3] clustered local events into groups and activities are modelled as sequential relationships among event groups using Dynamic Bayesian Networks. Their extended work was shown to have the capability of detecting suspicious behaviour in front of a secured entrance [4]. However, the types of activities modelled were restricted to a small set of events in a small local region without considering any true sense of global context. Brand and Kettnaker [5] attempted modelling scene activities from optical flows using a Multi-Observation-Mixture+Counter Hidden Markov Model (MOMC-HMM). A traffic circle at a crossroad is modelled as sequential states and each state is a mixture of multiple activities (observations). However, their anomaly detection is based only on how an individual behaves in isolation. How activities interact in a wider context is not considered. Wang et al [6] proposed hierarchical Bayesian models to learn visual interactions from low-level optical flow features. However, their framework is difficult to be extended to model behaviour correlation across different type of features, in which adding more features would significantly increase complexity of their models.

In our work, we model behaviour correlation by measuring the frequency of co-occurrence of any pairs of commonly occurred behaviours both locally and remotely over spatial locations. An accumulated concurrence matrix is constructed for a given training video set and matched with an instance of this matrix calculated for any testing video clip in order to detect irregular object correlations in the video clip both within the same region and across different regions in the scene. The proposed approach enables behaviour correlation to be modelled beyond a local spatial neighbourhood. Furthermore, representing visual behaviours using different dominant features at different spatial locations makes it possible

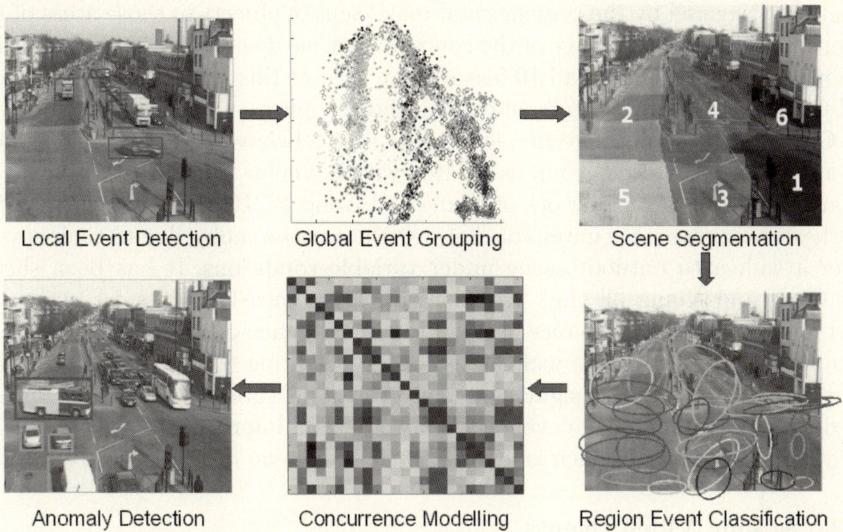

Fig. 1. Semantic scene segmentation and behaviour correlation for anomaly detection

to discover subtle unusual object behaviour correlations that either human prior knowledge is unaware of or it is difficult to be defined by human analysis. An overall data flow of the system is shown in Fig. 1.

2 Event Detection and Clustering

2.1 Image Events

We define an image event as a group of foreground neighbouring pixels detected using background subtraction. Different background models can be adopted. When only moving objects are of interest, we can use a dynamic Gaussian-Mixture background model [7]. As we also want to extract those long-staying objects, an alternative background model [8] is preferred.

Detected foreground pixels are grouped into blobs using connected components, with each blob corresponding to an image event given by a rectangular bounding box. An image event $\mathbf{v}_f$ is represented by a set of 10 features given the membership of a group as follows:

$$\mathbf{v}_f = [x, y, w, h, r_s, r_p, u, v, r_u, r_v], \tag{1}$$

where (x, y) and (w, h) are the centroid position and the width and height of the bounding box respectively, $r_s = w/h$ is the ratio between width and height, r_p is the percentage of foreground pixels in a bounding box, (u, v) is the mean optic flow vector for the bounding box, $r_u = u/w$ and $r_v = v/h$ are the scaling features between motion information and blob shape. Clearly, some of these features are more dominant for certain image events depending on their loci in a scene, as they are triggered by the presence and movement of objects in those areas of the scene. However, at this stage of the computation, we do not have any information about the scene therefore all 10 features are used at this initial step to represent all the detected image events across the entire scene.

Given detected image events, we wish to seek a behavioural grouping of these image events with each group associated with a similar type of behaviour. This shares the spirit with the work of Xiang and Gong [3]. However, direct grouping of these image events is unreliable because they are too noisy due to their spread over a wide-area outdoor scene under variable conditions. It has been shown by Gong and Xiang [9] that precision of feature measurements for events affects strongly the performance of event grouping. When processing video data of crowded outdoor scenes of wide-areas, variable lighting condition and occlusion can inevitably introduce significant noise to the feature measurement. Instead of directly grouping image events, we introduce an intermediate representation of atomic video event which is less susceptible to scene noise.

2.2   Atomic Video Events

Derived from image events, an atomic video event is defined as a spatio-temporal group of image events with similar features. To generate atomic video events, a video is cut into short non-overlapping clips and image events within a single clip are clustered into groups using K-means. Each group then corresponds to an atomic video event. In our system, we segment a video into clips of equal frame length N_f, where N_f is between 100 to 300 depending on the nature of a scene. For K-means clustering in each clip, the number of clusters is set to the average number of image event across all the frames in this clip. An atomic video event is represented by both the mean feature values of all the membership image events in its cluster, and their corresponding variances, resulting in a 20 components feature vector for each atomic video event, consisting of:

$$\mathbf{v} = [\bar{\mathbf{v}}_f, \bar{\mathbf{v}}_s], \tag{2}$$

where $\bar{\mathbf{v}}_f = mean(\mathbf{v}_f)$ and $\bar{\mathbf{v}}_s = var(\mathbf{v}_f)$, $\mathbf{v}_f$ given by Eqn. (1).

2.3   Event Grouping

We seek a behavioural grouping of all the atomic video events detected in the scene in a 20 dimensional feature space. Here we assume an atomic video event being a random variable following a Mixture of Gaussian (MoG) distribution. We need to determine both the number of Gaussian components in the mixture (model order selection) and their parameters. To automatically determine the model order, we adopt the Schwarz's Bayesian Information Criterion (BIC) model selection method [10]. Given the number of Gaussians K being determined, the Gaussian parameters and priors are computed using Expectation-Maximisation [11]. Each atomic video event is associated with the kth Gaussian representing a behaviour class in the scene, $1 \leq k \leq K$, which gives the maximum posterior probability.

3   Scene Segmentation

This behavioural grouping of atomic video events gives a concise and semantically more meaningful representation of a scene (top middle plot in Fig. 1). We consider that each group represents a behaviour type in the scene. However, such a behaviour representation is based on a global clustering of all the atomic video events detected in the entire scene without any spatial or temporal restriction. It thus does not provide a good model for capturing behaviour correlations more selectively, both in terms of spatial locality and temporal dependency. In order to impose more contextual constraints, we segment a scene semantically into regions according to event distribution with behaviour labelling, as follows.

We treat the problem similar to an image segmentation problem except that we represent each image position by a multivariate feature vector instead of RGB values. To that end, we introduce a mapping procedure transferring features from event domain to image domain. We assign each image pixel location of the scene a feature vector $\mathbf{p}$ with K components, where K is the number of groups of atomic video events estimated for a given scene, i.e. the number of behaviour types automatically determined by the BIC algorithm (Section 2.3). The value of the kth component p_k is given as the count of the kth behaviour type occurred at this image position throughout the video. In order to obtain reliable values of $\mathbf{p}$, we use the following procedure. First of all, the behavioural type label for an atomic video event is applied to all image events belonging to this atomic video event. Secondly, given an image event, its label is applied to all pixels within its rectangular bounding box. In other words, each image position is assigned with a histogram of different types of behaviours occurred at that pixel location for a given video. Moreover, because we perform scene segmentation by activities, those locations without or with few activities are to be removed from the segmentation procedure. For doing this, we apply a lower bound threshold TH_p to the number of events happened at each pixel location, i.e. the sum of component values of $\mathbf{p}$. Finally the value of this K dimensional feature vector $\mathbf{p}$ at each pixel location is scaled to $[0, 1]$ for scene segmentation.

With this normalised behavioural histogram representation in the image domain, we employ a spectral clustering technique modified from the method proposed by Zelnik-Manor and Perona [12]. Given a scene in which N locations with activities, an $N \times N$ affinity matrix $\mathbf{A}$ is constructed and the similarity between the features at the ith position and the jth position is computed according to Eqn. (3),

$$
\mathbf{A}(i,j) = \begin{cases} exp\left(-\dfrac{(d(\mathbf{p}_i, \mathbf{p}_j))^2}{\sigma_i \sigma_j}\right) exp\left(-\dfrac{(d(\mathbf{x}_i, \mathbf{x}_j))^2}{\sigma_x^2}\right), & \text{if } \|\mathbf{x}_i - \mathbf{x}_j\| \leq r \\ 0, & \text{otherwise} \end{cases}
\tag{3}
$$

where $\mathbf{p}_i$ and $\mathbf{p}_j$ are feature vectors at the ith and the jth locations, d represents Euclidean distance, σ_i and σ_j correspond to the scaling factors for the feature vectors at the ith and the jth positions, $\mathbf{x}_i$ and $\mathbf{x}_j$ are the coordinates and σ_x is the spatial scaling factor. r is the radius indicating a circle only within which, similarity is computed.

Proper computation of the scaling factors is a key for reliable spectral clustering. The original Zelnik-Perona's method computes σ_i using the distance between the current feature and the feature for a specific neighbour. This setting is very arbitrary and we will show that it suffers from under-fitting in our experiment. In order to capture more accurate statistics of local feature similarities, we compute σ_i as the standard deviation of feature distances between the current location and all locations within a given radius r. The scaling factor σ_x is computed as the mean of the distances between all positions and the circle center within the radius r. The affinity matrix is then normalised according to:

$$\bar{\mathbf{A}} = \mathbf{L}^{-\frac{1}{2}} \mathbf{A} \mathbf{L}^{-\frac{1}{2}} \tag{4}$$

where $\mathbf{L}$ is a diagonal matrix with $\mathbf{L}(s,s) = \sum_{t=1}^{N}(\mathbf{A}(s,t))$. $\bar{\mathbf{A}}$ is then used as the input to the Zelnik-Perona's algorithm which automatically determines the number of segments and performs segmentation. This procedure groups those pixel locations with activities into M regions for a given scene.

4   Behaviour Concurrence Modelling

4.1   Regional Event Classification

Recall that due to the lack of any prior information at the initial behavioural grouping stage for scene segmentation, all 10 features together with their corresponding variances were used to represent atomic video events. These settings are not necessarily optimal for accurately describing behaviours once the scene has been segmented semantically into regions. To address the problem, we re-classify behaviours in each region. Essentially, we follow the same procedure described in Section 2 but perform an additional computation to refine the grouping of atomic video events in each individual region as follows.

Given a segmented scene, we determine the most representative features in each region by computing entropy values for the features in $\mathbf{v}_f$ in each region and select the top five features with high entropy values. The selected features are then used for grouping image events in each video clip into atomic video events. When representing atomic video events, their corresponding 5 variances are also considered. This results in different and smaller set of features being selected for representing events in different regions. After atomic video event clustering, we obtain K_m regional event classes in each region m, where $1 \leq m \leq M$.

4.2   Behaviour Correlation Modelling

Suppose we have now obtained in total K_o clusters of atomic video events in all regions, i.e. $K_o = \sum_{m=1}^{M} K_m$, we wish to examine the frequency of concurrence among all pairs of behaviours happened in the scene throughout a video. Given a training video $\mathbf{F}$ which is segmented into N_c non-overlapping clips $\mathbf{F} = [\mathbf{f}_1, \cdots, \mathbf{f}_{N_c}]$, each atomic video event in a single clip $\mathbf{f}_n$, $1 \leq n \leq N_c$, has been clustered to a specific regional event class $\mathbf{b}_i$, where $1 \leq i \leq K_o$. To indicate the concurrence of a pair of regional event classes $\mathbf{b}_i$ and $\mathbf{b}_j$ occurred in clip n, we construct a $K_o \times K_o$ dimension binary matrix $\mathbf{C}_n$ so that

$$\mathbf{C}_n(i,j) = \begin{cases} 1, & \text{if } \mathbf{b}_i = TRUE \text{ and } \mathbf{b}_j = TRUE \\ 0, & \text{otherwise} \end{cases} \tag{5}$$

An accumulated concurrence matrix $\mathbf{C}$ over all the clips in the video is then computed as:

$$\mathbf{C} = \sum_{n=1}^{N_c} \mathbf{C}_n \tag{6}$$

It is clear that the diagonal components of $\mathbf{C}$ indicate the number of occurrence of event class $\mathbf{b}_i$ throughout the video and each other component $\mathbf{C}(i,j)$ corresponds to the total number of concurrence of event classes $\mathbf{b}_i$ and $\mathbf{b}_j$. To normalise the accumulated concurrence matrix $\mathbf{C}$, components in each row of $\mathbf{C}$ is divided by the diagonal component in this row. This results in a non-symmetric normalised matrix $\mathbf{C}_e$. The final symmetric concurrence matrix is computed as:

$$\mathbf{C}_f = \frac{1}{2}(\mathbf{C}_e + \mathbf{C}_e^T), \tag{7}$$

where T is transpose. After re-scaling the values in $\mathbf{C}_f$ to $[0,1]$, $\mathbf{C}_f$ is then used as the model to recognise irregular behaviour labelled atomic video event concurrence. It is worth pointing out that in practice, a measurement of concurrent frequency between a pair of atomic video event classes $\mathbf{b}_i$ and $\mathbf{b}_j$ is meaningful only when $\mathbf{b}_i$ and $\mathbf{b}_j$ individually occur sufficiently frequently. In order to remove those rarely occurred regional event classes from the concurrence matrix during training, we set a lower bound threshold TH_b to the diagonal components of accumulated concurrence matrix $\mathbf{C}$. If $\mathbf{C}(i,i) < TH_b$, the ith row and the ith column are removed from $\mathbf{C}$. The rectified matrix $\mathbf{C}$ is then used for generating the concurrence matrix $\mathbf{C}_f$.

4.3   Anomaly Detection

A test video is segmented into clips in the same way as the training video set. Image events are grouped into atomic video events using K-means. Each atomic video event is then assigned to a regional event class. In order to detect anomaly due to unexpected multi-object behaviour concurrence, we identify abnormal video clips as those with unexpected pairs of concurrences of regional event classes when compared with the concurrence matrix constructed from the training video set. More precisely, for a test video $\mathbf{Q}$ with N_q clips: $\mathbf{Q} = [\mathbf{q}_1, \cdots, \mathbf{q}_{N_q}]$, we generate a binary concurrence matrix $\mathbf{C}_t$ for each clip $\mathbf{q}_t$ by Eqn. (5). We then generate a matrix $\mathbf{CT}_t$ according to Eqn. (8).

$$\mathbf{CT}_t(i,j) = \begin{cases} 1 - \mathbf{C}_f(i,j), & \text{if } \mathbf{C}_t(i,j) = 1 \text{ and } \mathbf{C}_f(i,j) \leq TH_c \\ 0, & \text{otherwise} \end{cases} \tag{8}$$

where TH_c is a threshold. Given matrix $\mathbf{CT}_t$ for clip $\mathbf{q}_t$, a score S_t is computed as the mean of all the non-zero values in $\mathbf{CT}_t$. Based on the values of S_t, $t =$

$1, \cdots, N_q$, those clips with unexpected behavioural concurrence can be identified if the corresponding S_t values are higher than a threshold TH_s. In the identified irregular video clips, pairs of unexpected concurrent regional event classes can be further detected as the pairs whose values in $\mathbf{C}_f$ are lower than TH_c.

5   Experiments

We evaluated the performance of the proposed system using video data captured from two different public road junctions (Scene-1 and Scene-2). Example frames are shown in Fig. 2. Scene-1 is dominated by three types of traffic patterns: the vertical traffic, the leftward horizontal traffic and the rightward traffic, from multiple entry and exit points. In addition, vehicles are allowed to stop between the vertical traffic lanes waiting for turning right or left. In Scene-2, vehicles usually move in from the entrances near the left boundary and near the right bottom corner. They move towards the exits located on the top, at left bottom corner and near the right boundary. Both videos were recorded at 25Hz and have a frame size of 360×288 pixels.

Failure Mode For Tracking: We first highlight the inadequacy of tracking based representation for behaviour modelling in a crowded scene such as Scene-1. Fig. 3 (a) shows the trajectories extracted from a two-minute video clip. In (b), we plot a histogram of the durations of all the tracked object trajectories (red), 331 in total and compare it to that of the ground-truth (blue), which was exhaustively labelled manually for all the objects appeared in the scene (in total 114 objects). It is evident that inevitable and significant fragmentation of object trajectories makes a purely trajectory based representation unsuitable for accurate behaviour analysis in this type of scenes. Moreover, it is equally important to point out that monitoring object in isolation even over a prolonged period of time through tracking does not necessarily facilitate the detection and discovery of unexpected and previously unknown anomaly in a complex scene.

(a) Scene-1   (b) Scene-2

Fig. 2. Two public road scenarios for experiment

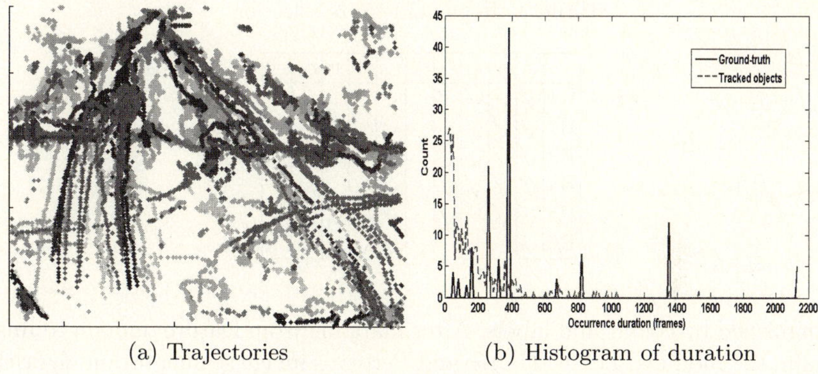

(a) Trajectories   (b) Histogram of duration

Fig. 3. Trajectory analysis

Event Clustering and Scene Segmentation: In this section, we show the performance of semantic event clustering and scene segmentation. In Scene-1, 22000 frames were used for training, in which 121583 image events were detected and grouped into 2117 atomic video events using K-means. In Scene-2, 415637 image events were detected from 45000 frames and grouped into 4182 atomic video events. The global atomic video events were automatically grouped into 13 and 19 clusters using the EM algorithm where the number of clusters in each scene was automatically determined by the BIC model selection method. The clustering results are shown in Fig. 4 (a) and (d) where clusters are

(a) Scene-1   (b) Proposed: $TH_p = 300$ (c) Original: $TH_p = 300$

(d) Scene-2   (e) Proposed: $TH_p = 200$ (f) Original: $TH_p = 200$

Fig. 4. Atomic video event classification and semantic scene segmentation

Table 1. Regional feature selection

	x	y	w	h	r_s	r_p	u	v	r_u	r_v
R1	√	√	√			√	√			
R2	√	√	√			√		√		
R3	√	√	√	√		√				
R4	√	√	√	√		√				
R5	√	√		√		√		√		
R6	√				√	√	√		√	

distinguished by colour and labels. After mapping from feature domain to image domain, the modified Zelnik-Manor and Perona's image segmentation algorithm was then used to segment Scene-1 and Scene-2 into 6 regions and 9 regions, respectively, as shown in Fig. 4 (b) and (e). For comparison, we also segmented the scenes using Zelnik-Manor and Perona's original algorithm (ZP) which resulted in 4 segments for Scene-1 and 2 segments for Scene-2 (Fig. 4 (c) and (f)). It is evident that Zelnik-Manor and Perona's original algorithm suffered from under-fitting severely and was not able to segment those scenes correctly according to expected traffic behaviours. In contrast, our approach provides a more meaningful semantic segmentation of both scenes.

Anomaly Detection: We tested the performance of anomaly detection using Scene-1. Comparing to Scene-2, Scene-1 contains more complex behaviour correlations that also subject to frequent deviations from normal correlations. Given the labelled scene segmentation shown in Fig. 4 (b), we re-classified atomic video events in each region. We performed a feature selection procedure which selected the 5 dominant features in each region with largest entropy values. The selected features in each region are shown in Table 1.

Atomic video events were then clustered in each region. From region 1 to region 6, the BIC determined 6, 5, 6, 4, 5 and 4 classes of events (behaviours) respectively. The clustering resulted in 30 local clusters of atomic video events in total (see Fig. 5 (a)). The number of concurrence for each pair of atomic event classes was then accumulated using the 73 clips in the training data to construct a 30×30 dimension accumulating concurrence matrix $\mathbf{C}$. By removing those behaviour which occurred less than 10 times (i.e. $TH_b = 10$), the dimension of the matrix $\mathbf{C}$ was reduced to 25×25. The concurrence matrix $\mathbf{C}_f$ was then computed by normalising and re-scaling $\mathbf{C}$ which is shown in Fig. 5 (b).

According to the scores shown in Fig. 5 (c), 7 clips had been picked out of a testing video consisting of 12000 frames (39 clips) as being abnormal with irregular concurrences shown in Fig. 6, in which objects with irregular concurrence are bounded by red and green boxes and the corresponding segments are highlighted using colour. Clip 4 detected a situation when a fire engine suddenly appeared and the surrounding moving vehicles had to stop unexpectedly. In Clip 28, another fire engine appeared. Although the fire engine did not significantly interrupt the normal traffic, it did caused a white van to stop in Region 3 which was not expected to be concurrent with horizontal traffic. A typical example was

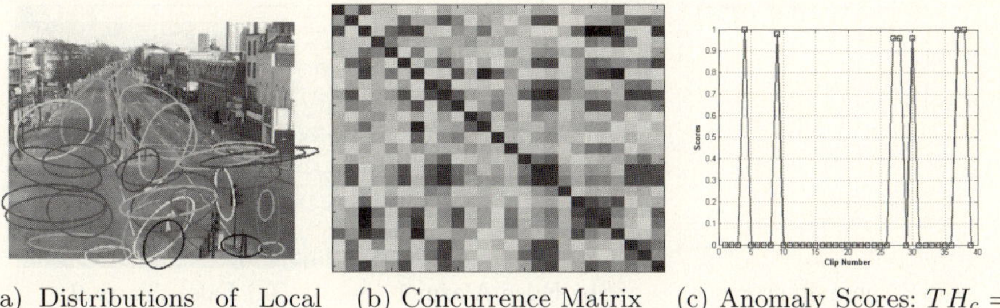

(a) Distributions of Local Behaviours   (b) Concurrence Matrix   (c) Anomaly Scores: $TH_c = 0.12$, $TH_s = 0.9$

Fig. 5. Local events classification and anomaly detection. In (a), the mean and covariance of the location of different classes of regional events are illustrated using ellipses in different colour.

(a) Clip 4    (b) Clip 9    (c) Clip 27

(d) Clip 28    (e) Clip 30    (f) Clip 37

(g) Clip 38    (h) Clip 38

Fig. 6. Detected irregular concurrences

(a) Anomaly scores	(b) False Alarm 1	(c) False Alarm 2

Fig. 7. False detections without scene segmentation

detected in Clip 30. Moreover, the second fire engine also caused strange driving behaviour for another car labelled in Clip 28 which strongly conflicted with the normal traffic. In Clip 9 and 37, two right-turn vehicles were detected in Region 2 and Region 5 respectively showing that they were quite close to each other which were not observed in the training data. Clip 27 indicates a false alarm mainly due to the imperfect blob detection which resulted in regional events being classified into wrong classes. In Clip 38, the irregular atomic events were detected in the same clip without frame overlapping (Fig. 6 (g) and (h)). This is an example that when the size of objects are large enough to cover two regions, error could also be introduced as most of vehicles in the training data have smaller size.

For comparison, we performed irregular concurrence detection without scene segmentation, i.e. only using globally clustered behaviours. The results are shown in Fig. 7. Compared with the proposed scheme, the scheme without scene segmentation gave much more false alarms (comparing (a) of Fig. 7 with (c) of Fig. 5). From the examples of false detections in Fig. 7 (b) and (c), it can be seen that using global behaviours without scene decomposition cannot accurately represent how objects behave locally. In other words, each of the global behaviour categories for the vehicles and pedestrians may not truly reflect the local behaviours of the objects and this would introduce more errors in detecting such abnormal correlations of subtle and short-duration behaviours. On the other hand, true irregular incidents were missed, e.g. the interruption from the fire engine was ignored. To summarise, when only using global classification, contextual constraints on local behaviour is not described accurately enough and general global correlation is too arbitrary. This demonstrates the advantage in behaviour correlation based on contextual constraint from semantic scene segmentation.

6   Conclusion

This paper presented a novel framework for detecting abnormal pedestrian and vehicle behaviour by modelling cross-correlation among different co-occurring objects both locally and globally in a given scene. Without tracking objects, the system was built based on local image events and atomic video events, which made the system more suitable for crowded scenes. Based on globally classified atomic video events, a scene was semantically segmented into regions and in each region, more detailed local events were re-classified. Local and global events correlations were learned by modelling event concurrence within the same region and across different regions. The correlation model was then used for detecting anomaly.

The experiments with public traffic data have shown the effectiveness of the proposed system on scene segmentation and anomaly detection. Compared with the scheme which identified irregularities only using atomic video events classified globally, the proposed system provided more detailed description of local behaviour, and showed more accurate anomaly detection and less false alarms. Furthermore, the proposed system is entirely unsupervised which ensures its generalisation ability and flexibility on processing video data with different scene content and complexity.

References

1. Hu, W., Xiao, X., Fu, Z., Xie, D., Tan, T., Maybank, S.: A system for learning statistical motion patterns. PAMI 28 (9), 1450–1464 (2006)
2. Johnson, N., Hogg, D.: Learning the distribution of object trajectories for event recognition. BMVC 2, 583–592 (1995)
3. Xiang, T., Gong, S.: Beyond tracking: Modelling activity and understanding behaviour. IJCV 67 (1), 21–51 (2006)
4. Xiang, T., Gong, S.: Video behavior profiling for anomaly detection. PAMI 30(5), 893–908 (2008)
5. Brand, M., Kettnaker, V.: Discovery and segmentation of activities in video. PAMI 22(8), 844–851 (2000)
6. Wang, X., Ma, X., Grimson, W.E.L.: Unsupervised activity perception by hierarchical bayesian models. In: CVPR, Minneapolis, USA, June 18-23, pp. 1–8 (2007)
7. Stauffer, C., Grimson, W.E.L.: Adaptive background mixture models for real-time tracking. In: CVPR, vol. 2, pp. 246–252 (1999)
8. Russell, D., Gong, S.: Minimum cuts of a time-varying background. In: BMVC, Edinburgh, UK, 1–10 (September 2006)
9. Gong, S., Xiang, T.: Scene event recognition without tracking. Special issue on visual surveillance, Acta Automatica Sinica 29(3), 321–331 (2003)
10. Schwarz, G.: Estimating the dimension of a model. Annals of Statistics 6(2), 461–464 (1978)
11. Dempster, A., Laird, N., Rubin, D.: Maximum likelihood from incomplete data via the em algorithm. Journal of the Royal Statistical Society, series B 39(1), 1–38 (1977)
12. Zelnik-Manor, L., Perona, P.: Self-tuning spectral clustering. In: NIPS (2004)

Robust Visual Tracking Based on an Effective Appearance Model

Xi Li[1], Weiming Hu[1], Zhongfei Zhang[2], and Xiaoqin Zhang[1]

[1] National Laboratory of Pattern Recognition, CASIA, Beijing, China
{lixi,wmhu,xqzhang}@nlpr.ia.ac.cn
[2] State University of New York, Binghamton, NY 13902, USA
zhongfei@cs.binghamton.edu

Abstract. Most existing appearance models for visual tracking usually construct a pixel-based representation of object appearance so that they are incapable of fully capturing both global and local spatial layout information of object appearance. In order to address this problem, we propose a novel spatial Log-Euclidean appearance model (referred as *SLAM*) under the recently introduced Log-Euclidean Riemannian metric [23]. *SLAM* is capable of capturing both the global and local spatial layout information of object appearance by constructing a block-based Log-Euclidean eigenspace representation. Specifically, the process of learning the proposed *SLAM* consists of five steps—appearance block division, online Log-Euclidean eigenspace learning, local spatial weighting, global spatial weighting, and likelihood evaluation. Furthermore, a novel online Log-Euclidean Riemannian subspace learning algorithm (*IRSL*) [14] is applied to incrementally update the proposed *SLAM*. Tracking is then led by the Bayesian state inference framework in which a particle filter is used for propagating sample distributions over the time. Theoretic analysis and experimental evaluations demonstrate the promise and effectiveness of the proposed *SLAM*.

1   Introduction

For visual tracking, handling appearance variations of an object is a fundamental and challenging task. In general, there are two types of appearance variations: intrinsic and extrinsic. Pose variation and/or shape deformation of an object are considered as the intrinsic appearance variations while the extrinsic variations are due to the changes resulting from different illumination, camera motion, camera viewpoint, and occlusion. Consequently, effectively modeling such appearance variations plays a critical role in visual tracking.

Hager and Belhumeur [1] propose a tracking algorithm which uses an extended gradient-based optical flow method to handle object tracking under varying illumination conditions. In [3], curves or splines are exploited to represent the appearance of an object to develop the Condensation algorithm for contour tracking. Due to the simplistic representation scheme, the algorithm is unable to handle the pose or illumination change, resulting in tracking failures under a varying lighting condition. Zhao *et al.*[18] present a fast differential EMD tracking method which is robust to illumination changes. Silveira and Malis [16] present a new algorithm for handling generic lighting changes.

D. Forsyth, P. Torr, and A. Zisserman (Eds.): ECCV 2008, Part IV, LNCS 5305, pp. 396–408, 2008.

Black *et al.*[4] employ a mixture model to represent and recover the appearance changes in consecutive frames. Jepson *et al.*[5] develop a more elaborate mixture model with an online EM algorithm to explicitly model appearance changes during tracking. Zhou *et al.*[6] embed appearance-adaptive models into a particle filter to achieve a robust visual tracking. Wang *et al.*[20] present an adaptive appearance model based on the Gaussian mixture model (GMM) in a joint spatial-color space (referred to as SMOG). SMOG captures rich spatial layout and color information. Yilmaz [15] proposes an object tracking algorithm based on the asymmetric kernel mean shift with adaptively varying the scale and orientation of the kernel. Nguyen *et al.*[17] propose a kernel-based tracking approach based on maximum likelihood estimation.

Lee and Kriegman [7] present an online learning algorithm to incrementally learn a generic appearance model for video-based recognition and tracking. Lim *et al.*[8] present a human tracking framework using robust system dynamics identification and nonlinear dimension reduction techniques. Black *et al.*[2] present a subspace learning based tracking algorithm with the subspace constancy assumption. A pre-trained, view-based eigenbasis representation is used for modeling appearance variations. However, the algorithm does not work well in the scene clutter with a large lighting change due to the subspace constancy assumption. Ho *et al.*[9] present a visual tracking algorithm based on linear subspace learning. Li *et al.*[10] propose an incremental PCA algorithm for subspace learning. In [11], a weighted incremental PCA algorithm for subspace learning is presented. Limy *et al.*[12] propose a generalized tracking framework based on the incremental image-as-vector subspace learning methods with a sample mean update. Chen and Yang [19] present a robust spatial bias appearance model learned dynamically in video. The model fully exploits local region confidences for robustly tracking objects against partial occlusions and complex backgrounds. In [13], li *et al.* present a visual tracking framework based on online tensor decomposition.

However, the aforementioned appearance-based tracking methods share a problem that their appearance models lack a competent object description criterion that captures both statistical and spatial properties of object appearance. As a result, they are usually sensitive to the variations in illumination, view, and pose. In order to tackle this problem, Tuzel *et al.* [24] and Porikli *et al.*[21] propose a covariance matrix descriptor for characterizing the appearance of an object. The covariance matrix descriptor, based on several covariance matrices of image features, is capable of fully capturing the information of the variances and the spatial correlations of the extracted features inside an object region. In particular, the covariance matrix descriptor is robust to the variations in illumination, view, and pose. Since a nonsingular covariance matrix is a symmetric positive definite (SPD) matrix lying on a connected Riemannian manifold, statistics for covariance matrices of image features may be computed through Riemannian geometry. Nevertheless, most existing algorithms for statistics on a Riemannian manifold rely heavily on the affine-invariant Riemannian metric, under which the Riemannian mean has no closed form. Recently, Arsigny *et al.*[23] propose a novel Log-Euclidean Riemannian metric for statistics on SPD matrices. Under this metric, distances and Riemannian means take a much simpler form than the widely used affine-invariant Riemannian metric.

Based on the Log-Euclidean Riemannian metric [23], we develop a tracking framework in this paper. The main contributions of the developed framework are as follows. First, the framework does not need to know any prior knowledge of the object, and only assumes that the initialization of the object region is provided. Second, a novel block-based spatial Log-Euclidean appearance model (*SLAM*) is proposed to fully capture both the global and local spatial properties of object appearance. In *SLAM*, the object region is first divided into several $p \times q$ object blocks, each of which is represented by the covariance matrix of image features. A low dimensional Log-Euclidean Riemannian eigenspace representation for each block is then learned online, and updated incrementally over the time. Third, we present a spatial weighting scheme to capture both the global and local spatial layout information among blocks. Fourth, while the Condensation algorithm [3] is used for propagating the sample distributions over the time, we develop an effective likelihood function based on the learned Log-Euclidean eigenspace model. Last, the Log-Euclidean Riemannian subspace learning algorithm (i.e., *IRSL*) [14] is applied to update the proposed *SLAM* as new data arrive.

2   The Framework for Visual Tracking

2.1   Overview of the Framework

The tracking framework includes two stages:(a) online *SLAM* learning; and (b)Bayesian state inference for visual tracking.

In the first stage, five steps are needed. They are appearance block division, online Log-Euclidean eigenspace learning, local spatial weighting, global spatial weighting, and likelihood evaluation, respectively. A brief introduction to these five steps is given as follows. First, the object appearance is uniformly divided into several blocks. Second, the covariance matrix feature from Eq. (2) in [14] is extracted for representing each block. After the Log-Euclidean mapping from Eq. (5) in [14], a low dimensional Log-Euclidean Riemannian eigenspace model is learned online. The model uses the incremental Log-Euclidean Riemannian subspace learning algorithm (*IRSL*) [14] to find the dominant projection subspaces of the Log-Euclidean unfolding matrices. Third, the block-specific likelihood between a candidate block and the learned Log-Euclidean eigenspace model is computed to obtain a block related likelihood map for object appearance. Fourth, the likelihood map is filtered by local spatial weighting into a new one. Fifth, the filtered likelihood map is further globally weighted by a spatial Gaussian kernel into a new one. Finally, the overall likelihood between a candidate object region and the learned *SLAM* is computed by multiplying all the block-specific likelihoods after local and global spatial weighting.

In the second stage, the object locations in consecutive frames are estimated by maximum a posterior (MAP) estimation within the Bayesian state inference framework in which a particle filter is applied to propagate sample distributions over the time. After MAP estimation, we just use the block related Log-Euclidean covariance matrices of image features inside the affinely warped image region associated with the highest weighted hypothesis to update the *SLAM*.

These two stages are executed repeatedly as time progresses. Moreover, the framework has a strong adaptability in the sense that when new image data arrive, the Log-

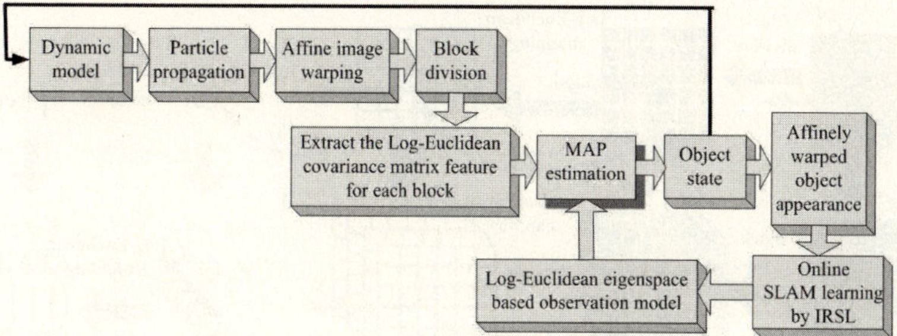

Fig. 1. The architecture of the tracking framework

Euclidean Riemannian eigenspace observation model follows the updating online. The architecture of the framework is shown in Fig. 1.

2.2   Spatial Log-Euclidean Appearance Model (*SLAM*)

The process of learning the SLAM consists of five steps—appearance block division, on-line Log-Euclidean eigenspace learning, local spatial weighting, global spatial weighting, and likelihood evaluation. The details of these five steps are given as follows.

(1) Appearance block division. Given an object appearance tensor $\mathcal{F} = \{F^t \in R^{m \times n}\}_{t=1,2,...,N}$, we divide the object appearance F^t at any time t into several $p \times q$ blocks ($m = n = 36$ and $p = q = 6$ in the paper), as illustrated in Figs. 2(a) and (b). For each block $F_{ij}^t \in R^{p \times q}$, the covariance matrix feature from Eq. (2) in [14] is extracted for representing F_{ij}^t, i.e., $\mathbf{C}_{ij}^t \in R^{d \times d}$. We call the covariance matrix $\mathbf{C}_{ij}^t$ as the block-(i,j) covariance matrix. In this case, the block-(i,j) covariance matrices $\{\mathbf{C}_{ij}^t\}_{t=1,2,...,N}$ constitute a block-(i,j) covariance tensor $\mathcal{A}_{ij} \in R^{d \times d \times N}$. If $\mathbf{C}_{ij}^t$ is a singular matrix, we replace $\mathbf{C}_{ij}^t$ with $\mathbf{C}_{ij}^t + \epsilon \mathbf{I}_d$, where ϵ is a very small positive constant ($\epsilon = 1e - 18$ in the experiments), and $\mathbf{I}_d$ is a $d \times d$ identity matrix. By the Log-Euclidean mapping from Eq. (5) in [14], as illustrated in Fig. 2(c), the block-(i,j) covariance subtensor $\mathcal{A}_{ij}$ is transformed into a new one:

$$\mathcal{L}\mathcal{A}_{ij} = \{\log(\mathbf{C}_{ij}^1), \ldots, \log(\mathbf{C}_{ij}^t), \ldots, , \log(\mathbf{C}_{ij}^N)\} \tag{1}$$

where ϵ is a very small positive constant, and $\mathbf{I}_d$ is a $d \times d$ identity matrix. We call $\mathcal{L}\mathcal{A}_{ij}$ as the block-(i,j) Log-Euclidean covariance subtensor, as illustrated in Fig. 2(d). Denote $[\cdot]$ as the rounding operator, m^* as $[\frac{m}{p}]$, and n^* as $[\frac{n}{q}]$. Consequently, all the Log-Euclidean covariance subtensors $\{(\mathcal{L}\mathcal{A}_{ij})_{m^* \times n^*}\}_{t=1,2,...,N}$ forms a Log-Euclidean covariance tensor $\mathcal{L}\mathcal{A}$ associated with the object appearance tensor $\mathcal{F} \in R^{m \times n \times N}$. With the emergence of new object appearance subtensors, $\mathcal{F}$ is extended along the time axis t (i.e., N increases gradually), leading to the extension of each Log-Euclidean covariance subtensor $\mathcal{L}\mathcal{A}_{ij}$ along the time axis t. Consequently, we need to track the changes of $\mathcal{L}\mathcal{A}_{ij}$, and need to identify the dominant projection subspace for a compact representation of $\mathcal{L}\mathcal{A}_{ij}$ as new data arrive.

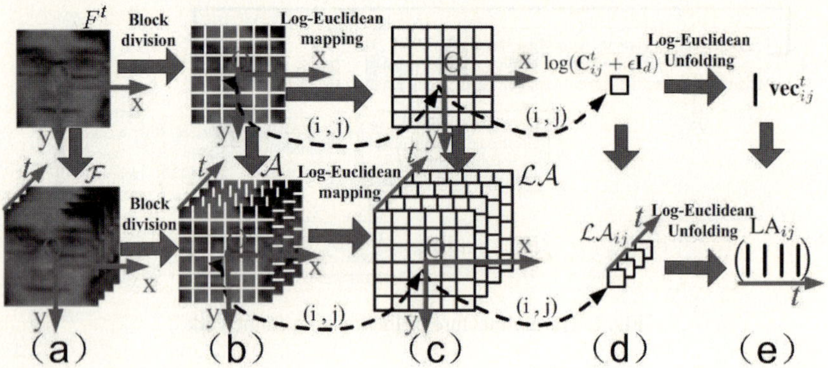

Fig. 2. Illustration of appearance block division, Log-Euclidean mapping, and Log-Euclidean unfolding. A face image F^t at time t is shown in the upper part of (a) while a 3-order face tensor $\mathcal{F} = \{F^t\}_{t=1,2,\ldots,N}$ (i.e., face image ensemble) is displayed in the lower one of (a). The results of appearance block division are exhibited in (b). The Log-Euclidean mapping results are shown in (c). An example of the block-(i,j) Log-Euclidean mapping is given in (d). (e) displays the results of Log-Euclidean unfolding.

Due to the vector space structure of $\log(\mathbf{C}^t_{ij})$ under the Log-Euclidean Riemannian metric, $\log(\mathbf{C}^t_{ij})$ is unfolded into a d^2-dimensional vector $\mathbf{vec}^t_{(i)}$ which is formulated as:

$$\mathbf{vec}^t_{(i)} = \mathrm{UT}(\log(\mathbf{C}^t_{ij})) = (c^t_1, c^t_2, \ldots, c^t_{d^2})^T \tag{2}$$

where $\mathrm{UT}(\cdot)$ is an operator unfolding a matrix into a column vector. The unfolding process can be illustrated by Figs. 2(e) and 3(a). In Fig. 3(a), the left part displays the covariance tensor $\mathcal{A}_{ij} \in \mathcal{R}^{d \times d \times N}$, the middle part corresponds to the Log-Euclidean covariance tensor $\mathcal{LA}_{ij}$, and the right part is associated with the Log-Euclidean unfolding matrix LA_{ij} with the t-th column being $\mathbf{vec}^t_{ij}$ for $1 \leq t \leq N$. As a result, LA_{ij} is formulated as:

$$\mathrm{LA}_{ij} = \left(\mathbf{vec}^1_{ij}\ \mathbf{vec}^2_{ij}\ \cdots\ \mathbf{vec}^t_{ij}\ \cdots\ \mathbf{vec}^N_{ij}\right). \tag{3}$$

The next step of the *SLAM* is to learn an online Log-Euclidean eigenspace model for $\mathcal{LA}_{ij}$. Specifically, we will introduce an incremental Log-Euclidean Riemannian subspace learning algorithm (*IRSL*) [14] for the Log-Euclidean unfolding matrix LA_{ij}. *IRSL* applies the online learning technique (R-SVD [12,27]) to find the dominant projection subspaces of LA_{ij}. Furthermore, a new operator $\mathrm{CVD}(\cdot)$ used in *IRSL* is defined as follows. Given a matrix $H = \{\mathrm{K}_1, \mathrm{K}_2, \ldots, \mathrm{K}_g\}$ and its column mean K, we let $\mathrm{CVD}(H)$ denote the SVD (i.e., singular value decomposition) of the matrix $\{\mathrm{K}_1 - \mathrm{K}, \mathrm{K}_2 - \mathrm{K}, \ldots, \mathrm{K}_g - \mathrm{K}\}$.

(2) Online Log-Euclidean eigenspace learning. For each Log-Euclidean covariance subtensor $\mathcal{LA}_{ij}$, *IRSL* [14] is used to incrementally learn a Log-Euclidean eigenspace model (i.e., LA_{ij}'s column mean $\bar{L}_{ij}$ and $\mathrm{CVD}(\mathrm{LA}_{ij}) = \mathrm{U}_{ij}\mathrm{D}_{ij}\mathrm{V}^T_{ij}$) for $\mathcal{LA}_{ij}$. For convenience, we call $\bar{L}_{ij}$ and $\mathrm{CVD}(\mathrm{LA}_{ij})$ as the block-(i,j) Log-Euclidean eigenspace

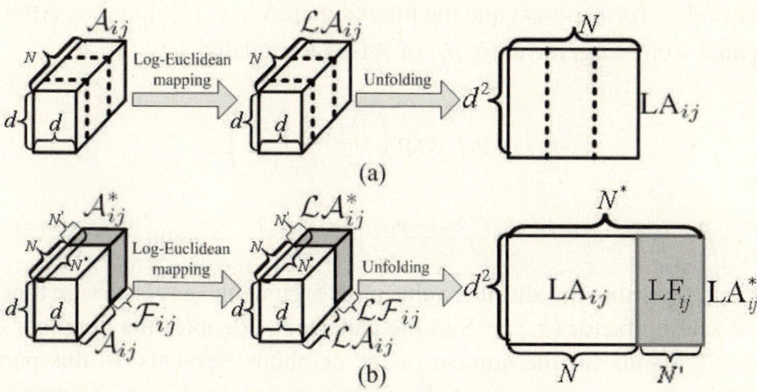

Fig. 3. Illustration of Log-Euclidean unfolding and *IRSL*. (a) shows the generative process of the Log-Euclidean unfolding matrix; (b) displays the incremental learning process of *IRSL*.

model. For a better understanding of *IRSL*, Fig. 3(b) is used to illustrate the incremental learning process of *IRSL*. Please see the details of *IRSL* in [14].

The distance between a candidate sample $B_{i,j}$ and the learned block-(i,j) Log-Euclidean eigenspace model (i.e. LA$_{ij}$'s column mean $\bar{L}_{ij}$ and CVD(LA$_{ij}$) $=$ $U_{ij}D_{ij}V_{ij}^T$) is determined by the reconstruction error norm:

$$RE_{ij} = \|(\mathbf{vec}_{ij} - \bar{L}_{ij}) - U_{(j)} \cdot U_{(j)}^T \cdot (\mathbf{vec}_{ij} - \bar{L}_{ij})\|^2 \tag{4}$$

where $\|\cdot\|$ is the Frobenius norm, and $\mathbf{vec}_{ij} = \mathrm{UT}(\log(B_{i,j}))$ is obtained from Eq. (2). Thus, the block-(i,j) likelihood p_{ij} is computed as: $p_{ij} \propto \exp(-RE_{ij})$. The smaller the RE_{ij}, the larger the likelihood p_{ij}. As a result, we can obtain a likelihood map $\mathcal{M} = (p_{ij})_{m^* \times n^*} \in R^{m^* \times n^*}$ for all the blocks.

(3) Local spatial weighting. In this step, the likelihood map $\mathcal{M}$ is filtered into a new one $\mathcal{M}_l \in R^{m^* \times n^*}$. The details of the filtering process are given as follows. Denote the

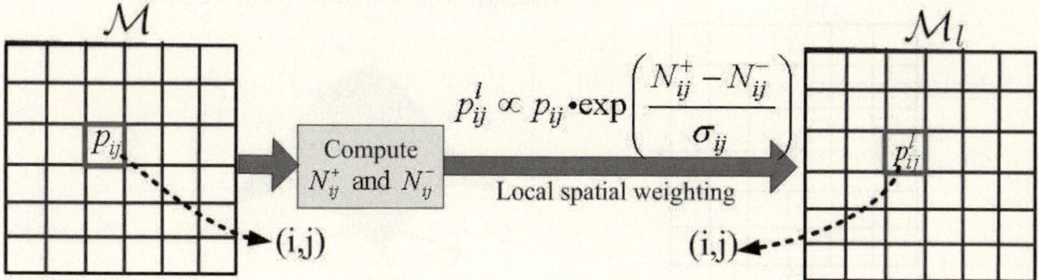

Fig. 4. Illustration of local spatial weighting for the i-th and j-th block. (a) shows the original likelihood map while (b) displays the filtered map by local spatial weighting for the i-th and j-th block.

original map $\mathcal{M} = (p_{ij})_{m^* \times n^*}$, and the filtered map $\mathcal{M}_l = (p_{ij}^l)_{m^* \times n^*}$. After filtering by local spatial weighting, the entry p_{ij}^l of $\mathcal{M}_l$ is formulated as:

$$p_{ij}^l \propto p_{ij} \cdot \exp\left(\frac{N_{ij}^+ - N_{ij}^-}{\sigma_{ij}}\right), \tag{5}$$

where $N_{ij}^+ = \frac{1}{k_{ij}} \sum_{u,v \in \mathbb{N}_{ij}} \mathrm{sgn}\left[\frac{|p_{uv}-p_{ij}|+(p_{uv}-p_{ij})}{2}\right]$, $N_{ij}^- = \frac{1}{k_{ij}} \sum_{u,v \in \mathbb{N}_{ij}} \mathrm{sgn}\left[\frac{|p_{uv}-p_{ij}|-(p_{uv}-p_{ij})}{2}\right]$, $|\cdot|$ is a function returning the absolute value of its argument, $\mathrm{sgn}[\cdot]$ is a sign function, σ_{ij} is a positive scaling factor ($\sigma_{ij} = 8$ in the paper), $\mathbb{N}_{ij}$ denotes the neighbor elements of p_{ij}, and $k_{i,j}$ stands for the number of the neighbor elements. In this paper, if all the 8-neighbor elements of p_{ij} exist, $k_{i,j} = 8$; otherwise, $k_{i,j}$ is the number of the valid 8-neighbor elements of p_{ij}. A brief discussion on the theoretical properties of Eq. (5) is given as follows. The second term of Eq. (5)(i.e., $\exp(\cdot)$) is a local spatial weighting factor. If N_{ij}^+ is smaller than N_{ij}^-, the factor will penalize p_{ij}; otherwise it will encourage p_{ij}. The process of local spatial weighting is illustrated in Fig. 4.

(4) Global spatial weighting. In this step, the filtered likelihood map $\mathcal{M}_l = (p_{ij}^l)_{m^* \times n^*}$ is further globally weighted by a spatial Gaussian kernel into a new one $\mathcal{M}_g = (p_{ij}^g) \in R^{m^* \times n^*}$. The global spatial weighting process is formulated as follows.

$$\begin{aligned}
p_{ij}^g &\propto p_{ij}^l \cdot \exp\left(-\|pos_{ij} - pos_o\|^2/2\sigma_{p_{ij}}^2\right) \\
&\propto p_{ij} \cdot \exp\left(-\|pos_{ij} - pos_o\|^2/2\sigma_{p_{ij}}^2\right) \cdot \exp\left(\frac{N_{ij}^+ - N_{ij}^-}{\sigma_{ij}}\right)
\end{aligned} \tag{6}$$

where pos_{ij} is the block-(i, j) positional coordinate vector, pos_o is the positional coordinate vector associated with the center O of the likelihood map $\mathcal{M}_l$, and $\sigma_{p_{ij}}$ is a scaling factor ($\sigma_{p_{ij}} = 3.9$ in the paper). The process of global spatial weighting can be illustrated by Fig. 5, where the likelihood map $\mathcal{M}_l$ (shown in Fig. 5(a)) is spatially weighted by the Gaussian kernel (shown in Fig. 5(b)).

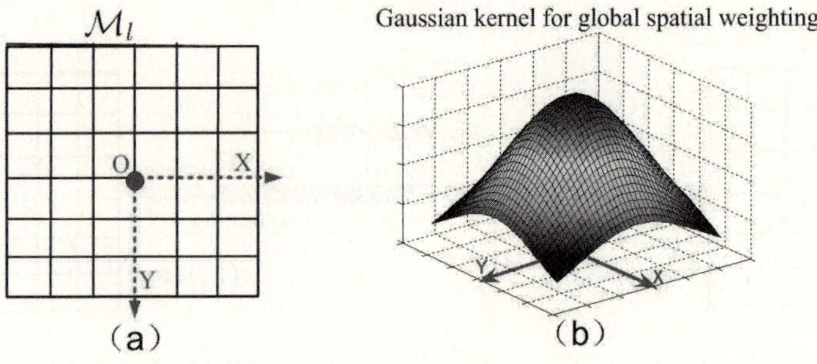

Fig. 5. Illustration of global spatial weighting. (a) shows the original likelihood map $\mathcal{M}_l$ while (b) exhibits the spatial weighting kernel for $\mathcal{M}_l$.

(5) Likelihood evaluation for *SLAM*. In this step, the overall likelihood between a candidate object region and the learned *SLAM* is computed by multiplying all the block-specific likelihoods after local and global spatial weighting. Mathematically, the likelihood is formulated as:

$$\mathbb{LIKI} \propto \prod_{1 \le i \le m^*} \prod_{1 \le j \le n^*} p_{ij}^g$$

$$\propto \prod_i \prod_j p_{ij} \cdot \exp\left(-\|pos_{ij} - pos_o\|^2 / 2\sigma_{p_{ij}}^2\right) \cdot \exp\left(\frac{N_{ij}^+ - N_{ij}^-}{\sigma_{ij}}\right) \tag{7}$$

2.3 Bayesian State Inference for Visual Tracking

For visual tracking, a Markov model with a hidden state variable is used for motion estimation. In this model, the object motion between two consecutive frames is usually assumed to be an affine motion. Let X_t denote the state variable describing the affine motion parameters (the location) of an object at time t. Given a set of observed images $\mathcal{O}_t = \{O_1, \ldots, O_t\}$, the posterior probability is formulated by Bayes' theorem as:

$$p(X_t|\mathcal{O}_t) \propto p(O_t|X_t) \int p(X_t|X_{t-1}) p(X_{t-1}|\mathcal{O}_{t-1}) dX_{t-1} \tag{8}$$

where $p(O_t|X_t)$ denotes the observation model, and $p(X_t|X_{t-1})$ represents the dynamic model. $p(O_t|X_t)$ and $p(X_t|X_{t-1})$ decide the entire tracking process. A particle filter [3] is used for approximating the distribution over the location of the object using a set of weighted samples.

In the tracking framework, we apply an affine image warping to model the object motion of two consecutive frames. The six parameters of the affine transform are used to model $p(X_t \mid X_{t-1})$ of a tracked object. Let $X_t = (x_t, y_t, \eta_t, s_t, \beta_t, \phi_t)$ where $x_t, y_t, \eta_t, s_t, \beta_t, \phi_t$ denote the x, y translations, the rotation angle, the scale, the aspect ratio, and the skew direction at time t, respectively. We employ a Gaussian distribution to model the state transition distribution $p(X_t|X_{t-1})$. Also the six parameters of the affine transform are assumed to be independent. Consequently, $p(X_t|X_{t-1})$ is formulated as:

$$p(X_t|X_{t-1}) = \mathcal{N}(X_t; X_{t-1}, \Sigma) \tag{9}$$

where Σ denotes a diagonal covariance matrix whose diagonal elements are $\sigma_x^2, \sigma_y^2, \sigma_\eta^2, \sigma_s^2, \sigma_\beta^2, \sigma_\phi^2$, respectively. The observation model $p(O_t \mid X_t)$ reflects the similarity between a candidate sample and the learned *SLAM*. In this paper, $p(O_t|X_t)$ is formulated as: $p(O_t|X_t) \propto \mathbb{LIKI}$, where $\mathbb{LIKI}$ is defined in Eq. (7). After maximum a posterior (MAP) estimation, we just use the block related Log-Euclidean covariance matrices of features inside the affinely warped image region associated with the highest weighted hypothesis to update the block related Log-Euclidean eigenspace model.

3 Experiments

In order to evaluate the performance of the proposed tracking framework, four videos are used in the experiments. The first three videos are recorded with moving cameras while the last video is taken from a stationary camera. The first two videos consist of 8-bit gray scale images while the last two are composed of 24-bit color images. Video 1 consists of dark gray scale images, where a man moves in an outdoor scene with drastically varying lighting conditions. In Video 2, a man walks from left to right in a bright road scene; his body pose varies over the time, with a drastic motion and pose change (bowing down to reach the ground and standing up back again) in the middle of the video stream. In Video 3, a girl changes her facial pose over the time in a color scene with varying lighting conditions. Besides, the girl's face is severely occluded by a man in the middle of the video stream. In the last video, a pedestrian moves along a corridor in a color scene. In the middle of the video stream, his body is severely occluded by the bodies of two other pedestrians.

During the visual tracking, the size of each object region is normalized to 36×36 pixels. Then, the normalized object region is uniformly divided into thirty-six 6×6 blocks. Further, a block-specific *SLAM* is online learned and online updated by *IRSL* every three frames. The maintained dimension r_{ij} of the block-(i, j) Log-Euclidean eigenspace model (i.e., U_{ij} referred in Sec. 2.2) learned by *IRSL* is obtained from the experiments. For the particle filtering in the visual tracking, the number of particles is set to be 200. The six diagonal elements $(\sigma_x^2, \sigma_y^2, \sigma_\eta^2, \sigma_s^2, \sigma_\beta^2, \sigma_\phi^2)$ of the covariance matrix Σ in Eq. (9) are assigned as $(5^2, 5^2, 0.03^2, 0.03^2, 0.005^2, 0.001^2)$, respectively.

Three experiments are conducted to demonstrate the claimed contributions of the proposed *SLAM*. In these four experiments, we compare tracking results of *SLAM* with those of a state-of-the-art Riemannian metric based tracking algorithm [21], referred here as *CTMU*, in different scenarios including drastic illumination changes, object pose variation, and occlusion. *CTMU* is a representative Riemannian metric based tracking algorithm which uses the covariance matrix of features for object representation. By using a model updating mechanism, *CTMU* adapts to the undergoing object deformations and appearance changes, resulting in a robust tracking result. In contrast to *CTMU*, *SLAM* constructs a block-based Log-Euclidean eigenspace representation to reflect the appearance changes of an object. Consequently, it is interesting and desirable to make a comparison between *SLAM* and *CTMU*. Furthermore, *CTMU* does not need additional parameter settings since *CTMU* computes the covariance matrix of image features as the object model. More details of *CTMU* are given in [21].

The first experiment is to compare the performances of the two methods *SLAM* and *CTMU* in handling drastic illumination changes using Video 1. In this experiment, the maintained eigenspace dimension r_{ij} in *SLAM* is set as 8. Some samples of the final tracking results are demonstrated in Fig. 6, where rows 1 and 2 are for *SLAM* and *CTMU*, respectively, in which five representative frames (140, 150, 158, 174, and 192) of the video stream are shown. From Fig. 6, we see that *SLAM* is capable of tracking the object all the time even in a poor lighting condition. In comparison, *CTMU* is lost in tracking from time to time.

The second experiment is for a comparison between *SLAM* and *CTMU* in the scenarios of drastic pose variation using Video 2. In this experiment, r_{ij} in *SLAM* is set as 6. Some samples of the final tracking results are demonstrated in Fig. 7, where rows 1 and 2 correspond to *SLAM* and *CTMU*, respectively, in which five representative frames (142, 170, 178, 183, and 188) of the video stream are shown. From Fig. 7, it is clear

Fig. 6. The tracking results of *SLAM* (row 1) and *CTMU* (row 2) over representative frames with drastic illumination changes

Fig. 7. The tracking results of *SLAM* (row 1) and *CTMU* (row 2) over representative frames with drastic pose variation

that *SLAM* is capable of tracking the target successfully even with a drastic pose and motion change while *CTMU* gets lost in tracking the target after this drastic pose and motion change.

The last experiment is to compare the tracking performance of *SLAM* with that of *CTMU* in the color scenarios with severe occlusions using Videos 3 and 4. The RGB color space is used in this experiment. r_{ij} for Videos 3 and 4 are set as 6 and 8, respectively. We show some samples of the final tracking results for *SLAM* and *CTMU* in Fig. 8, where the first and the second rows correspond to the performances of *SLAM* and *CTMU* over Video 3, respectively, in which five representative frames (158, 160, 162, 168, and 189) of the video stream are shown, while the third and the last rows correspond to the performances of *SLAM* and *CTMU* over Video 4, respectively, in which five representative frames (22, 26, 28, 32, and 35) of the video stream are shown. Clearly, *SLAM* succeeds in tracking for both Video 3 and Video 4 while *CTMU* fails.

In summary, we observe that *SLAM* outperforms *CTMU* in the scenarios of illumination changes, pose variations, and occlusions. *SLAM* constructs a block-based Log-Euclidean eigenspace representation to capture both the global and local spatial properties of object appearance. The spatial correlation information of object appearance is incorporated into *SLAM*. Even if the information of some local blocks is partially lost or drastically varies, *SLAM* is capable of recovering the information using the cues of the information from other local blocks. In comparison, *CTMU* only captures the statistical properties of object appearance in one mode, resulting in the loss of the local

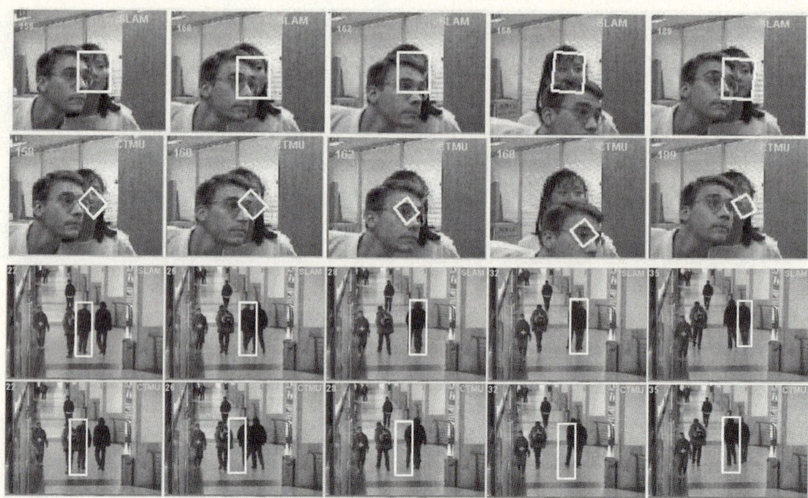

Fig. 8. The tracking results of *SLAM* and *CTMU* over representative frames in the color scenarios of severe occlusions. Rows 1 and 2 show the tracking results of *SLAM* and *CTMU* for Video 4, respectively. Rows 3 and 4 display the tracking results of *SLAM* and *CTMU* for Video 5, respectively.

spatial correlation information inside the object region. In particular, *SLAM* constructs a robust Log-Euclidean Riemannian eigenspace representation of each object appearance block. The representation fully explores the distribution information of covariance matrices of image features under the Log-Euclidean Riemannian metric, whereas *CTMU* relies heavily on an intrinsic mean in the Lie group structure without considering the distribution information of the covariance matrices of image features. Consequently, *SLAM* is an effective appearance model which performs well in modeling appearance changes of an object in many complex scenarios.

4 Conclusion

In this paper, we have developed a visual tracking framework based on the proposed spatial Log-Euclidean appearance model (*SLAM*). In this framework, a block-based Log-Euclidean eigenspace representation is constructed by *SLAM* to reflect the appearance changes of an object. Then, the local and global spatial weighting operations on the block-based likelihood map are performed by *SLAM* to capture the local and global spatial layout information of object appearance. Moreover, a novel criterion for the likelihood evaluation, based on the Log-Euclidean Riemannian subspace reconstruction error norms, has been proposed to measure the similarity between the test image and the learned subspace model during the tracking. *SLAM* is incrementally updated by the proposed online Log-Euclidean Riemannian subspace learning algorithm (*IRSL*). Experimental results have demonstrated the robustness and promise of the proposed framework.

Acknowledgment

This work is partly supported by NSFC (Grant No. 60520120099, 60672040 and 60705003) and the National 863 High-Tech R&D Program of China (Grant No. 2006AA01Z453). Z.Z. is supported in part by NSF (IIS-0535162). Any opinions, findings, and conclusions or recommendations expressed in this material are those of the authors and do not necessarily reflect the views of the NSF.

References

1. Hager, G., Belhumeur, P.: Real-time tracking of image regions with changes in geometry and illumination. In: Proc. CVPR, pp. 410–430 (1996)
2. Black, M.J., Jepson, A.D.: Eigentracking: Robust matching and tracking of articulated objects using view-based representation. In: Buxton, B.F., Cipolla, R. (eds.) ECCV 1996. LNCS, vol. 1064, pp. 329–342. Springer, Heidelberg (1996)
3. Isard, M., Blake, A.: Contour tracking by stochastic propagation of conditional density. In: Buxton, B.F., Cipolla, R. (eds.) ECCV 1996. LNCS, vol. 1065, pp. 343–356. Springer, Heidelberg (1996)
4. Black, M.J., Fleet, D.J., Yacoob, Y.: A framework for modeling appearance change in image sequence. In: Proc. ICCV, pp. 660–667 (1998)
5. Jepson, A.D., Fleet, D.J., El-Maraghi, T.F.: Robust Online Appearance Models for Visual Tracking. In: Proc. CVPR, vol. 1, pp. 415–422 (2001)
6. Zhou, S.K., Chellappa, R., Moghaddam, B.: Visual Tracking and Recognition Using Appearance-Adaptive Models in Particle Filters. IEEE Trans. on Image Processing 13, 1491–1506 (2004)
7. Lee, K., Kriegman, D.: Online Learning of Probabilistic Appearance Manifolds for Video-based Recognition and Tracking. In: Proc. CVPR, vol. 1, pp. 852–859 (2005)
8. Lim, H., Morariu3, V.I., Camps, O.I., Sznaier1, M.: Dynamic Appearance Modeling for Human Tracking. In: Proc. CVPR, vol. 1, pp. 751–757 (2006)
9. Ho, J., Lee, K., Yang, M., Kriegman, D.: Visual Tracking Using Learned Linear Subspaces. In: Proc. CVPR, vol. 1, pp. 782–789 (2004)
10. Li, Y., Xu, L., Morphett, J., Jacobs, R.: On Incremental and Robust Subspace Learning. Pattern Recognition 37(7), 1509–1518 (2004)
11. Skocaj, D., Leonardis, A.: Weighted and Robust Incremental Method for Subspace Learning. In: Proc. ICCV, pp. 1494–1501 (2003)
12. Limy, J., Ross, D., Lin, R., Yang, M.: Incremental Learning for Visual Tracking. In: NIPS, pp. 793–800. MIT Press, Cambridge (2005)
13. Li, X., Hu, W., Zhang, Z., Zhang, X., Luo, G.: Robust Visual Tracking Based on Incremental Tensor Subspace Learning. In: Proc. ICCV (2007)
14. Li, X., Hu, W., Zhang, Z., Zhang, X., Luo, G.: Visual Tracking Via Incremental Log-Euclidean Riemannian Subspace Learning. In: Proc. CVPR (2008)
15. Yilmaz, A.: Object Tracking by Asymmetric Kernel Mean Shift with Automatic Scale and Orientation Selection. In: Proc. CVPR (2007)
16. Silveira, G., Malis, E.: Real-time Visual Tracking under Arbitrary Illumination Changes. In: Proc. CVPR (2007)
17. Nguyen, Q.A., Robles-Kelly, A., Shen, C.: Kernel-based Tracking from a Probabilistic Viewpoint. In: Proc. CVPR (2007)
18. Zhao, Q., Brennan, S., Tao, H.: Differential EMD Tracking. In: Proc. ICCV (2007)

19. Chen, D., Yang, J.: Robust Object Tracking Via Online Dynamic Spatial Bias Appearance Models. IEEE Trans. on PAMI 29(12), 2157–2169 (2007)
20. Wang, H., Suter, D., Schindler, K., Shen, C.: Adaptive Object Tracking Based on an Effective Appearance Filter. IEEE Trans. on PAMI 29(9), 1661–1667 (2007)
21. Porikli, F., Tuzel, O., Meer, P.: Covariance Tracking using Model Update Based on Lie Algebra. In: Proc. CVPR, vol. 1, pp. 728–735 (2006)
22. Tuzel, O., Porikli, F., Meer, P.: Human Detection via Classification on Riemannian Manifolds. In: Proc. CVPR (2007)
23. Arsigny, V., Fillard, P., Pennec, X., Ayache, N.: Geometric Means in a Novel Vector Space Structure on Symmetric Positive-Definite Matrices. SIAM Journal on Matrix Analysis and Applications (2006)
24. Tuzel, O., Porikli, F., Meer, P.: Region Covariance: A Fast Descriptor for Detection and Classification. In: Leonardis, A., Bischof, H., Pinz, A. (eds.) ECCV 2006. LNCS, vol. 3952, pp. 589–600. Springer, Heidelberg (2006)
25. Pennec, X., Fillard, P., Ayache, N.: A Riemannian Framework for Tensor Computing. In: IJCV, pp. 41–66 (2006)
26. Rossmann, W.: Lie Groups: An Introduction Through Linear Group. Oxford Press (2002)
27. Levy, A., Lindenbaum, M.: Sequential Karhunen-Loeve Basis Extraction and Its Application to Images. IEEE Trans. on Image Processing 9, 1371–1374 (2000)

Key Object Driven Multi-category Object Recognition, Localization and Tracking Using Spatio-temporal Context

Yuan Li and Ram Nevatia

University of Southern California
Institute for Robotics and Intelligent Systems
Los Angeles, CA, USA
{yli8,nevatia}@usc.edu

Abstract. In this paper we address the problem of recognizing, localizing and tracking multiple objects of different categories in meeting room videos. Difficulties such as lack of detail and multi-object co-occurrence make it hard to directly apply traditional object recognition methods. Under such circumstances, we show that incorporating object-level spatio-temporal relationships can lead to significant improvements in inference of object category and state. Contextual relationships are modeled by a dynamic Markov random field, in which recognition, localization and tracking are done simultaneously. Further, we define human as the *key object* of the scene, which can be detected relatively robustly and therefore is used to guide the inference of other objects. Experiments are done on the CHIL meeting video corpus. Performance is evaluated in terms of object detection and false alarm rates, object recognition confusion matrix and pixel-level accuracy of object segmentation.

1   Introduction

Object recognition is a fundamental problem of computer vision. Its significance lies not only in the static image domain but also in video understanding and analysis, *e.g.*, is the man typing on a laptop or writing on a pad of paper? What objects have been put on the table and where are they? What is the motion of the passenger and his luggage if he is carrying any? Answering questions of this kind requires the ability to recognize, localize and even track different categories of objects from videos captured with a camera of usually broad view field.

There are a number of difficulties in this problem: background clutter, lack of image detail, occlusion, multi-object co-occurrence and motion. To enhance purely appearance-based approaches in the hope of overcoming these difficulties, we incorporate contextual information to aid object recognition and localization. There are three key notions in our approach: 1) spatial relationships between different object categories are utilized so that co-inference helps enhance accuracy; 2) temporal context is utilized to accumulate object evidence and to track objects continuously; 3) we borrow techniques from research efforts in single category object recognition to robustly detect *key objects* (such as humans) and use them to reduce inference space for other objects.

D. Forsyth, P. Torr, and A. Zisserman (Eds.): ECCV 2008, Part IV, LNCS 5305, pp. 409–422, 2008.

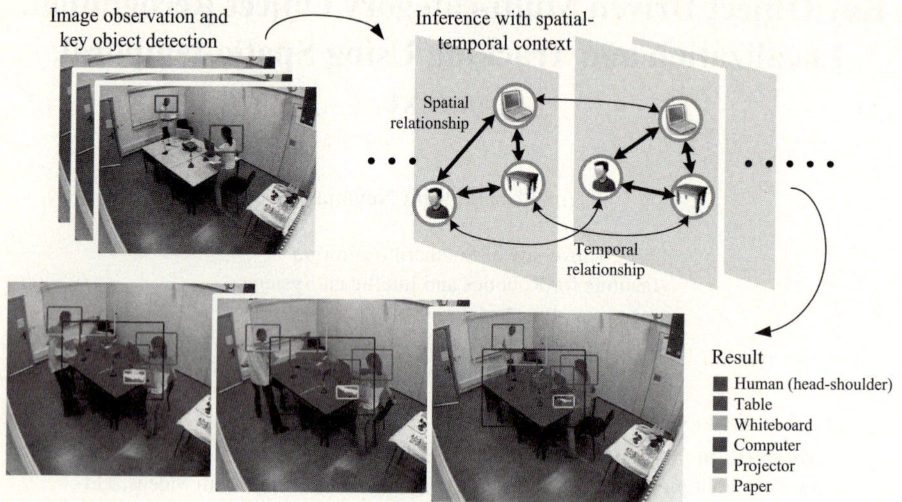

Fig. 1. Finding objects in spatio-temporal context

These concepts are modeled by a dynamic Markov random field (MRF). Figure 1 gives an illustration. Instead of letting each node represent a pixel or image blob in a pre-defined grid, as is commonly done in segmentation, in our model a node represents a hypothetical object in one frame, which enables integration of object-level information during inference. Spatial and temporal relationships are modeled by intra-frame and inter-frame edges respectively. Since objects are recognized on-the-fly and change with time, the structure of the MRF is also dynamic. To avoid building an MRF with excessive false hypothetical object nodes, key objects are detected first and provide contextual guidance for finding other objects. Inference over the resulting MRF gives an estimate of the states of all objects through the sequence. We apply our approach to meeting room scenes with humans as the key objects.

The rest of the paper is organized as follows: Section 2 summarizes related work by categories; Section 3 gives the formulation of the model; Section 4 defines the potential functions of the MRF and Section 5 describes the inference algorithm; Section 6 shows the experimental results; Section 7 discusses about future work and concludes the paper.

2   Related Work

Our approach uses elements from both object recognition and detection. Object recognition focuses on categorization of objects [1][2]; many approaches assume a close-up view of a single object in the input image. Object detection focuses on single category object classification and localization from the background [3][4][5]. Both have received intense research interest recently, bringing forward a large body of literature. While our approach assimilates several established ideas from the two, our emphasis is on integration of spatio-temporal context. We hereby focus on the recent growing effort in tackling object-related problems based on contextual relationships.

Object in the scene. Modeling object-scene relationship enables the use of prior knowledge regarding object category, position, scale and appearance. [6] learns a scene-specific prior distribution of the reference position of each object class to improve classification accuracy of image features. It assumes that a single reference position explains all observed features. [7] proposes a framework for placing local object detection in the 3D scene geometry of an image. Some other work seeks to classify the scene and objects at the same time [8][9]. [8] uses the recognized scene to provide strong prior of object position and scale. Inter-object relationship is not considered. [9] proposes an approach to recognize events and label semantic regions in images, but the focus is not on localizing individual objects.

Object categorization and segmentation in context. Object segmentation and categorization are often combined to enhance each other. When multiple categories are present, contextual knowledge fits in naturally [10][11][12]. [10] uses Conditional Random Field (CRF) to combine appearance, shape and context. *Shape filters* are used to classify each pixel, based on the appearance of a neighborhood; no object-level relationship is explicitly modeled. By counting the co-occurrence of every object pair, [11] exploits object-level context to refine the category label after each image segment has been categorized independently. While [11] does not model spatial relationship among objects, [12] captures spatial relationship by laying a grid-structured MRF on the image, with each node corresponding to the label of a rectangular image blob. Labeling of one blob is dependent on the labels of its neighbors. However, such relationship is constrained to adjacent image blobs.

Object and human action. There have been several attempts in collaborative recognition of object category and human action [13][14][15]. [13] uses the hand motion to improve the shape-based object classification from the top-down view of a desktop. In [14], objects such as chair, keyboards are recognized from surveillance video of an office scene. Bayesian classification of regions is done completely based on human pose and action signatures. Given estimated human upper body pose, [15] accomplishes human action segmentation and object recognition at the same time. All these approaches require the ability of tracking human poses or recognizing action, which is not a trivial task. But they have reflected the fact that many visual tasks are human centered. Namely the objects of most interest for recognition are those interacting closely with humans. This is also our motivation to choose human as the key object in our framework.

3   Model and Representation

In our approach, a dynamic MRF (Figure 2) is designed to integrate the relationship between the object state and its observation, the spatial relationships between objects, as well as the temporal relationships between the states of one object in successive frames. The MRF has the structure of an undirected graph $\mathcal{G}$, with a set of nodes $\mathcal{V}$ and a set of edges $\mathcal{E}$. Each node $v \in \mathcal{V}$ is associated with an unobserved state variable x_v and a observation y_v. Since we are considering a temporal sequence, each node belongs to exactly one time frame t.

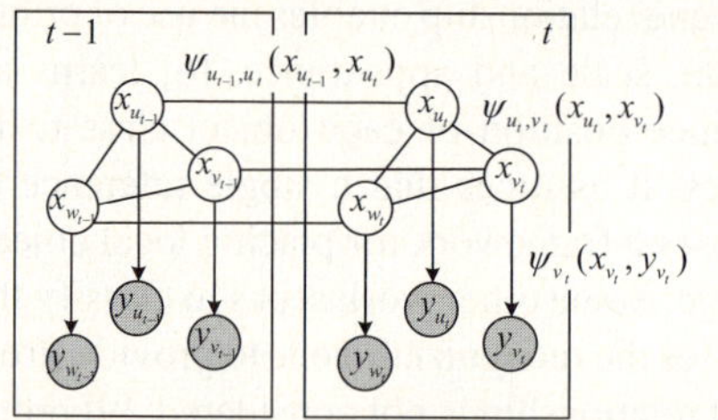 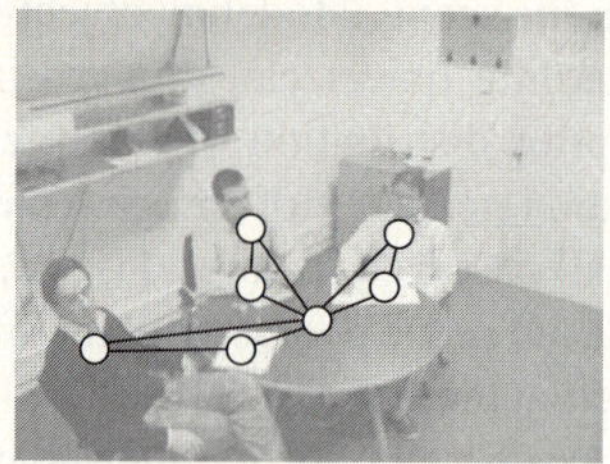

Fig. 2. The MRF defined in our problem (left) and an ideal graph structure for one input frame (right). Section 5 explains how to build such a graph.

We use a node v_t to represent a hypothetical object instance in frame t. Define

$$x_{v_t} = (c_{v_t}, p_{v_t}, s_{v_t}) \tag{1}$$

as the state of the object, where c_{v_t} stands for the object's category label, p_{v_t} for co-ordinates of its centroid and s_{v_t} for the logarithm of size[1]. y_{v_t} is defined as the image evidence of the object. There are two types of edges: intra-frame edges that represent the spatial relationships between different objects, and inter-frame edges that represent the temporal relationships between states of the same object in adjacent frames. Let the potential functions be pairwise, in which case the distribution of the MRF factorizes as

$$p(\mathbf{x}, \mathbf{y}) = \frac{1}{Z} \prod_{(v,u) \in \mathcal{E}} \psi_{v,u}(x_v, x_u) \prod_{v \in \mathcal{V}} \psi_v(x_v, y_v), \tag{2}$$

where $\mathbf{x} = \{x_v | v \in \mathcal{V}\}$ and $\mathbf{y} = \{y_v | v \in \mathcal{V}\}$, $\psi_{v,u}(x_v, x_u)$ models the spatio-temporal relationship, and $\psi_v(x_v, y_v)$ models the image observation likelihood. Given the structure of the MRF and the potential functions, the states of the objects can be inferred.

Note that rather than letting each node correspond to an image blob in a pre-defined image grid or a pixel, as is commonly done in segmentation literature [12][10], we let each node represent an object, which is similar to some tracking frameworks such as the Markov chain in Particle Filtering and MRF in collaborative tracking proposed by [16]. The reason is twofold: 1) object-based graph enables us to use object-level information during inference, while pixel- or grid-based graph can only model inter-object relationships locally along the boundary of objects; 2) object-based graph has fewer nodes and therefore the complexity of inference is much lower. However, one drawback of object-based graph is that accurate segmentation cannot be directly obtained. One new property of the object-based graph is that its structure is dynamic. In Section 5 we show that the nodes for new objects can be added to the graph online driven by detected key objects. Before that we first give our models for the potential functions.

4   Potential Functions

There are three types of edges in our model, each associated with one kind of potential function representing a specific semantic meaning.

---

[1] Logarithm is used because scale is multiplicative.

4.1   Observation Potential $\psi_v(x_v, y_v)$

We use two sources of observation evidence. The first is a single-category object detector for the *key objects*. For meeting room applications, we implement a patch-based cascade human upper body detector following the method in [17]. Let c^* stand for the category label of key objects, for each $x_v = (c_v, p_v, s_v)$ with $c_v = c^*$, we define the observation potential to be the likelihood output of the detector: $\psi_v(x_v, y_v) = p(c^*|x_v, y_v) = p(human|p_v, s_v)$. Please refer to [18] for deriving probability of an object class from a boosted classifier.

The second source of observation potential function targets all object categories of interest. We build our object classifier based on the Bag of Features [1] approach and combine it with image region. The motivation of our choice is the proven performance of Bag of Feature and the suggestion in recent literature that classification based on image segments provides better spatial coherence [2][11][14]. These ideas are tailored to our needs. Specifically, interest points are detected with the DoG and Harris corner detectors and at each interest point a 128d SIFT feature is extracted. During training, these features are used to build a code book by clustering. Also every input image is over-segmented by Mean Shift [19]; each segment is associated with interest points. Base on both the point features and the statistics of pixel intensity of the segments, a classifier is built to model $p(c|r_i)$, defined as the likelihood of any given segment r_i belonging to category c. This could be done by standard Bag of Feature categorization, or more sophisticated generative models such as [2]. We build a discriminative model by using AdaBoost to select and weight features from the code book. Given $p(c|r_i)$ for any segment r_i, the observation potential of object v is modeled as:

$$\psi_v(x_v, y_v) = \frac{\sum\limits_{r_i \in \mathcal{R}(x_v)} p(c_v|r_i)\zeta(r_i, x_v)}{\sum\limits_{r_i \in \mathcal{R}(x_v)} \zeta(r_i, x_v)}, \tag{3}$$

where $\mathcal{R}(x_v)$ stands for the set of segments that is associated with the object v; $\zeta(r_i, x_v)$ is a position weight for r_i, which allows the use of object shape prior. In our implementation we let $\mathcal{R}(x_v)$ include all segments which has at least 50% of its area within v's bounding box, and $\zeta(r_i, x_v)$ is defined as a Gaussian centered at p_v. Figure 3 shows an example for the category *paper*. We can see that it is hard to distinguish the paper from a bright computer screen or the white board by appearance (feature point and region).

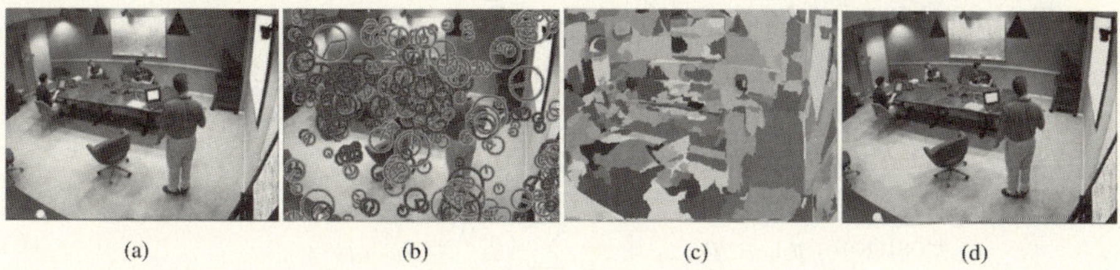

(a)   (b)   (c)   (d)

Fig. 3. An example of finding paper based on appearance. (a) Input image; (b) SIFT features (green: feature with positive weight in the classifier, red: feature with negative weight); (c) Segmentation; (d) observation likelihood $p(paper|r_i)$ for each region r_i (yellow: high likelihood).

Note that the observation potential here can be substituted by any object recognition method, possibly with a more complicated model and higher accuracy such as [2]. Here we do not elaborate on this since our emphasis is on the effect of introducing contextual relationship.

4.2 Spatial Potential $\psi_{v_t, u_t}(x_{v_t}, x_{u_t})$

Spatial potential function ψ_{v_t, u_t} is defined on edges between nodes within one frame but of different object categories. The purpose is to model inter-category correlation in terms of position and scale, *e.g.*, a person tends to sit on a chair beside a table and a laptop is often near a person and on a table. Such correlation generalizes well in our experience for the selected scenario.

When defining the form of the potential function, we want to avoid using very complicated models which introduce risk of over-fitting. In practice, we find that a single Gaussian function is sufficient for our amount of training data as well as the problem itself. Denote $N(\mu, \sigma, x)$ as a Gaussian function with mean μ, variance σ and x as the variable. Since nodes involved are from the same time frame, we suppress the subscript t in this subsection. Define

$$\psi_{v,u}(x_v, x_u) = N(\mu_p(c_u, c_v), \sigma_p(c_u, c_v), p_v - p_u) \cdot$$
$$N(\mu_s(c_u, c_v), \sigma_s(c_u, c_v), s_v - s_u), \tag{4}$$

where $\mu_p(c_u, c_v)$, $\sigma_p(c_u, c_v)$, $\mu_s(c_u, c_v)$ and $\sigma_s(c_u, c_v)$ are the model parameters that describes the relative position and size of two object depending on their category labels c_u and c_v. It is ideal to learn them by maximizing the sum of log likelihoods of all training samples $\{\mathbf{x}^{(i)}\}$. However, this is difficult because $\mathbf{x}^{(i)}$s of different training samples may have different dimensionalities (number of objects varies) and the graph structures also differ. Therefore potential functions are learned independently for each kind of edge in a piecewise manner [20]. The number of different spatial potential functions is $n(n-1)/2$ for n categories. The parameters of the spatial potential function between the categories c_1 and c_2 can be easily learned by maximizing

$$l = \sum_j \log \psi_{v,u}(x_v^{(j)}, x_u^{(j)}), \tag{5}$$

where $\{(x_v^{(j)}, x_u^{(j)})\}$ is the set of all pairs of objects that co-exist in a training sample and satisfy $c_v = c_1$ and $c_u = c_2$.

4.3 Temporal Potential $\psi_{v_{t-1}, v_t}(x_{v_{t-1}}, x_{v_t})$

To build the temporal potential function, feature points used in Section 4.1 are tracked by optical flow through frames. Let the positions of feature points associated with object v be $\{q_t^{(i)}\}_{i=1}^m$ at frame t, x_{v_t} can be estimated from x_{v_t-1} as:

$$\text{Position}: \tilde{p}_{v_t} = p_{v_{t-1}} + \frac{1}{m} \sum_{i=1}^m (q_t^{(i)} - q_{t-1}^{(i)}), \tag{6}$$

$$\text{Scale}: \tilde{s}_{v_t} = s_{v_{t-1}} + \log \left(\frac{\sum_{i=1}^m Dist(q_t^{(i)}, \tilde{p}_{v_t})}{\sum_{i=1}^m Dist(q_{t-1}^{(i)}, p_{v_{t-1}})} \right), \tag{7}$$

where $Dist(\cdot)$ is the distance between two points. The temporal potential is defined as a Gaussian distribution centered at the estimated position and scale with fixed variance:

$$\psi_{v_{t-1},v_t}(x_{v_{t-1}}, x_{v_t}) = N(\tilde{p}_{v_t}, \sigma_p, p_{v_t})N(\tilde{s}_{v_t}, \sigma_s, s_{v_t}). \tag{8}$$

5 Integration of Observation and Spatio-temporal Context

Given the graphical model defined above, there are two remaining issues in using it: how to build such a graph on-the-fly and how to do inference. We solve them in a unified manner by belief propagation (BP) [21][22]. *Augmenting nodes* are introduced as nodes that do not correspond to any specific object but are responsible for generating new object nodes by receiving belief messages from nodes of *key objects*. To distinguish *augmenting nodes* from the others, we refer to other nodes as *object nodes*. BP is then applied to compute the distribution $p(x_v|\mathbf{y})$ for all object nodes, from which the state of every object can be estimated. Since message passing in BP is essential for augmenting nodes, we first bring up the inference part and then introduce the augmenting nodes.

5.1 Inference

We choose BP as the inference method because of two main reasons. First, our graph has cycles and the structure is not fixed (due to addition and removal of object nodes, also inference is not done over the whole sequence but over a sliding window). Therefore it is inconvenient to use methods that require rebuilding the graph (such as the junction tree algorithm). While loopy BP is not guaranteed to converge to the true marginal, it has proven excellent empirical performance. Second, BP is based on local message passing and update, which is efficient and more importantly, gives us an explicit representation of the interrelationship between nodes (especially useful for the augmenting nodes).

At each iteration of BP, the message passing and update process is as follows. Define the neighborhood of a node $u \in V$ as $\Gamma(u) = \{v|(u,v) \in \mathcal{E}\}$, each node u send a message to its neighbor $v \in \Gamma(u)$:

$$m_{u,v}(x_v) = \alpha \int_{x_u} \psi_{u,v}(x_u, x_v)\psi_u(x_u, y_u) \prod_{w \in \Gamma(u)\backslash v} m_{w,u}(x_u)dx_u. \tag{9}$$

The marginal distribution of each object v is estimated by

$$p(x_v|\mathbf{y}) = \alpha\psi_v(x_v, y_v) \prod_{u \in \Gamma(v)} m_{u,v}(x_v). \tag{10}$$

In our problem x_v is a continuous variable whose distribution is non-Gaussian and hard to represent in an analytical form; also, the observation potential function can only be evaluated in a point-wise manner. Therefore we resort to nonparametric version of the BP algorithm [22]. Messages are represented by a nonparametric kernel density estimate. More details of this method can be found in [22]. As a result, a weighted sample set is obtained to approximate the marginal distribution of each object node v: $\{x_v^{(i)}, \omega_v^{(i)}\}_{i=1}^M \sim p(x_v|\mathbf{y})$. The sample set is generated by importance sampling; namely sample $\{x_v^{(i)}\} \sim \prod_{u \in \Gamma(v)} m_{u,v}(x_v)$ and let $\omega_v^{(i)} = \psi_v(x_v^{(i)}, y_v)$. We can then estimate the state of object v (except its category label) by MMSE: $\hat{x}_v = \sum_{i=1}^{M} \omega_v^{(i)} x_v^{(i)} / \sum_{i=1}^{M} \omega_v^{(i)}$.

5.2 Augmenting Nodes

Augmenting nodes find new objects by receiving "hints" (messages) from key object nodes. It is reasonable because we are more interested in finding objects that are closely related to key objects; by combining inter-category spatial relationships with detection techniques specially developed for key objects, other objects can be detected and recognized more robustly and efficiently.

Let the set of key objects in one frame be K, and consider finding new objects of category $c \neq c^*$. The ideal way is: for every subset K' of K, make the hypothesis that there is a new object a which is in context with K'. Based on the NBP paradigm, we estimate a's state by $p(x_a|\mathbf{y}) \propto \psi_u(x_a, y_a) \prod_{v \in K'} m_{v,a}(x_a)$. The number of such hypotheses is exponential of $|K|$, so we simplify it by letting K' contain only one key object (it is reasonable because if a new object is hinted by a set of key objects it is at least hinted by one in some extent). In this case $K' = \{v\}$, the distribution of a's state is estimated as $p(x_a|\mathbf{y}) \propto \psi_a(x_a, y_a) m_{v,a}(x_a)$. This is done for each v in K, each result in a weighted sample set of a hypothetic new object's state.

Further, if two hypotheses of the same category are close in position and scale, they should be the same new object. So for each category, Agglomerative Clustering is done on the union of the $|K|$ sample sets to avoid creating duplicated nodes. For each

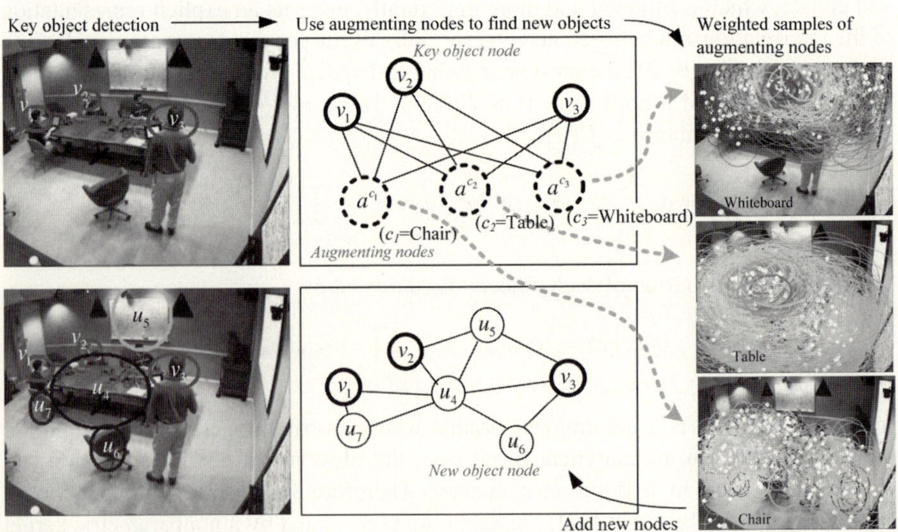

Fig. 4. Use of augmenting nodes to update graph structure. Augmenting nodes for each category are shown as one (dotted circle). For weighted samples, red indicates the highest possible weight, while blue indicates the lowest.

Table 1. Algorithm: inference over a sequence

Denote by $\mathcal{V}_t$ and $\mathcal{E}_t$ the sets of nodes and edges in frame t respectively.

With the graph $\mathcal{G}$ over a L-frame sliding window containing frame $(t - L)$ to $(t - 1)$, proceed as follows with the arrival of a new frame t:

- **Output** the estimated state $\hat{x}_v$ for each object node v of frame $(t - L)$. Remove sub-graph $(\mathcal{V}_{t-L}, \mathcal{E}_{t-L})$ from $\mathcal{G}$ and move the sliding window one frame forward.
- **Add** new sub-graph $(\mathcal{V}_t, \mathcal{E}_t)$ for frame t to $\mathcal{G}$ by algorithm in Table 2.
- **Inference**: perform the nonparametric BP algorithm over $\mathcal{G}$. For each object node v a weighted sample set is obtained: $\{x_v^{(i)}, \omega_v^{(i)}\}_{i=1}^M \sim p(x_v | \mathbf{y})$.
- **Evaluate** confidence of each object v by $W = \sum_{j=t-L+1}^{t} \sum_{i=1}^{M} \omega_{v_j}^{(i)}$. If $W < \gamma$, remove node v_j from frame j for each $j = (t - L + 1) \ldots t$. γ is an empirical threshold.

Table 2. Algorithm: build the sub-graph for a new frame t

Build the sub-graph $(\mathcal{V}_t, \mathcal{E}_t)$ for a new frame t as follows:

- For each object node $v_{t-1} \in \mathcal{V}_{t-1}$, let $\mathcal{V}_t \leftarrow \mathcal{V}_t \cup \{v_t\}$, $\mathcal{E} \leftarrow \mathcal{E} \cup \{(v_{t-1}, v_t)\}$. Pass message forward along edge (v_{t-1}, v_t) to get an approximation of $p(x_{v_t} | \mathbf{y}) \propto \psi_{v_t}(x_{v_t}, y_{v_t}) m_{v_{t-1}, v_t}(x_{v_t})$.
- Detect key object by applying $p(c^* | x)$ to all possible state x in the image. Cluster responses with confidence higher than τ_{c^*}. For each cluster non-overlapping with any existing node, create a new node v_t. Let the initial estimated state $\hat{x}_{v_t}$ be the cluster mean. Denote the set of all key object node as K.
- For each category $c \neq c^*$:
  - Create an augmenting node a for each key object node $v \in K$ and add an edge (v, a) between them.
  - For each such augmenting node and key object node pair $\{a, v\}$, sample $\{x_a^{(i)}, \omega_a^{(i)}\}_{i=1}^M \sim p(x_a | \mathbf{y}) \propto \psi_a(x_a, y_a) m_{v,a}(x_a)$.
  - Define the union of samples $S = \bigcup_a \{x_a^{(i)}, \omega_a^{(i)}\}_{i=1}^M$; let S' be the subset of S with samples whose weight are higher than τ_c.
  - Do clustering on S'; for each cluster non-overlapping with any existing node, create an object node u_t of category c. Let the initial estimated state $\hat{x}_{u_t}$ be the cluster mean.
  - $\mathcal{V}_t \leftarrow \mathcal{V}_t \cup \{u_t\}$. $\mathcal{E}_t \leftarrow \mathcal{E}_t \cup \{(u_t, v_t) | v_t \in \mathcal{V}_t, \psi_{u_t, v_t}(\hat{x}_{u_t}, \hat{x}_{v_t}) > \lambda\}$.
  - Remove augmenting nodes and corresponding edges.

---

high-weight cluster, a new object node is created. Figure 4 illustrates how to use augmenting nodes to update the graph.

More details of our overall algorithm and the algorithm of building sub-graph for each new frame are shown in Table 1 and Table 2.

6 Experiments

Experiments are done on the CHIL meeting video corpus [23]. Eight categories of objects are of interest: *human, table, chair, computer, projector, paper, cup* and *whiteboard* (or *projection screen*).

For testing we use 16 videos captured from three sites (IBM, AIT and UPC) and three camera views for each site. Each sequence has about 400 frames. One frame out of every

60 is fully annotated for evaluation. For training the parameters of spatial potential function, we selected 200 images from two views of the IBM and UPC site (no intersection between training images and test videos), and manually annotated the object size and position. Observation models for objects are trained with object instances from images of various meeting room and office scenes including a training part from CHIL.

We design our experiments to compare three methods with different levels of context: 1) no context, *i.e.* object observation model is directly applied to each frame; 2) spatial context only, *i.e.* a MRF without the temporal edges is applied in a frame-by-frame manner; 3) spatio-temporal context, *i.e.* the full model with both spatial and temporal edges is applied to the sequence.

6.1   Quantitative Analysis

Quantitative analysis is performed with metrics focusing on three different aspects: object detection and tracking, image segment categorization and pixel-level segmentation accuracy.

Object-level detection and tracking. The overall object detection rate and false alarm rate is shown in Figure 6(left). Two methods are compared: frame-based method with only spatial context, and the spatio-temporal method using the complete model. For the spatial-only method, an ROC curve is obtained by changing the threshold τ_c for creating new object nodes. The result shows that integrating temporal information helps improve detection rate and reduce false alarms, which is the effect of temporal smoothing and evidence accumulation. In object-level evaluation we do not include the non-contextual method, because the object observation model is based on classifying image segments, and we find that applying exhaustive search using such a model does not give a meaningful result. Some visual results of these methods can be found in Figure 5(a)(c)(d) respectively.

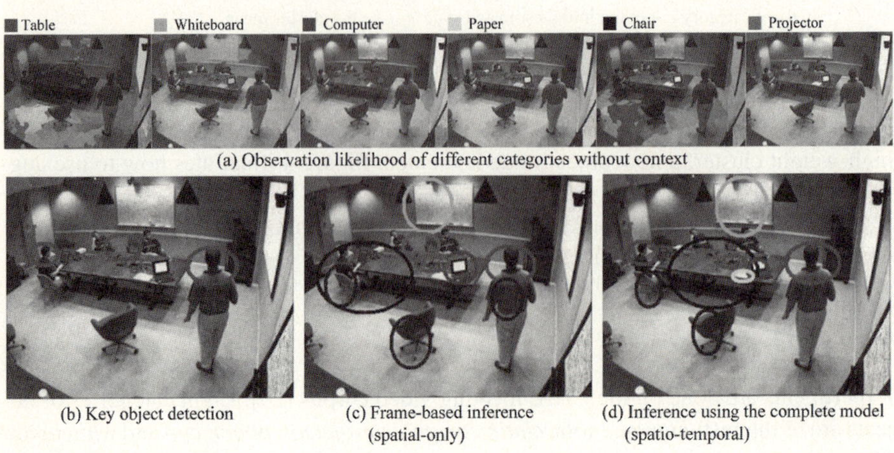

(a) Observation likelihood of different categories without context

(b) Key object detection   (c) Frame-based inference   (d) Inference using the complete model
                                        (spatial-only)   (spatio-temporal)

Fig. 5. Comparison among observation with no context, inference using spatial relationship only and inference using spatio-temporal relationship

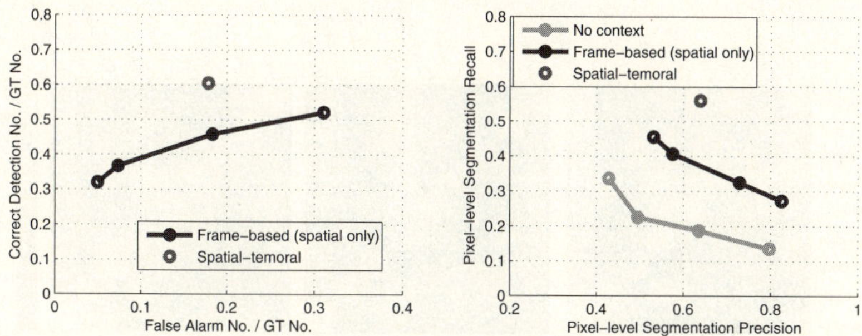

Fig. 6. Object detection rate and false alarm rate (left); pixel-level segmentation precision and recall (right)

Table 3. Object tracking evaluation

Category	Ground truth trajectories	Mostly tracked trajectories (%GT)	Partially tracked trajectories (%GT)	Fragments
Human	64	46 (71.9%)	12 (18.8%)	12
Chair	30	10 (33.3%)	4 (13.3%)	2
Paper	40	21 (52.5%)	7 (17.5%)	0
Cup	11	2 (18.2%)	0 (0%)	0
Computer	24	10 (41.7%)	3 (12.5%)	2
Table	16	14 (87.5%)	0 (0%)	0
Screen	14	12 (85.7%)	1 (7.1%)	2
Projector	13	7 (53.8%)	2 (15.4%)	0
All	212	122 (57.5%)	29 (13.7%)	8

For the spatio-temporal method, we further evaluate its performance by the number of objects that are consistently tracked through the sequence, as shown in Table 3. All the numbers stand for trajectories, where *mostly tracked* is defined as at least 80% of the trajectory is tracked, and *partially tracked* defined as at least 50% is tracked. When a trajectory is broken into two, a *fragment* is counted. We can see that small objects such as *cups* and *computers* are harder to detect and track. *Paper* has a high false alarm rate, probably due to lack of distinct interior features (Figure 8(h) shows segments of human clothes detected as *paper*). Most fragments belong to human trajectories, because humans exhibit much more motion than other objects.

Image segment-level categorization. To compare with the result of applying object observation without contextual information, we compute the categorization accuracy of all the image segments in the form of a confusion matrix (Figure 7). The matrix shows that incorporating context helps reduce the confusion between different object categories, such as *paper* versus *whiteboard*. It is also observed that many objects are easily confused with *table*, mainly because they are often on top of or adjacent to the *table*.

Pixel-level segmentation. We obtain segmentation of each object based on the likelihood $p(c|r_i)$ of each segment r_i classified as category c. Pixel-level precision and recall rates of the three methods are shown in Figure 6(right). Similar to the previous two evaluations, the spatio-temporal method gives the best result. The segmentation

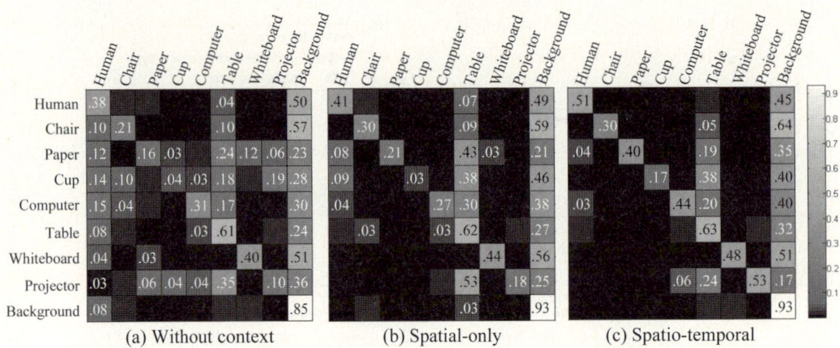

(a) Without context   (b) Spatial-only   (c) Spatio-temporal

Fig. 7. Confusion matrix of image region categorization by different methods. The value at (i, j) stands for the proportion of segments of category i classified as category j.

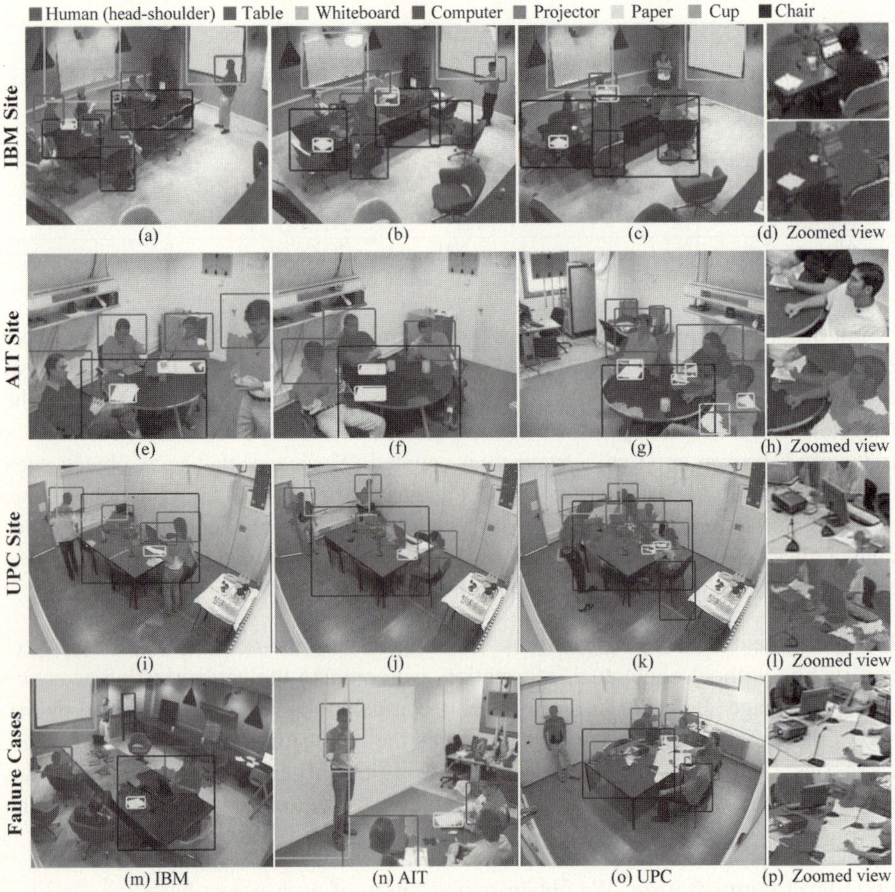

(a)   (b)   (c)   (d) Zoomed view

(e)   (f)   (g)   (h) Zoomed view

(i)   (j)   (k)   (l) Zoomed view

(m) IBM   (n) AIT   (o) UPC   (p) Zoomed view

Fig. 8. Sample results of our method

accuracy is not high, since this is only a simple post-process. But from the sample results of the spatio-temporal method in Figure 8 we can see that most detected objects are reasonably segmented when the object position and scale are correctly inferred.

6.2 Scenario Analysis

Figure 8 shows some sample results of our method on data from different meeting room sites and views. Objects that are in close interaction with key objects (humans) are detected more accurately. The method also has a tolerance to missed detection of key objects, *e.g.*, for the IBM site, although human detection rate is not high due to complex background, most objects are reasonably detected (Figure 8(a)-(c)). However such tolerance is to certain extent: Figure 8(m) shows a case when missed detections of key objects cause failure in detecting other objects.

Partial occlusions are frequently encountered and handled, such as occlusions of tables, whiteboards and laptops. But there is a bigger chance of failure when only a small part of an object is visible, such as in Figure 8(a)-(d) the table is broken into two; in Figure 8(n)(o) the table or part of it is missing from detection. This is also true for small objects, *e.g.* in Figure 8(g)(h) the paper occluded by the hand is broken into two. But in such case the result is still correct in the image segment level.

The bottleneck of performance is the observation model for objects other than key objects. As in Figure 8(p) the computer and projector are missing simply because observation likelihood is low. Although contextual information improves the overall result, the observation model in our current implementation is relatively simple compared with the complexity of the object recognition problem.

7 Conclusion

In this paper we address the problem of recognizing, localizing and tracking multiple categories of objects in a certain type of scenes. Specifically, we consider eight categories of common objects in meeting room videos. Given the difficulty of approaching this problem by purely appearance-based methods, we propose the integration of spatio-temporal context through a dynamic MRF, in which each node represents an object and the edges represent inter-object relationships. New object hypotheses are proposed online by adding *augmenting nodes*, which receive belief messages from the detected *key objects* of the scene (humans in our case). Experimental results show that the performance is greatly enhanced by incorporating contextual information.

There are many open problems and promising directions regarding the topic of object analysis in video. First, a stronger object observation model is needed, and our current training and testing sets are very limited. Second, we made no assumption of a fixed camera, but it can be a strong cue for inference, *e.g.* the position and scale of the stationary objects (such as tables) can be inferred from the activity area of the moving objects (such as humans). Third, 3D geometry of the scene or depth information should be useful for modeling occlusions. Last but not least, object recognition and tracking can be combined with action recognition [14][15] so as to better understand the semantics of human activities.

Acknowledgments. This research is supported, in part, by the U.S. Government VACE program. Yuan Li is funded, in part, by a Provost's Fellowship from USC.

References

1. Fergus, R., Perona, P., Zisserman, A.: Object class recognition by unsupervised scale-invariant learning. In: CVPR (2003)
2. Cao, L., Fei-Fei, L.: Spatially coherent latent topic model for concurrent object segmentation and classification. In: ICCV (2007)
3. Viola, P., Jones, M.: Rapid object detection using a boosted cascade of simple features. In: CVPR (2001)
4. Dalal, N., Triggs, B.: Histograms of oriented gradients for human detection. In: CVPR (2005)
5. Wu, B., Nevatia, R.: Cluster boosted tree classifier for multi-view, multi-pose object detection. In: ICCV (2007)
6. Sudderth, E.B., Torralba, A., Freeman, W.T., Willsky, A.S.: Learning hierarchical models of scenes, objects, and parts. In: ICCV (2005)
7. Hoiem, D., Efros, A.A., Hebert, M.: Putting objects in perspective. In: CVPR (2006)
8. Torralba, A., Murphy, K., Freeman, W., Rubin, M.: Context-based vision system for place and object recognition. In: ICCV (2003)
9. Li, L.-J., Fei-Fei, L.: What, where and who? classifying events by scene and object recognition. In: ICCV (2007)
10. Shotton, J., Winn, J., Rother, C., Criminisi, A.: Textonboost: Joint appearance, shape and context modeling for multi-class object recognition and segmentation. In: ECCV (2006)
11. Rabinovich, A., Vedaldi, A., Galleguillos, C., Wiewiora, E., Belongie, S.: Objects in context. In: ICCV (2007)
12. Carbonetto, P., de Freitas, N., Barnard, K.: A statistical model for general contextual object recognition. In: Pajdla, T., Matas, J(G.) (eds.) ECCV 2004. LNCS, vol. 3021, pp. 350–362. Springer, Heidelberg (2004)
13. Moore, D.J., Essa, I.A., Heyes, M.H.: Exploiting human actions and object context for recognition tasks. In: ICCV (1999)
14. Peursum, P., West, G., Venkatesh, S.: Combining image regions and human activity for indirect object recognition in indoor wide-angle views. In: ICCV (2005)
15. Gupta, A., Davis, L.S.: Objects in action: an approach for combining action understanding and object perception. In: CVPR (2007)
16. Yu, T., Wu, Y.: Collaborative tracking of multiple targets. In: CVPR (2004)
17. Wu, B., Nevatia, R.: Tracking of multiple humans in meetings. In: V4HCI (2006)
18. Friedman, J., Hastie, T., Tibshirani, R.: Additive logistic regression: a statistical view of boosting. Annals of Statistics 28(2), 337–407 (2000)
19. Comaniciu, D., Meer, P.: Mean shift: A robust approach toward feature space analysis. IEEE Transaction on Pattern Analysis and Machine Intelligence 24(5), 603–619 (2002)
20. Sutton, C., McCallum, A.: Piecewise training for undirected models. In: Conference on Uncertainty in Artificial Intelligence (2005)
21. Pearl, J.: Probabilistic Reasoning in Intelligent Systems. Morgan Kaufman, San Mateo (1988)
22. Sudderth, E.B., Ihler, A.T., Freeman, W.T., Willsky, A.S.: Nonparametric belief propagation. In: CVPR (2003)
23. CHIL: The chil project, http://chil.server.de/

A Pose-Invariant Descriptor for Human Detection and Segmentation

Zhe Lin and Larry S. Davis

Institute of Advanced Computer Studies
University of Maryland, College Park, MD 20742
{zhelin,lsd}@umiacs.umd.edu

Abstract. We present a learning-based, sliding window-style approach for the problem of detecting humans in still images. Instead of traditional concatenation-style image location-based feature encoding, a global descriptor more invariant to pose variation is introduced. Specifically, we propose a principled approach to learning and classifying human/non-human image patterns by simultaneously segmenting human shapes and poses, and extracting articulation-insensitive features. The shapes and poses are segmented by an efficient, probabilistic hierarchical part-template matching algorithm, and the features are collected in the context of poses by tracing around the estimated shape boundaries. Histograms of oriented gradients are used as a source of low-level features from which our pose-invariant descriptors are computed, and kernel SVMs are adopted as the test classifiers. We evaluate our detection and segmentation approach on two public pedestrian datasets.

1   Introduction

Human detection is a widely-studied problem in vision. It still remains challenging due to highly articulated body postures, viewpoint changes, varying illumination conditions, and background clutter. Combinations of these factors result in large variability of human shapes and appearances in images. We present an articulation-insensitive feature extraction method and apply it to machine learning-based human detection. Our research goal is to robustly and efficiently detect and segment humans under varying poses.

Numerous approaches have been developed for human detection in still images or videos. Most of them use shape information as the main discriminative cue. These approaches can be roughly classified into two categories. The first category models human shapes globally or densely over image locations, *e.g.* shape template hierarchy in [1], an over-complete set of haar wavelet features in [2], rectangular features in [3], histograms of oriented gradients (HOG) in [4] or locally deformable Markov models in [5]. Global schemes such as [4, 6] are designed to tolerate certain degrees of occlusions and shape articulations with a large number of samples and have been demonstrated to achieve excellent performance with well-aligned, more-or-less fully visible training data. The second category of approaches uses local feature-based approaches to learn body part

D. Forsyth, P. Torr, and A. Zisserman (Eds.): ECCV 2008, Part IV, LNCS 5305, pp. 423–436, 2008.

and/or full-body detectors based on sparse interest points and descriptors as in [7, 8], from predefined pools of local curve segments [9, 10], k-adjacent segments [11], or edgelets [12]. In [13], several part detectors are trained separately for each body part, and combined with a second-level classifier. Compared to the global schemes, part (or local feature)-based approaches [12, 8, 14] are more adept in handling partial occlusions, and flexible in dealing with shape articulations. Shape cues are also combined with motion cues for human detection in [15, 16], simultaneous detection and segmentation in [17].

Dalal and Triggs [4] introduced a powerful image descriptor - HOG, and provided an extensive experimental evaluation using linear and gaussian-kernel SVMs as the test classifiers. Later, Zhu *et al.* [18] improved its computational efficiency significantly by utilizing a boosted cascade of rejectors. Recently, Tuzel *et al.* [6] reported better detection performance than [4] on the INRIA dataset. They use covariant matrices as image descriptors and classify patterns on Riemannian manifolds. Similarly, Maji *et al.* [19] also demonstrate promising results using multi-level HOG descriptors and faster (histogram intersection) kernel SVM classification. In [20], two-fold adaboost classifiers are adopted for simultaneous part selection and pedestrian classification. Ref. [21] combines different features in a single classification framework.

Previous discriminative learning-based approaches mostly train a binary classifier on a large number of positive and negative samples where humans are roughly center-aligned. These approaches represent appearances by concatenating information along 2D image coordinates for capturing spatially recurring local shape events in training data. However, due to highly articulated human poses and varying viewing angles, a very large number of (well-aligned) training samples are required; moreover, the inclusion of information from whole images inevitably make them sensitive to biases in training data (in the worst case, significant negative effects can occur from arbitrary image regions), consequently the generalization capability of the trained classifier can be compromised. Motivated by these limitations, we extract features adaptively in the local context of poses, i.e. we propose a pose-invariant feature extraction method for simultaneous human detection and segmentation. The intuition is that pose-adapted features produce much better spatial repeatability and recurrence of local shape events. Specifically, we segment human poses on both positive and negative samples[1] and extract features adaptively in local neighborhoods of pose contours, *i.e.* in the pose context. The set of all possible pose instances are mapped to a canonical pose such that points on an arbitrary pose contour have one-to-one correspondences to points in the canonical pose. This ensures that our extracted feature descriptors correspond well to each other, and also invariant to varying poses. Our main contributions are summarized as follows:

- An extended part-template tree model and an automatic learning algorithm are introduced for simultaneous human detection and pose segmentation.

---

[1] For negative samples, pose estimation is forced to proceed even though no person in them.

- A fast hierarchical part-template matching algorithm is used to estimate human shapes and poses based on both gradient magnitude and orientation matching. Human shapes and poses are represented by parametric models, and the estimation problem is formulated and optimized in a probabilistic framework.
- Estimated optimal poses are used to impose spatial priors (for possible humans) to encode pose-invariant features in nearby local pose contexts. One-to-one correspondence is established between sets of contour points of an arbitrary pose and a canonical pose.

The paper is organized as follows. Section 2 gives an overview; Section 3 describes the details of our pose-invariant feature extraction method; Section 4 introduces our learning and classification schemes and demonstrates experiments and evaluations; finally, Section 5 concludes the paper and discusses possible future extensions.

2   Overview of the Approach

We illustrate and evaluate our approach mainly using the INRIA person dataset[2] [4] and the MIT-CBCL pedestrian dataset[3] [2, 13]. In these datasets, training and testing samples all consist of 128×64 image patches. Negative samples are randomly selected from raw (person-free) images, positive samples are cropped (from annotated images) such that persons are roughly aligned in location and scale.

For each training or testing sample, we first compute a set of histograms of (gradient magnitude-weighted) edge orientations for non-overlapping 8×8 rectangular regions (or cells) evenly distributed over images. Motivated by the success of HOG descriptors [4] for object detection, we employ coarse-spatial and fine-orientation quantization to encode the histograms, and normalization is performed on groups of locally connected cells, *i.e.* blocks. Then, given the orientation histograms, a probabilistic hierarchical part-template matching technique is used to estimate shapes and poses based on an efficient part-based synthesis approach under a probabilistic framework. A fast k-fold greedy search algorithm is used for the likelihood optimization. The part-template tree model in [14] used for the hierarchical matching is learned from a set of annotated silhouette images. Given the pose and shape estimates, block features closest to each pose contour point are collected; finally, the histograms of the collected blocks are concatenated in the order of pose correspondence to form our feature descriptor. As in [4], each block (consisting of 4 histograms) is normalized before collecting features to reduce sensitivity to illumination changes. The one-to-one point correspondence from an arbitrary pose model to the canonical one reduces sensitivity of extracted descriptors to pose variations. Figure 1 shows an illustration of our feature extraction process.

---

[2] http://lear.inrialpes.fr/data
[3] http://cbcl.mit.edu/software-datasets/PedestrianData.html

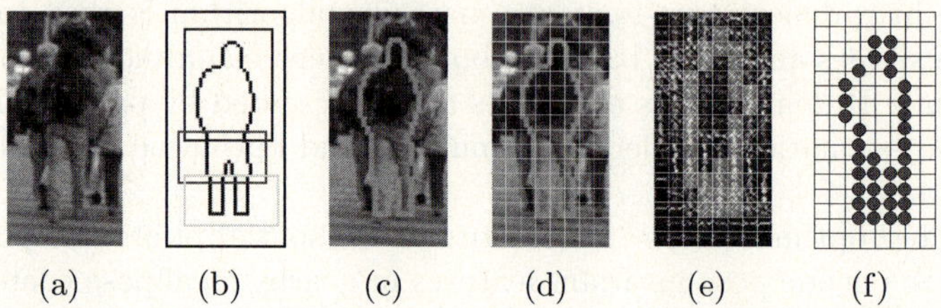

(a)   (b)   (c)   (d)   (e)   (f)

Fig. 1. Overview of our feature extraction method. a) A training or testing image, b) Part-template detections, c) Pose and shape segmentation, d) Cells overlaid onto pose contours, e) Orientation histograms and cells overlapping with the pose boundary, f) Block centers relevant to the descriptor.

3   Pose-Invariant Descriptors

3.1   Low-Level Feature Representation

For pedestrian detection, histograms of oriented gradients (HOG) [4] exhibited superior performance in separating image patches into human/non-human. These descriptors ignore spatial information locally, hence are very robust to small alignment errors. We use a very similar representation as our low-level feature description, $i.e.$ (gradient magnitude-weighted) edge orientation histograms.

Given an input image $\mathbf{I}$, we calculate gradient magnitudes $|G_{\mathbf{I}}|$ and edge orientations $O_{\mathbf{I}}$ using simple difference operators $(-1, 0, 1)$ and $(-1, 0, 1)^t$ in horizontal-x and vertical-y directions, respectively. We quantize the image region into local 8×8 non-overlapping cells, each represented by a histogram of (unsigned) edge orientations (each surrounding pixel contributes a gradient magnitude-weighted vote to the histogram bins). Edge orientations are quantized into $N_b = 9$ orientation bins $[k\frac{\pi}{N_b}, (k+1)\frac{\pi}{N_b})$, where $k = 0, 1 ... N_b - 1$. For reducing aliasing and discontinuity effects, we also use trilinear interpolation as in [4] to vote for the gradient magnitudes in both spatial and orientation dimensions. Additionally, each set of neighboring 2×2 cells form a block. This results in overlapping blocks where each cell is contained in multiple blocks. For reducing illumination sensitivity, we normalize the group of histograms in each block using L_2 normalization with a small regularization constant ϵ to avoid dividing-by-zero. Figure 2 shows example visualizations of our low-level HOG descriptors.

The above computation results in our low-level feature representation consisting of a set of raw (cell) histograms (gradient magnitude-weighted) and a set of normalized block descriptors indexed by image locations. As will be explained in the following, both unnormalized cell histograms and block descriptors are used for inferring poses and computing final features for detection.

3.2   Part-Template Tree Model

For highly articulated objects like humans, part-based detection approaches (*e.g.* [7, 12]) have been shown to be capable of handling partial object/

Fig. 2. Examples of two training samples and visualization of corresponding (unnormalized and L_2-normalized) edge orientation histograms

inter-occlusions and are flexible in modeling shape articulations. In contrast, global shape template-based approaches are capable of simultaneously detecting and segmenting human shapes, *e.g.* [1] and its generalization [22] using a Bayesian inference. The merits of these two schemes are combined in a unified (top-down and bottom-up) optimization framework for simultaneous detection and segmentation in [14]. Specifically, it extends the hierarchical template matching method in [1] by decomposing the global shape models into parts and constructing a part-template tree for matching it to images hierarchically.

In order to more efficiently and reliably estimate human shapes and poses in the image, we learn the part-template tree model [14] and extend the matching algorithm in a probabilistic optimization framework. We train the part-template tree on a set of annotated silhouette images to learn the distribution of part models in each of the tree layers and to handle a wider range of articulations of people. The structure of our learned part-template tree model is roughly shown in Figure 3. The part-template tree was initially constructed using a simple pose generator and body-part decomposer. Each part in the tree can be viewed as a parametric model, where part location and sizes are the model

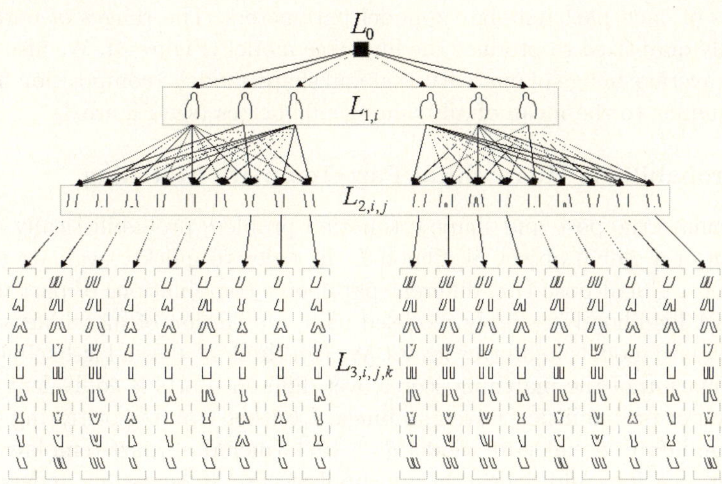

Fig. 3. An illustration of the extended part-template tree

parameters. As shown in the figure, the tree consists of 186 part-templates, *i.e.* 6 head-torso (ht) models, 18 upper-leg (ul) models, 162 lower-leg (ll) models, and organized hierarchically based on the layout of human body parts in a top-to-bottom manner. Due to the tree structure, a fast hierarchical shape (or pose) matching scheme can be applied using the model. For example, using hierarchical part-template matching (which will be explained later), we only need to match 24 part-templates to account for the complexity of matching 486 global shape models using the method in [1], so it is extremely fast. For the details of the tree model construction method, readers are referred to [14].

Learning the Part-template Tree. We learn the part-template tree model based on a training set which consists of 404 (plus mirrored versions) manually annotated binary silhouette images (white foreground and black background). Those silhouettes are chosen from a subset of positive image patches of the IN-RIA person database. Each of the training silhouette images is sent through the tree from the root node to leaf nodes and the degree of coverage (both foreground and background) consistency between each part template $T_{\theta_j}, j \in \{ht, ul, ll\}$ and the observation is measured. Here, each part-template is considered to be covered by a binary rectangular image patch M (see Figure 1(b) for an example). The degree of coverage consistency $\rho(\theta_j | S)$ between a part-template T_{θ_j} and a silhouette image S is defined as the pixel-wise similarity of the part-template coverage image $M(\theta_j)$ and the binary sub-silhouette S_j (corresponding to the same region as the part-template), *i.e.* $\rho(\theta_j | S) = 1 - \frac{\sum_{\mathbf{x}} |S_i(\mathbf{x}) - M(\theta_j, \mathbf{x})|}{n}$, where n is the total number of pixels in the rectangular part-template region. Then, we can estimate the best set of part models $\theta^* = \{\theta_j^*\}$ for the training silhouette S by maximum likelihood estimation: $\theta_j^* = \arg\max_{\theta_j \in \Theta_j} \rho(\theta_j | S)$, where Θ_j denotes the set of all possible part template parameters. This process is repeated for all training silhouettes and the ranges of part template models are estimated based on the statistics of each part-template's model parameters. The ranges of parameters are evenly quantized to produce the final tree model (Figure 3). We also verified that the average image of our learned global shape models (composition of parts) is very similar to the mean of all training silhouettes (see Figure 4).

3.3   Probabilistic Hierarchical Part-Template Matching

We formulate the pose and shape estimation problem probabilistically as maximization of a global object likelihood L. In order to quickly evaluate the likelihood for a global pose (*i.e.* different parameter combinations of part models), the object likelihood is simply modeled as a *summation* of matching scores of part-template models in all tree layers. We can think of L as a log-likelihood and the summation of the matching scores over different parts is equivalent to multiplication of probabilities. Given an image I (either training or testing sample) and a candidate global pose model $\theta = \{\theta_j\}$ (including part-template indices and their locations and scales), in the simplest case, if we assume independence between part-template models θ_j in different layers, the object likelihood can be simply represented as follows:

$$L(\theta|I) = L(\theta_{ht}, \theta_{ul}, \theta_{ll}|I) = \sum_{j \in \{ht,ul,ll\}} L(\theta_j|I). \tag{1}$$

For the purpose of pose estimation, we should jointly consider different parts θ_j for optimization of L. Hence, based on the layer structure of the tree in Figure 3, the likelihood L is decomposed into conditional likelihoods as follows:

$$\begin{aligned}
L(\theta|I) &= L(\theta_{ht}|I) + L(\theta_{ul}|\theta_{ht}, I) + L(\theta_{ll}|\theta_{ht}, \theta_{ul}, I) \\
&= L(\theta_{ht}|I) + L(\theta_{ul}|\theta_{ht}, I) + L(\theta_{ll}|\theta_{ul}, I),
\end{aligned} \tag{2}$$

where the decomposition is performed in a top-to-bottom order of the layers, and independence is assumed between the two non-joining layers, ht and ll. We used Eq. 2 as our optimization model.

Part-Template Likelihood. A part template T_{θ_j} (defined by model parameters θ_j) is characterized by its boundary curve segments (see Figure 3) and edge orientations of points along the segment. We match individual part-templates using a method similar to Chamfer matching [1]. Instead of using distance transforms, we collect matching scores (magnitudes of corresponding orientation bins in the map of edge orientation histograms) along the part-template contour. The matching scores are measured using look-up tables for speed. Magnitudes from neighboring histogram bins are weighted to reduce orientation biases and to regularize the matching scores of each template point.

More formally, the likelihood $L(\theta_j(\mathbf{x}, s)|I)$ of a part template-T_{θ_j} at location $\mathbf{x}$ and scale s is modeled as follows:

$$L(\theta_j(\mathbf{x}, s)|I) = \frac{1}{|T_{\theta_j}|} \sum_{\mathbf{t} \in T_{\theta_j}} d'_I(\mathbf{x} + s\mathbf{t}), \tag{3}$$

where $|T_{\theta_j}|$ denotes the length of the part-template, and $\mathbf{t}$ represents individual contour points along the template. Suppose the edge orientation of contour point $\mathbf{t}$ is $O(\mathbf{t})$, its corresponding orientation bin index $B(\mathbf{t})$ is computed as: $B(\mathbf{t}) = [O(\mathbf{t})/(\pi/9)]$ ($[x]$ denotes the maximum integer less-or-equal to x), and the unnormalized (raw) orientation histogram at location $(\mathbf{x} + s\mathbf{t})$ is $H = \{h_i\}$. Then, the individual matching score d'_I at contour point $\mathbf{t}$ is expressed as:

$$d'_I(\mathbf{x} + s\mathbf{t}) = \sum_{b=-\delta}^{\delta} w(b) h_{B(\mathbf{t})+b}, \tag{4}$$

where δ is a neighborhood range, and $w(b)$ is a symmetric weight distribution[4].

Optimization. The structure of our part-template model and the form (summation) of the global object likelihood L suggest that the optimization problem

---

[4] For simplicity, we use $\delta = 1$, and $w(1) = w(-1) = 0.25, w(0) = 0.5$ in our experiments.

can be solved by dynamic programming or belief propagation [23] to achieve globally optimal solutions. But, these algorithms are computationally too expensive for dense scanning of all windows for detection. For efficiency, we perform the optimization, *i.e.* the maximization of L, by a fast k-fold greedy search procedure. Algorithm 1 illustrates the overall matching (optimization) process. We keep scores for all nodes ($k = 1, 2...K$) in the second layer (*i.e.* the torso layer) instead of estimating the best k in step 1 of the algorithm. In the following steps, a greedy procedure is individually performed for each of those K nodes (or threads).

Algorithm 1. Probabilistic Hierarchical Part-Template Matching

---

1) For a set of locations $\mathbf{x}$ and scales s, match all K head-torso part-templates in layer L_1 with the image and compute their part-template likelihoods $L(\theta_{ht}^k(\mathbf{x}, s)|I), k = 1, 2...K$.

2) For $k = 1...K$, repeat the following steps (3)-(4), and select $k = k^*$ and $\theta = \theta^*$ with the maximum $L(\theta|I)$.

3) According to the part-template model θ_{ht}^k of Layer L_1, estimate the maximum conditional-likelihood leg models $\theta_{ul}^*|\theta_{ht}^k$ in L_2 and $\theta_{ll}^*|\theta_{ul}^*, \theta_{ht}^k$ in L_3 using a greedy search algorithm along the tree.

4) Given the above part-template's model estimates, compute the current global object likelihood based on Eq. 2.

5) Return the global pose model estimates $\theta^* = \{\theta_{ht}^k, \theta_{ul}^*, \theta_{ll}^*\}$.

Pose model parameters estimated by the hierarchical part-template matching algorithm are directly used for pose segmentation by part-synthesis (region connection). Figure 4 shows the process of global pose (shape) segmentation by the part-template synthesis.

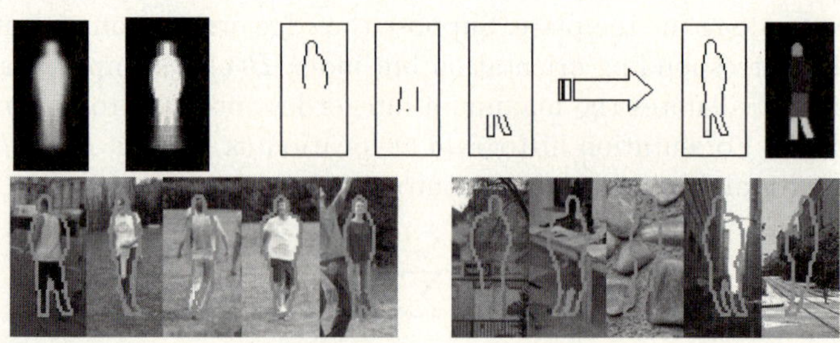

Fig. 4. An illustration of pose segmentation. Top-Left: Mean image of training silhouettes and our 486 learned global shape models; Top-Right: Best part-templates (three images on the left side) in each layer are combined to produce final global shape and pose estimates (two images on the right side); Bottom: example pose (shape) segmentation on positive/negative samples.

2 Texture-Consistent Shadow Removal

In this paper, we provide a brush tool for users to mark the shadow boundary. As illustrated in Fig. 1(c), users can select a brush with much larger size than the boundary, and do not need to delineate the boundary precisely. The brush strokes divide an image into three areas: definite umbra area, definite lit area, and boundary, which consists of penumbra area as well as parts of the umbra and lit area. Our algorithm precisely locates the penumbra area from the user specified boundary, and removes the shadow seamlessly. A working example of our algorithm is illustrated in Fig. 1.

This paper aims to remove shadow effects such that the resulting shadow-free image has consistent texture between the shadow and lit area. We first construct a new image gradient field that removes the gradients induced by the shadow effect and has consistent gradient characteristics between the shadow and lit area. Then we can reconstruct the shadow-free image from the new gradient field through 2D integration by solving a Poisson equation similar to previous work (c.f. [2,6,13]). The major challenge is to construct the new image gradient field G^n given only the rough shadow boundary from users. In § 2.1, we describe a novel algorithm to estimate the illumination change curves across the shadow boundary and cancel the effect of illumination change on the gradient field in the penumbra area. In the § 2.2, we describe a method to estimate the shadow effect on the texture characteristics in the shadow area and transform the characteristics of gradients there to be compatible with that in the lit area.

2.1 Estimate Illumination Change in Penumbra Area

Properly handling the shadow boundary or the penumbra area is a challenge for shadow removal. The ambiguity of the shadow boundary often makes automatic shadow boundary detection methods fail. Relying on users to provide the precise shadow boundary casts a heavy burden on them. To relieve users' burden, Mohan *et al.* [10] presented a piece-wise model where users only need to specify connected line segments to delineate the boundary. However, when dealing with complex shadow boundaries like the eagle's right wing in Fig. 1(c), their method will still require users to specify a large number of key points. To further reduce users' burden, we only require a rough specification of the shadow boundary from users using brush tools as illustrated in Fig. 1(c).

Given an inaccurate shadow boundary specification, our method simultaneously locates the shadow boundary precisely and estimates the illumination change $C(x, y)$ in Equation 2 in the penumbra area. The complex shape of the shadow boundary makes devising a parametric model of $C(x, y)$ difficult. However, we observe that any line segment crossing the boundary has an easily parameterizable illumination profile. Therefore, we model $C(x, y)$ by sampling line segments across the boundary and estimating a parametric model for each as illustrated in Fig. 3(a). Since the user provided-boundary usually is not accurate enough, unlike [3], we do not sample $C(x, y)$ using line segments perpendicular to the boundary. Instead, like [10], we use a vertical/horizontal sampling line per

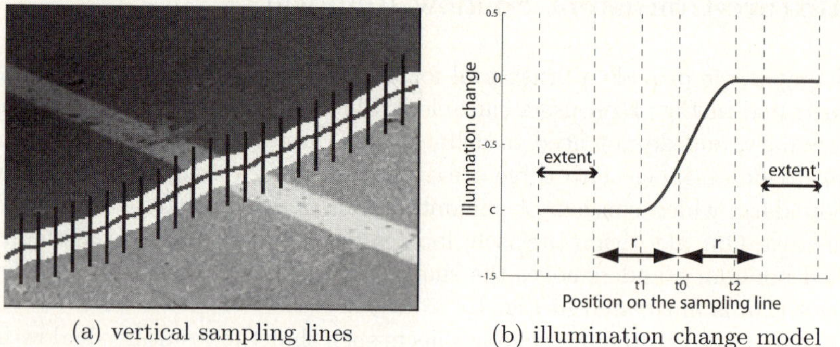

(a) vertical sampling lines   (b) illumination change model

Fig. 3. Sampling illumination change surface using line segments. (a): vertical sampling lines. (b): t_0 and r are the brush center and brush radius. $[t_1, t_2]$ is the penumbra area. *extent* is the range in the umbra and lit area, used to estimate the gradient characteristics.

pixel along the boundary and use the estimated illumination change to cancel the shadow effect on the gradient in Y/X direction. We estimate horizontal and vertical illumination change sampling lines independently.

We model the illumination change along each line segment as the following C^1 continuous piece-wise polynomial as illustrated in Fig. 3(b):

$$C_l(t) = \begin{cases} c, & t < t_1; \\ f(t), & t_1 \leq t \leq t_2; \\ 0, & \text{else.} \end{cases} \tag{3}$$

This piece-wise polynomial model can be parameterized by 3 parameters, denoted as $M_l(c, t_1, t_2)$. Here t_1 and t_2 define the penumbra area along the sampling line. (Without losing generality, we assume $t < t_1$ lies in the umbra area and $t > t_2$ lies in the lit area.) $c(\leq 0)$ is the reduction of the illumination in the umbra area. $f(t)$ is a cubic curve determined by the two boundary points, (t_1, c) and $(t_2, 0)$, and the derivatives at these two points, $f'(t_1) = 0$ and $f'(t_2) = 0$. This illumination change model is determined by both the location of the penumbra area and the characteristics how the illumination changes from c in the umbra area to 0 in the lit area. Due to these combined properties, our method estimates the penumbra area location and the illumination change simultaneously by estimating the above piece-wise polynomial model.

Because we assume that the illumination change surface is smooth, neighboring illumination change models along the shadow boundary should be similar to each other. So we solve for all these models simultaneously instead of fitting each model separately. We formulate the problem of finding illumination change models as an optimization problem, aiming to balance the fitness of the models to the shadow image and the smoothness between neighboring models.

$$E = \sum_{li} E_{fit}(M_{li}, \tilde{I}) + \lambda \sum_{li} \sum_{lj \in N(li)} E_{sm}(M_{li}, M_{lj}) \tag{4}$$

where $E_{fit}(M_{li}, \tilde{I})$ measures the fitness error of the illumination change model M_{li} to the original shadow image $\tilde{I}$, $E_{sm}(M_{li}, M_{lj})$ measures the similarity between M_{li} and M_{lj}, and $N(li)$ denotes the neighborhood of sampling line li. λ is a parameter, with a default value 10.

We measure $E_{fit}(M_{li}, \tilde{I})$, the fitness error of the model M_{li} to the shadow image $\tilde{I}$, as how well the gradient in the penumbra area fits into its neighborhood along the sampling line after shadow effect compensation according to M_{li}.

$$E_{fit}(M_{li}, \tilde{I}) = -\Pi_{t \in [t_{i0}-r_i, t_{i0}+r_i]} \varphi(\hat{G}_{li}(t), T_{li}^{tex}) \tag{5}$$

$$\hat{G}_{li}(t) = \tilde{G}_{li}(t) - C'_{li}(t) \tag{6}$$

where C_{li} is the illumination change curve of M_{li} as defined in Equation 3, C'_{li} is its first derivative, $\tilde{G}_{li}$ is the gradient along li, and $\hat{G}_{li}(t)$ is the gradient after canceling the shadow effect. T_{li}^{tex} is the texture distribution along li. $\varphi(,)$ measures the fitness of the gradient to the distribution T_{li}^{tex}. We model the texture distribution along li as a normal distribution $N(\mu_i, \sigma_i^2)$ of the gradients, which can be estimated explicitly from the umbra and lit extension along li as illustrated in Fig. 3(b). Accordingly, we define the fitness measure as follows:

$$\varphi(G_{li}(t), T_{li}^{tex}) = \frac{exp(-(G_{li}(t) - \mu_i)^2 / 2\sigma_i^2)}{\sqrt{2\pi\sigma_i^2}} \tag{7}$$

We define $E_{sm}(M_{li}, M_{lj})$, the smoothness cost between neighboring illumination change models as follows:

$$E_{sm}(M_{li}, M_{lj}) = \gamma(c_i - c_j)^2 + (1 - \gamma)((t_{1i} - t_{1j})^2 + (t_{2i} - t_{2j})^2)$$

where the first term measures the difference between the illumination steps from the umbra to lit area, and the second term measures the difference between the location of the penumbra area along sampling lines. We emphasize the fact that the illumination change inside the umbra area is mostly uniform by weighting the first term significantly. The default value for γ is 0.9.

Directly solving the minimization problem in Equation 4 is time-consuming. We approximate the optimal solution in two steps:

1. For each sampling line li, we find an optimal illumination change model M_{li}^o which fits the shadow image most by minimizing the fitness error defined in Equation 5. Since the extent of the penumbra area is small, we use a brute-force search method.
2. With the optimal illumination change model M_{li}^o of each sampling line, we approximate the fitness error term in Equation 4 using the difference between the illumination change model M_{li} and M_{li}^o as follows:

$$E = \sum_{li} E_{sm}(M_{li}, M_{li}^o) + \lambda \sum_{li} \sum_{lj \in N(li)} E_{sm}(M_{li}, M_{lj})$$

The above energy minimization is a quadratic minimization problem. We solve it using a Preconditioned Conjugate Gradient method [14].

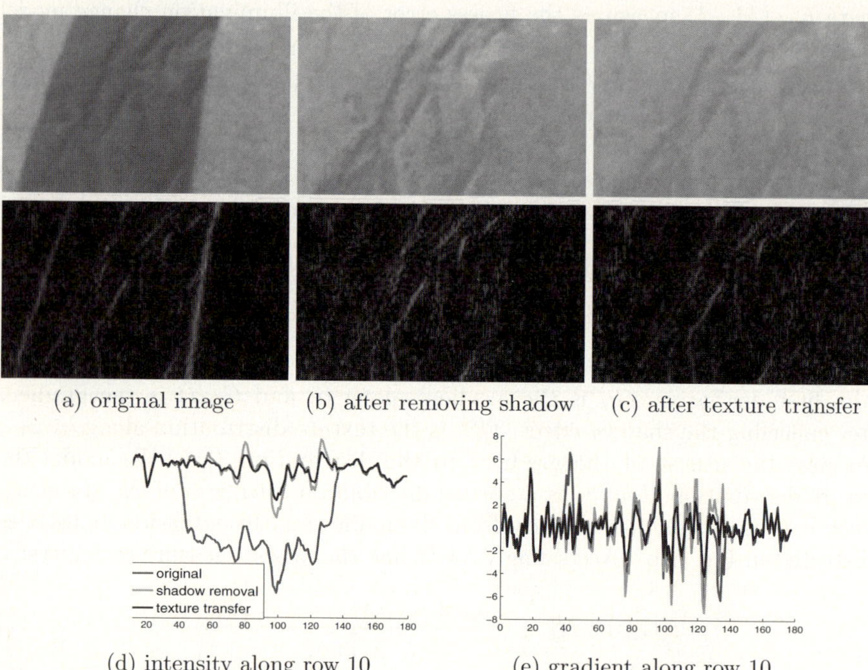

(a) original image   (b) after removing shadow   (c) after texture transfer

(d) intensity along row 10   (e) gradient along row 10

Fig. 4. Reconstruct the gradient field for shadow removal. (a) shows the original image and its gradient field along X direction. For the sake of illustration, we encode the negative and positive gradient values using the GREEN and RED channels respectively. From the original gradient field, we can see the shadow effect on the gradient field by noticing the strong edges along the shadow boundary. By estimating the illumination change across the penumbra area, the shadow effect on the gradient field is canceled as illustrated in (b) and (d). However, as we can see in (b) and (e) right, the shadow area is more contrasty than the lit area, causing inconsistent texture characteristics. This inconsistency is removed after gradient transformation as shown in (c) and (e).

After obtaining the illumination change model along each sampling line, we apply it to the gradient field to cancel the shadow effect according to Equation 6. An example of canceling the shadow effect on the gradients in the penumbra area is shown in Fig. 4(a) and (b).

2.2   Estimate Shadow Effect on Texture Characteristics

Canceling the shadow effect on the gradients in the penumbra area can effectively match the illumination in the shadow area (including penumbra and umbra area) to that in the lit area. However, as illustrated in Fig. 4(b) and (c), it cannot guarantee the texture consistency between the shadow and lit area since the shadow can also affect the texture characteristics in the whole shadow area (§ 1). Our method estimates the shadow effect on the gradient characteristics and transfers the shadow-effect free gradient characteristics to the shadow area to make it compatible with the lit area.

Like transferring color between images [15], where the global color characteristics of an image is parameterized using its sampling mean and deviation, we model the texture characteristics using the sampling mean and deviation of the gradient field. So if given the target mean and deviation, we transform the gradient field in the shadow area as follows:

$$G^s(x,y) = \hat{\mu}^t + \frac{(\hat{G}^s(x,y) - \hat{\mu}^s) * \hat{\sigma}^t}{\hat{\sigma}^s} \tag{8}$$

where $\hat{G}^s$ and G^s are the gradients in the shadow area before and after transformation respectively, and $\hat{\mu}^s$ and $\hat{\sigma}^s$ are the mean and deviation of $\hat{G}^s$. $\hat{\mu}^t$ and $\hat{\sigma}^t$ are the target mean and deviation.

Like transferring color [15], using the characteristics parameters of the lit area as the target parameters can achieve consistent texture characteristics between the shadow and lit area. However, this scheme works well only if the texture distribution is globally homogeneous in the image. Otherwise it can destroy local textures in the shadow area. We calculate the target characteristics parameters by estimating the shadow effect on the gradient distribution and canceling this effect from the original gradient field. Assuming the gradient distribution around the shadow boundary is homogenous and the shadow effect is independent of the shadow-free image, we estimate the shadow effect parameters from gradients around the boundary as follows:

$$\begin{cases} \mu_{se} = \mu_b^s - \mu_b^l \\ \sigma_{se}^2 = \sigma_b^{s\,2} - \sigma_b^{l\,2} \end{cases} \tag{9}$$

where μ_{se} and σ_{se} are the mean and deviation of the shadow effect on gradients in the shadow area. μ_b^s and σ_b^s are the mean and deviation of the gradients in the umbra side along the shadow boundary(the *extent* parts as illustrated in Fig. 3(b)) , and μ_b^l and σ_b^l are those in the lit area side. Accordingly, the target mean and deviation can be calculated by canceling the shadow effect as follows:

$$\begin{cases} \hat{\mu}^t = \hat{\mu}^s - \mu_{se} \\ \hat{\sigma}^t = \sqrt{\hat{\sigma}^{s\,2} - \sigma_{se}^2} \end{cases} \tag{10}$$

Fig. 4(b) and (c) shows that the gradient field transformation leads to consistent texture characteristics between the shadow and lit area. Please refer to the whole image in Fig. 6(a) to examine the consistency of the texture.

3   Results

We have experimented with our method on photos with shadows from *Flickr*. These photos have different texture characteristics. We report some representative ones together with the results in Fig. 1, Fig. 2, Fig. 6, Fig. 7 and Fig. 8, as well as comparison to many representative works [2,6,4,3,10]. (**Please refer to**

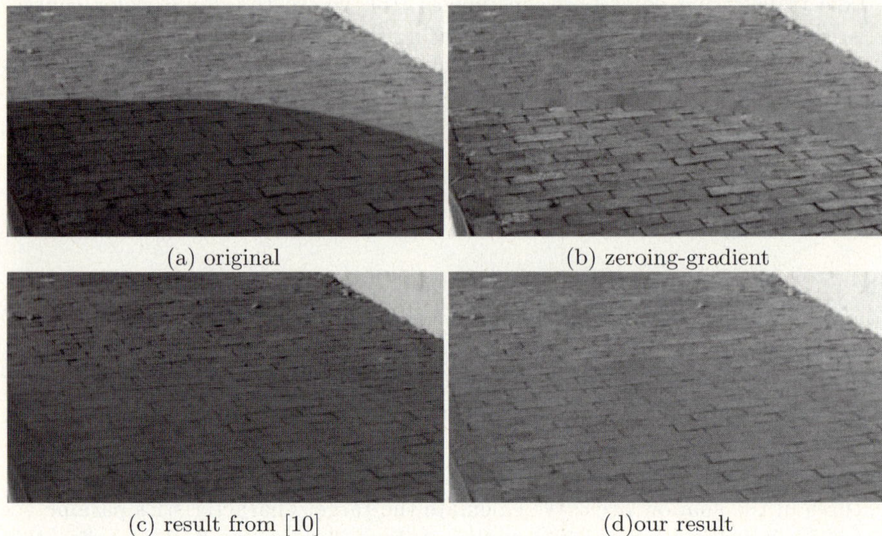

(a) original   (b) zeroing-gradient

(c) result from [10]   (d)our result

Fig. 5. Images in (a) and (c) are from [10]. (b) shadow removed by nullifying the gradients in the boundary [2,6]. (c) shadow removed using the method from [10]. There, not only the illuminance level in the lit area is changed, but also the shadow area is not as contrasty as the lit area. Our method creates a texture-consistent result.

the electronic version of this paper to examine the results. Zooming in on the images will be helpful for the examination.)

For all the experiments, users specify the shadow boundaries with a brush tool. Users do not need to delineate the boundary precisely as shown in Fig. 1(c) (notice the eagle's right wing). They can pick a brush with much larger size than the real shadow boundary area to cover the boundary as shown in the second column of Fig. 6. Given the user specified shadow boundary, our system can automatically perform shadow removal efficiently. The majority of the time is spent on solving the Poisson equation, whose complexity is dependent on the number of pixels in the shadow region. It takes about 3 seconds to remove a shadow region with about 60,000 colored pixels on a 2.2GHz Athlon machine.

Fig. 2 and Fig. 5 compare our method to other representative methods. Methods [2,6] cancel the shadow effect by zeroing the gradients in the boundary area. In this way, the textures there are nullified as shown in Fig. 2(c). While in-painting [4] can partially solve this problem, it sometimes destroys the continuity of the texture as shown in Fig. 2(e). The recent method from [3] can effectively remove shadow, however the texture in the original shadow area is not consistent with that in the lit area as shown in Fig. 2(d). Our method can not only remove the shadows, but also keep the texture consistency between the shadow and lit area as shown in Fig. 2(f). Fig. 5 compares our method to the recent work from [10]. While the illuminance between the lit and the original shadow area is balanced in the result from [10], the illuminance level in the lit area is changed. More overall, the lit and the original shadow area have different

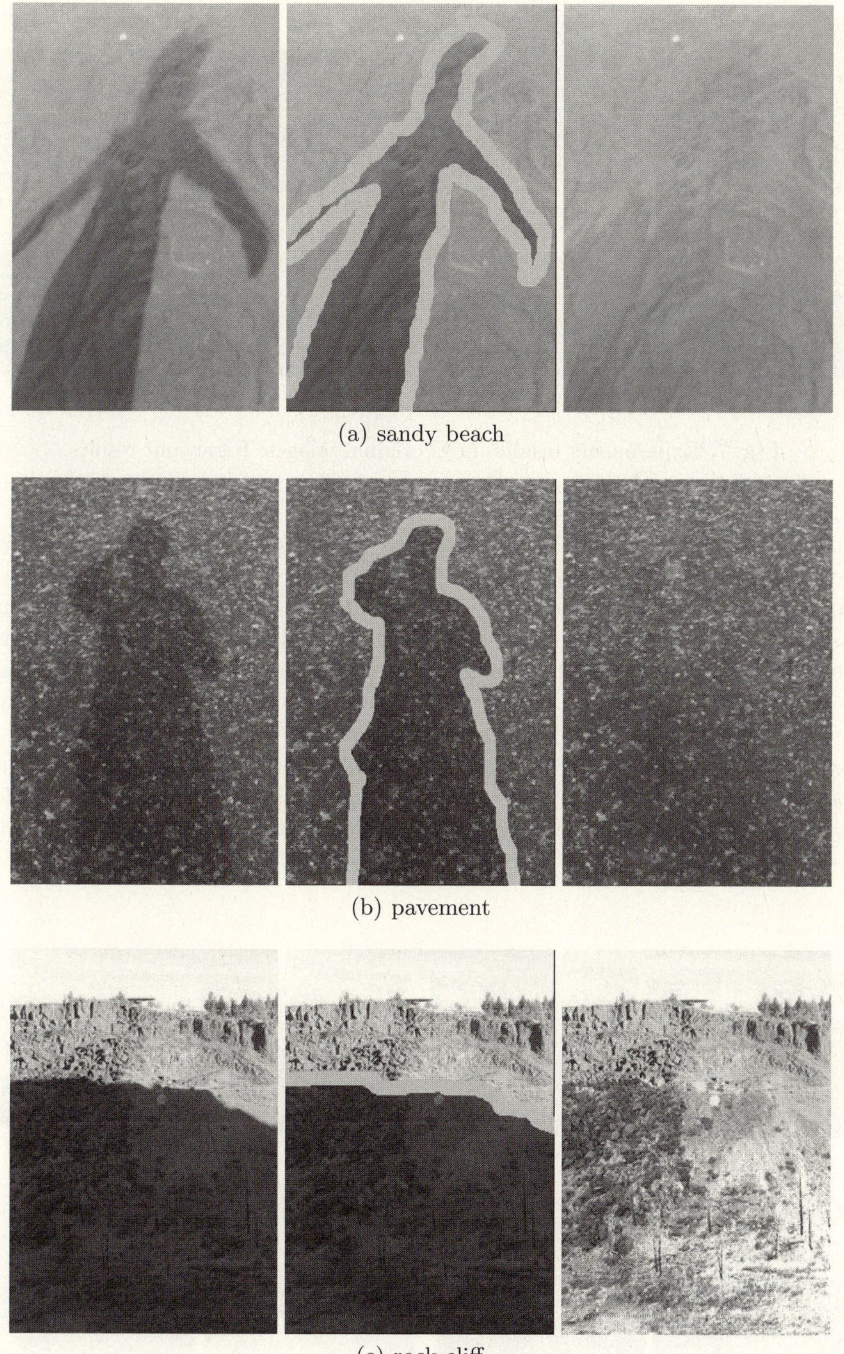

(a) sandy beach

(b) pavement

(c) rock cliff

Fig. 6. Experiments results. Left: original images; Middle: shadow boundaries; Right: our results.

(a) tree in hill   (b) sandy beach

(c) desert sand dawn break   (d) bridge over river

Fig. 7. Experiments results. Left: original images; Right: our results.

contrasty levels as shown in Fig. 5(c). Our method effectively removes the shadow as well as keeps the consistent texture characteristics across the whole image as shown in Fig. 5(d) and other examples. For instance, in the Fig. 7(b), the texture of small shell grains in the shadow area and in the lit area is consistent. For the desert example in Fig. 7(c), the highlights across the original shadow boundary are consistent between the shadow and lit area. For the river surface example in Fig. 7(d), the ripples in the shadow area are consistent with that in the lit area. Particularly, the wavefront in the middle is continuous across the original shadow boundaries. For the tree example in Fig. 7(a), the soil inside the shadow region is consistent with the lit area surrounding it. The hill example in Fig. 8(a) is similar.

(a) rock cliffs   (b) mountain above clouds

(c) volcano above clouds   (d) cast shadow of semi-transparent object

Fig. 8. Experiments results. Left: original images; Right: our results.

From the results in Fig. 6, 7 and 8, we can see that the proposed algorithm can seamlessly remove shadows in images with various texture characteristics. For example, the shadows are on the beach (Fig. 6(a)), on the road surfaces (Fig. 6(b)), on the sands (Fig. 7(b)), on the desert (Fig. 7(c)), on the river surface (Fig. 7(d)), on the hills (Fig. 7(a) and Fig. 8(a)), etc. Our method works well on specular surfaces such as Fig. 6(a), as well as Lambertian surfaces, such as examples in Fig. 7.

Examples in Fig. 8(b) and (c) are very interesting. Noticing the mountains in these examples, shadow removal reveals the beautiful texture details in the original dark shadow areas, which are concealed in the original shadow images. What is particularly interesting is that shadow removal recovers the *blue glacier ice* phenomenon[1] in the Fig. 8(b) (Notice the blue-cyan area of the snow in the left bottom.).

We found from the experiments that our method does not work well on some images. Taking Fig. 8(d) as an example, the shadow area in the original image looks more reddish than its surrounding lit area. This is because when the lighting is blocked by the semi-transparent red leaf, its red component can still pass through. For this kind of cast shadow, the general shadow model in Equation 2 used in previous work (including ours) does not hold. Noticing the original shadow region in the resulting image, we can still sense the reddish component there. In future, analyzing the caustics of shadow from its context may help solve this problem. However, our current method is effective for many images.

4   Conclusion

In this paper, we presented a texture-consistent shadow removal method. Specifically, we construct a shadow-effect free and texture-consistent gradient field between the shadow and lit area and recover the shadow-free image from it by solving a Poisson equation. The experiments on shadow images from *Flickr* demonstrate the effectiveness of the proposed method.

Currently, our method provides users with a brush tool to specify the shadow boundary. The brush tool is very popular in digital photography software. As illustrated in the examples in previous sections, our method does not require a precise shadow boundary. We envision our method a convenient tool for interactive photo editing. Of course, integrating an automatic shadow detection algorithm can make our method even easier to use.

We characterize texture characteristics using the sampling mean and deviation of the gradient field. Based on our current experiments on photos from *Flickr*, this global model works well. An important reason for its success is that a global transformation on an image or its various representations usually preserves important properties of the original image. In fact, similar models work pretty well in other applications like color transfer [15] as well.

---

[1] `http://www.northstar.k12.ak.us/schools/joy/denali/OConnor/`
   `colorblue.html`

Acknowledgements. We would like to thank reviewers for their constructive suggestions. The *Flickr* images are used under a Creative Commons license from *Flickr* users: etamil, viktoria_s, 82684220@N00, el_chupacabrito, magnusvk, 30201239@N00, erikogan, 24342028@N00, mishox, gandhu, hamedmasoumi and lexnger. This research was sponsored in part by NSF grant IIS-0416284.

References

1. Barrow, H., Tenenbaum, J.: Recovering intrinsic scene characteristics from images. In: Computer Vision Systems. Academic Press, London (1978)
2. Weiss, Y.: Deriving intrinsic images from image sequences. In: IEEE ICCV, pp. 68–75 (2001)
3. Arbel, E., Hel-Or, H.: Texture-preserving shadow removal in color images containing curved surfaces. In: IEEE CVPR (2007)
4. Finlayson, G.D., Hordley, S.D., Lu, C., Drew, M.S.: On the removal of shadows from images. IEEE Trans. Pattern Anal. Mach. Intell. 28(1), 59–68 (2006)
5. Liu, Z., Huang, K., Tan, T., Wang, L.: Cast shadow removal combining local and global features. In: The 7th International Workshop on Visual Surveillance (2007)
6. Finlayson, G.D., Hordley, S.D., Drew, M.S.: Removing shadows from images. In: 7th European Conference on Computer Vision, pp. 823–836 (2002)
7. Salvador, E., Cavallaro, A., Ebrahimi, T.: Cast shadow segmentation using invariant color features. Comput. Vis. Image Underst. 95(2), 238–259 (2004)
8. Levine, M.D., Bhattacharyya, J.: Removing shadows. Pattern Recognition Letters 26(3), 251–265 (2005)
9. Wu, T.P., Tang, C.K., Brown, M.S., Shum, H.Y.: Natural shadow matting. ACM Trans. Graph. 26(2), 8 (2007)
10. Mohan, A., Tumblin, J., Choudhury, P.: Editing soft shadows in a digital photograph. IEEE Comput. Graph. Appl. 27(2), 23–31 (2007)
11. Baba, M., Mukunoki, M., Asada, N.: Shadow removal from a real image based on shadow density. ACM SIGGRAPH 2004 Posters, 60 (2004)
12. Fredembach, C., Finlayson, G.D.: Hamiltonian path based shadow removal. In: BMVC, pp. 970–980 (2005)
13. Pérez, P., Gangnet, M., Blake, A.: Poisson image editing. ACM Trans. Graph. 22(3), 313–318 (2003)
14. Barrett, R., Berry, M., Chan, T.F., Demmel, J., Donato, J., Dongarra, J., Eijkhout, V., Pozo, R., Romine, C., der Vorst, H.V.: Templates for the Solution of Linear Systems: Building Blocks for Iterative Methods. SIAM, Philadelphia (1994)
15. Reinhard, E., Ashikhmin, M., Gooch, B., Shirley, P.: Color transfer between images. IEEE Comput. Graph. Appl. 21(5), 34–41 (2001)

large visual variability. This is due to the fact that these images depict objects that tend to appear together. The algorithm also generalizes well: when the clusters were transfered to the test set it still produced a good output (see figure 4).

Word prediction. Our approach to scene discovery is based on the internal representation of the word classifier, so these promising results suggest a good word annotation prediction performance. Table 1 shows the precision, recall and F1-measure of our word prediction model is competitive with the best state-of-the-art methods using this dataset. Changing the value of ϵ in equation 3 traces out the precision-recall curve; we show the equal error rate ($P = R$) result. It is remarkable that the kernelized classifier does not provide a substantial improvement over the linear classifier. The reason for this may lie in the high dimensionality of the feature space, in which all points are roughly at the same distance. In fact, using a standard RBF kernel produced significantly lower results; thus the sigmoid kernel, with a broarder support, performed much better. Because to this and the higher computational complexity of the kernelized classifier, we will use the linear classifier for the rest of the experiments.

The **influence of the tracenorm regularization** is clear when the results are compared to independent linear SVMs on the same features (that corresponds to using the Frobenius norm regularization, equation 2). The difference in performance indicates

Table 1. Comparison of the performance of our word annotation prediction method with that of Co-occurance model (Co-occ), Translation Model (Trans), Cross-Media Relevance Model (CMRM), Text space to image space (TSIS), Maximum Entropy model (MaxEnt), Continuous Relevance Model (CRM), 3×3 grid of color and texture moments (CT-3×3), Inference Network (InfNet), Multiple Bernoulli Relevance Models (MBRM), Mixture Hierarchies model (MixHier), PicSOM with global features, and linear independent SVMs on the same features. The performance of our model is provided for the linear and kernelized (sigmoid) classifiers.* *Note: the results of the PicSOM method are not directly comparable as they limit the annotation length to be at most five (we do not place this limit as we aim to complete the annotations for each image).*

Method	P	R	F1	Ref
Co-occ	0.03	0.02	0.02	[16]
Trans	0.06	0.04	0.05	[7]
CMRM	0.10	0.09	0.10	[9]
TSIS	0.10	0.09	0.10	[5]
MaxEnt	0.09	0.12	0.10	[10]
CRM	0.16	0.19	0.17	[11]
CT-3×3	0.18	0.21	0.19	[25]
CRM-rect	0.22	0.23	0.23	[8]
InfNet	0.17	0.24	0.23	[15]
Independent SVMs	0.22	0.25	0.23	
MBRM	0.24	0.25	0.25	[8]
MixHier	0.23	0.29	0.26	[4]
This work (Linear)	0.27	0.27	0.27	
This work (Kernel)	0.29	0.29	0.29	
PicSOM	0.35*	0.35*	0.35*	[23]

Fig. 5. Example **word completion results**. Correctly predicted words are below each image in blue, predicted words not in the annotations ("False Positives") are *italic red*, and words not predicted but annotated ("False Negatives") are in green. Missing annotations are not uncommon in the Corel dataset. Our algorithm performs scene clustering by predicting all the words that *should* be present on an image, as it learns correlated words (e. g. images with *sun* and *plane* usually contain *sky*, and images with *sand* and *water* commonly depict *beaches*). Completed word annotations are a good guide to scene categories while original annotations might not be; this indicates visual information really matters.

the sharing of features among the word classifiers is beneficial. This is specially true for words that are less common.

Annotation completion. The promising performance of the approach results from its generalization ability; this in turn lets the algorithm predict words that are not annotated in the training set but *should* have been. Figure 5 shows some examples of word completion results. It should be noted that performance evaluation in the Corel dataset is delicate, as missing words in the annotation are not uncommon.

Discriminative scene prediction. The Corel dataset is divided into sets (CDs) that do not necessarily depict different scenes. As it can be observed in figure 3, some correctly clustered scenes are spread among different CD labels (e. g. *sunsets, people*). In order to evaluate our unsupervised scene discovery, we selected a subset of 10 out of the 50 CDs from the dataset so that the CD number can be used as a reliable proxy for scene labels. The subset consists of CDs: 1 (sunsets), 21 (race cars), 34 (flying airplanes), 130 (african animals), 153 (swimming), 161 (egyptian ruins), 163 (birds and nests), 182 (trains), 276 (mountains and snow) and 384 (beaches). This subset has visually very disimlar pictures with the same labels and visually similar images (but depicting different objects) with different labels. The train/test split of [7] was preserved.

To evaluate the performance of the *unsupervised* scene discovery method, we label each cluster with the most common CD label in the **training** set and then evaluate the scene detection performance in the **test** set. We compare our results with the same clustering thechnique on the image features directly. In this space the cosine distance losses

Table 2. Comparison of the performance of our scene discovery on the latent space with another unsupervised method and four supervised methods on image features directly. Our model produced significantly better results that the unsupervised method on the image features, and is only surpassed by the supervised kernelized SVM. For both unsupervised methods, clustering is done on the train set and performance is measured on the test set (see text for details).

Method	Accuracy
Unsupervised Latent space (this work)	0.848
Unsupervised Image features clustering	0.697
Supervised Image features KNN	0.848
Supervised Image features SVM (linear)	0.798
Supervised Image features SVM (kernel)	0.948
Supervised "structural learning" [2,18]	0.818

its meaning and thus we use the euclidean distance. We also computed the performance of two *supervised* approaches on the image features: k nearest neighbors (KNN), support vector machines (SVM), and "structural learning" (introduced in [2] and used in a vision application -Reuters image classification- in [18]). We use a one-vs-all approach for the SVMs. Table 2 show the the latent space is indeed a suitable space for scene detection: it clearly outperforms clustering on the original space, and only the supervised SVM using a kernel provides an improvement over the performance of our method.

The difference with [18] deserves further exploration. Their algorithm classifies *topics* (in our case scenes) by first learning a classification of *auxiliary tasks* (in this case words), based in the framework introduced in [2]. [18] starts by building *independent*

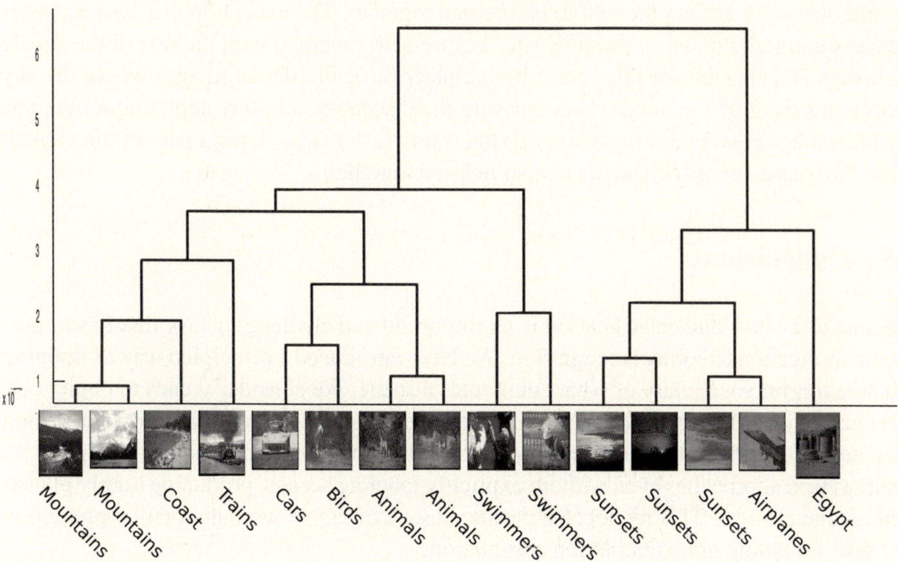

Fig. 6. Dendrogram for our clustering method. Our scene discovery model produces 1.5 *protoscenes* per scene. Clusters belonging to the same scene are among the first to be merged

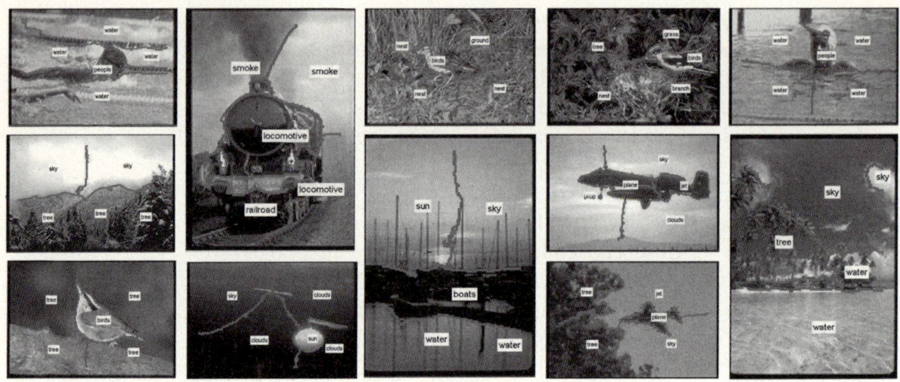

Fig. 7. Future work includes unsupervised **region annotation**. Example images show promising results for region labeling. Images are presegmented using normalized cuts (red lines), features are computed in each region and fed to our classifier as if they were whole image features.

SVM classifiers on the auxiliary tasks/words. As we showed in table 1, this leads to lower performance in word classification when compared to our *correlated* classifiers. On top of this [18] runs an SVD to correlate the output of the classifiers. It is remarkable that our algorithm provides a slight performance advantage despite the fact [18] is supervised and learns the topic classifier directly, whereas our formulation is unsupervised and does not use topic labels.

Figure 4 depicts a dendrogram of the complete-link clustering method applied to the clusters found by our scene discovery algorithm. As expected clusters belonging to the same scene are among the first to be merged together. The exception is a *sunset* cluster that is merged with an *airplane* cluster before being merged with the rest of the *sunset* clusters. The reason for this is that both cluster basically depict images where the sky occupies most of the image. Is is pleasing that "scenery" clusters depicting mountains and beaches are merged together with the train cluster (also depicts panoramic views); the *birds* and *animals* clusters are also merged together.

5   Conclusions

Scene discovery and classification is an important and challenging task that has important applications in object recognition. We have introduced a principled way of defining a meaningful vocabulary of what constitutes a scene. We consider scenes to depict correlated objects and present visual similarity. We introduced a max-margin factorization model to learn these correlations. The algorithm allows for scene discovery on par with supervised approaches even without explicitly labeling scenes, producing highly plausible scene clusters. This model also produced state of the art word annotation prediction results including good annotation completion.

Future work will include using our classifier for weakly supervised region annotation/labeling. For a given image, we use normalized cuts to produce a segmentation.

Using our classifier, we know what words describe the image. We then *restrict* our classifier to these word subsets and to the features in each of the regions. Figure 7 depicts examples of such annotations. These are promising preliminary results; since quantitative evaluation of this procedure requires having a ground truth labels for each segment, we only show qualitative results.

Acknowledgements

The authors would like to thank David Forsyth for helpful discussions.

This work was supported in part by the National Science Foundation under IIS - 0534837 and in part by the Office of Naval Research under N00014-01-1-0890 as part of the MURI program. Any opinions, findings and conclusions or recommendations expressed in this material are those of the author(s) and do not necessarily reflect those of the National Science Foundation or the Office of Naval Research.

References

1. Amit, Y., Fink, M., Srebro, N., Ullman, S.: Uncovering shared structures in multiclass classification. In: ICML, pp. 17–24 (2007)
2. Ando, R.K., Zhang, T.: A high-performance semi-supervised learning method for text chunking. In: ACL (2005)
3. Bosch, A., Zisserman, A., Munoz, X.: Scene classification via plsa. In: Leonardis, A., Bischof, H., Pinz, A. (eds.) ECCV 2006. LNCS, vol. 3954, pp. 517–530. Springer, Heidelberg (2006)
4. Carneiro, G., Vasconcelos, N.: Formulating semantic image annotation as a supervised learning problem. In: CVPR, vol. 2, pp. 163–168 (2005)
5. Celebi, E., Alpkocak, A.: Combining textual and visual clusters for semantic image retrieval and auto-annotation. In: 2nd European Workshop on the Integration of Knowledge, Semantics and Digital Media Technology, 30 November - 1 December 2005, pp. 219–225 (2005)
6. Chapelle, O., Haffner, P., Vapnik, V.: SVMs for histogram-based image classification. IEEE Transactions on Neural Networks, special issue on Support Vectors (1999)
7. Duygulu, P., Barnard, K., de Freitas, J.F.G., Forsyth, D.A.: Object recognition as machine translation: Learning a lexicon for a fixed image vocabulary. In: Heyden, A., Sparr, G., Nielsen, M., Johansen, P. (eds.) ECCV 2002. LNCS, vol. 2353, pp. 97–112. Springer, Heidelberg (2002)
8. Feng, S.L., Manmatha, R., Lavrenko, V.: Multiple bernoulli relevance models for image and video annotation. In: CVPR, vol. 02, pp. 1002–1009 (2004)
9. Jeon, J., Lavrenko, V., Manmatha, R.: Automatic image annotation and retrieval using cross-media relevance models. In: SIGIR, pp. 119–126 (2003)
10. Jeon, J., Manmatha, R.: Using maximum entropy for automatic image annotation. In: Enser, P.G.B., Kompatsiaris, Y., O'Connor, N.E., Smeaton, A.F., Smeulders, A.W.M. (eds.) CIVR 2004. LNCS, vol. 3115, pp. 24–32. Springer, Heidelberg (2004)
11. Lavrenko, V., Manmatha, R., Jeon, J.: A model for learning the semantics of pictures. In: NIPS (2003)
12. Lazebnik, S., Schmid, C., Ponce, J.: Beyond bags of features: Spatial pyramid matching for recognizing natural scene categories. In: CVPR, pp. 2169–2178 (2006)
13. Li, F.-F., Perona, P.: A bayesian hierarchical model for learning natural scene categories. In: CVPR, vol. 2, pp. 524–531 (2005)

14. Liu, J., Shah, M.: Scene modeling using co-clustering. In: ICCV (2007)
15. Metzler, D., Manmatha, R.: An inference network approach to image retrieval. In: Enser, P.G.B., Kompatsiaris, Y., O'Connor, N.E., Smeaton, A.F., Smeulders, A.W.M. (eds.) CIVR 2004. LNCS, vol. 3115, pp. 42–50. Springer, Heidelberg (2004)
16. Mori, Y., Takahashi, H., Oka, R.: Image-to-word transformation based on dividing and vector quantizing images with words. In: Proc. of the First International Workshop on Multimedia Intelligent Storage and Retrieval Management (1999)
17. Oliva, A., Torralba, A.B.: Modeling the shape of the scene: A holistic representation of the spatial envelope. International Journal of Computer Vision 42(3), 145–175 (2001)
18. Quattoni, A., Collins, M., Darrell, T.: Learning visual representations using images with captions. In: CVPR (2007)
19. Quelhas, P., Odobez, J.-M.: Natural scene image modeling using color and texture visterms. Technical report, IDIAP (2006)
20. Rabinovich, A., Vedaldi, A., Galleguillos, C., Wiewiora, E., Belongie, S.: Objects in context. In: ICCV (2007)
21. Rennie, J.D.M., Srebro, N.: Fast maximum margin matrix factorization for collaborative prediction. In: ICML, pp. 713–719 (2005)
22. van Gemert, J.C., Geusebroek, J.-M., Veenman, C.J., Snoek, C.G.M., Smeulders, A.W.M.: Robust scene categorization by learning image statistics in context. In: CVPRW Workshop (2006)
23. Viitaniemi, V., Laaksonen, J.: Evaluating the performance in automatic image annotation: Example case by adaptive fusion of global image features. Image Commun. 22(6), 557–568 (2007)
24. Vogel, J., Schiele, B.: Natural scene retrieval based on a semantic modeling step. In: CIVR, pp. 207–215 (2004)
25. Yavlinsky, A., Schofield, E., Rger, S.: Automated image annotation using global features and robust nonparametric density estimation. In: Leow, W.-K., Lew, M., Chua, T.-S., Ma, W.-Y., Chaisorn, L., Bakker, E.M. (eds.) CIVR 2005. LNCS, vol. 3568, pp. 507–517. Springer, Heidelberg (2005)

observe that the results obtained with our RH and an OAR approach (see results obtained by Griffin et al. [10]) are comparable.

As to be expected, our approach does not depend on the image representation. Best results on Caltech-256 dataset in a similar setup (53% average accuracy for 10 training images) where achieved by Varma [26] using a combination of multiple channels. Our method could be combined with this multi-representation approach. Note that it could even be applied to different data types, but this is beyond the scope of this paper. In the following we use the image representation described in Sect. 4.1 as it is fast to compute and does not impact the evaluation of our class hierarchy construction.

Figure 6 compares the complexity in the number of categories. The complexity in the OAR setup is linear (red squares). The complexity of our Relaxed Hierarchy method is confirmed to be sublinear. The exact gain depends on the parameter α, see the datapoints along the right edge. Note that α is expressed here as r—the number of relaxed training samples per class, i.e., $\alpha = r/15$. For 250 categories and a setting of $\alpha = 3/15 = 0.2$ (blue diamonds) which corresponds to minor performance loss, we observe a reduction of the computation time by $1/3$. This ratio will further increase with the number of categories.

Figure 7 demonstrates the speed-for-accuracy trade-off (green circles) that can be tuned with the α parameter. As shown in Sect. 3, with the increase of the parameter value the set of classes is more willingly treated as separable. Greater α values lead to better computational gain, but could degrade the classification accuracy. Note that the complexity is sublinear independently of the parameter setting (see Fig. 6), but for the smaller number of classes one may choose to accept a small loss in accuracy for a significant gain in computation time. For instance, for Caltech-256 we find the setting of $\alpha = 0.2$ ($r = 3$) reasonable, as the absolute loss in the accuracy is only about 2%, while the computational gain

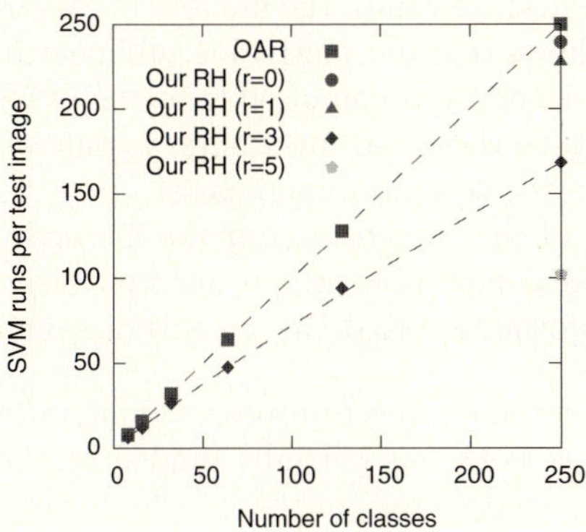

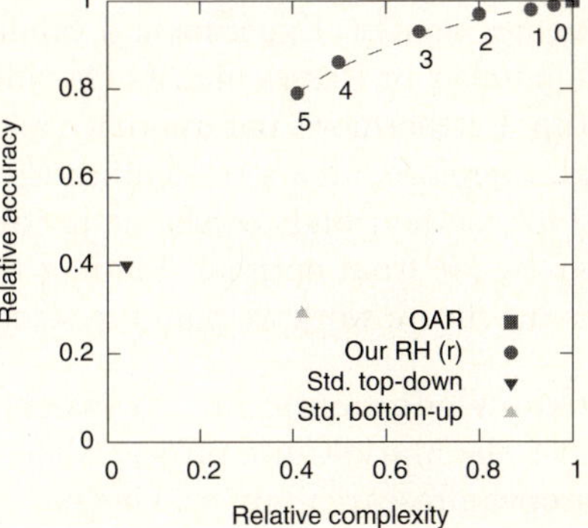

Fig. 6. Complexity in the number of classes. The α relaxation parameter expressed in the number of per-class training samples r (i.e., $\alpha = r/15$) is given in parenthesis for our method.

Fig. 7. Speed-for-accuracy trade-off and comparison with existing approaches. Next to the RH datapoints the α relaxation parameter expressed in the number of samples r is shown ($\alpha = r/15$).

of 1/3 is noticeable. Setting $\alpha = 0.33$ ($r = 5$) leads to the computational gain of 3/5, but in exchange for another 2% of accuracy.

Figure 7 compares the results obtained with our class hierarchies for different α values (green circles) to two existing methods for class hierarchy construction (triangles). The baseline top-down method follows the approach of Liu et al. [12,14], but we use normalized cuts instead of k-means. This makes it more comparable to our method and is also similar to the approach of Chen et al. [11]. The baseline bottom-up method follows the agglomerative clustering based approach of Zhigang et al. [13], but uses the same inter-class similarity measure as the top-down approach. Note that the only difference between the compared methods is the algorithm used for class hierarchy construction, i.e., we keep the same image representation and settings of the Support Vector Machines. Still, the magnitude of the difference is surprising. The standard bottom-up method seems to fail completely. The standard top-down approach has low computational complexity, but the loss in terms of classification accuracy is enormous. This confirms our claim, see Subsect. 2.3, that popular *disjoint* approaches for construction of class hierarchies fail when dealing with a large number of visual object categories. Note that the cited methods were evaluated on visual data and performed well. However, the number of categories never exceeded 14 classes and was usually kept below 10.

5 Summary

We have shown that existing approaches for constructing class hierarchies for visual recognition do not scale well with the number of categories. Methods that perform disjoint class-set partitioning assume good class separability and thus fail to achieve good performance on visual data when the number of categories becomes large. Thus, we have proposed a method that detects classes at the partitioning boundary and postpones uncertain decisions until the number of classes becomes smaller. Experimental validation shows that our method is sublinear in the number of classes and its classification accuracy is comparable to the OAR setup. Furthermore, our approach allows to tune the speed-for-accuracy trade-off and, therefore, allows to significantly reduce the computational costs.

Our method finds a reliable partitioning of the categories, but the hierarchy may be far from optimal. Finding the optimal partitioning is a hard problem. For the future work we plan use semantic information to drive the optimization.

Acknowledgments. M. Marszałek is supported by the European Community under the Marie-Curie project VISITOR. This work was partially funded by the European research project CLASS.

References

1. Everingham, M., van Gool, L., Williams, C., Winn, J., Zisserman, A.: Overview and results of classification challenge. In: The PASCAL VOC 2007 Challenge Workshop, in conj. with ICCV (2007)

2. Mikolajczyk, K., Schmid, C.: Scale and affine invariant interest point detectors. IJCV (2004)
3. Lindeberg, T.: Feature detection with automatic scale selection. IJCV (1998)
4. Lowe, D.: Distinctive image features form scale-invariant keypoints. IJCV (2004)
5. Willamowski, J., Arregui, D., Csurka, G., Dance, C.R., Fan, L.: Categorizing nine visual classes using local appearance descriptors. In: IWLAVS (2004)
6. Schölkopf, B., Smola, A.: Learning with Kernels: Support Vector Machines, Regularization, Optimization and Beyond (2002)
7. Lazebnik, S., Schmid, C., Ponce, J.: Beyond bags of features: Spatial pyramid matching for recognizing natural scene categories. In: CVPR (2006)
8. Zhang, J., Marszalek, M., Lazebnik, S., Schmid, C.: Local features and kernels for classification of texture and object categories: A comprehensive study. IJCV (2007)
9. Fei-Fei, L., Fergus, R., Perona, P.: One-shot learning of object categories. PAMI (2007)
10. Griffin, G., Holub, A., Perona, P.: Caltech-256 object category dataset. Technical report (2007)
11. Chen, Y., Crawford, M., Ghosh, J.: Integrating support vector machines in a hierarchical output space decomposition framework. In: IGARSS (2004)
12. Liu, S., Yi, H., Chia, L.T., Deepu, R.: Adaptive hierarchical multi-class SVM classifier for texture-based image classification. In: ICME (2005)
13. Zhigang, L., Wenzhong, S., Qianqing, Q., Xiaowen, L., Donghui, X.: Hierarchical support vector machines. In: IGARSS (2005)
14. Yuan, X., Lai, W., Mei, T., Hua, X., Wu, X., Li, S.: Automatic video genre categorization using hierarchical SVM. In: ICIP (2006)
15. Zweig, A., Weinshall, D.: Exploiting object hierarchy: Combining models from different category levels. In: ICCV (2007)
16. He, X., Zemel, R.: Latent topic random fields: Learning using a taxonomy of labels. In: CVPR (2008)
17. Griffin, G., Perona, P.: Learning and using taxonomies for fast visual category recognition. In: CVPR (2008)
18. Nister, D., Stewenius, H.: Scalable recognition with a vocabulary tree. In: CVPR (2006)
19. Philbin, J., Chum, O., Isard, M., Sivic, J., Zisserman, A.: Object retrieval with large vocabularies and fast spatial matching. In: CVPR (2007)
20. Casasent, D., Wang, Y.C.: A hierarchical classifier using new support vector machines for automatic target recognition. Neural Networks (2005)
21. Shi, J., Malik, J.: Normalized cuts and image segmentation. PAMI (2000)
22. Rahimi, A., Recht, B.: Clustering with normalized cuts is clustering with a hyperplane. In: SLCV (2004)
23. Fowlkes, C., Belongie, S., Chung, F., Malik, J.: Spectral grouping using the Nyström method. PAMI (2004)
24. Everingham, M., Zisserman, A., Williams, C., van Gool, L.: The PASCAL visual object classes challenge 2006 (VOC 2006) results. Technical report (2006)
25. Marszałek, M., Schmid, C.: Semantic hierarchies for visual object recognition. In: CVPR (2007)
26. Perona, P., Griffin, G., Spain, M.: The Caltech 256 Workshop. In: Conj. with ICCV (2007)

Sample Sufficiency and PCA Dimension for Statistical Shape Models

Lin Mei*, Michael Figl, Ara Darzi, Daniel Rueckert, and Philip Edwards*

Dept. of Biosurgery and Surgical Technology Imperial College London, UK
{l.mei,eddie.edwards}@imperial.ac.uk

Abstract. Statistical shape modelling(SSM) is a popular technique in computer vision applications, where the variation of shape of a given structure is modelled by principal component analysis (PCA) on a set of training samples. The issue of sample size sufficiency is not generally considered. In this paper, we propose a framework to investigate the sources of SSM inaccuracy. Based on this framework, we propose a procedure to determine sample size sufficiency by testing whether the training data stabilises the SSM. Also, the number of principal modes to retain (PCA dimension) is usually chosen using rules that aim to cover a percentage of the total variance or to limit the residual to a threshold. However, an ideal rule should retain modes that correspond to real structural variation and discard those that are dominated by noise. We show that these commonly used rules are not reliable, and we propose a new rule that uses bootstrap stability analysis on mode directions to determine the PCA dimension.

For validation we use synthetic 3D face datasets generated using a known number of structural modes with added noise. A 4-way ANOVA is applied for the model reconstruction accuracy on sample size, shape vector dimension, PCA dimension, and the noise level. It shows that there is no universal sample size guideline for SSM, nor is there a simple relationship to the shape vector dimension (with p-Value=0.2932). Validation of our rule for retaining structural modes showed it detected the correct number of modes to retain where the conventional methods failed. The methods were also tested on real 2D (22 points) and 3D (500 points) face data, retaining 24 and 70 modes with sample sufficiency being reached at approximately 50 and 150 samples respectively. We provide a foundation for appropriate selection of PCA dimension and determination of sample size sufficiency in statistical shape modelling.

1   Introduction

Statistical shape modelling (SSM) is a technique for analysing variation of shape and generating or inferring unseen shapes. A set of sample shapes is collected and PCA is performed to determine the principal modes of shape variation. These modes can be optimised to fit the model to a new individual, which is the familiar active shape model (ASM) [1,2,3]. Further information, such as texture, can be included to create an active appearance model [4] or morphable model [5].

---

* We would like to thank Tyco Healthcare for funding Lin Mei's PhD studentship. We are also grateful to many other members of the Department of Computing and the Department of Biosurgery and Surgical Technology at Imperial College.

D. Forsyth, P. Torr, and A. Zisserman (Eds.): ECCV 2008, Part IV, LNCS 5305, pp. 492–503, 2008.

Despite its popularity, PCA-based SSMs are normally trained from datasets for which the issue of sufficiency is not considered. The PCA dimension for an SSM is often chosen by rules that assume either a given percentage or level of noise. As will be shown later in this paper that these two methods are highly dependent on sample size.

In this paper, we review the discussions on sample size sufficiency for a closely related field, common factor analysis (CFA), and design a mathematical framework to investigate the source of PCA model error. This framework provides a theoretical evaluation of the conventional rules for retaining PCA modes, and enables analysis of sample size sufficiency for PCA. We then propose a rule for retaining only stable PCA modes that uses a t-test between the bootstrap stability of mode directions from the training data and those from pure Gaussian noise. The convergence of the PCA dimension can then be used as an indication of sample sufficiency.

We verify our framework by a 4-way ANOVA for reconstruction accuracy is applied to the models trained from synthetic datasets generated under different conditions. Our PCA dimension rule and procedure for sample sufficiency determination are validated on the synthetic datasets and demonstrated on real data.

2   Background

2.1   Minimum Sample Size for CFA

There is little literature on the issue of minimum sample size for PCA. In the related field of CFA, however, this issue has been thoroughly discussed. CFA is commonly used to test or discover common variation shared by different test datasets. Guidelines for minimum sample size in CFA involve either a universal size regardless of the data dimension or a ratio to the data dimension. Recommendations for minimum size neglecting the sample dimension and the number of expected factors vary from 100 to 500 [6]. Such rules are not supported by tests on real data. Doubts have been raised about a universal sample size guideline since it neglects the data dimension. Size-variable ratios (SVR) may be more appropriate and values of between 2:1 to 20:1 have been suggested [7]. There have been a number of tests using real data, but no correlation was found between SVR and the mode stability [8], nor has any minimum value for SVR emerged [9]. The minimum sample size needed in these real tests is not consistent either, varying from 50 [8], to 78-100 [9], 144 [10], 400 [11] and 500 or more [12].

The inconsistency among these results shows that the minimum size depends on some nature of the data other than its dimension. MacCallum et al. [13,14] proposed a mathematical framework for relating the minimum sample size for CFA with its communality and overdetermination level. They then designed an experiment using 4-way ANOVA to study the effects of communality, overdetermination level, model error and sample size on the accuracy in recovering the genuine factors from synthetic data. The results showed that communality had the dominant effect on the accuracy regardless of the model error. The effect of overdetermination level was almost negligible when communality is high. In low communality tests, accuracy improves with larger sample size and higher accuracy was found in tests with lower overdetermination levels.

There is no equivalent to communality and overdetermination level for PCA. Instead, the factors we consider are the data dimension and the number of genuine structural modes that are retained.

2.2 Number of Modes to Retain for SSM

Many rules choosing the number of modes to retain for SSM and PCA have been proposed [6,15,16,17]. The most popular rule used in SSM for structural mode determination is simply to take the leading modes covering a percentage of the total variance in the sample set. The percentage is arbitrarily set, which equivalent to simply assuming a corresponding percentage of noise.

Another popular rule is to discard the least principal modes until the sum of total variance, which is the model residual on the training data, reaches a certain threshold. This threshold is normally set according to the error tolerance of the application.

Stability measurements for PCA have been proposed to determine the number of modes. Given two shape models trained from different sample sets, Daudin et al [18] used a sum of correlation coefficients between pairs of principal components; Besse et al [19] used a loss function derived from an Euclidean distance between orthogonal projectors; Babalola et al [20] used the Bhattacharya Metric to measure the similarity of PCA models from different sample sets. Resampling techniques such as boot-strapping [18] and jackknifing [19] can be used. The distribution of PCA modes across the replicates reflects their distribution in the population, allowing stability analysis to be performed. The selected principal modes span a subspace. Besse et al. proposed a framework for choosing the number of modes based on their spanned-space stability [21]. This method differentiates structural modes and noise-dominated modes when the sample set is large. However, as will be shown in the section 4.3, this method can only provide a estimation of the number of modes when the sample size is sufficient.

3 Theories

3.1 Sources of PCA Model Inaccuracy

We propose the following mathematical framework to examine the characteristics affecting the sufficiency of a sample set drawn from a population with genuine modes of variation, listed in the column of A. Due to the presence of noise, we have $\widehat{X}$ instead of X, and the PCA modes from $\widehat{X}$ are $\widehat{A}$. The model inaccuracy can be expressed as the difference between the covariance matrices $\Delta = \widehat{X}\widehat{X}^T - XX^T$.

Let $X = AW$ and $\widehat{X} = \widehat{AW}$, we have:

$$\widehat{X} = \widehat{AW} = AA^T\widehat{AW} + (I - AA^T)\widehat{AW} \tag{1}$$

Since A is orthonormal, $(I - AA^T)$ is a diagonal matrix with only 1s and 0s. Hence $(I - AA^T) = NN^T$. Equation 1 becomes:

$$\widehat{X} = AA^T\widehat{AW} + NN^T\widehat{AW} = A\widehat{W}_A + N\widehat{W}_N \tag{2}$$

Applying equation 2 on the covariance matrix of $\boldsymbol{X}$:

$$\widehat{X}\widehat{X}^T = A\widehat{W_A}\widehat{W_A^T}A^T + A\widehat{W_A}\widehat{W_N^T}N^T + N\widehat{W_N}\widehat{W_A^T}A^T + N\widehat{W_N}\widehat{W_N^T}N^T$$
$$= A\widehat{\Sigma_{AA}}A^T + A\Sigma_{AN}N^T + N\Sigma_{NA}A^T + N\Sigma_{NN}N^T \qquad (3)$$

The model inaccuracy becomes:

$$\Delta = \widehat{X}\widehat{X}^T - XX^T$$
$$= A(\widehat{\Sigma_{AA}} - \Sigma_{AA})A^T + A\Sigma_{AN}N^T + N\Sigma_{NA}A^T + N\Sigma_{NN}N^T$$
$$= (A(\Sigma_{EE})A^T + A\Sigma_{AN}N^T + N\Sigma_{NA}A^T) + N\Sigma_{NN}N^T \qquad (4)$$

A PCA model error consists of two parts:

$E_N = N\Sigma_{NN}N^T$, the error introduced by sampling noise modes that are orthogonal to $\mathbb{A}$. E_N depends only on the noise level introduced by human interaction or measurement error during the process of building an SSM. Increasing sample size would cause little reduction in this error if noise level remains the same.

$E_l = A(\Sigma_{EE})A^T + A\Sigma_{AN}N^T + N\Sigma_{NA}A^T$, the error along the subspace spanned by structural modes, $\mathbb{A}$. This is due to noise affecting the sample coefficients and insufficient coverage of the dimensions in $\mathbb{A}$. Therefore, E_l increases with PCA dimension, $rank(A)$. It also affected by noise level because at high noise level some structural modes with small variances may be swamped by noise.

Rather counter-intuitively, E_l is not dependent on the shape vector dimension, as will be shown in the section 4.2. However, higher E_N can result from higher shape vector dimension, which therefore increases Δ.

3.2   Sample Size Requirement

According to the framework in section 3.1, the sample size requirement for PCA only depends on two factors: number of structural modes in the dataset, and the level of noise. Hence we propose the following procedure for sample sufficiency determination. For a sample set, X, of n samples:

PCA Sample Size Sufficiency Test
1) Apply PCA on X, to get a set of modes B.
2) Starting with a reasonably small number, n^*, construct a set X_j^* of n^* samples randomly drawn, allowing repeats, from X.
3) Apply PCA to X_j^* to get a set of modes B_j^* and resolve mode correspondence with respect to B.
4) Find the number of structural modes in B_j^*, k.
5) Repeat 2-4 with an increased n^*. If k converges before n^* reaches n, we have sufficient samples. Otherwise, further sample data is required.

Step 4 in this procedure requires determination of the number of structural modes, which is a common question for PCA. These rules are sometimes is called *stopping rules*, since they determine where to stop including the principal modes.

3.3 PCA Dimension by the Stability of Mode Direction

It is generally assumed that most of the structural variation is modelled by the leading principal modes. Noise affects the lower ranking modes and dominates those after a cut-off point further down the rank. Going back to our framework, a cut-off point is chosen for retaining principal modes in order to reduce E_N. However, since genuine structural variation may still be present in the least principal modes, discarding them would increase E_l. There is trade-off between E_N and E_l. Stopping rules should aim at discarding only modes that are sufficiently dominated by noise.

Assuming noise is randomly spread across all the dimensions, mode instability can be a good indication of the point where noise begins to dominate. There is a risk with tests using the magnitude of the variance that stopping rules will be dominated by the first few modes and fail to identify the correct cut-off point. Also, it is the mode directions that define the basis of a shape model for fitting or synthetic shape generation. Therefore we propose a stopping rule based on the stability of the mode direction only.

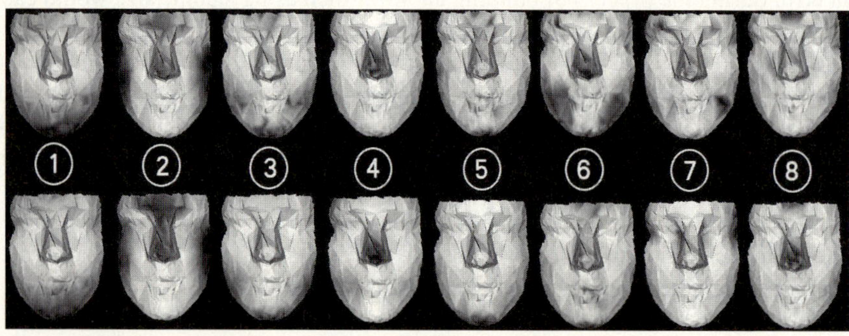

Fig. 1. Comparison of Leading 8 Eigenmodes from two mutually exclusive sets of 50 samples from our 3D face mesh database, aligned according to eigenvalue ranks. Darker texture implies larger variation, showing many mismatched after the 4th mode.

3.3.1 Establishing Mode Correspondence

Examining individual modes requires mode correspondence. Normally, this is done by matching those with the same eigenvalue ranks. Significant variation can be found between individual modes drawn from different sample sets with the same ranking, as shown in figure 1. Although leading modes may correspond, mode 5 on the top seems to correspond with mode 6 on the bottom, and modes after 6 on the top seems to correspond to none at the bottom. However, the combined modes from different sample sets may still span similar subspaces. Mode alignment can be achieved by minimising the distance between these subspaces.

For the leading PCA modes $\{(a_i, \lambda_i) \,|\, |a_i| = 1\}$ of an n-dimensional distribution, we define the principal spanned space (PSS) as the subspace $\mathbb{S}^k$ spanned by $\{a_i\}$, where the distance measure used by Besse et al.[19] can be applied:

$$d(\mathbb{A}^k, \mathbb{B}^k) = k - trace(AA^T BB^T) \tag{5}$$

where the columns of A and B are the modes spanning PSS $\mathbb{A}^k$ and $\mathbb{B}^k$.

For two sets of PCA modes, a_i and b_i, trained from different sample sets of a common distribution, the following rule can be used to establish correspondence. The first mode in a_i corresponds to the mode of a replicate that minimises $d(\$_a{}^1, \$_b{}^1)$, and we proceed iteratively. Assume we have already aligned $\$_a{}^k$, the PSS from the first k modes in a_i, to the spanned space $\$_b{}^k$ from k modes in the replicate b_i. The mode in b_i that corresponds to the $k+1^{\text{th}}$ mode in a_i will be the one that minimises $d(\$_a{}^{k+1}, \$_b{}^{k+1})$.

3.3.2 Bootstrap Stability of PCA Modes

Bootstrap stability analysis can be used to analyse mode stability. We use the angles between mode directions as the measurement of distance between corresponding modes from different replicates. The instability, ξ, of mode a_i is given by:

$$\xi(a_i) = \frac{\sum_{j=1}^{m} arccos(a_{i'_j} \cdot \widehat{\alpha_i})\cdot}{m\pi} \qquad (6)$$

where $\widehat{\alpha_i}$ is the mean mode vector and m is the number of bootstrap replicates.

3.3.3 Stopping Rule Based on a *t*-Test against Synthetic Gaussian Noise

Since noise-dominated modes should have higher instability than structural modes, a threshold on ξ can be used to differentiate them from structural modes. However, the choice for the threshold is arbitrary and is found to be sensitive to the size of replicates. Instead, assuming the distribution of angles between corresponding modes is Gaussian, a one-tailed *t*-test can be used to establish whether a mode is dominated by noise to a given significance level.

We generate a pure Gaussian noise dataset to compare with the test dataset. All conditions must be the same – the dimensionality, the number of samples in the dataset, the number of replicates, and the number of samples in each replicate. Since we are only interested in mode directions, the level of noise is not important. Let the angle for the first pure noise mode to be α_1 and the angle for the test samples to be a_i, The null hypothesis of the *t*-test is $H_0 : \xi(\alpha_1) > \xi(a_i)$. By rejecting H_0 at a given confidence level, one can safely conclude that a mode is not dominated by noise.

4 Experiments

We demonstrate the correctness of our theories with three sets of experiments. First a 4-way ANOVA is performed on synthetic datasets to show how the PCA model accuracy is affected by different features as it is discussed in section 3.1. Then we show that our stopping rule is able to identify the correct number of modes in the synthetic samples for which commonly used rules fail. This shows that our rule can be used to determine PCA sample sufficiency, by following the procedure presented in section 3.2. This is applied to two different sets of real samples.

4.1 Real Datasets

Two real shape datasets are used in the experiments. The first one comprises 150 samples of 3D faces with 5090 points each from University of Notre Dame [22] preprocessed using Papatheodorou's method [23], and the second one consists of 135 samples

498   L. Mei et al.

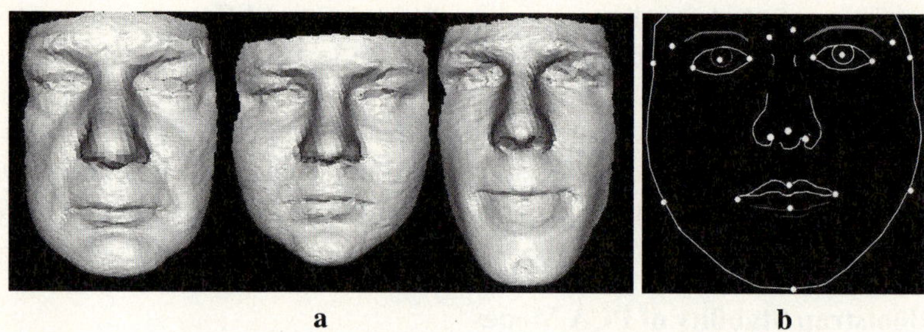

a   b

Fig. 2. Examples from real 3D Face database (a) and landmarks of 2D AR Face database

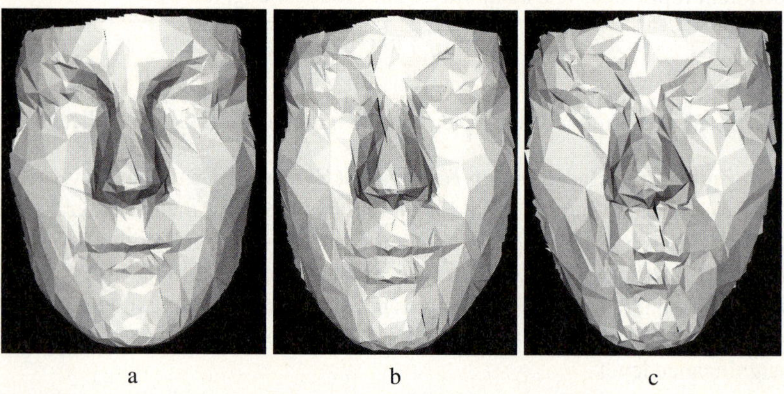

a   b   c

Fig. 3. Examples of three synthetic faces generated with 70 modes with shape vector dimension being 2100. Different noise levels are applied: 0.1mm (a), 0.25mm (b) and 0.5mm (c). Noise starts to become visible in (b) and (c).

from the landmarks (22 points) [24] of 2D AR face database [25]. Examples from these two datasets are shown in figure 2.

4.2   ANOVA Results

For validation of our framework, we generate a dataset consists of 8960 subsets, each having different combinations of: sample sizes, numbers of modes to generate, levels of Gaussian noise and decimated to different number of points. A list of choices for different characteristics are shown as follows

Sample Sizes (SS): 50, 100, 150, 200, 250, 300, 350, 400
450, 500, 550, 600, 650, 700, 750, 800
Shape Vector Dimension: 300, 600, 900, 1200, 1500, 1800, 2100
Number of Genuine Modes: 10, 20, 30, 40, 50, 60, 70, 80
Gaussian Noise Levels (in mm): 0.05, 0.1, 0.15, 0.2, 0.25, 0.3, 0.35, 0.4, 0.45, 0.5
Examples of faces generated under different conditions are given in figure 3.

PCA is applied to each of the 8960 subsets of the first synthetic dataset. Results are compared to the original modes that used to generate the data. Measurement described in equation 5 is used to calculate the error of the models trained from the subsets. A 4-way ANOVA was performed to find out which characteristics influence the model accuracy. As shown in table 1, the results confirm the correctness of our framework introduced in section 3.1. Sample size and number of genuine modes in the dataset act as the major source of influence on the model accuracy. Noise also has a significant but small influence. Also the result showed that the effect of sample dimension is negligible.

Table 1. Result of 4-Way ANOVA

Source	Sum of Squares	DoF	Mean Squares	F-Statistic	p-Value
Sample Size	194.534	15	12.9689	11315.66	¡0.03
Sample Dimension	0.008	6	0.0014	1.22	0.2932
Number of Genuine Modes	83.055	7	11.865	10352.48	¡0.03
Gaussian Noise Level	0.513	9	0.057	49.74	¡0.03

4.3 Number of Modes to Retain for SSM

We have validated previous stopping rules and our method using synthetic data generated with a known number of background modes. These shapes are generated using the leading 80 modes of the model built from all the 150 3D Faces, decimated to 500 points for faster experiments. Gaussian noise with 1mm standard deviation is added to each element of the shape vector. Example faces from the synthetic set are shown in figure 4. Stopping rules applied to this dataset should not retain more than 80 modes.

We validated the rule which retains 95% of the cumulative variance using synthetic datasets sized from 100 to 600. Compactness plots are shown in figure 5(a). With increasing sample size, the number of modes retained by this rule increases beyond 80, where the noise dominates the variance. These noise modes contribute to an increasing proportion of the total variance with increasing sample size, and the number of modes covering 95% of the total variance increases accordingly. A similar trend was also found

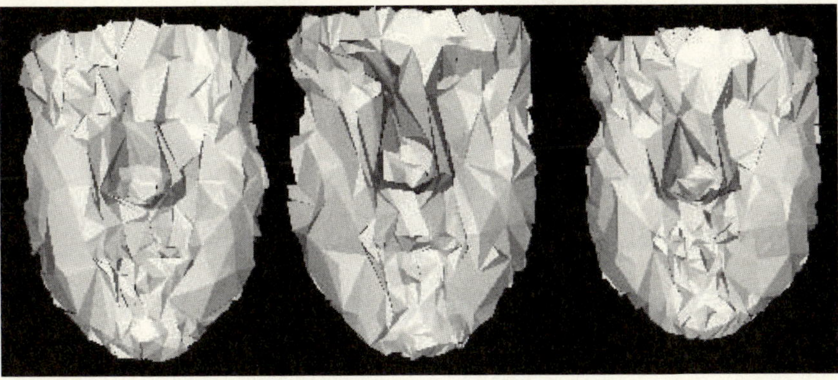

Fig. 4. Synthetic faces generated using 80 modes, added 1mm Gaussian noise on to each element of the shape vector with dimension 1500

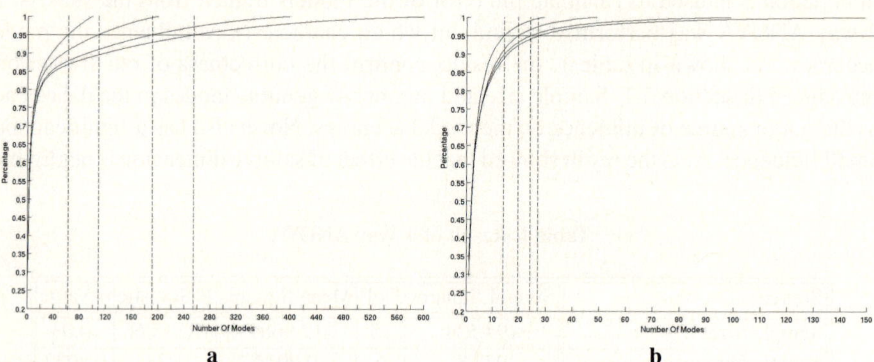

a   b

Fig. 5. 95% thresholded compactness plots of synthetic 3D face datasets (a) with 100, 200, 400 and 600 samples and real 3D face datasets (b) with 30, 50, 100 and 150 samples. The number of retained modes is clearly dependent on sample size.

Table 2. Number of modes to keep the point error below 1mm

Number of Samples	50	100	150	200	250	300	350	400	450	500
Number of Modes	32	60	95	108	120	140	169	186	204	219

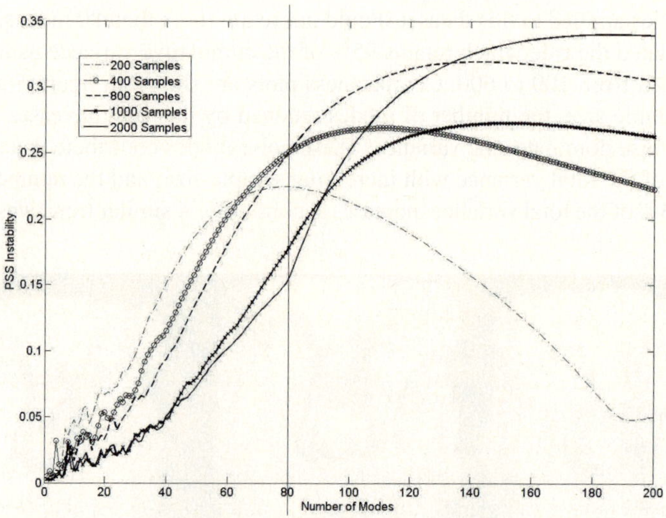

Fig. 6. Instability of PSS for synthetic datasets for synthetic datasets sized from 200 to 2000

for the real data as shown in figure 5(b), which strongly suggests that this rule is unreliable and should not be used. A similar effect, as shown in table 2, was found for the stopping rule that discards the least principal modes until the average error of each point reaches 1mm.

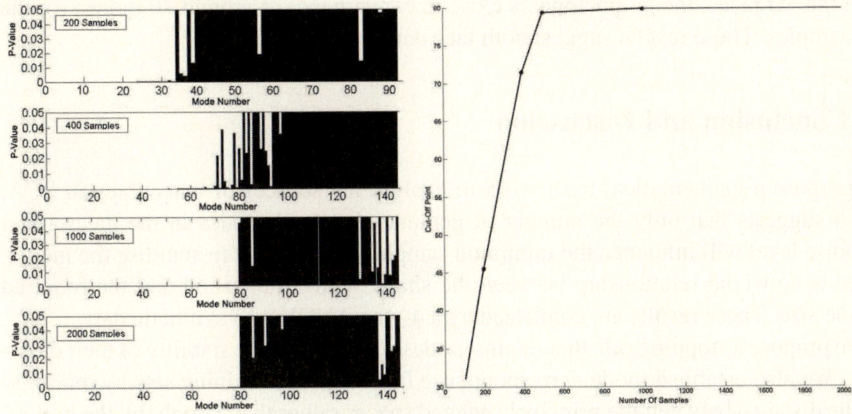

Fig. 7. *t*-Test Based stopping rule on synthetic datasets

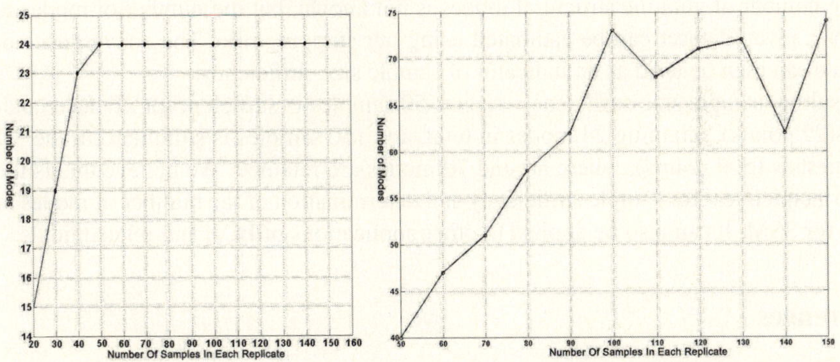

Fig. 8. Result of real datasets sufficiency test. Left: 2D faces; Right: 3D faces.

The method of Besse et al [21] was validated with synthetic datasets sized from 200 to 400. A plot of instability, measured as the distance between subspaces spanned by different replicates, is shown in figure 6. Although this method provides a visible indication of the correct number of modes to retain when the sample size is sufficiently large, it cannot identify the lower number of modes that should be retained when the sample size is insufficient.

Our method was validated with synthetic datasets sized from 100 to 2000. Figure 7 shows the number of modes to retain versus the sample size is also shown. Our stopping rule does not have the tendency to go beyond 80 with large sample sizes. It also identifies a lower number of stable modes to retain for smaller sample sizes. It appears a sample size of around 500 is sufficient.

4.4   Sample Size Sufficiency Test for SSM

Figure 8 shows the results of sample size sufficiency tests on the three real datasets we have. For the 2D dataset, the plot obviously converges at 24 modes with 50 samples.

With the 3D faces, the graph appears close to convergence at around 70 modes for the 150 samples. These results suggest both face datasets are sufficient.

5   Conclusion and Discussion

We propose a mathematical framework to analyse the sources of inaccuracy in PCA, which suggests that only the number of genuine structural modes in the dataset and the noise level will influence the minimum sample size required to stabilise the model. There is no trivial relationship between the shape vector dimension and the required sample size. These results are confirmed by a 4-way ANOVA on synthetic data.

We propose a stopping rule that retains modes according to the stability of their directions. We also establish mode correspondence from different training sets by minimising the distance between the principal spanned spaces rather than simply by the rank of their eigenvalues. For a synthetic dataset generated with known structural modes plus added noise, our method converges correctly where conventional methods did not.

The number of genuine structural modes is not known, but the number of modes to use for a given dataset can be estimated using our stopping rule. The convergence of this rule can then be used as an indicator of sample size sufficiency.

Resulting sample size sufficiency suggest 50 samples is sufficient for 2D face landmarks(22 points), retaining 24 modes in total, and 150 samples is sufficient for the 3D face meshes (500 points), where around 70 modes are retained. We believe this is the first principled test for sample sufficiency and determination of the number of modes to retain for SSM. It can also be applied to other applications of PCA and related fields.

References

1. Cootes, F., Hill, A., Taylor, C., Haslam, J.: The use of active shape models for locating structures in medical images. In: Proc. IPMI, pp. 33–47 (1993)
2. Cootes, T., Taylor, C., Cooper, D., Graham, J.: Active shape models and their training and application. Comput. Vis. Image Underst. 61(1), 38–59 (1995)
3. Sukno, F.M., Ordas, S., Butakoff, C., Cruz, S.: Active shape models with invariant optimal features: Application to facial analysis. IEEE Trans. Pattern Anal. Mach. Intell. 29(7), 1105–1117 (2007) (Senior Member-Alejandro F. Frangi)
4. Cootes, T., Edwards, G., Taylor, C.: Active appearance models. IEEE Transactions on Pattern Analysis and Machine Intelligence 23(6), 681–685 (2001)
5. Blanz, V., Vetter, T.: Face recognition based on fitting a 3D morphable model. IEEE Transactions On Pattern Analysis And Machine Intelligence 25, 1063–1074 (2003)
6. Osborne, J., Costello, A.: Sample size and subject to item ratio in principal components analysis. Practical Assessment, Research and Evaluation 9(11) (2004)
7. Guadagnoli, E., Velicer, W.: Relation of sample size to the stability of component patterns. Psychological Bulletin 103, 265–275 (1988)
8. Barrett, P., Kline, P.: The observation to variable ratio in factor analysis. Personality Study and Group Behavior 1, 23–33 (1981)
9. Arrindell, W., van der Ende, J.: An empirical test of the utility of the observations-to-variables ratio in factor and components analysis. Applied Psychological Measurement 9(2), 165–178 (1985)

10. Velicer, W., Peacock, A., Jackson, D.: A comparison of component and factor patterns: A monte carlo approach. Multivariate Behavioral Research 17(3), 371–388 (1982)
11. Aleamoni, L.: Effects of size of sample on eigenvalues, observed communalities, and factor loadings. Journal of Applied Psychology 58(2), 266–269 (1973)
12. Comfrey, A., Lee, H.: A First Course in Factor Analysis. Lawrence Erlbaum, Hillsdale (1992)
13. MacCallum, R., Widaman, K., Zhang, S., Hong, S.: Sample size in factor analysis. Psychological Methods 4, 84–99 (1999)
14. MacCallum, R., Widaman, K., Hong, K.P.S.: Sample size in factor analysis: The role of model error. Multivariate Behavioral Research 36, 611–637 (2001)
15. Jackson, D.: Stopping rules in principal components analysis: a comparison of heuristical and statistical approaches. Ecology 74, 2204–2214 (1993)
16. Jolliffe, I.: Principal Component Analysis, 2nd edn. Springer, Heidelberg (2002)
17. Sinha, A., Buchanan, B.: Assessing the stability of principal components using regression. Psychometrika 60(3), 355–369 (2006)
18. Daudin, J., Duby, C., Trecourt, P.: Stability of principal component analysis studied by the bootstrap method. Statistics 19, 341–358 (1988)
19. Besse, P.: PCA stability and choice of dimensionality. Statistics& Probability 13, 405–410 (1992)
20. Babalola, K., Cootes, T., Patenaude, B., Rao, A., Jenkinson, M.: Comparing the similarity of statistical shape models using the bhattacharya metric. In: Larsen, R., Nielsen, M., Sporring, J. (eds.) MICCAI 2006. LNCS, vol. 4190, pp. 142–150. Springer, Heidelberg (2006)
21. Besse, P., de Falguerolles, A.: Application of resampling methods to the choice of dimension in PCA. In: Hardle, W., Simar, L. (eds.) Computer Intensive Methods in Statistics, pp. 167–176. Physica-Verlag, Heidelberg (1993)
22. University of Notre Dame Computer Vision Research Laboratory: Biometrics database distribution (2007), `http://www.nd.edu/~cvrl/UNDBiometricsDatabase.html`
23. Papatheodorou, T.: 3D Face Recognition Using Rigid and Non-Rigid Surface Registration. PhD thesis, VIP Group, Department of Computing, Imperial College, London University (2006)
24. Cootes, T.: The AR face database 22 point markup (N/A),
    `http://www.isbe.man.ac.uk/~bim/data/tarfd_markup/`
    `tarfd_markup.html`
25. Martinez, A., Benavente, R.: The AR face database (2007),
    `http://cobweb.ecn.purdue.edu/~aleix/aleix_face_DB.html`

Locating Facial Features with an Extended Active Shape Model

Stephen Milborrow and Fred Nicolls

Department of Electrical Engineering
University of Cape Town, South Africa
`www.milbo.users.sonic.net`

Abstract. We make some simple extensions to the Active Shape Model of Cootes et al. [4], and use it to locate features in frontal views of upright faces. We show on independent test data that with the extensions the Active Shape Model compares favorably with more sophisticated methods. The extensions are (i) fitting more landmarks than are actually needed (ii) selectively using two- instead of one-dimensional landmark templates (iii) adding noise to the training set (iv) relaxing the shape model where advantageous (v) trimming covariance matrices by setting most entries to zero, and (vi) stacking two Active Shape Models in series.

1 Introduction

Automatic and accurate location of facial features is difficult. The variety of human faces, expressions, facial hair, glasses, poses, and lighting contribute to the complexity of the problem.

This paper focuses on the specific application of locating features in unobstructed frontal views of upright faces. We make some extensions to the Active Shape Model (ASM) of Cootes et al. [4] and show that it can perform well in this application.

2 Active Shape Models

This section describes Active Shape Models [8].

A *landmark* represents a distinguishable point present in most of the images under consideration, for example, the location of the left eye pupil (Fig. 1). We locate facial features by locating landmarks.

A set of landmarks forms a shape. Shapes are represented as vectors: all the x- followed by all the y-coordinates of the points in the shape. We align one shape to another with a similarity transform (allowing translation, scaling, and rotation) that minimizes the average euclidean distance between shape points. The *mean shape* is the mean of the aligned training shapes (which in our case are manually landmarked faces).

The ASM starts the search for landmarks from the mean shape aligned to the position and size of the face determined by a global face detector. It then

D. Forsyth, P. Torr, and A. Zisserman (Eds.): ECCV 2008, Part IV, LNCS 5305, pp. 504–513, 2008.

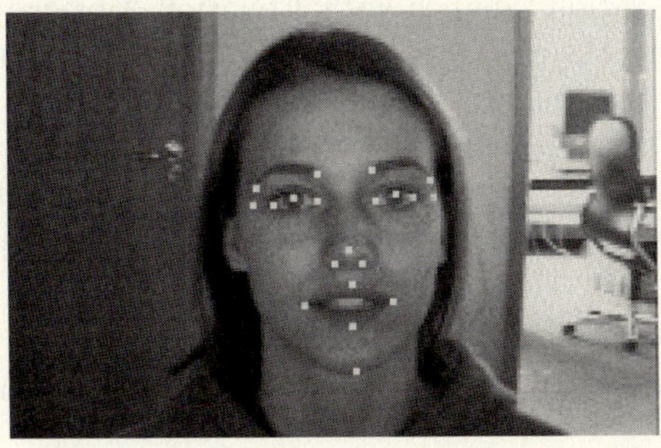

Fig. 1. A face with correctly positioned landmarks. This image is from the BioID set [15].

repeats the following two steps until convergence (i) suggest a tentative shape by adjusting the locations of shape points by template matching of the image texture around each point (ii) conform the tentative shape to a global shape model. The individual template matches are unreliable and the shape model pools the results of the weak template matchers to form a stronger overall classifier. The entire search is repeated at each level in an image pyramid, from coarse to fine resolution.

It follows that two types of submodel make up the ASM: the *profile model* and the *shape model.*

The profile models (one for each landmark at each pyramid level) are used to locate the approximate position of each landmark by template matching. Any template matcher can be used, but the classical ASM forms a fixed-length normalized gradient vector (called the *profile*) by sampling the image along a line (called the *whisker*) orthogonal to the shape boundary at the landmark. During training on manually landmarked faces, at each landmark we calculate the mean profile vector $\bar{\mathbf{g}}$ and the profile covariance matrix $\mathbf{S_g}$. During searching, we displace the landmark along the whisker to the pixel whose profile $\mathbf{g}$ has lowest Mahalanobis distance from the mean profile $\bar{\mathbf{g}}$:

$$MahalanobisDistance = (\mathbf{g} - \bar{\mathbf{g}})^{\mathbf{T}}\mathbf{S_g^{-1}}(\mathbf{g} - \bar{\mathbf{g}}). \tag{1}$$

The shape model specifies allowable constellations of landmarks. It generates a shape $\hat{\mathbf{x}}$ with

$$\hat{\mathbf{x}} = \bar{\mathbf{x}} + \mathbf{\Phi b} \tag{2}$$

where $\bar{\mathbf{x}}$ is the mean shape, $\mathbf{b}$ is a parameter vector, and $\mathbf{\Phi}$ is a matrix of selected eigenvectors of the covariance matrix $\mathbf{S_s}$ of the points of the aligned training shapes. Using a standard principal components approach, we model as much variation in the training set as we want by ordering the eigenvalues λ_i of $\mathbf{S}_s$ and keeping an appropriate number of the corresponding eigenvectors in $\mathbf{\Phi}$. We use a single shape model for the entire ASM but scale it for each pyramid level.

We can generate various shapes with Equation 2 by varying the vector parameter $\mathbf{b}$. By keeping the elements of $\mathbf{b}$ within limits (determined during model building) we ensure that generated face shapes are lifelike.

Conversely, given a suggested shape $\mathbf{x}$, we can calculate the parameter $\mathbf{b}$ that allows Equation 2 to best approximate $\mathbf{x}$ with a model shape $\hat{\mathbf{x}}$. Cootes and Taylor [8] describe an iterative algorithm that gives the $\mathbf{b}$ and $\mathbf{T}$ that minimizes

$$distance(\mathbf{x},\ \mathbf{T}(\bar{\mathbf{x}} + \mathbf{\Phi}\mathbf{b})) \tag{3}$$

where $\mathbf{T}$ is a similarity transform that maps the model space into the image space.

3   Related Work

Active Shape Models belong to the class of models which after a shape is situated near an image feature interact with the image to warp the shape to the feature. They are deformable models likes snakes [16], but unlike snakes they use an explicit shape model to place global constraints on the generated shape. ASMs were first presented by Cootes et al. [3]. Cootes and his colleagues followed with a succession of papers cumulating in the classical ASM described above [8] [4].

Many modifications to the classical ASM have been proposed. We mention just a few. Cootes and Taylor [6] employ a shape model which is a mixture of multivariate gaussians, rather than assuming that the shapes come from the single gaussian distribution implicit in the shape model of the classical ASM. Romdhani et al. [22] use Kernel Principal Components Analysis [23] and a Support Vector Machine. Their software trains on 2D images, but models non-linear changes to face shapes as they are rotated in 3D. Rogers and Graham [21] robustify ASMs by applying robust least-squares techniques to minimize the residuals between the model shape and the suggested shape. Van Ginneken et al. [12] take the tack of replacing the 1D normalized first derivative profiles of the classical ASM with local texture descriptors calculated from "locally orderless images" [17]. Their method automatically selects the optimum set of descriptors. They also replace the classical ASM profile model search (using Mahalanobis distances) with a k-nearest-neighbors classifier. Zhou et al. [25] estimate shape and pose parameters using Bayesian inference after projecting the shapes into a tangent space. Li and Ito [24] build texture models with AdaBoosted histogram classifiers. The Active Appearance Model [5] merges the shape and profile model of the ASM into a single model of appearance, and itself has many descendants. Cootes et al. [7] report that landmark localization accuracy is better on the whole for ASMs than AAMs, although this may have changed with subsequent developments to the AAM.

4   Extensions to the ASM

We now look at some extensions to the classical ASM. Figure 3 (Sec. 5.1) shows the increase in performance for each of these extensions.

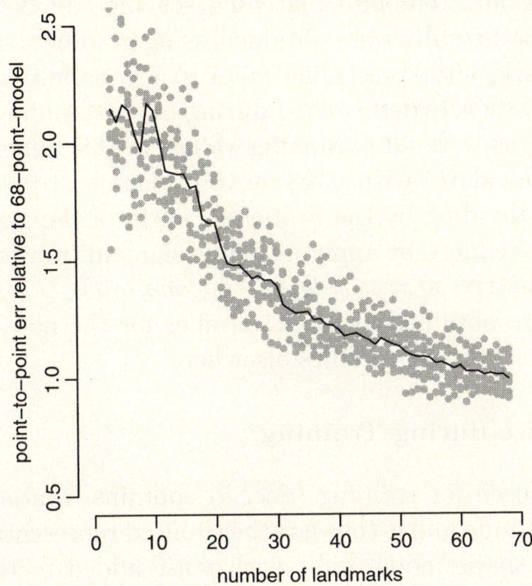

Fig. 2. Mean error versus number of landmarks

4.1   Number of Landmarks

A straightforward way to improve the mean fit is to increase the number of landmarks in the model (Fig. 2). Fitting a landmark tends to help fitting other landmarks, so results are improved by fitting more landmarks than are actually needed. Search time increases roughly linearly with the number of landmarks.

Fig. 2 was constructed as follows from the XM2VTS [19] set of manually landmarked faces. For a given number (from 3 to 68) of landmarks, that number of landmarks was chosen randomly from the 68 in the XM2VTS test. With the chosen landmarks, a model was built and tested to give one gray dot. This was repeated ten times for each number of landmarks. The black line shows the mean error for each number of landmarks.

4.2   Two Dimensional Profiles

The classical ASM uses a one-dimensional profile at each landmark, but using two-dimensional "profiles" can give improved fits. Instead of sampling a one-dimensional line of pixels along the whisker, we sample a square region around the landmark. Intuitively, a 2D profile area captures more information around the landmark and this information if used wisely should give better results.

During search we displace the sampling region in both the "x" and "y" directions, where x is orthogonal to the shape edge at the landmark and y is tangent to the shape edge. We must rely on the face being approximately upright because 2D profiles are aligned to the edges of the image. The profile covariance matrix $\mathbf{S_g}$ of a set of 2D profiles is formed by treating each 2D profile matrix as a long vector (by appending the rows end to end), and calculating the covariance of the vectors.

Any two dimensional template matching scheme can be used, but the authors found that good results were obtained using gradients over a 13x13 square around the landmark, after prescaling faces to a constant width of 180 pixels. The values 13 and 180 were determined during model building by measurements on a validation set, as were all parameter values in this paper (Sec. 5).

Gradients were calculated with a 3x3 convolution mask $((0,0,0),(0,-2,1),(0,1,0))$ and normalized by dividing by the Frobenius norm of the gradient matrix. The effect of outliers was reduced by applying a mild sigmoid transform to the elements x_i of the gradient matrix: $x_i' = x_i/(abs(x_i) + constant)$.

Good results were obtained using 2D profiles for the nose and eyes and surrounding landmarks, with 1D profiles elsewhere.

4.3   Adding Noise during Training

The XM2VTS set used for training (Sec. 5) contains frontal images of mostly caucasian working adults and is thus a rather limited representation of the variety of human faces. A shape model built with noise added to the training shapes helps the trained model generalize to a wider variety of faces. Good results can be obtained with the following techniques:

1. Add gaussian noise with a standard deviation of 0.75 pixels to the x- and y-positions of each training shape landmark. In effect, this increases variability in the training set face shapes.
2. Randomly choose the left or the right side each face. Generate a stretching factor ϵ for each face from a gaussian distribution with a standard deviation of 0.08. Stretch or contract the chosen side of the face by multiplying the x position (relative to the face center) of each landmark on that side by $1 + \epsilon$. This is roughly equivalent to rotating the face slightly.

4.4   Loosening Up the Shape Model

In Equation 2, the constraints on the generated face shape are determined by the number of eigenvectors n_{eigs} in Φ and the maximum allowed values of elements in the parameter vector $\mathbf{b}$. When conforming the shape suggested by the profile models to the shape model, we clip each element b_i of $\mathbf{b}$ to $b_{max}\sqrt{\lambda_i}$ where λ_i is the corresponding eigenvalue The parameters n_{eigs} and b_{max} are global constants determined during model building by parameter selection on a validation set. See [8] for details.

The profile models are most unreliable when starting the search (for example, a jaw landmark can snag on the collar), but become more reliable as the search progresses. We can take advantage of this increase in reliability with two modifications to the standard ASM procedure described above. The first modification sets n_{eigs} and b_{max} for the final pyramid level (at the original image scale) to larger values. The second sets n_{eigs} and b_{max} for the final iteration at each pyramid level to larger values. In both cases the landmarks at that stage of the search tend to be already positioned fairly accurately, for the given pyramid level. It is therefore less likely that the profile match at any landmark is grossly mispositioned, allowing the shape constraints to be weakened.

These modifications are effective for 2D but not for 1D profiles. The 1D profile matches are not reliable enough to allow the shape constraints to be weakened.

4.5   Trimming the Profile Covariance Matrices

For 2D profiles, calculation of the Mahalanobis distances dominates the overall search time. We can reduce this time (with little or no effect on landmark location accuracy) by "trimming" the covariance matrix.

The covariance between two pixels in a profile tends to be much higher for pixels that are closer together. This means that we can ignore covariances for pixels that are more than 3 pixels apart, or equivalently clear them to 0. Clearing elements of a covariance matrix may result in a matrix that is no longer positive definite (which is necessary for a meaningful Mahalanobis distance calculation in Equation 1). We therefore adjust the trimmed matrix to a "nearby" positive definite matrix. This can be done by iterating the following procedure a few times: perform a spectral decomposition of the trimmed covariance matrix $A = Q\Lambda Q^T$, set zero or negative eigenvalues in Λ to a small positive number, reconstruct the matrix from the modified Λ, and re-trim. A suitable "small positive number" is $iter\_nbr \times abs(min(eig\_vals(A)))$. More rigorous ways of forcing positive definiteness are presented in Gentle [11] and in Bates and Maechler [1].

Trimming the covariance matrices in conjunction with a sparse matrix multiplication routine roughly halves the overall search time.

4.6   Stacking Models

Accurate positioning of the start shape is crucial — it is unlikely that an ASM search will recover completely from a bad start shape. One way of better positioning the start shape is to run two ASM searches in series, using the results of the first search as the start shape for the second search. In practice is suffices to use 1D profiles for the first model and to start the second model at pyramid level 1, one level below full size. Stacking helps the worst fits, where the start shape is often badly mis-positioned, but has little effect where the start shape is already well positioned.

5   Experimental Results

Before giving experimental results we briefly review model assessment in more general terms [13]. The overall strategy for selecting parameters is

1. for each model parameter
2.   for each parameter value
3.   train on a set of faces
4.   evaluate the model by using it to locate landmarks
5.   select the value of the parameter that gives the best model
6. test the final model by using it to locate landmarks.

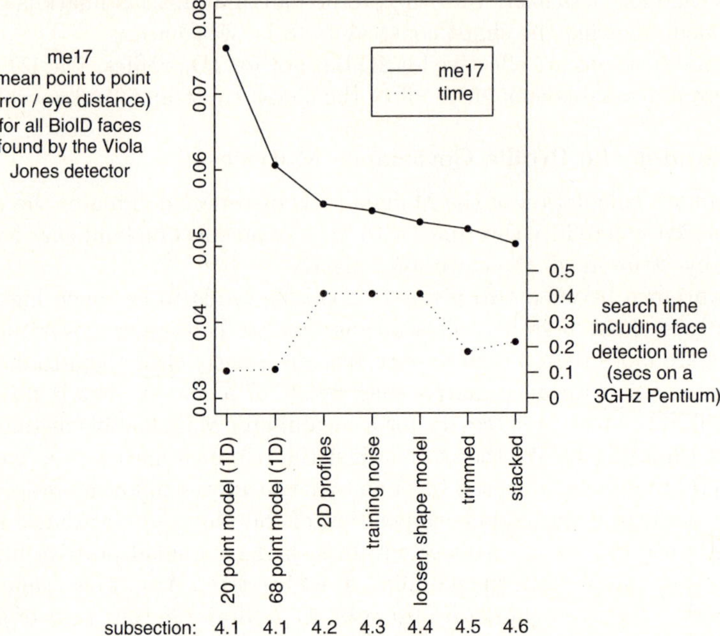

Fig. 3. Relative performance of various models

Two processes are going on here: model selection which estimates the performance of different models in order to choose one (steps 2-5 above), and model assessment which estimates the final model's performance on new data (step 6 above). We want to measure the generalization ability of the model, not its ability on the set it was trained on, and therefore need three independent datasets (i) a *training set* for step 3 above (ii) a parameter selection or *validation set* for step 4 above, and (iii) a *test set* for step 6 above.

For the training set we used the XM2VTS [19] set. We effectively doubled the size of the training set by mirroring images, but excluded faces that were of poor quality (eyes closed, blurred, etc.).

For the validation set we used the AR [18] set. So, for example, we used the AR set for choosing the amount of noise discussed in section 4.3. We minimized overfitting to the validation set by using a different subset of the AR data for selecting each parameter. Subsets consisted of 200 randomly chosen images.

For the test set we used the BioID set [15]. More precisely, the test set is those faces in the BioID set that were successfully found by the OpenCV [14] implementation of the Viola-Jones face detector (1455 faces, which is 95.7% of the total 1521 BioID faces).

We used manual landmarks for these three sets from the FGNET project [9].

Cross validation on a single data set is another popular approach. We did not use cross validation because three datasets were available and because of the many instances of near duplication of images within each dataset.

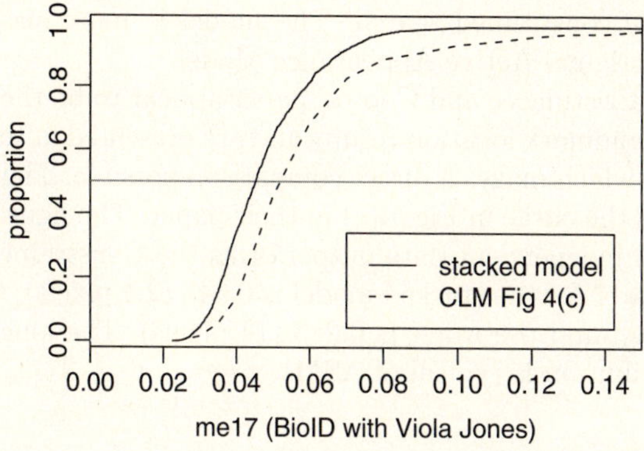

Fig. 4. Comparison to Constrained Local Model [10]

Following Cristinacce [10], we present results in terms of the *me17* measure. The me17 is calculated by taking the mean of the euclidean distances between each of the 17 internal face points located by the search and the corresponding manually landmarked point. This mean is normalized by dividing by the distance between the manually landmarked eye pupils. We use only 17 of the 20 manually landmarked BioID points because the 3 points near the sides of the face have a high variability across human landmarkers.

5.1   Relative Performance

Fig. 3 summarizes and compares results from applying each of the modifications described in this paper. Each graph point represents the me17 averaged over all faces in the test set, for the given model. Each model incorporates the improvements of the models to its left but not to its right.

For example, the entry labeled **4.2 2D profiles** shows results for the model described in section 4.2. The model uses the 2D profiles described in that section and incorporates the techniques prior to but not subsequent to section 4.2. The graph shows that using 2D profiles decreases the me17 from 0.061 to 0.055 but increases the search time from 110 to 410 ms.

The mean me17 of the final stacked model is 66% of the initial 20 point model. The biggest single improvement comes from adding more points to the model, followed by using 2D profiles, followed by stacking. A different test set or different evaluation order would give somewhat different results, but the graph is representative of the relative performance of the various modifications.

5.2   Comparison to Previously Published Results

Fig. 4 compares the best model in this paper, the stacked model (section 4.6), to the Constrained Local Model presented in Cristinacce and Cootes [10]. Briefly, the Constrained Local Model is similar to an Active Appearance Model [5], but instead of modeling texture across the whole face it models a set of local feature templates. During search, the feature templates are matched to the image using an efficient shape constrained search. The model is more accurate and more robust than the original Active Appearance Model.

The results in Cristinacce and Cootes' paper appear to be the best previously published facial landmark location results and are presented in terms of the me17 on the BioId set, which makes a direct comparison possible. The dotted curve in Fig. 4 reproduces the curve in Fig. 4(c) in their paper. The figure shows that the stacked model on independent data outperforms the Constrained Local Model.

The median me17 for the stacked model is 0.045 (2.4 pixels), the best me17 is 0.0235 (1.4 pixels), and the worst is 0.283 (14 pixels). The long right hand tail of the error distribution is typical of ASMs.

6   Conclusion and Future Work

This paper presented some modifications to the Active Shape Model which make it competitive with more sophisticated methods of locating features in frontal views of upright faces.

A few simple rules of thumb for improving ASMs became apparent. You can get better fits by adding more landmarks. You can discard most elements of the covariance matrices for increased speed without loss of quality. You get better results with a better start shape, and you can do this by running two models in series.

The techniques used in this paper are fairly standard. Perhaps the main contribution of the paper is assembling them together in a sound fashion. Advantages of the techniques are their simplicity and applicability for use in conjunction with other methods. For example, extra landmarks and stacked models would possibly improve the performance of the Constrained Local Model shown in Fig. 4.

The results are still not as good as manual landmarks. Further work will investigate combining multiple profiling techniques at each landmark with a decision tree [2] or related method. Here the training process would try different profiling techniques at each landmark and build a decision tree (for each landmark) that would select or combine techniques during searching.

Additional documentation and source code to reproduce the results in this paper can be found at this project's web site [20].

References

1. Bates, D., Maechler, M.: Matrix: A Matrix package for R. See the nearPD function in this R package for methods of forcing positive definiteness (2008),
   http://cran.r-project.org/web/packages/Matrix/index.html
2. Breiman, Friedman, Olshen, Stone: Classification and Regression Trees. Wadsworth (1984)
3. Cootes, T.F., Cooper, D.H., Taylor, C.J., Graham, J.: A Trainable Method of Parametric Shape Description. BMVC 2, 54–61 (1991)
4. Cootes, T.F., Taylor, C.J., Cooper, D.H., Graham, J.: Active Shape Models — their Training and Application. CVIU 61, 38–59 (1995)

5. Cootes, T.F., Edwards, G.J., Taylor, C.J.: Active Appearance Models. In: Burkhardt, H., Neumann, B. (eds.) ECCV 1998. LNCS, vol. 1407, pp. 484–498. Springer, Heidelberg (1998)
6. Cootes, T.F., Taylor, C.J.: A Mixture Model for Representing Shape Variation. Image and Vision Computing 17(8), 567–574 (1999)
7. Cootes, T.F., Edwards, G.J., Taylor, C.J.: Comparing Active Shape Models with Active Appearance Models. In: Pridmore, T., Elliman, D. (eds.) Proc. British Machine Vision Conference, vol. 1, pp. 173–182 (1999)
8. Cootes, T.F., Taylor, C.J.: Technical Report: Statistical Models of Appearance for Computer Vision. The University of Manchester School of Medicine (2004), www.isbe.man.ac.uk/~bim/refs.html
9. Cootes, T.F., et al.: FGNET manual annotation of face datasets (2002), www-prima.inrialpes.fr/FGnet/html/benchmarks.html
10. Cristinacce, D., Cootes, T.: Feature Detection and Tracking with Constrained Local Models. BMVC 17, 929–938 (2006)
11. Gentle, J.E.: Numerical Linear Algebra for Applications in Statistics. Springer, Heidelberg (1998); See page 178 for methods of forcing positive definiteness
12. van Ginneken, B., Frangi, A.F., Stall, J.J., ter Haar Romeny, B.: Active Shape Model Segmentation with Optimal Features. IEEE-TMI 21, 924–933 (2002)
13. Hastie, T., Tibshirani, R., Friedman, J.: The Elements of Statistical Learning: Data Mining, Inference, and Prediction. Springer, Heidelberg (2003); See chapter 7 for methods of model assessment
14. Intel: Open Source Computer Vision Library. Intel (2007)
15. Jesorsky, O., Kirchberg, K., Frischholz, R.: Robust Face Detection using the Hausdorff Distance. AVBPA 90–95 (2001)
16. Kass, M., Witkin, A., Terzopoulos, D.: Snakes: Active Contour Models. IJCV 1, 321–331 (1987)
17. Koenderink, J.J., van Doorn, A.J.: The Structure of Locally Orderless Images. IJCV 31(2/3), 159–168 (1999)
18. Martinez, A.M., Benavente, R.: The AR Face Database: CVC Tech. Report 24 (1998)
19. Messer, K., Matas, J., Kittler, J., Luettin, J., Maitre, G.: XM2VTS: The Extended M2VTS Database. AVBPA (1999)
20. Milborrow, S.: Stasm software library (2007), http://www.milbo.users.sonic.net/stasm
21. Rogers, M., Graham, J.: Robust Active Shape Model Search. In: Heyden, A., Sparr, G., Nielsen, M., Johansen, P. (eds.) ECCV 2002. LNCS, vol. 2353, pp. 517–530. Springer, Heidelberg (2002)
22. Romdhani, S., Gong, S., Psarrou, A.: A Multi-view Non-linear Active Shape Model using Kernel PCA. BMVC 10, 483–492 (1999)
23. Scholkopf, S., Smola, A., Muller, K.: Nonlinear Component Analysis as a Kernel Eigenvalue Problem. Neural Computation 10(5), 1299–1319 (1998)
24. Li, Y., Ito, W.: Shape Parameter Optimization for AdaBoosted Active Shape Model. ICCV 1, 251–258 (2005)
25. Zhou, Y., Gu, L., Zhang, H.J.: Bayesian Tangent Shape Model: Estimating Shape and Pose Parameters via Bayesian Inference. In: CVPR (2003)

Dynamic Integration of Generalized Cues for Person Tracking

Kai Nickel and Rainer Stiefelhagen

Universität Karlsruhe (TH), InterACT
Am Fasanengarten 5, 76131 Karlsruhe, Germany

Abstract. We present an approach for the dynamic combination of multiple cues in a particle filter-based tracking framework. The proposed algorithm is based on a combination of democratic integration and layered sampling. It is capable of dealing with deficiencies of single features as well as partial occlusion using the very same dynamic fusion mechanism. A set of simple but fast cues is defined, which allow us to cope with limited computational resources. The system is capable of automatic track initialization by means of a dedicated attention tracker permanently scanning the surroundings.

1 Introduction

Visual person tracking is a basic prerequisite for applications in fields like surveillance, multimodal man-machine interaction or smart spaces. Our envisioned scenario is that of an autonomous robot with limited computational resources operating in a common space together with its users. The tracking range varies from close distance, where the portrait of the user spans the entire camera image, to far distance, where the entire body is embedded in the scene. In order to tackle the problem, we present a multi-cue integration scheme within the framework of particle filter-based tracking. It is capable of dealing with deficiencies of single features as well as partial occlusion by means of the very same dynamic fusion mechanism. A set of simple but fast cues is defined, allowing to cope with limited on-board resources.

The choice of cues is a crucial design criterion for a tracking system. In real-world applications, each single cue is likely to fail in certain situations such as occlusion or background clutter. Thus, a dynamic integration mechanism is needed to smooth over a temporary weakness of certain cues as long as there are other cues that still support the track. In [1], Triesch and Von Der Malsburg introduced the concept of *democratic integration* that weights the influence of the cues according to their agreement with the joint hypothesis. The competing cues in [1] were based on different feature types such as color, motion, and shape. In this paper, we use the principle of democratic integration in a way that also includes the competition between different regions of the target object. We show that this allows us to deal with deficiencies of single feature types as well as with partial occlusion using one joint integration mechanism.

D. Forsyth, P. Torr, and A. Zisserman (Eds.): ECCV 2008, Part IV, LNCS 5305, pp. 514–526, 2008.

The combination of democratic integration and particle filters has been approached before by Spengler and Schiele [2]. In their work, however, the integration weights were held constant, thus falling short behind the real power of democratic integration. This has also been pointed out by Shen et al. [3], who did provide a cue quality criterion for dynamic weight adaptation. This criterion is formulated as the distance of the tracking hypothesis based on all cues and the hypothesis based on the cue alone. The problem with this formulation is that, due to resampling, the proposal distribution is generally strongly biased toward the final hypothesis. Thus, even cues with uniformly mediocre scores tend to agree well with the joint mean of the particle set. We therefore propose a new quality criterion based on weighted MSE that prefers cues which actually focus their probability mass around the joint hypothesis.

Democratic integration combines cues in the form of a weighted sum. In a particle filter framework, this means that all cues have to be evaluated simultaneously for all particles. As pointed out by Pérez et al. [4], this can be alleviated by *layered sampling*, if the cues are ordered from coarse to fine. In the proposed algorithm, we therefore combine two-stage layered sampling with democratic integration on each stage to increase efficiency by reducing the required number of particles.

For each object to be tracked, we employ one dedicated Condensation-like tracker [5]. By using separate trackers instead of one single tracker running in a joint state space, we accept the disadvantage of potentially not being able to find the global optimum. On the other hand, however, we thereby avoid the exponential increase in complexity that typically prevents the use of particle filters in high-dimensional state spaces. There are a number of approaches dealing with this problem, such as Partitioned Sampling [6], Trans-dimensional MCMC [7], or the Hybrid Joint-Separable formulation [8]. Although these approximations reduce the complexity of joint state space tracking significantly, they still require noticeably more computational power than the separate tracker approach.

The remainder of this paper is organized as follows: In section 2, we briefly describe the concept of particle filters and layered sampling. In section 3 we present our multi-cue integration scheme, which is the main contribution of this paper. It is followed, in section 4, by the definition of the cues that we actually use in the live tracking system. In section 5, the multi-person tracking logic including automatic track initialization and termination is described. Finally, section 6 shows the experiments and results.

2   Particle Filter-Based Tracking

Particle filters represent a generally unknown probability density function by a set of random samples $s_t^{(1..n)}$ and associated weights $\pi_t^{(1..n)}$ with $\sum \pi_t^{(i)} = 1$. In one of the simplest cases, the Condensation algorithm [5], the evolution of the particle set is a two-stage process which is guided by the observation and the state evolution model:

1. The prediction step (including resampling): randomly draw n new particles from the old set with a likelihood proportional to the particle weights. Propagate the new particles by applying the state evolution model $p(\mathbf{s}_t|\mathbf{s}_{t-1})$.
2. The measurement step: adjust the weights of the new particles with respect to the current observation $\mathbf{z}_t$: $\pi_t^{(i)} \propto p(\mathbf{z}_t|\mathbf{s}_t^{(i)})$.

The final tracking hypothesis for the current time instance $\hat{\mathbf{s}}_t$ can be obtained from the sample set as

$$\hat{\mathbf{s}}_t = \sum_{i=0..n} \pi_t^{(i)} \mathbf{s}_t^{(i)} \tag{1}$$

2.1   Layered Sampling

Assuming that $\mathbf{z}$ is made up of M conditionally independent measurement sources, i.e. different cues, the observation likelihood of a particle $\mathbf{s}$ can be factorized as follows[1]:

$$p(\mathbf{z}|\mathbf{s}) = \prod_{m=1..M} p(\mathbf{z}^m|\mathbf{s}) \tag{2}$$

According to [4], the state evolution can then be decomposed into M successive intermediate steps:

$$p(\mathbf{s}_t|\mathbf{s}_{t-1}) = \int p_M(\mathbf{s}_t|\mathbf{s}^{M-1}) \cdots p_1(\mathbf{s}^1|\mathbf{s}_{t-1}) d\mathbf{s}^1 \cdots d\mathbf{s}^{M-1} \tag{3}$$

where $\mathbf{s}^1 \cdots \mathbf{s}^{M-1}$ are auxiliary state vectors[2]. In case of a Gaussian evolution model, this corresponds to a fragmentation into M successive steps with lower variances. Then, [4] make the approximation that the likelihood for the m-th cue $p(\mathbf{z}^m|\mathbf{s})$ can be incorporated after applying the m-th state evolution model $p_m(\mathbf{s}^m|\mathbf{s}^{m-1})$. This leads to a layered sampling strategy, where at the m-th stage new samples are simulated from a Monte Carlo approximation of the distribution $p_m(\mathbf{s}^m|\mathbf{s}^{m-1})\pi^{m-1}$ with an associated importance weight $\pi^m \propto p(\mathbf{z}^m|\mathbf{s}^m)$. As [4] point out, the benefit of layered sampling arises in cases where the cues can be ordered from coarse to fine, e.g. the first cue produces a reliable but rough estimation for the state, while the second cue produces a sharp and peaky estimation. Then, the layered sampling approach will effectively guide the search in the state space, with each stage refining the result from the previous stage. We will apply layered sampling in section 5 in combination with the multi-cue integration scheme described in the following.

3   Dynamic Multi-cue Integration

In the Bayesian tracking formulation used in this work, cues have the function of scoring the match between a state vector $\mathbf{s}$ and the observation $\mathbf{z}$. A joint score combining the cues from the set of all cues C can be formulated as a weighted sum

---

[1] The time index t is omitted for the sake of brevity wherever possible.
[2] We omit the according formula for splitting the proposal distribution, because in Condensation, the proposal distribution is identical to the evolution model.

$$p(\mathbf{z}|\mathbf{s}) = \sum_{c \in C} r_c p_c(\mathbf{z}|\mathbf{s}), \tag{4}$$

where $p_c(\mathbf{z}|\mathbf{s})$ is the the single-cue observation model, and r_c is the mixture weight for cue c, with $\sum_c r_c = 1$.

Democratic integration [1] is a mechanism to dynamically adjust the mixture weights r_c, termed reliabilities, with respect to the agreement of the single cue c with the joint result. For each cue, a quality measure q_c is defined that quantifies the agreement, with values close to zero indicating little agreement and values close to one indicating good agreement. The reliabilities are updated after each frame by a leaky integrator using the normalized qualities:

$$r_c^{t+1} = (1 - \tau)r_c^t + \tau \frac{q_c}{\sum_c q_c} \tag{5}$$

with the parameter τ controlling the speed of adaptation.

3.1   Cue Quality Measure

In the original paper [1], tracking is implemented as an exhaustive search over a support map, and the quality measure is defined over a single cue's support map. In [3], a different quality measure dedicated to particle filters is proposed: Based on the current particle set $\mathbf{s}^{(1..n)}$ and an auxiliary set of weights $\pi_c^{(1..n)} \propto p_c(\mathbf{z}|\mathbf{s}^{(1..n)})$, a tracking hypothesis $\hat{\mathbf{s}}_c$ is generated according to eq. 1 and compared to the joint hypothesis $\hat{\mathbf{s}}$. The L_2-norm distance $|\hat{\mathbf{s}}_c - \hat{\mathbf{s}}|_2$ is normalized by means of a sigmoid function and then taken as quality measure.

Although this formulation looks straightforward, there is a problem associated with it: Imagine the common situation where a cue finds little or no support at all, and therefore assigns uniform likelihood values to all of the particles. Let's assume further that the state of the target has not changed for a while, so that in consequence, due to resampling, the particle distribution is equally spread around the actual state. In this case, the cue-based hypothesis $\hat{\mathbf{s}}_c$ will be close to $\hat{\mathbf{s}}$ resulting in a high quality value q_c despite the fact that the cue is actually not at all able to locate the target.

To address this problem, we need a quality measure that quantifies how well the probability mass agglomerates around the joint hypothesis $\hat{\mathbf{s}}$. The inverse mean-square error $(\sum_i \pi_c^{(i)} |\mathbf{s}^{(i)} - \hat{\mathbf{s}}|_2^2)^{-1}$ of the particle set weighted with the respective cue's weights π_c meets this requirement, but is dependent on the actual location of the particles. We eliminate this dependency by relating the cue's MSE to the MSE of a hypothetical baseline cue which assigns uniform weights $\frac{1}{n}$ to each particle. Because a good cue is not only supposed to converge to the target location but also to assign high values to the target, we multiply the term with the cue's non-normalized response at the joint hypothesis $p_c(\mathbf{z}|\hat{\mathbf{s}})$. Thus, we come to the following formulation for a universal cue quality measure in the context of particle-filter based tracking:

$$q_c = \frac{\sum_{i=1..n} \frac{1}{n} |\mathbf{s}^{(i)} - \hat{\mathbf{s}}|^\lambda}{\sum_{i=1..n} \pi_c^{(i)} |\mathbf{s}^{(i)} - \hat{\mathbf{s}}|^\lambda} p_c(\mathbf{z}|\hat{\mathbf{s}}) \tag{6}$$

The exponent $\lambda > 0$ can be used to tweak the volatility of the quality measure: high values of λ emphasize the quality difference between cues whereas low values produce more similar qualities for all cues.

3.2 Generalized Cue Competition

In order to allow for a fruitful combination, the set of cues should be orthogonal in the sense that different cues tend to fail under different circumstances. One way to reduce the chances of co-occurrence of failure is to use different cue-specific feature transformations $\mathcal{F}(\mathbf{z})$ like motion, color, or shape. Failure of one feature can thus more likely be compensated by other features.

$$p_c(\mathbf{z}|\mathbf{s}) = p_c(\mathcal{F}(\mathbf{z})|\mathbf{s}) \tag{7}$$

The other option to generate orthogonal cues is to use different state model transformations $\mathcal{A}(\mathbf{s})$:

$$p_c(\mathbf{z}|\mathbf{s}) = p_c(\mathbf{z}|\mathcal{A}(\mathbf{s})) \tag{8}$$

This is motivated by the fact that cues relying on certain aspects of the state vector may still be used while other aspects of the state are not observable. In our implementation, $\mathcal{A}(\mathbf{s})$ represents a certain projection from state space to image space, i.e. a certain image sub-region of the target. This is useful in a situation, where due to partial occlusion one region of the target object can be observed, while another region cannot.

In this work, we aim at combining the advantages of both strategies, i.e. dynamically combining cues that are based on different feature types as well as dynamically weighting cues that focus on different regions of the target but are based on the same feature type. Therefore, we use a generalized definition of the cues $c = (\mathcal{F}, \mathcal{A})$ that comprises different feature types $\mathcal{F}(\mathbf{z})$ and different state transformations $\mathcal{A}(\mathbf{s})$:

$$p_c(\mathbf{z}|\mathbf{s}) = p_{\mathcal{F},\mathcal{A}}(\mathcal{F}(\mathbf{z})|\mathcal{A}(\mathbf{s})), \tag{9}$$

All cues in this unified set will then compete equally against each other, guided by the very same integration mechanism. Thus, the self-organizing capabilities of democratic integration can be used to automatically select the specific feature types as well as the specific regions of the target that are most suitable in the current situation.

3.3 Cue Model Adaptation

Certain cues, such as color models or templates, allow for online adaptation of their internal parameters to better match the current target appearance. In [1], this adaptation is described as a continuous update process with a fixed time constant τ_c:

$$P_c^{t+1} = (1 - \tau_c)P_c^t + \tau_c \hat{P}_c, \tag{10}$$

with P_c being the internal parameters of cue c, and $\hat{P}_c$ being new parameters acquired from the image region given by the joint hypothesis $\hat{\mathbf{s}}$.

One of the issues with adaptation is due to the fact that after an update step, the cue is not guaranteed to perform better than before. Although the update step always results in a higher score for the prototype region at $\hat{\mathbf{s}}$, it can happen that the updated model produces higher scores also for other regions than the correct one. This actually reduces the cue's discriminative power and, in consequence, its reliability r_c. We therefore propose the following test to be carried out before accepting an update:

1. Calculate q_c' (eq. 6) using the new parameters $\hat{P}_c$
2. Perform the update step (eq. 10) only if $q_c' > q_c$

4 Fast Cues for 3D Person Tracking

In the targeted application, one or more people are to be tracked in the vicinity of an autonomous robot featuring a calibrated stereo camera. As the on-board computational resources are strictly limited, cues have to be found that rely on features that can be evaluated rapidly. Our proposed cues are based on the following well-known feature types: difference image, color histogram back-projection, Haar-feature cascades and stereo correlation.

As motivated in section 3.2, we use different transformations of the state vector in order to handle partial occlusion: some cues focus on the human head region only, whereas other cues concentrate on the torso and legs region respectively. These regions are determined using the "3-box model" of the human body depicted in Fig. 1. The real-world extensions of the 3 cuboids are geared to model an average human being; their relative positions depend on the height of the head above the ground plane.

By combining the feature types motion, color and stereo with the 3 different body parts, and by using 4 different detectors, we obtain a total number of 13 cues that will be described in the following. Fig. 2 shows the different feature types as a snapshot from a test sequence.

In the following, we will use $\mathcal{F}(\mathbf{z})$ to denote a feature map, i.e. an image in which the intensity of a pixel is proportional to the presence of a feature, such as color or motion. An image region corresponding to a state vector $\mathbf{s}$ will be denoted as $\mathcal{A}(\mathbf{s})$ (see Fig. 1), $|\mathcal{A}(\mathbf{s})|$ is the size of the region, and $\sum_{\mathcal{A}(\mathbf{s})} \mathcal{F}(\mathbf{z})$ is the sum of pixel values of $\mathcal{F}(\mathbf{z})$ inside region $\mathcal{A}(\mathbf{s})$. All regions in our system are rectilinear bounding boxes, so the sum can be calculated efficiently by means of 4 table lookups in the integral image [9].

4.1 Motion Cues

The difference image $\mathcal{M}(\mathbf{z})$ is generated by pixel-wise thresholding the absolute difference of the current frame's and the previous frame's intensity images. For a moving object, we can expect high values of $\mathcal{M}(\mathbf{z})$ in the region $\mathcal{A}(\mathbf{s})$ around object's current location $\mathbf{s}$. The motion cue's observation likelihood is given as:

$$p_{\mathcal{M},\mathcal{A}}(\mathbf{z}|\mathbf{s}) = \frac{\sum_{\mathcal{A}(\mathbf{s})} \mathcal{M}(\mathbf{z})}{|\mathcal{A}(\mathbf{s})|} \cdot \frac{\sum_{\mathcal{A}(\mathbf{s})} \mathcal{M}(\mathbf{z})}{\sum \mathcal{M}(\mathbf{z})} \tag{11}$$

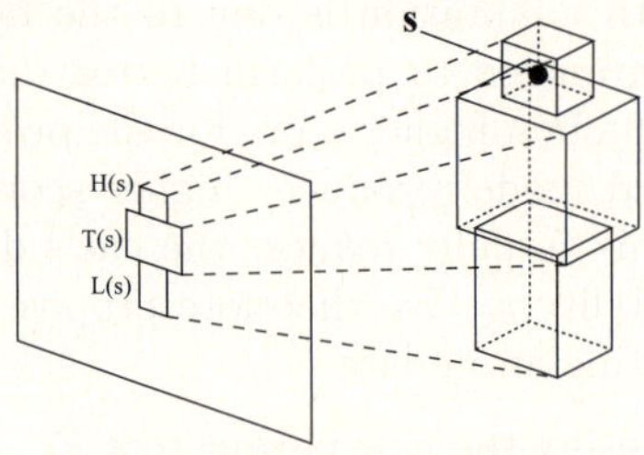

Fig. 1. The 3-box model of the human body: the state vector **s** is transformed into the image space as the projection of a cuboid representing either the head, torso, or leg region. The projection of the cuboid is approximated by a rectilinear bounding box.

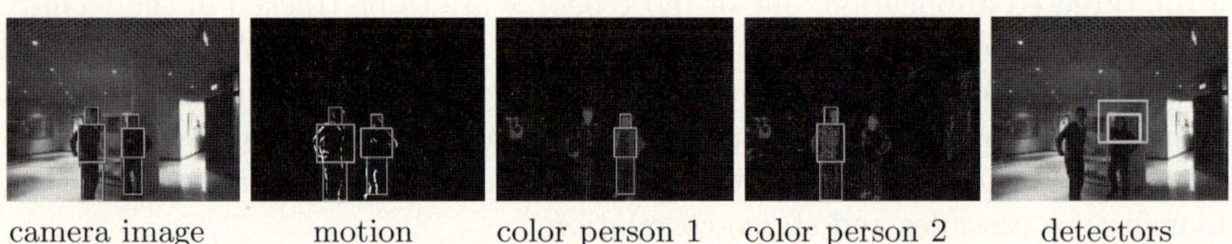

Fig. 2. Snapshot from a test sequence showing the different feature types. In this visualization, the color support maps for head, torso and legs of the respective person are merged into the RGB-channels of the image. The tracking result is superimposed.

The left factor seeks to maximize the amount of foreground within the region. The right factor seeks to cover all foreground pixels in the image. It prevents the motion cue from preferring tiny regions filled with motion, while ignoring the rest.

We employ 3 motion cues, termed M-H, M-T and M-L, dedicated to either the head, torso or legs region as depicted in Fig. 1. We rely on the ability of the integration mechanism (see section 3) to automatically cancel the influence of the motion cues in case of camera motion. This is justified by the fact that the agreement of the motion cues with the final tracking hypothesis will drop whenever large portions of the image exceed the threshold.

4.2   Color Cues

We employ three adaptive color cues C-H, C-T, C-L for the three body regions. For each of the cues, we use a 3-dimensional histogram with 16 bins per channel in RGB color space that automatically adapts to the target region using the mechanism described in section 3.3. A second histogram is built from the entire image; it acts as a model for the background color distribution. The quotient histogram of the target histogram and the background histogram is back-projected and forms the support map $\mathcal{C}(\mathbf{z})$ for a color cue. The observation likelihood is given analogous to eq. 11 as:

$$p_{C,\mathcal{A}}(\mathbf{z}|\mathbf{s}) = \frac{\sum_{\mathcal{A}(\mathbf{s})} \mathcal{C}(\mathbf{z})}{|\mathcal{A}(\mathbf{s})|} \cdot \frac{\sum_{\mathcal{A}(\mathbf{s})} \mathcal{C}(\mathbf{z})}{\sum \mathcal{C}(\mathbf{z})} \tag{12}$$

4.3 Detector Cues

For each particle, the head region $\mathcal{A}(\mathbf{s})$ is projected to the image plane, and the bounding box of the projection is being classified with a single run of the detector proposed by [9]. The detectors are organized stages that need to be passed one by one in order to produce a positive response. The ratio $m(\mathcal{A}(\mathbf{s})) = \left(\frac{\text{stages passed}}{\text{stages total}}\right)^{\omega}$ can be interpreted as a confidence value for the detection, with the exponent ω controlling the steepness of decay for each stage that is not being passed.

In order to smooth the scores of nearby particles, we define the score of a particle $\mathbf{s}$ as the highest overlap between its region $\mathcal{A}(\mathbf{s})$ and all the positively classified regions $\mathcal{A}' \in \{\mathcal{A}(\mathbf{s}^{(i)}) | \mathcal{A}(\mathbf{s}^{(i)})\text{is face}\}_{i=1..n}$ by any of the other particles:

$$p_{\mathcal{D},\mathcal{A}}(\mathbf{z}|\mathbf{s}) = \max_{\mathcal{A}'} m(\mathcal{A}') \cdot d(\mathcal{A}', \mathcal{A}(\mathbf{s})), \tag{13}$$

with d being a distance metric based on rectangle overlap.

We use four detector cues in total: one for frontal faces (D-F), one for left (D-L) and one for right (D-R) profile faces, and one for upper bodies (D-U). Implementation and training of the detectors is based on [10,11] as provided by the OpenCV library.

4.4 Stereo Correlation Cues

In traditional stereo processing [12], a dense disparity map is generated by exhaustive area correlation followed by several post-filtering steps. Apart from the computational effort of generating a dense disparity map, there is another, more fundamental problem, namely the choice of the size of the area correlation window. If a windows is too large, it smoothes over fine details, if it is too small, it tends to produce noisy results. In our approach, we can avoid these issues: we use the entire target region $\mathcal{A}(\mathbf{s})$ as correlation window and search for optimal correlation along the epipolar lines. The adaptive correlation window is thus as large as possible and as small as necessary given the current size of the target.

The response of the stereo cue is given by the distance of the discovered disparity $\hat{d}(\mathcal{A}(\mathbf{s}))$ and the hypothesized disparity $d(\mathcal{A}(\mathbf{s}))$:

$$p_{\mathcal{S},\mathcal{A}}(\mathbf{z}|\mathbf{s}) = \left(1 + |\hat{d}(\mathcal{A}(\mathbf{s})) - d(\mathcal{A}(\mathbf{s}))|^{\kappa}\right)^{-1}, \tag{14}$$

with κ being a parameter to control the volatility of the cue. The complexity of the local search for the disparity $\hat{d}(\mathcal{A}(\mathbf{s}))$ is scale-invariant because it can be implemented efficiently by means of integral images, as proposed by [13] for dense disparity calculation. We employ 3 stereo cues, one for the head (S-H), torso (S-T), and legs (S-L).

5 Multi-person Tracking Logic

As motivated in the introduction, we run one dedicated particle filter for each person to be tracked. The state space consists of the location and velocity of the person's head centroid in 3-dimensional space: $\mathbf{s}^{(i)} = (x, y, z, \dot{x}, \dot{y}, \dot{z})$. The state evolution $p(\mathbf{s}_t|\mathbf{s}_{t-1})$ is implemented as a 1st-order motion model with additive Gaussian noise on the velocity components.

5.1   Democratic Integration and Layered Sampling

Multi-cue integration as described by eq. 4 is suitable for all kinds of cues that are *optional* for the target, which means that the target may or may not have the property implied by the cue at the moment. There are, however, cues that are indispensable as track foundation and therefore must not be ruled out by the fusion mechanism. In our application, this applies to the stereo cues: a track should not be able to exist if it is not supported by at least one of the stereo cues as these represent strict geometrical constraints. One way of ensuring this would be to multiply the response of the stereo cues with the response of the regular cues. A more efficient way is layered sampling as described in section 2.1. We use it to evaluate the stereo cues $C_S \subset C$ before the regular cues $C_R \subset C$, as shown in Fig. 3. By evaluating the mandatory stereo cues first, followed by a resampling step, the resulting particle set $\mathbf{s}_t^{1,(1..n)}$ clusters only in those regions of the state space that are well supported by the stereo cues. The particles on the second stage can now more efficiently evaluate the regular cues.

1st layer:

- resample $\mathbf{s}_{t-1}^{(1..n)}$ wrt. $\pi_{t-1}^{(1..n)}$
- propagate with partial evolution model (cf. eq. 3)
$$\mathbf{s}_t^{1,(1..n)} \longleftarrow p_1(\mathbf{s}_t^{1,(i)}|\mathbf{s}_{t-1}^{(i)})$$
- evaluate stereo cues: $\pi_t^{1,(i)} \propto \sum_{c \in C_S} r_c p_c(\mathbf{z}|\mathbf{s}_t^{1,(i)})$
- apply collision penalty: $\pi_t^{1,(i)} \longleftarrow \pi_t^{1,(i)} - v(\mathbf{s}_t^{1,(i)})$

2nd layer:

- resample $\mathbf{s}_t^{1,(1..n)}$ wrt. $\pi_t^{1,(1..n)}$
- propagate with partial evolution model (cf. eq. 3)
$$\mathbf{s}_t^{(1..n)} \longleftarrow p_2(\mathbf{s}_t^{(i)}|\mathbf{s}_t^{1,(i)})$$
- evaluate regular cues: $\pi_t^{(i)} \propto \sum_{c \in C_R} r_c p_c(\mathbf{z}|\mathbf{s}_t^{(i)})$

Dem. integration:

- calculate track hypothesis $\hat{\mathbf{s}}_t = \sum_i \pi_t^{(i)} \mathbf{s}_t^{(i)}$
- update reliabilities (cf. eqs. 5 and 6)
$$r_{c \in C_S} \longleftarrow \hat{\mathbf{s}}_t, \mathbf{s}_t^{1,(1..n)}, \pi_t^{1,(1..n)}$$
$$r_{c \in C_R} \longleftarrow \hat{\mathbf{s}}_t, \mathbf{s}_t^{(1..n)}, \pi_t^{(1..n)}$$

Fig. 3. Two-stage layered sampling algorithm with democratic cue integration

Apart from the geometrical constraints implied by the stereo cues, there is another strict constraint, namely the collision penalty, which is enforced in the 1st layer of the algorithm in Fig. 3. The function $v(\mathbf{s})$ penalizes particles that are close to those tracks with a higher track quality than the current track (see following section). Thereby, we guarantee mutual exclusion of tracks.

5.2   Automatic Track Initialization

The question of when to spawn a new tracker and when to terminate a tracker that has lost its target is of high importance, and can become more difficult than the actual tracking problem. We define the quality measure for a tracker to be the joint response from both stereo and regular cues at the tracker's hypothesis $\hat{\mathbf{s}}$:

$$Q(\hat{\mathbf{s}}) = \sum_{c \in C_S} r_c p_c(\mathbf{z}|\hat{\mathbf{s}}) \cdot \sum_{c \in C_R} r_c p_c(\mathbf{z}|\hat{\mathbf{s}}) \tag{15}$$

The final quality measure Q is a result of temporal filtering with a time constant ν:

$$Q^{t+1} = (1 - \nu)Q^t + \nu Q(\hat{\mathbf{s}}) \tag{16}$$

Trackers falling below a certain threshold $Q < \Theta$ for a certain amount of time Γ will be discarded.

In order to discover potential targets, we employ an additional tracker termed *attention tracker*. The attention tracker permanently scans the state space, searching for promising regions. It is, however, repelled by existing tracks by means of the collision penalty $v(\mathbf{s})$. Unlike regular trackers, 50% of the attention tracker's particles are not propagated by means of the state evolution model, but are drawn randomly from the state space. This guarantees good coverage of the state space and still allows some clustering around interesting regions. As the attention tracker must remain general, its cues' parameters are not allowed to adapt. After each frame, the distribution of the attention tracker's particles is clustered with a k-means algorithm. If one of the clusters exceeds the threshold Θ, a new regular tracker is initialized at that location.

6   Experiments

We evaluated the algorithm on 11 test sequences, some of them including camera motion. The head's bounding box was manually labeled in 3 of the 15 frames per second to obtain the ground truth. In total, 2312 frames were labeled. From the 3D tracking output, a head-sized box was projected to the image and compared to the manually labeled box. If there was no overlap between the boxes, the frame was counted as a miss and a false positive. As the tracker was free to output 0, 1 or more tracks, the number of misses and false positives do not need to be identical.

Overall, the tracker showed solid performance throughout the experiments. Critical situations for track loss – although it occurred rarely – were periods in which the user rested virtually motionless either at far distance or in a turned-away position, so that in consequence the detectors did not respond. Then, the tracker had to rely solely on the automatically initialized color models, which were not always significant enough. Another issue were phantom tracks that were triggered by non-human motion or false detections. They were sometimes kept alive by the color models which adapted to the false positive region. In most

Table 1. Tracking results on the evaluation set

	misses	false pos.
Fixed reliabilities (baseline)	10.2%	8.1%
Dynamic integration (Shen et al.)	11.1%	8.8%
Dynamic integration (equation 6)	4.6%	4.6%

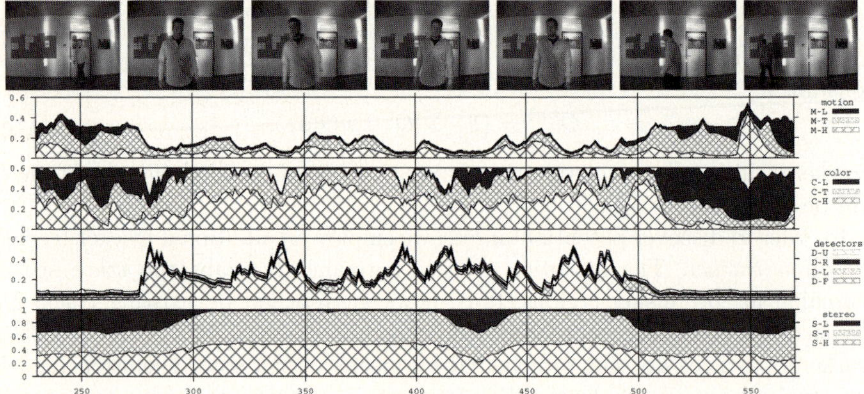

Fig. 4. Evolution of cue reliabilities in an example sequence. The three stereo cues constitute the first layer of the algorithm, their reliabilities sum up to 1. The remaining ten cues are used in layer 2 and sum up to 1 likewise. In the beginning of the interval, the subject approaches the camera. While he is walking (frame 250), the motion cues for legs and torso (M-L,M-T) contribute significantly to the track. At around frame 300, the subject's legs disappear, and in consequence the reliabilities of all leg-related cues (M-L, C-L, S-L) drop automatically. While the subject is standing in front of the camera (frames 300-500), the frontal face detection cue D-F and the the head color cue C-H dominate the track. The influence of the head color cue C-H drops dramatically, when the subject turns around (frame 520) and walks in front of the wooden pinboard, which has a skin-color like appearance.

cases, however, this could be avoided by the adaptation control mechanisms described in section 3.3.

Table 1 shows the results of the evaluation. The proposed algorithm was compared to a baseline system with static reliabilities, and to a system using the dynamic cue quality formulation by Shen et al. [3]. The proposed algorithm clearly outperforms the two other systems both in the number of misses and false positives. Figure 4 discusses the evolution of cue reliabilities for an example sequence.

6.1 Implementation Details

In the implementation, we made the following additions to the algorithm: The color cue for the head region (C-H) is expected to converge to general skin color;

its model is therefore shared among all trackers. An new box-type for the upper body detector was used; it comprises head and upper half of the torso. To avoid dominance, we limited the range for a cue's influence to $0.03 \leq r_c \leq 0.6$. We found, however, that these situations rarely occur. Boxes that get projected outside the visible range or that are clipped to less than 20% of their original size, are scored with a minimum score of 0.001. The approximate runtime of the algorithm was $30ms$ per frame for an empty scene, plus another $10ms$ per person being tracked. These values are based on an image size of 320×240 pixels, and a 2.4GHz Pentium CPU. The most important parameter values are given in Table 2.

Table 2. Parameters of the algorithm

# of particles per tracker	$n = 150$
Track threshold / timeout	$\Theta = 0.25,\ \Gamma = 2s$
Track quality time constant	$\nu = 0.33$
Cue reliability time constant	$\tau = 0.25$
Color update time constant	$\tau_c = 0.01$
Cue tweaking factors	$\lambda = 4, \kappa = 4, \omega = 10$

7 Conclusion

We have presented a new approach for dynamic cue combination in the framework of particle filter-based tracking. It combines the concepts of democratic integration and layered sampling and enables a generalized kind of competition among cues. With this method, cues based on different feature types compete directly with cues based on different target regions. In this way, the self-organizing capabilities of democratic integration can be fully exploited. In an experimental validation, the proposed new cue quality measure has been shown to improve the tracking performance significantly.

Acknowledgments

This work has been funded by the German Research Foundation (DFG) as part of the Sonderforschungsbereich 588 "Humanoid Robots".

References

1. Triesch, J., Malsburg, C.V.D.: Democratic integration: Self-organized integration of adaptive cues. Neural Comput. 13(9), 2049–2074 (2001)
2. Spengler, M., Schiele, B.: Towards robust multi-cue integration for visual tracking. Machine Vision and Applications 14, 50–58 (2003)
3. Shen, C., Hengel, A., Dick, A.: Probabilistic multiple cue integration for particle filter based tracking. In: International Conference on Digital Image Computing - Techniques and Applications, pp. 309–408 (2003)

4. Pérez, P., Vermaak, J., Blake, A.: Data fusion for visual tracking with particles. Proceedings of the IEEE 92(3), 495–513 (2004)
5. Isard, M., Blake, A.: Condensation–conditional density propagation for visual tracking. International Journal of Computer Vision 29(1), 5–28 (1998)
6. MacCormick, J., Blake, A.: A probabilistic exclusion principle for tracking multiple objects. International Journal of Computer Vision 39(1), 57–71 (2000)
7. Smith, K., Gatica-Perez, D., Odobez, J.M.: Using particles to track varying numbers of interacting people. In: IEEE Conf. on Computer Vision and Pattern Recognition, Washington, DC, USA, pp. 962–969 (2005)
8. Lanz, O.: Approximate bayesian multibody tracking. IEEE Transactions on Pattern Analysis and Machine Intelligence 28(9), 1436–1449 (2006)
9. Viola, P., Jones, M.: Robust real-time object detection. In: ICCV Workshop on Statistical and Computation Theories of Vision (July 2001)
10. Lienhart, R., Maydt, J.: An extended set of haar-like features for rapid object detection. In: ICIP, vol. 1, pp. 900–903 (September 2002)
11. Kruppa, H., Castrillon-Santana, M., Schiele, B.: Fast and robust face finding via local context. In: IEEE Intl. Workshop on Visual Surveillance and Performance Evaluation of Tracking and Surveillance (October 2003)
12. Scharstein, D., Szeliski, R.: A taxonomy and evaluation of dense two-frame stereo correspondence algorithms. IJCV 47(1/2/3), 7–42 (2002)
13. Veksler, O.: Fast variable window for stereo correspondence using integral images. In: IEEE Conf. on Computer Vision and Pattern Recognition, pp. 556–561 (2003)

Extracting Moving People from Internet Videos

Juan Carlos Niebles[1,2], Bohyung Han[3], Andras Ferencz[3], and Li Fei-Fei[1]

[1] Princeton University, Princeton NJ, USA
[2] Universidad del Norte, Colombia
[3] Mobileye Vision Technologies, Princeton NJ, USA

Abstract. We propose a fully automatic framework to detect and extract arbitrary human motion volumes from real-world videos collected from *YouTube*. Our system is composed of two stages. A person detector is first applied to provide crude information about the possible locations of humans. Then a constrained clustering algorithm groups the detections and rejects false positives based on the appearance similarity and spatio-temporal coherence. In the second stage, we apply a top-down pictorial structure model to complete the extraction of the humans in arbitrary motion. During this procedure, a density propagation technique based on a mixture of Gaussians is employed to propagate temporal information in a principled way. This method reduces greatly the search space for the measurement in the inference stage. We demonstrate the initial success of this framework both quantitatively and qualitatively by using a number of *YouTube* videos.

1 Introduction

Human motion analysis is notoriously difficult because human bodies are highly articulated and people tend to wear clothing with complex textures that obscure the important features needed to distinguish poses. Uneven lighting, clutter, occlusions, and camera motions cause significant variations and uncertainties. Hence it is no surprise that the most reliable person detectors are built for upright walking pedestrians seen in typically high quality images or videos.

Our goal in this work is to be able to *automatically* and *efficiently* carve out spatio-temporal volumes of human motions from arbitrary videos. In particular, we focus our attention on videos that are typically present on internet sites such as *YouTube*. These videos are representative of the kind of real-world data that is highly prevalent and important. As the problem is very challenging, we do not assume that we can find every individual. Rather, our aim is to enlarge the envelope of upright human detectors by tracking detections from typical to atypical poses. Sufficient data of this sort will allow us in the future to learn even more complex models that can reliably detect people in arbitrary poses. Two example sequences and the system output are shown in Fig. 1.

Our first objective is to find moving humans automatically. In contrast to much of the previous work in tracking and motion estimation, our framework does not rely on manual initialization or a strong *a priori* assumption on the

D. Forsyth, P. Torr, and A. Zisserman (Eds.): ECCV 2008, Part IV, LNCS 5305, pp. 527–540, 2008.

Fig. 1. Two example outputs. Our input videos are clips downloaded from *YouTube* and thus are often low resolution, captured by hand-held moving cameras, and contain a wide range of human actions. In the top sequence, notice that although the boundary extraction is somewhat less accurate in the middle of the jump, the system quickly recovers once more limbs become visible.

number of people in the scene, the appearance of the person or the background, the motion of the person or that of the camera. To achieve this, we improve a number of existing techniques for person detection and pose estimation, leveraging on temporal consistency to improve both the accuracy and speed of existing techniques. We initialize our system using a state-of-the-art upright pedestrian detection algorithm [1]. While this technique works well on average, it produces many false positive windows and very often fails to detect. We improve this situation by building an appearance model and applying a two-pass constrained clustering algorithm [2] to verify and extend the detections.

Once we have these basic detections, we build articulated models following [3,4,5] to carve out arbitrary motions of moving humans into continuous spatio-temporal volumes. The result can be viewed as a segmentation of the moving person, but we are not aiming to achieve pixel-level accuracy for the extraction. Instead, we offer a relatively efficient and accurate algorithm based on the prior knowledge of the human body configuration. Specifically, we enhance the speed and potential accuracy of [4,5] by leveraging temporal continuity to constrain the search space and applying semi-parametric density propagation to speed up evaluation.

The paper is organized as follows. After reviewing previous work in the area of human motion analysis in Section 1.1, we describe the overall system architecture in Section 2. Two main parts of our system, person detection/clustering and extraction of moving human boundaries, are presented in Sections 3 and 4, respectively. Finally, implementation details and experimental results are described in Section 5.

1.1   Related Work

Body Tracking. The most straightforward method to track humans is to consider them as blobs and use generic object tracking methods such as [6,7]. More complex methods attempt to model the articulation of the body

[8,9,10,11,12,13,14,15]. Most of these methods rely on a manual initialization, strong priors to encode the expected motion, a controlled or very simple environment with good foreground/background separation, and/or seeing the motion from multiple cameras.

Pedestrian Detection and Pose Estimation. Several fairly reliable pedestrian detection algorithms have been developed recently [1,16,17,18,19,20]. However, these methods typically deal with upright persons only, and the detection accuracy is significantly reduced by even moderate pose variations. Furthermore, these algorithms offer little segmentation of the human, providing only a bounding box of the body.

To model body configurations, tree shaped graphical models have shown promising results [3,4,5]. These generative models are often able to find an accurate pose of the body and limbs. However, they are less adept at making a discriminative decision: is there a person or not? They are typically also very expensive computationally in both the measurement and inference steps.

We build on these models and address the discrimination problem by initializing detections with an upright person detector. To improve computational efficiency, our algorithm exploits temporal information and uses more efficient semi-parametric (Gaussian mixture) representations of the distributions.

Based on similar intuitions, [21] uses temporal information to reduce the search space progressively in applying pictorial structures to videos. Ren et al. [22] takes another approach to human pose estimation in videos by casting the figure tracking task into a foreground/background segmentation problem using multiple cues, though the algorithm seems to rely on objects having a high contrast with the background.

2   System Architecture

Our system consists of two main components. The first component generates object-level hypotheses by coupling a human detector with a clustering algorithm. In this part, the state of each person, including location, scale and trajectory, is obtained and used to initialize the body configuration and appearance models for limb-level analysis. Note that in this step two separate problems – detection and data association – are handled simultaneously, based on the spatio-temporal coherence and appearance similarity.

The second component extracts detailed human motion volumes from the video. In this stage, we further analyze each person's appearance and spatio-temporal body configuration, resulting in a probability map for each body part. We have found that we can improve both the robustness and efficiency of the algorithm by limiting the search space of the measurement and inference around the modes of the distribution. To do this, we model the density function as a mixture of Gaussians in a sequential Bayesian filtering framework [23,24,25].

The entire system architecture is illustrated in Fig. 2. More details about each step are described in the following two sections.

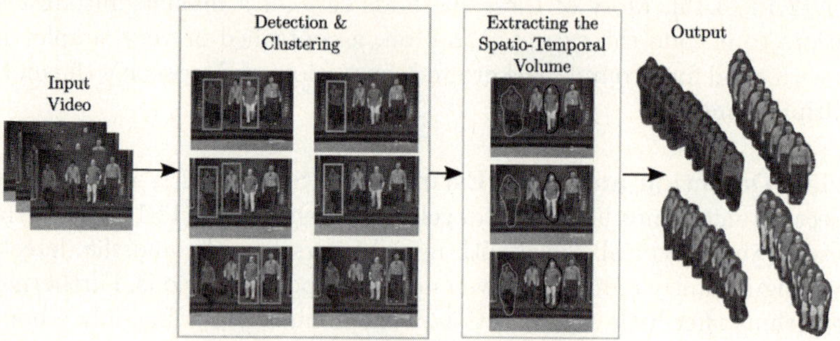

Fig. 2. Overall system

The focus of our work is to extract arbitrarily complex human motions from *YouTube* videos that involve a large degree of variability. We face several difficult challenges, including:

1. Compression artifacts and low quality of videos
2. Multiple shots in a video
3. Unknown number of people in each shot or sequence
4. Unknown human motion and poses
5. Unknown camera parameters and motion
6. Background clutter, motion and occlusions

We will refer back to these points in the rest of the paper as we describe how the components try to overcome them.

3   People Detection and Clustering

As Fig. 2 shows, our system begins with a step to estimate location, scale, and trajectories of moving persons. This step is composed of the following two parts.

3.1   Initial Hypothesis by Detection

We first employ an human detection algorithm [1] to generate a large number of hypotheses for persons in a video. This method, which trains a classifier cascade using boosting of HOG features to detect upright standing or walking people, has serious limitations. It only detects upright persons and cannot handle arbitrary poses (challenge 4). The performance is degraded in the presence of compression artifacts (challenge 1). Moreover, since it does not use any temporal information, the detection is often inconsistent and noisy, especially in scale. It is, therefore, difficult to reject false positives and recover miss-detections effectively. The complexity increases dramatically when multiple people are involved (challenge 3). This step, therefore, serves only as an initial hypotheses proposal stage. Additional efforts are required to handle various exceptions.

3.2   People Clustering

The output of the person detector is a set of independent bounding boxes; there are no links for the same individual between detections. The detections also have significant noise, false alarms and miss-detections especially due to the low quality of the video (challenge 1). In order to recover from these problems, we incorporate a clustering algorithm based on the temporal and appearance coherence of each person. The goal of clustering in our system is to organize all correct detections into groups, where each corresponds to a single person in the sequence (challenge 3), while throwing away false alarms. To achieve this, we apply a constrained clustering paradigm [2] in two hierarchical stages, adding both positive (should link) edges and negative (can not link) constraints between the detections. See Fig. 3 for an example.

Stage 1. In the first stage, we focus on exploiting the temporal-coherence cue by associating detections from multiple frames with the help of a low-level tracking algorithm [7]. When the first detection is observed, a low-level tracker is initialized with the detected bounding box. A new detection in a consequent frame is assigned to an existing track if it coherently overlaps with the tracker predictions. In this case, we reinitialize the tracker with the associated detection bounding box. When no existing track can explain the new detection, a new track is created. Due to the complexity of the articulated human body, a low-level tracker is susceptible to drift from the person. We thus limit the temporal life of the tracker by counting the number of frames after the last detection and terminating the track at the last detection if the maximum gap (e.g. 100 frames) is surpassed. Very small clusters with few detections are discarded. The clusters produced in this first stage are almost always correct but over-segmented tracks (see Fig. 3 (b)). This is because the person detector often fails to detect a person in the video for many frames in a row – especially when the person performs some action that deviates from an upright pose.

Stage 2. The stage 2 agglomerative constrained clustering views the stage 1 clusters as atomic elements, and produces constraints between them with positive weights determined by appearance similarity and negative constraints determined by temporal/positional incompatibility.

For the appearance similarity term, we select multiple high-scoring detection windows for each stage 1 cluster, and generate probability maps for the head and torso locations using a simple two-part pictorial structure [4]. We use these results to (1) remove false detections by rejecting clusters that have unreliable head/torso estimation results (e.g., high uncertainty in the estimated head and torso locations), and (2) generate a weighted mask for computing color histogram descriptors for both the head and the torso. The appearance of the person in each cluster is then modeled with the color distributions of head and torso.

After the second pass of our hierarchical clustering, we obtain one cluster per person in the sequence. Fig. 3 (c) illustrates the final clustering result, which shows that three different persons and their trajectories are detected correctly, despite the fact that the appearance of these individuals are very similar (Fig. 3 (d)).

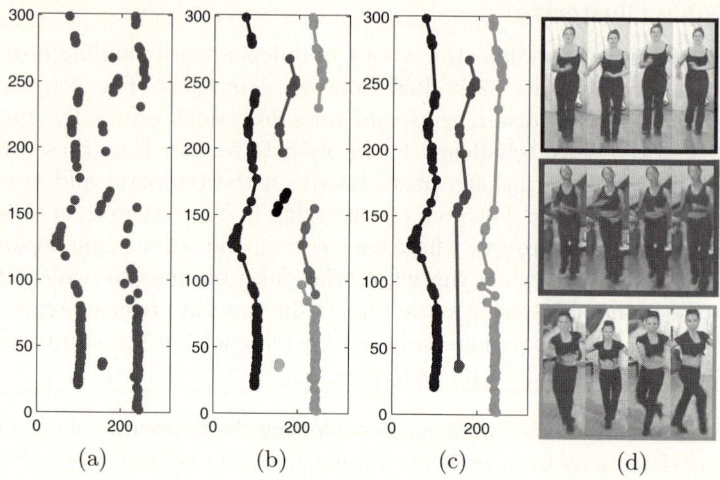

Fig. 3. Human detection and clustering result. From noisy detections, three tracks of people are identified successfully by filling gaps and removing outliers. (In this figure, the horizontal and vertical axis are the x locations and frame numbers, respectively.) (a) Original detection (b) Initial clusters after step 1 (c) Final clusters (d) Example images of three similar people that correctly clustered into different groups.

4   Extracting Spatio-temporal Human Motion Volume

We now have a cluster for each person, with a detection bounding box giving the location, scale, and appearance in some subset of the frames. Our goal is to find the body configuration for all the frames of the cluster (challenge 4), both where we have detections and where we do not. In this section, we discuss how to extract human body pose efficiently in every frame.

The existing algorithms for human motion analysis based on belief propagation such as [3,5] typically require exhaustive search of the input image because minimal (or no) temporal information is employed for the inference. Our idea is to propagate the current posterior to the next frame for the future measurement.

4.1   Overview

We summarize here the basic theory for the belief propagation and inference in [3,4]. Suppose that each body part p_i is represented with a 4D vector of $(x_i, y_i, s_i, \theta_i)$ – location, scale and orientation. The entire human body B is composed of m parts, i.e. $B = \{p_1, p_2, \ldots, p_m\}$. Then, the log-likelihood given the measurement from the current image I is

$$L(B|I) \propto \sum_{(i,j) \in E} \Psi(p_i - p_j) + \sum_i \Phi(p_i) \tag{1}$$

where $\Psi(p_i - p_j)$ is the relationship between two body parts p_i and p_j, and $\Phi(p_i)$ is the observation for body part p_i. E is a set of edges between directly connected body parts. Based on the given objective function, the inference procedure by message passing is characterized by

$$M_i(p_j) \propto \sum_{p_j} \Psi(p_i - p_j)O(p_i) \tag{2}$$

$$O(p_i) \propto \Phi(p_i) \prod_{k \in C_i} M_k(p_i) \tag{3}$$

where $M_i(p_j)$ is the message from part p_i to p_j, $O(p_i)$ is the measurement of part p_i, and C_i is a set of children of part p_i. The top-down message from part p_j to p_i for the inference is defined by

$$P(p_i|I) \propto \Phi(p_i) \sum_{p_j} \Psi(p_i - p_j)P(p_j|I), \tag{4}$$

which generates the probability map of each body part in the 4D state.

Based on this framework, we propose a method to propagate the density function in the temporal domain in order to reduce search space and temporally consistent results. The rest of the section describes the details of our algorithm.

4.2   Initialization

The first step for human body extraction is to estimate an initial body configuration and create a reliable appearance model. The initial location of the human is given by the method presented in Section 3. Note that the bounding box produced by the detection algorithm does not need to be very accurate since most of the background area will be removed by further processing. Once a potential human region is found, we apply a pose estimation technique [4] based on the same pictorial structure and obtain the probability map of the configuration of each body part through the measurement and inference step. In other words, the output of this algorithm is the probability map $P_p(u, v, s, \theta)$ for each body part p, where (u, v) is location, s is scale and θ is orientation. A sample probability map is presented in Fig. 4 (b)-(d). Although this method creates accurate probability maps for each human body part, it is too computationally expensive to be used in video processing. Thus, we adopt this algorithm only for initialization.

4.3   Representation of Probability Map

The original probability map P_p is represented by a discrete distribution in 4D space for each body part. There are several drawbacks of the discrete density function. First of all, it requires a significant amount of memory space, which is proportional to the image size and granularity of the orientations and scales, even if most of the pixels in the image have negligible probabilities. Second, the propagation of a smooth distribution is more desirable for the measurement in the next step since a spiky discrete density function may lose a significant number of potentially good candidates by sampling.

Instead of using the non-parametric and discrete probability map, we employ a parametric density function. However, finding a good parametric density function is not straightforward, especially when the density function is highly multi-modal as in human body. In our problem, we observe that the probability map for each orientation is mostly uni-modal and close to a Gaussian distribution[1]. We employ a mixture of N Gaussians for the initialization of human body configuration, where N is the number of different orientations.

Denote by $\mathbf{x}_i^{(k)}$ and $\omega_i^{(k)}$ ($i = 1, \ldots, n$) the location and weight of each point in the k-th orientation probability map. Let $\theta^{(k)}$ be the orientation corresponding the k-th orientation map. The mean ($\mathbf{m}^{(k)}$), covariance ($\mathbf{P}^{(k)}$) and weight ($\kappa^{(k)}$) of the Gaussian distribution for the k-th orientation map is then given by

$$\mathbf{m}^{(k)} = \begin{pmatrix} \mathbf{x}^{(k)} \\ \theta^{(k)} \end{pmatrix} = \begin{pmatrix} \sum_i \omega_i^{(k)} \mathbf{x}_i^{(k)} \\ \theta^{(k)} \end{pmatrix} \tag{5}$$

$$\mathbf{P}^{(k)} = \begin{pmatrix} \mathbf{V_x} & \mathbf{0} \\ \mathbf{0}^\top & V_\theta \end{pmatrix} = \begin{pmatrix} \sum_i \omega_i^{(k)} (\mathbf{x}_i^{(k)} - \mathbf{m}^{(k)})(\mathbf{x}_i^{(k)} - \mathbf{m}^{(k)})^\top & \mathbf{0} \\ \mathbf{0}^\top & V_\theta \end{pmatrix} \tag{6}$$

$$\kappa^{(k)} = \sum_i \mathbf{x}_i^{(k)} / \sum_k \sum_i \mathbf{x}_i^{(k)} \tag{7}$$

where $\mathbf{V_x}$ and V_θ are (co)variance matrices in spatial and angular domain, respectively. The representation of the combined density function based on the entire orientation maps is given by

$$\hat{f}(\mathbf{x}) = \frac{1}{(2\pi)^{d/2}} \sum_{i=1}^{N} \frac{\kappa^{(k)}}{\mid \mathbf{P}^{(k)} \mid^{1/2}} \exp\left(-\frac{1}{2} D^2\left(\mathbf{x}, \mathbf{x}^{(k)}, \mathbf{P}^{(k)}\right) \right) \tag{8}$$

where $D^2\left(\mathbf{x}, \mathbf{x}^{(k)}, \mathbf{P}^{(k)}\right)$ is the Mahalanobis distance from $\mathbf{x}$ to $\mathbf{x}^{(k)}$ with covariance $\mathbf{P}^{(k)}$.

Although we simplify the density functions for each orientation as a Gaussian, it is still difficult to manage them in an efficient way especially because the number of components will increase exponentially when we propagate the density to the next time step. We therefore adopt Kernel Density Approximation (KDA) [26] to further simplify the density function with little sacrifice in accuracy. KDA is a density approximation technique for a Gaussian mixture. The algorithm finds the mode locations of the underlying density function by an iterative procedure, such that a compact mixture of Gaussians based on the detected mode locations is found.

Fig. 4 presents the original probability map and our approximation using a mixture of Gaussians for each body part after the pose estimation. Note that the approximated density function is very close to the original one and that the multi-modality of the original density function is well preserved.

---

[1] Arms occasionally have significant outliers due to their flexibility. A uni-modal Gaussian fitting may result in more error here.

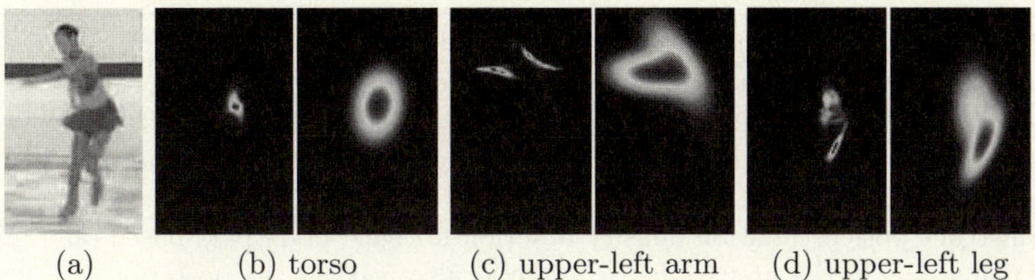

(a)   (b) torso   (c) upper-left arm   (d) upper-left leg

Fig. 4. Comparison between the true probability map for the pose estimation (left in each sub-figure) and its Gaussian mixture approximation (right) for each body part. The approximated density functions are propagated for the measurement in the next time step. Note that our approximation results look much wider since different scales in the color palette are applied for better visualization.

4.4   Measurement, Inference and Density Propagation

Fast and accurate measurement and inference are critical in our algorithm. As shown in Eq. (2) and (3), the bottom-up message is based on all the information up to the current node as well as the relative configuration with the parent node. Exhaustive search is good for generating the measurement information at all possible locations. However, it is very slow and, more importantly, the performance for the inference may be affected by spurious observations; noisy measurement incurred by an object close to or moderately far from the real person may corrupt the inference process. A desirable reduction of search space not only decreases computation time, but also improves the accuracy. The search space for measurement and inference is determined by a probability density function characterizing potential state of human body, where a mixture of Gaussians are propagated in sequential Bayesian filtering framework [23,24,25].

In our method, we perform local search based on the spatio-temporal information. We first diffuse the posterior density function from the previous frame, which is done analytically thanks to the Gaussian mixture representation. Based on the diffused density, locally dense samples are drawn to make measurements and a discrete density function is constructed. Note that inference is performed using the discrete density function. But a parametric representation of density function is propagated to the next time step for the measurement. After the inference, the pose estimation density function is converted to a mixture of Gaussians by the method described in Section 4.3. The posterior is given by the product of the diffused density and the pose estimation density function in the current frame. This step is conceptually similar to the integration of the measurement and inference history (temporal smoothing). We denote by $\mathbf{X}$ and $\mathbf{Z}$ the state and observation variable in the sequential Bayesian filtering framework, respectively. The posterior at the time step t of the state is given by the product of two Gaussian mixture as follows:

$$p(\mathbf{X}_t|\mathbf{Z}_{1:t}) \propto p(\mathbf{Z}_t|\mathbf{X}_t)p(\mathbf{X}_t|\mathbf{Z}_{1:t-1}) \tag{9}$$

$$= \left(\sum_{i=1}^{N_1} \mathcal{N}(\kappa_i, \mathbf{x}_i, \mathbf{P}_i) \right) \left(\sum_{j=1}^{N_2} \mathcal{N}(\tau_j, \mathbf{y}_j, \mathbf{Q}_j) \right), \tag{10}$$

Algorithm 1. Moving human body extraction

1: Apply human detection algorithm to a sequence
2: Apply clustering algorithm based on the detection. Create the initial body configuration and appearance at the first detection. Also, obtain the number of people in the video.
3: Construct pose estimation density function for each body part based on a mixture of Gaussians in the first frame, where it is also used as the posterior.
4: **while** not the end of sequence **do**
5:     Go to the next frame
6:     Diffuse the posterior of the previous frame
7:     Perform the measurement and inference with the locally dense samples
8:     Create a Gaussian mixture with the discrete pose estimation distribution
9:     Compute the posterior by multiplying diffusion and pose estimation density
10:     **if** there exists the detection of the same person **then**
11:         Reinitialize the appearance and body configuration of the person (optional)
12:     **end if**
13: **end while**

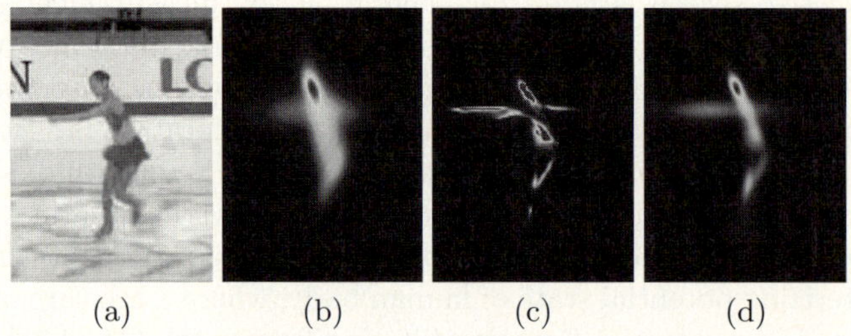

(a)   (b)   (c)   (d)

Fig. 5. Density functions in one step of the human motion extraction. (a) Original frame (cropped for visualization) (b) Diffused density function (c) Measurement and inference results (d) Posterior (Note that the probability maps for all orientations are shown in a single image by projection.)

where $\mathcal{N}(\cdot)$ represents a Gaussian distribution with parameters of weight, mean, and covariance. The first and second terms in the right hand side represent diffusion and pose estimation density function, respectively. Note that the product of two Gaussian mixtures is still a Gaussian mixture, but it causes the exponential increase of the number of components. So KDA is required again to maintain a compact representation of the density function.

The density propagation algorithm for inference is summarized in Algorithm 1, and illustrated in Fig. 5.

5   Experiments

In order to evaluate our proposed approach, we have collected a dataset of 50 sequences containing moving humans downloaded from *YouTube*. The sequences contain natural and complex human motions and various challenges mentioned

Multiple Instance Boost Using Graph Embedding Based Decision Stump for Pedestrian Detection

Junbiao Pang[1,2,3], Qingming Huang[1,2,3], and Shuqiang Jiang[2,3]

[1] Graduate university of Chinese Academy of Sciences, Beijing, 100190, China
[2] Key Lab. of Intelligent Information Processing, Chinese Academy of Sciences (CAS)
[3] Institute of Computing Technology, CAS, Beijing, 100190, China
{jbpang,qmhuang,sqjiang}@jdl.ac.cn

Abstract. Pedestrian detection in still image should handle the large appearance and stance variations arising from the articulated structure, various clothing of human as well as viewpoints. In this paper, we address this problem from a view which utilizes multiple instances to represent the variations in multiple instance learning (MIL) framework. Specifically, logistic multiple instance boost (LMIBoost) is advocated to learn the pedestrian appearance model. To efficiently use the histogram feature, we propose the graph embedding based decision stump for the data with non-Gaussian distribution. First the topology structure of the examples are carefully designed to keep between-class far and within-class close. Second, K-means algorithm is adopted to fast locate the multiple decision planes for the weak classifier. Experiments show the improved accuracy of the proposed approach in comparison with existing pedestrian detection methods, on two public test sets: INRIA and VOC2006's person detection subtask [1].

1   Introduction

Pedestrian detection is a practical requirement of many today's automated surveillance, vehicle driver assistance systems and robot vision systems. However, the issue of large appearance and stance variations accompanied with different viewpoints makes pedestrian detection very difficult. The reasons can be multifold, such as variable human clothing, articulated human structure and illumination change, etc. The variations bring various challenges including miss-alignment problem, which is often encountered in non-rigid object detection.

There exist a variety of pedestrian detection algorithms from the different perspectives, directly template matching [2], unsupervised model [3], *traditional* supervised model [4,5,6] and so on. Generally, these approaches cope with "mushroom" shape – the torso is wider than the legs, which dominates the frontal pedestrian, and deal with "scissor" shape – the legs are switching in walk, which dominates the lateral pedestrian. However, for some uncommon stances, such as mounting on bike, they incline to fail. In these conditions, the variations often impair the performance of these conventional approaches. Fig. 1 shows some false negatives generated by Dalal et al [4]. These false negatives are typically non-"mushroom" or non-"scissor" shape, and have large variations between each other.

D. Forsyth, P. Torr, and A. Zisserman (Eds.): ECCV 2008, Part IV, LNCS 5305, pp. 541–552, 2008.
© Springer-Verlag Berlin Heidelberg 2008

Fig. 1. Some detection results in our method producing fewer false negatives than Dalal et al do [4]

The key notion of our solution is that the variations are represented within multiple instances, and the "well" aligned instances are automatically selected to train a classifier via multiple instance learning (MIL) [7,8]. In MIL, a training example is not singletons, but is represented as a "bag" where all of the instances in a bag share the bag's label. A positive bag means that at least one instance in the bag is positive, while a negative bag means that all instances in the bag are negative. To pedestrian detection, the standard scanning window is considered as the "bag", a set of sub-images in window are treated as instances. If one instance is classified as pedestrian, the pedestrian is located in detection stage. The logistic multiple instance boost (LMIBoost) [9] is utilized to learn the pedestrian appearance, which assumes the average relationship between bag's label and instance's label.

Considering the non-Gaussian distribution (which dominates the positive and negative examples) and aims of detection (which are accurate and fast), a graph embedding based weak classifier is proposed for histogram feature in boosting. The graph embedding can effectively model the non-Gaussian distribution, and maximally separate the pedestrians from negative examples in low dimension space [10]. After feature is projected onto discriminative one dimension manifold, K-means is utilized to fast locate the multiple decision planes for the decision stump. The proposed weak classifier has the following advantages: 1) it handles training examples with any distribution; and 2) it not only needs less computation cost, but also results in robust boosting classifier. The main contributions of the proposed algorithm are summarized as following:

– The pose variations are handled by multiple instance learning. The variations between examples are represented within the instances, and are automatically reduced during learning stage.
– Considering the boost setting, graph embedding based decision stump is proposed to handle training data with non-Gaussian distribution.

In the next section, related work is briefly summarized. Section 3 introduces the LMIBoost for solving the variations. Section 4 first introduces the graph embedding based discriminative analysis, and then presents the multi-channel decision stump. In section 5, we describe the experimental settings for pedestrian detection. Finally the experiment and conclusion sections are provided, respectively.

2   Related Work

Generally, the "mushroom" or "scissor" shape encourages the use of template matching and traditional machine learning approach as discussed in section 1. The contour templates are hierarchically matched via Chamfer matching [2]. A polynomial support vector machine (SVM) is learned with Haar wavelets as human descriptor [5] (and variants are described in [11]). Similar to still images, a real-time boosted cascade detector also uses Haar wavelets descriptor but extracted from space-time differences in video [6]. In [4], an excellent pedestrian detector is described by training a linear SVM classifier using densely sampled histogram of oriented gradients (HOG) feature (this is a variant of Lowe's SIFT descriptor [12]). In a similar approach [13], the near real-time detection performance is achieved by training a cascade detector using SVM and HOG feature in AdaBoost. However, their "fixed-template-style" detectors are sensitive to pose variations. If the pose or appearance of the pedestrian has large change, the "template"-like methods are doomed to fail. Therefore, more robust feature is proposed to withstand translation and scale transformation [14].

Several existing publications have been aware of the pose variation problem, and have handled it by "divide and conquer"– the parts based approach. In [15], the body parts are explicitly represented by co-occurrences of local orientation features. The separate detector is trained for each part using AdaBoost. Pedestrian location is determined by maximizing the joint likelihood of the part occurrences according to the geometric relations. Codebook approach avoids explicitly modeling the body segments or the body parts, and instead uses unsupervised methods to find part decompositions [16]. Recently, the body configuration estimation is exploited to improve pedestrian detection via structure learning [17]. However, parts based approaches have two drawbacks. First, different part detector has to be applied to the same image patch. This reduces the detection speed. Second, labeling and aligning the local parts are tedious and time-costing work in supervised learning. Therefore, the deformable part model supervised learns the holistic classifier to coarsely locate the person, and then utilizes part filters to refine body parts in unsupervised method [18].

The multiple instance learning(MIL) problem is first identified in [8], which represents ambiguously labeled examples using axis-paralled hyperrectangles. Previous applications of MIL in vision have focused on image retrieval [19]. The seemingly most similar work to ours may be the upper-body detection [20]. Viola et al use Noisy-OR boost which assumes that only sparse instances are upper-body in a positive bag. However, in our pedestrian detection setting, the instances in a positive bag are all positive, and this facilitates to simply assume that every instance in a bag contributes equally to the bag's class label.

In pedestrian detection, the histogram feature (such as SIFT, HOG) is typically used. The histogram feature can be computed rapidly using an intermediate data representation called "Integral Histogram" [21]. However, the efficient use of the histogram feature is not well discussed. In [13], the linear SVM and HOG feature is used as weak classifier. Kullback-Leibler (K-L) Boost uses the log-ratio between the positive and negative projected histograms as weak classifier. The projection function is optimized by maximizing the K-L divergence between the positive and negative features [22]. SVM has high computational cost and hence reduces the detection speed. Optimizing the projection function in K-L Boost is also computationally costly and numerically unstable. Fisher linear discriminative analysis (FLDA) is used as weak classifier for histogram feature [23]. Despite the success of FLDA for building weak classifier, it still has the following limitations: it is optimal only in the case that the data for each class are approximate Gaussian distribution with equal covariance matrix.

Although the histogram feature is projected into one dimension manifold using the projection functions, the learned manifold does not directly supply classification ability. The widely used decision stump is a kind of threshold-type weak classifier, but a lot of discriminative information is lost [24]. Therefore, the single-node, multi-channel split decision tree is introduce to exploit the discriminative ability. In face detection [25], Huang et al use the histogram to approximate the distributions of the real value feature by dividing the feature into many sub-regions with equal width in RealBoost. Then a weak classifier based on a look up table (LUT) function is built by computing the log-ratio on each sub-bins. However, the equal regions unnecessarily waste decision stump in low discriminative region. In [26], the unequal regions are obtained by exhaustively merging or splitting the large number of histogram bins via Bayes decision rule. In this paper, we avoid exhaustive searching and emphasize on fast designing the multi-channel decision stump via K-means clustering.

3   Logistic Multiple Instance Boost

If pedestrian have uncommon stance, human-centering normalization often produces miss-aligned examples as illustrated in Fig. 1. Intuitively, some parts of human can be aligned by shifting the normalization window. Therefore, we augment the training set by perturbing the training examples. The created instances can take advantage of all information of the "omega" heads and the rectangle bodies. Moreover, the augmented training set should cover the possible pose variations for MIL. Fig. 2 illustrates the proposed approach.

Compared with traditionally supervised learning, an instance in MIL is indexed with two indices: i which indexes the bag, and j which indexes the instance within the bag. Given a bag $\mathbf{x}_i$, the conditional probability of the bag-level class $\mathbf{y}_i$ is

$$p(\mathbf{y}_i|\mathbf{x}_i) = \frac{1}{n_i} \sum_{j=1}^{n_i} p(y_{ij}|x_{ij}),\tag{1}$$

where n_i is the number of the instances in the i-th bag, y_{ij} is the instance-level class label for the instance x_{ij}. Equation.(1) indicates that every instance contributes equally to the bag's label. This simple assumption is suitable for the instances generated by perturbing around the person. Because the generated every instance is positive pedestrian image.

The instance-level class probability is given as $p(y|x) = 1/(1 + e^{\beta x})$, where β is the parameter to be estimated. Controlling the parameter β gives different

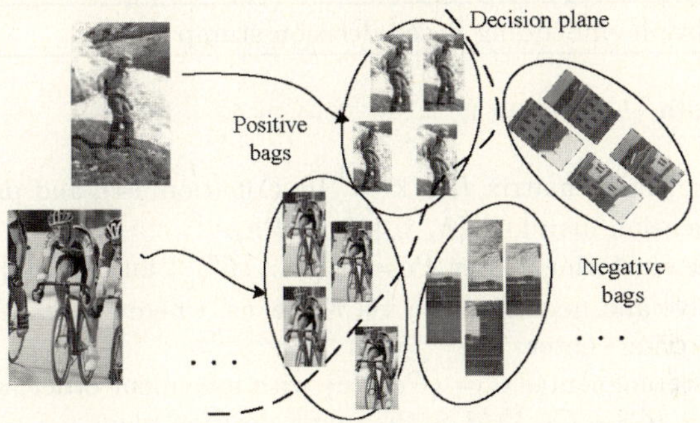

Fig. 2. Overview of the multiple instance learning process. The training example is first converted into a bag of instances. Note that we only generate the instances spatially, and the instances can also be generated at different scales. Therefore, the resulting classifier will withstand the translation and scale transformation.

instance-level class probability, which gives different contribution to bag-level probability. Ideally, the "well" aligned instances should be assigned higher probability than the non-aligned. Given a collection of N *i.i.d* bags $\mathbf{x}_1, \ldots, \mathbf{x}_N$, the parameter β can be estimated by maximizing the bag-level binomial log-likelihood function

$$L = \sum_i^N \left[\mathbf{y}_i \mathrm{log} p(\mathbf{y}_i = 1 | \mathbf{x}_i) + (1 - \mathbf{y}_i) \mathrm{log} p(\mathbf{y}_i = 0 | \mathbf{x}_i) \right]. \tag{2}$$

Equation.(2) can not be solved analytically. Xu el al [9] propose an boosting method to maximize the log-likelihood function. We need to learn a bag-level function $\mathbf{F}(\mathbf{x}) = \sum_m c_m \mathbf{f}_m(\mathbf{x})$ and the corresponding *strong* classifier $H = \mathrm{sign}(\mathbf{F}(\mathbf{x}))$, where weights $c_1, \ldots, c_M \in \mathbb{R}$, the $\mathbf{f}$ is the bag-level weak classifier. The expected empirical loss is

$$E[I(\mathbf{F}(\mathbf{x}) \neq \mathbf{y})] = -\frac{1}{N} \sum_{i=1}^N \mathbf{y}_i \mathbf{F}(\mathbf{x}_i), \tag{3}$$

where $I(\cdot)$ is the indicator function. We are interesting in wrapping the bag-level weak classifier $\mathbf{f}$ with the instance-level weak classifier f. Using the Equation.(1), Equation.(3) is converted into the instance-level's exponential loss $E_{\mathbf{x}} E_{\mathbf{y}|\mathbf{x}} [e^{-yf}]$ as $e^{-\mathbf{y}\mathbf{H}} \geq I(H(\mathbf{x}) \neq \mathbf{y}), \forall M$. One searches for the optimal update $c_m f_m$ such that minimizes

$$E_{\mathbf{x}} E_{\mathbf{y}|\mathbf{x}} \left[e^{-y_{ij} F_{m-1}(x_{ij}) - c_m y_{ij} f_m(x_{ij})} \right] = \sum_i w_i e^{[(2\epsilon_i - 1)c_m]}, \tag{4}$$

where $\epsilon_i = \sum_j \mathbf{1}_{f_m(x_{ij}) \neq y_{ij}} / n_i$, w_i is the example's weight. The error ϵ_i describes the discrepancy between the bag's label and instance's label. The instance in positive bags with higher score $f(x_{ij})$ gives higher confidence to the bag's label, even though there are some negative instances occurring in the positive bag.

Algorithm. 1 Graph embedding based decision stump

Input:
  The training data $\{h_i, y_i\}, i = 1, \ldots, n$
Training:
  1. Learn the projection matrix $P \in \mathbb{R}^{1 \times D}$ by Equation. (4), and project the data into one dimension manifold $\{\hat{h}_i, y_i\}, \hat{h}_i = Ph_i$.
  2. Calculate the clustering center $Pc = \{C_1^p, \ldots, C_{N_p}^p\}$ and $Nc = \{C_1^n, \ldots, C_{N_n}^n\}$ for the positive and negative data via K-means, where N_p and N_n is the number of clustering center.
  3. Sort the clustering center $C = \{Pc, Nc\}$ with ascendent order, and find the middle value $r_k = (C_k + C_{k+1})/2$ as the rough decision plane.
  4. Generate the histogram with the intervals $\sigma_k = (r_k, r_{k+1}]$, and produce the class label ω_c for each interval via Bayesian decision rule.
  5. Iteratively merge adjacent intervals with same decision label ω_c to produce a set of consistent intervals $\hat{\sigma}_k$.
Output:
A LUT function $\text{lup}(k)$ on the merged intervals $\hat{\sigma}_k, k = 1, \ldots, K$.

Therefore, the final classifier often classifies these bags as positive. The variations problem in training examples will be reduced.

4   Graph Embedding Based Decision Stump

4.1   Supervised Graph Embedding

Let $h_i \in \mathbb{R}^D (i = 1, 2, \ldots, n)$ be the D-dimensional histogram feature and $y_i \in \{\omega_c\}_{c=1}^2$ be the associated class label. The feature is written as matrix form: $H = (h_1 | h_2 | \ldots | h_n)$. Let $G = \{\{h_i\}_{i=1}^n, S\}$ be an undirected weighted graph with vertex set $\{h_i\}_{i=1}^n$ and the similarity matrix $S \in \mathbb{R}^{n \times n}$. The element $s_{i,j}$ of matrix S measures the similarity of vertex pair i and j. The unsupervised graph embedding is defined as the optimal low dimension vector representations for the vertices of graph G

$$P^* = \min_{P^T H M H^T P = I} \sum_{i,j} ||Ph_i - Ph_j||^2 s_{i,j} = \min_{P^T H M H^T P = I} 2\text{tr}(P^T H L H^T P), \quad (5)$$

where projection $P \in \mathbb{R}^{d \times D}, (d < D)$ maps feature h from high dimension space $\mathbb{R}^D$ to low dimension space $\mathbb{R}^d$. The elements in the diagonal matrix M is $m_{i,j} = \sum_{i \neq j} s_{i,j}$, and the Laplacian matrix L is $M - S$.

The similarity $s_{i,j}$ connects the relationship between high dimension and low dimension space. If two vertexes h_i and h_j are close, $s_{i,j}$ will be large, and vice versa. To classification, the projection P should keep the between-class far and within-class close. The similarity matrix S should reflect the separable ability. The between-class similarity $s_{i,j}^b$ and within-class similarity $s_{i,j}^w$ can be defined as[1]

---

[1] We refer the interested reader to [10] for more details.

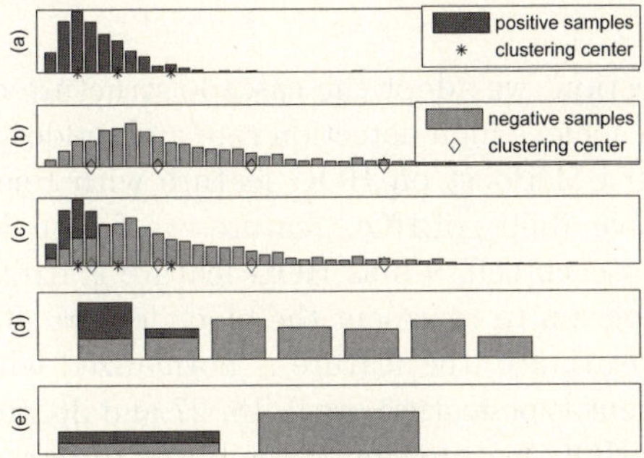

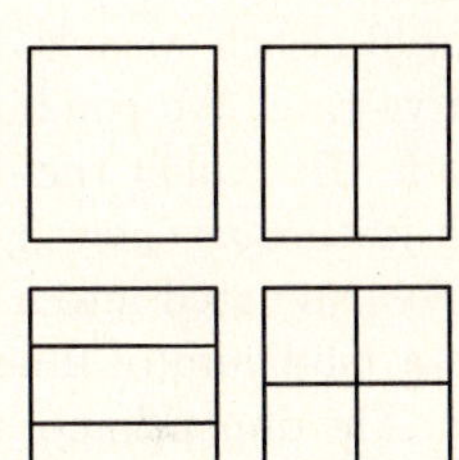

Fig. 3. A demonstration of generating the multi-channel decision stump. (a)-(b) Cluster on positive and negative examples, respectively. (d)Generate the decision stumps via histogram. (e)Merge the consistent decision stumps.

Fig. 4. 4 type block feature

$$s^b_{i,j} = \begin{cases} 1/n - 1/n_c \text{ if } & y_i = y_j = \omega_c, \\ 1/n \quad \text{ if } & y_i \neq y_j, \end{cases} \qquad s^w_{i,j} = \begin{cases} 1/n_c \text{ if } & y_i = y_j = \omega_c, \\ 0 \text{ if } & y_i = y_j, \end{cases} \tag{6}$$

where n_c is the cardinality of the ω_c class. The pairwise $s^b_{i,j}$ and $s^w_{i,j}$ try to keep within-class sample close (since $s^w_{i,j}$ is positive and $s^b_{i,j}$ is negative if $y_i = y_j$) and between-class sample pairs apart (since $s^b_{i,j}$ is positive if $y_i \neq y_j$). The projection matrix P can be calculated by Fisher criterion

$$P^* = \max_{P^T H(M_w - S_w)H^T P = I} \text{tr}(P^T H(M_b - S_b)H^T P). \tag{7}$$

The projection matrix $P = [p_1, p_2, \ldots, p_l]$ are solved by generalized eigenvectors corresponding to the l largest eigenvalues p in $H(M_w - S_w)H^T p_l = \lambda H(M_b - S_b)H^T p_l$.

4.2 Multi-channel Decision Stump

According to Bayesian decision theory, if class conditional probability $p(\omega_1|x) > p(\omega_2|x)$ we would naturally incline to decide that the true label of x is ω_1, and vice versa. Using Bayes rule $p(\omega|x) = p(x|\omega)p(\omega)$, the optimal decision plane is located at where $p(x|\omega_1) = p(x|\omega_2)$ with $p(\omega_1) = p(\omega_2)$. We obtain the Bayes error $p(error|x) = \int min[p(x|\omega_1), p(x|\omega_2)]dx$. However, the $p(x|\omega_c)$ is not directly available. To accurately estimate the $p(x|\omega_c)$, histogram needs large numbers of bins via uniform sampling in [25,26]. We avoid estimating the $p(x|\omega_c)$ with uniform sampling or rejection sampling. As demonstrated in Fig. 3(c), we consider the local region of feature space, and the location at the middle of two modal is a natural decision plane. The decision plane would approximately minimize Bayes error, if $p(\omega_1) = p(\omega_2)$. Algorithm. 1 shows the graph embedding based decision stump. Note that the number of decision planes is automatically decided.

5   Pedestrian Detection

To achieve the fast pedestrian detection, we adopt the cascade structure of detector [6]. Each stage is designed to achieve high detection rate and modest false positive rate. We combine $K = 30$ LMIBoost on HOG feature with rejection cascade. To exploit the discriminative ability of HOG feature, we design 4 type block feature as showed in Fig.4. In each cell, 9-bins HOG feature is extracted and concatenated into a single histogram to represent the block feature. To obtains a modicum of illumination invariance, the feature is normalized with L2 norm. The dimension of the 4 different type feature are 9, 18, 27 and 36, respectively. The 453×4 number of block HOG feature can be computed from a single detection window.

Assuming that the i-th cascade stage is trained, we classify all the possible detection window on the negative training images with the cascade of the previous k-1 LMIBoost classifiers. The examples which are misclassified in scanning window form the possible new negative training set. While, the positive training samples do not change during bootstrap. Let N_{pi} and N_{ni} be cardinality of the positive and negative training examples at i-th stage. Considering the influence of asymmetric training data on the classifier and computer RAM limitations, we constrain N_{pi} and N_{ni} to be approximately equal.

According to "There is no free lunch" theorem, it is very important to choose suitable number of instances in a bag for training and detection. More instances in a bag will represent more variations and improve the detection results, but will also reduce the training and detection speed. We experimentally set 4 instances for training and detection, respectively. Each level of cascade classifier is optimized to correctly detect at least 99% of the positive bags, while reject at least 40% of the negative bags.

6   Experiments

To test our method, we perform the experiments on two public dataset: INRIA [4] and VOC2006 [1]. The INRIA dataset contains 1239 pedestrian images (2478 with their left-right reflections) and 1218 person-free images for training. In the test set, there are 566 images containing pedestrians. The pedestrian images provided by INRIA dataset have large variations (but most of them have standing pose), different clothing and urban background. This dataset is very close to real-life setting. The VOC2006's *person* detection subtask supplies 319 images with 577 person as training set, and 347 images with 579 person as validation set. 675 images with 1153 person is supplied as test data. Note that the VOC2006's person detection dataset contains various human activities, different stances and clothing. Some examples of the two different datasets are showed in Fig. 8.

6.1   Performance Comparisons on Multiple Datasets

We plot the detection error tradeoff curves on a log-log scale for INRIA dataset. The y-axis corresponds to the miss rate, $FalseNeg/(FalseNeg+TruePos)$ and the

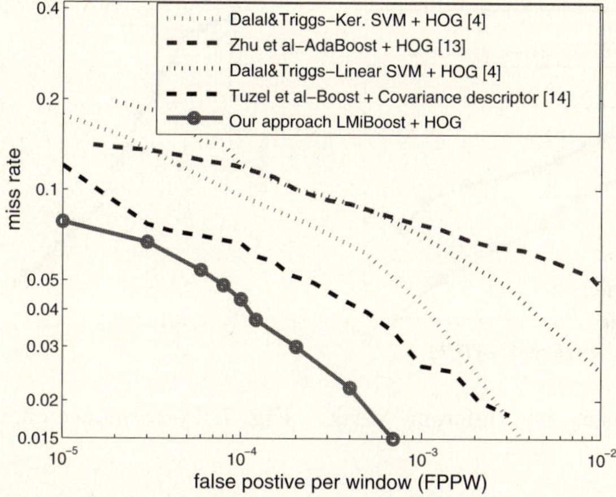

Fig. 5. Comparison results on INRIA dataset. Note that the curve of our detector is generated by changing the number of cascade stage used.

x-axis corresponds to false positives per window(FPPW), $FalsePos/(TrueNeg + FalsePos)$. We compare ours results with [4,13,14] on INRIA dataset. Although it has been noted that kernel SVM is computationally expensive, we consider both the kernel and linear SVM method of [4]. Only the best performing result, the L2-norm in HOG feature, is considered. Covariance descriptor [14] is also compared. Fig. 5 shows that the performance of our method is comparable to the state-of-art approaches. We achieve 4.3% miss rate at 10^{-4} FPPW. Notice that all the results by other methods are quoted directly from the original papers, since we perform the same separation of training-testing sets.

The Fig.7 shows the precision-recall cure on VOC2006 person detection subtask for comp3 [1]. The protocol of the comp3 is that the training data is composed of the training set and validation set. The non-normalized examples are first approximately aligned, and then be converted into a bag of instances. Some truncated and difficult examples in training data are discarded. The standard scanning window technique is adopted for detection, although the scanning window may be not suitable for VOC2006 detection subtask. The average precision scores is 0.23, which is better than the best results 0.164 reported by INRIA_Douze [1]. In Fig. 8, several detection results are showed for different scenes with human having variable appearance and pose. Significantly overlapping detection windows are averaged into a single window.

6.2 Analysis of the Weak Classifiers

For our next experiment, we conduct experiments to compare the performance of different weak classifiers. A common set of parameters (such as, false positive rate for every stage)are controlled equally for cascade training. Two detectors are trained with different weak classifiers, including FLDA and graph embedding based decision stump.

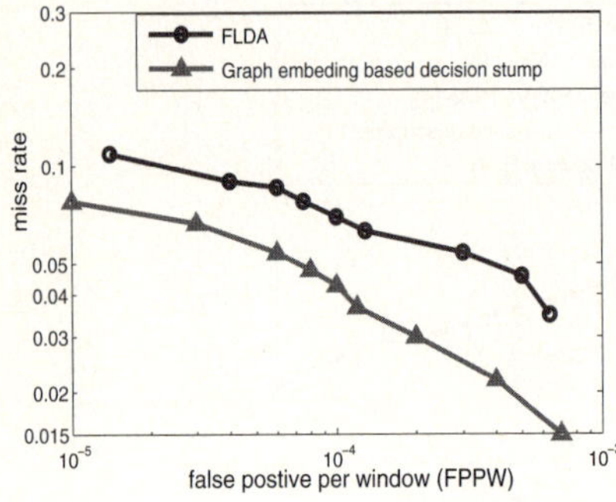

Fig. 6. Comparison on different weak classifier

Fig. 7. Performance on VOC2006 dataset

(a) INRIA dataset

(b) VOC2006 dataset

Fig. 8. Some detection samples on INRIA and VOC2006 datasets

The performance results on INRIA show that the detectors based on graph embedding decision stump outperforms the detector based on FLDA in Fig. 6. Unlike the other LUT weak classifier [26,25], the bins of decision stumps are automatically decided by algorithm.

6.3 Analysis of the Detection Speed

There are 90% of the negative examples are rejected at first five stage. The speed of the cascaded detector is directly related to the number of feature evaluated per scanned sub-window. For INRIA dataset, on average our method requires to evaluate 10.05 HOG feature per negative detection window. Densely scanning at 0.8 scale and 4 pixel step in a 320×240 image needs average 150ms under PC with 2.8GHz CPU and 512RAM. While, 250ms for 320×240 image is reported in Zhu et al's detector [13].

7 Conclusion and Future Work

We introduce the multiple instance learning into the pedestrian detection for solving pose variations. The training example does not need to be well aligned, but to be represented as a bag of instances. To efficiently utilizing histogram feature, a graph embedding based decision stump is proposed. The weak classifier guarantees the fast detection and better discriminative ability. The promising performances of the approach are shown on INRIA and VOC2006's person detection subtask.

Using multiple instance learning has enabled detector robust to the pose and appearance variations. Theoretically, the more instances are supplied, the more variations would be learned. Modeling the average relationship between the instance's label and bag's label may be unsuitable when there are large numbers of instances in a positive bag. In future, more experiments will be carried out to compare the different way to model the relationship.

Acknowledgements

This work was supported in part by National Natural Science Foundation of China under Grant 60773136 and 60702035, in part by National Hi-Tech Development Program (863 Program) of China under Grant 2006AA01Z117 and 2006AA010105. We would also thank the anonymous reviewers for their valuable comments.

References

1. Everingham, M., Zisserman, A., Williams, C.K.I., Gool, L.V.: The PASCAL Visual Object Classes Challenge (VOC 2006) Results (2006),
   http://www.pascal-network.org/challenges/VOC/voc2006/results.pdf
2. Gavrila, D.M.: Pedestrian detection from a moving vehicle. In: Vernon, D. (ed.) ECCV 2000. LNCS, vol. 1843, pp. 37–49. Springer, Heidelberg (2000)
3. Bissacco, A., Yang, M., Soatto, S.: Detection human via their pose. In: Proc. NIPS (2006)
4. Dalal, N., Triggs, B.: Histograms of oriented gradients for human detection. In: Proc. CVPR, vol. I, pp. 886–893. IEEE, Los Alamitos (2005)
5. Papageorgiou, P., Poggio, T.: A trainable system for object detection. IJCV, 15–33 (2000)
6. Viola, P., Jones, M., Snow, D.: Detecing pedestrians using patterns of motion and appearance. In: Proc. ICCV (2003)
7. Maron, O., Lozanno-Perez, T.: A framework for multiple-instance learning. In: Proc. NIPS, pp. 570–576 (1998)
8. Dietterich, T., Lathrop, R., Lozano-Perez, T.: Solving the multiple instance problem with axis-parallel rectangles. Artifical intelligence, 31–71 (1997)
9. Xu, X., Frank, E.: Logistic regression and boosting for labeled bags of instances. In: Dai, H., Srikant, R., Zhang, C. (eds.) PAKDD 2004. LNCS (LNAI), vol. 3056, pp. 272–281. Springer, Heidelberg (2004)

10. Sugiyama, M.: local fisher discriminat analysis for supervised dimensionality reduction. In: Proc. ICML (2006)
11. Monhan, A., Papageorgiou, C., Poggio, T.: Example-based object detection in images by componets. IEEE Trans. PAMI 23, 349–360 (2001)
12. Lowe, D.G.: Distinctive image features from scale-invariant keypoints, 91–110 (2004)
13. Zhu, Q., Avidan, S., Yeh, M.C., Cheng, K.T.: Fast human detection using a cascade of histograms of oriented gradients. In: Proc. CVPR, vol. 2, pp. 1491–1498. IEEE, Los Alamitos (2006)
14. Tuzel, O., Porikli, F., Meer, P.: Human detection via classification on riemannina manifolds. In: Proc. CVPR. IEEE, Los Alamitos (2007)
15. Zisserman, A., Schmid, C., Mikolajczyk, K.: Human detection based on a probabilistic assembly of robust part detectors. In: Pajdla, T., Matas, J(G.) (eds.) ECCV 2004. LNCS, vol. 3021, pp. 69–82. Springer, Heidelberg (2004)
16. Leibe, B., Seemann, E., Schiele, B.: Pedestrian detection in crowed scenes. In: Proc. CVPR, pp. 878–885. IEEE, Los Alamitos (2005)
17. Tran, D., Forsyth, D.A.: Configuration estimates improve pedestrian finding. In: Proc. NIPS (2007)
18. Felzenszwalb, P., Mcallester, D., Ramanan, D.: A discriminatively trained, multiscale, deformable part model. In: Proc. CVPR. IEEE, Los Alamitos (2008)
19. Maron, O., Ratan, A.: Multiple-instance learning for natural scene classification. In: Proc. ICML (1998)
20. Viola, P., Platt, J.C., Zhang, C.: Multiple instance boosting for object detection. In: Proc. NIPS (2006)
21. Porikli, F.M.: Integral histogram: a fast way to extract histogram in cartesian space. In: Proc. CVPR, pp. 829–836. IEEE, Los Alamitos (2005)
22. Liu, C., Shum, H.Y.: Kullback-leibler boosting. In: Proc. CVPR, pp. 587–594 (2003)
23. Laptev, I.: Improvements of object detection using boosted histograms. In: BMVC (2006)
24. Viola, P., Jones, M.: Rapid object detection using a boosted cascade of simple features. In: Proc. CVPR. IEEE, Los Alamitos (2001)
25. Huang, C., Ai, H., Wu, B., Lao, S.: Boosting nested cascade detector for multi-view face detection. In: Proc. ICPR. IEEE, Los Alamitos (2004)
26. Xiao, R., Zhu, H., Sun, H., Tang, X.: Dynamic cascades for face detection. In: Proc. ICCV. IEEE, Los Alamitos (2007)

Object Detection from Large-Scale 3D Datasets Using Bottom-Up and Top-Down Descriptors

Alexander Patterson IV, Philippos Mordohai, and Kostas Daniilidis

University of Pennsylvania
{aiv,mordohai,kostas}@seas.upenn.edu

Abstract. We propose an approach for detecting objects in large-scale range datasets that combines bottom-up and top-down processes. In the bottom-up stage, fast-to-compute local descriptors are used to detect potential target objects. The object hypotheses are verified after alignment in a top-down stage using global descriptors that capture larger scale structure information. We have found that the combination of spin images and Extended Gaussian Images, as local and global descriptors respectively, provides a good trade-off between efficiency and accuracy. We present results on real outdoors scenes containing millions of scanned points and hundreds of targets. Our results compare favorably to the state of the art by being applicable to much larger scenes captured under less controlled conditions, by being able to detect object classes and not specific instances, and by being able to align the query with the best matching model accurately, thus obtaining precise segmentation.

1   Introduction

Object detection and recognition in images or videos is typically done based on color and texture properties. This paradigm is very effective for objects with characteristic appearance, such as a stop sign or the wheel of a car. There are, however, classes of objects for which 3D shape and not appearance is the most salient feature. Cars are an object category, whose appearance varies a lot within the class, as well as with viewpoint and illumination changes. Instead of representing these objects with a collection of appearance models, specific to each viewpoint, several researchers have used range scanners and addressed object recognition in 3D. Range as an input modality offers the advantages of using the full dimensionality of an object and avoiding any scale ambiguity due to projection. In addition, figure-ground segmentation is easier in 3D than in 2D images since separation in depth provides powerful additional cues. On the other hand, range sensors have significantly lower resolution compared to modern cameras and alignment between the query and the database models still has to be estimated. The challenges associated with object detection in 3D are due to intra-class shape variations, different sampling patterns due to different sensors or different distance and angle between the sensor and the object, targets that are almost always partial due to self-occlusion and occlusion.

D. Forsyth, P. Torr, and A. Zisserman (Eds.): ECCV 2008, Part IV, LNCS 5305, pp. 553–566, 2008.

Fig. 1. Cars detection results from real LIDAR data. Cars have false colors.

In this paper, we present an approach for detecting and recognizing objects characterized by 3D shape from large-scale datasets. The input is a point cloud acquired by range sensors mounted on moving vehicles. A part of the input is used as training data to provide manually labeled exemplars of the objects of interest, as well as negative exemplars where objects of interest are not present. Our algorithm automatically detects potential locations for the target objects in a bottom-up fashion. These locations are then processed by the top-down module that verifies the hypothesized objects by aligning them with models from the training dataset. We show results on a very large-scale dataset which consists of hundreds of millions of points. To the best of our knowledge no results have been published for datasets of this size. State-of-the-art 3D recognition systems on real data [1,2,3,4,5] have shown very high recognition rates, but on high-resolution scenes containing a few small objects captured in controlled environments.

We believe that our research is a first step towards fully automatic annotation of large scenes. Recent advances in sensor technology have made the acquisition and geo-registration of the data possible. Detailed 3D models can be generated very efficiently to provide high-quality visualization [6], but their usefulness for everyday applications is limited due to the absence of semantic annotation. Much like image-based visualizations, such as the Google Street View, these representations cannot answer practical questions, such as "where is the nearest gas station, mailbox or phonebooth". Automatic methods for scene annotation would dramatically increase the benefits users can derive from these large collections of data. While our methods are not currently capable of addressing the problem in its full extend, this paper introduces a framework for object detection from range data that makes a step towards automatic scene annotation. Some results on car detection can be seen in Fig. 1.

The main technical contribution of our work is the combination of a bottom-up and a top-down process to efficiently detect and verify the objects of interest. We use spin images [7] as local descriptors to differentiate between the target objects and clutter and Extended Gaussian Images (EGIs) [8] to ascertain the presence of a target at the hypothesized locations. This scheme enables us to process very large datasets with high precision and recall. Training requires little effort, since the user has to click one point in each target object, which is then automatically segmented from the scene. The remaining points are used as negative examples. Spin images are computed on both positive and negative examples. EGIs only need to be computed for the positive exemplars of the training set, since they are used to align the bottom-up detection with the model database. Accurate alignment estimates between similar but not identical objects enable us to segment the target objects from the clutter. The contributions of our work can be summarized as follows:

- The combination of bottom-up and top-down processing to detect potential targets efficiently and verify them accurately.
- The capability to perform training on instances that come from the same object category as the queries, but are not necessarily identical to the queries.
- Minimal user efforts during training.
- Object detection for large-scale datasets captured in uncontrolled environments.
- Accurate segmentation of target objects from the background.

2   Related Work

In this section, we briefly overview related work on local and global 3D shape descriptors and 3D object recognition focusing only on shape-based descriptors. Research on appearance-based recognition has arguably been more active recently, but is not directly applicable in our experimental setup.

Global shape descriptors include EGIs [8], superquadrics [9], complex EGIs [10], spherical attribute images [11] and the COSMOS [12]. Global descriptors are more discriminative since they encapsulate all available information. On the other hand, they are applicable to single segmented objects and they are sensitive to clutter and occlusion. A global representation in which occlusion is explicitly handled is the spherical attribute image proposed by Hebert et al. [11].

A method to obtain invariance to rigid transformations was presented by Osada et al. [13] who compute shape signatures for 3D objects in the form of shape statistics, such as the distance between randomly sampled pairs of points. Liu et al. [14] introduced the directional histogram model as a shape descriptor and achieved orientation invariance by computing the spherical harmonic transform. Kazhdan et al. [15] proposed a method to make several types of shape descriptors rotationally invariant, via the use of spherical harmonics. Makadia et al. [16] compute the rotational Fourier transform [17] to efficiently compute the correlation between EGIs. They also propose the constellation EGI, which we use in Section 5 to compute rotation hypotheses.

Descriptors with local support are more effective than global descriptors for partial data and data corrupted by clutter. Stein and Medioni [18] combined surface and contour descriptors, in the form of surface splashes and super-segments, respectively. Spin images were introduced by Johnson and Hebert [7] and are among the most popular such descriptors (See Section 4). Ashbrook et al. [19] took a similar approach based on the pairwise relationships between triangles of the input mesh. Frome et al. [20] extended the concept of shape contexts to 3D. Their experiments show that 3D shape contexts are more robust to occlusion and surface deformation than spin images but incur significantly higher computational cost. Huber et al. [21] propose a technique to divide range scans of vehicles into parts and perform recognition under large occlusions using spin images as local shape signatures.

Local shape descriptors have been used for larger scale object recognition. Johnson et al.[1] use PCA-compressed spin images and nearest neighbor search to find the most similar spin images to the query. Alignment hypotheses are estimated using these correspondences and a variant of the ICP algorithm [22] is used for verification. Shan et al. [23] proposed the shapeme histogram projection algorithm which can match partial object by projecting the descriptor of the query onto the subspace of the model database. Matei et al. [3] find potential matches for spin images using locality sensitive hashing. Geometric constraints are then used to verify the match. Ruiz-Correa et all. [5] addressed deformable shape recognition via a two-stage approach that computes numeric signatures (spin images) to label components of the data and then computes symbolic signatures on the labels. This scheme is very effective, but requires extensive manual labeling of the training data. Funkhouser and Shilane [24] presented a shape matching system that uses multi-scale, local descriptors and a priority queue that generates the most likely hypotheses first.

In most of the above methods, processing is mostly bottom-up, followed in some cases by a geometric verification step. A top-down approach was proposed by Mian et al. [4] who represent objects by 3D occupancy grids which can be matched using a 4D hash table. The algorithm removes recognized objects from the scene and attempts to recognize the remaining data until no additional library object can be found.

Our method can detect cars in real scenes in the presence of clutter and sensor noise. Very few of the papers mentioned above ([1,2,3,4,5]) present results on real data. Among the ones that do, Matei et al. [3] classified cars that had been previously segmented. Johnson et al. [1], Carmichael et al. [2] and Mian et al. [4] show object detection from real scenes containing multiple objects. It should be noted, however, that the number of objects in the scene is small and that all objects were presented to the algorithm during training. Ruiz-Correa et. all [5] are able to handle intra-class variation, at the cost of large manual labeling effort. The goal of our work is more ambitious than [1,2,3,4,20,23] in order to make more practical applications possible. Our algorithm is not trained on exemplars identical to the queries, but on other instances from the same class. This enables us to deploy the system on very large-scale datasets with moderate training efforts, since we only have to label a few instances from the object categories we are interested in.

3   Algorithm Overview

Our algorithm operates on 3D point clouds and entails a bottom-up and a top-down module. The steps for annotation and training are the following:

1. The user selects one point on each target object.
2. The selected target objects are automatically extracted from the background.
3. Compute surface normals for all points[1] in both objects and background.
4. Compute spin images on a subset of the points for both objects and background and insert into spin image database DB_{SI} (Section 4).
5. Compute an EGI for each object (not for the background). Compute constellation EGI and density approximation. Insert into EGI database DB_{EGI} (Section 5).

Processing on test data is performed as follows:

1. Compute normals for all points and spin images on a subset of the points.
2. Classify spin images as positive (object) or negative (background) according to their nearest neighbors in DB_{SI}.
3. Extract connected components of neighboring positive spin images. Each connected component is a query (object hypothesis).
4. Compute an EGI and the corresponding constellation EGI for each query.
5. For each query and model in DB_{EGI} (Section 5):
   (a) Compute rotation hypothesis using constellation EGIs.
   (b) For each rotation hypothesis with low distance according to Section (5.3), compute translation in frequency domain.
   (c) Calculate the overlap between query and model.
6. If the overlap is above the threshold, declare positive detection (Section 5).
7. Label all points that overlap with each of the models of DB_{EGI} after alignment as object points to obtain segmentation.

4   Bottom-Up Detection

The goal of the bottom-up module is to detect potential target locations in the point cloud with a bias towards high recall to minimize missed detections. Since detection has to be performed on very large point clouds, we need a representation that can be computed and compared efficiently. To this end we use spin images [7]. A spin image is computed in a cylindrical coordinate system defined by a reference point and its corresponding normal. All points within this region are transformed by computing α, the distance from the reference normal ray and β the height above the reference normal plane. Finally a 2D histogram of α and β is computed and used as the descriptor. Due to integration around the normal of the reference point, spin images are invariant to rotations about the normal. This is not the case with 3D shape contexts [20] or EGIs (Section 5) for which several rotation hypotheses have to be evaluated to determine a match. Since

---

[1] During normal computation, we also estimate the reliability of the normals, which is used to select reference points for the spin images.

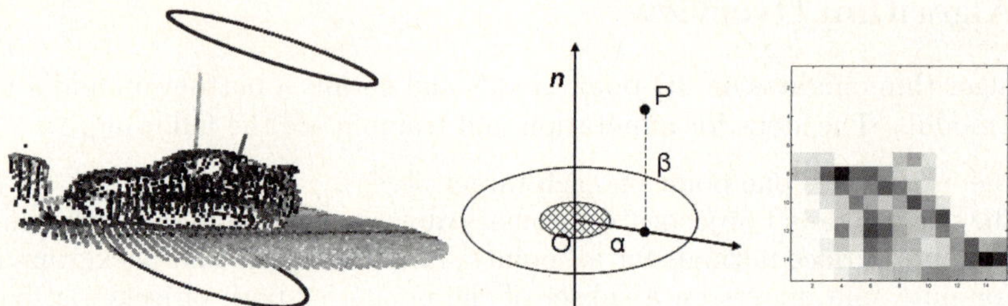

Fig. 2. Left: spin image computation on real data. The blue circles delineate the cylindrical support region and the red vector is the normal at the reference point. Middle: illustration of spin image computation. O is the reference point and n its normal. A spin image is a histogram of points that fall into radial (α) and elevation (β) bins. Right: the spin image computed for the point on the car.

spin image comparison is a simple distance between vectors, their comparisons are computationally cheaper, but less discriminative.

Johnson and Hebert [7] computed spin images on meshes. This can compensate for undesired effects due to varying sample density, since triangles contribute to each bin of the histogram with their area. Triangulating the point cloud to obtain a mesh is not trivial in our case, not only because of the computational cost, but also due to noise, sparsity and sampling patterns of the data. Similar to [20], we compute spin images directly from the point clouds and weigh the contribution of each point by its inverse density to account for sampling differences. Local density is computed in balls centered at every point. Isolated points are removed. Accounting for variations in point density is important for point clouds captured by range sensors since the density of samples on a surface is a function of sensor type, as well as distance and angle to the sensor.

Given a point cloud, regardless of whether it contains training or test data, normals for all points are computed using tensor voting [25]. We then need to select reference points for the spin images. Our experiments have shown that spin images vary smoothly as long as the reference point is on the same surface and the normal is accurate. Therefore, reference points need to be dense enough to capture all surfaces of the object, but higher density is redundant. For cars, a distance of $0.4m$ was found to offer a good trade-off between coverage and computational efficiency. We obtain such a sampling by placing a 3D grid of the desired resolution in the dataset and dropping vertices that have no scanned points in their voxel. Since the reference points need to be among the points sampled by the scanner, the retained vertices are moved to the median of the nearest points to account for noisy samples. (The grid can be seen in the two rightmost images of Fig. 3.) A spin image is computed for each of these points unless the eigenvalues of the tensor after tensor voting indicate that the estimated normal is unreliable [25]. Our spin images have 15 radial and 15 elevation bins resulting in a 225-D descriptor.

For the training data, the user has to specify the targets, which are assumed to be compact objects lying on the ground, by clicking one point on each. Then, an

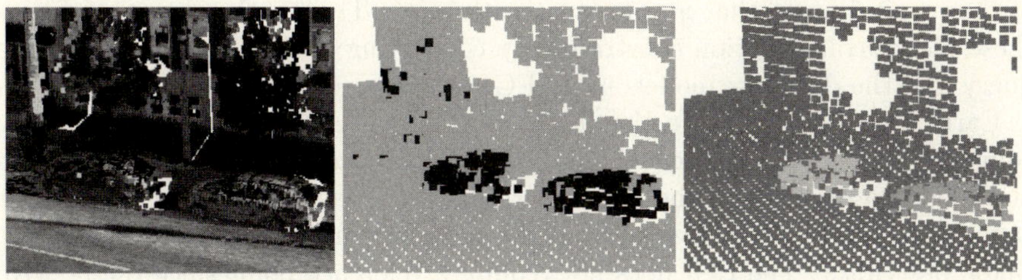

Fig. 3. Left: input point cloud. Middle: Classification of spin images as target (blue) and background (cyan). (Only the reference points are shown.) Right: target spin image centers clustered into object hypotheses. Isolated target spin images are rejected.

automatic algorithm segments the object as a connected component protruding from the ground. The ground can be reliably estimated in a small neighborhood around the selected point as the lowest smooth surface that bounds the data. Spin images computed for points on the targets are inserted into the spin image database DB_{SI} as positive exemplars, while spin images from the background are inserted as negative exemplars. We have implemented the database using the Approximate Nearest Neighbor (ANN) k-d tree [26].

During testing, query spin images are computed on reference points on a grid placed on the test data as above. Each query spin image is classified according to the nearest neighbor retrieved from DB_{SI}. Some results on real data can be seen in Fig. 3. Potential locations of the target objects can be hypothesized in areas of high density of positive detections, while isolated false positives can be easily pruned from the set of detections.

Object hypotheses (queries) are triggered by spin images that have been classified as positive (target). Target spin images are grouped into clusters by a simple region growing algorithm that starts from a spin image reference point and connects it to all neighboring target spin images within a small radius. When the current cluster cannot be extended any further, the algorithm initializes a new cluster. Raw points that are within a small distance from a cluster of spin images are also added to it to form a query. Since neighboring spin images overlap, the bottom-up portion of our algorithm is robust to some miss-classifications.

5   Top-Down Alignment and Verification

The second stage of processing operates on the queries (clustered points with normals) proposed by the bottom-up stage and verifies whether targets exist at those locations. Spin images without geometric constraints are not discriminative enough to determine the presence of a target with high confidence. Spin image classification is very efficient, but only provides local evidence for the presence of a potential part of a target and not for a configuration of parts consistent with a target. For instance a row of newspaper boxes can give rise to a number of spin images that are also found in cars, but cannot support a configuration of those spin images that is consistent with a car. The top-down stage enforces these global configuration constraints by computing an alignment between the query and the database models using EGI descriptors.

Early research has shown that there is a unique EGI representation for any convex object [27], which can be obtained by computing the density function of all surface normals on the unit sphere. If the object is not convex, its shape cannot be completely recovered from the EGI, but the latter is still a powerful shape descriptor. The EGI does not require a reference point since the relative positions of the points are not captured in the representation. This property makes EGIs effective descriptors for our data in which a reference point cannot be selected with guaranteed repeatability due to occlusion, but the distribution of normals is fairly stable for a class of objects.

5.1   Computing EGIs

EGIs are computed for the positive object examples in the training set. Objects are segmented with assistance from the user, as described in Section 4. For the test data, an EGI is computed for each object hypothesis extracted according to the last paragraph of Section 4. Each EGI contains the normals of all input points of the cluster, oriented so that they point outwards, towards the scanner. These orientations can be computed since the trajectory of the sensor is available to us. The majority of objects are scanned only from one side and, as a result, the normals typically occupy at most a hemisphere of the EGI. This viewpoint dependence occurs for both the queries and database objects and thus requires no special treatment. If necessary, database models can be mirrored to increase the size of the database without additional manual labeling since model symmetry is modeled by the EGI.

5.2   Constellation EGIs

Unlike spin images, comparing two EGIs requires estimating a rotation that aligns them before a distance can be computed. One can compute the rotational Fourier transform [17] to efficiently compute all correlations between EGIs [16]. This technique is efficient if all rotations need to be computed, but it is sensitive to clutter, missing parts and quantization. Our experiments have shown that quantization can have adverse effects on rotation and distance computations

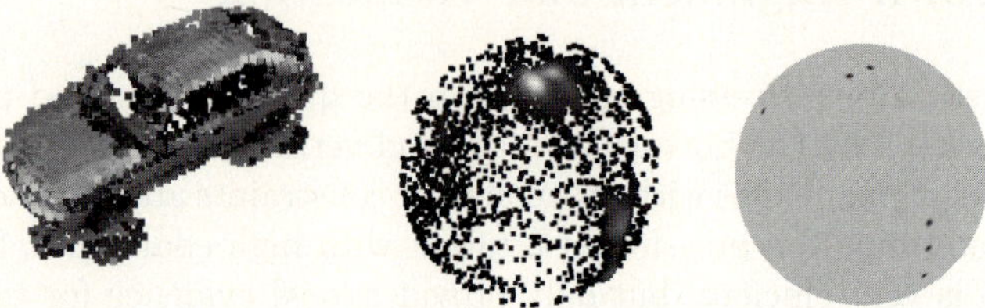

Fig. 4. Left: a database model of a car. Middle: illustration of an EGI in which points are color-coded according to their density. Right: the corresponding constellation EGI.

using EGIs. We can use the constellation EGI to cue a more efficient distance computation. Therefore, we avoid quantizing the orientations of the normals in an EGI and do not treat it as an orientation histogram.

Instead of an exhaustive search using either spatial or Fourier methods, we use a technique that generates discrete alignment hypotheses, which was originally proposed in [16]. A *constellation EGI* records the locations of local maxima in the distribution of normals in the EGI. We call these maxima *stars*, since they resemble stars in the sky. An EGI and the corresponding constellation EGI for an object can be seen in Fig. 4. Two constellation EGIs can be matched by sampling pairs of stars that subtend the same angle on the sphere. Each sample generates matching hypotheses with two stars of the other EGI. If the angles between each pair are large enough and similar, a rotation hypothesis for the entire descriptor is generated. Note that a correspondence between two pairs of stars produces two possible rotations. Similar rotations can be clustered to reduce the number of hypotheses that need to be tested. The resulting set of rotations are evaluated based on the distance between the entire EGIs and not just the stars.

5.3   Hypothesis Verification

Conceptually, the rotation hypothesis that achieves the best alignment of the descriptors is the one that maximizes the cross-correlation between all normal vectors of the first and the second EGI. This computation is exact, but computationally expensive since models and queries consist of thousands of points each. To reduce the computational complexity, we select a smaller set of normals and compute the weights of kernels which are centered on this set, thus closely approximating the original EGI via interpolation. This computation is performed once per EGI and significantly reduces the cost of distance computation. Specifically, to create an approximation of the EGI for a set of input normals, we compute the density at all input normals on the sphere. We then select a subset of samples by greedily choosing a predetermined number of points. Each choice is made by computing the current interpolation via nearest neighbor, and then adding the normal with the largest deviation between approximated and actual values. Our method is similar to [28], but operates on the sphere. Once we have a set of kernel centers N_s which is a subset of all normals N, the weights of the kernels are computed as follows:

$$v_{ij} = \frac{max(d_{max} - \arccos(\hat{n}_i^T \hat{n}_j), 0)}{\sum_j (max(d_{max} - \arccos(\hat{n}_i^T \hat{n}_j), 0))}, \quad i \in N, j \in N_s$$

$$D_j = V^\dagger D_i, \tag{1}$$

Where D_j are the coefficients at the sparse set of normals N_s, and d_{max} is the range of the kernel function. Using this new representation, we can compute the distance between two EGIs, using a sparse set of samples, after applying a rotation hypothesis. If the two shapes are identical, the density values should be equal over the entire sphere. We measure the deviation from an ideal match by predicting the density on the samples of one EGI using the interpolation

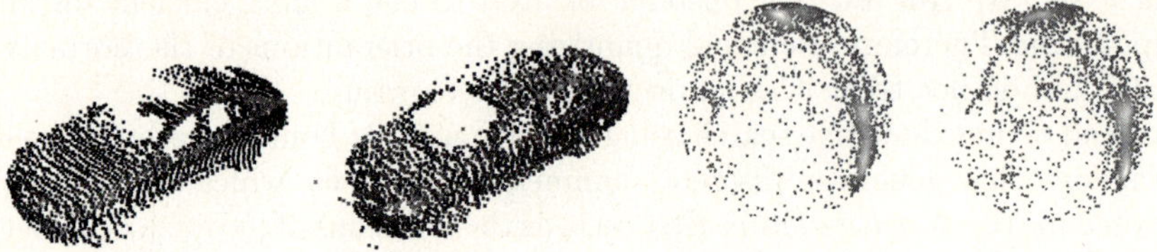

Fig. 5. Alignment of a database model (left car and left EGI) and a query (right car and right EGI) that have been aligned. The car models are shown separately for clarity of the visualization. Notice the accuracy of the rotation estimation. The query has been segmented by the positive spin image clustering algorithm and the model by removing the ground after the user specified one point.

function of the other EGI and comparing them with the original density values. Specifically we use the l_1 distance computed at the query points which we can now interpolate once the normals N_s are rotated according to each hypothesized rotation. The minimum distance provides an estimate of the best rotation to align the two objects, but no estimate of translation and most importantly no indication of whether the objects actually match. Typically, 1-5 rotations are close enough to the minimum distance. For these, we estimate the translation and compute the final distance in the following section.

5.4   Alignment and Distance Computation

Given the few best rotation hypotheses based in section 5.3, we compute the translation that best aligns the two models in the frequency domain. We adopt the translation estimation method of [16] in which translation is estimated using a Fourier transform in $\mathbb{R}^3$. This is less sensitive to noise in the form of missing parts or clutter than global alignment methods that estimate complete rigid transformations in the Fourier domain. We begin by voxelizing the model and the query to obtain binary occupancy functions in 3D. We then compute their convolution efficiently using the fft and take the maximum as our translation.

Finally, we need a measure of distance to characterize the quality of the alignment that is flexible enough to allow for deformation between the query and the model. We experimented with the ICP distance [22], without performing ICP iterations, but found the overlap between the query and model to be more effective because the quantization in the translation estimation caused large ICP distance errors even though the models were similar. The overlap is computed as the inlier ratio over all points of the model and query, where an inlier is a point with a neighboring point from the other model that is closer than a threshold distance and whose normal is similar to that of the point under consideration. Figure 5 shows an alignment between a query and a database object and their corresponding EGIs. Selecting the overlap points after alignment results in precise segmentation of the object from the background.

Fig. 6. Left: The precision-recall curve for car detection on 200 million points containing 1221 cars. (Precision is the x-axis and recall the y-axis.) Right: Screenshot of detected cars. Cars are in random colors and the background in original colors.

Fig. 7. Screenshots of detected cars, including views from above. (There is false negative at the bottom of the left image.) **Best viewed in color.**

6   Experimental Results

We processed very large-scale point clouds captured by a moving vehicle equipped with four range scanners and precise Geo-location sensors. The dataset consists of about 200 million points, 2.2 million of which were used for training. The training set included 17 cars which were selected as target objects. We compute 81,172 spin images for the training set (of which 2657 are parts of cars) and 6.1 million for the test set. Each spin image has a 15×15 resolution computed in a cylindrical support region with height and radius both set to $2m$. Reference points for the spin images are selected as in Section 4 with an average distance between vertices of $0.4m$. The spin images of the training set are inserted into DB_{SI}.

EGIs are computed for each target object in the training set, approximated by picking a smaller set of 200 normals, that minimize the interpolation error on all samples. The approximated EGIs are inserted into DB_{EGI}, which is a simple list with 17 entries. Since our method only requires very few representatives from each class, we were able to perform the experiments using a few sedans, SUVs and vans as models.

The query trouping threshold is set to $1m$ (Section 4). This groups points roughly up to two grid positions away. The EGI matching thresholds (Section 5.2) are set as follows: Each pair must of stars must subtend an angle of at least $30°$ and the two angles must not differ by more than $5°$. Rotations that meet these requirements are evaluated according to Section 5.3. For the best rotation hypotheses, the metric used to make the final decision is computed: the percentage of inliers on both models after alignment. For a point to be an inlier there has to be at least one other point from the other model that is within $30cm$ and whose normal deviates by at most $35°$ from the normal of the current point. We have found the inlier fraction to be more useful than other distance metrics.

Results on an test area comprising 220 million points and 1221 cars are shown in Figs. 6 and 7. After bottom-up classification there were approximately 2200 detections of which about 1100 were correct. The top-down step removes about 1200 false positives and 200 true positives. The precision-recall curve as the inlier threshold varies for the full system is shown in Fig. 6. For the point marked with a star, there are 905 true positives, 74 false positives and 316 false negatives (missed detections) for a precision of 92.4% and a recall of 74.1%.

7 Conclusion

We have presented an approach for object detection from 3D point clouds that is applicable to very large datasets and requires limited training efforts. Its effectiveness is due to the combination of bottom-up and top-down mechanisms to hypothesize and test locations of potential target objects. An application of our method on car detection has achieved very satisfactory precision and recall on an area far larger than the test area of any previously published method. Moreover, besides a high detection rate, we are able to accurately segment the objects of interest from the background. We are not aware of any other methodology that obtains comparable segmentation accuracy without being trained on the same instances that are being segmented.

A limitation of our approach we intend to address is that search is linear in the number of objects in the EGI database. We are able to achieve satisfactory results with a small database, but sublinear search is a necessary enhancement to our algorithm.

Acknowledgments

This work is partially supported by DARPA under the Urban Reasoning and Geospatial ExploitatioN Technology program and is performed under National Geospatial-Intelligence Agency (NGA) Contract Number HM1582-07-C-0018. The ideas expressed herein are those of the authors, and are not necessarily endorsed by either DARPA or NGA. This material is approved for public release; distribution is unlimited.

The authors are also grateful to Ioannis Pavlidis for his help in labeling the ground truth data.

backgrounds. Many improvements have been published since, such as a recent method that dynamically selects the appropriate number of components for each pixel [1]. We will use it as a benchmark against which we compare our approach because it is representative of this whole class of techniques.

Introducing a GMM is not the only way to model a dynamic background. Elgammal et al. proposed to model both background and foreground pixel intensities by a nonparametric kernel density estimation [6]. In [7], Sheikh and Shah proposed to model the full background with a single distribution, instead of one distribution per pixel, and to include location into the model.

Because these methods do not decouple illumination from other causes of background changes, they are more sensitive to drastic light effects than our approach.

Shadows cast by moving objects cause illumination changes that follow them, thereby hindering the integration of shadowed pixels into the background model. This problem can be alleviated by explicitly detecting the shadows [8]. Most of them consider them as binary [8], with the notable exception of [9] that also considers penumbra by using the ratio between two images of a planar background. Our approach also relies on image ratios, but treats shadows as a particular *illumination effect*, a wider class that also include the possibility of switching lights on.

Another way to handle illumination changes is by using illumination invariant features, such as edges. Edge information alone is not sufficient, because some part of the background might be uniform. Thus, Jabri et al. presented an approach to detect people fusing colour and edge information [10]. More recently, Heikkilä and Pietikäinen modelled the background using histograms of local binary patterns [11]. The bilayer segmentation of live video presented in [12] fuses colour and motion clues in a probabilistic framework. In particular, they observe in a labeled training set the relation between the image features and their target segmentation. We follow here a similar idea by training beforehand histograms of correlation and amount of texture, allowing us to fuse illumination, colour and texture clues.

3   Method

Our method can serve in two different contexts. For background subtraction, where both the scene and the camera are static. For augmented reality applications, where an object is moving in the camera field and occlusions have to be segmented for realistic augmentation.

Let us assume that we are given an unoccluded *model image* of a background scene or an object. Our goal is to segment the pixels of an *input image* in two parts, those that belong to the same object in both images and those that are occluded. If we are dealing with a moving object, we first need to register the input image and create an image that can be compared to the model image pixelwise. In this work, we restrict ourselves to planar objects and use publicly available software [13] for registration. If we are dealing with a static scene and camera, that is, if we are performing standard background subtraction, registration is not necessary. It is the only difference between both contexts, and the rest of the method is common. In both cases, the intensity and colour of individual pixels are affected mostly by illumination changes and the presence of occluding objects.

Changes due to illumination effects are highly correlated across large portions of the image and can therefore be represented by a low dimensional model that accounts for variations across the whole image. In this work, we achieve this by representing the ratio of intensities between the stored *background image* and an input image in all three channels as a Gaussian Mixture Model (GMM) that has very few components—2 in all the experiments shown in this paper. This is in stark contrast with more traditional background subtraction methods [2,3,4,1] that introduce a model for each pixel and do not explicitly account for the fact that inter-pixel variations are correlated.

Following standard practice [14], we model the pixel colours of occluding objects, such as people walking in front of the camera, as a mixture of Gaussian and uniform distributions.

To fuse these clues, we model the whole image — background, foreground and shadows — with a single mixture of distributions. In our model, each pixel is drawn from one of five distributions: Two Gaussian kernels account for illumination effects, and two more Gaussians, completed by a uniform distribution, represent the foreground. An Expectation Maximization algorithm assigns pixels to one of the five distributions (E-step) and then optimizes the distributions parameters (M-step).

Since illumination changes preserve texture whereas occluding objects radically change it, the correlation between image patches in the model and input images provides a hint as to whether pixels are occluded or not in the latter, especially where there is enough texture.

In order to lower the computational burden, we assume pixel independence. Since this abusive assumption entails the loss of the relation between a pixel and its neighbors, it makes it impossible to model texture. However, to circumvent this issue, we characterize each pixel of the input image by a five dimensional feature vector: The usual red, green, and blue values plus the normalized cross-correlation and texturedness values. Feature vectors are then assumed independent, allowing an efficient maximization of a global image likelihood, by optimizing the parameters of our mixture. In the remainder of this section, we introduce in more details the different components of our model.

3.1   Illumination Likelihood Model

First, we consider the background model, which is responsible for all pixels that have a counterpart in the *model image* m. If a pixel u_i of the *input image* u shows the occlusion free target object, the luminance measured by the camera depends on the light reaching the surface (the irradiance e_i) and on its albedo. Irradiance e_i is function of visible light sources and of the surface normal. Under the lambertian assumption, the pixel value u_i is: $u_i = e_i a_i$, where a_i is the albedo of the target object at the location pointed by u_i. Similarly, we can write: $m_i = e_m a_i$,with e_m assumed constant over the surface. This assumption is correct if the model image m has been taken under uniform illumination, or if a textured model free of illumination effects is available. Combining the above equations yields:

$$l_i = \frac{u_i}{m_i} = \frac{e_i}{e_m},$$

which does not depend on the surface albedo. It depends on the surface orientation and on the illumination environment. In the specific case of a planar surface lit by distant light sources and without cast shadows, this ratio can be expected to be constant for all i [9]. In the case of a 3 channel colour camera, we can write the function l_i that computes a colour illumination ratio for each colour band:

$$l_i = \left[\begin{array}{ccc} \frac{u_{i,r}}{m_{i,r}} & \frac{u_{i,g}}{m_{i,g}} & \frac{u_{i,b}}{m_{i,b}} \end{array} \right]^T ,$$

where the additional indices r, g, b denotes the red, green and blue channel of pixel u_i, recpectively.

In our background illumination model we suppose that the whole scene can be described by K different illumination ratios, that correspond to areas in u_i with different orientations and/or possible cast shadows. Each area is modelled by a Gaussian distribution around the illumination ratio μ_k and with full covariance Σ_k. Furthermore we introduce a set of binary latent variables $x_{i,k}$ that take the value 1 iff pixel i belongs to Gaussian k and 0 otherwise. Then, the probability of the ratio l_i is given by:

$$p(l_i \mid x_i, \mu, \Sigma) = \prod_{k=1}^{K} \pi_k^{x_{i,k}} \mathcal{N} \left(l_i; \mu_k, \Sigma_k \right)^{x_{i,k}} , \tag{1}$$

where μ, Σ denote all parameters of the K Gaussians. π_k weights the relative importance of the different mixture components. Even though the ratios l_i are not directly observed, this model has much in common with a generative model for illumination ratios.

So far we described the background model. The foreground model is responsible for all pixels that do not correspond to the model image m. These pixels are assumed to be generated by sampling the foreground distribution, which we model as a mixture of $\bar{K}$ Gaussians and a uniform distribution. By this choice, we implicitly assume that the foreground object is composed of $\bar{K}$ colours μ_k, handled by the normal distributions $\mathcal{N}(u_i; \mu_k, \Sigma_k)$, and some suspicious pixels that occur with probability $1/256^3$. Again, as in the background model, the latent variables are used to select a specific Gaussian or the uniform distribution. The probability of observing a pixel value u_i given the state of the latent variable x_i and the parameters μ, Σ is given by:

$$p(u_i \mid x_i, \mu, \Sigma) = \left(\frac{\pi_{K+\bar{K}+1}}{256^3} \right)^{x_{i,K+\bar{K}+1}} \prod_{k=K+1}^{K+\bar{K}} \pi_k^{x_{i,k}} \mathcal{N} \left(u_i; \mu_k, \Sigma_k \right)^{x_{i,k}} . \tag{2}$$

The overall model consist of the background (Eq. 1) and the foreground (Eq. 2) model. Our latent variables x_i select the one distribution among the total $K+\bar{K}+1$

components which is active for pixel i. Consider figures 2(a) and 2(b) for example: The background pixels could be explained by $K = 2$ illumination ratios, one for the cast shadow and one for all other background pixels. The hand in the foreground could be modelled by the skin colour and the black colour of the shirt ($\bar{K} = 2$). The example in Fig. 2 shows clearly that the importance of the latent variable components is not equal. In practice, there is often one Gausssian which models a global illimination change, *i.e.* most pixels are assigned to this model by the latent variable component $x_{i,k}$. To account for the possibly changing importance, we have introduced π_k that globally weight the contribution of all Gaussian mixtures $k = 1 \ldots \bar{K}$ and the uniform distribution $k = \bar{K} + 1$.

A formal expression of our model requires combining the background pdf of Eq. 1 and the foreground pdf of Eq. 2. However, one is defined over illumination, whereas the other over pixel colour, making direct probabilities incompatible. We therefore express the background model as a function of pixel colour instead of illumination:

$$p(u_i \,|\, x_i, \boldsymbol{\mu}, \boldsymbol{\Sigma}) = \frac{1}{|\,J_i\,|} p(l_i \,|\, x_i, \boldsymbol{\mu}, \boldsymbol{\Sigma}) \,, \tag{3}$$

where $|\,J_i\,|$ is the determinant of the Jacobian of function $l_i(u_i)$. Multiplying this equation with Eq. 2 composes the complete colour pdf.

Some formulations define an appropriate prior model on the latent variables $\boldsymbol{x}$. Such a prior model would incorporate the prior belief that the model selection $\boldsymbol{x}$ shows spatial [14] and spatio-temporal [12] correlations. These priors on the latent variable $\boldsymbol{x}$ have shown to improve the performance of many vision algorithms [15]. However, they increase the complexity and slow down the computation substantially. To circumvent this, we propose in the next section a spatial likelihood model, which can be seen as a model to capture the spatial nature of pixels and which allows real-time performance.

3.2   Spatial Likelihood Model

In this section, we present an image feature and a way to learn off-line its relationship with our target segmentation. Consider an extended image patch around pixel i for which we extract a low dimensional features vector $f_i = [f_i^1, f_i^2]$. The basic idea behind our spatial likelihood model is to capture texture while keeping a pixel independence assumption. To achieve real-time performance we use two features that can be computed very fast and model their distribution independently for the background and for the foreground, by histograms of the discretized feature values. We use the normalized cross-correlation (NCC) between input and model image as one feature and a measure of the amount of texture as the other feature. f_i^1 is given by:

$$f_i^1 = \frac{\sum_{j \in w_i} (u_j - \bar{u}_i)(m_j - \bar{m}_i)}{\sqrt{\sum_{j \in w_i} (u_j - \bar{u}_i)^2 \sum_{j \in w_i} (m_j - \bar{m}_i)^2}} \,,$$

where w_i denotes a window around pixel i, and $\bar{u}_i = \frac{1}{|w_i|} \sum_{j \in w_i} u_j$ is the average over w_i. The correlation is meaningful only in windows containing texture. Thus, the texturedness of window i is quantified by:

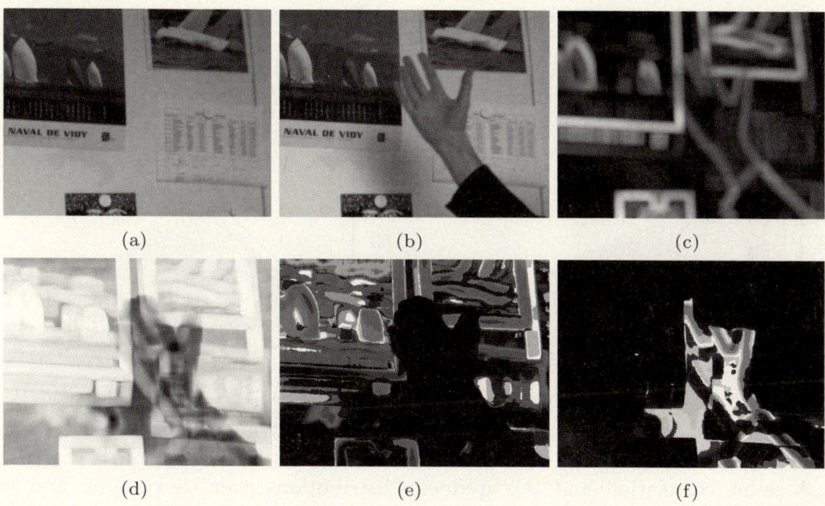

Fig. 2. Elements of the approach. (a) Background image m. (b) Input image u. (c) Textureness image f^2. (d) Correlation image f^1. (e) Probability of observing f on the background, according to the histogram $h(f_i \mid v_i)$ (f) Probability of observing f on the foreground, according to the histogram $\bar{h}(f_i \mid \bar{v}_i)$.

$$f_i^2 = \sqrt{\sum_{j \in w_i} (u_j - \bar{u}_i)^2} + \sqrt{\sum_{j \in w_i} (m_j - \bar{m}_i)^2}.$$

We denote the background and foreground distributions by $h(f_i \mid v_i)$ and $\bar{h}(f_i \mid \bar{v}_i)$, respectively. They are trained from a set of manually segmented image pairs. Since joint correlation and amount of texture is modelled, the histograms remain valid for new illumination conditions and for new backgrounds. Therefore, the training is done only once, off-line. Once normalized, these histograms model the probability of observing a feature f_i on the backgound or on the foreground. Fig. 3 depicts both distributions. One can see that both distributions are dissociate, especially in highly textured areas.

Figure 2 shows a pair of model and input images, the corresponding texture and correlation images f_i^2 and f_i^1, and the results of applying the histograms to f. It is obvious that the correlation measure is only meaningful in textured areas. In uniform areas, because NCC is invariant to illumination, it can not make the difference between a background with some uniform illumination or a uniform foreground.

Both histograms are learnt in the two cases of background and foreground which are related to the latent variable x_i designing one of the distributions of our model. Therefore, h can be used together with all background distributions corresponding to $\{x_{i,1}, ..., x_{i,K}\}$ and $\bar{h}$ with all foreground ones, corresponding to $\{x_{i,K+1}, ..., x_{i,K+\bar{K}+1}\}$.

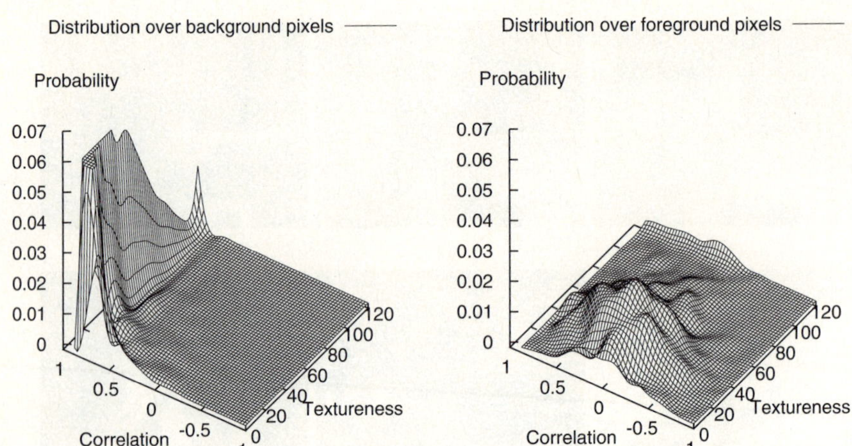

Fig. 3. Joint correlation and texturedness distributions over background and foreground pixels

3.3   Maximum Likelihood Estimation

Having defined the illumination and the spatial likelihood model we are now in the position to describe the Maximum Likelihood (ML) estimation of the combined model. Let $\theta = \{\mu, \Sigma, \pi\}$ denote the vector of all unknowns. The ML estimate $\tilde{\theta}$ is given by:

$$\tilde{\theta} = \arg\max_{\theta} \left\{ \log \sum_{x} p(u, f, x \mid \theta) \right\} \tag{4}$$

where $p(u, f, x \mid \theta) = p(u, x \mid \theta)p(f, x \mid \theta)$ represents the combined pdf of the illumination and the spatial likelihood models given by the product of eqs. 3, 2 and the histrogram distributions $h(f_i \mid v_i)$, $\bar{h}(f_i \mid \bar{v})$. Since the histogram distributions are computed over an image patch, the pixel contributions are not independent. However, in order to reach the real-time constraints, we assume the factorisation over all pixels i in Eq. 4 to be approximately true. We see this problem as a trade-off between *(i)* a prior model on x, that models spatial interactions [12,15] with a higher computational complexity and *(ii)* a more simple, real time model for which the independence assumption is violated, in the hope that the spatially dependent feature description f account for pixel dependence.

The pixel independence assumption simplifies the ML estimate to:

$$\tilde{\theta} = \arg\max_{\theta} \left\{ \log \prod_{i} \sum_{x_i} p(u_i, l_i, f_i, x_i \mid \theta) \right\} \tag{5}$$

The expectation-maximization (EM) algorithm can maximize equation 5. It alternates the computation between an expectation step (E-step), and a maximization step (M-step).

E-Step. On the $(t+1)^{th}$ iteration the conditional expectation $\boldsymbol{b}^{t+1}$ of the log-likelihood *w.r.t.* the posterior $p(\boldsymbol{x}\,|\,\boldsymbol{u},\boldsymbol{\theta})$ is computed in the E-step. By construction, *i.e.* by the pixel independence, this leads to a closed-form solution for the latent variable expectations b_i, which are often called beliefs. Note, that in other formulations, where the spatial correlation is modelled explicitly, the E-step requires graph-cut optimisation [14] or other iterative approximations like mean field [15]. The update equations for the expected values $b_{i,k}$ of $x_{i,k}$ are given by:

$$b_{i,k=1...K}^{t+1} = \frac{1}{N}\pi_k\frac{1}{|\,J_i\,|}\mathcal{N}(l_i;\mu_k^t,\Sigma_k^t)h(f_i\,|\,v_i) \tag{6}$$

$$b_{i,k=K+1...\bar{K}}^{t+1} = \frac{1}{N}\pi_k\mathcal{N}(u_i;\mu_k^t,\Sigma_k^t)\bar{h}(f_i\,|\,\bar{v}_i) \tag{7}$$

$$b_{i,\bar{K}+1}^{t+1} = \frac{1}{N}\pi_{K+\bar{K}+1}\frac{1}{256^3}\bar{h}(f_i\,|\,\bar{v}_i)\ ,$$

where $N = \sum_k b_{i,k}^{t+1}$ normalises the beliefs $b_{i,k}^{t+1}$ to one. The first line in Eq. 6 corresponds to the beliefs that the kth normal distribution of the illumination background model is active for pixel i. Similarly, the other two lines (Eq. 7) correspond to the beliefs *w.r.t.* for the foreground illumination model.

M-Step. Given the beliefs $b_{i,k}^{t+1}$, the M-step maximises the log-likelihood by replacing the binary latent variables $x_{i,k}$ by their expected value $b_{i,k}^{t+1}$.

$$\mu_k^{t+1} = \begin{cases} \frac{1}{N_k}\sum_{i=1}^{N}b_{i,k}^{t+1}l_i & \text{if } k \leq K \\ \frac{1}{N_k}\sum_{i=1}^{N}b_{i,k}^{t+1}u_i & \text{otherwise} \end{cases} , \tag{8}$$

where $N_k = \sum_{i=1}^{N}b_{ik}^{t+1}$. Similarly, we obtain:

$$\Sigma_k^{t+1} = \begin{cases} \frac{1}{N_k}\sum_{i=1}^{N}b_{i,k}^{t+1}(l_i-\mu_k)(l_i-\mu_k)^{\mathrm{T}} & \text{if } k \leq K \\ \frac{1}{N_k}\sum_{i=1}^{N}b_{i,k}^{t+1}(u_i-\mu_k)(u_i-\mu_k)^{\mathrm{T}} & \text{otherwise} \end{cases} \tag{9}$$

$$\pi_k^{t+1} = \frac{N_k}{\sum_k N_k} \tag{10}$$

Alternating E and M steps ensure convergence to a local minimum. After convergence, we can compute the segmentation by summing the beliefs corresponding to the foreground and the background model. The probability of a pixel beeing described by the background model is therefore given by:

$$p(v_i\,|\,\tilde{\boldsymbol{\theta}},\boldsymbol{u}) = \sum_{k=1}^{K}b_{i,k}\ . \tag{11}$$

In the next section, we discuss implementation and performance issues.

3.4   Implementation Details

Our algorithm can be used in two different manners. First, it can run on-line, with a single E-M iteration at each frame, which allows fast computation. On very abrupt illumination changes, convergence is reached after a few frames (rarely more than 6). Second, the algorithm can run offline, with only two images as input instead of a video history. In this case, several iterations, typically 5 to 10, are necessary before convergence.

Local NCC can be computed efficiently with integral images, with a complexity linear with respect to the number of pixels and constant with respect to the window size. Thus, the complexity of the complete algorithm is also linear with the number of pixels, and the full process of acquiring, segmenting, and displaying images is achieved at a rate of about 2.3×10^6 pixels per second, using a single core of a 2.0GHz CPU. This is about 18 fps for half PAL (360x288), 12 FPS for 512x384, and 5-6 FPS for 720x576 images.

Correlation and texturedness images, as presented in section 3.2, are computed from single channel images. We use the green channel only, because it is more represented on a Bayer pattern. The correlation window is a square of 25×25 pixels, cropped at image borders.

For all experiments presented in the paper, $K = 2$ and $\bar{K} = 2$. The histograms h and $\bar{h}$ have been computed only once, from 9 pairs of images (about 2×10^6 training pixels). Training images do not contain any pattern or background used in test experiments.

The function l_i as presented in previous section is sensitive to limited dynamic range and to limited precision in low intensity values. Both following functions assume the same role with more robustness and give good result:

$$l_i^a(u_i) = \left[\arctan\left(\frac{u_{i,r}}{m_{i,r}}\right) \ \arctan\left(\frac{u_{i,g}}{m_{i,g}}\right) \ \arctan\left(\frac{u_{i,b}}{m_{i,b}}\right) \right]^T$$

$$l_i^c(u_i) = \left[\frac{u_{i,r} + c}{m_{i,r} + c} \ \frac{u_{i,g} + c}{m_{i,g} + c} \ \frac{u_{i,b} + c}{m_{i,b} + c} \right]^T$$

where c is an arbitrary positive constant. In our experiments, we use $c = 64$.

4   Results

In this section, we show results on individual frames of video sequences that feature both sudden illumination changes and shadows cast by occluding objects. We also compare those results to those produced by state-of-the-art techniques [1,11].

4.1   Robustness of Illumination Changes and Shadows

We begin by the sequence of Fig. 5 in which an arm is waved in front of a cluttered wall. The arm casts a shadow, which affects the scene's radiosity and causes the camera to automatically adapt its luminosity settings. With default parameters, the algorithm of [1] reacts to this by slowly adapting its background model. However, this adaptation cannot cope with the rapidly moving shadow and produces the poor result of Fig. 5(a). This can be prevented by increasing

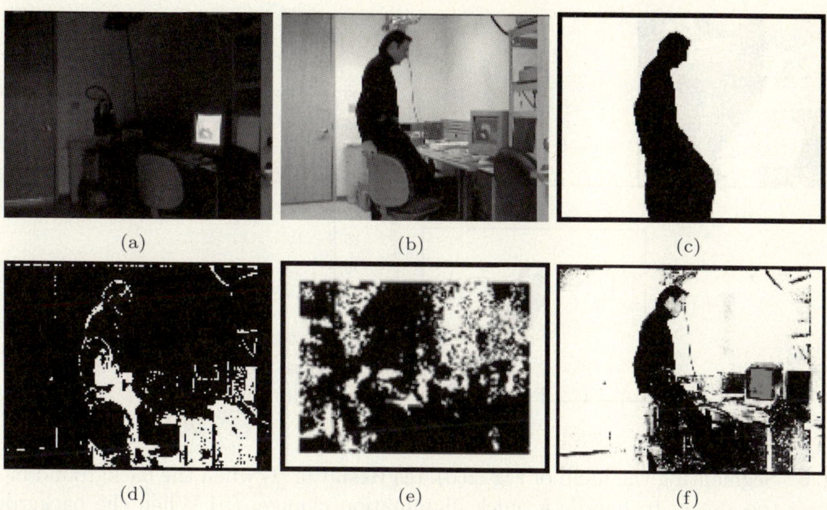

Fig. 4. Segmenting the *light switch* test images from [16]. (a) Background model. (b) Test image. (c) Manually segmented ground truth. (d) The output of Zivkovic's method [1]. (e) Result published in [11], using an approach based on local binary patterns. (f) Our result, obtained solely by comparing (a) and (b). Unlike the other two methods, we used no additional video frames.

the rate at which the background adapts, but, as shown in Fig. 5(b), it results in the sleeve being lost. By contrast, by explicitly reevaluating the illumination parameters at every frame, our algorithm copes much better with this situation, as shown in Fig. 5(c). To compare these two methods independently of specific parameter choices, we computed the ROC curve of Fig. 5(d). We take precision to be the number of pixels correctly tagged as foreground divided by the total number of pixels marked as foreground and recall to be the number of pixels tagged as foreground divided by the number of foreground pixels in the ground truth. The curve is obtained by binarizing using different thresholds for the probability of Eq. 11. We also represent different runs of [1] by crosses corresponding to different choices of its learning rate and the decision threshold. As expected, our method exhibits much better robustness towards illumination effects.

Fig. 1 depicts a sequence with even more drastic illumination changes that occur when the subject turns on one light after the other. The GMM based-method [1] immediately reacts by classifying most of the image as foreground. By contrast, our algorithm correctly compares the new images with the background image, taken to be the average of the first 25 frames of the sequence.

Fig. 4 shows the *light switch* benchmark of [16]. We again built the background representation by averaging 25 consecutive frames showing the room with the light switched off. We obtain good results when comparing it to an image where the light is turned on even though, unlike the other algorithms [1,11], we use a single frame instead of looking at the whole video. To foreground recall of 82% that appears in [11] entails a precision of only 25%, whereas our method achieves

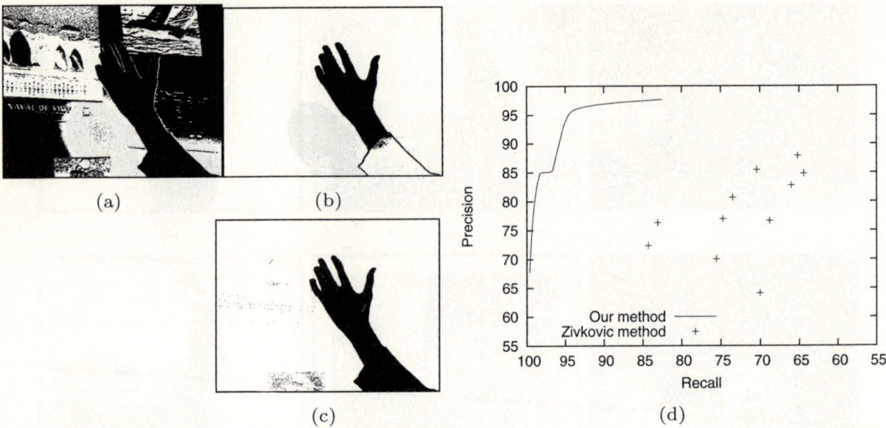

Fig. 5. Segmenting the hand of Fig. 2(b). (a) Result of [1] when the background model adjusts too slowly to handle a quick illumination change. (b) When the background model adjusts faster. (d) ROC curve for our method obtained by varying a threshold on the probability of Eq. 11. The crosses represent results obtained by [1] for different choices of learning rate and decision threshold.

49% for the same recall. With default parameters, the algorithm of [1] cannot handle this abrupt light change and yields a precision of 13% for a recall of 70%.

Finally, as shown in Fig. 6, we ran our algorithm on one of the PETS 2006 video sequences that features an abandoned luggage to demonstrate that our technique is indeed appropriate for surveillance applications because it does not lose objects by unduly merging them in the background.

4.2   Augmented Reality

Because our approach is very robust to abrupt illumination changes, it is a perfect candidate for occlusion segmentation in augmented reality. The task is

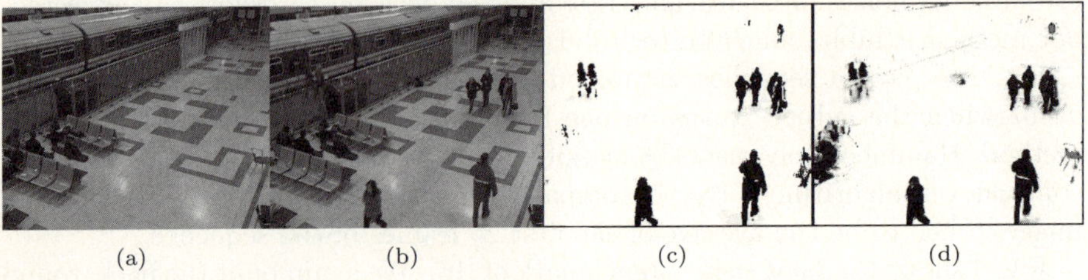

Fig. 6. PETS 2006 Dataset. (a) Initial frame of the video, used as background model. (b) Frame number 2800. (c) The background subtraction of [1]: The abandoned bag in the middle of the scene has mistakenly been integrated into the background. (d) Our method correctly segment the bag, the person who left after sitting on the bottom left corner, and the chair that has been removed on the right.

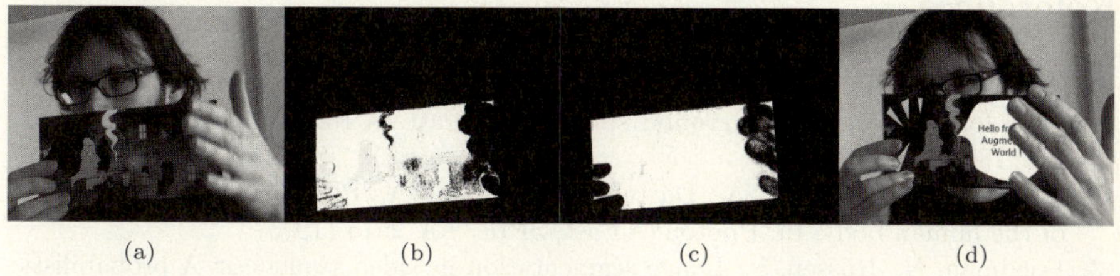

(a)   (b)   (c)   (d)

Fig. 7. Occlusion segmentation on a moving object. (a) Input frame in which the card is tracked. (b) Traditional background subtraction provide unsatisfying results because of the shadow cast by the hand, and because it learned the fingers hiding the bottom left corner as part of the background. (c): Our method is far more robust and produces a better segmentation. (d) We use its output as an alpha channel to convincingly draw the virtual text and account for the occluding hand.

the following: A user holds an object that is detected and augmented. If the detected pattern is occluded by a real object, the virtual object should also be occluded. In order to augment only the pixels actually showing the pattern, a visibility mask is required. Technically, any background subtraction technique could produce it, by unwarping the input images in a reference frame, and by rewarping the resulting segmentation back to the input frame.

The drastic illumination changes produced by quick rotation of the pattern might hinder a background subtraction algorithm that has not been designed for such conditions. That is why the Gaussian mixture based background subtraction method of [1] has difficulties to handle our test sequence illustrated by figure 7. On the other hand, the illumination modeling of our approach is able to handle this situation well and, unsurprisingly, shows superior results. The quality of the resulting segmentation we obtain allows convincing occluded augmented reality, as illustrated by figure 7(d).

5   Conclusion

We presented a fast background subtraction algorithm that handles heavy illumination changes by relying on a statistical model, not of the pixel intensities, but of the illumination effects. The optimized likelihood also fuses texture correlation clues by exploiting histograms trained off-line.

We demonstrated the performance of our approach under drastic light changes that state-of-the-art technique have trouble to handle.

Moreover, our technique can be used to segment the occluded parts of a moving planar object and therefore allows occlusion handling for augmented reality applications.

Although we do not explicitly model spatial consistency, the learnt histograms of correlation captures texture. Similarly, we could easily extend our method by integrating temporal dependence using temporal features.

References

1. Zivkovic, Z., van der Heijden, F.: Efficient adaptive density estimation per image pixel for the task of background subtraction. Pattern Recognition Letters 27(7), 773–780 (2006)
2. Wren, C., Azarbayejani, A., Darrell, T., Pentland, A.: Pfinder: Real-time tracking of the human body. In: Photonics East, SPIE, vol. 2615 (1995)
3. Friedman, N., Russell, S.: Image segmentation in video sequences: A probabilistic approach. In: Annual Conference on Uncertainty in Artificial Intelligence, pp. 175–181 (1997)
4. Stauffer, C., Grimson, W.: Adaptive background mixture models for real-time tracking. In: CVPR, pp. 246–252 (1999)
5. Bishop, C.: Pattern Recognition and Machine Learning. Springer, Heidelberg (2006)
6. Elgammal, A., Duraiswami, R., Harwood, D., Davis, L.: Background and foreground modeling using nonparametric kernel density for visual surveillance. Proceedings of the IEEE 90, 1151–1163 (2002)
7. Sheikh, Y., Shah, M.: Bayesian modeling of dynamic scenes for object detection. PAMI 27, 1778–1792 (2005)
8. Prati, A., Mikic, I., Trivedi, M., Cucchiara, R.: Detecting moving shadows: Algorithms and evaluation. PAMI 25, 918–923 (2003)
9. Stauder, J., Mech, R., Ostermann, J.: Detection of moving cast shadows for object segmentation. IEEE Transactions on Multimedia 1(1), 65–76 (1999)
10. Jabri, S., Duric, Z., Wechsler, H., Rosenfeld, A.: Detection and location of people in video images using adaptive fusion of color and edge information. In: International Conference on Pattern Recognition, vol. 4, pp. 627–630 (2000)
11. Heikkila, M., Pietikainen, M.: A texture-based method for modeling the background and detecting moving objects. PAMI 28(4), 657–662 (2006)
12. Criminisi, A., Cross, G., Blake, A., Kolmogorov, V.: Bilayer segmentation of live video. In: CVPR, pp. 53–60 (2006)
13. Lepetit, V., Pilet, J., Geiger, A., Mazzoni, A., Oezuysal, M., Fua, P.: Bazar, `http://cvlab.epfl.ch/software/bazar`
14. Rother, C., Kolmogorov, V., Blake, A.: Grabcut: Interactive foreground extraction using iterated graph cuts. ACM SIGGRAPH (2004)
15. Fransens, R., Strecha, C., Van Gool, L.: A mean field EM-algorithm for coherent occlusion handling in map-estimation problems. In: CVPR (2006)
16. Toyama, K., Krumm, J., Brumitt, B., Meyers, B.: Wallflower: principles and practice of background maintenance. In: International Conference on Computer Vision, vol. 1, pp. 255–261 (1999)

Closed-Form Solution to Non-rigid 3D Surface Registration*

Mathieu Salzmann, Francesc Moreno-Noguer,
Vincent Lepetit, and Pascal Fua

EPFL - CVLab,
1015 Lausanne, Switzerland

Abstract. We present a closed-form solution to the problem of recovering the 3D shape of a non-rigid inelastic surface from 3D-to-2D correspondences. This lets us detect and reconstruct such a surface by matching individual images against a reference configuration, which is in contrast to all existing approaches that require initial shape estimates and track deformations from image to image.

We represent the surface as a mesh, and write the constraints provided by the correspondences as a linear system whose solution we express as a weighted sum of eigenvectors. Obtaining the weights then amounts to solving a set of quadratic equations accounting for inextensibility constraints between neighboring mesh vertices. Since available closed-form solutions to quadratic systems fail when there are too many variables, we reduce the number of unknowns by expressing the deformations as a linear combination of modes. The overall closed-form solution then becomes tractable even for complex deformations that require many modes.

1   Introduction

3D shape recovery of deformable surfaces from individual images is known to be highly ambiguous. The standard approach to overcoming this is to introduce a deformation model and to recover the shape by optimizing an objective function [1,2,3,4,5,6,7,8] that measures the fit of the model to the data. However, in practice, this objective function is either non-convex or involves temporal consistency. Thus, to avoid being trapped in local minima, these methods require initial estimates that must be relatively close to the true shape. As a result, they have been shown to be effective for tracking, but not for registration without *a priori* shape knowledge.

By contrast, we propose here a solution to detecting and reconstructing inelastic 3D surfaces from correspondences between an individual image and a reference configuration, in closed-form, and without any initial shape estimate.

More specifically, we model flexible inelastic surfaces as triangulated meshes whose edge lengths cannot change. Given an image of the surface in a known

---

* This work was supported in part by the Swiss National Science Foundation and in part by the European Commission under the IST-project 034307 DYVINE (Dynamic Visual Networks).

D. Forsyth, P. Torr, and A. Zisserman (Eds.): ECCV 2008, Part IV, LNCS 5305, pp. 581–594, 2008.

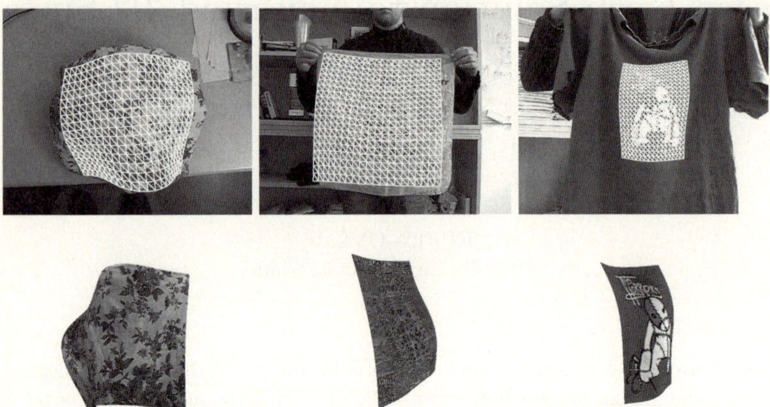

Fig. 1. 3D Reconstruction of non-rigid objects from and individual image and a reference configuration. Results were obtained in closed-form, without any initial estimate. Top: Recovered mesh overlaid on the original image. Bottom: Re-textured side view of the retrieved surface.

3D configuration, and correspondences between that *model image* and an *input image* in which the shape is unknown, retrieving the mesh's vertex coordinates involves solving a rank-deficient linear system encoding the projection equations. Taking our inspiration from our recent paper on rigid object pose estimation [9], we express the solution of this linear system as a weighted sum of the corresponding matrix's eigenvectors associated with the smallest eigenvalues. We compute these weights by using Extended Linearization [10] to solve a set of quadratic constraints that preserve edge lengths. In its simplest form, this method is only directly applicable to very small meshes because, for larger ones, the number of unknowns after Extended Linearization grows fast, thus yielding an intractable problem. We overcome this difficulty by expressing the surface deformations as a linear combination of deformation modes. This preserves the linear formulation of the correspondence problem, but dramatically reduces the size of the corresponding linear system, while improving its conditioning. Therefore, the quadratic constraints required to guarantee inextensibility are also expressed in terms of a smaller number of variables, making Extended Linearization practical. As a result, we can solve our problem in closed-form even when using enough modes to model complex deformations such as those of Fig. 1, which yields a 3D reconstruction that jointly minimizes edge length variations and reprojects correctly on the input image.

2   Related Work

3D reconstruction of non-rigid surfaces from images has attracted increasing attention in recent years. It is a severely under-constrained problem and many different kinds of prior models have been introduced to restrict the space of possible shapes to a manageable size.

Most of the models currently in use trace their roots to the early physics-based models that were introduced to delineate 2D shapes [11] and reconstruct relatively simple 3D ones [12].

As far as 2D problems are concerned, their more recent incarnations have proved effective for image registration [13,14] and non-rigid surface detection [15,16]. Many variations of these models have also been proposed to address 3D problems, including superquadrics [1], triangulated surfaces [2], or thin-plate splines [17]. Additionally, dimensionality reduction was introduced through modal analysis [3,18], where shapes are represented as linear combinations of deformation modes. Finally, a very recent work [19] proposes to set bounds on distances between feature points, and use them in conjunction with a thin-plate splines model to reconstruct inextensible surfaces.

One limitation of the physics-based models is that they rarely describe accurately the non-linear physics of large deformations. In theory, this could be remedied by introducing more sophisticated finite-element modeling. However, in practice, this often leads to vastly increased complexity without a commensurate gain in performance. As a result, in recent years, there has been increasing interest in statistical learning techniques that build surface deformation models from training data. Active Appearance Models [20] pioneered this approach by learning low-dimensional linear models for 2D face tracking. They were quickly followed by Active Shape Models [5] and Morphable Models [4] that extended it to 3D. More recently, linear models have also been learned for structure-from-motion applications [6,21] and tracking of smoothly deforming 3D surfaces [7].

There has also been a number of attempts at performing 3D surface reconstruction without resorting to a deformation model. One approach has been to use lighting information in addition to texture clues to constrain the reconstruction process [8], which has only been demonstrated under very restrictive assumptions on lighting conditions and is therefore not generally applicable. Other approaches have proposed to use motion models over video sequences. The reconstruction problem was then formulated either as solving a large linear system [22] or as a Second Order Cone Programming problem [23]. These formulations, however, rely on tightly bounding the vertex displacements from one frame to the next, which makes them applicable only in a tracking context where the shape in the first frame of the sequence is known.

In all the above methods, shape recovery entails minimizing an objective function. In most cases, the function is non convex, and therefore, one can never be sure to find its global minimum, especially if the initial estimate is far from the correct answer. In the rare examples formulated as convex problems [23], the solution involves temporal consistency, which again requires a good initialization.

By contrast, many closed-form solutions have been proposed for pose estimation of rigid objects [24,25,26]. In fact, the inspiration for our method came from our earlier work [9] in that field. However, reconstructing a deformable surface involves many more variables than the 6 rigid motion degrees of freedom. In the remainder of this paper, we show that this therefore requires a substantially different approach.

3   Closed-Form 3D Reconstruction

In this section, we show that recovering the 3D shape of a flexible surface from 3D-to-2D correspondences can be achieved by solving a set of quadratic equations accounting for inextensibility, which can be done in closed-form.

3.1   Notations and Assumptions

We represent our surface as a triangulated mesh made of n_v vertices $\mathbf{v}_i = [x_i, y_i, z_i]^T$, $1 \leq i \leq n_v$ connected by n_e edges. Let $\mathbf{X} = [\mathbf{v}_1^T, \cdots, \mathbf{v}_{n_v}^T]^T$ be the vector of coordinates obtained by concatenating the $\mathbf{v}_i$.

We assume that we are given a set of n_c 3D-to-2D correspondences between the surface and an image. Each correspondence relates a 3D point on the mesh, expressed in terms of its barycentric coordinates in the facet to which it belongs, and a 2D feature in the image.

Additionally, we assume the camera to be calibrated and, therefore, that its matrix of intrinsic parameters $\mathbf{A}$ is known. To simplify our notations without loss of generality, we express the vertex coordinates in the camera referential.

3.2   Linear Formulation of the Correspondence Problem

We first show that, given a set of 3D-to-2D correspondences, the vector of vertex coordinates $\mathbf{X}$ can be found as the solution of a linear system.

Let $\mathbf{x}$ be a 3D point belonging to facet f with barycentric coordinates $[a_1, a_2, a_3]$. Hence, we can write it as $\mathbf{x} = \sum_{i=1}^{3} a_i \mathbf{v}_{f,i}$, where $\{\mathbf{v}_{f,i}\}_{i=1,2,3}$ are the three vertices of facet f. The fact that $\mathbf{x}$ projects to the 2D image location (u, v) can now be expressed by the relation

$$\mathbf{A} \left(a_1 \mathbf{v}_{f,1} + a_2 \mathbf{v}_{f,2} + a_3 \mathbf{v}_{f,3} \right) = k \begin{bmatrix} u \\ v \\ 1 \end{bmatrix} , \tag{1}$$

where k is a scalar accounting for depth. Since, from the last row of Eq. 1, k can be expressed in terms of the vertex coordinates, we have

$$\begin{bmatrix} a_1 \mathbf{B} & a_2 \mathbf{B} & a_3 \mathbf{B} \end{bmatrix} \begin{bmatrix} \mathbf{v}_{f,1} \\ \mathbf{v}_{f,2} \\ \mathbf{v}_{f,3} \end{bmatrix} = \mathbf{0} , \text{ with } \mathbf{B} = \mathbf{A}_{2 \times 3} - \begin{bmatrix} u \\ v \end{bmatrix} \mathbf{A}_3 , \tag{2}$$

where $\mathbf{A}_{2 \times 3}$ are the first two rows of $\mathbf{A}$, and $\mathbf{A}_3$ is the third one. n_c such correspondences between 3D surface points and 2D image locations therefore provide $2n_c$ linear constraints such as those of Eq. 2. They can be jointly expressed by the linear system

$$\mathbf{MX} = \mathbf{0} , \tag{3}$$

where $\mathbf{M}$ is a $2n_c \times 3n_v$ matrix obtained by concatenating the $\begin{bmatrix} a_1 \mathbf{B} & a_2 \mathbf{B} & a_3 \mathbf{B} \end{bmatrix}$ matrices of Eq. 2.

Although solving this system yields a surface that reprojects correctly on the image, there is no guarantee that its 3D shape corresponds to reality. This stems from the fact that, for all practical purposes, $\mathbf{M}$ is rank deficient. More specifically, even where there are many correspondences, one third, i.e. n_v, of the eigenvalues of $\mathbf{M}^T\mathbf{M}$ are very close to zero [22], as illustrated by Fig. 2(c). As a result, even small amounts of noise produce large instability in the recovered shape.

This suggests that additional constraints have to be added to guarantee a unique and stable solution. In most state-of-the-art approaches, these constraints are provided by deformation models and are enforced via an iterative method. By contrast, we will argue that imposing inextensibility of the surface yields a closed-form solution to the problem.

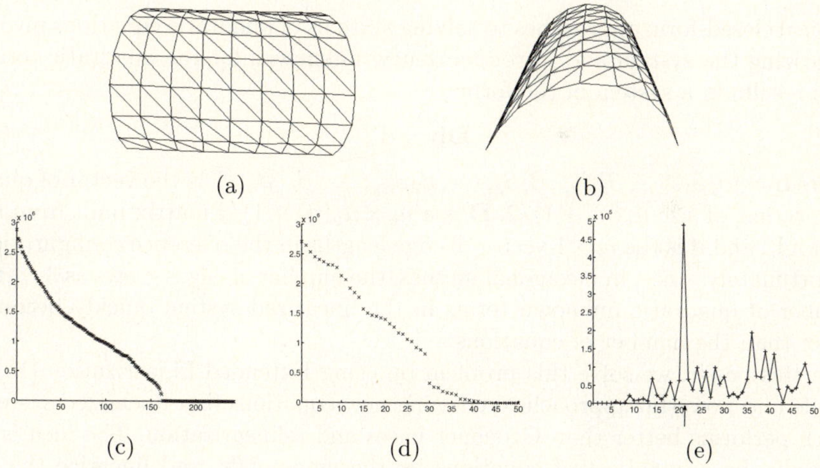

Fig. 2. (a,b) Original and side views of a surface used to generate a synthetic sequence. The 3D shape was reconstructed by an optical motion capture system. (c,d) Eigenvalues of the linear system written from correspondences randomly established for the synthetic shape of (a). (c) The system was written in terms of 243 vertex coordinates. One third of the eigenvalues are close to zero. (d) The system was written in terms of 50 PCA modes. There are still a number of near zero eigenvalues. (e) First derivative of the curve (d) (in reversed x-direction). We take the maximum value of n_l to be the one with maximum derivative, which corresponds to the jump in (d).

3.3 Inextensible Meshes

Following the idea introduced in [9], we write the solution of the linear system of Eq. 3 as a weighted sum of the eigenvectors $\mathbf{l}_i$, $1 \leq i \leq n_v$ of $\mathbf{M}^T\mathbf{M}$, which are those associated with the eigenvalues that are almost zero. Therefore we write

$$\mathbf{X} = \sum_{i=1}^{n_v} \beta_i \mathbf{l}_i \,, \tag{4}$$

since any such linear combination of $\mathbf{l}_i$ is in the kernel of $\mathbf{M}^T\mathbf{M}$ and produces a mesh that projects correctly on the image. Our problem now becomes finding appropriate values for the β_i, which are the new unknowns.

We are now in a position to exploit the inextensibility of the surface by choosing the β_i so that edge lengths are preserved. Such β_i can be expressed as the solution of a set of quadratic equations of the form

$$\| \sum_{i=1}^{n_v} \beta_i \mathbf{l}_i^j - \sum_{i=1}^{n_v} \beta_i \mathbf{l}_i^k \|^2 = \| \mathbf{v}_j^{ref} - \mathbf{v}_k^{ref} \|^2 \,, \tag{5}$$

where $\mathbf{l}_i^j$ is the 3×1 sub-vector of $\mathbf{l}_i$ corresponding to the coordinates of vertex $\mathbf{v}_j$, and $\mathbf{v}_j^{ref}$ and $\mathbf{v}_k^{ref}$ are two neighboring vertices in the reference configuration.

3.4   Extended Linearization

Typical closed-form approaches to solving systems of quadratic equations involve linearizing the system and introducing new unknowns for the quadratic terms. This results in a system of the form

$$\mathbf{Db} = \mathbf{d} \,, \tag{6}$$

where $\mathbf{b} = [\beta_1\beta_1, \cdots, \beta_1\beta_{n_v}, \beta_2\beta_2, \cdots, \beta_2\beta_{n_v}, \cdots, \beta_{n_v}\beta_{n_v}]^T$ is the vector of quadratic terms, of size $n_v(n_v+1)/2$. $\mathbf{D}$ is a $n_e \times n_v(n_v+1)/2$ matrix built from the known $\mathbf{l}_i$, and $\mathbf{d}$ is the $n_e \times 1$ vector of edge lengths in the reference configuration. Unfortunately, since, in hexagonal meshes, the number of edges grows as $3n_v$, the number of quadratic unknown terms in the linearized system quickly becomes larger than the number of equations.

In this paper, we solve this problem by using Extended Linearization [10], a simple and powerful approach to creating new equations in a linearized system, which performs better than Groebner bases and relinearization. The idea is to multiply the original set of equations by the monomials, and linearize the resulting system. In our particular case, we can, for example, multiply the existing quadratic equations by each of the linear terms, thus creating new equations of the form

$$\beta_1 \left(\| \sum_{i=1}^{n_v} \beta_i \mathbf{l}_i^j - \sum_{i=1}^{n_v} \beta_i \mathbf{l}_i^k \|^2 \right) = \beta_1 \left(\| \mathbf{v}_j^{ref} - \mathbf{v}_k^{ref} \|^2 \right) \,,$$

$$\vdots$$

$$\beta_{n_v} \left(\| \sum_{i=1}^{n_v} \beta_i \mathbf{l}_i^j - \sum_{i=1}^{n_v} \beta_i \mathbf{l}_i^k \|^2 \right) = \beta_{n_v} \left(\| \mathbf{v}_j^{ref} - \mathbf{v}_k^{ref} \|^2 \right) \,.$$

Let $\mathbf{b}^c = [\beta_1\beta_1\beta_1, \cdots, \beta_1\beta_1\beta_{n_v}, \beta_1\beta_2\beta_2, \cdots, \beta_1\beta_2\beta_{n_v}, \beta_2\beta_2\beta_2, \cdots, \beta_{n_v}\beta_{n_v}\beta_{n_v}]^T$, and $\mathbf{b}^l = [\beta_1, \cdots, \beta_{n_v}]^T$. The resulting system can be written as

$$\begin{bmatrix} 0 & \cdots & 0 & \mathbf{D}_1^{1,1} & \cdots & \mathbf{D}_1^{1,n_v} & \mathbf{D}_1^{2,2} & \cdots & \mathbf{D}_1^{n_v,n_v} & 0 & \cdots & \cdots & \cdots & 0 \\ \cdots & \cdots & \cdots & \cdots & \cdots & \cdots & \cdots & \cdots & \cdots & \cdots & \cdots & \cdots & \cdots & \cdots \\ -\mathbf{d}_1 & 0 & \cdots & 0 & \cdots & \cdots & \cdots & \cdots & 0 & \mathbf{D}_1^{1,1} & \cdots & \mathbf{D}_1^{n_v,n_v} & 0 & \cdots \\ \cdots & \cdots & \cdots & \cdots & \cdots & \cdots & \cdots & \cdots & \cdots & \cdots & \cdots & \cdots & \cdots & \cdots \end{bmatrix} \begin{bmatrix} \mathbf{b}^l \\ \mathbf{b} \\ \mathbf{b}^c \end{bmatrix} = \begin{bmatrix} \mathbf{d}_1 \\ \vdots \\ \mathbf{0} \\ \vdots \end{bmatrix} \,, \tag{7}$$

where we only show the first line of the original system of Eq. 6 and its product with β_1, and where $\mathbf{D}_1^{i,j}$ stands for the coefficient on the first line of $\mathbf{D}$ corresponding to the product $\beta_i\beta_j$.

It can be shown that multiplying the inextensibility equations by all the β_i only yields a sufficient number of equations for very small meshes, i.e. less than 12 vertices for a hexagonal mesh. In theory, one could solve this problem by applying Extended Linearization iteratively by re-multiplying the new equations by the linear terms. However, in practice, the resulting system quickly becomes so large that it is intractable, i.e. for a 10×10 mesh, the number of equations only becomes larger than the number of unknowns when the size of the system is of the order 10^{10}. In other words, Extended Linearization cannot deal with a problem as large as ours and we are not aware of any other closed-form approach to solving systems of quadratic equations that could. We address this issue in the next section.

3.5   Linear Deformation Model

As discussed above, to solve the set of quadratic equations that express edge length preservation, we need to reduce its size to the point where Extended Linearization becomes a viable option. Furthermore, we need to do this in such a way that the solution of the correspondence problem can still be expressed as the solution of a system of linear equations, as discussed in Section 3.2. To this end, we model the plausible deformations of the mesh as a linear combination of n_m deformation modes [6,7], much in the same spirit as those the morphable models used to represent face deformations [4]. We write

$$\mathbf{X} = \mathbf{X_0} + \sum_{i=1}^{n_m} \alpha_i \mathbf{p}_i = \mathbf{X_0} + \mathbf{P}\alpha \, , \tag{8}$$

where the $\mathbf{p}_i$ are the deformation modes and the α_i their associated weights. In our implementation, modes were obtained by applying Principal Component Analysis to a matrix of registered training meshes in deformed configurations, from which the mean shape $\mathbf{X_0}$ was subtracted [7]. The $\mathbf{p}_i$ therefore are the eigenvectors of the data covariance matrix. Nonetheless, they could also have been derived by modal analysis, which amounts to computing the eigenvectors of a stiffness matrix, and is a standard approach in physics-based modeling [3].

In this formulation, recovering the shape amounts to computing the weights α. Since the shape must satisfy Eq. 3, α must then satisfy

$$\mathbf{M}(\mathbf{X_0} + \mathbf{P}\alpha) = \mathbf{0} \ . \tag{9}$$

When solving this system, to ensure that the recovered weights do not generate shapes exceedingly far from our training data, we introduce a regularization term by penalizing α_i with the inverse of the corresponding eigenvalue σ_i of the data covariance matrix. We therefore solve

$$\begin{bmatrix} \mathbf{MP} & \mathbf{MX_0} \\ w_r\mathbf{S} & \mathbf{0} \end{bmatrix} \begin{bmatrix} \alpha \\ 1 \end{bmatrix} = \mathbf{0} \, , \tag{10}$$

where $\mathbf{S}$ is an $n_m \times n_m$ diagonal matrix whose elements are the σ_i^{-1} and w_r is a regularization weight that only depends on the maximum σ_i, and whose precise value has only little influence on the results.

As shown in Fig. 2(d), we have considerably reduced the number of near-zero eigenvalues. The system of Eq. 10 is therefore better conditioned than the one of Eq. 3, but still does not yield a well-posed problem that would have a unique solution. This is attributable to the fact that, because the solution is expressed as a sum of deformation modes, inextensibility constraints, which are non linear, are not enforced.

Nonetheless, we can follow the same procedure as in Sections 3.3 and 3.4. We write the solution of the linear system of Eq. 10 as a weighted sum of the eigenvectors $\tilde{\mathbf{l}}_i$, $1 \leq i \leq n_l \ll n_m$ associated with the smallest eigenvalues of its matrix, and find the weights $\tilde{\beta}_i$ as the solution of the linearized system of quadratic equations

$$\tilde{\mathbf{D}}\tilde{\mathbf{b}} = \tilde{\mathbf{d}} \, , \tag{11}$$

where $\tilde{\mathbf{b}} = [\tilde{\beta}_1, \cdots, \tilde{\beta}_{n_l}, \tilde{\beta}_1\tilde{\beta}_1, \cdots, \tilde{\beta}_1\tilde{\beta}_{n_l}, \tilde{\beta}_2\tilde{\beta}_2, \cdots, \tilde{\beta}_2\tilde{\beta}_{n_l}, \cdots, \tilde{\beta}_{n_l}\tilde{\beta}_{n_l}]^T$ now also contains the linear terms arising in the quadratic equations from the mean shape $\mathbf{X_0}$. Furthermore, the system also encodes the additionnal linear equation that constrains the $\tilde{\beta}_i\tilde{\mathbf{l}}_{i,n_m+1}$ to sum up to 1, where $\tilde{\mathbf{l}}_{i,n_m+1}$ is the last element of $\tilde{\mathbf{l}}_i$.

Since in practice $n_l \ll n_m \ll n_v$, the system is now much smaller. Therefore a single iteration of Extended Linearization is sufficient to constrain its solution while keeping it tractable, even for relatively large numbers of modes—in practice up to 60—thus allowing complex deformations.

In this formulation, the number n_l of eigenvectors strongly depends on the number n_m of modes used for the recovery. However, as shown in Fig. 2(e), we can easily set the maximum number $\hat{n}_l$ of eigenvectors to use by picking the number corresponding to the maximum first derivative of the ordered eigenvalues curve. We then simply test for all $n_l \leq \hat{n}_l$ and pick the optimal value as the one that, for a small enough reprojection error, gives the smallest mean edge length variation. In practice, $\hat{n}_l$ was typically about 25 when using 60 deformation modes.

4   Experimental Results

In this section we show that our method can be successfully applied to reconstructing non-rigid shapes from individual images and a reference configuration. We present results on both synthetic data and real images.

4.1   Synthetic Data

We first applied our method to images, such as those of Fig. 2(a), synthesized by projecting known deformed shapes using a virtual camera. The deformed shapes were obtained by recovering the 3D locations of reflective markers stuck on a 200×200mm piece of cardboard with an optical motion capture system. This allowed us to randomly create n_{cf} perfect correspondences per facet to

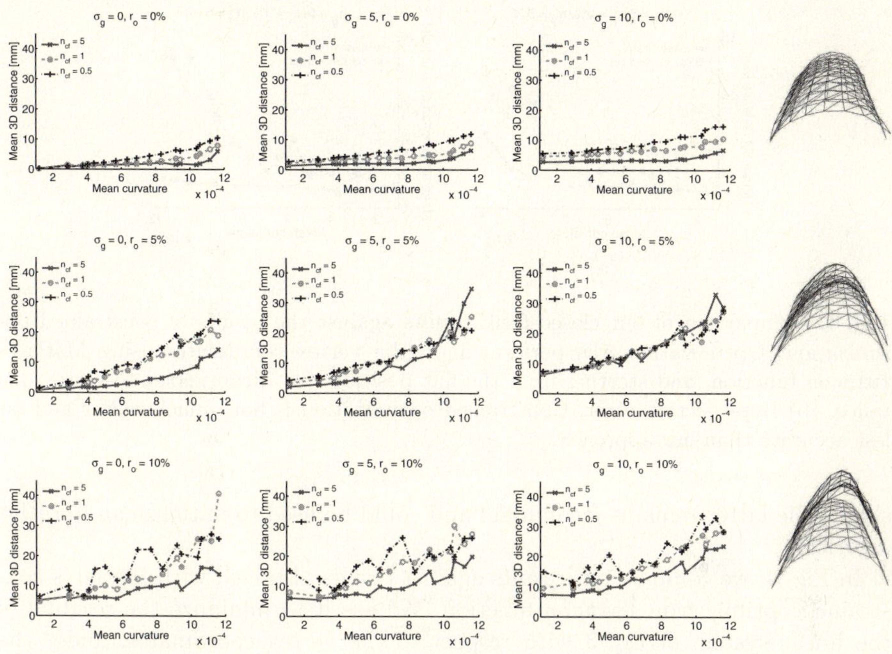

Fig. 3. Shape recovery of a 200×200mm synthetic mesh imaged by a virtual camera placed 20cm away from it. Each plot shows the mean vertex-to-vertex 3D distance between the recovered surface and the ground-truth as a function of its mean curvature. The three different curves in each graph correspond to a varying number of correspondences per facet. Left to right, the gaussian noise added to the correspondences increases. Top to bottom, the number of outliers grows. For each experiments, we plot the average over 40 trials. The rightmost column shows in blue recovered shapes for the ground-truth surface of Fig. 2(a,b), shown in red. The corresponding mean vertex-to-vertex distances are 9mm, 19mm and 38mm. This highlights the fact that even for distances around 40mm, the recovered shape remains meaningful.

which we added zero mean gaussian noise of variance σ_g. Finally, we simulated outliers by setting the image coordinates of r_o percents of the correspondences to uniformly and randomly distributed values.

In Fig. 3, we show results as a function of the surface's mean curvature, the maximum one being that of Fig. 2(a). Each plot includes three curves corresponding to $n_{cf} = \{5, 1, 1/2\}$, which depict the mean vertex-to-vertex 3D distance between the recovered mesh and ground-truth. The plots are ordered on a grid whose x-direction corresponds to $\sigma_g = \{0, 5, 10\}$ and y-direction to $r_o = \{0\%, 5\%, 10\%\}$. Each experiment was repeated 40 times, and we show the average results. Note that the error grows with the mean curvature of the shape, which is natural since the shape becomes more ambiguous when seen from the viewpoint shown in Fig. 2(a). In the rightmost column, we display three shapes reconstructed from the image of Fig. 2(a) with their corresponding ground-truth. Note that even for average distances of 40mm between the true and recovered

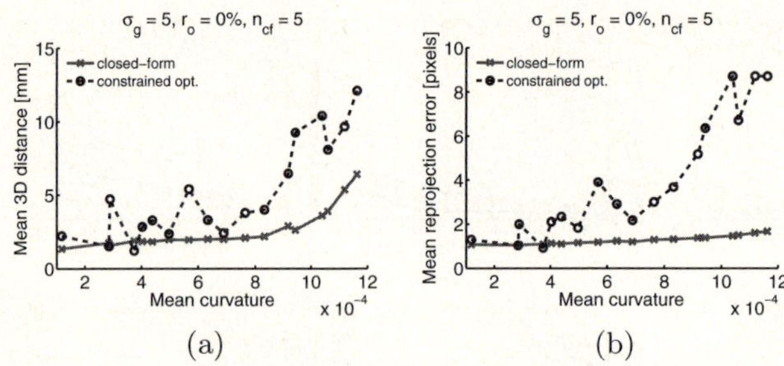

Fig. 4. Comparison of our closed-form results against the results of constrained optimization. Optimization was performed on the vertex coordinates using Matlab's `fmincon` function, and starting from the flat position. (a) Mean vertex-to-vertex distance. (b) Reprojection error. Constrained optimization is both much slower and far less accurate than our approach.

shape, the latter remains meaningful and could be used to initialize an iterative algorithm.

In Fig. 4, we compare our results against results obtained with Matlab's constrained optimization `fmincon` function. We use it to minimize the residual of the linear system of Eq. 3 with respect to the vertex coordinates, under the constraints that edge lengths must remain constant. We first tried to use the similar representation in terms of modes. However, since the constraints could never be truly satisfied, the algorithm would never converge towards an acceptable solution. This forced us to directly use the vertex coordinates. To improve convergence and prevent the surface from crumpling, we added a smoothness term [11]. For all the frames, the initialization was set to the flat position. In Fig. 4(a), we show the mean 3D vertex-to-vertex distance for the case where $\sigma_g = 5$, $r_o = 0$, and $n_{cf} = 5$. The red curve corresponds to our closed-form solution and the blue one to constrained optimization. Note that our approach gives much better results. Furthermore, it is also much faster, requiring only 1.5 minutes per frame as opposed to 1.5 hours for constrained optimization. Fig. 4(b) shows the reprojection errors for the same cases.

4.2 Real Images

We tested our method on a folded bed-sheet, a piece of cloth and a t-shirt deforming in front of a 3-CCD DV-camera. In all these cases, we first established SIFT [27] correspondences between the reference image and the input one. We then detected the surface in 2D, which can be done in closed-form by simply solving the linear system built from SIFT matches, augmented with linear smoothing equations [11]. For each facet, we then warped the reference image to best match the input one based on the retrieved 2D shape, and finally established dense correspondences by sampling the barycentric coordinates of the facet, and matching small regions between the input image and the warped reference one using normalized cross-correlation. Note that, even when we show results on

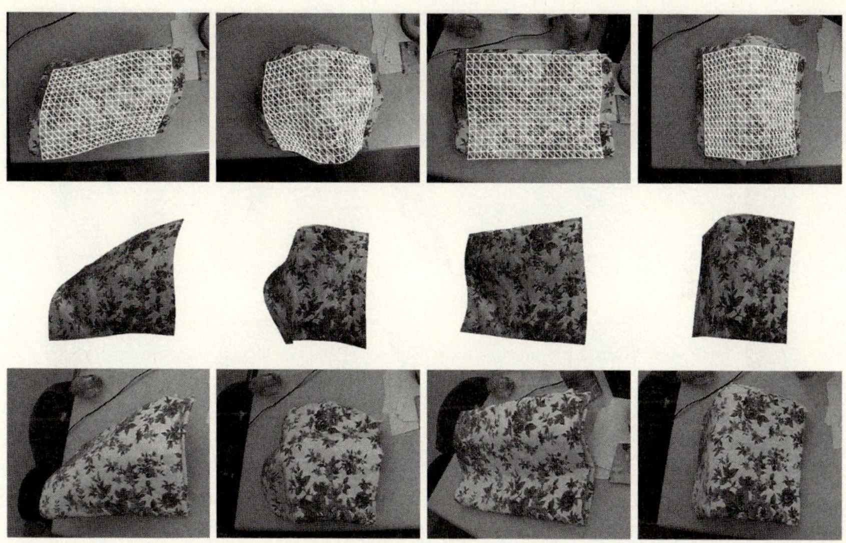

Fig. 5. 3D registration of a folded bed-sheet to an individual image given a reference configuration. Top Row: Recovered mesh overlaid on the original image. Middle Row: Synthesized textured view using the recovered shape. Bottom Row: Real side view of the sheet from similar viewpoints. Despite lighting changes, the synthetic images closely match the real ones.

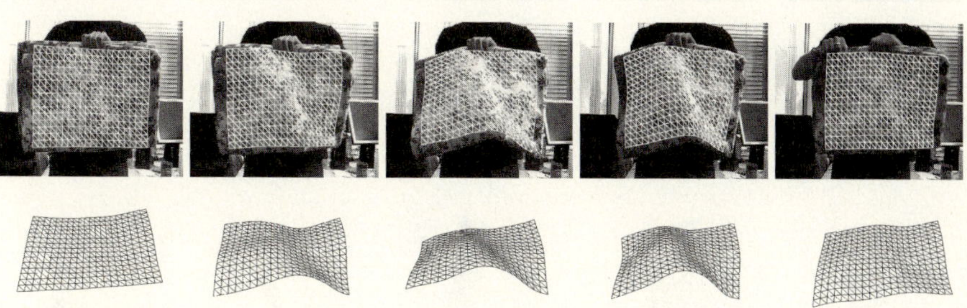

Fig. 6. Shape recovery of a bed-sheet. Top Row: Recovered mesh overlaid on the original image. Bottom Row: Mesh seen from a different viewpoint.

video sequences, nothing links one frame to the next, and no initialization is required. Corresponding videos are given as supplementary material.

In the case of the sheet, we deformed it into several unrelated shapes, took pictures from 2 different views for each deformation, and reconstructed the surface from a single image and a reference configuration. In Fig. 5, we show the results on four different cases. From our recovered shape, we generated synthetic textured images roughly corresponding to the viewpoint of the second image. As can be seen in the two bottom rows of Fig. 5, our synthetic images closely match the real side views. Additionally, we also reconstructed the same sheet

Fig. 7. Shape recovery of a piece of cloth. From Top to Bottom: Mesh computed in closed-form overlaid on the input image, side view of that mesh, refined mesh after 5 Gauss-Newton iterations.

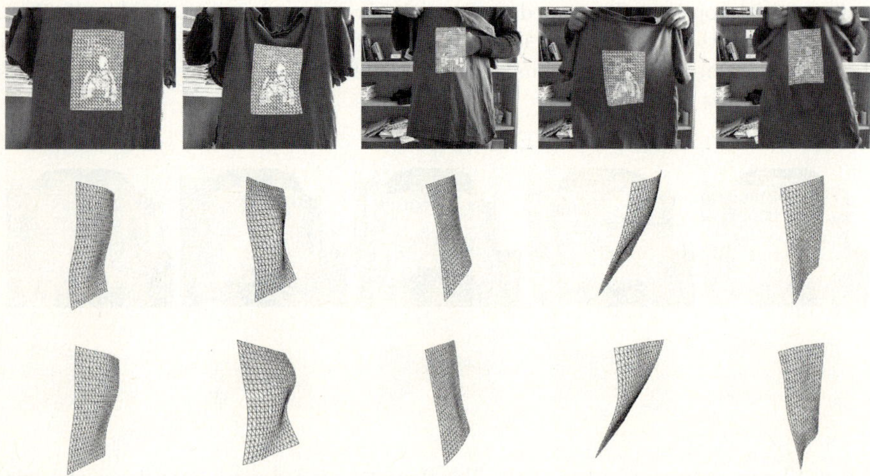

Fig. 8. Shape recovery of the central part of a t-shirt. From Top to Bottom: Mesh computed in closed-form overlaid on the input image, side view of that mesh, refined mesh after 5 Gauss-Newton iterations.

from the images of a video sequence, and show the results in Fig. 6. Note that no initialization was required, and that nothing links one frame to the next.

In Figs. 7 and 8, we show results for images of a piece of cloth and of a t-shirt waved in front of the camera. Note that in both cases, the closed-form solution closely follows what we observe in the videos. To further refine it, we implemented a simple Gauss-Newton optimization technique, and minimize the residual $\|\tilde{\mathbf{D}}\tilde{\mathbf{b}} - \tilde{\mathbf{d}}\|$ corresponding to Eq. 11 with respect to the $\tilde{\beta}_i$. In the third row of the figures, we show the refined mesh after 5 iterations this scheme. This proved sufficient to recover finer details at a negligible increase in overall computation time.

5   Conclusion

In this paper, we presented a closed-form solution to the problem of recovering the shape of a non-rigid inelastic surface from an individual image and a reference configuration. We showed that the reconstruction could be obtained by solving a system of quadratic equations representing distance constraints between neighboring mesh vertices.

In future work, we intend to investigate what additional quadratic constraints could be introduced to the current system of distance constraints. They could come from additional sources of image information, such as lighting. Having a larger number of quadratic equations would hopefully relieve the need for Extended Linearization, and result in smaller, and therefore faster to solve, linear systems.

References

1. Metaxas, D., Terzopoulos, D.: Constrained deformable superquadrics and nonrigid motion tracking. PAMI 15, 580–591 (1993)
2. Cohen, L., Cohen, I.: Deformable models for 3-d medical images using finite elements and balloons. In: CVPR, pp. 592–598 (1992)
3. Pentland, A.: Automatic extraction of deformable part models. IJCV 4, 107–126 (1990)
4. Blanz, V., Vetter, T.: A Morphable Model for The Synthesis of 3–D Faces. ACM SIGGRAPH, 187–194 (1999)
5. Matthews, I., Baker, S.: Active Appearance Models Revisited. IJCV 60, 135–164 (2004)
6. Torresani, L., Hertzmann, A., Bregler, C.: Learning non-rigid 3d shape from 2d motion. In: NIPS (2003)
7. Salzmann, M., Pilet, J., Ilić, S., Fua, P.: Surface Deformation Models for Non-Rigid 3–D Shape Recovery. PAMI 29, 1481–1487 (2007)
8. White, R., Forsyth, D.: Combining cues: Shape from shading and texture. In: CVPR (2006)
9. Moreno-Noguer, F., Lepetit, V., Fua, P.: Accurate Non-Iterative $O(n)$ Solution to the PnP Problem. In: ICCV (2007)
10. Courtois, N., Klimov, A., Patarin, J., Shamir, A.: Efficient algorithms for solving overdefined systems of multivariate polynomial equations. In: Preneel, B. (ed.) EUROCRYPT 2000. LNCS, vol. 1807. Springer, Heidelberg (2000)
11. Kass, M., Witkin, A., Terzopoulos, D.: Snakes: Active Contour Models. IJCV 1, 321–331 (1988)
12. Terzopoulos, D., Witkin, A., Kass, M.: Symmetry-seeking Models and 3D Object Reconstruction. IJCV 1, 211–221 (1987)

13. Bartoli, A., Zisserman, A.: Direct Estimation of Non-Rigid Registration. In: BMVC (2004)
14. Gay-Bellile, V., Bartoli, A., Sayd, P.: Direct estimation of non-rigid registrations with image-base self-occlusion reasoning. In: ICCV (2007)
15. Pilet, J., Lepetit, V., Fua, P.: Real-Time Non-Rigid Surface Detection. In: CVPR (2005)
16. Zhu, J., Lyu, M.R.: Progressive finit newton approach to real-time nonrigid surface detection. In: ICCV (2007)
17. McInerney, T., Terzopoulos, D.: A Finite Element Model for 3D Shape Reconstruction and Nonrigid Motion Tracking. In: ICCV (1993)
18. Delingette, H., Hebert, M., Ikeuchi, K.: Deformable surfaces: A free-form shape representation. Geometric Methods in Computer Vision (1991)
19. Perriollat, M., Hartley, R., Bartoli, A.: Monocular Template-based Reconstruction of Inextensible Surfaces. In: BMVC (2008)
20. Cootes, T., Edwards, G., Taylor, C.: Active Appearance Models. In: Burkhardt, H., Neumann, B. (eds.) ECCV 1998. LNCS, vol. 1407. Springer, Heidelberg (1998)
21. Llado, X., Bue, A.D., Agapito, L.: Non-rigid 3D Factorization for Projective Reconstruction. In: BMVC (2005)
22. Salzmann, M., Lepetit, V., Fua, P.: Deformable Surface Tracking Ambiguities. In: CVPR (2007)
23. Salzmann, M., Hartley, R., Fua, P.: Convex Optimization for Deformable Surface 3–D Tracking. In: ICCV (2007)
24. Quan, L., Lan, Z.: Linear N-Point Camera Pose Determination. PAMI 21, 774–780 (1999)
25. Fiore, P.D.: Efficient linear solution of exterior orientation. PAMI 23, 140–148 (2001)
26. Ansar, A., Daniilidis, K.: Linear pose estimation from points or lines. PAMI 25, 578–589 (2003)
27. Lowe, D.: Distinctive Image Features from Scale-Invariant Keypoints. IJCV 20, 91–110 (2004)

Implementing Decision Trees and Forests on a GPU

Toby Sharp

Microsoft Research, Cambridge, UK
`toby.sharp@microsoft.com`

Abstract. We describe a method for implementing the evaluation and training of decision trees and forests entirely on a GPU, and show how this method can be used in the context of object recognition.

Our strategy for evaluation involves mapping the data structure describing a decision forest to a 2D texture array. We navigate through the forest for each point of the input data in parallel using an efficient, non-branching pixel shader. For training, we compute the responses of the training data to a set of candidate features, and scatter the responses into a suitable histogram using a vertex shader. The histograms thus computed can be used in conjunction with a broad range of tree learning algorithms.

We demonstrate results for object recognition which are identical to those obtained on a CPU, obtained in about 1% of the time.

To our knowledge, this is the first time a method has been proposed which is capable of evaluating or training decision trees on a GPU. Our method leverages the full parallelism of the GPU.

Although we use features common to computer vision to demonstrate object recognition, our framework can accommodate other kinds of features for more general utility within computer science.

1   Introduction

1.1   Previous Work

Since their introduction, randomized decision forests (or random forests) have generated considerable interest in the machine learning community as new tools for efficient discriminative classification [1,2]. Their introduction in the computer vision community was mostly due to the work of Lepetit et al in [3,4]. This gave rise to a number of papers using random forests for: object class recognition and segmentation [5,6], bilayer video segmentation [7], image classification [8] and person identification [9].

Random forests naturally enable a wide variety of visual cues (e.g. colour, texture, shape, depth etc.). They yield a probabilistic output, and can be made computationally efficient. Because of these benefits, random forests are being established as efficient and general-purpose vision tools. Therefore an optimized implementation of both their training and testing algorithms is desirable.

D. Forsyth, P. Torr, and A. Zisserman (Eds.): ECCV 2008, Part IV, LNCS 5305, pp. 595–608, 2008.
© Springer-Verlag Berlin Heidelberg 2008

This work is complementary to that of Shotton et al [6] in which the authors demonstrate a fast recognition system using forests. Although they demonstrate real-time CPU performance, they evaluate trees sparsely at 1% of pixels and still achieve only 8 frames per second, whereas our evaluations are dense and considerably quicker.

At the time of writing, the premium desktop CPU available is the Intel Core 2 Extreme QX9775 3.2 GHz quad-core. This chip has a theoretical peak performance of 51.2 Gflops using SSE instructions (12.8 Gflops without SSE). With DDR3 SDRAM at 200 MHz, system memory bandwidth peaks at 12.8 GB/s. In contrast, the premium desktop GPU is the nVidia GeForce GTX 280. With its 240 stream processors it has a theoretical peak of 933 Gflops and a memory bandwidth of 141 GB/s.

In [10], the authors demonstrate a simple but effective method for performing plane-sweep stereo on a GPU and achieve real-time performance. In [11], the authors present belief propagation with a chequerboard schedule based on [12]. We follow in similar fashion, presenting no new theory but a method for realizing the GPU's computational power for decision trees and forests.

To our knowledge, this is the first time a method has been proposed which is capable of evaluating or training decision trees on a GPU. In [13], the authors explored the implementation of neural networks for machine learning on a GPU, but did not explore decision trees.

1.2   Outline

Algorithm 1 describes how a binary decision tree is conceptually evaluated on input data. In computer vision techniques, the input data typically correspond to feature values at pixel locations. Each parent node in the tree stores a binary function. For each data point, the binary function at the root node is evaluated on the data. The function value determines which child node is visited next. This continues until reaching a leaf node, which determines the output of the procedure. A forest is a collection of trees that are evaluated independently.

In §2 we describe the features we use in our application which are useful for object class recognition. In §3, we show how to map the evaluation of a decision

Fig. 1. *Left:* A 320 × 213 image from the Microsoft Research recognition database [14] which consists of 23 labeled object classes. *Centre:* The mode of the pixelwise distribution given by a forest of 8 trees, each with 256 leaf nodes, trained on a subset of the database. This corresponds to the `ArgMax` output option (§3.3). This result was generated in 7 ms. *Right:* The ground truth labelling for the same image.

Algorithm 1. Evaluate the binary decision tree with root node N on input x

---

1. **while** N has valid children **do**
2.   **if** $TestFeature(N, x) = true$ **then**
3. $N \leftarrow N.RightChild$
4.   **else**
5. $N \leftarrow N.LeftChild$
6.   **end if**
7. **end while**
8. **return** data associated with N

---

forest to a GPU. The decision forest data structure is mapped to a *forest texture* which can be stored in graphics memory. GPUs are highly data parallel machines and their performance is sensitive to flow control operations. We show how to evaluate trees with a *non-branching* pixel shader. Finally, the training of decision trees involves the construction of histograms – a *scatter* operation that is not possible in a pixel shader. In §4, we show how new GPU hardware features allow these histograms to be computed with a combination of pixel shaders and vertex shaders. In §5 we show results with speed gains of 100 times over a CPU implementation.

Our framework allows clients to use any features which can be computed in a pixel shader on multi-channel input. Our method is therefore applicable to more general classification tasks within computer science, such as multi-dimensional approximate nearest neighbour classification. We present no new theory but concentrate on the highly parallel implementation of decision forests. Our method yields very significant performance increases over a standard CPU version, which we present in §5.

We have chosen Microsoft's Direct3D SDK and High Level Shader Language (HLSL) to code our system, compiling for Shader Model 3.

2 Visual Features

2.1 Choice of Features

To demonstrate our method, we have adopted visual features that generalize those used by many previous works for detection and recognition, including [15,16,3,17,7]. Given a single-channel input image I and a rectangle R, let σ represent the sum $\sigma(I, R) = \sum_{\mathbf{x} \in R} I(\mathbf{x})$.

The features we use are differences of two such sums over rectangles R_0, R_1 in channels c_0, c_1 of the input data. The response of a multi-channel image I to a feature $F = \{R_0, c_0, R_1, c_1\}$ is then $\rho(I, F) = \sigma(I[c_0], R_0) - \sigma(I[c_1], R_1)$. The Boolean test at a tree node is given by the threshold function $\theta_0 \leq \rho(I, F) < \theta_1$. This formulation generalizes the Haar-like features of [15], the summed rectangular features of [16] and the pixel difference features of [3]. The generalization of features is important because it allows us to execute the same code for all the nodes in a decision tree, varying only the values of the parameters. This will enable us to write a non-branching decision evaluation loop.

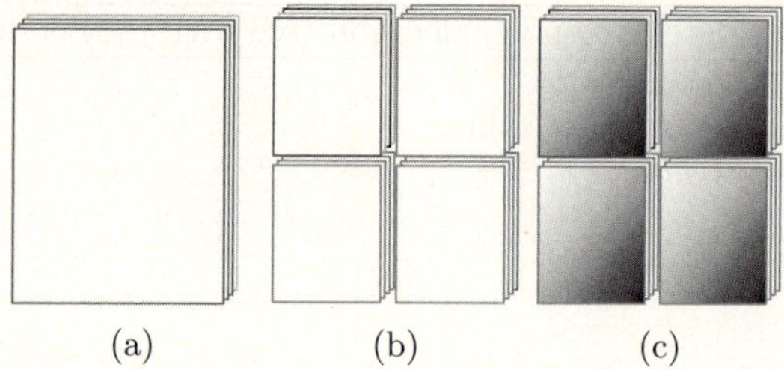

(a)   (b)   (c)

Fig. 2. Processing images for feature computation. Each group of four rectangles represents a four-component (`ARGB`) texture, and each outline in the group represents a single component (channel) of the texture. *(a)* An original sRGB image. *(b)* The image is convolved with 16 filters to produce 16 data channels in 4 four-component textures (§2.3). *(c)* The filtered textures are then integrated (§2.4).

The rectangular sums are computed by appropriately sampling an integral image [15]. Thus the input data consists of multiple channels of integral images. The integration is also performed on the GPU (§2.4).

2.2 Input Data Channels

Prior to integration, we pre-filter sRGB images by applying the bank of separable 2D convolution filters introduced in [14] to produce a 16-channel result. This over-complete representation incorporates local texture information at each pixel. The convolution is also performed on the GPU (§2.3). The pipeline for preparing input textures is shown in Figure 2.

2.3 Convolution with the Filter Bank

For object class recognition, we pre-filter images by convolving them with the 17-filter bank introduced in [14] to model local texture information. Whereas the authors of that work apply their filters in the CIE Lab colour space, we have found it sufficient to apply ours only to the non-linear R, G, B and Y channels. The Gaussians are applied to the RGB channels, and the derivative and Laplacian filters to the luma.

To perform the separable convolution on the GPU, we use the two-pass technique of [18].

Since the pixel shader operates on the four texture components in parallel, up to four filters can be applied in one convolution operation. All 17 filters can therefore be applied in 5 convolutions. In practice we prefer to omit the largest scale Laplacian, applying 16 filters in 4 convolutions.

2.4 Image Integration

The sums over rectangular regions are computed using integral images [15]. Integral images are usually computed on the CPU using an intrinsically serial method, but they can be computed on the GPU using prefix sums [19]. This algorithm is also known as parallel scan or recursive doubling. For details on how this can be implemented on the GPU, see [20].

```
bool TestFeature(sampler2D Input, float2 TexCoord, Parameters Params)
{    // Evaluate the given Boolean feature test for the current input pixel
     float4 Sum1 = AreaSum(Input, TexCoord, Params.Rect1);
     float4 Sum2 = AreaSum(Input, TexCoord, Params.Rect2);
     float Response = dot(Sum1, Params.Channel1) - dot(Sum2, Params.Channel2);
     return Params.Thresholds.x <= Response && Response < Params.Thresholds.y;
}
```

Fig. 3. HLSL code which represents the features used to demonstrate our system. These features are suitable for a wide range of detection and recognition tasks.

2.5 Computation of Features

Figure 3 shows the HLSL code which is used to specify our choice of features (§2.1). The variables for the feature are encoded in the **Parameters** structure. The Boolean test for a given node and pixel is defined by the **TestFeature** method, which will be called by the evaluation and training procedures as necessary.

We would like to stress that, although we have adopted these features to demonstrate our implementation and show results, there is nothing in our framework which requires us to use a particular feature set. We could in practice use any features that can be computed in a pixel shader independently at each input data point, e.g. pixel differences, dot products for BSP trees or multi-level forests as in [6].

3   Evaluation

Once the input data textures have been prepared, they can be supplied to a pixel shader which performs the evaluation of the decision forest at each pixel in parallel.

3.1 Forest Textures

Our strategy for the evaluation of a decision forest on the GPU is to transform the forest's data structure from a list of binary trees to a 2D texture (Figure 4). We lay out the data associated with a tree in a four-component **float** texture, with each node's data on a separate row in breadth-first order.

In the first horizontal position of each row we store the texture coordinate of the corresponding node's left child. Note that we do not need to store the right child's position as it always occupies the row after the left child. We also store all the feature parameters necessary to evaluate the Boolean test for the node. For each leaf node, we store a unique index for the leaf and the required output – a distribution over class labels learned during training.

To navigate through the tree during evaluation, we write a pixel shader that uses a local 2D *node coordinate* variable in place of a pointer to the current node (Figure 5). Starting with the first row (root node) we read the feature parameters

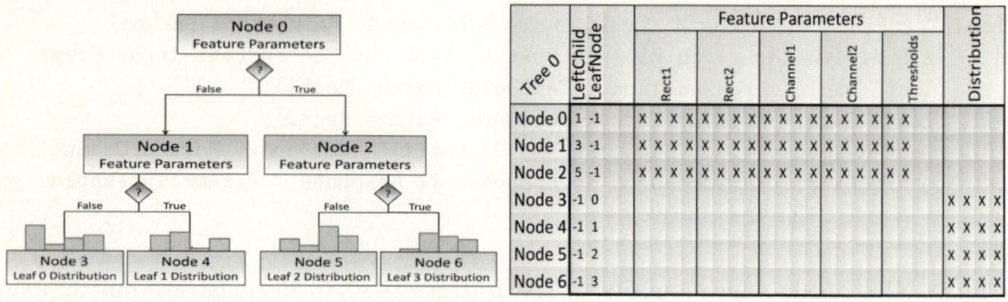

Tree 0	LeftChild	LeafNode	Feature Parameters					Distribution
			Rect1	Rect2	Channel1	Channel2	Thresholds	
Node 0	1	-1	X X X X	X X X X	X X X X	X X X X	X X	
Node 1	3	-1	X X X X	X X X X	X X X X	X X X X	X X	
Node 2	5	-1	X X X X	X X X X	X X X X	X X X X	X X	
Node 3	-1	0						X X X X
Node 4	-1	1						X X X X
Node 5	-1	2						X X X X
Node 6	-1	3						X X X X

Fig. 4. *Left*: A decision tree structure containing parameters used in a Boolean test at each parent node, and output data at each leaf node. *Right*: A 7×5 *forest texture* built from the tree. Empty spaces denote unused values.

```
float4 Evaluate(uniform sampler2D Forest, uniform sampler2D Input,
        uniform float2 PixSize, in float2 TexCoord : TEXCOORD0) : COLOR0
{
    float2 NodeCoord = PixSize * 0.5f;
    // Iterate over the levels of the tree, from the root down...
    [unroll] for (int nLevel = 1; nLevel < MAX_DEPTH; nLevel++)
    {
        float LeftChild = tex2D(Forest, NodeCoord).x;
        // Read the feature parameters for this node...
        Parameters Params = ReadParams(Forest, NodeCoord, PixSize);
        // Perform the user-supplied Boolean test for this node...
        bool TestResult = TestFeature(Input, TexCoord, Params);
        // Move the node coordinate according to the result of the test...
        NodeCoord.y = LeftChild + TestResult * PixSize.y;
    }
    // Read the output distribution associated with this leaf node...
    return Distribution(Forest, NodeCoord);
}
```

Fig. 5. An HLSL pixel shader which evaluates a decision tree on each input point in parallel without branching. Here we have omitted evaluation on multiple and unbalanced trees for clarity.

and evaluate the Boolean test on the input data using texture-dependent reads. We then update the vertical component of our node coordinate based on the result of the test and the value stored in the child position field. This has the effect of walking down the tree according to the computed features. We continue this procedure until we reach a row that represents a leaf node in the tree, where we return the output data associated with the leaf.

For a forest consisting of multiple trees, we tile the tree textures horizontally. An outer loop then iterates over the trees in the forest; we use the horizontal component of the node coordinate to address the correct tree, and the vertical component to address the correct node within the tree. The output distribution for the forest is the mean of the distributions for each tree.

5   Results

Our test system consists of a dual-core Intel Core 2 Duo 2.66 GHz and an nVidia GeForce GTX 280. (Timings on a GeForce 8800 Ultra were similar.) We have coded our GPU version using Direct3D 9 with HLSL shaders, and a CPU version using C++ for comparison only. We have not developed an SSE version of the CPU implementation which we believe may improve the CPU results somewhat (except when performance is limited by memory bandwidth). Part of the appeal of the GPU implementation is the ability to write shaders using HLSL which greatly simplifies the adoption of vector instructions.

In all cases identical output is attained using both CPU and GPU versions. Our contribution is a method of GPU implementation that yields a considerable speed improvement, thereby enabling new real-time recognition applications.

We separate our results into timings for pre-processing, evaluation and training. All of the timings depends on the choice of features; we show timings for our generalized recognition features. For reference, we give timings for our feature pre-processing in Figure 8.

5.1   Tree Training

Training time can be prohibitively long for randomized trees, particularly with large databases. This leads to pragmatic short-cuts such as sub-sampling training data, which in turn has an impact on the discrimination performance of learned trees.

Our training procedure requires time linear in the number of training examples, the number of trees, the depth of the trees and the number of candidate features evaluated.

Description	Resolution (pixels)	CPU (ms)	GPU (ms)	Speed-up ($\times$)
Convolution with 16 separable filters (§2.3)	320×240	94	8.5	11.1
	640×480	381	15.5	24.6
Integration of 16 data channels (§2.4)	320×240	9.2	7.0	1.31
	640×480	31	16.6	1.87

Fig. 8. Timings for feature pre-processing

Operation	CPU (s)	CPU (%)	GPU (s)	GPU (%)	Speed-up ($\times$)
$100\times$ Leaf image computation (§4.2)	6.0	6	0.2	2	30
$10^4 \times$ Feature responses (§4.3)	39.5	41	0.2	2	198
$10^4 \times$ Histogram accumulations (§4.4)	52.1	53	11.8	96	4.4
Total	97.6	100	12.2	100	8.0

Fig. 9. Breakdown of time spent during one training round with 100 training examples and a pool of 100 candidate features. Note the high proportion of time spent updating the histogram.

To measure training time, we took 100 images from the labeled object recognition database of [14] with a resolution of 320×213. This data set has 23 labeled classes. We used a pool of 100 candidate features for each training round. The time taken for each training round was 12.3 seconds. With these parameters, a balanced tree containing 256 leaf nodes takes 98 seconds to train. Here we have used every pixel of every training image.

Training time is dominated by the cost of evaluating a set of candidate features on a training image and aggregating the feature responses into a histogram. Figure 9 shows a breakdown of these costs. These figures are interesting as they reveal two important insights:

First, the aggregation of the histograms on the GPU is comparatively slow, dominating the training time significantly. We experimented with various different method for accumulating the histograms, maintaining the histogram in system memory and performing the incrementation on the CPU. Unfortunately, this did not substantially reduce the time required for a training round. Most recently, we have begun to experiment with using CUDA [24] for this task and we anticipate a significant benefit over using Direct3D.

Second, the computation of the rectangular sum feature responses is extremely fast. We timed this operation as able to compute over 10 million rectangular sums per ms on the GPU. This computation time is insignificant next to the other timings, and this leads us to suggest that we could afford to experiment with more arithmetically complex features without harming training time.

5.2 Forest Evaluation

Our main contribution is the fast and parallel evaluation of decision forests on the GPU. Figure 10 show timings for the dense evaluation of a decision forest, with various different parameters.

Our method maps naturally to the GPU, exploiting its parallelism and cache, and this is reflected in the considerable speed increase over a CPU version by around two orders of magnitude.

Resolution (pixels)	Output Mode	Trees	Classes	CPU (ms)	GPU (ms)	Speed-up ($\times$)
320×240	TreeLeaf	1	N/A	70.5	0.75	94
320×240	ForestLeaves	4	N/A	288	3.0	96.1
320×240	Distribution	8	4	619	5.69	109
320×240	ArgMax	8	23	828	6.85	121
640×480	TreeLeaf	1	N/A	288	2.94	97.8
640×480	ForestLeaves	4	N/A	1145	12.1	95.0
640×480	Distribution	8	4	2495	23.1	108
640×480	ArgMax	8	23	3331	25.9	129

Fig. 10. Timings for evaluating a forest of decision trees. Our GPU implementation evaluates the forest in about 1% of the time required by the CPU implementation.

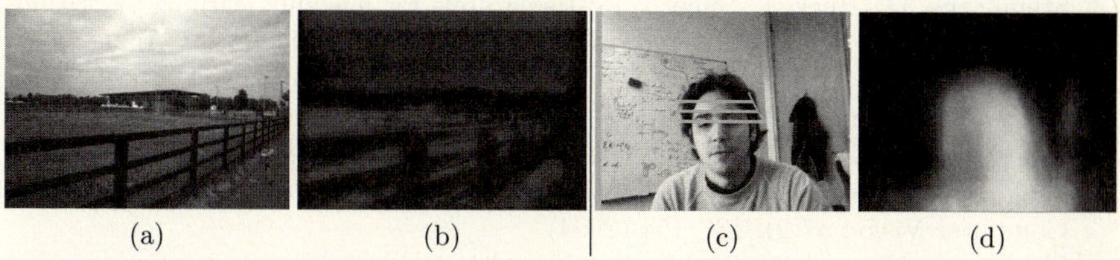

(a)   (b)   (c)   (d)

Fig. 11. (a)-(b) Object class recognition. A forest of 8 trees was trained on labelled data for grass, sky and background labels. *(a)* An outdoor image which is not part of the training set for this example. *(b)* Using the `Distribution` output option, the blue channel represents the probability of sky and the green channel the probability of grass at each pixel (5 ms). **(c)-(d) Head tracking in video.** A random forest was trained using spatial and temporal derivative features instead of the texton filter bank. *(c)* A typical webcam video frame with an overlay showing the detected head position. This frame was not part of the training set for this example. *(d)* The probability that each pixel is in the foreground (5 ms).

5.3   Conclusion

We have shown how it is possible to use GPUs for the training and evaluation of general purpose decision trees and forests, yielding speed gains of around 100 times.

References

1. Amit, Y., Geman, D.: Shape quantization and recognition with randomized trees. Neural Computation 9(7), 1545–1588 (1997)
2. Breiman, L.: Random forests. ML Journal 45(1), 5–32 (2001)
3. Lepetit, V., Fua, P.: Keypoint recognition using randomized trees. IEEE Trans. Pattern Anal. Mach. Intell. 28(9), 1465–1479 (2006)
4. Ozuysal, M., Fua, P., Lepetit, V.: Fast keypoint recognition in ten lines of code. In: IEEE CVPR (2007)
5. Winn, J., Criminisi, A.: Object class recognition at a glance. In: IEEE CVPR, video track (2006)
6. Shotton, J., Johnson, M., Cipolla, R.: Semantic texton forests for image categorization and segmentation. In: IEEE CVPR, Anchorage (2008)
7. Yin, P., Criminisi, A., Winn, J.M., Essa, I.A.: Tree-based classifiers for bilayer video segmentation. In: CVPR (2007)
8. Bosh, A., Zisserman, A., Munoz, X.: Image classification using random forests and ferns. In: IEEE ICCV (2007)
9. Apostolof, N., Zisserman, A.: Who are you? - real-time person identification. In: BMVC (2007)
10. Yang, R., Pollefeys, M.: Multi-resolution real-time stereo on commodity graphics hardware. In: CVPR, vol. (1), pp. 211–220 (2003)
11. Brunton, A., Shu, C., Roth, G.: Belief propagation on the gpu for stereo vision. In: CRV, p. 76 (2006)
12. Felzenszwalb, P.F., Huttenlocher, D.P.: Efficient belief propagation for early vision. International Journal of Computer Vision 70(1), 41–54 (2006)

13. Steinkraus, D., Buck, I., Simard, P.: Using gpus for machine learning algorithms. In: Proceedings of Eighth International Conference on Document Analysis and Recognition, 2005, 29 August-1 September 2005, vol. 2, pp. 1115–1120 (2005)
14. Winn, J.M., Criminisi, A., Minka, T.P.: Object categorization by learned universal visual dictionary. In: ICCV, pp. 1800–1807 (2005)
15. Viola, P.A., Jones, M.J.: Robust real-time face detection. International Journal of Computer Vision 57(2), 137–154 (2004)
16. Shotton, J., Winn, J.M., Rother, C., Criminisi, A.: TextonBoost: Joint appearance, shape and context modeling for multi-class object recognition and segmentation. In: Leonardis, A., Bischof, H., Pinz, A. (eds.) ECCV 2006. LNCS, vol. 3951, pp. 1–15. Springer, Heidelberg (2006)
17. Deselaers, T., Criminisi, A., Winn, J.M., Agarwal, A.: Incorporating on-demand stereo for real time recognition. In: CVPR (2007)
18. James, G., O'Rorke, J.: Real-time glow. In: GPU Gems: Programming Techniques, Tips and Tricks for Real-Time Graphics, pp. 343–362. Addison-Wesley, Reading (2004)
19. Blelloch, G.E.: Prefix sums and their applications. Technical Report CMU-CS-90-190, School of Computer Science, Carnegie Mellon University (November 1990)
20. Hensley, J., Scheuermann, T., Coombe, G., Singh, M., Lastra, A.: Fast summed-area table generation and its applications. Comput. Graph. Forum 24(3), 547–555 (2005)
21. Quinlan, J.R.: Induction of decision trees. Machine Learning 1(1), 81–106 (1986)
22. Quinlan, J.: C4.5: Programs for Machine Learning. Morgan Kaufmann, California (1992)
23. Scheuermann, T., Hensley, J.: Efficient histogram generation using scattering on gpus. In: SI3D, pp. 33–37 (2007)
24. http://www.nvidia.com/cuda

General Imaging Geometry
for Central Catadioptric Cameras

Peter Sturm[1] and João P. Barreto[2]

[1] INRIA Rhône-Alpes and Laboratoire Jean Kuntzmann, Grenoble, France
[2] ISR/DEEC, University of Coimbra, 3030 Coimbra, Portugal

Abstract. Catadioptric cameras are a popular type of omnidirectional imaging system. Their imaging and multi-view geometry has been extensively studied; epipolar geometry for instance, is geometrically speaking, well understood. However, the existence of a bilinear matching constraint and an associated fundamental matrix, has so far only been shown for the special case of para-catadioptric cameras (consisting of a paraboloidal mirror and an orthographic camera). The main goal of this work is to obtain such results for all central catadioptric cameras. Our main result is to show the existence of a general 15×15 fundamental matrix. This is based on and completed by a number of other results, e.g. the formulation of general catadioptric projection matrices and plane homographies.

1   Introduction and Previous Work

The geometry of single and multiple images has been extensively studied in computer vision and photogrammetry [1]. The picture is rather complete for perspective cameras and many results have been obtained for other camera models too, e.g. catadioptric [2,3,4,5], fisheyes [6,7], pushbroom [8], x-slit [9], oblique [10,11], non-central mosaics [12,11]; this list is not intended to be exhaustive.

Besides perspective cameras, the most studied case is probably that of catadioptric ones. Baker and Nayar have shown which catadioptric devices have a single effective viewpoint, i.e. are central cameras [13]. Among those, the most useful ones are the para-catadioptric and the hyper-catadioptric models, using a mirror of paraboloidal/hyperboloidal shape, coupled with an orthographic/perspective camera. The epipolar geometry of these devices has been studied by Svoboda and Pajdla [2] who showed the existence of epipolar *conics*. Geyer and Daniilidis have shown the existence of a fundamental matrix for para-catadioptric cameras [4,14]; Sturm has extended this to fundamental matrices and trifocal tensors for mixtures of para-catadioptric and perspective images [5]. Barreto showed that the framework can also be extended to cameras with lens distortion due to the similarities between the para-catadioptric and division models [15,16]

However, no such results have so far been obtained for the general catadioptric camera model, i.e. including hyper-catadioptric cameras. In this paper, we present a number of novel results concerning (bi-) linear formulations for single and two-view geometry, valid for all central catadioptric cameras. First, we show that the projection of a 3D point can be modeled using a projection matrix of

D. Forsyth, P. Torr, and A. Zisserman (Eds.): ECCV 2008, Part IV, LNCS 5305, pp. 609–622, 2008.
© Springer-Verlag Berlin Heidelberg 2008

size 6×10 and how this may be used for calibrating catadioptric cameras with a straightforward DLT approach, something which has not been possible up to now. We then give analogous results for the backprojection of image points and the projection of quadrics and conics. These are the basis for our main result, the general fundamental matrix for catadioptric cameras. It is of size 15×15 and an explicit compact expression is provided. Finally, we also show the existence of plane homographies, again of size 15×15, that relate sets of matching image points that are the projections of coplanar scene points.

Our results, like those cited above for para-catadioptric cameras, are based on the use of so-called *lifted coordinates* to represent geometric objects. For example, 2D points are usually represented by 3-vectors of homogeneous coordinates; their lifted coordinates are 6-vectors containing all degree-2 monomials of the original coordinates. Lifted coordinates have also been used to model linear pushbroom cameras [8] and to perform multi-body structure from motion [17,18].

Organization. We describe our results in two ways, geometrically and algebraically. In the next section, we immediately describe all main results in a purely geometrical way, which we find rather intuitive and which hopefully guides the reader through the more technical later sections. In section 3, we introduce notations and background, mostly associated to Veronese maps which are extensively used in this work. In sections 4 to 8, we develop algebraic formulations for the projection of 3D points, backprojection of image points, projection of quadrics and conics, epipolar geometry, and plane homographies.

2   Geometrical Description of Our Results

We use the sphere based model for catadioptric projection introduced by Geyer and Daniilidis [19]. All central catadioptric cameras can be modeled by a unit sphere and a perspective camera, such that the projection of 3D points can be performed as follows in two steps. First, one projects the point onto the sphere, to an intersection of the sphere and the line joining its center and the 3D point. There are two such intersection points; it is usually assumed that the only physically feasible one can be singled out. That point is then projected into the perspective camera. This model covers all central catadioptric cameras, the type of which is encoded by the distance between the perspective camera and the center of the sphere, e.g. 0 for perspective, 1 for para-catadioptric, $0 < \xi < 1$ for hyper-catadioptric. Note that even for the para-catadioptric case, where the true camera is an affine one, the camera in the sphere based model is still perspective. An algebraic formulation of the model is given in section 3. We now give geometrical and intuitive descriptions of our main results.

Projection of a 3D point. In the first step of the projection as described above, the intersections of the sphere with the line spanned by its center and the 3D point, are computed. There are two mathematical solutions which when projected to the perspective camera, give the two mathematical image points, cf. figure 1. For most real catadioptric cameras, only one of them can be observed;

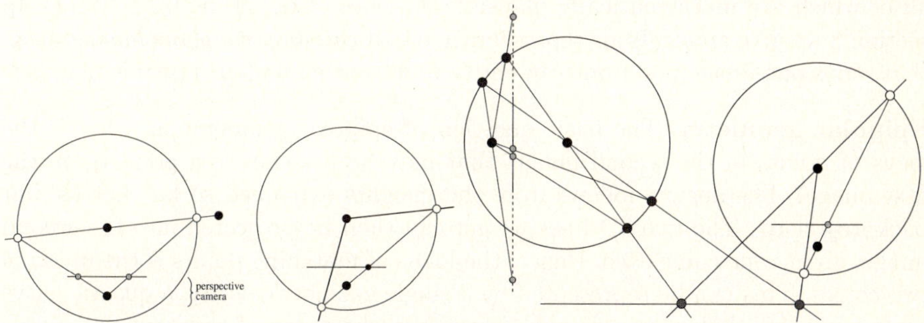

Fig. 1. *Left:* Projection of a 3D point to two image points (in cyan). *Middle:* Backprojection of an image point to two 3D lines. *Right:* Illustration of plane homography. The image point in cyan in the right camera is backprojected to the scene plane, giving the two red points. Each one of them is projected to the unit sphere of the second camera on two blue points. The four image points in the second camera are shown in cyan.

the second intersection point with the sphere is hidden from the 3D point by the mirror. An exception is the case of an elliptical mirror where a 3D point may actually be seen twice in the image. Although for the most useful catadioptric cameras, a 3D point is in reality visible in only one image point, it turns out that in order to obtain multi-linear expressions for epipolar geometry, plane homographies etc., both mathematical image points have to be considered, see later. So, from now on we consider two image points per 3D point and similar considerations will be done for backprojection etc. in the following. The algebraic formulation of the projection of 3D points, in the form of a 6×10 general *catadioptric projection matrix*, is given in section 4.

Projection of a 3D line. It is well known that a central catadioptric image of a 3D line, is a conic [19,20]. In this work, we do not explicitly require the projection of 3D lines, but we still mention this result since it is helpful to understand the epipolar geometry, see below.

Backprojection of an image point. First, the image point is backprojected relative to the perspective camera, giving rise to a 3D line. Then, its two intersection points with the sphere are computed. Finally, the two lines spanned by the sphere center and these intersection points, are generated. These are the backprojection lines of the image point, cf. figure 1 (middle). In section 5 we show how to represent the union of two 3D lines by a single algebraic object, leading to the formulation of a 6×6 *backprojection matrix*.

Two images of a plane. Consider $\mathbf{q}_1$, the image of a 3D point on a scene plane Π, in the first image. What are the possible matching points in the second image? These can be determined as follows. Let us first backproject $\mathbf{q}_1$ to 3D; as we have seen just before, this gives two 3D lines. Their intersections with Π are the two points on that plane that may be observed in $\mathbf{q}_1$. Let us project both of them into the second image. This gives a total of four points in the second image, all of which are mathematically plausible matches of $\mathbf{q}_1$, cf. figure 1 (right). In section 8 we give an algebraic representation of a *catadioptric plane homography*, that maps one image point onto an entity representing its four possible matches.

Epipolar geometry. The basic question of epipolar geometry is: what is the locus of points in the second image, that may be matches of a point $\mathbf{q}_1$ in the first image? The answer follows from the insights explained so far. Let us first backproject $\mathbf{q}_1$. The two 3D lines we get can then be projected into the second image, giving one conic each. Hence, the locus of matching points is the union of two conics. This can be represented by a single geometric entity, a quartic curve (note of course that not every quartic curve is the union of two conics).

Hence, if a multi-linear fundamental matrix exists that represents this epipolar geometry, it must map an image point into some representation of a quartic curve. The equation of a planar quartic curve depends on 15 coefficients (defined up to scale), one per 4-th order monomial of a 2D point's homogeneous coordinates. Hence, we may expect the fundamental matrix to be of size $15 \times \cdots$. Like for perspective images, we may expect that the transpose of the fundamental matrix gives the fundamental matrix going from the second to the first image. The fundamental matrix for catadioptric images should thus intuitively be of size 15×15. This is indeed the case, as is shown in section 7.

3   Background

Notations. We do not distinguish between a projective transformation and the matrix representing it. Matrices are represented by symbols in sans serif font, e.g. M and vectors by bold symbols, e.g. $\mathbf{Q}$. Equality of matrices or vectors up to a scalar factor is written as $\sim$. $[\mathbf{a}]_\times$ denotes the skew-symmetric matrix associated with the cross product of a 3-vector $\mathbf{a}$.

Camera model. As mentioned before, we use the sphere based model [19]. Without loss of generality, let the unit sphere be located at the origin and the optical center of the perspective camera, at the point $\mathbf{C}_p = (0, 0, -\xi)^\mathsf{T}$. The perspective camera is modeled by the projection matrix $\mathsf{P} \sim \mathsf{A}_p \mathsf{R}_p (\mathsf{I} - \mathbf{C}_p)$. For full generality, we include a rotation R_p; this may encode an actual rotation of the true camera looking at the mirror, but may also simply be a projective change of coordinates in the image plane, like for para-catadioptric cameras, where the true camera's rotation is fixed, modulo rotation about the mirror axis. Note that all parameters of the perspective camera, i.e. both its intrinsic and extrinsic parameter sets, are intrinsic parameters for the catadioptric camera. Hence, we replace $\mathsf{A}_p \mathsf{R}_p$ by a generic projective transformation K from now on. The **intrinsic parameters of the catadioptric camera** are thus ξ and K.

The *projection* of a 3D point $\mathbf{Q}$ goes as follows (cf. section 2). The two intersection points of the sphere and the line joining its center and $\mathbf{Q}$, are $\left(Q_1, Q_2, Q_3, \pm\sqrt{Q_1^2 + Q_2^2 + Q_3^2} \right)^\mathsf{T}$. Their images in the perspective camera are

$$\mathbf{q}_{\pm} \sim \mathsf{K}\mathbf{r}_{\pm} \sim \mathsf{K}\begin{pmatrix} Q_1 \\ Q_2 \\ Q_3 \pm \xi\sqrt{Q_1^2 + Q_2^2 + Q_3^2} \end{pmatrix}$$

In the following, we usually first work with the intermediate image points $\mathbf{r}_{\pm} \sim \mathsf{K}^{-1}\mathbf{q}_{\pm}$, before giving final results for the actual image points $\mathbf{q}_{\pm}$.

Plücker line coordinates. 3D lines may be represented by 6-vectors of so-called Plücker coordinates. Let $\mathbf{A}$ and $\mathbf{B}$ be the non-homogeneous coordinates of two generic 3D points. Let us define the line's Plücker coordinates as the 6-vector $\mathbf{L} = \left(\mathbf{A}^\mathsf{T} - \mathbf{B}^\mathsf{T}, (\mathbf{A} \times \mathbf{B})^\mathsf{T}\right)^\mathsf{T}$.

All lines satisfy the Plücker constraint $\mathbf{L}^\mathsf{T}\mathsf{W}\mathbf{L} = 0$ where W is

$$\mathsf{W} = \begin{pmatrix} 0 & \mathsf{I} \\ \mathsf{I} & 0 \end{pmatrix}$$

Two lines $\mathbf{L}$ and $\mathbf{L}'$ cut one another if and only if $\mathbf{L}^\mathsf{T}\mathsf{W}\mathbf{L}' = 0$. Consider a rigid transformation for points

$$\begin{pmatrix} \mathsf{R} & \mathbf{t} \\ \mathbf{0}^\mathsf{T} & 1 \end{pmatrix}$$

Lines are mapped accordingly using the transformation

$$\mathsf{T} = \begin{pmatrix} \mathsf{R} & 0_{3\times 3} \\ [\mathbf{t}]_{\times}\mathsf{R} & \mathsf{R} \end{pmatrix}$$

Second order line complexes. A second order line complex is a set of 3D lines that satisfy a quadratic equation in the Plücker coordinates [21]. It can be represented by a symmetric 6×6 matrix C such that exactly the lines on the complex satisfy $\mathbf{L}^\mathsf{T}\mathsf{C}\mathbf{L} = 0$. Note that C is only defined up to adding multiples of W. Henceforth we call second order line complexes shortly *line complexes*. In this paper, we use line complexes to represent the union of two 3D lines. Rigid displacements of line complexes are carried out as

$$\mathsf{T}^{-\mathsf{T}}\mathsf{C}\mathsf{T}^{-1} \quad \text{with} \quad \mathsf{T}^{-1} = \begin{pmatrix} \mathsf{R}^\mathsf{T} & 0_{3\times 3} \\ -\mathsf{R}^\mathsf{T}[\mathbf{t}]_{\times} & \mathsf{R}^\mathsf{T} \end{pmatrix} \tag{1}$$

Lifted coordinates from symmetric matrix equations. The derivation of (multi-) linear relations for catadioptric imagery requires the use of lifted coordinates. The Veronese map $V_{n,d}$ of degree d maps points of $\mathcal{P}^n$ into points of an m dimensional projective space $\mathcal{P}^m$, with $m = \begin{pmatrix} n+d \\ d \end{pmatrix} - 1$.

Consider the second order Veronese map $V_{2,2}$, that embeds the projective plane into the 5D projective space, by lifting the coordinates of point $\mathbf{q}$ to

$$\hat{\mathbf{q}} = \left(q_1^2 \; q_1q_2 \; q_2^2 \; q_1q_3 \; q_2q_3 \; q_3^2\right)^\mathsf{T}$$

Vector $\hat{\mathbf{q}}$ and matrix $\mathbf{q}\mathbf{q}^\mathsf{T}$ are composed by the same elements. The former can be derived from the latter through a suitable re-arrangement of parameters.

Define $\mathbf{v}(\mathsf{U})$ as the vector obtained by stacking the columns of a generic matrix U [22]. For the case of $\mathbf{q}\mathbf{q}^{\mathsf{T}}$, $\mathbf{v}(\mathbf{q}\mathbf{q}^{\mathsf{T}})$ has several repeated elements because of matrix symmetry. By left multiplication with a suitable permutation matrix S that adds the repeated elements, it follows that

$$\hat{\mathbf{q}} = \mathsf{D}^{-1} \underbrace{\begin{pmatrix} 1 & 0 & 0 & 0 & 0 & 0 & 0 & 0 & 0 \\ 0 & 1 & 0 & 1 & 0 & 0 & 0 & 0 & 0 \\ 0 & 0 & 0 & 0 & 1 & 0 & 0 & 0 & 0 \\ 0 & 0 & 1 & 0 & 0 & 0 & 1 & 0 & 0 \\ 0 & 0 & 0 & 0 & 0 & 1 & 0 & 1 & 0 \\ 1 & 0 & 0 & 0 & 0 & 0 & 0 & 0 & 1 \end{pmatrix}}_{\mathsf{S}} \mathbf{v}(\mathbf{q}\mathbf{q}^{\mathsf{T}}), \tag{2}$$

with D a diagonal matrix, $D_{ii} = \sum_{j=1}^{9} S_{ij}$. This process of computing the lifted representation of a point $\mathbf{q}$ can be extended to any second order Veronese map $V_{n,2}$ independently of the dimensionality of the original space. It is also a mechanism that provides a compact representation for square symmetric matrices. If U is symmetric, then it is uniquely represented by $\mathbf{v}_{sym}(\mathsf{U})$, the column-wise vectorization of its upper right triangular part:

$$\mathbf{v}_{sym}(\mathsf{U}) = \mathsf{D}^{-1}\mathsf{S}\mathsf{U} = (U_{11}, U_{12}, U_{22}, U_{13}, \cdots, U_{nn})^{\mathsf{T}}$$

Let us now discuss the lifting of linear transformations. Consider A such that $\mathbf{r} = \mathsf{A}\mathbf{q}$. The relation $\mathbf{r}\mathbf{r}^{\mathsf{T}} = \mathsf{A}(\mathbf{q}\mathbf{q}^{\mathsf{T}})\mathsf{A}^{\mathsf{T}}$ can be written as a vector mapping

$$(\mathbf{r}\mathbf{r}^{\mathsf{T}}) = (\mathsf{A} \otimes \mathsf{A})(\mathbf{q}\mathbf{q}^{\mathsf{T}}),$$

with $\otimes$ denoting the Kronecker product [22]. Using the symmetric vectorization, we have $\hat{\mathbf{q}} = \mathbf{v}_{sym}(\mathbf{q}\mathbf{q}^{\mathsf{T}})$ and $\hat{\mathbf{r}} = \mathbf{v}_{sym}(\mathbf{r}\mathbf{r}^{\mathsf{T}})$, thus:

$$\hat{\mathbf{r}} = \underbrace{\mathsf{D}^{-1}\mathsf{S}(\mathsf{A} \otimes \mathsf{A})\mathsf{S}^{\mathsf{T}}}_{\hat{\mathsf{A}}} \hat{\mathbf{v}}$$

We have just derived the expression for lifting linear transformations. A has a lifted counterpart $\hat{\mathsf{A}}$ such that $\mathbf{r} = \mathsf{A}\mathbf{q}$ *iff* $\hat{\mathbf{r}} = \hat{\mathsf{A}}\hat{\mathbf{q}}$. For the case of a second order Veronese map, the lifting of a 2D projective transformation A is $\hat{\mathsf{A}}$ of size 6×6. This lifting generalizes to any projective transformation, independently of the dimensions of its original and target spaces, i.e. it is also applicable to rectangular matrices. We summarize a few useful properties [22].

$$\widehat{\mathsf{AB}} = \hat{\mathsf{A}}\hat{\mathsf{B}} \quad \widehat{\mathsf{A}^{-1}} = \hat{\mathsf{A}}^{-1} \quad \widehat{\mathsf{A}^{\mathsf{T}}} = \mathsf{D}^{-1}\hat{\mathsf{A}}^{\mathsf{T}}\mathsf{D} \tag{3}$$

Also, for symmetric matrices U and M, we have the following property:

$$\mathsf{U} = \mathsf{A}\mathsf{M}\mathsf{A}^{\mathsf{T}} \Rightarrow \mathbf{v}_{sym}(\mathsf{U}) = \hat{\mathsf{A}}\,\mathbf{v}_{sym}(\mathsf{M}) \tag{4}$$

Also, note that $\hat{\mathsf{A}}$ is non-singular *iff* A is non-singular. In this paper, we use the following liftings: 3-vectors $\mathbf{q}$ to 6-vectors $\hat{\mathbf{q}}$, 4-vectors $\mathbf{Q}$ to 10-vectors $\hat{\mathbf{Q}}$, and 6-vectors $\mathbf{u}$ to 21-vectors $\hat{\mathbf{u}}$. Analogously, 3×3 matrices W are lifted to 6×6 ones $\hat{\mathsf{W}}$ and 3×4 matrices to 6×10 ones. We also use the fourth order Veronese map of $\mathcal{P}^2$, mapping 3-vectors $\mathbf{q}$ to 15-vectors $\hat{\hat{\mathbf{q}}}$ containing the quartic monomials of

q. We call this *double lifting*; it applies analogously to 3×3 matrices W, which are doubly lifted to 15×15 matrices $\hat{\hat{\mathsf{W}}}$. Finally, note that applying two second order Veronese maps in succession, is **not** equivalent to applying one fourth order Veronese map: for a 3-vector $\mathbf{q}$, $\hat{\hat{\mathbf{q}}}$ is a 15-vector, whereas $\hat{\mathbf{w}}$ where $\mathbf{w}$ is the 6-vector $\mathbf{w} = \hat{\mathbf{q}}$, is of length 21. We thus denote successive application of two second order liftings by $\,\mathring{}\,$, e.g. for a 3×3 matrix E, we get a 21×21 matrix $\mathring{\mathsf{E}}$.

4   Projection of 3D Points

As explained in section 2, a 3D point is mathematically projected to two image points. How to represent two 2D points via a single geometric entity? One way is to compute the degenerate dual conic generated by them, i.e. the dual conic containing exactly the lines going through at least one of the two points. Let the two image points be $\mathbf{q}_+$ and $\mathbf{q}_-$ (see section 3). The dual conic is given by

$$\Omega \sim \mathbf{q}_+\mathbf{q}_-^{\mathsf{T}} + \mathbf{q}_-\mathbf{q}_+^{\mathsf{T}} \sim \mathsf{K} \begin{pmatrix} Q_1^2 & Q_1Q_2 & Q_1Q_3 \\ Q_1Q_2 & Q_2^2 & Q_2Q_3 \\ Q_1Q_3 & Q_2Q_3 & Q_3^2 - \xi^2(Q_1^2 + Q_2^2 + Q_3^2) \end{pmatrix} \mathsf{K}^{\mathsf{T}}$$

This can be written as a linear mapping of the 3D point's lifted coordinates, onto the vectorized matrix of the conic:

$$\mathbf{v}_{sym}(\Omega) \sim \widehat{\mathsf{K}}_{6\times 6} \underbrace{\begin{pmatrix} 1 & 0 & 0 & 0 & 0 & 0 \\ 0 & 1 & 0 & 0 & 0 & 0 \\ 0 & 0 & 1 & 0 & 0 & 0 \\ 0 & 0 & 0 & 1 & 0 & 0 \\ 0 & 0 & 0 & 0 & 1 & 0 \\ -\xi^2 & 0 & -\xi^2 & 0 & 0 & 1-\xi^2 \end{pmatrix}}_{\mathsf{X}_\xi} \begin{pmatrix} \mathsf{I}_6 & \mathsf{0}_{6\times 4} \end{pmatrix} \hat{\mathbf{Q}}$$

So far, we have projected a 3D point given in the catadioptric camera's local coordinate system. If we introduce extrinsic parameters of the camera, i.e. a pose matrix $\mathsf{T} = \mathsf{R} \begin{pmatrix} \mathsf{I} & -\mathbf{t} \end{pmatrix}$ then we can write the projection operation as

$$\mathbf{v}_{sym}(\Omega) \sim \widehat{\mathsf{K}}_{6\times 6}\mathsf{X}_{\xi,6\times 6}\widehat{\mathsf{R}}_{6\times 6} \begin{pmatrix} \mathsf{I}_6 & \mathsf{T}'_{4\times 6} \end{pmatrix} \hat{\mathbf{Q}}_{10}$$

where T' depends only on $\mathbf{t}$.

We have thus derived a 6×10 **catadioptric projection matrix** P_{cata}. Furthermore, it can be decomposed, like the projection matrix of a perspective camera, into two matrices containing either intrinsic or extrinsic parameters:

$$\mathsf{P}_{cata} = \underbrace{\widehat{\mathsf{K}}\mathsf{X}_\xi}_{\mathsf{A}_{cata}} \underbrace{\widehat{\mathsf{R}}_{6\times 6} \begin{pmatrix} \mathsf{I}_6 & \mathsf{T}'_{4\times 6} \end{pmatrix}}_{\mathsf{T}_{cata}}$$

We find here a 6×6 **catadioptric calibration matrix** A_{cata}. Note that the restriction of the projection to points lying in a plane in 3D, leads straightforwardly to a 6×6 homography, analogous to the plane-to-image homographies used e.g. for perspective camera calibration.

Camera Calibration. Let us consider how to set up equations for calibrating a camera. In the perspective case, a 2D–3D point correspondence allows to write $\mathbf{q} \sim \mathsf{P}\mathbf{Q}$. One way to set up linear equations on P is to write $[\mathbf{q}]_\times \mathsf{P}\mathbf{Q} = \mathbf{0}_3$. What is the analogous expression in the catadioptric case? Let again $\mathbf{q}$ and $\mathbf{Q}$ be a 2D–3D point correspondence. Since each 3D point is projected to two 2D points, one may not directly be able to compare $\mathbf{q}$ to the image of $\mathbf{Q}$, unlike in the perspective case. Instead, as mentioned above, the projection matrix maps $\mathbf{Q}$ (rather, its lifted version) onto the coefficients of a degenerate dual conic Ω. The point $\mathbf{q}$ must be one of the two generators of Ω. This implies that all lines through $\mathbf{q}$ must lie on Ω. Hence: $\forall \mathbf{p} : \mathbf{p} \times \mathbf{q} \in \Omega$, which gives $\forall \mathbf{p} : \mathbf{p}^\mathsf{T} [\mathbf{q}]_\times \Omega [\mathbf{q}]_\times \mathbf{p} = 0$. Thus, $[\mathbf{q}]_\times \Omega [\mathbf{q}]_\times = \mathbf{0}_{3\times 3}$. This gives 6 constraints that can be written as

$$\left(\widehat{[\mathbf{q}]_\times} \right)_{6\times 6} \mathbf{v}_{sym}(\Omega) = \widehat{[\mathbf{q}]_\times}\, \mathsf{P}_{cata}\, \hat{\mathbf{Q}} = \mathbf{0}_6$$

We thus find an expression that is very similar to that for perspective cameras and that may be directly used for calibrating catadioptric cameras using e.g. a standard DLT like approach. While a 3×3 skew symmetric matrix has rank 2, its lifted counterpart is rank 3. Therefore, each 3D-to-2D match provides 3 linear constraints on the 59 parameters of P_{cata}, and DLT calibration can be done with a minimum of 20 matches.

5   Backprojection of Image Points

This is essential for deriving the proposed expression of the fundamental matrix. Similarly to the case of projection, we want to express the backprojection function of a catadioptric camera as a linear mapping. Recall from section 2, that the backprojection of an image point gives **two** 3D lines. How to represent two 3D lines via a single geometric entity? Several possibilities may exist; the one that seems appropriate is to use a second order line complex: consider two 3D lines $\mathbf{L}_+$ and $\mathbf{L}_-$. All lines that cut at least one of them, form a second order line complex, represented by a 6×6 matrix C such that lines on the complex satisfy $\mathbf{L}^\mathsf{T}\mathsf{C}\mathbf{L} = 0$. The matrix C is given as (with W as defined in section 3)

$$\mathsf{C} \sim \mathsf{W} \left(\mathbf{L}_+\mathbf{L}_-^\mathsf{T} + \mathbf{L}_-\mathbf{L}_+^\mathsf{T} \right) \mathsf{W}$$

The backprojection lines $\mathbf{L}_\pm$ of an image point $\mathbf{q}$ are spanned by the origin (center of the sphere) and points $(\mathbf{b}_\pm^\mathsf{T}, 1)^\mathsf{T}$, thus $\mathbf{L}_\pm \sim (\mathbf{b}_\pm^\mathsf{T}\ \mathbf{0}^\mathsf{T})^\mathsf{T}$. Here,

$$\mathbf{b}_\pm = (\mathbf{r}^\mathsf{T}\mathbf{r})\mathbf{C}_p + \left(\xi r_3 \pm \sqrt{\xi^2 r_3^2 - (\mathbf{r}^\mathsf{T}\mathbf{r})(\xi^2 - 1)} \right) \mathbf{r}$$

with $\mathbf{r} \sim \mathsf{K}^{-1}\mathbf{q}$ and $\mathbf{C}_p$ the center of the perspective camera (cf. section 3). The line complex C generated by the two lines, is

$$\mathsf{C} \sim \begin{pmatrix} 0 & 0 \\ 0 & \mathbf{b}_+\mathbf{b}_-^\mathsf{T} + \mathbf{b}_-\mathbf{b}_+^\mathsf{T} \end{pmatrix} \sim \begin{pmatrix} 0 & 0 \\ 0 & \xi^2(\mathbf{r}^\mathsf{T}\mathbf{r})\mathbf{e}_3\mathbf{e}_3^\mathsf{T} - \xi^2 r_3(\mathbf{e}_3\mathbf{r}^\mathsf{T} + \mathbf{r}\mathbf{e}_3^\mathsf{T}) + (\xi^2 - 1)\mathbf{r}\mathbf{r}^\mathsf{T} \end{pmatrix}$$

where $\mathbf{e}_3 = (0,0,1)^{\mathsf{T}}$. C is by construction symmetric and of rank 2 and it has 9 non-zero coefficients. Let M be the lower right 3×3 submatrix of C and the 6-vector $\mathbf{m}$ its vectorized version: $\mathbf{m} = \mathbf{v}_{sym}(\mathsf{M})$. We have the following linear backprojection equation:

$$\mathbf{m} \sim \mathsf{B}_\xi \hat{\mathbf{r}} = \mathsf{B}_\xi \hat{\mathsf{K}}^{-1} \hat{\mathbf{q}} \tag{5}$$

with

$$\mathsf{B}_\xi = \begin{pmatrix} \xi^2 - 1 & 0 & 0 & 0 & 0 & 0 \\ 0 & \xi^2 - 1 & 0 & 0 & 0 & 0 \\ 0 & 0 & \xi^2 - 1 & 0 & 0 & 0 \\ 0 & 0 & 0 & -1 & 0 & 0 \\ 0 & 0 & 0 & 0 & -1 & 0 \\ \xi^2 & 0 & \xi^2 & 0 & 0 & -1 \end{pmatrix}$$

We call $\mathsf{B}_{cata} = \mathsf{B}_\xi \hat{\mathsf{K}}^{-1}$ the **backprojection matrix**.

6   Projection of Line Complexes and Quadrics

To the best of our knowledge the projection of general quadric surfaces in catadioptric cameras has never been studied. The existing literature concerns only the projection of spheres, for calibration purposes [23]. The problem can be conveniently addressed by considering a line-based representation of quadrics, via line complexes. The set of 3D lines tangent to a quadric form a line complex [21]. A conic on a 3D scene plane can also be represented by a line complex [24,25], and the following results apply thus to the projection of both quadrics and conics. As discussed in the previous section, a line complex can be represented by a 6×6 symmetric matrix C. Let C be split in 3×3 blocks:

$$\mathsf{C} \sim \begin{pmatrix} \mathsf{U} & \mathsf{N} \\ \mathsf{N}^{\mathsf{T}} & \mathsf{M} \end{pmatrix}$$

The image of C consists of all points $\mathbf{q}$ such that at least one of their backprojection rays lies on C. Let $\mathbf{L}_{\pm} \sim (\mathbf{b}_{\pm}^{\mathsf{T}} \ \mathbf{0}^{\mathsf{T}})^{\mathsf{T}}$ be the Plücker coordinates of the two backprojections of $\mathbf{q}$ (cf. section 5). Hence, $\mathbf{q}$ lies on the image of C *iff*

$$(\mathbf{L}_{+}^{\mathsf{T}} \mathsf{C} \mathbf{L}_{+})(\mathbf{L}_{-}^{\mathsf{T}} \mathsf{C} \mathbf{L}_{-}) = (\mathbf{b}_{+}^{\mathsf{T}} \mathsf{U} \mathbf{b}_{+})(\mathbf{b}_{-}^{\mathsf{T}} \mathsf{U} \mathbf{b}_{-}) = 0$$

By replacing $\mathbf{b}_{\pm}$ with its definition (section 5) and developing this equation, we get the following constraint on the doubly lifted coordinates of $\mathbf{q}$:

$$\hat{\hat{\mathbf{q}}}^{\mathsf{T}} \underbrace{\hat{\hat{\mathsf{K}}}^{-\mathsf{T}} \mathsf{X}_{lc}}_{\mathsf{P}_{lc}} \hat{\mathbf{u}} = 0$$

where $\mathbf{u} = \mathbf{v}_{sym}(\mathsf{U})$, $\hat{\mathbf{u}}$ is a 21-vector and the 15×21 matrix X_{lc} only depends on the intrinsic parameter ξ and is highly sparse (not shown due to lack of space). Since the coefficients of $\hat{\hat{\mathbf{q}}}$ are 4th order monomials of $\mathbf{q}$, we conclude that a central catadioptric image of any line complex, and thus of any quadric or conic, is a quartic curve. We may call the 15×21 matrix P_{lc} the **line complex projection matrix** for catadioptric cameras. It maps the lifted coefficients of the line complex to the 15 coefficients of the quartic curve in the image.

7   The General Catadioptric Fundamental Matrix

We are now ready to derive an analytical expression for the fundamental matrix for any pair of catadioptric images. As suggested in section 2, we perform the following steps: *(i)* Backproject a point $\mathbf{q}$ from the first image to its two backprojection rays, represented by a line complex. *(ii)* Map the line complex from the coordinate system of the first camera, to that of the second one, via a rigid transformation (rotation and translation). *(iii)* Project the transformed line complex into the second camera.

We already know from the previous sections that the result has to be a *quartic epipolar curve* since it is the image of a line complex. In our case, the line complex is degenerate (the "envelope" of just two lines – the backprojection rays – not a full quadric). Hence, and as described in section 2, the quartic epipolar curve is indeed the union of two conics.

Let us now derive the full expression of the catadioptric fundamental matrix. The only remaining missing piece is that the backprojection of an image point (step *(i)*) gives the coefficients of a line complex, but that the projection of the line complex (step *(iii)*) requires its *lifted* coefficients. Hence, we need to insert that lifting between steps *(ii)* and *(iii)*.

Recall from section 5 that the backprojection line complex is obtained as:

$$\mathsf{C} \sim \begin{pmatrix} 0 & 0 \\ 0 & \mathsf{M} \end{pmatrix} \quad \text{with} \quad \mathbf{v}_{sym}(\mathsf{M}) = \mathsf{B}_{cata}\hat{\mathbf{q}}$$

The rigid transformation of step *(ii)* gives (cf. equation (1) in section 3)

$$\mathsf{C}' \sim \mathsf{T}^{-\mathsf{T}}\mathsf{C}\mathsf{T}^{-1} \sim \begin{pmatrix} [\mathbf{t}]_\times \mathsf{R}\mathsf{M}\mathsf{R}^{\mathsf{T}}[\mathbf{t}]_\times^{\mathsf{T}} & -[\mathbf{t}]_\times^{\mathsf{T}}\mathsf{R}\mathsf{M}\mathsf{R}^{\mathsf{T}} \\ -\mathsf{R}\mathsf{M}\mathsf{R}^{\mathsf{T}}[\mathbf{t}]_\times & \mathsf{R}\mathsf{M}\mathsf{R}^{\mathsf{T}} \end{pmatrix}$$

Recall from section 6 that the projection of a line complex, when expressed in the local camera coordinate system, only involves its upper left 3×3 submatrix:

$$\mathsf{U} = [\mathbf{t}]_\times \mathsf{R}\mathsf{M}\mathsf{R}^{\mathsf{T}}[\mathbf{t}]_\times^{\mathsf{T}} = \mathsf{E}\mathsf{M}\mathsf{E}^{\mathsf{T}}$$

where we encounter the well-known essential matrix $\mathsf{E} = [\mathbf{t}]_\times \mathsf{R}$.

Since U and M are symmetric, property (4) from section 3 allows to write the folllwing relation between their vectorized versions $\mathbf{u} = \mathbf{v}_{sym}(\mathsf{U}), \mathbf{m} = \mathbf{v}_{sym}(\mathsf{M})$:

$$\mathbf{u} = \hat{\mathsf{E}}_{6\times6}\mathbf{m}$$

Finally, the required lifted coefficients of the line complex are obtained as:

$$\hat{\mathbf{u}} = \mathring{\mathsf{E}}_{21\times21}\hat{\mathbf{m}}$$

The last remaining detail is to express $\hat{\mathbf{m}}$ in terms of the image point $\mathbf{q}$. From $\mathbf{m} = \mathsf{B}_{cata}\,\hat{\mathbf{q}}$, we deduce

$$\hat{\mathbf{m}} = \hat{\mathsf{B}}_{cata,21\times15}\,\hat{\hat{\mathbf{q}}}$$

We can now introduce the **catadioptric fundamental matrix**:

$$\mathsf{F}_{cata,15\times15} \sim \mathsf{P}_{lc}\,\mathring{\mathsf{E}}\,\hat{\mathsf{B}}_{cata} \tag{6}$$

We have already explained that $\mathsf{F}_{cata}\ \hat{\hat{\mathbf{q}}}$ gives a quartic epipolar curve. The **epipolar constraint** can thus be written as

$$\hat{\hat{\mathbf{q}}}_2^{\mathsf{T}}\ \mathsf{F}_{cata}\ \hat{\hat{\mathbf{q}}}_1 = 0$$

which has the familiar form known for perspective cameras. F_{cata} has rank 6 and its left/right null space has dimension 9. While a perspective view has a single epipole, in an omnidirectional view there are a pair of epipoles, $\mathbf{e}_+$ and $\mathbf{e}_-$, corresponding to the two antipodal intersections of the baseline with the sphere, cf. section 4. The nullspace of F_{cata} comprises the doubly lifted coordinates vectors of both epipoles. We conjecture that they are the only doubly lifted 3-vectors in the nullspace of F_{cata}, but this has to be proven in future work.

We have no space to discuss it, but for mixtures involving one hyper-catadioptric and one other camera, the size of F_{cata} is smaller (15×6 for para–hyper and 6×6 for hyper–perspective). Other special cases are already known [14,5].

8   The General Catadioptric Plane Homography

We give an algebraic formulation of the different steps involved in the plane homography operation, as described in section 2. This section omits many details due to lack of space. Let $\Pi = (\mathbf{n}^{\mathsf{T}}, d)^{\mathsf{T}}$ be the plane and $\mathbf{q}_1$ a point in the first image. We start by backprojecting the point to the plane, giving two 3D points

$$\mathbf{Q}_\pm \sim \underbrace{\begin{pmatrix} d\mathsf{I}_3 \\ -\mathbf{n}^{\mathsf{T}} \end{pmatrix}}_{\mathsf{Y}_{4\times 3}} \mathbf{b}_\pm$$

with $\mathbf{b}_\pm$ as in section 5. We project them to the second image using the projection matrix given in section 4. To do so, we first have to lift the coordinates of these 3D points: $\hat{\mathbf{Q}}_\pm \sim \hat{\mathsf{Y}}_{10\times 6}\ \hat{\mathbf{b}}_\pm$. The projection then gives two dual conics in the second image (cf. section 4), represented by 6-vectors $\omega_\pm \sim \mathsf{P}_{cata}\ \hat{\mathsf{Y}}\ \hat{\mathbf{b}}_\pm$.

Let us compute the following symmetric 6×6 matrix:

$$\Gamma \sim \omega_+\omega_-^{\mathsf{T}} + \omega_-\omega_+^{\mathsf{T}} \sim \mathsf{P}_{cata}\hat{\mathsf{Y}}\underbrace{\left(\hat{\mathbf{b}}_+\hat{\mathbf{b}}_-^{\mathsf{T}} + \hat{\mathbf{b}}_-\hat{\mathbf{b}}_+^{\mathsf{T}}\right)}_{\mathsf{Z}}\hat{\mathsf{Y}}^{\mathsf{T}}\mathsf{P}_{cata}^{\mathsf{T}} \tag{7}$$

What does Γ represent? It is the "union" of two degenerate dual conics, ω_+ and ω_-. Hence, it represents a dual quartic curve (this can also be proven more formally). Further, each of the two dual conics represents two image points; it contains exactly the lines going through at least one of them. Hence, Γ contains exactly the lines going through at least one of the total of four considered image points. These four points are nothing else than the four points explained in section 2, i.e. the possible four matches of $\mathbf{q}_1$.

The expression of the dual quartic curve represented by Γ can be written as $\hat{\hat{\mathbf{l}}}^{\mathsf{T}}\gamma = 0$, where $\hat{\hat{\mathbf{l}}}$ denotes the fourth order Veronese map of a generic line $\mathbf{l}$, and

γ is a 15-vector containing sums of the 21 coefficients of Γ (this is analogous to the reduction explained in equation (2)).

When developing the expression for Z in (7), it can be seen that its coefficients are linear in $\hat{\mathbf{q}}_1$. The coefficients of γ may thus be computed via a linear mapping of $\hat{\mathbf{q}}_1$; that mapping is given by a 15×15 matrix:

$$\gamma \sim \mathsf{H}_{cata,15\times15}\,\hat{\hat{\mathbf{q}}}_1$$

The matrix H_{cata} is the **catadioptric plane homography**. Its explicit form is omitted due to lack of space.

By the same approach as in section 4, we can derive the following constraint equation:

$$\widehat{[\mathbf{q}_2]}_\times\,\mathsf{H}_{cata}\,\hat{\hat{\mathbf{q}}}_1 = \mathbf{0}_{15}$$

Of the 15 constraints contained in the above equation, only five are linearly independent. Hence, in order to estimate the $15^2 = 225$ coefficients of H_{cata}, we need at least 45 matches.

In the special case of para-catadioptric cameras, the homography is of size 6×6 and each match gives 6 equations, 3 of which are linearly independent. Hence, 12 matches are needed to estimate the 36 coefficients of that plane homography.

9   Conclusions, Discussion, and Perspectives

Our motivation for this work is to get a complete picture of the imaging and multi-view geometry of catadioptric (and other) cameras. We have shown that the basic concepts – projection, backprojection, epipolar geometry, or plane homography – can all be written as (multi-) linear mappings. These results are first of all of conceptual value, and we consider them as theoretical contributions.

Concerning potential practical applications, we note that a linear estimation of the catadioptric fundamental matrix requires 224 matches... We thus do not currently believe that F_{cata} will be of practical use.

However, the catadioptric projection matrix and the plane-to-image homography described in section 4 may indeed prove useful for calibrating catadioptric and possibly other omnidirectional cameras. We show this by an illustrative experiment, cf. Figure 2. Corner extraction for calibration grids, despite being trivial for perspective cameras, is still problematic for images with strong non-linear distortions [26]. In the perspective case we typically indicate the area of interest by manually clicking 4 corners; they enable the estimation of an homography and the projection of the grid into the image. The final position of all corners is accurately determined by refining the initial estimate using image processing techniques. Such a procedure has not been possible until now for non-conventional imagery with non-perspective distortions.

From section 4 it follows that the homography from a plane to any catadioptric image is represented by a 6×6 matrix H. We estimated it from the required minimum of 12 manually selected matches by the DLT procedure suggested in section 4 (left side of Fig. 2). All corners were then projected into the image and refined

Fig. 2. Estimation of the homography mapping a planar grid into a catadioptric image. It was determined from 12 clicked points (left side). Each corner is mapped into a pair of image projections. The lines joining corresponding pairs form a pencil going through the principal point which confirms the correctness of the estimation.

using a corner detector. From this initial step only 7 out of 91 points were missed. The procedure was repeated a second time using all the good points and 6 more points were correctly detected. The estimated homography maps each plane point to a pair of antipodal image points (right side of Fig. 2). The shown result suggests that the plane-to-image homography can be well estimated and that it is useful for extracting and matching corners of a planar grid. Current work deals with calibrating the camera from such homographies, from multiple images of the grid.

There are several perspectives for our work. The shown results can be specialized to e.g. para-catadioptric cameras, leading to simpler expressions. It may also be possible that due to the coordinate liftings used, some of the results hold not only for catadioptric cameras, but also for other models, e.g. classical radial distortion; this will be investigated. Current work is concerned with developing practical DLT like calibration approaches for catadioptric cameras, using 3D or planar calibration grids. Promising results have already been obtained.

Acknowledgements. João P. Barreto is grateful to the Portuguese Science Foundation by generous funding through grant PTDC/EEA-ACR/68887/2006. Peter Sturm acknowledges support by the French ANR project CAVIAR.

References

1. Hartley, R., Zisserman, A.: Multiple View Geometry in Computer Vision, 2nd edn. Cambridge University Press, Cambridge (2004)
2. Svoboda, T., Pajdla, T.: Epipolar geometry for central catadioptric cameras. IJCV 49, 23–37 (2002)
3. Kang, S.: Catadioptric self-calibration. In: CVPR, pp. 1201–1207 (2000)

4. Geyer, C., Daniilidis, K.: Structure and motion from uncalibrated catadioptric views. In: CVPR, pp. 279–286 (2001)
5. Sturm, P.: Mixing catadioptric and perspective cameras. In: OMNIVIS, pp. 37–44 (2002)
6. Mičušík, B., Pajdla, T.: Structure from motion with wide circular field of view cameras. PAMI 28(7), 1135–1149 (2006)
7. Claus, D., Fitzgibbon, A.: A rational function for fish-eye lens distortion. In: CVPR, pp. 213–219 (2005)
8. Gupta, R., Hartley, R.: Linear pushbroom cameras. PAMI 19, 963–975 (1997)
9. Feldman, D., Pajdla, T., Weinshall, D.: On the epipolar geometry of the crossed-slits projection. In: ICCV, pp. 988–995 (2003)
10. Pajdla, T.: Stereo with oblique cameras. IJCV 47, 161–170 (2002)
11. Seitz, S., Kim, J.: The space of all stereo images. IJCV 48, 21–38 (2002)
12. Menem, M., Pajdla, T.: Constraints on perspective images and circular panoramas. In: BMVC (2004)
13. Baker, S., Nayar, S.: A theory of catadioptric image formation. In: ICCV, pp. 35–42 (1998)
14. Geyer, C., Daniilidis, K.: Properties of the catadioptric fundamental matrix. In: Tistarelli, M., Bigun, J., Jain, A.K. (eds.) ECCV 2002. LNCS, vol. 2359, pp. 140–154. Springer, Heidelberg (2002)
15. Barreto, J.: A unifying geometric representation for central projection systems. CVIU 103 (2006)
16. Barreto, J.P., Daniilidis, K.: Epipolar geometry of central projection systems using veronese maps. In: CVPR, pp. 1258–1265 (2006)
17. Wolf, L., Shashua, A.: Two-body segmentation from two perspective views. In: CVPR, pp. 263–270 (2001)
18. Vidal, R., Ma, Y., Soatto, S., Sastry, S.: Two-view multibody structure from motion. IJCV 68, 7–25 (2006)
19. Geyer, C., Daniilidis, K.: A unifying theory for central panoramic systems. In: Vernon, D. (ed.) ECCV 2000. LNCS, vol. 1843, pp. 445–461. Springer, Heidelberg (2000)
20. Barreto, J., Araujo, H.: Geometric properties of central catadioptric line images and its application in calibration. PAMI 27, 1237–1333 (2005)
21. Semple, J., Kneebone, G.: Algebraic Projective Geometry. Claredon Press (1998)
22. Horn, R., Johnson, C.: Topics in Matrix Analysis. Cambridge University Press, Cambridge (1991)
23. Ying, X., Zha, H.: Using sphere images for calibrating fisheye cameras under the unified imaging model of the central catadioptric and fisheye cameras. In: ICPR, pp. 539–542 (2006)
24. Ponce, J., McHenry, K., Papadopoulo, T., Teillaud, M., Triggs, B.: On the absolute quadratic complex and its application to autocalibration. In: CVPR, pp. 780–787 (2005)
25. Valdés, A., Ronda, J., Gallego, G.: The absolute line quadric and camera autocalibration. IJCV 66, 283–303 (2006)
26. Mei, C., Rives, P.: Single view point omnidirectional camera calibration from planar grids. In: ICRA, pp. 3945–3950 (2007)

Estimating Radiometric Response Functions from Image Noise Variance

Jun Takamatsu[1,2], Yasuyuki Matsushita[3], and Katsushi Ikeuchi[1,4,⋆]

[1] Microsoft Institute for Japanese Academic Research Collaboration (MS-IJARC)
[2] Nara Institute of Science and Technology
`j-taka@is.naist.jp`
[3] Microsoft Research Asia
`yasumat@microsoft.com`
[4] Institute of Industrial Science, the University of Tokyo
`ki@cvl.iis.u-tokyo.ac.jp`

Abstract. We propose a method for estimating radiometric response functions from observation of image noise variance, not profile of its distribution. The relationship between radiance intensity and noise variance is affine, but due to the non-linearity of response functions, this affinity is not maintained in the observation domain. We use the non-affinity relationship between the observed intensity and noise variance to estimate radiometric response functions. In addition, we theoretically derive how the response function alters the intensity-variance relationship. Since our method uses noise variance as input, it is fundamentally robust against noise. Unlike prior approaches, our method does not require images taken with different and known exposures. Real-world experiments demonstrate the effectiveness of our method.

1 Introduction

Many computer vision algorithms rely on the assumption that image intensity is linearly related to scene radiance recorded at the camera sensor. However, this assumption does not hold with most cameras; the linearity is not maintained in the actual observation due to non-linear *camera response functions*. Linearization of observed image intensity is important for many vision algorithms to work, therefore the estimation of the response functions is needed.

Scene radiance intensity (input) I and observed image intensity (output) O are related by the response function f as $O = f(I)$. Assuming it is continuous and monotonic, the response function can be inverted to obtain the *inverse response function* g $(= f^{-1})$, and measured image intensities can be linearized using $I = g(O)$. Since only observed output intensities O are usually available, most estimation methods attempt to estimate the inverse response functions.

⋆ Institute of Industrial Science, University of Tokyo.

D. Forsyth, P. Torr, and A. Zisserman (Eds.): ECCV 2008, Part IV, LNCS 5305, pp. 623–637, 2008.

1.1   Prior Work

Radiometric calibration methods assume known characteristics of radiance at a camera to estimate unknown response functions. One class of these methods uses information about the ratio of input radiance intensities. The Macbeth color checker is used for estimating the response functions by several research groups such as [1]. Nayar and Mitsunaga [2] use an optical filter with spatially varying transmittance; the variation corresponds to the ratio. Instead of using such special equipment, some methods use a set of input images taken with different exposures from a fixed viewpoint so that the ratio becomes available. These methods are divided into parametric and non-parametric approaches.

In the parametric framework, Mann and Picard [3] propose a method that assumes the response functions can be approximated by gamma correction functions. Mitsunaga and Nayar use a polynomial function for the representation [4]. Grossberg and Nayar apply principal component analysis (PCA) to a database of real-world response functions (DoRF) and show that the space of response functions can be represented by a small number of basis functions [5]. Mann [6] uses the *comparametric* equation that defines the relationship between two images taken with different exposure times.

Debevec and Malik estimate response functions with a non-parametric representation using a smoothness constraint [7]. Tsin *et al.* estimate non-parametric forms of response functions using the statistical analysis [8]. Pal *et al.* propose to use a Bayesian network consisting of probabilistic imaging models and prior models of response functions [9]. While non-parametric approaches have greater descriptive power, the large number of unknown variables often lead to computationally expensive or unstable algorithms.

A few authors have proposed more general estimation methods that allow camera motion or scene motion between input images. Grossberg and Nayar [10] estimate the function from images with scene motion using brightness histograms, which avoid difficult pixelwise image registration. Kim and Pollefeys [11] propose a method that allows free camera motion by finding the correspondence among images. These methods can handle more general cases, but still require images taken with multiple exposures.

Another class of the estimation methods is based on the physics of the imaging process. Lin *et al.* estimate the function from the color blending around edges in a single image [12]. For grayscale images, a 1D analogue of the 3D color method is presented in [13]. Wilburn *et al.* propose to use the temporal blending of irradiance [14]. Matsushita and Lin propose an estimation method from profiles of image noise distributions [15]; their method relies on the assumption that profiles of noise distributions are symmetric. More recently, Takamatsu *et al.* [16] propose to use a probabilistic intensity similarity measure [17] for the estimation. Unlike these approaches, our method only uses *noise variance*, which is considered a lower level of information.

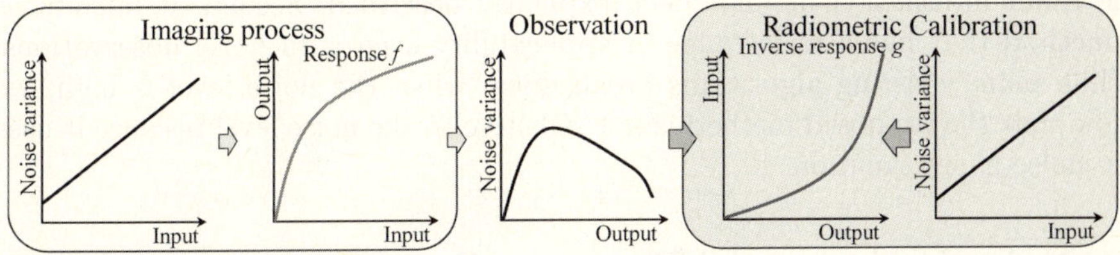

Fig. 1. The noise variance in the input domain has an affine relationship with input intensity level. Due to the non-linearity of the response function, the affine relationship is lost in the output domain. The proposed method estimates the response functions by recovering the affinity of the measured noise variances.

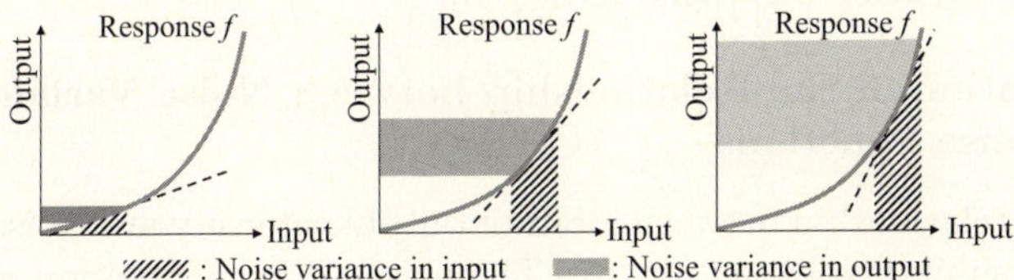

Fig. 2. The relationship between response function and noise variances in input and output domains. The magnitude of output noise variance (height of the filled region) varies with the slope of the response function with a fixed input noise variance.

1.2   Proposed Method

In this paper, we propose a method for estimating camera response functions from observation of image noise variance. Figure 1 illustrates the overview of the proposed method. As shown in the figure, noise variance in the input domain has an affine relationship with input intensity. This relationship, however, is not maintained in the output domain due to the non-linearity of camera response functions. By estimating a function that projects the non-affine intensity-variance relationship to an affine one, our method derives the response functions. Figure 2 depicts the relationship between the response function and noise variances in input and output domains. The ratio of the input/output noise variance is closely related to the slope of the response function.

Unlike Matsushita and Lin's work [15], our method does not rely on profiles of noise distributions which are more difficult to obtain in practice. The proposed method has the following two advantages over their method. First, it uses the more reasonable assumption; the affinity of intensity-variance relationship in the input domain. We derive this relationship from the nature of photon (electron) noise [18,19]. Second, the proposed method uses only noise variances rather than noise profiles, therefore it only requires less amount of information about noise. Practically, it makes the procedure of noise measurement easier.

This paper has two major contributions. It first theoretically derives how the radiometric response function alters the intensity-variance relationship, and vice versa. While it has been pointed out that these two quantities are related [8,20], the exact alternation has not been explicitly described. Second, it introduces a method that has a wide range of applicability even with noisy observations. While many existing algorithms break down when the noise level is high, we show that the proposed method is not sensitive to the noise level because it uses the noise as information.

2 Noise Variance and Response Functions

This section provides a theoretical relationship between the response function and noise variances in the input and output domains. We show that the levels of noise variance in the input and output domains are related by the square of the first derivative of the camera response function. This result is later used to develop the estimation algorithm in Section 3.

2.1 Derivation of the Relationship between Noise Variance and Response Functions

We treat the relationship between input or output intensity and noise variance as the function similar to Liu $et\ al.$ [20]. The noise variance function $\sigma_O^2(\tilde{O})$ in the output domain, where the noise-free output intensity is $\tilde{O}$, can be described as

$$\sigma_O^2(\tilde{O}) = \int (O - \mu_O)^2 p(O|\tilde{O})dO = \int (f(I) - \mu_O)^2 p(I|\tilde{I})dI.$$

The conditional density function (cdf) $p(O|\tilde{O})$ represents the noise distribution in the output domain, i.e., the probability that the output intensity becomes O when the noise-free intensity is $\tilde{O}$. Likewise, the cdf $p(I|\tilde{I})$ represents the noise distribution in the input domain when the noise-free input intensity is $\tilde{I}$. The function f is the response function, and $\tilde{O}$ and $\tilde{I}$ are related by f as $\tilde{O} = f(\tilde{I})$. $\mu_O(= \mu_O(\tilde{O}))$ is the expectation of the output intensity with the cdf $p(O|\tilde{O})$.

Using the Taylor series of $f(I)$ around $\tilde{I}$ and assuming that the second- and higher-degree terms of the series are negligible (discussed in Section 2.3), we obtain

$$\sigma_O^2(\tilde{O}) \simeq f'^2(\tilde{I})\sigma_I^2(\tilde{I}), \tag{1}$$

where $\sigma_I^2(\tilde{I})$ is the function of noise variance in the input domain when the noise-free input intensity is $\tilde{I}$. This equation shows the relationship between the noise variance and the response functions. This relationship has been pointed out in [8,20], however, the correctness of the equation was not thoroughly discussed.

To simplify the notation in derivation of Eq. (1), we define:

$$\mu_I = \mu_I(\tilde{I}) = \int Ip(I|\tilde{I})dI, \ I_d = \mu_I - \tilde{I},$$

$$M_n = M_n(\tilde{I}) = \int (I - \mu_I)^n p(I|\tilde{I})dI \ (n \geq 0). \tag{2}$$

μ_I is the expectation, or mean, of input intensity I with the cdf $p(I|\tilde{I})$. I_d denotes the difference between the expectation μ_I and the noise-free intensity $\tilde{I}$. M_n is the n-th moment about the mean. Note that we do not put any assumptions about the profile and model of the noise distributions.

The expectation of the output intensity μ_O where the noise-free output intensity is $\tilde{O}$ can be written as

$$\mu_O = \mu_O(\tilde{O}) = \int f(I)p(I|\tilde{I})dI = \tilde{O} + \sum_{j=1}^{\infty} N_j \frac{f^{(j)}(\tilde{I})}{j!}, \tag{3}$$

where N_j is the j-th moment about $\tilde{I}$ defined as follows.

$$N_j = \int (I - \tilde{I})^j p(I|\tilde{I})dI = \int ((I - \mu_I) + I_d)^j p(I|\tilde{I})dI = \sum_{k=0}^{j} \binom{j}{k} I_d^{j-k} M_k. \tag{4}$$

Note that the response function f is represented using its Taylor series.

From the definition, the noise variance function $\sigma_O^2(\tilde{O})$ where the noise-free output intensity is $\tilde{O}$ can be derived as

$$\sigma_O^2(\tilde{O}) = \sum_{j=1}^{\infty} \left(\frac{f^{(j)}(\tilde{I})}{j!} \right)^2 L_{j,j} + 2 \sum_{j=1}^{\infty} \sum_{k>j}^{\infty} \frac{f^{(j)}(\tilde{I})}{j!} \frac{f^{(k)}(\tilde{I})}{k!} L_{j,k}, \tag{5}$$

where $L_{j,k}$ is defined as

$$L_{j,k} = \int ((I - \mu_I)^j - N_j)((I - \mu_I)^k - N_k)p(I|\tilde{I})dI = N_{j+k} - N_j N_k. \tag{6}$$

As can be seen in the above equation, $L_{j,k}$ is commutative ($L_{j,k} = L_{k,j}$).

A detailed calculation gives us $L_{1,1} = M_2 = \sigma_I^2(\tilde{I})$, $L_{1,2} = 2I_d\sigma_I^2(\tilde{I}) + M_3, \cdots$. Substituting these terms into Eq. (5), we obtain:

$$\begin{aligned} \sigma_O^2(\tilde{O}) &= f'^2(\tilde{I})L_{1,1} + f'(\tilde{I})f''(\tilde{I})L_{1,2} + \cdots \\ &= f'^2(\tilde{I})\sigma_I^2(\tilde{I}) + f'(\tilde{I})f''(\tilde{I})(2I_d\sigma_I^2(\tilde{I}) + M_3) + \cdots . \end{aligned} \tag{7}$$

Eq. (7) is the exact form of the relationship between the response function and noise variances in the input and output domains. By discarding the second- and higher-degree terms of Eq. (7), Eq. (1) is obtained. We discuss the validity of this approximation in Section 2.3.

2.2 Noise Variance Function

Eq. (1) shows the relationship between the response function and noise variance functions in the input and output domains. The input to our method is the measured noise variance in the output domain. This section models the noise variance function $\sigma_I^2(\tilde{I})$ in the input domain so that the estimation algorithm for the inverse response function g can be developed.

Input intensity I with camera noise can be written as

$$I = aP + N_{DC} + N_S + N_R, \tag{8}$$

where a is a factor of photon-to-electron conversion efficiency with amplification, and P is the number of photons. N_{DC}, N_S, and N_R indicate dark current noise, shot noise, and readout noise, respectively [21][1]. The noise-free input intensity $\tilde{I}$ equals to aP.

Now we consider the noise variance function in the input domain. We assume the different noise sources are independent. From Eq. (8), the noise variance function in the input domain, $\sigma_I^2(\tilde{I})$, can be written as

$$\sigma_I^2(\tilde{I}) = \tilde{I}\sigma_S^2 + \sigma_{DC}^2 + \sigma_R^2, \tag{9}$$

where σ_*^2 denotes the variances of the different noise sources [8]. Eq. (9) can be written in a simplified form as

$$\sigma_I^2(\tilde{I}) = A\tilde{I} + B, \tag{10}$$

where $A = \sigma_S^2$ and $B = \sigma_{DC}^2 + \sigma_R^2$. This equation clearly shows the affine relationship between the noise-free input intensity $\tilde{I}$ and the noise variance $\sigma_I^2(\tilde{I})$.

2.3 Validity of the Approximation

It is important to consider the validity of the approximation in Eq. (7). In this section, we show it in the following steps. First, we show that $L_{i,j}$ becomes exponentially smaller as $i+j$ increases. Second, the second largest term in Eq. (7), $L_{1,2}$, is small enough to be negligible compared with $L_{1,1}$ through a detailed calculation. Hereafter, we normalize the input and output intensity ranges from 0 to 1.

Relationship between $L_{i,j}$ and $i+j$. By assuming independence of different noise sources, the second- and higher-order moments can be computed by summing up the moments of noises from different sources. Three types of noise sources must be considered: dark current noise, shot noise, and readout noise [21]. We do not consider low-light conditions [22], so the effect of the dark current noise becomes small.

The probability density of the readout noise can be considered to have a normal distribution with a mean value equal to the noise-free input intensity $\tilde{I}$. The moment can be written as

$$N_{R_i} = \begin{cases} 0 & (i \text{ is odd}) \\ \prod_{j=1}^{i/2}(2j-1)\sigma_R^2 & (i \text{ is even}). \end{cases} \tag{11}$$

In Eq. (11), $(2j-1)\,\sigma_R^2 \ll 1$, so the i-th moment of the readout noise N_{R_i} about the noise-free input intensity becomes exponentially smaller as i increases.

Shot noise is modeled as Poisson distribution [21]. From the theory of generalized Poisson distribution [23], the moment M_{S_i} about the mean of the distribution is defined as

$$M_{S_i} \simeq a^{i-2}\big(\sigma_{S_{\tilde{I}}}^2 + O(\sigma_{S_{\tilde{I}}}^4)\big)\ (i \geq 2), \tag{12}$$

---

[1] The effect of the fixed-pattern noise is included in the term P.

since the minimum unit of the distribution equals to a (See Eq. (8)). $\sigma^2_{S_{\tilde{I}}}$ is the variance of shot noise where the noise-free input intensity is $\tilde{I}$. By substituting this into Eq. (4) yields

$$N_{S_i} \simeq I_d^i + \sum_{j=2}^{i} \binom{i}{j} I_d^{i-j} a^{j-2} \sigma^2_{S_{\tilde{I}}}. \tag{13}$$

Even in the worst case where $\binom{i}{j}$ is overestimated as 2^i, N_{S_i} becomes exponentially smaller, since Eq. (13) is rewritten as

$$N_{S_i} \leq I_d^i + \sum_{j=2}^{i} (2I_d)^{i-j} (2a)^{j-2} (2\sigma_{S_{\tilde{I}}})^2, \tag{14}$$

and we know that $2I_d \ll 1$, $2a \ll 1$, and $(2\sigma_{S_{\tilde{I}}})^2 \ll 1$. This equation shows that N_{S_i} exponentially decreases as i increases.

The term $L_{i,j}$ is defined as $L_{i,j} = N_{i+j} - N_i N_j$ in Eq. (6). Because the i-th moment of image noise N_i can be computed as the sum of the readout and shot noise as $N_i = N_{R_i} + N_{S_i}$, it also becomes exponentially smaller as i increases. From these results, we see that the term $L_{i,j}$ becomes exponentially smaller as $i + j$ increases.

Ratio of $L_{1,1}$ to $L_{1,2}$. Now we show that $L_{1,2}$ is small enough to be negligible compared with $L_{1,1}$. A detailed calculation gives us $L_{1,2} = 2I_d M_2 + M_3$. The third moment of shot noise M_{S_3} can be computed from Eq. (12). Also, the third moment of readout noise can be obtained using Eq. (4) as

$$M_{R_3} = N_{R_3} - 3I_d N_{R_2} - I_d^3 = -3I_d \sigma_R^2 - I_d^3. \tag{15}$$

From these results, the following equation is obtained:

$$L_{1,2} = 2I_d M_2 - 3I_d \sigma_R^2 - I_d^3 + a\sigma^2_{S_{\tilde{I}}}. \tag{16}$$

Since $M_2 \simeq \sigma_R^2 + \sigma^2_{S_{\tilde{I}}}$, $a \ll I_d$, if $M_2 \not\ll I_d^2$, the order of $L_{1,2}$ is roughly the same as the order of $I_d M_2$. Since I_d is the difference between the noise-free input intensity and the mean which can be naturally considered very small, it is implausible to have cases where $M_2 \ll I_d^2$.

From these results, the order of $L_{1,2}$ is roughly equivalent to the order of $I_d L_{1,1}$, and I_d is small because it is computed in the normalized input domain, e.g., in the order of 10^{-2} ($\simeq 1/2^8$) in 8-bit image case. Therefore, $L_{1,2}$ is about 10^{-2} times smaller than $L_{1,1}$.

To summarize, $L_{1,2}$ is sufficiently small compared with $L_{1,1}$, and $L_{i,j}$ decreases exponentially as $i + j$ increases. Also, because response functions are smooth, $f'(\tilde{I}) \not\ll f''(\tilde{I})$. Therefore, Eq. (7) can be well approximated by Eq. (1).

3   Estimation Algorithm

This section designs an evaluation function for estimating inverse response functions g, using the result of the previous section.

3.1 Evaluation Function

From Eqs. (1) and (10), the noise variance $\sigma_O^2(\tilde{O})$ of the output intensity O is

$$\sigma_O^2(\tilde{O}) \simeq f'^2(\tilde{I})\sigma_I^2(\tilde{I}) + \sigma_Q^2 = f'^2(\tilde{I})(A\tilde{I} + B) + \sigma_Q^2. \tag{17}$$

σ_Q^2 is the variance of the quantization noise, which affects after applying the response function. Using the inverse response function g, Eq. (17) can be rewritten as

$$\sigma_O^2(\tilde{O}) = \frac{1}{g'(\tilde{O})^2}(Ag(\tilde{O}) + B) + \sigma_Q^2. \tag{18}$$

The variance of the quantization noise σ_Q^2 becomes $\sigma_Q^2 = l^2/12$, where l is the quantization interval. Since its distribution is uniform, the following equation holds:

$$\sigma_Q^2 = \int_{-\frac{l}{2}}^{\frac{l}{2}} x^2 p(x)dx = \frac{1}{l}\int_{-\frac{l}{2}}^{\frac{l}{2}} x^2 dx = \frac{l^2}{12}. \tag{19}$$

In the following, we use $\sigma_{O_m}^2(\tilde{O})$ to represent the measured noise variance to discriminate from the analytic form of the noise variance $\sigma_O^2(\tilde{O})$. Using Eq. (18) and the measured noise variances $\sigma_{O_m}^2(\tilde{O})$, our method estimates the inverse response function g that minimizes the following evaluation function:

$$E_1(g; \sigma_{O_m}^2(\tilde{O})) = \min_{A,B} \sum_{\tilde{O}} \left(\sigma_O^2(\tilde{O}) - \sigma_{O_m}^2(\tilde{O})\right)^2. \tag{20}$$

Eq. (20) involves the estimation of A and B, which can be simply solved by linear least square fitting, given g.

To make the algorithm robust against the measuring errors, namely the erroneous component in the measured noise, we use weighting factors. Eq. (20) is changed to

$$E_2(g; \sigma_{O_m}^2(\tilde{O})) = \min_{A,B} \frac{1}{\sum w(\tilde{O})} \sum_{\tilde{O}} w(\tilde{O})\left(\sigma_O^2(\tilde{O}) - \sigma_{O_m}^2(\tilde{O})\right)^2, \tag{21}$$

where the weight function $w(\tilde{O})$ controls the reliability on the measured noise variance $\sigma_{O_m}^2(\tilde{O})$ at the intensity level $\tilde{O}$. We use a Cauchy distribution (Lorentzian function) for computing the weight function $w(\tilde{O})$:

$$w(\tilde{O}) = \frac{1}{e^2 + \rho}, \tag{22}$$

where e is defined as $e = \sigma_O^2(\tilde{O}) - \sigma_{O_m}^2(\tilde{O})$. A damping factor ρ controls the relationship between the difference e and weight $w(\tilde{O})$. As ρ becomes smaller, the weight $w(\tilde{O})$ decreases more rapidly as the difference e increases.

We also add a smoothness constraint to the evaluation function, and the evaluation function becomes

$$E_3(g; \sigma_{O_m}^2(\tilde{O})) = \frac{1}{\sum_{\tilde{O}} \sigma_{O_m}^2(\tilde{O})} E_2 + \lambda_s \frac{1}{n_{\tilde{O}}} \sum_{\tilde{O}} g''(\tilde{O})^2, \tag{23}$$

where $n_{\tilde{O}}$ is the number of possible noise-free output intensity levels, e.g., 256 in 8-bit case. λ_s is a regularization factor that controls the effect of the smoothness constraint. $1/\sum_{\tilde{O}} \sigma^2_{O_m}(\tilde{O})$ is a normalization factor that makes E_2 independent of the degree of noise level.

Our method estimates the inverse response function $\hat{g}$ by minimizing Eq. (23) given the measured noise variance $\sigma^2_{O_m}(\tilde{O})$:

$$\hat{g} = \operatorname*{argmin}_{g} E_3\left(g; \sigma^2_{O_m}(\tilde{O})\right). \tag{24}$$

3.2 Representation of Inverse Response Functions

To reduce the computational cost, we represent the inverse response functions using a parametric model proposed by Grossberg and Nayar [5]. In their method, principal component analysis (PCA) is performed on the database of real-world response functions (DoRF) to obtain a small number of eigenvectors that can represent the space of the response functions. As done by Lin *et al.* [12,13], we compute the principal components of the inverse response functions using the DoRF. Using the principal components, we represent the inverse response function g as $\mathbf{g} = \mathbf{g}_0 + \mathbf{H}\mathbf{c}$, where $\mathbf{g}_0$ is the mean vector of all the inverse response functions, $\mathbf{H}$ is a matrix in which a column vector represents an eigenvector, and $\mathbf{c}$ is a vector of PCA coefficients. Following Lin *et al.* [12,13], we use the first five eigenvectors. Using this representation, the number of unknown variables is significantly decreased, e.g., from 256 to 5 in the case of 8-bit images.

3.3 Implementation

In our implementation, we set the damping factor ρ to the variance of the difference e in Eq. (22). The regularization factor λ_s is set to 5×10^{-7} from our empirical observation. Minimization is performed in an alternating manner. We perform the following steps until convergence:

1. minimize the evaluation function in Eq. (23) with fixing the weight function $w(\tilde{O})$
2. recompute the values of the weight function $w(\tilde{O})$ using the current estimation result

We use the Nelder-Mead Simplex method [24] as the minimization algorithm implemented in Matlab as a function `fminsearch`. The values of the weight function $w(\tilde{O})$ are set to one for every $\tilde{O}$ at the beginning. During the experiments, we used five initial guesses for the inverse response function g as the input to the algorithm. The converged result that minimizes the energy score is finally taken as the global solution.

4 Experiments

We used two different setups to evaluate the performance of the proposed algorithm; one is with multiple images taken by a fixed video camera, the other is using a single image. The two setups differ in the means for collecting noise variance information.

4.1   Multiple-Images Case

In this experiment, the measurements of noise variances are obtained by capturing multiple shots of a static scene from a fixed viewpoint with fixed camera parameters. From multiple images, a histogram of output intensities is created for each pixel. From the histogram, the noise-free output intensity $\tilde{O}$ is determined by taking the mode of the distribution, assuming that the noise-free intensity should correspond to the most frequently observed signal. The pixelwise histograms are then merged together to form the histogram $h(O, \tilde{O})$ for each output intensity level $\tilde{O}$. Finally, the noise distribution $p(O|\tilde{O})$ is computed by normalizing the histogram $h(O, \tilde{O})$ as

$$p(O|\tilde{O}) = \frac{h(O, \tilde{O})}{\sum_O h(O, \tilde{O})}. \tag{25}$$

Results. We used three different video cameras for this experiment: *Sony DCR-TRV9E* (Camera A), *Sony DCR-TRV900 NTSC* (Camera B), and *Sony DSR-PD190P* (Camera C). To obtain the ground truth of Camera C, we used Mitsunaga and Nayar's method [4], and the Macbeth color checker-based method [1], and combined these results by taking the mean. For Camera A and B, we used only the Macbeth color checker-based method [1] to obtain the ground truth because the exposure setting was not available in these cameras. The results obtained by the proposed method are compared with the ground truth curves.

Figure 3 shows the results of our algorithm. The top row shows the plot of the estimated inverse response functions with the corresponding ground truth

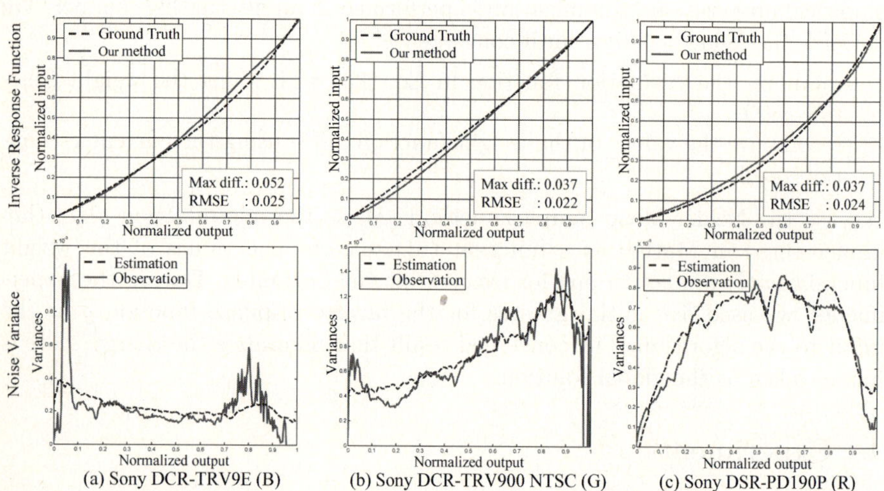

Fig. 3. Results of our estimation method. Top row: comparison of inverse response functions. Bottom row: measured noise variance and fitting result.

curves. The bottom row shows the estimated and measured distributions of noise variances; the horizontal axis is the normalized output, and the vertical axis corresponds to the noise variance. Figure 4 shows the scenes used to obtain these results.

Figure 3 (a) shows an estimation result using the blue channel of Camera A. The maximum difference is 0.052 and the RMSE is 0.025 in terms of normalized input. As shown in the bottom of (a), the noise variances in lower output levels contain severe measured errors. Our algorithm is robust against such errors because of the use of adaptive weighting factors. Figure 3 (b) shows the result of Camera B (green channel). The maximum difference is 0.037 and the RMSE is 0.022. Figure 3 (c) shows the estimation result of Camera C (red channel). The input frames are obtained by setting the camera gain to 12 db which causes high noise level. The maximum difference is 0.037 and the RMSE is 0.024.

Table 1 summarizes all the experimental results. For each camera, three different scenes are used. The algorithm is applied to RGB-channels independently, therefore 9 datasets for each camera are used. Disparity represents the mean of maximum differences in normalized input. From these results, the proposed method performs well even though the algorithm only uses the noise variance as input.

4.2   Comparison with Another Noise-Based Estimation Method

Figure 5 shows the comparison between our method and Matsushita and Lin's method [15]. Unlike other estimation methods, these two methods take noise as input. We use Camera B for the comparison.

As shown in the result, the estimation results are equivalent when the number of images is relatively large. However, Matsushita and Lin's method breaks down when the number of samples becomes small, and our method shows significant superiority. In statistics, it is known that variance of measured from samples' variance is inversely proportional to the number of the samples. Therefore, the measured variance becomes more stable than the profile of noise distribution does, as the number of samples increases. In addition, Matsushita and Lin's symmetry criterion naturally requires large number of samples to make the noise profiles smooth, while it does not hold in the lower number of samples in Figure 5. These are why our method works well when the number of samples is relatively small.

Table 1. Mean RMSE and disparity of the estimated inverse response functions in terms of normalized input. Three different scenes were used for each camera.

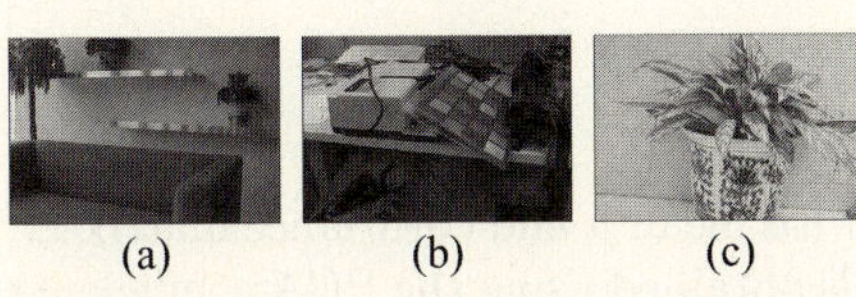

(a)   (b)   (c)

Fig. 4. Recorded scenes corresponding to the results in Figure 3 (a-c)

Camera	Mean RMSE	Disparity
A. DCR-TRV9E	0.026	0.053
B. DCR-TRV900	0.024	0.040
C. DSR-PD190P	0.033	0.055

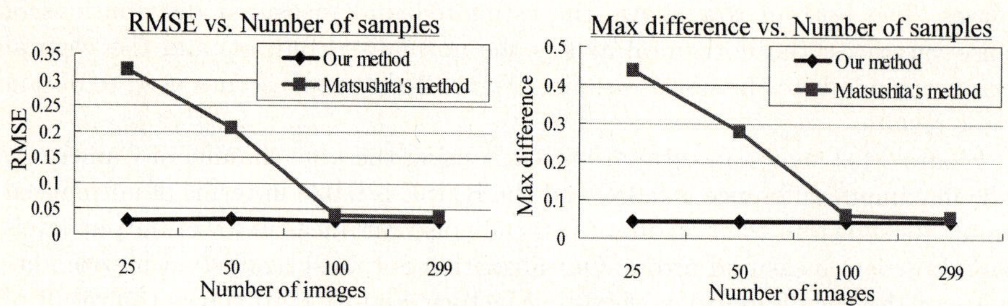

Fig. 5. Comparison between our method and Matsushita and Lin's method [15]. Our method uses noise variance, but not profiles of noise distributions. Our method works well even when the sampling number is relatively small.

4.3   Single-Image Case

We describe a single-image case where only one shot of the scene is available. In this setup, the distribution of noise variances are collected from uniformly colored image regions. However, the measured noise distribution is often insufficient to determine the inverse response functions because the limited measurements do not span the entire range of output levels. To better constrain the problem, we use a prior model $p(g)$ of the inverse response functions obtained from the DoRF as done in [12] and [15].

Using the prior model $p(g)$, the MAP (maximum a posteriori) estimation is performed by maximizing the cdf $p(g|\sigma^2_{O_m}(\tilde{O}))$ which represents the probability of the inverse response function being g when the measured noise variances are $\sigma^2_{O_m}(\tilde{O})$ as

$$\hat{g} = \underset{g}{\operatorname{argmax}}\, p(g|\sigma^2_{O_m}(\tilde{O})) = \underset{g}{\operatorname{argmax}}\big(\log p(\sigma^2_{O_m}(\tilde{O})|g) + \log p(g)\big). \qquad (26)$$

The likelihood $p(\sigma^2_{O_m}(\tilde{O})|g)$ is defined as

$$p(\sigma^2_{O_m}(\tilde{O})|g) = \frac{1}{Z}\exp\big(-\lambda_p E_3(g; \sigma^2_{O_m}(\tilde{O}))\big), \qquad (27)$$

where Z is the normalization factor, and λ_p is a regularization coefficient that determines the weight on the evaluation function E_3. We empirically set λ_p to 2×10^4 in the experiments. The prior model $p(g)$ is formed using a multivariate Gaussian mixture model as

$$p(g) = \sum_{i=1}^{K} \alpha_i \mathcal{N}(g; \mu_i, \mathbf{\Sigma}_i), \qquad (28)$$

where $\mathcal{N}$ represents a normal distribution with mean μ and covariance matrix $\mathbf{\Sigma}$, and α_i is a weight factor. The prior model is obtained using the PCA coefficients of the inverse response functions in the DoRF by applying the cross-entropy method [25]. The number of normal distributions K is set to 5 in our experiments.

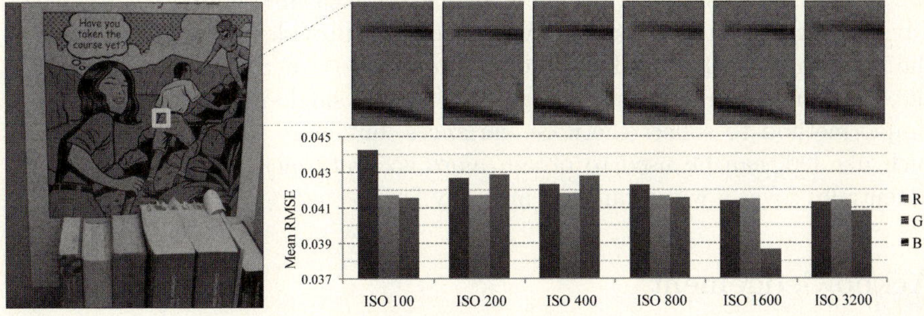

Fig. 6. Relationship between the noise level and mean RMSE of the estimates. Left image shows one of the photographed scenes. Top row shows magnification of a part of the image at different ISO levels. Bottom row shows the mean RMSE of RGB channels at each ISO gain level, and demonstrates that our estimation method is independent of noise levels.

Results. We used a *Canon EOS-20D* camera for the experiment. To obtain the ground truth, we used Mitsunaga and Nayar's method [4] using images taken with different exposures. Since our focus is on estimating the inverse response functions from the measured noise variances, we photographed a scene composed of relatively flat and uniformly colored surfaces, so that the noise variances can be easily obtained. The left image in Figure 6 shows one of two scenes used for the experiment. We photographed them five times each at six different camera gains (ISO 100 $\sim$ 3200). We manually selected 21 homogeneous image regions to obtain the noise variances as input. In total, we ran our estimation algorithm 60 times ($=$ 2 scenes $\times$ 5 shots $\times$ 6 ISO levels) for each RGB color channel.

Figure 6 summarizes the results of estimation at different ISO levels. The noise level increases with the ISO gain level, as shown by the cropped images on the top. The results indicate that the estimation is unaffected by the greater noise level. The mean RMSE is almost constant across the different ISO levels, which verifies that our method is not sensitive to the noise level.

5   Conclusions

In this paper, we have proposed the method for estimating a radiometric response function using noise variance, not noise distribution, as input. The relationship between the radiometric response function and noise variances in input and output domains is explicitly derived, and this result is used to develop the estimation algorithm. The experiments are performed for two different scenarios; one is with multiple shots of the same scene, and the other is only from a single image. These experiments quantitatively demonstrate the effectiveness of the proposed algorithm, especially its robustness against noise. With our method, either special equipment or images taken with multiple exposures are not necessary.

Limitations. It is better for our method that the measured noise variances cover a wide range of intensity levels. Wider coverage provides more information to the algorithm, so the problem becomes more constrained. This becomes an issue, particularly in the single-image case. In the single-image case, we used a simple method to collect the noise variances, but more sophisticated methods such like [20] can be used to obtain more accurate measurements that could potentially cover a wider range of intensity levels.

Acknowledgement

The authors would like to thank Dr. Bennett Wilburn for his useful feedback on this research.

References

1. Chang, Y.C., Reid, J.F.: Rgb calibration for color image analysis in machine vision. IEEE Trans. on Image Processing 5, 1414–1422 (1996)
2. Nayar, S.K., Mitsunaga, T.: High dynamic range imaging: Spatially varying pixel exposures. In: Proc. of Comp. Vis. and Patt. Recog. (CVPR), pp. 472–479 (2000)
3. Mann, S., Picard, R.: Being 'undigital' with digital cameras: Extending dynamic range by combining differently exposed pictures. In: Proc. of IS & T 48th Annual Conf., pp. 422–428 (1995)
4. Mitsunaga, T., Nayar, S.K.: Radiometric self-calibration. In: Proc. of Comp. Vis. and Patt. Recog. (CVPR), pp. 374–380 (1999)
5. Grossberg, M.D., Nayar, S.K.: What is the space of camera response functions? In: Proc. of Comp. Vis. and Patt. Recog. (CVPR), pp. 602–609 (2003)
6. Mann, S.: Comparametric equations with practical applications in quantigraphic image processing. IEEE Trans. on Image Processing 9, 1389–1406 (2000)
7. Debevec, P.E., Malik, J.: Recovering high dynamic range radiance maps from photographs. Proc. of ACM SIGGRAPH, 369–378 (1997)
8. Tsin, Y., Ramesh, V., Kanade, T.: Statistical calibration of ccd imaging process. In: Proc. of Int'l Conf. on Comp. Vis. (ICCV), pp. 480–487 (2001)
9. Pal, C., Szeliski, R., Uyttendale, M., Jojic, N.: Probability models for high dynamic range imaging. In: Proc. of Comp. Vis. and Patt. Recog. (CVPR), pp. 173–180 (2004)
10. Grossberg, M.D., Nayar, S.K.: What can be known about the radiometric response function from images? In: Heyden, A., Sparr, G., Nielsen, M., Johansen, P. (eds.) ECCV 2002. LNCS, vol. 2353, pp. 189–205. Springer, Heidelberg (2002)
11. Kim, S.J., Pollefeys, M.: Radiometric alignment of image sequences. In: Proc. of Comp. Vis. and Patt. Recog. (CVPR), pp. 645–651 (2004)
12. Lin, S., Gu, J., Yamazaki, S., Shum, H.Y.: Radiometric calibration from a single image. In: Proc. of Comp. Vis. and Patt. Recog. (CVPR), pp. 938–945 (2004)
13. Lin, S., Zhang, L.: Determining the radiometric response function from a single grayscale image. In: Proc. of Comp. Vis. and Patt. Recog. (CVPR), pp. 66–73 (2005)
14. Wilburn, B., Xu, H., Matsushita, Y.: Radiometric calibration using temporal irradiance mixtures. In: Proc. of Comp. Vis. and Patt. Recog. (CVPR) (2008)

15. Matsushita, Y., Lin, S.: Radiometric calibration from noise distributions. In: Proc. of Comp. Vis. and Patt. Recog. (CVPR) (2007)
16. Takamatsu, J., Matsushita, Y., Ikeuchi, K.: Estimating camera response functions using probabilistic intensity similarity. In: Proc. of Comp. Vis. and Patt. Recog. (CVPR) (2008)
17. Matsushita, Y., Lin, S.: A probabilistic intensity similarity measure based on noise distributions. In: Proc. of Comp. Vis. and Patt. Recog. (CVPR) (2007)
18. Janesick, J.R.: Photon Transfer. SPIE Press (2007)
19. Schechner, Y.Y., Nayar, S.K., Belhumeur, P.N.: Multiplexing for optimal lighting. IEEE Trans. on Patt. Anal. and Mach. Intell. 29, 1339–1354 (2007)
20. Liu, C., Freeman, W.T., Szeliski, R., Kang, S.B.: Noise estimation from a single image. In: Proc. of Comp. Vis. and Patt. Recog. (CVPR), pp. 901–908 (2006)
21. Healey, G.E., Kondepudy, R.: Radiometric ccd camera calibration and noise estimation. IEEE Trans. on Patt. Anal. and Mach. Intell. 16, 267–276 (1994)
22. Alter, F., Matsushita, Y., Tang, X.: An intensity similarity measure in low-light conditions. In: Leonardis, A., Bischof, H., Pinz, A. (eds.) ECCV 2006. LNCS, vol. 3954, pp. 267–280. Springer, Heidelberg (2006)
23. Consul, P.C.: Generalized Poisson Distributions: Properties and Applications. Marcel Dekker Inc., New York (1989)
24. Nelder, J.A., Mead, R.: A simplex method for function minimization. Computer Journal 7, 308–312 (1965)
25. Botev, Z., Kroese, D.: Global likelihood optimization via the cross-entropy method with an application to mixture models. In: Proc. of the 36th Conf. on Winter simul., pp. 529–535 (2004)

Solving Image Registration Problems Using Interior Point Methods

Camillo Jose Taylor and Arvind Bhusnurmath

GRASP Laboratory, University of Pennsylvania

Abstract. This paper describes a novel approach to recovering a parametric deformation that optimally registers one image to another. The method proceeds by constructing a global convex approximation to the match function which can be optimized using interior point methods. The paper also describes how one can exploit the structure of the resulting optimization problem to develop efficient and effective matching algorithms. Results obtained by applying the proposed scheme to a variety of images are presented.

1   Introduction

Image registration is a key problem in computer vision that shows up in a wide variety of applications such as image mosaicing, medical image analysis, face tracking, handwriting recognition, stereo matching and motion analysis. This paper considers the problem of recovering the parameters of a deformation that maps one image onto another. The main contribution is a novel approach to this problem wherein the image matching problem is reformulated as a Linear Program (LP) which can be solved using interior point methods. The paper also describes how one can exploit the special structure of the resulting LP to derive efficient implementations which can effectively solve problems involving hundreds of thousands of pixels and constraints.

One of the principal differences between the proposed approach and other approaches that have been developed [1,2,3] is that the scheme seeks to construct a *global* convex approximation to the matching function associated with the registration problem as opposed to constructing a *local* convex model around the current parameter estimate. The approach is intended for situations where the displacements between frames are large enough that local matches at the pixel level are likely to be ambiguous. For example, in the experiments we consider images that are 320 pixels on side where individual pixels may be displaced by up to 40 pixels along each dimension. The approximation procedure is designed to capture the uncertainties inherent in matching a given pixel to a wide swath of possible correspondents.

One common approach to solving image matching problems proceeds by extracting feature points in the two images, establishing correspondences between the frames, and then using a robust estimation procedure to recover the parameters of the transformation. This approach is exemplified by the work of

D. Forsyth, P. Torr, and A. Zisserman (Eds.): ECCV 2008, Part IV, LNCS 5305, pp. 638–651, 2008.

Mikolajczyk and Schmid [4] who proposed a very effective scheme for detecting and matching interest points under severe affine deformations. This approach works best when the interframe motion is close to affine since more complicated deformation models can distort the feature points beyond recognition. Further, it becomes increasingly difficult to apply robust estimation methods as the complexity of the deformation model increases since an ever increasing number of reliable point matches are required.

Belongie and Malik [5] proposed an elegant approach to matching shapes based on information derived from an analysis of contour features. This approach is similar to [4] in that it revolves around feature extraction and pointwise correspondence. The method described in this work is very different from these in that it avoids the notion of features altogether, instead it proceeds by constructing a matching function based on low level correlation volumes and allows every pixel in the image to constrain the match to the extent that it can.

Shekhovstov Kovtun and Hlavac [6] have developed a novel method for image registration that uses Sequential Tree-Reweighted Message passing to solve a linear program that approximates a discrete Markov Random Field optimization problem. Their work also seeks to construct a globally convex approximation to the underlying image matching problem but the approach taken to formulating and solving the optimization problem differ substantially from the method discussed in this paper.

Linear programming has been previously applied to motion estimation [7,8]. The work by Jiang $et\ al.$. [7] on matching feature points is similar to ours in that the data term associated with each feature is approximated by a convex combination of points on the lower convex hull of the match cost surface. However, their approach is formulated as an optimization over the interpolating coefficients associated with these convex hull points which is quite different from the approach described in this paper. Also their method uses the simplex method for solving the LP while the approach described in this paper employs an interior point solver which allows us to exploit the structure of the problem more effectively.

2   Image Registration Algorithm

The objective of the algorithm is to recover the deformation that maps a base image onto a target image. This deformation is modeled in the usual manner by introducing two scalar functions $D_x(x, y, \mathbf{p_x})$ and $D_y(x, y, \mathbf{p_y})$ which capture the displacement of a pixel at location (x, y) along the horizontal and vertical directions respectively [9,5,10]. Here $\mathbf{p_x}$ and $\mathbf{p_y}$ represent vectors of parameters that are used to model the deformation. Consider for example an affine deformation where the horizontal displacements are given by $D_x(x, y) = c_1 + c_2 x + c_2 y$, then $p_x = [c_1, c_2, c_3]$ would capture the parameters of this transformation. In the sequel we will restrict our consideration to models where the displacements can be written as a linear function of the parameters. That is, if we let D_x and

D_y represent vectors obtained by concatenating the displacements at all of the pixels then $D_x = C\mathbf{p_x}$ and $D_y = C\mathbf{p_y}$ for some matrix C. Here the columns of the matrix C constitute the basis vectors of the displacement field [9].

2.1   Formulating Image Matching as an LP

The problem of recovering the deformation that maps a given base image onto a given target image can be phrased as an optimization problem. For every pixel in the target image one can construct an objective function, e_{xy}, which captures how similar the target pixel is to its correspondent in the base image as a function of the displacement applied at that pixel.

Figure 1(a) shows an example of one such function for a particular pixel in one of the test images. This particular profile was constructed by computing the ℓ_2 difference between the RGB value of the target pixel and the RGB values of the pixels in the base image for various displacements up to ± 10 pixels in each direction.

Our goal then is to minimize an objective function $E(\mathbf{p_x}, \mathbf{p_y})$ which models how the discrepancy between the target and base images varies as a function of the deformation parameters, $\mathbf{p_x}$ and $\mathbf{p_y}$.

$$E(\mathbf{p_x}, \mathbf{p_y}) = \sum_{\mathbf{x}} \sum_{\mathbf{y}} \mathbf{e_{xy}}(\mathbf{D_x}(\mathbf{x}, \mathbf{y}, \mathbf{p_x}), \mathbf{D_y}(\mathbf{x}, \mathbf{y}, \mathbf{p_y})) \tag{1}$$

In general, since the component e_{xy} functions can have arbitrary form the landscape of the objective function $E(\mathbf{p_x}, \mathbf{p_y})$ may contain multiple local minima

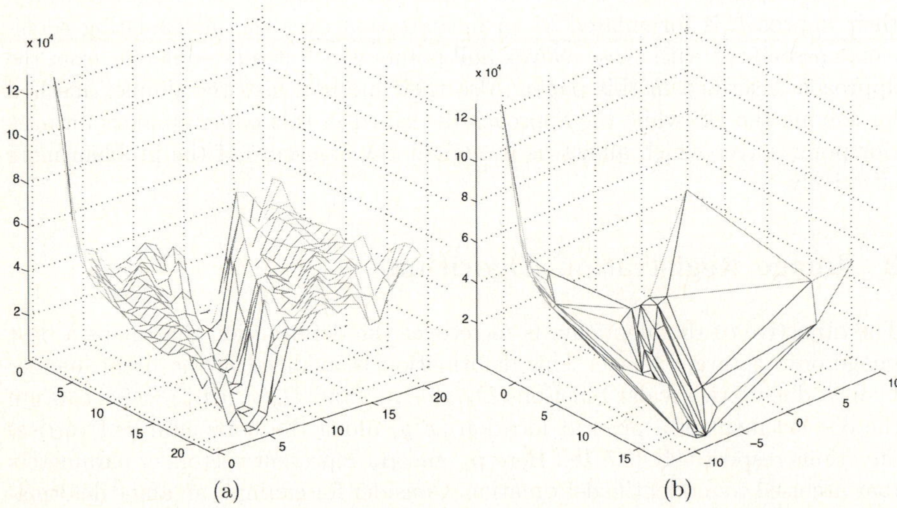

(a)   (b)

Fig. 1. (a) Error surface associated with particular pixel in the target image that encodes how compatible that pixel is with various x, y displacements (b) Piecewise planar convex approximation of the error surface

which can confound most standard optimization methods that proceed by constructing local approximations of the energy function.

The crux of the proposed approach is to introduce a convex approximation for the individual objective functions e'_{xy}. This leads directly to an approximation of the global objective function $E'(\mathbf{p_x}, \mathbf{p_y})$ which is convex in the deformation parameters. Once this has been done, one can recover estimates for the deformation parameters and, hence, the deformation by solving a convex optimization problem which is guaranteed to have a unique minimum.

The core of the approximation step is shown in Figure 1(b), here the original objective function is replaced by a convex lower bound which is constructed by considering the convex hull of the points that define the error surface. This convex lower hull is bounded below by a set of planar facets.

In order to capture this convex approximation in the objective function we introduce one auxiliary variable $z(x, y)$ for every pixel in the target image. There are a set of linear constraints associated with each of these variables which reflect the constraint that this value must lie above all of the planar facets that define the convex lower bound.

$$z(x, y) \geq a_x^i(x, y)D_x(x, y, p_x) + a_y^i(x, y)D_y(x, y, p_y) - b^i(x, y) \ \ \forall i \qquad (2)$$

Here the terms a_x^i, a_y^i and b^i denote the coefficients associated with each of the facets in the approximation.

The problem of minimizing the objective function $E'(\mathbf{p_x}, \mathbf{p_y})$ can now be rephrased as a linear program as follows:

$$\min_{p_x, p_y, z} \sum_x \sum_y z(x, y) \qquad (3)$$
$$\text{st } z(x, y) \geq a_x^i(x, y)D_x(x, y, p_x) + a_y^i D_y(x, y, p_y) - b^i(x, y) \ \forall x, y, i \qquad (4)$$

This can be written more compactly in matrix form as follows:

$$\min_{p_x, p_y, z} \mathbf{1^T z} \qquad (5)$$
$$\text{st } A_x D_x + A_y D_y - I_z \mathbf{z} \leq \mathbf{b}$$
$$D_x = C \mathbf{p_x}$$
$$D_y = C \mathbf{p_y}$$

where A_x and A_y are I_z are sparse matrices obtained by concatenating the constraints associated with all of the planar facets and z and b are vectors obtained by collecting the $z(x, y)$ and $b^i(x, y)$ variables respectively.

Note that the A_x, A_y and I_z matrices all have the same fill pattern and are structured as shown in equation 6, the non zero entries in the I_z matrix are all 1. In this equation M denotes the total number of pixels in the image and S_i refers to the number of planar facets associated with pixel i.

$$A = \begin{bmatrix} a_{11} & 0 & \cdots\cdots & 0 \\ a_{21} & 0 & \cdots\cdots & 0 \\ \vdots & 0 & \cdots\cdots & 0 \\ a_{S_1 1} & 0 & \cdots\cdots & 0 \\ 0 & a_{12} & 0 \cdots & 0 \\ 0 & a_{22} & 0 \cdots & 0 \\ 0 & \vdots & 0 \cdots & 0 \\ 0 & a_{S_2 2} & 0 \cdots & 0 \\ 0 & 0 & \ddots \ \ddots & 0 \\ 0 & \cdots\cdots & 0 & a_{1M} \\ 0 & \cdots\cdots & 0 & \vdots \\ 0 & \cdots\cdots & 0 & a_{S_M M} \end{bmatrix} \tag{6}$$

The linear program shown in Equation 5 can be augmented to include constraints on the displacement entries, D_x, D_y and the z values as shown in Equation 7. Here the vectors b_{lb} and b_{ub} capture the concatenated lower and upper bound constraints respectively. It would also be a simple matter to include bounding constraints on the parameter values at this stage. Alternatively one could easily add a convex regularization term to reflect a desire to minimize the bending energy associated with the deformation.

$$\min_{p_x, p_y, z} \mathbf{1^T z} \tag{7}$$

$$\begin{bmatrix} A_x & A_y & -I_z \\ & -I & \\ & I & \end{bmatrix} \begin{pmatrix} C & 0 & 0 \\ 0 & C & 0 \\ 0 & 0 & I \end{pmatrix} \begin{pmatrix} \mathbf{p_x} \\ \mathbf{p_y} \\ z \end{pmatrix} \leq \begin{pmatrix} b \\ b_{lb} \\ b_{ub} \end{pmatrix}$$

Note that the proposed approximation procedure increases the ambiguity associated with matching any individual pixel since the convex approximation is a lower bound which may significantly under estimate the cost associated with assigning a particular displacement to a pixel. What each pixel ends up contributing is a set of convex terms to the global objective function. The linear program effectively integrates the convex constraints from tens of thousands of pixels, constraints which are individually ambiguous but which collectively identify the optimal parameters. In this scheme each pixel contributes to constraining the deformation parameters to the extent that it is able. Pixels in homogenous regions may contribute very little to the global objective while well defined features may provide more stringent guidance. There is no need to explicitly identify distinguished features since local matching ambiguities are handled through the approximation process.

2.2   Solving the Matching LP

Once the image registration problem has been reformulated as the linear program given in equation 7 the barrier method [11] can be employed to solve the problem. In this method, a convex optimization problem of the following form

$$\min \quad f_0(x)$$
$$\text{st } f_i(x) \leq 0, \, i = 1, \ldots, m \tag{8}$$

is solved by minimizing $\phi(x, t) = t f_0(x) - \sum_{i=1}^{m} \log(-f_i(x))$ for increasing values of t until convergence. At each value of t a local step direction, the Newton step, needs to be computed. This involves the solution of a system of linear equations involving the Hessian and the gradient of $\phi(x, t)$. The Hessian can be computed from the following expression $H = [A^T \text{diag}(s^{-2})A]$ where $s = b - Ax$ and s^{-2} denotes the vector formed by inverting and squaring the elements of s. Similarly the gradient of the $\phi(x, t)$ can be computed from the following expression:

$$g = -tw - A^T s^{-1} \tag{9}$$

Then the Newton step is computed by solving

$$[A^T \text{diag}(s^{-2})A]\delta x = g \tag{10}$$

For our matching problem, it can be shown that this Newton step system can be written in the following form:

$$\begin{bmatrix} H_p & H_z^T \\ H_z & D_6 \end{bmatrix} \begin{pmatrix} \delta p \\ \delta z \end{pmatrix} = \begin{pmatrix} g_p \\ g_z \end{pmatrix} \tag{11}$$

where

$$H_p = \begin{bmatrix} (C^T D_1 C) & (C^T D_2 C) \\ (C^T D_2 C) & (C^T D_3 C) \end{bmatrix}$$
$$H_z = \begin{bmatrix} (D_4 C) & (D_5 C) \end{bmatrix} \tag{12}$$

δp and δz denote proposed changes in the deformation parameters and the z variables respectively and $D_1, D_2, D_3, D_4, D_5, D_6$ are all diagonal matrices.

At this point we observe that since the matrix D_6 is diagonal we can simplify the linear system in Equation 11 via the Schur complement. More specifically we can readily solve for δz in terms of δp as follows: $\delta z = D_6^{-1}(g_z - H_z \delta p)$. Substituting this expression back into the system yields the following expression where all of the auxiliary z variables have been elided.

$$(H_p - H_z^T D_6^{-1} H_z)\delta p = (g_p - H_z^T D_6^{-1} g_z) \tag{13}$$

This can be written more concisely as follows:

$$H_p' \delta p = g_p' \tag{14}$$

In short, computing the Newton Step boils down to solving the linear system in Equation 14. Note that the size of this system depends only on the dimension of the parameter vector, p. For example if one were interested in fitting an affine model which involves 6 parameters, 3 for $\mathbf{p_x}$ and 3 for $\mathbf{p_y}$, one would only end up solving a linear system with six degrees of freedom. Note that the computational complexity of this key step *does not depend on the number of pixels being considered or on the number of constraints that were used to construct the convex approximation.* This is extremely useful since typical matching problems will involve hundreds of thousands of pixels and a similar number of constraint equations. Even state of the art LP solvers like MOSEK and TOMLAB would have difficulty solving problems of this size.

2.3   Deformation Models

Experiments were carried out with two classes of deformation models. In the first class the displacements at each pixel are computed as a polynomial function of the image coordinates. For example for a second order model:

$$D_x(x,y) = c_1 + c_2 x + c_3 y + c_4 xy + c_5 x^2 + c_6 y^2 \tag{15}$$

These deformations are parameterized by the coefficients of the polynomials. The complexity of the model can be adjusted by varying the degree of the polynomial. A number of interesting deformation models can be represented in this manner include affine, bilinear, quadratic and bicubic.

Another class of models can be represented as a combination of an affine deformation and a radial basis function. That is

$$D_x(x,y) = c_1 + c_2 x + c_3 y + \sum_i k_i \phi(\|(x,y) - (x_i, y_i)\|) \tag{16}$$

Once again the deformation model is parameterized by the coefficients c_1, c_2, c_3, k_i and the function ϕ represents the interpolating kernel. Two different variants of this kernel were considered in the experiments, a Gaussian kernel, $\phi(r) = \exp(-(r/\sigma)^2)$ and a thin plate spline kernel $\phi(r) = r^2 \log r$. In the sequel we will refer to the former as the Gaussian deformation model and the latter as the Thin Plate Spline model.

In the experiments the coordinates of the kernel centers, (x_i, y_i) were evenly distributed in a grid over the the image. The complexity of the model can be varied by varying the number of kernel centers employed. All of the experiments that used this model employed 16 kernel centers arranged evenly over the image in a four by four grid.

2.4   Coarse to Fine

It is often advantageous to employ image registration algorithms in a coarse to fine manner [1]. In this mode of operation the base and target images are downsampled to a lower resolution and then matched. The deformation recovered from this stage is used to constrain the search for matches at finer scales. With this scheme, gross deformations are captured at the coarser scales while the finer scales fill in the details. It also serves to limit the computational effort required since one can effectively constrain the range of displacements that must be considered at the finer scales which limits the size of the correlation volumes that must be constructed.

In the experiments described in section 3.1 the images are first downsampled by a factor of 4 and then matched. The deformations computed at this scale inform the search for correspondences at the next finer scale which is downsampled from the originals by a factor of 2.

Note that as the approach proceeds to finer scales, the convex approximation is effectively being constructed over a smaller range of disparities which means that it increasingly approaches the actual error surface.

3   Experimental Results

Two different experiments were carried out to gauge the performance of the registration scheme quantitatively. In the first experiment each of the images in our data set was warped by a random deformation and the proposed scheme was employed to recover the parameters of this warp. The recovered deformation was compared to the known ground truth deformation to evaluate the accuracy of the method.

In the second set of experiments the registration scheme was applied to portions of the Middlebury stereo data set. The disparity results returned by the method were then compared to the ground truth disparities that are provided for these image pairs.

3.1   Synthetic Deformations

In these experiments the proposed scheme was applied to a number of different images. In each case, a random deformation was constructed using a particular motion model. The base image was warped by the deformation to produce the target image and the registration algorithm was employed to recover this deformation. In these experiments each of the base images was at most 320 pixels on side. The deformations that were applied were allowed to displace the pixels in the base image by up to $\pm 12.5\%$ of the image size. Hence for an image 320 pixels on side each pixel in the image can be displaced by ± 40 pixels along each dimension. The random deformations were specifically constructed to fully exercise the range of displacements so the maximum allowed displacement values are achieved in the applied warps. In order to recover such large deformations, the registration scheme is applied in a coarse to fine manner as described in Section 2.4.

The underlying matching functions associated with each of the pixels in the target image, e_{xy}, are constructed by simply comparing the pixel intensity in the target image to the pixels in a corresponding range in the base image. This is equivalent to conducting sum of squared difference (SSD) matching for each pixel using a 1×1 matching window.

In order to provide a quantitative evaluation of the scheme, the recovered deformation field, $(D_x(x, y), D_y(x, y))$ was compared to the known ground truth deformation field $(D_x^t(x, y), D_y^t(x, y))$ and the mean, median and maximum discrepancy between these two functions over the entire image was computed. The

Table 1. This table details the deformation applied to each of the images in the data set and reports the discrepancy between the deformation field returned by the method and the ground truth displacement field

Image	Deformation Model	no. of parameter	error in pixels		
			mean	median	max
Football	Gaussian	38	0.1524	0.1306	0.5737
Hurricane	Gaussian	38	0.1573	0.1262	0.7404
Spine	Affine	6	0.1468	0.1314	0.4736
Peppers	Gaussian	38	0.1090	0.0882	0.7964
Cells	Thin Plate Spine	38	0.1257	0.1119	0.8500
Brain	Gaussian	38	0.1190	0.0920	0.8210
Kanji	third degree polynomial	20	0.1714	0.0950	2.5799
Aerial	bilinear	8	0.0693	0.0620	0.2000
Face1	Gaussian	38	0.1077	0.0788	0.6004
Face2	Gaussian	38	0.5487	0.3095	4.6354

results are tabulated in Table 1. This table also indicates what type of deformation model was applied to each of the images along with the total number of parameters required by that model.

Note that in every case the deformed result returned by the procedure is almost indistinguishable from the given target. More importantly, the deformation fields returned by the procedure are consistently within a fraction of a pixel of the ground truth values. The unoptimized Matlab implementation of the matching procedure takes approximately 5 minutes to proceed through all three scales and produce the final deformation field for a given image pair.

3.2   Stereo Data Set

The image registration scheme was applied to regions of the image pairs taken from the Middlebury stereo data set. This data set was chosen because it included ground truth data which allows us to quantitatively evaluate the deformation results returned by the registration scheme. Here the vertical displacement between the two images is zero and the horizontal displacement field $D_x(x, y)$ is modeled as an affine function.

The correlation volume was computed using sum of squared difference matching with a five by five correlation window. For the teddy image, the correlation volume was constructed by considering displacements between 12 and 53 pixels while for the venus image the displacement range was 3 to 20 pixels. In this case, the convex lower bound approximations to the individual score functions degenerates to a piecewise linear profile along the horizontal dimension.

In each of the images two rectangular regions were delineated manually and an affine displacement model was fit to the pixels within those regions using the proposed method.

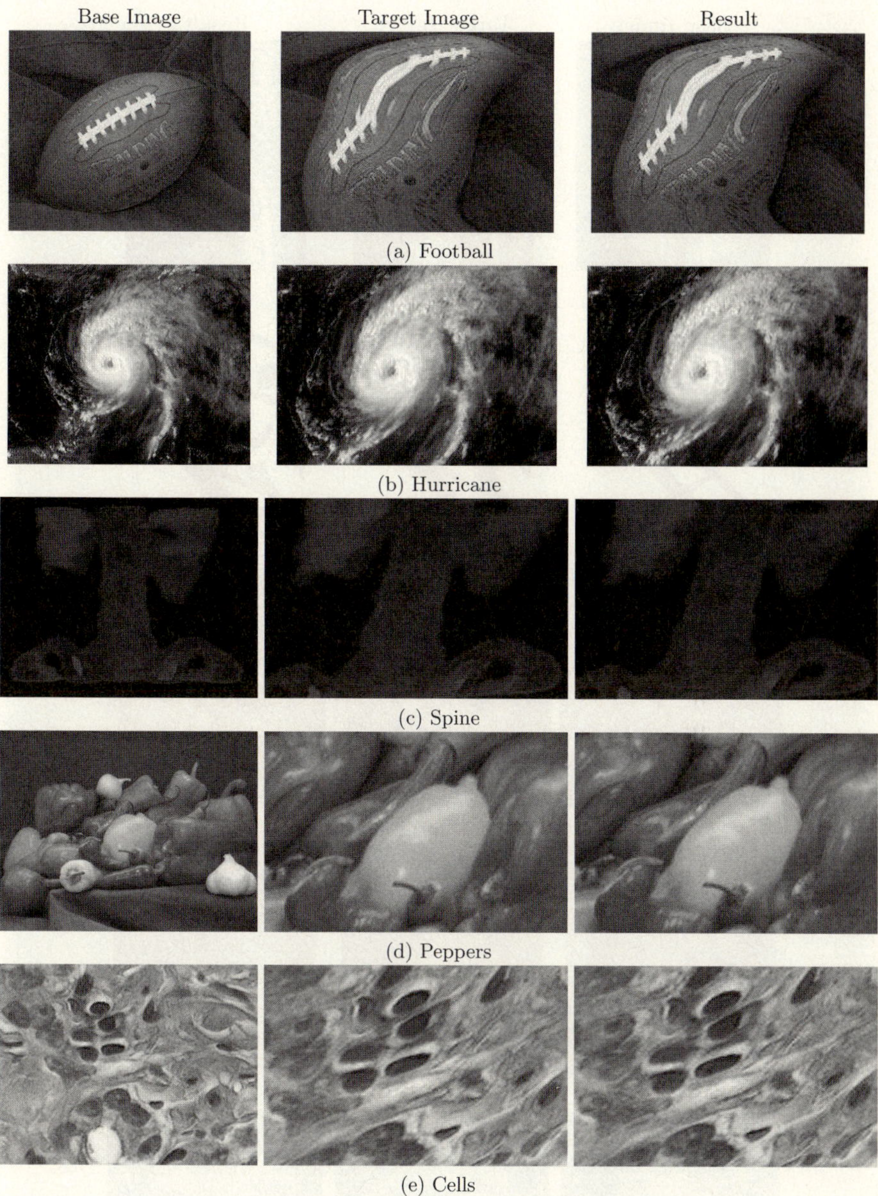

Fig. 2. Results obtained by applying the proposed method to actual image pairs. The first two columns correspond to the input base and target images respectively while the last column corresponds to the result produced by the registration scheme.

The first column of Figure 4 shows the left image in the pair, the second column shows what would be obtained if one used the raw SSD stereo results and the final column shows the ground truth disparities.

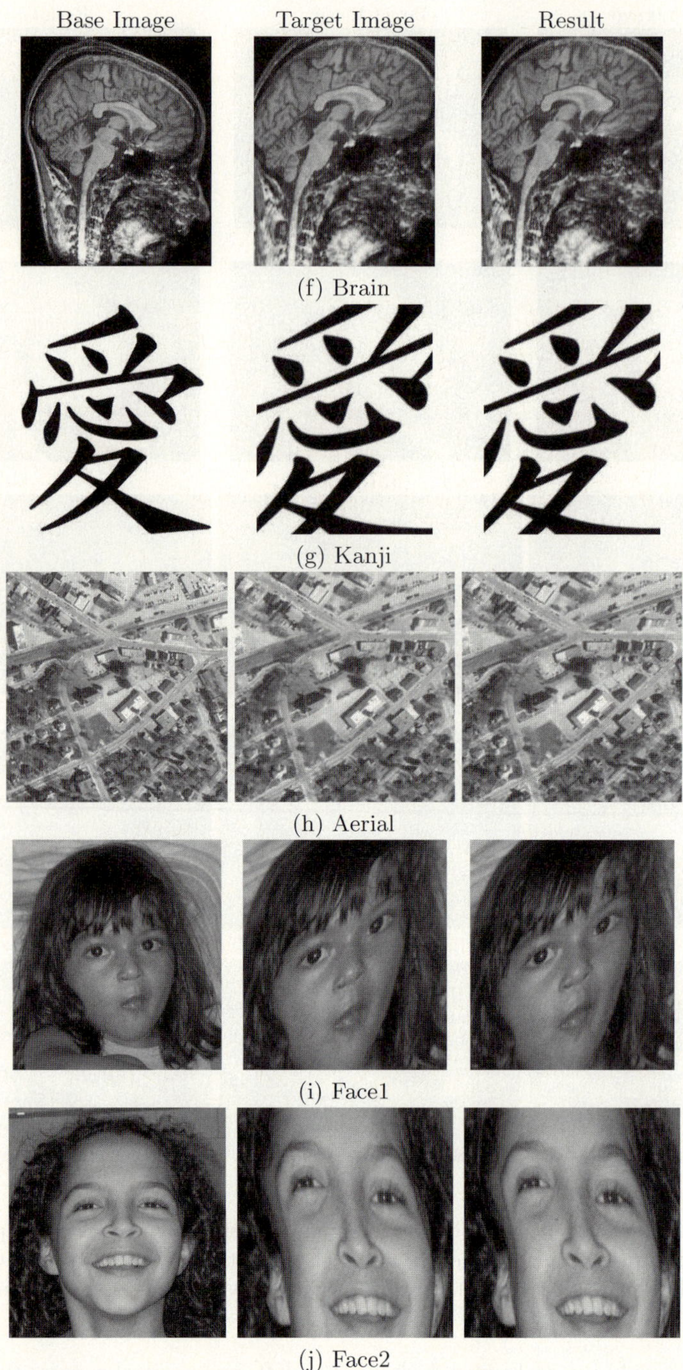

Fig. 3. More Registration Results

Left Image   SSD Disparity Solution   Ground Truth Disparity

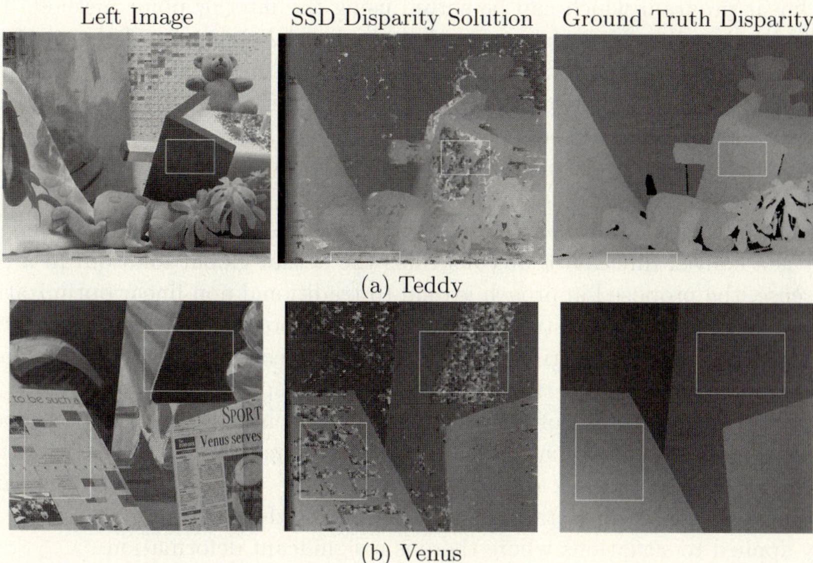

(a) Teddy

(b) Venus

Fig. 4. The proposed image registration scheme was applied to the delineated regions in the Middlebury Stereo Data Set. The first column shows the left image, the second column the raw results of the SSD correlation matching and the last column the ground truth disparity.

Table 2. This table reports the discrepancy between the affine deformation field returned by the method and the ground truth disparities within each region

		error in pixels	
Image	Region	mean	median
teddy	bird house roof	0.2558	0.2245
teddy	foreground	0.9273	0.8059
venus	left region	0.0317	0.0313
venus	right region	0.0344	0.0317

The selected rectangles are overlaid on each of the images. These regions were specifically chosen in areas where there was significant ambiguity in the raw correlation scores to demonstrate that the method was capable of correctly integrating ambiguous data. Table 2 summarizes the results of the fitting procedure. The reconstructed disparity fields within the regions were compared to the ground truth disparities and the mean and median discrepancy between these two fields is computed over all of the pixels within the region.

4 Conclusion

This paper has presented a novel approach to tackling the image registration problem wherein the original image matching objective function is approximated by a linear program which can be solved using the interior point method. The paper also describes how one can exploit the special structure of the resulting linear program to develop efficient algorithms. In fact the key step in the resulting resulting procedure only involves inverting a symmetric matrix whose dimension reflects the complexity of the model being recovered.

While the convex approximation procedure typically increases the amount of ambiguity associated with any individual pixels, the optimization procedure effectively aggregates information from hundreds of thousands of pixels so the net result is a convex function that constrains the actual global solution. In a certain sense, the proposed approach is dual to traditional non-linear optimization schemes which seek to construct a *local* convex approximation to the objective function. The method described in this work proceeds by constructing a *global* convex approximation over the specified range of displacements.

A significant advantage of the approach is that once the deformation model and displacement bounds have been selected, the method is insensitive to initialization since the convex optimization procedure will converge to the same solution regardless of the start point. This means that the method can be directly applied to situations where there is a significant deformation.

The method does not require any special feature detection or contour extraction procedure. In fact all of the correlation volumes used in the experiments were computed using nothing more than pointwise pixel comparisons. Since the method does not hinge on the details of the scoring function more sophisticated variants could be employed as warranted. The results indicate the method produces accurate results on a wide range of image types and can recover fairly large deformations.

References

1. Bajcsy, R., Kovacic, S.: Multiresolution elastic matching. Computer Vision, Graphics and Image Processing 46(1), 1–21 (1989)
2. Cootes, T., Edwards, G., Taylor, C.: Active appearance models. IEEE Transactions on Pattern Analysis and Machine Intelligence 23(6), 681–685 (2001)
3. Baker, S., Matthews, I.: Equivalence and efficiency of image alignment algorithms. In: IEEE Conference on Computer Vision and Pattern Recognition, pp. 1090–1097 (2001)
4. Mikolajczyk, K., Schmid, C.: Scale and affine invariant interest point detectors. International Journal of Computer Vision 60(1), 63–86 (2004)
5. Belongie, S., Malik, J.: Shape matching and object recognition using shape contexts. IEEE Transactions on Pattern Analysis and Machine Intelligence 24(24), 509 (2002)
6. Shekhovstov, A., Kovtun, I., Hlavac, V.: Efficient mrf deformation model for non-rigid image matching. In: IEEE Conference on Computer Vision and Pattern Recognition (2007)
7. Jiang, H., Drew, M., Li, Z.N.: Matching by linear programming and successive convexification. PAMI 29(6) (2007)
8. Ben-Ezra, M., Peleg, S., Werman, M.: Real-time motion analysis with linear programming. In: ICCV (1999)

9. Friston, K.J., Ashburner, J., Frith, C.D., Poline, J.B., Heather, J.D., Frackowiak, R.S.J.: Spatial registration and normalization of images. Human Brain Mapping 2, 165–189 (1995)
10. Modersitzki, J.: Numerical Methods for Image Registration. Oxford University Press, Oxford (2004)
11. Boyd, S., VandenBerghe, L.: Convex Optimization. Cambridge University Press, Cambridge (2004)

3D Face Model Fitting for Recognition

Frank B. ter Haar and Remco C. Veltkamp

Department of Information and Computing Sciences, Utrecht University, the Netherlands

Abstract. This paper presents an automatic efficient method to fit a statistical deformation model of the human face to 3D scan data. In a global to local fitting scheme, the shape parameters of this model are optimized such that the produced instance of the model accurately fits the 3D scan data of the input face. To increase the expressiveness of the model and to produce a tighter fit of the model, our method fits a set of predefined face components and blends these components afterwards. Quantitative evaluation shows an improvement of the fitting results when multiple components are used instead of one. Compared to existing methods, our fully automatic method achieves a higher accuracy of the fitting results. The accurately generated face instances are manifold meshes without noise and holes, and can be effectively used for 3D face recognition: We achieve 97.5% correct identification for 876 queries in the UND face set with 3D faces. Our results show that contour curve based face matching outperforms landmark based face matching.

1   Introduction

The use of 3D scan data for face recognition purposes has become a popular research area. With high recognition rates reported for several large sets of 3D face scans, the 3D shape information of the face proved to be a useful contribution to person identification. The major advantage of 3D scan data over 2D color data, is that variations in scaling and illumination have less influence on the appearance of the acquired face data. However, scan data suffers from noise and missing data due to self-occlusion. To deal with these problems, 3D face recognition methods should be invariant to noise and missing data, or the noise has to be removed and the holes interpolated. Alternatively, data could be captured from multiple sides, but this requires complex data acquisition. In this work we propose a method that produces an accurate fit of a statistical 3D shape model of the face to the scan data. The 3D geometry of the generated face instances, which are without noise and holes, are effectively used for 3D face recognition.

Related work. The task to recognize 3D faces has been approached with many different techniques as described in surveys of Bowyer et al. [1] and Scheenstra et al. [2]. Several of these 3D face recognition techniques are based on 3D geodesic surface information, such as the methods of Bronstein et al. [3] and Berretti et al. [4]. The geodesic distance between two points on a surface is the length of the shortest path between two points. To compute accurate 3D geodesic distances for face recognition purposes, a 3D face without noise and without holes is desired. Since this is typically not the case with laser range scans, the noise has to be removed and the holes in the 3D surface interpolated. However, the success of basic noise removal techniques, such as Laplacian smoothing is

D. Forsyth, P. Torr, and A. Zisserman (Eds.): ECCV 2008, Part IV, LNCS 5305, pp. 652–664, 2008.

very much dependent on the resolution of the scan data. Straightforward techniques to interpolate holes using curvature information or flat triangles often fail in case of complex holes, as pointed out in [5]. The use of a deformation model to approximate new scan data and interpolate missing data is a gentle way to regulate flaws in scan data.

A well known *statistical deformation model* specifically designed for surface meshes of 3D faces, is the 3D morphable face model of Blanz and Vetter [6]. This statistical model was built from 3D face scans with dense correspondences to which Principal Component Analysis (PCA) was applied. In their early work, Blanz and Vetter [6] fit this 3D morphable face model to 2D color images and cylindrical depth images from the $Cyberware^{TM}$ scanner. In each iteration of their fitting procedure, the model parameters are adjusted to obtain a new 3D face instance, which is projected to 2D cylindrical image space allowing the comparison of its color values (or depth values) to the input image. The parameters are optimized using a stochastic Newton algorithm. More recently, Blanz et al. [7] proposed a method to fit their 3D morphable face model to more common textured depth images. The fitting process is similar to their previous algorithm, but now the cost function is minimized using both color and depth values after the projection of the 3D model to 2D cylindrical image space. To initialize their fitting process, they manually select seven corresponding face features on their model and in the depth scan. A morphable model of expressions was proposed by Lu et al. [8]. Starting from an existing neutral scan, they use their expression model to adjust the vertices in a small region around the nose to obtain a better fit of the neutral scan to a scan with a certain expression.

Non-statistical deformation models were proposed as well. Huang et al. [9] proposed a global to local deformation framework to deform a shape with an arbitrary dimension (2D, 3D or higher) to a new shape of the same class. They show their framework's applicability to 3D faces, for which they deform an incomplete source face to a target face. Kakadiaris et al. [10] deform an annotated face model to scan data. Their deformation is driven by triangles of the scan data attracting the vertices of the model. The deformation is restrained by a stiffness, mass and damping matrix, which control the resistance, velocity and acceleration of the model's vertices. The advantage of such deformable faces is that they are not limited to the statistical changes of the input shapes, so the deformation has less restrictions. However, this is also their disadvantage, because these models cannot rely on statistics in case of noise and missing data.

Contribution. First, we propose a fully automatic algorithm to efficiently optimize the parameters of the morphable face model, creating a new face instance that accurately fits the 3D geometry of the scan data. Unlike other methods, ours needs no manual initialization, so that batch processing of large data sets has become feasible. Second, we quantitatively evaluate our fitted face models and show that the use of multiple components improves the fitting process. Thirdly, we show that our model fitting method is more accurate than existing methods. Fourthly, we show that the accurately generated face instances can be effectively used for 3D face recognition.

2   Morphable Face Model

In this work we fit the morphable face model of the USF Human ID 3D Database [11] to 3D scan data to obtain a clean model of the face scan, that we use to identify

3D faces. This statistical point distribution model (PDM) was built from 100 cylindrical 3D face scans with neutral expressions from which n=75,972 correspondences were selected using an optic flow algorithm. Each face shape S_i was described using the set of correspondences $S = (x_1, y_1, z_1, ..., x_n, y_n, z_n)^T \in \Re^{3n}$ and a mean face $\bar{S}$ was determined. PCA was applied to these 100 sets S_i to obtain the m=99 most important eigenvectors of the PDM. The mean face $\bar{S}$, the eigenvectors $s_i = (\Delta x_1, \Delta y_1, \Delta z_1, ..., \Delta x_n, \Delta y_n, \Delta z_n)^T$, the eigenvalues λ_i ($\sigma_i^2 = \lambda_i$) and weights w_i are used to model new face instances according to $S_{inst} = \bar{S} + \sum_{i=1}^{m} w_i \sigma_i s_i$. Weight w_i represents the number of standard deviations a face instance morphs along eigenvector ev_i. Since the connectivity of the n correspondences in the PDM is known, each instance is a triangular mesh with proper topology and without holes.

3 Face Scans

We fit the morphable face model to the 3D frontal face scans of the University of Notre Dame (UND) Biometrics Database [12]. This set contains 953 range scans and a corresponding 2D color texture from 277 different subjects. All except ten scans were used in the Face Recognition Grand Challenge (FRGC v.1). Because the currently used morphable model is based on faces with neutral expressions only, it makes no sense to use collections containing many non-neutral scans such as the FRGC v.2. Nevertheless, our proposed method performs well for the small expression variations of the UND set. Throughout this work, we have only used the 3D scan data and neglected the available 2D color information.

We aim at 3D face recognition, so we need to segment the face from each scan. For that, we employ our pose normalization method [13] that normalizes the pose of the face and localizes the tip of the nose. Before pose normalization was applied to the UND scan data, we applied a few basic preprocessing steps to the scan data: the 2D depth images were converted to triangle meshes by connecting the adjacent depth samples with triangles, slender triangles and singularities were removed, and only considerably large components were retained.

The cleaned surface meshes were randomly sampled, such that every $\approx$2.0 mm^2 of the surface is approximately sampled once. The pose normalization method uses these locations in combination with their surface normal as initial placements for a nose tip template. To locations where this template fits well, a second template of global face

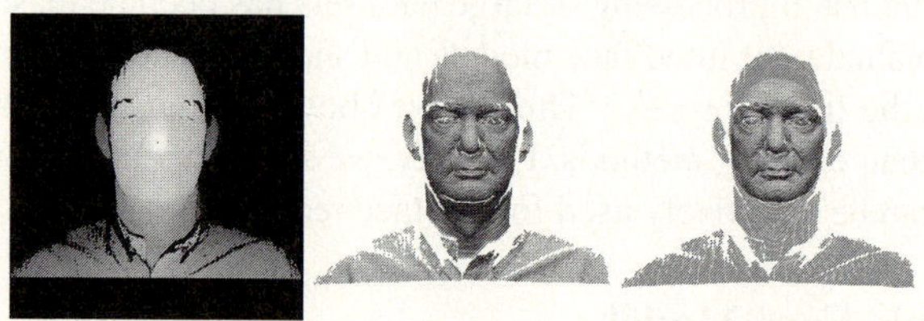

Fig. 1. Face segmentation. The depth image (left) is converted to a surface mesh (middle). The surface mesh is cleaned, the tip of the nose is detected and the face segmented (right, in pink).

features is fitted to normalize the face's pose and to select the tip of the nose. The face is then segmented by removing the scan data with a Euclidean distance larger than 100 mm from the nose tip. These face segmentation steps are visualized in Fig. 1.

4   Face Model Fitting

In general, 3D range scans suffer from noise, outliers, and missing data and their resolution may vary. The problem with single face scans, such as the UND face scans, is that large areas of the face are missing, which cannot be filled by simple hole filling techniques. When the morphable face model is fitted to a 3D face scan, a model is obtained that has no holes, has a proper topology, and has an assured resolution. By adjusting the $m=99$ weights w_i for the eigenvectors, the morphable model creates a new face instance. To fit the morphable model to 3D scan data, we need to find the optimal set of m weights w_i. In this section, we describe a fully automatic method that efficiently finds a proper model of the face scan in the m-dimensional space.

4.1   Distance Measure

To evaluate if an instance of the morphable face model is a good approximation of the 3D face scan, we use the Root Mean Square (RMS) distance of the instance's vertices to their closest points in the face scan. For each vertex point (p) from the instance (M_1), we find the vertex point (p') in the scan data (M_2) with the minimal Euclidean distance

$$e_{min}(p, M_2) = min_{p' \in M_2} d(p, p') \ , \tag{1}$$

using a kD-tree. The RMS distance is then measured between M_1 and M_2 as:

$$d_{rms}(M_1, M_2) = \sqrt{\frac{1}{n} \sum_{i=1}^{n} e_{min}(p_i, M_2)^2} \ , \tag{2}$$

using n vertices from M_1. Closest point pairs (p,p') for which p' belongs to the boundary of the face scan, are not used in the distance measure.

The morphable face model has $n=75,972$ vertices that cover the face, neck and ear regions and its resolution in the upward direction is three times higher than in its sideways direction. Because the running time of our measure is dependent on the number of vertices, we recreated the morphable face model such that it contains only the face (data within 110 mm from the tip of the nose) and not the neck and ears. To obtain a more uniform resolution of for the model, we reduced the upward resolution to one third of the original model. The number of vertices of this adjusted morphable mean face is now $n=12,964$ vertices, a sample every ≈ 2.6 mm^2 of the face area.

4.2   Iterative Face Fitting

With the defined distance measure for an instance of our compressed morphable face model, the m-dimensional space can be searched for the optimal instance. The fitting is done by choosing a set of m weights w_i, adjusting the position of the instance's vertices according to $S_{inst} = \bar{S} + \sum_{i=1}^{m} w_i \sigma_i s_i$, measuring the RMS-distance of the new instance to the scan data, selecting new weights and continue until the optimal instance is found. Knowing that each instance is evaluated using a large number of vertices, an exhaustive search for the optimal set of m weights is too computationally expensive.

A common method to solve large combinatorial optimization problems is *simulated annealing* (SA) [14]. In our case, random m-dimensional vectors could be generated which represent different *morphs* for a current face instance. A morph that brings the current instance closer to the scan data is accepted (downhill), and otherwise it is either accepted (uphill to avoid local minima) or rejected with a certain probability. In each iteration, the length of the m-dimensional morph vector can be reduced as implementation of the "temperature" scheme. The problem with such a naive SA approach is that most random m-dimensional morph vectors are uphill. In particular close to the optimal solution, a morph vector is often rejected, which makes it hard to produce an accurate fit. Besides this inefficiency, it doesn't take the eigensystem of the morphable face model into account.

Instead, we propose an *iterative downhill walk* along the consecutive eigenvectors from a current instance towards the optimal solution. Starting from the mean face $\bar{S}$ ($\forall_{i=1}^{m} w_i = 0$), try new values for w_1 and keep the best fit, then try new values for w_2 and keep the best fit, and continue until the face is morphed downhill along all m eigenvectors. Then iterate this process with a *smaller search space* for w_i. The advantage in computation costs of this method is twofold. First, the discrete number of morphs in the selected search space directly defines the number of rejected morphs per iteration. Second, optimizing one w_i at a time means only a one (instead of m) dimensional modification of the current face instance $S_{new} = S_{prev} + (w_{new} - w_{prev})\sigma_i s_i$.

Because the first eigenvectors induce the fitting of global face properties (e.g. face height and width) and the last eigenvectors change local face properties (e.g. nose length and width), each iteration follows a global to local fitting scheme (see Fig. 2). To avoid local minima, two strategies are applied. (1) The selected w_i in one iteration is not evaluated in the next iteration, forcing a new (similar) path through the m-dimensional space. (2) The vertices of the morphable face model are uniformly divided over three

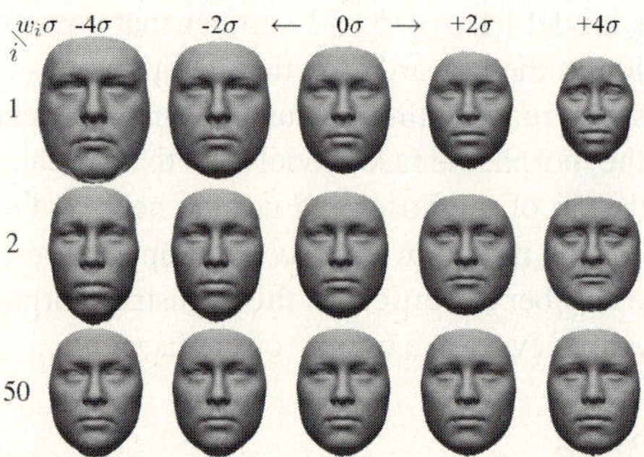

Fig. 2. Face morphing along eigenvectors starting from the mean face (center column). Different weights for the principal eigenvectors (e.g. i=1,2) changes the global face shape. For latter eigenvectors the shape changes locally (e.g. i=50).

sets and in each iteration a different set is modified and evaluated. Only in the first and last iteration all vertices are evaluated. Notice that this also reduces the number of vertices to fit and thus the computation costs.

The fitting process starts with the mean face and morphs in place towards the scan data, which means that the scan data should be well aligned to the mean face. To do so, the segmented and pose normalized face is placed with its center of mass on the center of mass of the mean face, and finely aligned using the Iterative Closest Point (ICP) algorithm [15]. The ICP algorithm iteratively minimizes the RMS distance between vertices. To further improve the effectiveness of the fitting process, our approach is applied in a *coarse fitting* and a *fine fitting* step.

4.3   Coarse Fitting

The more the face scan differs from the mean face $\bar{S}$, the less reliable the initial alignment of the scan data to the mean face is. Therefore, the mean face is coarsely fitted to the scan data by adjusting the weights of the first ten principal eigenvectors (m_{max}=10) in a single iteration (k_{max}=1) with 10 different values for w_{new}=[-1.35, -1.05, ..., 1.05, 1.35] as in Algorithm `ModelFitting`($\bar{S}$,*scan*). Fitting the model by optimizing the first ten eigenvectors results in the face instance S_{coarse}, with global face properties similar to those of the scan data. After that, the alignment of the scan to S_{coarse} is further improved with the ICP algorithm.

4.4   Fine Fitting

Starting with the improved alignment, we again fit the model to the scan data. This time the model fitting algorithm is applied using all eigenvectors (m_{max}=m) and multiple iterations (k_{max}=9). In the first iteration of Algorithm `ModelFitting`($\bar{S}$,*scan*), 10 new weight values w_{new} are tried for each eigenvector, to cover a large range of facial variety. The best w_{new} for every sequential eigenvector is used to morph the instance closer to the face scan. In the following k_{max}-1 iterations only four new weight values w_{new} are tried around w_i with a range w_{range} equal to w_{incr} of the previous iteration. By iteratively searching for a better w_i in a smaller range, the weights are continuously optimized. Local minima are avoided as described in Sect. 4.2. The range of the first iteration and the number of new weights tried in each next iteration were empirically selected as good settings.

4.5   Multiple Components

Knowing that the morphable model was generated from 100 3D face scans, an increase of its expressiveness is most likely necessary to cover a large population. To increase the expressiveness, also Blanz and Vetter [6] proposed to independently fit different components of the face, namely the eyes, nose, mouth, and the surrounding region. Because each component is defined by its own linear combination of shape parameters, a larger variety of faces can be generated with the same model. The fine fitting scheme from the previous section was developed to be applicable to either the morphable face model as a whole, but also to individual components of this model.

Algorithm 1. ModelFitting(S_{inst} to $scan$)

1: $w_{range} = 1.5$, $w_{incr} = 0.3$
2: **for** $k \leftarrow 1$ **to** k_{max} **do**
3:     select vertices (uniform subset of component)
4:     **for** $i \leftarrow 1$ **to** m_{max} **do**
5:         $w_{min} = w_i - w_{range} + \frac{1}{2} w_{incr}$
6:         $w_{max} = w_i + w_{range} - \frac{1}{2} w_{incr}$
7:         **for** $w_{new} \leftarrow w_{min}$ **to** w_{max} **do**
8:             morph S_{inst} with w_{new}
9:             $d_{rms}(S_{inst}, scan)$ smaller $\rightarrow$ keep w_{new}
10:             undo morph
11:             $w_{new} = w_{new} + w_{incr}$
12:         morph S_{inst} with $w_i \leftarrow$ best w_{new}
13:     $w_{range} = w_{incr}$, $w_{incr} = \frac{1}{2} w_{incr}$
14: **return** S_{inst}

Component selection. All face instances generated with the morphable model are assumed to be in correspondence, so a component is simply a subset of vertices in the mean shape $\bar{S}$ (or any other instance). We define seven components in our adjusted morphable face model (see Fig. 3). Starting with the improved alignment, we can individually fit each of the components to the scan data using the fine fitting scheme, obtaining a higher precision of the fitting process (as shown in Sect. 6.1). Individual components for the left and right eyes and cheeks were selected, so that our method applies to non-symmetric faces as well. The use of multiple components has no influence on the fitting time, because the total number of vertices remains the same and only the selected vertices are modified and evaluated.

Component blending. A drawback of fitting each component separately is that inconsistencies may appear at the borders of the components. During the fine fitting, the border triangles of two components may start to intersect, move apart, or move across (Fig. 3). The connectivity of the complete mesh remains the same, so two components moving apart remain connected with elongated triangles at their borders. We solve these inconsistencies by means of a post-processing step, as described in more detail below.

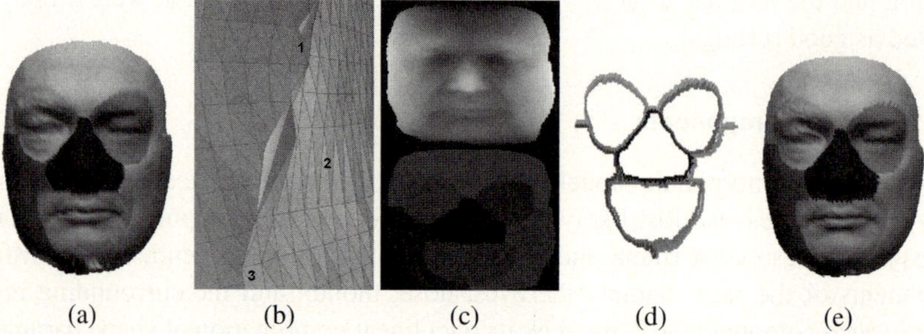

(a)   (b)   (c)   (d)   (e)

Fig. 3. Multiple components (a) may intersect (b1), move apart (b2), or move across (b3). Simulating a cylindrical scan (c) and smoothing the new border vertices (d) solves these problems (e).

Knowing that the morphable face model is created from cylindrical range scans and that the position of the face instance doesn't change, it is easy to synthetically rescan the generated face instance. Each triangle of the generated face instance S_{fine} is assigned to a component (Fig. 3a). A cylindrical scanner is simulated, obtaining a cylindrical depth image $d(\theta, y)$ with a surface sample for angle θ, height y with radius distance d from the y-axis through the center of mass of $\bar{S}$ (Fig. 3c). Basically, each sample is the intersection point of a horizontal ray with its closest triangle, so we still know to which component it belongs. The cylindrical depth image is converted to a 3D triangle mesh by connecting the adjacent samples and projecting the cylindrical coordinates to 3D. This new mesh S'_{fine} has a guaranteed resolution depending on the step sizes of θ and y, and the sampling solves the problem of intersecting and stretching triangles. However, ridges may still appear at borders where components moved across. Therefore, Laplacian smoothing is applied to the border vertices and their neighbors (Fig. 3d). Finally, data further then 110 mm from the tip of the nose is removed to have the final model S_{final} (Fig. 3e) correspond to the segmented face. In Sect. 6.1, we evaluate both the single and multiple component fits.

5   Face Recognition

Our model fitting algorithm provides a clean model of a 3D face scan. In this section, we use this newly created 3D geometry as input for two 3D face matching methods. One compares facial landmarks and the other compares extracted contour curves.

Landmarks. All vertices of two different instances of the morphable model are assumed to have a one-to-one correspondence. Assuming that facial landmarks such as the tip of the nose, corners of the eyes, etc. are morphed towards the correct position in the scan data, we can use them to match two 3D faces. So, we assigned 15 anthropomorphic landmarks to the mean face and obtain their new locations by fitting the model to the scan data. To match two faces $\mathbf{A}$ and $\mathbf{B}$ we use the sets of $c=15$ corresponding landmark locations:

$$d_{corr}(\mathbf{A}, \mathbf{B}) = \sum_{i=1}^{c} d_p(a_i, b_i) \ , \tag{3}$$

where distance d_p between two correspondences a_i and b_i is the squared difference in Euclidean distance e to the nose tip landmark p_{nt}:

$$d_p(a_i, b_i) = (e(a_i, p_{nt}) - e(b_i, p_{nt}))^2 \ . \tag{4}$$

Contour curves. Another approach is to fit the model to scans $\mathbf{A}$ and $\mathbf{B}$ and use the new clean geometry as input for a more complex 3D face recognition method. To perform 3D face recognition, we extract from each fitted face instance three 3D facial contour curves, and match only these curves to find similar faces. The three curves were extracted and matched as described by ter Haar and Veltkamp [13].

In more detail, after pose normalization and the alignment of the face scan to both $\bar{S}$ and S_{coarse}, a correct pose of the face scan is assumed and thus a correct pose of the final face instance S_{final}. Starting from the nose tip landmark p_{nt}, 3D profile curves can

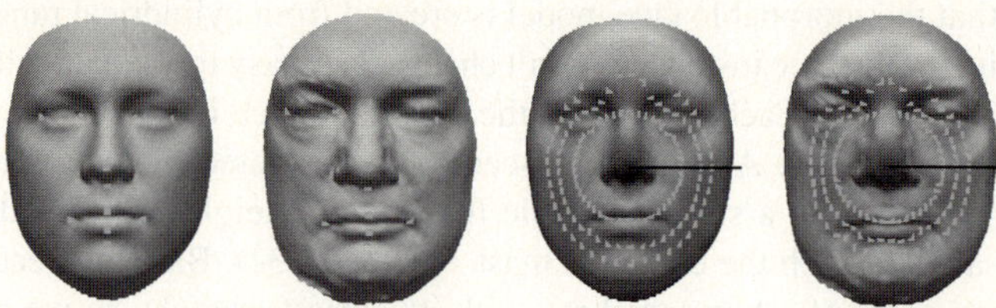

Fig. 4. The similarity of two 3D faces is determined using one-to-one correspondences, with on the left 15 corresponding landmarks and on the right 135 corresponding contour samples. The optimal XY-, C-, and G-contour curves (inner to outer) were extracted, for which the G-contour uses the (colored) geodesic distances. The line shown in black is one of the N_p profiles.

be extracted by walking the 3D surface in different directions (radii in the XY-plane). Samples along a profile from one face should correspond to samples along the same profile on another face. In case two faces are identical, these samples have the same Euclidean distance to the tip of the nose p_{nt}. For different faces, these samples cause a dissimilarity. The 3D face matching algorithm extracts N_p=45 profiles curves and extracts from each profile curve:

- One XY-sample, the location where the distance $\sqrt{(x^2 + y^2)}$ to p_{nt} equals r.
- One C-sample, the location where the curve length of the profile to p_{nt} equals r.
- One G-sample, the location on the profile where the length of the shortest geodesic path over the entire surface to p_{nt} equals r.

The shortest geodesic paths were computed using the fast marching method [16]. The combination of N_p=45 XY-samples at the same distance r builds a XY-contour, similarly a C-contour and a G-contour are constructed. Based on a training set of morphable face instances, the curves that were found most distinctive were selected, namely the XY-contour at r=34 mm, the C-contour at r=68 mm, and the G-contour at r=77 mm. The information of each 3D face instance is now reduced to a set of 135 ($3 \times N_p$) 3D sample points, with one-to-one correspondence to the same set of 135 samples in a different face instance. The similarity of faces **A** and **B** is again defined by d_{corr}, with c=135 correspondences.

6 Results

The results described in this section are based on the UND face scans. For each of the 953 scans we applied our face segmentation method (Sect. 3). Our face segmentation method correctly normalized the pose of all face scans and adequately extracted the tip of the nose in each of them. The average distance and standard deviation of the 953 automatically selected nose tips to our manually selected nose tips was 2.3 ±1.2 mm.

Model fitting was applied to the segmented faces, once using only a single component and once using multiple components. Both instances are quantitatively evaluated in Sect. 6.1, and both instances were used for 3D face recognition in Sect. 6.2.

6.1 Face Model Fitting

In this section we evaluate the face model fitting as follows. Each segmented face was aligned to $\bar{S}$ and the coarse fitting method of Sect. 4.3 was applied. After the improved alignment of the scan data to S_{coarse}, the fine fitting method of Sect. 4.4 was applied to either the entire face (one component) or to each of the individual components (multiple components). For a fair comparison the same post-processing steps (Sect. 4.5) were applied to both S_{fine} instances. Fig. 5 shows qualitative better fits when multiple components are used instead of a single component. Globally, by looking at the more frequent surface interpenetration of the fitted model and face scan, which means a tighter fit. Locally, by looking at facial features, such as the nose, lips and eyes. Note that our fitting method correctly neglects facial hair, which is often a problem for 3D face recognition methods.

To quantitatively evaluate the produced fits, we determined the RMS distance (Eq. 2) for each of the fitted models to their face scan $d_{rms}(S_{final}, scan)$ and for the scan data to the face instance $d_{rms}(scan, S_{final})$. Points paired with boundary points are not included, so that results report merely the measurements in overlapping face regions. Results are reported over all 953 scans in Table 1. They show that our face morphing method provides accurate alignments of the morphable face model to the scan data for both the single component and multiple components. All results are in favor of multiple component morphs. Important to know is that the segmented UND faces have approximately twice the number of vertices compared to the fitted face model. Therefore, the closest point distances are higher for the scan to model case.

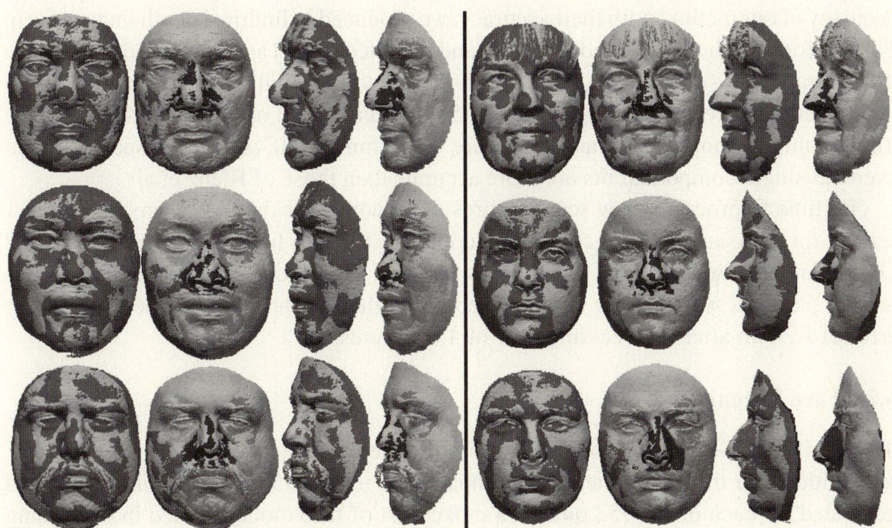

Fig. 5. Fitted face models S_{final} based on a single component (1st and 3rd column) and multiple components (2nd and 4rd column) to scan data in blue. Results from the front and side view, show a qualitative better fit of the multiple components to the scan data. The last two subjects on the right were also used in [7].

Table 1. The quantitative evaluation (in mm) of our face fitting method

measure	outliers	$M_1 \rightarrow M_2$	min	max	mean	sd
d_{rms}	yes	1 component $\rightarrow$ scan	0.478	3.479	0.776	0.176
d_{rms}	yes	7 components $\rightarrow$ scan	0.374	2.076	0.608	0.123
d_{rms}	yes	scan $\rightarrow$ 1 component	0.787	7.236	1.115	0.561
d_{rms}	yes	scan $\rightarrow$ 7 components	0.696	6.269	0.935	0.503
$d_{avr.depth}$	yes	scan $\leftrightarrow$ 1 component	0.393	4.704	0.692	0.290
$d_{avr.depth}$	yes	scan $\leftrightarrow$ 7 components	0.254	2.542	0.444	0.197
$d_{avr.depth}$	no	scan $\leftrightarrow$ 1 component	0.393	2.379	0.656	0.183
$d_{avr.depth}$	no	scan $\leftrightarrow$ 7 components	0.254	1.818	0.423	0.120

Table 2. Recognition rates and mean average precisions based on landmarks and contour curves for single and multiple component fits.

features	model fit	RR	MAP
landmarks	1 component	85.8%	0.872
landmarks	7 components	85.2%	0.862
contours	1 component	96.3%	0.952
contours	7 components	97.5%	0.967

Comparison. Blanz et al. [7] reported the accuracy of their model fitting method using the average depth error between the cylindrical depth images of the input scan and the output model. The mean depth error over 300 FRGC v.1 scans was 1.02 mm when they neglected outliers (distance > 10 mm) and 2.74 mm otherwise. To compare the accuracy of our method with their accuracy, we produced cylindrical depth images (as in Fig. 3c) for both the segmented face scan and the fitted model and computed the average depth error $|d_{scan}(\theta, y) - d_{final}(\theta, y)|$ without and with the outliers. For the fitted single component these errors $d_{avr.depth}$ are 0.656 mm and 0.692 mm, respectively. For the fitted multiple components these errors are 0.423 mm and 0.444 mm, respectively. So even our single component fits are more accurate then those of Blanz et al.

Our time to process a raw scan requires $\approx$3 seconds for the face segmentation, $\approx$1 second for the coarse fitting, and $\approx$30 seconds for the fine fitting on a Pentium IV 2.8 GHz. Blanz method reported $\approx$4 minutes on a 3.4 GHz Xeon processor, but includes texture fitting as well. Huang et al. [9] report for their deformation model a matching error of 1.2 mm after a processing time of 4.6 minutes.

6.2   Face Recognition

As described in Sect. 5, we can use the 953 morphed face instances to perform 3D face recognition. For this experiment, we computed the 953×953 dissimilarity matrix and generated for each of the 953 queries a ranked list of face models sorted in decreasing similarity. From these ranked lists, we computed the recognition rate (RR) and the mean average precision (MAP). A person is recognized (or identified) when the face retrieved on top of the ranked list (excluding the query) belongs to the same subject as the query. For 77 subjects only a single face instance is available which cannot be identified, so the RR is based on the remaining 876 queries. The mean average precision (MAP) of the ranked lists are reported, to elaborate on the retrieval of all relevant faces, i.e. all faces from the same subject.

Four 3D face recognition experiments were conducted, namely face recognition based on landmark locations from the fitted single component and the fitted multiple components, and based on contour curves from the fitted single component and the fitted multiple components. Results in Table 2 show that the automatically selected anthropomorphic landmarks are not reliable enough for effective 3D face recognition with 85.8% and 85.2% recognition rates (RR). Notice that the landmarks obtained from the single component fit perform better than those from the multiple component fit. This is probably caused by three landmarks (outer eye corners and Sellion) lying close to component boundaries, where the fitting can be less reliable.

The fitted face model is an accurate representation of the 3D scan data. This accuracy allows the contour based method to achieve high recognition rates (see Table 2). For the single component fits, the contour matching achieves a RR of 96.3% and for multiple component fits even 97.5%. For a high recognition rate, only one of the relevant faces in the dataset is required on top of each ranked list. The reported MAPs show that most of the other relevant faces are retrieved before the irrelevant ones. Some of the queries that were not identified, have a non-neutral expression (happy, angry, biting lips, etc.) while its relevant faces have a neutral expression. A face recognition method invariant to facial expressions, will most likely increase the performance even further.

Comparison. Blanz et al. [7] achieved a 96% RR for 150 queries in a set of 150 faces (from the FRGC v.1). To determine the similarity of two face instances, they computed the scalar product of the 1000 obtained model coefficients. Using a set of facial depth curves, Samir et al. [17] reported a 90.4% RR for 270 queries in a set of 470 UND scans. Mian et al. [18] reported a 86.4% RR for 277 queries in a set of 277 UND scans.

7   Concluding Remarks

Where other methods need manual initialization, we presented a fully automatic 3D face morphing method that produces a fast and accurate fit for the morphable face model to 3D scan data. Based on a global to local fitting scheme the face model is coarsely fitted to the automatically segmented 3D face scan. After the coarse fitting, the face model is either finely fitted as a single component or as a set of individual components. Inconsistencies at the borders are resolved using an easy to implement post-processing method. Our results show that the use of multiple components produces a tighter fit of the face model to the face scan, but assigned anthropomorphic landmarks may lose their reliability for 3D face identification. Face matching using facial contours, shows higher recognition rates based on the multiple component fits then for the single component fits. This means that the obtained 3D geometry after fitting multiple components has a higher accuracy. With a recognition rate of 97.5% for a large dataset of 3D faces, our model fitting method proves to produce highly accurate fits usable for 3D face recognition.

Acknowledgements

This research was supported by the FP6 IST Network of Excellence 506766 AIM@-SHAPE and partially supported by FOCUS-K3D FP7-ICT-2007-214993. The authors thank the University of South Florida for providing the USF Human ID 3D Database.

References

1. Bowyer, K.W., Chang, K., Flynn, P.: A survey of approaches and challenges in 3D and multi-modal 3D + 2D face recognition. CVIU 101(1), 1–15 (2006)
2. Scheenstra, A., Ruifrok, A., Veltkamp, R.C.: A Survey of 3D Face Recognition Methods. In: Kanade, T., Jain, A., Ratha, N.K. (eds.) AVBPA 2005. LNCS, vol. 3546, pp. 891–899. Springer, Heidelberg (2005)
3. Bronstein, A.M., Bronstein, M.M., Kimmel, R.: Three-dimensional face recognition. IJCV 64(1), 5–30 (2005)
4. Berretti, S., Del Bimbo, A., Pala, P., Silva Mata, F.: Face Recognition by Matching 2D and 3D Geodesic Distances. In: Sebe, N., Liu, Y., Zhuang, Y.-t., Huang, T.S. (eds.) MCAM 2007. LNCS, vol. 4577, pp. 444–453. Springer, Heidelberg (2007)
5. Davis, J., Marschner, S.R., Garr, M., Levoy, M.: Filling holes in complex surfaces using volumetric diffusion. 3DPVT, 428–861 (2002)
6. Blanz, V., Vetter, T.: A morphable model for the synthesis of 3D faces. SIGGRAPH, 187–194 (1999)
7. Blanz, V., Scherbaum, K., Seidel, H.P.: Fitting a Morphable Model to 3D Scans of Faces. In: ICCV, pp. 1–8 (2007)
8. Lu, X., Jain, A.: Deformation Modeling for Robust 3D Face Matching. PAMI 30(8), 1346–1356 (2008)
9. Huang, X., Paragios, N., Metaxas, D.N.: Shape Registration in Implicit Spaces Using Information Theory and Free Form Deformations. PAMI 28(8), 1303–1318 (2006)
10. Kakadiaris, I., Passalis, G., Toderici, G., Murtuza, N., Theoharis, T.: 3D Face Recognition. In: BMVC, pp. 869–878 (2006)
11. Sarkar.S.: USF HumanID 3D Face Database. University of South Florida
12. Chang, K.I., Bowyer, K.W., Flynn, P.J.: An Evaluation of Multimodal 2D+3D Face Biometrics. PAMI 27(4), 619–624 (2005)
13. ter Haar, F.B., Veltkamp, R.C.: A 3D Face Matching Framework. In: Proc. Shape Modeling International (SMI 2008), pp. 103–110 (2008)
14. Kirkpatrick, S., Gelatt, C.D., Vecchi, M.P.: Optimization by Simulated Annealing. Science 220, 4598, 671–680 (1983)
15. Besl, P.J., McKay, N.D.: A method for registration of 3D shapes. PAMI 14(2), 239–256 (1992)
16. Kimmel, R., Sethian, J.: Computing geodesic paths on manifolds. Proc. of National Academy of Sciences 95(15), 8431–8435 (1998)
17. Samir, C., Srivastava, A., Daoudi, M.: Three-Dimensional Face Recognition Using Shapes of Facial Curves. PAMI 28(11), 1858–1863 (2006)
18. Mian, A.S., Bennamoun, M., Owens, R.: Matching Tensors for Pose Invariant Automatic 3D Face Recognition. IEEE A3DISS (2005)

as possible, but there is, to our knowledge, no criterion for defining an optimal distribution of control points.

3   Multiscale Vector Splines

3.1   Parametric Spline Model

Thin-plate vector splines minimize the 2nd order div-curl regularity, but are inappropriate for multiscale estimation as they are defined from a harmonic basis function. A multiscale scheme actually requires using a basis function that provides a local representation, hence locally supported or rapidly decaying. The contribution of this paper is to formulate a multiscale model, based on a spline parameterized by the scale value and on a pyramidal representation of images at different scales.

We consider the spline approximation problem with the 2nd order div-curl norm and either the luminance or the mass conservation equation, through the observation operators $\mathcal{L}_i$ assessed on the n control points $\mathbf{x}_i$:

$$\min J(\mathbf{w}) = \left\{ \sum_{i=1}^{n} (\mathcal{L}_i \mathbf{w} - \mathbf{w}_i)^2 + \lambda \int_{\Omega} \alpha \|\nabla \operatorname{div} \mathbf{w}\|^2 + \beta \|\nabla \operatorname{curl} \mathbf{w}\|^2 \right\} \quad (8)$$

Rather than exactly solving equation (8), which would lead to the thin-plate spline, the minimum is searched for among a set of spline functions suitable for the multiscale formalism and satisfying the two following properties. (1) The spline is defined from a unique bell-shaped radial basis function of unit support. The choice of this function is not critical as long as it is positive, decreasing and at least three times continuously differentiable in order to compute the 2nd order div-curl semi-norm. We make use of the basis function ψ proposed by [14] and defined as $\psi(r) = (1 - r)^6 (35r^2 + 18r + 3)$ for $|r| \leq 1$. (2) The spline is a linear combination of translates of the basis function over a regular lattice of m grid points, whose sampling defines the scale parameter h. These translates are dilated by a factor γ proportional to h. The parameters defining the spline are the m weights $\mathbf{q} = (\mathbf{q}_j)$ (each weight $\mathbf{q}_j$ being homogeneous to a motion vector with u and v components) applied to the translates of the basis function. The parametric expression of the vector spline is thus:

$$\mathbf{w}_{\mathbf{q},h}(x) = \sum_{\mathbf{v}_j \in \mathbb{Z}^2, h\mathbf{v}_j \in \Omega} \mathbf{q}_j \psi(\|\frac{\mathbf{x} - h\mathbf{v}_j}{\gamma}\|) \quad (9)$$

where $\mathbf{v}_j$ spans a regular lattice of unit spacing in the image domain Ω.

A new expression of the functional J is defined by substituting, in equation (8), $\mathbf{w}$ by its parametric form $\mathbf{w}_{\mathbf{q},h}$ (9). Let us first consider the first term of J. If the observation operator is based on the luminance conservation equation, its new expression becomes:

$$\|\mathbf{I_x}\Psi\mathbf{q^u} + \mathbf{I_y}\Psi\mathbf{q^v} - \mathbf{I_t}\|^2 = \|A_l\mathbf{q} - \mathbf{I_t}\|^2 \quad (10)$$

$\mathbf{I_t}$ being the n-dimensional vector of the temporal derivatives at the control points; Ψ being the $n \times m$ matrix of general term $\psi((x_i - k_j)/\gamma, (y_i - l_j)/\gamma)$ with i indexing the n control points and j the m grid points (k, l); $\mathbf{I_x}$ and $\mathbf{I_y}$ are the $n \times n$ diagonal matrices of the image spatial derivatives at the control points. In the case of mass conservation, the first term of J becomes:

$$\|\mathbf{I_x}\Psi\mathbf{q^u} + \mathbf{I_y}\Psi\mathbf{q^v} + \mathbf{ID_x}\Psi + \mathbf{ID_y}\Psi - \mathbf{I_t}\|^2 = \|A_m\mathbf{q} - \mathbf{I_t}\|^2 \tag{11}$$

where $\mathbf{I}$ is the $n \times n$ diagonal matrix formed by the image values at control points, $\mathbf{D_x}\Psi$ and $\mathbf{D_y}\Psi$ are the matrices of the spatial derivatives of Ψ. Whatever the conservation equation, the first term of J is then rewritten as a quadratic function of $\mathbf{q}$.

Let us now analyze the second term of J. By introducing the matrix of differential operators $Q(D)$:

$$Q(D) = \begin{pmatrix} \sqrt{\alpha}\partial_{xx} + \sqrt{\beta}\partial_{yy} & (\sqrt{\alpha} - \sqrt{\beta})\partial_{xy} \\ (\sqrt{\alpha} - \sqrt{\beta})\partial_{xy} & \sqrt{\alpha}\partial_{yy} + \sqrt{\beta}\partial_{xx} \end{pmatrix} \tag{12}$$

J then factorizes as:

$$\alpha \int \|\nabla \operatorname{div} \mathbf{w}\|^2 + \beta \int \|\nabla \operatorname{curl} \mathbf{w}\|^2 = \int \|Q(D)\mathbf{w}\|^2 \tag{13}$$

The second term of J is finally rewritten as the quadratic expression $\|R\mathbf{q}\|^2$, with:

$$R = \begin{pmatrix} \sqrt{\alpha}\partial_{xx}\Psi + \sqrt{\beta}\partial_{yy}\Psi & (\sqrt{\alpha} - \sqrt{\beta})\partial_{xy}\Psi \\ (\sqrt{\alpha} - \sqrt{\beta})\partial_{xy}\Psi & \sqrt{\alpha}\partial_{yy}\Psi + \sqrt{\beta}\partial_{xx}\Psi \end{pmatrix}. \tag{14}$$

The substitution of $\mathbf{w}$ by the parametric expression $\mathbf{w_{q,}}_h$ allows J to be rewritten as a quadratic function of $\mathbf{q}$:

$$J(\mathbf{q}) = \|A\mathbf{q} - I_t\|^2 + \lambda\|R\mathbf{q}\|^2 \tag{15}$$

with A being either A_l or A_m depending on the conservation equation chosen. Finding the minimum of J with respect to $\mathbf{q}$ is now a linear optimization problem.

The matrices A and R, in (15), have a band structure since ψ has a compact support of size γ. The width of the band depends on the ratio of γ to the scale parameter h. If γ is smaller than h, the matrices A and R are diagonal and the vector spline is zero everywhere except in the vicinity of the grid points. If γ is large compared to h, the resulting vector spline can accurately approximate the thin-plate spline, but the A and R matrices are dense and require a heavy computational load. $\gamma = 3h$ has been empirically chosen as a good compromise between the computational speed and the accuracy of the spline. The band structure allows an efficient numerical solving to be implemented.

3.2   Hierarchical Motion Estimation

A multiscale scheme is required for two main reasons. (1) The parametric spline allows the image motion to be assessed, given a spatial scale parameter h, and provided that the conservation equation can be computed. On satellite image sequences, a too strong motion and/or a too coarse time sampling cause large displacements between successive frames, preventing the linearization of the conservation equation. (2) Turbulent flows are associated with a large spectrum of spatial and temporal scales. We therefore make use of a pyramidal scheme, in which motion is hierachically computed from the coarsest to the finest scale.

Let I_0 and I_1 be two successive images of the sequence. Both are represented using a pyramid, from the full resolution $I_0(0)$ and $I_1(0)$ to the coarsest scale $I_0(p_{max})$ and $I_1(p_{max})$. To each index p corresponds a scale parameter $h(p)$. The motion is initially computed at the coarsest scale with the parametric spline at scale $h(p_{max})$, yielding the motion field $\mathbf{w}(p_{max})$. This initial coarse motion field is then progresively refined at each scale $h(p)$ by first compensating the image $I_0(p)$ with $\mathbf{w}(p+1)$ and computing the motion increment $\delta\mathbf{w}(p)$ between the compensated image and $I_1(p)$. The finest scale motion ($p = 0$) is thus expressed as the sum of the coarse scale motion $\mathbf{w}(p_{max})$ and of the increments describing the finer resolutions:

$$\mathbf{w}(0) = \mathbf{w}(p_{max}) + \sum_{p=p_{max}-1}^{0} \delta\mathbf{w}(p) \tag{16}$$

The link between the scale parameter $h(p)$ and the real spatial scale of the evolving image structures is not obvious: at one level of the pyramid, the motion is computed using a scale parameter $h(p)$ corresponding to a basis function of support $\gamma = 3h(p)$. The basis function is thus able to represent motion patterns with spatial size less than $3h(p)$; but there is no guarantee that all motion patterns of that size will be represented: this will occur only if enough control points have been selected in the existing patterns.

4   Results

The first result intends to demonstrate the efficiency of accounting for the conservation only at control points. For this purpose, the motion is computed using

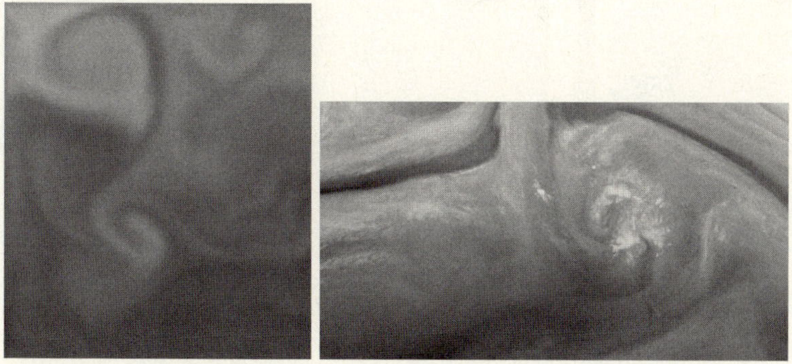

Fig. 1. Extract of the test sequences: left, OPA; right: Meteosat

the multiscale vector spline and compared to the result of Corpetti's method [15]. Both methods minimize the second order div-curl regularity constraint, make use of either luminance or mass conservation and are solved in a multiscale scheme. The two methods differ in the data confidence term of the minimized energy (computed on control points selected by double thresholding for the multiscale spline, on the whole image domain for Corpetti's method) and in the numerical minimization scheme (multiscale vector spline vs variational minimization). Two comparisons are displayed. First, the motion is computed using the luminance conservation equation on the synthetic 'OPA' sequence (on the left in figure 1), obtained by numerical simulation with the OPA ocean circulation model[1]. The

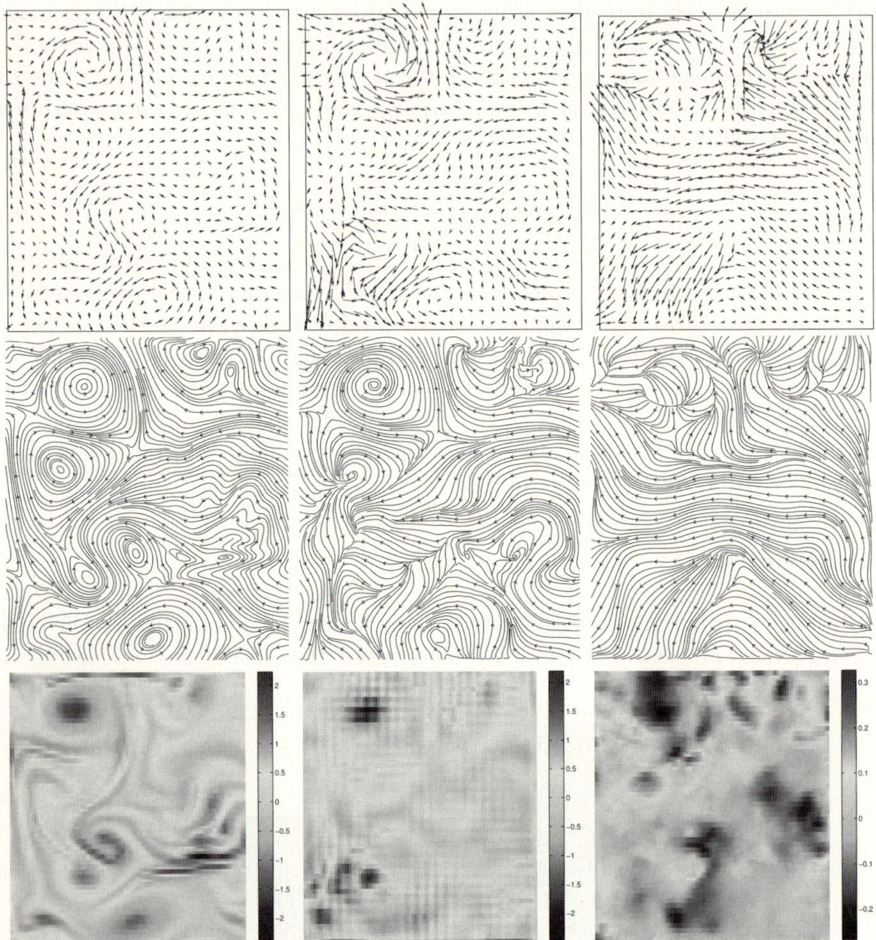

Fig. 2. Motion fields estimated on the OPA sequence using luminance conservation. Left to right: reference motion, multiscale spline, Corpetti and Mémin. Top to bottom: motion field, streamlines, vorticity.

---

[1] Thanks to Marina Levy, LOCEAN, IPSL, France.

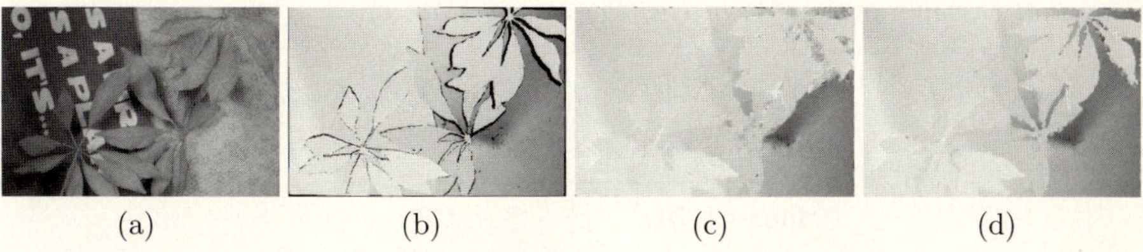

(a)   (b)   (c)   (d)

Fig. 1. (a) input image 0 of the Middlebury "Schefflera" sequence; (b) color-coded ground truth flow (hue = direction, intensity = magnitude, black = unknown); color-coded flows, estimated using: (c) a continuous TV-L^1 flow model; (d) the proposed optimization strategy (AAE = $2.91°$)

where the small, positive constant θ ensures a tight coupling of $\boldsymbol{u}$ and $\boldsymbol{v}$. In contrast to the original energy (6), this convex approximation can be solved using a dual approach. Our proposed optimization strategy can now easily be applied by replacing $\boldsymbol{v}$ with a binary fusion of two flow proposals $\boldsymbol{\alpha}$ and $\boldsymbol{\beta}$:

$$\min_{\boldsymbol{u},\phi} \left\{ \sum_{d=1}^{2} \int_{\Omega} |\nabla u_d| \, d\boldsymbol{x} + \frac{1}{2\theta} \sum_{d=1}^{2} \int_{\Omega} (u_d - [(1-\phi)\,\alpha_d + \phi\beta_d])^2 \, d\boldsymbol{x} \right.$$
$$\left. + \lambda \int_{\Omega} (1-\phi)\,\rho(\boldsymbol{\alpha},\boldsymbol{x}) + \phi\rho(\boldsymbol{\beta},\boldsymbol{x}) \, d\boldsymbol{x} \right\} \ . \qquad (8)$$

The relaxed version of (8) is a minimization problem in two variables, $\boldsymbol{u}$ and ϕ. We therefore have to perform an alternating minimization procedure:

1. For ϕ fixed, solve for every u_d:

$$\min_{u_d} \left\{ \int_{\Omega} |\nabla u_d| \, d\boldsymbol{x} + \frac{1}{2\theta} \int_{\Omega} (u_d - [(1-\phi)\,\alpha_d + \phi\beta_d])^2 \, d\boldsymbol{x} \right\} \qquad (9)$$

2. For $\boldsymbol{u}$ fixed, solve for ϕ:

$$\min_{\phi} \left\{ \frac{1}{2\theta} \sum_{d=1}^{2} \int_{\Omega} (u_d - [(1-\phi)\,\alpha_d + \phi\beta_d])^2 \, d\boldsymbol{x} \right.$$
$$\left. + \lambda \int_{\Omega} (1-\phi)\,\rho(\boldsymbol{\alpha},\boldsymbol{x}) + \phi\rho(\boldsymbol{\beta},\boldsymbol{x}) \, d\boldsymbol{x} \right\} \qquad (10)$$

The subproblem (9) is the well understood image denoising model of Rudin, Osher, and Fatemi [19]. For this model, Chambolle proposed an efficient and globally convergent numerical scheme, based on a dual formulation [20]. In practice, a gradient descent/reprojection variant of this scheme performs better [9], although there is no proof for convergence. In order to make this paper self-contained, we reproduce the relevant results from [20,9]:

Proposition 1. *The solution of (9) is given by*

$$u_d = [(1-\phi)\,\alpha_d + \phi\beta_d] - \theta\nabla \cdot \boldsymbol{p}_d \ . \qquad (11)$$

The dual variable $\boldsymbol{p}_d$ is obtained as the steady state of

$$\tilde{\boldsymbol{p}}_d^{k+1} = \boldsymbol{p}_d^k + \frac{\tau}{\theta}\nabla\left(\theta\nabla\cdot\boldsymbol{p}_d^k - [(1-\phi)\,\alpha_d + \phi\beta_d]\right) \ ,$$

$$\boldsymbol{p}_d^{k+1} = \frac{\tilde{\boldsymbol{p}}_d^{k+1}}{\max\left\{1,|\tilde{\boldsymbol{p}}_d^{k+1}|\right\}} \ , \tag{12}$$

where k is the iteration number, $\boldsymbol{p}_d^0 = \boldsymbol{0}$, and $\tau \le 1/4$.

The subproblem (10) permits a direct solution.

Proposition 2. *The solution of (10) is given by clamping*

$$\tilde{\phi} = \begin{cases} \dfrac{(\boldsymbol{u}-\boldsymbol{\alpha})^T\,(\boldsymbol{\beta}-\boldsymbol{\alpha}) + \lambda\theta\,(\rho(\boldsymbol{\beta},\boldsymbol{x}) - \rho(\boldsymbol{\alpha},\boldsymbol{x}))}{(\boldsymbol{\beta}-\boldsymbol{\alpha})^T\,(\boldsymbol{\alpha}-\boldsymbol{\beta})} & \text{where } \boldsymbol{\alpha}\ne\boldsymbol{\beta} \\ 0 & \text{elsewhere} \end{cases} \tag{13}$$

to the range $[0,1]$:

$$\phi = \max\left\{0, \min\left\{1,\tilde{\phi}\right\}\right\} \ . \tag{14}$$

Proof: Starting with the Euler-Lagrange equation of (10)

$$\frac{1}{\theta}\left(\boldsymbol{u} - [(1-\phi)\,\boldsymbol{\alpha} + \phi\boldsymbol{\beta}]\right)^T(\boldsymbol{\alpha}-\boldsymbol{\beta}) + \lambda\left[\rho(\boldsymbol{\beta},\boldsymbol{x}) - \rho(\boldsymbol{\alpha},\boldsymbol{x})\right] \ , \tag{15}$$

we try to solve for ϕ, yielding

$$\phi\,(\boldsymbol{\beta}-\boldsymbol{\alpha})^T(\boldsymbol{\alpha}-\boldsymbol{\beta}) = (\boldsymbol{u}-\boldsymbol{\alpha})^T(\boldsymbol{\beta}-\boldsymbol{\alpha}) + \lambda\theta\,(\rho(\boldsymbol{\beta},\boldsymbol{x}) - \rho(\boldsymbol{\alpha},\boldsymbol{x})) \ . \tag{16}$$

Wherever $\boldsymbol{\alpha} = \boldsymbol{\beta}$, $\rho(\boldsymbol{\alpha},\boldsymbol{x}) = \rho(\boldsymbol{\beta},\boldsymbol{x})$, hence ϕ can arbitrarily be chosen in $[0,1]$. Everywhere else, we can divide by $(\boldsymbol{\beta}-\boldsymbol{\alpha})^T(\boldsymbol{\alpha}-\boldsymbol{\beta})$, yielding (13). $\square$

Please note that the data residuals $\rho(\boldsymbol{\alpha},\boldsymbol{x})$ and $\rho(\boldsymbol{\beta},\boldsymbol{x})$ are just constants. They have to be calculated only once per fusion step, and the sole requirement is that their range is $\mathbb{R}_0^+$, i.e. almost any cost function can be used. Once the relaxed version of problem (8) is solved, a final thresholding of ϕ is required to obtain the binary fusion of the two flow proposals $\boldsymbol{\alpha}$ and $\boldsymbol{\beta}$. Since the continuous solution of ϕ is already close to binary, the threshold μ is not critical. Our heuristic solution is to evaluate the energy of the original TV-L^1 energy (8) for $\boldsymbol{\alpha}$, $\boldsymbol{\beta}$, and a few different thresholds $\mu \in (0,1)$. Finally, we select the threshold yielding the flow field with the lowest energy.

3.1   Extension to a Second-Order Prior

The presented approach is by no means limited to Total Variation regularization. As an example, we will apply the proposed technique to an optical flow model with a prior based on decorrelated second-order derivatives [21]. This second-order prior has the intrinsic property to penalize only deviations from piecewise affinity. Since spatial second-order derivatives are not orthogonal and the local information of orientation and shape are entangled, a decorrelation is necessary. In [22], Danielsson et al. used circular harmonic functions to map the the second-order derivative operators into an orthogonal space. In two spatial dimensions, the decorrelated operator is given by

$$\Diamond = \sqrt{\frac{1}{3}} \left(\frac{\partial^2}{\partial_x^2} + \frac{\partial^2}{\partial_y^2}, \ \sqrt{2} \left(\frac{\partial^2}{\partial_x^2} - \frac{\partial^2}{\partial_y^2} \right), \ \sqrt{8} \frac{\partial^2}{\partial_x \partial_y} \right)^T . \tag{17}$$

The magnitude of this operator, defined as the Euclidean vector norm

$$\|\Diamond u\| = \sqrt{\frac{1}{3}} \sqrt{ \left(\frac{\partial^2 u}{\partial_x^2} + \frac{\partial^2 u}{\partial_y^2} \right)^2 + 2 \left(\frac{\partial^2 u}{\partial_x^2} - \frac{\partial^2 u}{\partial_y^2} \right)^2 + 8 \left(\frac{\partial^2 u}{\partial_x \partial_y} \right)^2 } , \tag{18}$$

measures the local deviation of a function u from being affine. Adapting the TV-L^1 flow model (6) to the new prior is a matter of replacing the TV regularization in (6–8) with the Euclidean norm of the new operator (18). Instead of minimizing the ROF energy (9), step 1 of the alternate optimization of u and ϕ now amounts to solving

$$\min_{u_d} \left\{ \int_\Omega \|\Diamond u_d\| \, d\boldsymbol{x} + \frac{1}{2\theta} \int_\Omega \left(u_d - [(1-\phi)\,\alpha_d + \phi\beta_d] \right)^2 d\boldsymbol{x} \right\} \tag{19}$$

for every u_d, while keeping ϕ fixed.

Proposition 3. *The solution of (19) is given by*

$$u_d = [(1-\phi)\,\alpha_d + \phi\beta_d] - \theta\Diamond \cdot \boldsymbol{p}_d . \tag{20}$$

The dual variable $\boldsymbol{p}_d$ is obtained as the steady state of

$$\tilde{\boldsymbol{p}}_d^{\,k+1} = \boldsymbol{p}_d^k + \frac{\tau}{\theta} \left[\Diamond \left([(1-\phi)\,\alpha_d + \phi\beta_d] - \theta\Diamond \cdot \boldsymbol{p}_d^k \right) \right] ,$$

$$\boldsymbol{p}_d^{\,k+1} = \frac{\tilde{\boldsymbol{p}}_d^{\,k+1}}{\max\left\{ 1, |\tilde{\boldsymbol{p}}_d^{\,k+1}| \right\}} , \tag{21}$$

where k is the iteration number, $\boldsymbol{p}_d^0 = \boldsymbol{0}$, and $\tau \leq 3/112$. For a proof and further details please refer to [21].

Moreover, we employ the following standard finite differences approximation of the $\Diamond$ operator:

$$(\Diamond u)_{i,j} = \begin{pmatrix} \sqrt{\frac{1}{3}} \left(u_{i,j-1} + u_{i,j+1} + u_{i-1,j} + u_{i+1,j} - 4u_{i,j} \right) \\ \sqrt{\frac{2}{3}} \left(u_{i-1,j} + u_{i+1,j} - u_{i,j-1} - u_{i,j+1} \right) \\ \sqrt{\frac{8}{3}} \left(u_{i,j} + u_{i+1,j+1} - u_{i,j+1} - u_{i+1,j} \right) \end{pmatrix} , \tag{22}$$

where (i, j) denote the indices of the discrete image domain, enforcing Dirichlet boundary conditions on $\partial\Omega$. For details on the discretization of $\Diamond \cdot \boldsymbol{p}$, please consult [21].

In the continuous setting, such an extension requires minor adaptions of the solution procedure and incurs only a small increase of time and memory requirements. Most combinatorial optimization approaches, however, are limited to unary and pairwise clique potentials, hence such a second-order prior can not be used. Extending combinatorial optimization algorithms to higher-order cliques (e.g. as proposed in [5,23]) is either expensive in time and space or imposes restrictions on the potentials, e.g. [23] restricts the potentials to the Potts model.

4   Experiments

In this section, we first restrict the optical flow model to a single dimension (rectified stereo) in order to analyze its behavior in a simplified setting. In Section 4.2 we will use image sets from the Middlebury optical flow database [17] to illustrate that the proposed algorithm yields state-of-the-art flow estimates.

Most of the algorithm has been implemented in C++, with the exception of the numerical schemes of the solvers, which have been implemented using CUDA 1.0. All subsequent experiments have been performed on an Intel Core 2 Quad CPU at 2.66 GHz (the host code is single-threaded, so only one core was used) with an NVidia GeForce 8800 GTX graphics card, running a 32 bit Linux operating system and recent NVidia display drivers. Unless noted otherwise, in all our experiments the parameters were set to $\lambda = 50$ and $\theta = 0.1$.

4.1   Illustrative Stereo Experiment

A restriction of the optical flow model (8) to a single displacement u permits a direct comparison of the estimated solutions to the global optimum (cf. [24] for details on calculating the global optimum of this multi-label problem). Since the approach presented in [24] is based on a discrete solution space, we further restrict our method by using only constant disparity proposals in 0.5 pixel increments.

(a)   (b)   (c)

Fig. 2. (a) im2 of the Middlebury "Teddy" stereo pair; (b) the corresponding ground truth disparity map; (c) a mask for the pixels, which are also visible in im6

Moreover, this (rectified) stereo setting simplifies discussing the effects caused by the relaxation and the seemingly asymmetric formulation.

All experiments in this section use the grayscale version of the "Teddy" stereo pair [25] and a set of constant disparity proposals in the range 0 to 59 pixels in 0.5 pixel increments. Figure 2(a) shows `im2` of the stereo pair, Fig. 2(b) the corresponding ground truth disparity, and Fig. 2(c) is the mask of non-occluded regions.

Relaxation and Thresholding. For every fusion step (4), the binary function ϕ : $\Omega \to \{0, 1\}$ has to be optimized. Since this is a non-convex problem, we proposed to relax ϕ to the range $[0, 1]$, solve the continuous problem, and finally threshold ϕ. Obviously, this only leads to reasonable fusions of α and β, if the optimal ϕ is close to binary. Figure 3 shows, how a 64-bin histogram of the relaxed function ϕ evolves during a typical optimization procedure. Since ϕ is initialized with 0, in the beginning proposal α is chosen over β in the whole image. However, the "traces" in Fig. 3 illustrate that in several image regions the value of ϕ flips to 1, i.e. in these regions proposal β is energetically favored. Once the algorithm converged, the histogram of ϕ is close to binary, just as expected.

Symmetry. Since we initialize $\phi(\boldsymbol{x}) = 0$, the formulation appears to be asymmetric with respect to the flow proposals α and β, but in practice the effects of swapping the proposals is negligible. Figure 4(a) shows some intermediate solution of the "Teddy" disparity map, which is used as proposal α in the following experiment. The energy of α is 325963 and there clearly is room for improvement. The disparity map resulting from a fusion with the constant proposal $\beta(\boldsymbol{x}) = 20$ has an energy of 284748 and can be seen in Fig. 4(b). To highlight the image

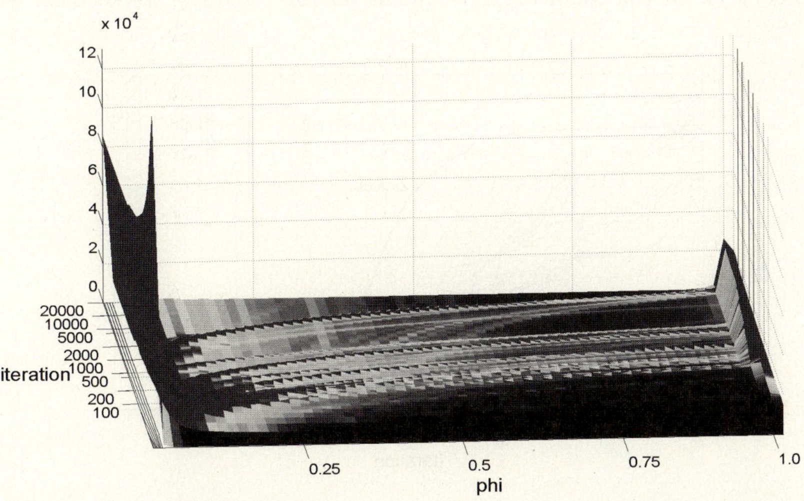

Fig. 3. The evolution of a 64-bin histogram of ϕ during the solution procedure. Starting at $\phi(\boldsymbol{x}) = 0$, i.e. with proposal $\boldsymbol{\alpha}$, several image regions flip to the alternative proposal $\boldsymbol{\beta}$. It is clearly apparent that the converged histogram is close to binary. Please note that a logarithmic scale is used for the "iteration" axis.

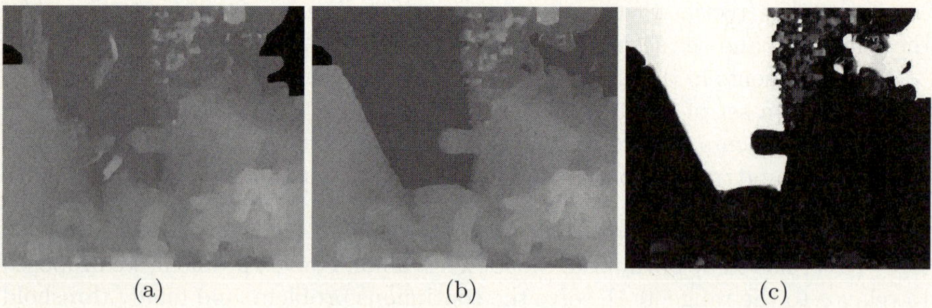

Fig. 4. Two intermediate disparity maps, before (a) and after (b) a binary fusion with the proposal $\beta(\boldsymbol{x}) = 20$. (c) shows the corresponding continuous optimum of ϕ.

regions that were switched to proposal β during this fusion step, Fig. 4(c) shows the continuous optimum of ϕ (before thresholding). Please note that ϕ is mostly binary, except for regions where neither α nor β are close to the true disparity. Repeating the experiment with α and β switched leads to visually indistinguishable results and an energy of 284727, i.e. the order of the proposals does *not* matter in the binary fusion step.

Disparity Estimation via Randomized Sweeping. In the discrete α-expansion algorithm [5], the minimization of a multi-label problem is performed by "agglomerating" an increasingly accurate solution by repeatedly allowing the current labeling to switch to an alternative label. One sweep through the label space is called a *cycle*. In the following experiment on the "Teddy" stereo pair, we

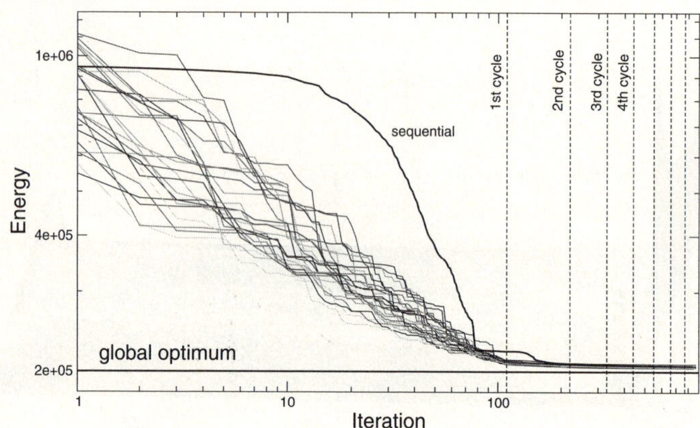

Fig. 5. Decrease of the energy of the flow field with every successful fusion. Results of randomized runs are shown as thin, colored lines; the thick, black line shows the progress for a sequential run. The dashed vertical lines delimit fusion cycles, the thick horizontal line marks the global optimum for the TV-L^1 flow model (6).

emulate this behavior by sweeping through a set of constant disparity proposals in the range 0 to 59 in 0.5 pixel increments. Figure 5 depicts, how the energy of the flow field decreases with every successful fusion step. Sweeping the "label space" in consecutive order of the displacements results in the thick, black line labeled "sequential". The results for 25 distinct runs with a randomized order of the proposals are illustrated using thin, colored lines. The dashed vertical lines delimit fusion cycles, i.e. between two of those vertical lines every possible disparity is tested exactly once. The thick horizontal line represents the energy of the global optimum ($E = 199891$) for this model at $\lambda = 50$, see Fig. 6(c) for a disparity map of this solution. It is clearly apparent that after three or more cycles the particular sweeping order does not have a significant influence. After eight cycles, the mean energy of all runs is 205114, with the best run being roughly 2 % better than the worst. Two exemplary disparity maps are shown in Figs. 6(a) and 6(b) – they differ from the global optimum (shown in Fig. 6(c)) mainly in occluded areas, i.e. areas, where Fig. 2(c) is black. The mean energy of the disparity maps, estimated using this approximated minimization technique, is 2.6% higher than the global optimum, which is consistent with empirical results reported for the discrete α-expansion algorithm.

Depending on the quality of the model, reducing the energy of a solution might not result in a lower error on the true problem. Hence, in order to compare the true errors of the global optimum and the approximated disparity fields, the results have been evaluated on the Middlebury stereo vision benchmark [25]. At an error threshold of 0.5, the global optimum mislabels 16.9 % of the pixels in non-occluded regions; the estimated solution shown in Fig. 6(b) has a slightly larger error of 17.2 %, which is consistent with the energy differences.

4.2   Optical Flow Estimation

For optical flow estimation, merely fusing a set of constant proposals, just like we did in the stereo case, is not feasible. One of the main reasons is that the expected range of motion is not known beforehand. Simply assuming very large ranges to be on the safe side either results in a coarse sampling of the solution space or in a huge number of constant flow proposals. Such a brute force method could be made tractable by using a scale pyramid, but this would still be inelegant and slow. However, the presented optimization strategy is not limited to fusing constant proposals – any flow field can be used as proposed solution. Thus, one obvious solution to the problem is to estimate a set of flow fields using a standard optical flow algorithm and fuse them. Therefore, before starting any fusion experiments, a set of 27 TV-regularized flows ($\lambda \in \{10, 25, 40, 65, 100, 150, 200, 500, 1000\}$, $\theta \in \{0.05, 0.1, 0.15\}$) and a set of 24 second-order prior regularized flows ($\lambda \in \{10, 25, 40, 55, 90, 200, 500, 1000\}$, $\theta \in \{0.05, 0.1, 0.15\}$) have been estimated using the algorithms described in [16,21]. Since the proposed optimization strategy is not limited to convex data terms, we used a truncated data term $\rho = \min\{1, 1 - r\}$, where r is the normalized cross-correlation, calculated on 3×3 patches across all color channels.

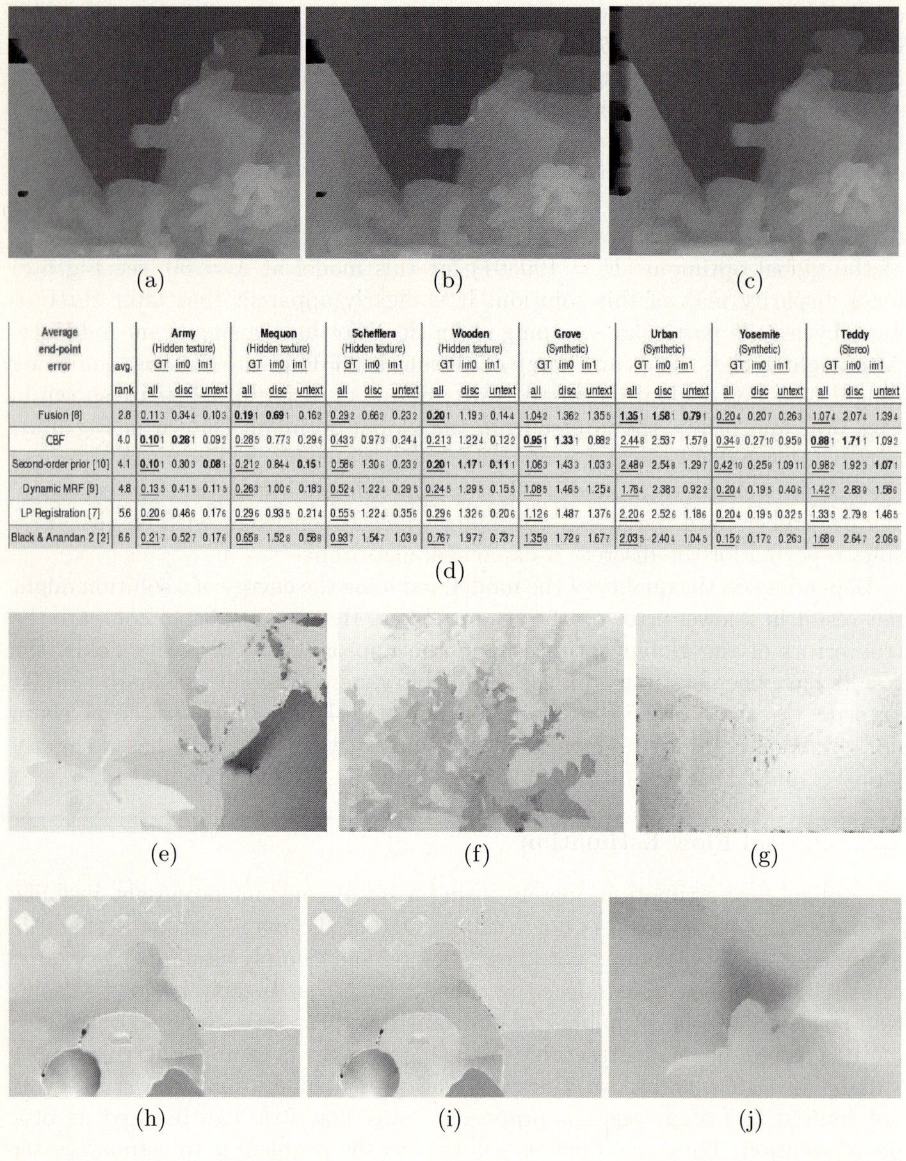

Fig. 6. First row: (a), (b) disparity maps, estimated by repeatedly fusing constant proposals using the proposed optimization strategy; (c) global optimum for the TV-L^1 model. Second row: average end-point error results for the Middlebury benchmark dataset; the proposed method, labeled "CBF", was ranked 2nd at the time of submission. Third row: (e–g) show color-coded flow fields for the sequences "Schefflera", "Grove", and "Yosemite". Last row: (h) shows a color-coded flow field for the Middlebury "RubberWhale" sequence, estimated using the second-order prior (AAE = $3.14°$); (i) and (j) show the color-coded flow fields for a TV-regularized fusion of TV and second-order prior flows for the "RubberWhale" (AAE = $2.87°$) and the "Dimetrodon" (AAE = $3.24°$) sequences.

Figure (d), the average end-point error table, is transcribed below. Small subscripts after each value are its per-column rank.

Average end-point error	avg. rank	Army (Hidden texture) GT all	im0 disc	im1 untext	Mequon (Hidden texture) GT all	im0 disc	im1 untext	Schefflera (Hidden texture) GT all	im0 disc	im1 untext	Wooden (Hidden texture) GT all	im0 disc	im1 untext
Fusion [8]	2.8	0.11_3	0.34_4	0.10_3	$\mathbf{0.19}_1$	$\mathbf{0.69}_1$	0.16_2	0.29_2	0.66_2	0.23_2	$\mathbf{0.20}_1$	1.19_3	0.14_4
CBF	4.0	$\mathbf{0.10}_1$	$\mathbf{0.28}_1$	0.09_2	0.28_5	0.77_3	0.29_6	0.43_3	0.97_3	0.24_4	0.21_3	1.22_4	0.12_2
Second-order prior [10]	4.1	$\mathbf{0.10}_1$	0.30_3	$\mathbf{0.08}_1$	0.21_2	0.84_4	$\mathbf{0.15}_1$	0.58_6	1.30_6	0.23_2	$\mathbf{0.20}_1$	$\mathbf{1.17}_1$	$\mathbf{0.11}_1$
Dynamic MRF [9]	4.8	0.13_5	0.41_5	0.11_5	0.26_3	1.00_6	0.18_3	0.52_4	1.22_4	0.29_5	0.24_5	1.29_5	0.15_5
LP Registration [7]	5.6	0.20_6	0.46_6	0.17_6	0.29_6	0.93_5	0.21_4	0.55_5	1.22_4	0.35_6	0.29_6	1.32_6	0.20_6
Black & Anandan 2 [2]	6.6	0.21_7	0.52_7	0.17_6	0.65_8	1.52_8	0.58_8	0.93_7	1.54_7	1.03_9	0.76_7	1.97_7	0.73_7

Average end-point error	Grove (Synthetic) GT all	im0 disc	im1 untext	Urban (Synthetic) GT all	im0 disc	im1 untext	Yosemite (Synthetic) GT all	im0 disc	im1 untext	Teddy (Stereo) GT all	im0 disc	im1 untext
Fusion [8]	1.04_2	1.36_2	1.35_5	$\mathbf{1.35}_1$	$\mathbf{1.58}_1$	$\mathbf{0.79}_1$	0.20_4	0.20_7	0.26_3	1.07_4	2.07_4	1.39_4
CBF	$\mathbf{0.95}_1$	$\mathbf{1.33}_1$	0.88_2	2.44_8	2.53_7	1.57_9	0.34_9	0.27_{10}	0.95_9	$\mathbf{0.88}_1$	$\mathbf{1.71}_1$	1.09_2
Second-order prior [10]	1.06_3	1.43_3	1.03_3	2.48_9	2.54_8	1.29_7	0.42_{10}	0.25_9	1.09_{11}	0.98_2	1.92_3	$\mathbf{1.07}_1$
Dynamic MRF [9]	1.08_5	1.46_5	1.25_4	1.76_4	2.38_3	0.92_2	0.20_4	0.19_5	0.40_6	1.42_7	2.83_9	1.58_6
LP Registration [7]	1.12_6	1.48_7	1.37_6	2.20_6	2.52_6	1.18_6	0.20_4	0.19_5	0.32_5	1.33_5	2.79_8	1.46_5
Black & Anandan 2 [2]	1.35_9	1.72_9	1.67_7	2.03_5	2.40_4	1.04_5	0.15_2	0.17_2	0.26_3	1.68_9	2.64_7	2.06_9

A TV-regularized fusion of all 27 TV flows has been submitted to the Middlebury optical flow evaluation site, where at the time of submission this method (labeled "CBF") was ranked second or better for 14 out of 16 error measures. Figure 6(d) shows the average end-point error of the six top-ranked algorithms, and Figs. 6(e)–6(g) show some color-coded results. Please visit the evaluation page at http://vision.middlebury.edu/flow/eval/ for other error measures and further images. Due to the challenging diversity of the data sets, this was the only experiment where we used $\lambda = 90$ to improve the results on "Schefflera" and "Grove" at the cost of a rather noisy "Yosemite" result. On the "Urban" sequence (640×480 pixels), the outlined flow estimation procedure took 218 s (138 s for precalculating 27 TV-regularized flows and 80 s for 6 fusion cycles).

Comparing the Figs. 6(e) and 1(c) indicates a slight improvement of the estimated flow, but since we only fuse $TV\text{-}L^1$ proposals, the correct solution is never "offered." For further improvements, we have to resort to the brute force strategy of fusing a number of constant flow proposals, but since we already have a good estimate of the flow, the solution space is small and it quickly converges. The final result has an average angular error (AAE) of $2.91°$ (see Fig. 1(d)).

Furthermore, the second-order prior regularized optical flow algorithm (see Section 3.1) was used to fuse the 24 precalculated second-order flows. Figure 6(h) shows a color-coded result for the "RubberWhale" sequence of the Middlebury training dataset (AAE $= 3.14°$). Using *all* precalculated flows and a TV-regularized fusion algorithm yields even better results: Figs. 6(i) and 6(j) show the color-coded flow fields for the "RubberWhale" (AAE $= 2.87°$) and "Dimetrodon" (AAE $= 3.24°$) sequences, respectively.

5   Conclusion

The presented optimization strategy permits large optimization moves in a variational context, by restating the minimization problem as a sequence of binary subproblems. After verifying that the introduced approximations are reasonable, we showed that typical solutions for a stereo problem are within a few percent of the global optimum (in energy as well as in the true error measure). Finally, we showed that applying this optimization strategy to optical flow estimation yields state-of-the-art results on the challenging Middlebury optical flow dataset.

References

1. Mumford, D.: Bayesian rationale for energy functionals. In: Geometry-driven diffusion in Computer Vision, pp. 141–153. Kluwer Academic Publishers, Dordrecht (1994)
2. Szeliski, R., Zabih, R., Scharstein, D., Veksler, O., Kolmogorov, V., Agarwala, A., Tappen, M., Rother, C.: A comparative study of energy minimization methods for Markov random fields with smoothness-based priors. IEEE Trans. Pattern Anal. Mach. Intell. 30(6), 1068–1080 (2008)
3. Pearl, J.: Probabilistic Reasoning in Intelligent Systems: Networks of Plausible Inference. Morgan Kaufmann, San Francisco (1988)

4. Kolmogorov, V.: Convergent tree-reweighted message passing for energy minimization. IEEE Trans. Pattern Anal. Mach. Intell. 28(10), 1568–1583 (2006)
5. Boykov, Y., Veksler, O., Zabih, R.: Fast approximate energy minimization via graph cuts. IEEE Trans. Pattern Anal. Mach. Intell. 23, 1222–1239 (2001)
6. Veksler, O.: Graph cut based optimization for MRFs with truncated convex priors. In: Proc. of the CVPR (June 2007)
7. Lempitsky, V., Rother, C., Blake, A.: LogCut – efficient graph cut optimization for Markov random fields. In: Proc. of the ICCV (October 2007)
8. Komodakis, N., Tziritas, G., Paragios, N.: Fast, approximately optimal solutions for single and dynamic MRFs. In: Proc. of the CVPR (June 2007)
9. Chambolle, A.: Total variation minimization and a class of binary MRF models. Energy Minimization Methods in Comp. Vision and Pattern Rec. 136–152 (2005)
10. Nikolova, M., Esedoglu, S., Chan, T.F.: Algorithms for finding global minimizers of image segmentation and denoising models. SIAM J. on App. Math. 66 (2006)
11. Horn, B.K.P., Schunck, B.G.: Determining optical flow. Artificial Intelligence 17, 185–203 (1981)
12. Black, M.J., Anandan, P.: A framework for the robust estimation of optical flow. In: Proc. of the ICCV, pp. 231–236 (May 1993)
13. Aubert, G., Deriche, R., Kornprobst, P.: Computing optical flow via variational techniques. SIAM Journal on Applied Mathematics 60(1), 156–182 (2000)
14. Papenberg, N., Bruhn, A., Brox, T., Didas, S., Weickert, J.: Highly accurate optic flow computation with theoretically justified warping. International Journal of Computer Vision 67(2), 141–158 (2006)
15. Bruhn, A., Weickert, J., Kohlberger, T., Schnörr, C.: A multigrid platform for real-time motion computation with discontinuity-preserving variational methods. International Journal of Computer Vision 70(3), 257–277 (2006)
16. Zach, C., Pock, T., Bischof, H.: A duality based approach for realtime TV-L1 optical flow. In: Hamprecht, F.A., Schnörr, C., Jähne, B. (eds.) DAGM 2007. LNCS, vol. 4713, pp. 214–223. Springer, Heidelberg (2007)
17. Baker, S., Scharstein, D., Lewis, J.P., Roth, S., Black, M., Szeliski, R.: A database and evaluation methodology for optical flow. In: Proc. of the ICCV (2007)
18. Aujol, J.F., Gilboa, G., Chan, T.F., Osher, S.: Structure-texture image decomposition – modeling, algorithms, and parameter selection. International Journal of Computer Vision 67(1), 111–136 (2006)
19. Rudin, L., Osher, S., Fatemi, E.: Nonlinear total variation based noise removal algorithms. Physica D 60, 259–268 (1992)
20. Chambolle, A.: An algorithm for total variation minimization and applications. Journal of Mathematical Imaging and Vision 20, 89–97 (2004)
21. Trobin, W., Pock, T., Cremers, D., Bischof, H.: An unbiased second-order prior for high-accuracy motion estimation. In: Rigoll, G. (ed.) DAGM 2008. LNCS, vol. 5096, pp. 396–405. Springer, Heidelberg (2008)
22. Danielsson, P.E., Lin, Q.: Efficient detection of second-degree variations in 2D and 3D images. Journal of Visual Comm. and Image Representation 12, 255–305 (2001)
23. Kohli, P., Kumar, P., Torr, P.H.: P3 & beyond: Solving energies with higher order cliques. In: Proc. of the CVPR (June 2007)
24. Pock, T., Schoenemann, T., Cremers, D., Bischof, H.: A convex formulation of continuous multi-label problems. In: Forsyth, D., Torr, P., Zisserman, A. (eds.) ECCV 2008. LNCS, vol. 5304, pp. 792–805. Springer, Heidelberg (2008)
25. Scharstein, D., Szeliski, R.: High-accuracy stereo depth maps using structured light. In: Proc. of the CVPR, vol. 1, pp. 195–202 (June 2003)

Unified Crowd Segmentation

Peter Tu, Thomas Sebastian, Gianfranco Doretto, Nils Krahnstoever,
Jens Rittscher, and Ting Yu

GE Global Research, Niskayuna, NY USA
tu@crd.ge.com

Abstract. This paper presents a unified approach to crowd segmentation. A global solution is generated using an Expectation Maximization framework. Initially, a head and shoulder detector is used to nominate an exhaustive set of person locations and these form the person hypotheses. The image is then partitioned into a grid of small patches which are each assigned to one of the person hypotheses. A key idea of this paper is that while whole body monolithic person detectors can fail due to occlusion, a partial response to such a detector can be used to evaluate the likelihood of a single patch being assigned to a hypothesis. This captures local appearance information without having to learn specific appearance models. The likelihood of a pair of patches being assigned to a person hypothesis is evaluated based on low level image features such as uniform motion fields and color constancy. During the E-step, the single and pairwise likelihoods are used to compute a globally optimal set of assignments of patches to hypotheses. In the M-step, parameters which enforce global consistency of assignments are estimated. This can be viewed as a form of occlusion reasoning. The final assignment of patches to hypotheses constitutes a segmentation of the crowd. The resulting system provides a global solution that does not require background modeling and is robust with respect to clutter and partial occlusion.

1 Introduction

The segmentation of crowds into individuals continues to be a challenging research problem in computer vision [1,2,3,4,5]. The automation of video surveillance systems in public venues such as airports, mass-transit stations and sports stadiums requires the ability to detect and track individuals through complex sites. We identify three challenges that make this problem particularly difficult: *(i) Partial occlusion.* In many crowded scenes people can be partially occluded by others. Monolithic detectors [2,6,7] that model the shape and appearance of an entire person typically fail in such situations and hence cannot reliably detect people in crowded environments. *(ii) Dynamic backgrounds.* When cameras are fixed, statistical background models are commonly used to identify foreground regions [8]. However, this approach fails when the background is dynamic. Further, background modeling is not applicable for moving cameras, such as those mounted on pan tilt devices or mobile platforms. *(iii) Foreground clutter.* The presence of moving non-person objects such as luggage carts, shopping trolleys

D. Forsyth, P. Torr, and A. Zisserman (Eds.): ECCV 2008, Part IV, LNCS 5305, pp. 691–704, 2008.

and cleaning equipment can clutter the foreground of the scene. A robust crowd segmentation algorithm should be immune to foreground clutter without having to explicitly model the appearance of every non-person object.

This paper presents a unified approach to crowd segmentation that effectively addresses these three challenges. The proposed system combines bottom-up and top-down approaches in a unified framework to create a robust crowd segmentation algorithm. We first review a number of relevant approaches.

Low level feature grouping has been used to segment crowds [5,9]. These approaches take advantage of the fact that the motion field for an individual is relatively uniform and hence tracked corners with common trajectories can be grouped together to form individuals. However, difficulties arise when multiple individuals have similar trajectories. *Monolithic classifiers* capture the shape and appearance space for the whole body using relatively simple learning methods [10,6,7]. The direct application of these classifiers to non-crowded scenes generates reasonable segmentations, however failure modes can occur when partial occlusions are encountered. *Part based constellation models* [11,12,13] construct boosted classifiers for specific body parts such as the head, the torso and the legs, and each positive detection generates a Hough-like vote in a parametrized person space. The detection of local maxima in this space constitutes a segmentation. A similar approach [2] uses interest operators to nominate image patches which are mapped to a learned code book. A drawback of these approaches is that the identification of local maxima in the Hough space can be problematic under crowded and cluttered environments - a global approach is required.

The previous approaches can be considered to be bottom-up methods where local context is used. On the other hand, *global approaches* that rely on background segmentation has been proposed in [14,4]. In [14], Markov Chain Monte Carlo (MCMC) algorithms are used to nominate various crowd configurations which are then compared with foreground silhouette images. However, this form of random search can be computationally expensive. To address this issue an Expectation Maximization (EM) based approach has been developed [4]. In this framework, a hypothesis nomination scheme generates a set of possible person locations. Image features are then extracted from foreground silhouettes and a global search for the optimal assignment of features to hypotheses is performed. The set of hypotheses that receive a significant number of assignments constitute the final segmentation. Reliance on accurate foreground background segmentation is a weakness of both of these approaches.

1.1   Overview of the Unified Approach

In this paper we extend the global EM crowd segmentation framework [4] to use appearance-based features that do not rely on background segmentation. A head and shoulder classifier is used to generate an initial set of hypothesized person locations, a grid of patches are then superimposed on the image. A globally optimal assignment of patches to hypotheses defines the final segmentation. The likelihood of a single patch to hypothesis assignment is evaluated based on local appearance. However, instead of learning an appearance and spatial distribution

iterative fashion so as to emphasize samples that were mis-classified during previous iteration. In this application an iterative site-specific approach is used for learning. Initial training data from the site of interest is manually labeled and a classifier is constructed. It is then applied to additional imagery taken from the site, and the resulting false positives are incorporated into the training data, while correctly classified negative training samples are removed. A new classifier is then constructed and this process is repeated until no new false positive responses are generated. In this manner a series of strong classifiers are constructed which are combined to form a cascaded classifier. The type of weak classifiers that are appropriate for this application is now considered.

3.2   Weak Classifiers

A particular type of weak classifier can be characterized as follows:

$$wc(s; \mathcal{R}(s)), \tag{4}$$

where $\mathcal{R}$ is a region of interest defined relative to the sample bounding box associated with a sample s. If the average values for a set of image statistics are above (or below) a given set of thresholds, then the weak classifier produces a positive or negative response accordingly. Once the type of image statistics have been selected, the weak classifier is essentially parametrized by the relative location and dimensions of its region of interest $\mathcal{R}$. In general the threshold values for selected weak classifiers are determined during the learning phase of the boosting process. If we restrict our hypothesis space to this type of weak classifier, then a patch specific partial response for a whole body classifier can be generated.

Based on the boosting algorithm a strong whole body classifier for the sample s is defined as:

$$sc(s) = \sum_{i=1}^{M} \alpha_i wc(s; \mathcal{R}_i(s)) \tag{5}$$

The basic idea for generating patch specific responses is that each weak classifier will only collect statistics over the intersection of $\mathcal{R}(s)$ and the patch z. Since average statics are used, the thresholds learned during boosting remain valid. However, instead of having a $1/-1$ response, each weak classifier will have its response modulated by the ratio of the areas of $\mathcal{R}(s) \cap z$ and $\mathcal{R}(s)$. Based on this idea, the partial response for a strong classifier with respect to a given patch z and sample s is defined as:

$$sc(s, z) = \sum_{i=1}^{M} \alpha_i wc_i(s, z) \frac{\int_{\mathcal{R}_i(s) \cap z} dx}{\int_{\mathcal{R}_i(s)} dx}, \tag{6}$$

where

$$wc_i(s, z) = wc(s; \mathcal{R}_i(s) \cap z). \tag{7}$$

Note that if the region of interest associated with a particular weak classifier does not intersect with the patch z, then this weak classifier will have no effect on the strong classifier decision.

For a given person hypothesis c_k, a sample s_k can be constructed so that for a particular patch z_i, the direct patch to hypothesis affinity measure can be defined as:

$$g_k(z_i) = sc(s_k, z_i) \tag{8}$$

Figure 2 shows a set of cascaded classifiers that were used to construct the whole body classifier. In this application the image statistic used is the magnitude of the edge responses for pixels that exhibited an orientation similar to the preferred orientation of the weak classifier. Edge magnitude and orientation are calculated using the Sobel operator. Given such a whole body classifier the next question is to determine the appropriate patch size. If the patch is too large, then there is risk of contamination by occlusion. On the other hand, if the patch is too small the ability to discriminate between correct and incorrect patch assignments diminishes. To understand this tradeoff a training set of positive and negative whole body samples was collected. Patches with widths ranging from 0.25W to 1.0W (W = person width) were evaluated across the entire bounding box for each training sample. For each relative patch location, the average number of positively responding strong classifiers from the cascaded whole body classifier was recorded. As shown in Figure 3, when the patch width was reduced below 0.5W the ability to discriminate between positive and negative samples was reduced significantly. Thus for this application a nominal patch width of 0.5W is chosen.

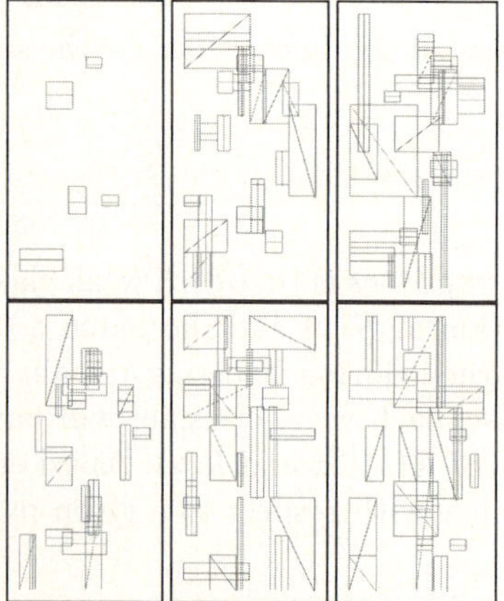 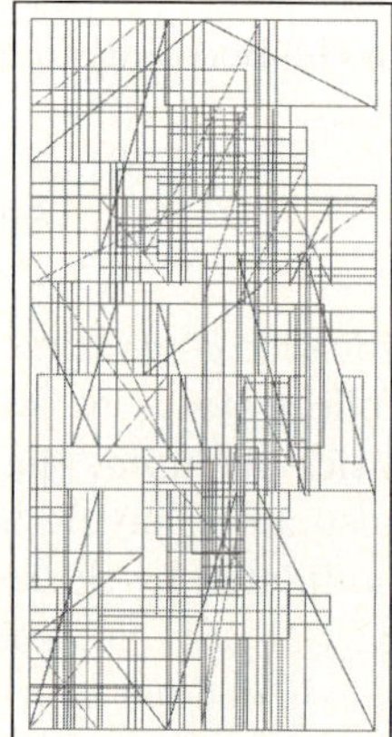

Fig. 2. This figure shows the six strong classifiers that were constructed for the whole body classifier plus all six cascades shown together. Each pink box represents the region of interest for a weak classifier. The line interior to each region of interest depict the weak classifier's preferred orientation. Green lines represent positive features (the average statistic must be above its threshold) and red lines are for negative features (average statistic must be below its threshold).

Positive Response	Negative Response	Patch Width	Positive Response	Negative Response	Patch Width

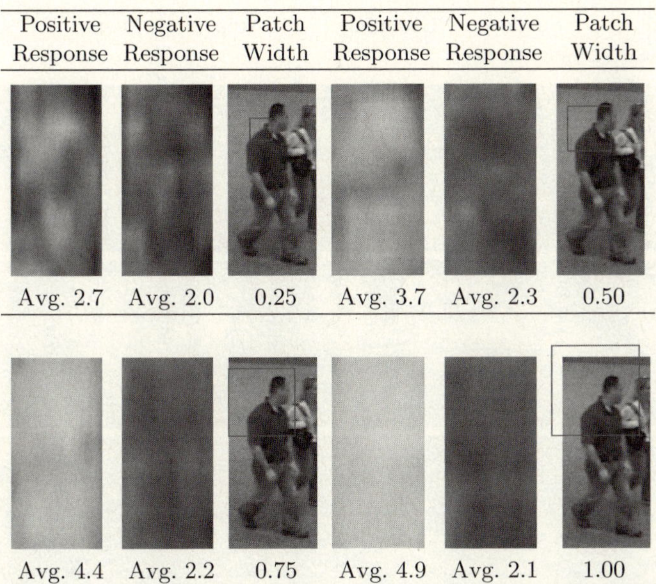

Avg. 2.7	Avg. 2.0	0.25	Avg. 3.7	Avg. 2.3	0.50
Avg. 4.4	Avg. 2.2	0.75	Avg. 4.9	Avg. 2.1	1.00

Fig. 3. This figure shows the effect of changing the patch size. Patch size is varied as a function of W = person width. In each case the average number of positively responding strong classifiers from the whole body cascaded classifier is shown as a function of patch location for both positive (person) and negative (non-person) images. Note that when the patch width is reduced below 0.5W the ability to discriminate between positive and negative samples is significantly reduced.

3.3   Hypothesis Nomination

For this application, hypothesis nomination as described in section 2 is achieved using a scanning window approach. For every possible sample, the partial response for the whole body classifier is evaluated based on a patch covering the hypothesized head and shoulder regions. The set of positive responses constitute the initial hypotheses set C.

4   Experiments

Unrehearsed imagery acquired at a mass transit site serves as the source of test imagery for this paper. A whole body classifier was trained for this site. We first illustrate the intermediate steps of our approach on a few representative frames (see Figure 4). The "Initial Hypothesis" column of figure 4 shows the initial set of hypotheses generated by the head and shoulders classifier. Note that while an appropriate hypothesis was generated for each person in each image, several false hypotheses were also generated. The "Single Assignment" column of figure 4 illustrates the direct affinity between each patch and each hypothesis as computed using equation 8. Each patch is color coded based on the hypothesis

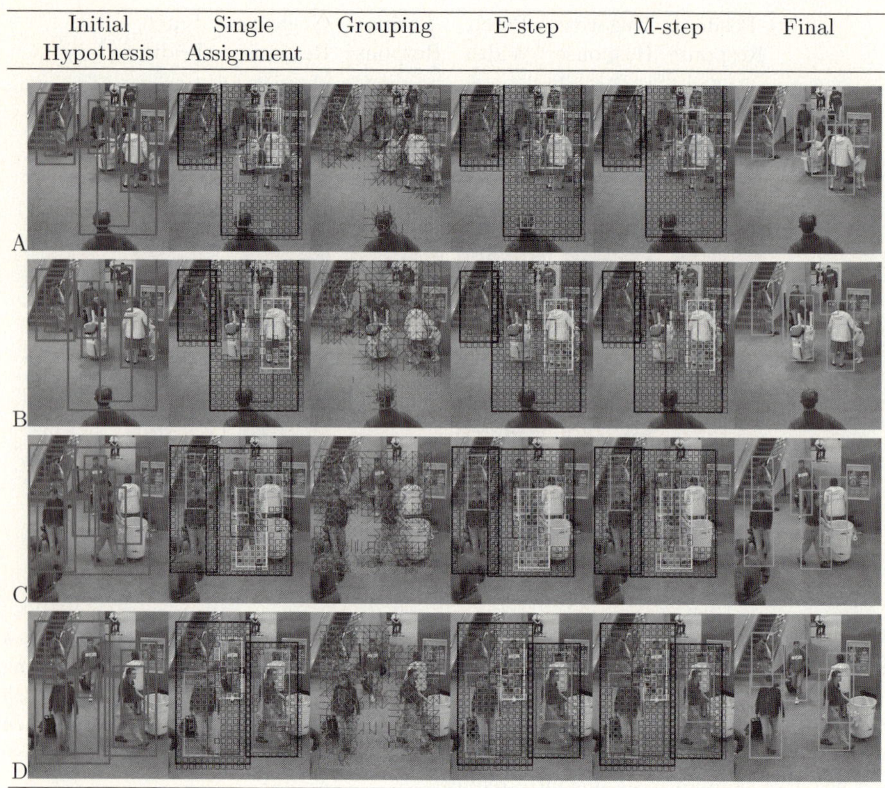

Initial Hypothesis   Single Assignment   Grouping   E-step   M-step   Final

Fig. 4. Six stages of the crowd segmentation process are illustrated for four test images. Overlapping patches of 0.5W are used. However, for clarity smaller patches are shown. The initial hypotheses generated by the head and shoulder classifier are shown in the first column. In the second column, the patches are shown color coded based on their strongest direct assignment as calculated by the whole body classifier. The null hypothesis is shown in black. In the third column, neighboring patches with strong similarity measures based on color constancy are connected by green line segments. The assignment of patches to hypotheses based on the first E-step is shown in the fourth column. The assignment after multiple rounds of both the E and M steps are shown in the fifth column. The final segmentation is shown in the last column.

for which it has the highest direct affinity. Patches that are black have the greatest affinity for the null hypothesis. A significant number of patches have the greatest direct affinity for their true hypothesis, however confusion occurs when multiple hypotheses overlap. An example of this can be seen in row A of the Single Assignment column. In addition, patches that are only associated with false detections tend to have a greater affinity for the null hypothesis.

The "Grouping" column of figure 4 illustrates the effectiveness of the pairwise assignment criteria. For purposes of clarity, only neighboring patches with high similarity measures are shown to be linked in green (blue otherwise). Note that

Fig. 5. This figure shows an assortment of crowd segmentation results. Note that the algorithm produces the correct segmentation in case of severe partial occlusion (right column), and in presence of cleaning equipment (bottom left) and a variety of suitcases and bags.

patches associated with the same article of clothing tend to be grouped together. Also, the background often exhibits continuity in appearance and such patches tend to be grouped together.

The "E-step" column of figure 4 shows the patch assignment after the first iteration of the "E-step". Most of the false hypotheses have received very few patch assignments, while the true hypotheses have been assigned patches in an appropriate manner. However, inconsistencies have also been generated. For example, in row A of the "E-step" column, a number of patches have been assigned to the bottom of the green hypothesis. As seen in the "M-step" column, these inconsistent assignments have been correctly removed.

The "Final" column of figure 4 shows the final segmentation. Figure 5 shows similar results from a variety of images. Note that the algorithm is successful when confronted with partial occlusion and clutter such as the janitor's equipment and various suitcases.

The algorithm was also applied to a video sequence (see supplemental material). To measure overall performance 117 frames were processed. The initial hypothesis generator produced 480 true detections, 32 false detections and 79 missed detections. After application of the crowd segmentation algorithm, the number of false detections were reduced by 72 percent at a cost of falsely rejecting 2 percent of the true detections. For purposes of comparison, we applied the Histogram of Oriented Gradients (HOG) [6] to this dataset. Our implementation uses camera calibration information for automatic scale selection. The performance tabulated in Table 1 shows that our crowd segmentation outperformed HOG, arguably this is due to partial occlusion.

Table 1. Comparison of HOG [6] person detector to the proposed crowd segmentation algorithm

	True Detects	Missed Detects	False Alarms
Crowd Segmentation	470	89	9
HOG [6]	387	172	20

frame 1	frame 30	frame 80	frame 117

Fig. 6. Four example frames from tracking the results of the crowd segmentation process

The purpose of crowd segmentation algorithms in general is to reliably track the location of people over time. The final segmentation of the sequence previously described was processed by a general-purpose person tracking algorithm. At every time step, locations and estimates of the location uncertainties are projected into the scene ground-plane via an unscented transform. Our tracker processes these detections in a scene ground-plane reference frame, where the dynamical models are intuitively formulated. Our approach to tracking is similar to [17] and [18]. We follow an efficient detect and track approach [19] using a JPDAF filter [20], which has excellent performance in high degrees of clutter while being efficient in the presence of many targets. The tracking results in Figure 6 show the trajectories of all people in the scene.

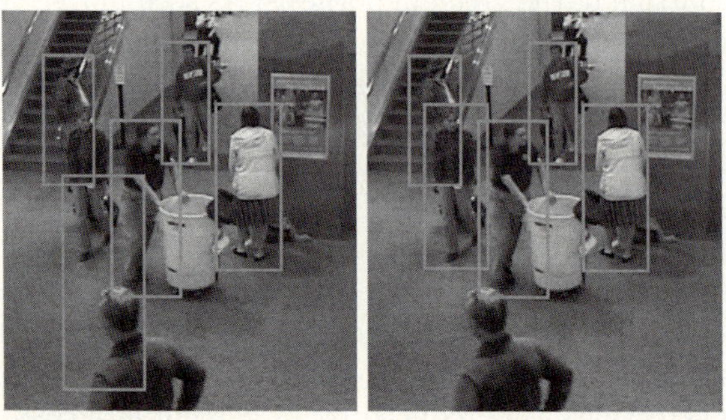

Fig. 7. This figure illustrates the effect of using motion fields in the pairwise patch assignment. The example on the left shows a frame where the algorithm results in both a false positive and a false negative. However, when the motion information from dense optical flow is used the correct segmentation results, as shown on the right.

The results thus far used pairwise patch similarity function based on color constancy as defined in Equation 1. However, this is not always enough as shown in the left image of Figure 7 where the crowd segmentation algorithm resulted in both a false and a missed detection. An experiment was performed where the pairwise patch similarity measures were augmented by the use of a motion consistency measure based on dense optical flow. As can be seen from the right image in Figure 7, this results in a correct segmentation.

5   Discussion

The framework presented in this paper has incorporated many of the strengths of previously proposed crowd segmentation methods into a single unified approach. A novel aspect of this paper is that monolithic whole body classifiers were used to analyze partially occluded regions by considering partial responses associated with specific image patches. In this way appearance information is incorporated into a global optimization process alleviating the need for foreground background segmentation. The EM framework was also able to consider low level image cues such as color histograms and thus take advantage of the potential color constancy associated with clothing and the background. Parametrization of the likelihood function allowed for the enforcement of global consistency of the segmentation. It was shown that these parameters can be estimated during the *M-step* and that this facilitates consistency based on occlusion reasoning.

In the course of experimentation it was found that at various times, different aspects of the crowd segmentation system proved to be the difference between success and failure. For example when confronted with clutter, the appearance based classifiers provide the saliency required to overcome these challenges. However, when multiple people having similar clothing are encountered, the motion field can become the discriminating factor. A robust system must be able to take advantage of its multiple strengths and degrade gracefully when confronted by their weaknesses.

References

1. Munder, S., Gavrila, D.: An experimental study on pedestrian classification. IEEE Trans. on Pattern Analysis and Machine Intelligence 28(11), 1863–1868 (2006)
2. Leibe, B., Seemann, E., Schiele, B.: Pedestrian detection in crowded scenes. IEEE Computer Vision and Pattern Recognition, 878–885 (2005)
3. Leibe, B., Cornelis, N., Cornelis, K., Gool, L.V.: Dynamic 3d scene analysis from a moving vehicle. IEEE Computer Vision and Pattern Recognition, 1–8 (2007)
4. Rittscher, J., Tu, P.H., Krahnstoever, N.: Simultaneous estimation of segmentation and shape. IEEE Computer Vision and Pattern Recognition 2, 486–493 (2005)
5. Brostow, G.J., Cipolla, R.: Unsupervised bayesian detection of independent motion in crowds. IEEE Computer Vision and Pattern Recognition I, 594–601 (2006)
6. Dalal, N., Triggs, B.: Histograms of oriented gradients for human detection. IEEE Computer Vision and Pattern Recognition, 886–893 (2005)

7. Tuzel, O., Porikli, F., Meer, P.: Pedestrian detection via classification on riemannian manifolds. IEEE Computer Vision and Pattern Recognition (2007)
8. Stauffer, C., Grimson, W.: Adaptive background mixture models for real-time tracking. IEEE Computer Vision and Pattern Recognition 2, 246–252 (1998)
9. Rabaud, V., Belongie, S.: Counting crowded moving objects. IEEE Computer Vision and Pattern Recognition, 705–711 (2006)
10. Viola, P., Jones, M., Snow, D.: Detecting pedestrians using patterns of motion and appearance. International Journal of Computer Vision 2, 734–741 (2003)
11. Fergus, R., Perona, P., Zisserman, A.: A visual category filter for Google images. In: European Conference on Computer Vision, vol. 1, pp. 242–256 (2004)
12. Mikolajczyk, K., Schmid, C., Zisserman, A.: Human detection based on a probabilistic assembly of robust part detectors. In: European Conference on Computer Vision (2004)
13. Wu, B., Nevatia, R.: Detection and tracking of multiple partially occluded humans by bayesian combination of edgelet based part detectors. International Journal of Computer Vision 75(2), 247–266 (2007)
14. Zhao, T., Nevatia, R.R.: Bayesian human segmentation in crowded situations. IEEE Computer Vision and Pattern Recognition 2, 459–466 (2003)
15. Chui, H., Rangarajan, A.: A new point matching algorithm for non-rigid registration. Computer Vision and Image Understanding 89(3), 114–141 (2003)
16. Viola, P., Jones, M.J.: Robust real-time face detection. International Journal of Computer Vision 57(2), 137–154 (2004)
17. Krahnstoever, N., Tu, P., Sebastian, T., Perera, A., Collins, R.: Multi-view detection and tracking of travelers and luggage in mass transit environments. In: Proc. Ninth IEEE International Workshop on Performance Evaluation of Tracking and Surveillance (PETS) (2006)
18. Leibe, B., Schindler, K., Gool, L.V.: Coupled detection and trajectory estimation for multi-object tracking. In: International Conference on Computer Vision (ICCV 2007), Rio de Janeiro, Brasil (October 2007)
19. Blackman, S., Popoli, R.: Design and Analysis of Modern Tracking Systems. Artech House Publishers (1999)
20. Rasmussen, C., Hager, G.: Joint probabilistic techniques for tracking multi-part objects. In: Proc. IEEE Conference on Computer Vision and Pattern Recognition, pp. 16–21 (1998)

Quick Shift and Kernel Methods
for Mode Seeking

Andrea Vedaldi and Stefano Soatto

University of California, Los Angeles
Computer Science Department
{vedaldi,soatto}@ucla.edu

Abstract. We show that the complexity of the recently introduced medoid-shift algorithm in clustering N points is $O(N^2)$, with a small constant, if the underlying distance is Euclidean. This makes medoid shift considerably faster than mean shift, contrarily to what previously believed. We then exploit kernel methods to extend both mean shift and the improved medoid shift to a large family of distances, with complexity bounded by the effective rank of the resulting kernel matrix, and with explicit regularization constraints. Finally, we show that, under certain conditions, medoid shift fails to cluster data points belonging to the same mode, resulting in over-fragmentation. We propose remedies for this problem, by introducing a novel, simple and extremely efficient clustering algorithm, called quick shift, that explicitly trades off under- and over-fragmentation. Like medoid shift, quick shift operates in non-Euclidean spaces in a straightforward manner. We also show that the accelerated medoid shift can be used to initialize mean shift for increased efficiency. We illustrate our algorithms to clustering data on manifolds, image segmentation, and the automatic discovery of visual categories.

1   Introduction

Mean shift [9,3,5] is a popular non-parametric clustering algorithm based on the idea of associating each data point to a mode of the underlying probability density function. This simple criterion has appealing advantages compared to other traditional clustering techniques: The structure of the clusters may be rather arbitrary and the number of clusters does not need to be known in advance.

Mean shift is not the only "mode seeking" clustering algorithm. Other examples include earlier graph-based methods [13] and, more recently, *medoid shift* [20]. Unlike mean shift, medoid shift extends easily to general metric spaces (i.e. spaces endowed with a distance). In fact, mean shift is essentially a gradient ascent algorithm [3,5,24] and the gradient may not be defined unless the data space has additional structure (e.g. Hilbert space or smooth manifold structure). While there have been recent efforts to generalize mean shift to non-linear manifolds [21], medoid shift does not require any additional steps to be used on curved spaces. Moreover, the algorithm is non-iterative and there is no need for a stopping heuristic. Its biggest disadvantage is its computational complexity [20].

D. Forsyth, P. Torr, and A. Zisserman (Eds.): ECCV 2008, Part IV, LNCS 5305, pp. 705–718, 2008.

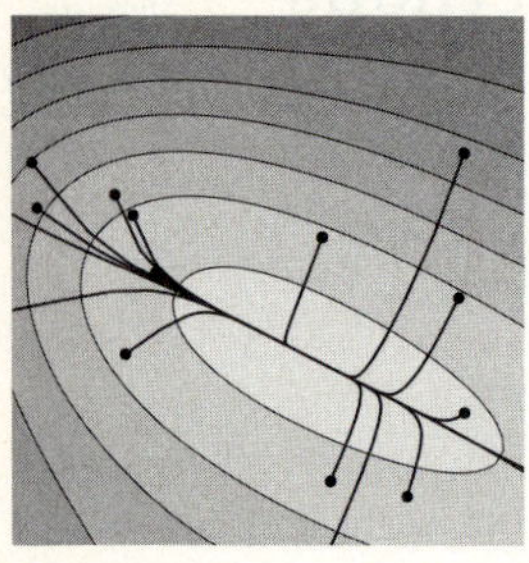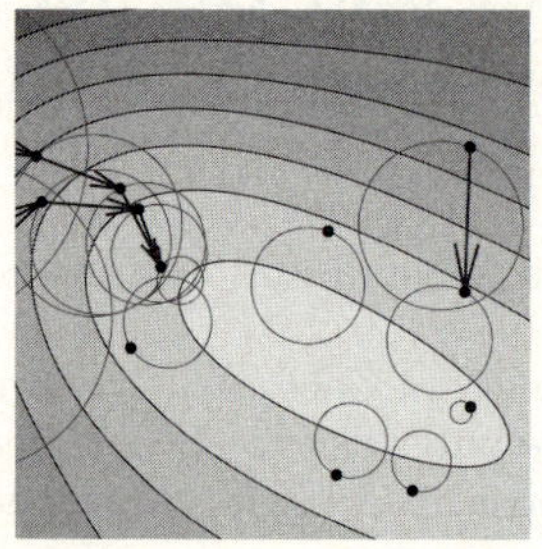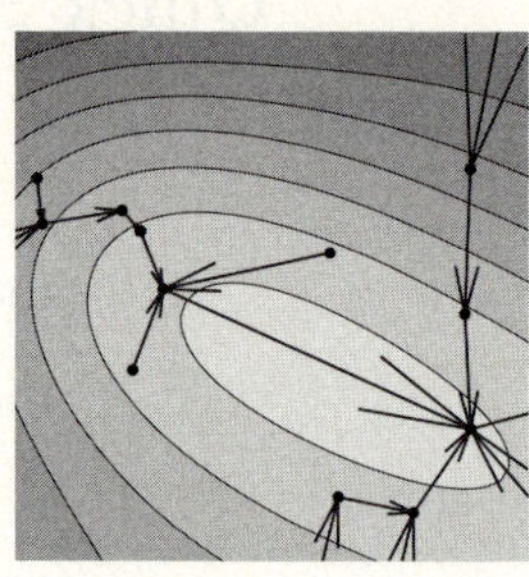

Fig. 1. Mode seeking algorithms. Comparison of different mode seeking algorithms (Sect. 2) on a toy problem. The black dots represent (some of) the data points $x_i \in \mathcal{X} \subset \mathbb{R}^2$ and the intensity of the image is proportional to the Parzen density estimate $P(x)$. **Left.** Mean shift moves the points uphill towards the mode approximately following the gradient. **Middle.** Medoid shift approximates mean shift trajectories by connecting data points. For reason explained in the text and in Fig. 2, medoid shifts are constrained to connect points comprised in the red circles. This disconnects portions of the space where the data is sparse, and can be alleviated (but not solved) by iterating the procedure (Fig. 2). **Right.** Quick shift (Sect. 3) seeks the energy modes by connecting nearest neighbors at higher energy levels, trading-off mode over- and under-fragmentation.

Depending on the implementation, medoid shift requires between $O(dN^2 + N^3)$ and $O(dN^2 + N^{2.38})$ operations to cluster N points, where d is the dimensionality of the data. On the other hand, mean shift is only $O(dN^2T)$, where T is the number of iterations of the algorithm, and clever implementations yield $dT \ll N$.

Contributions. In this paper we show that the computational complexity of Euclidean medoid shift is only $O(dN^2)$ (with a small constant), which makes it faster (not slower!) than mean shift (Sect. 3). We then generalize this result to a large family of non-Euclidean distances by using kernel methods [18], showing that in this case the complexity is bounded by the effective dimensionality of the kernel space (Sect. 3). Working with kernels has other advantages: First, it extends to mean shift (Sect. 4); second, it gives an explicit interpretation of non-Euclidean medoid shift; third, it suggests why such generalized mode seeking algorithms skirt the *curse of dimensionality*, despite estimating a density in complex spaces (Sect. 4). In summary, we show that kernels extend mode seeking algorithms to non-Euclidean spaces in a simple, general and efficient way.

Can we conclude that medoid shift should replace mean shift? Unfortunately, not. We show that the weak point of medoid shift is its inability to identify consistently all the modes of the density (Sect. 2). This fact was addressed implicitly by [20] who reiterate medoid shift on a simplified dataset (similar to [2]). However, this compromises the non-iterative nature of medoid shift and changes the underlying density function (which may be undesirable). Moreover, we show that this fix does not always work (Fig. 2).

We address this issue in two ways. First, we propose using medoid shift to simplify the data and initialize the more accurate mean shift algorithm (Sect. 5.2 and Sect. 5.3). Second, we propose an alternative mode seeking algorithm that can trade off mode over- and under-fragmentation (Sect. 3). This algorithm, related to [13], is particularly simple and fast, yields surprisingly good segmentations, and returns a one parameter family of segmentations where model selection can be applied.

We demonstrate these algorithms on three tasks (Sect. 5): Clustering on a manifold (Sect. 5.1), image segmentation (Sect. 5.2), and clustering image signatures for automatic object categorization (Sect. 5.3). The relative advantages and disadvantages of the various algorithms are discussed.

2   Mode Seeking

Given N data points $x_1, \ldots, x_N \in \mathcal{X} = \mathbb{R}^d$, a *mode seeking* clustering algorithm conceptually starts by computing the *Parzen density estimate*

$$P(x) = \frac{1}{N} \sum_{i=1}^{N} k(x - x_i), \quad x \in \mathbb{R}^d \tag{1}$$

where $k(x)$ can be a Gaussian or other window.[1] Then each point x_i is moved towards a mode of $P(x)$ evolving the trajectory $y_i(t)$, $t > 0$ uphill, starting from $y_i(0) = x_i$ and following the gradient $\nabla P(y_i(t))$. All the points that converge to the same mode form a cluster.

A mode seeking algorithm needs (i) a numerical scheme to evolve the trajectories $y_i(t)$, (ii) a halting rule to decide when to stop the evolution and (iii) a clustering rule to merge the trajectory end-points. Next, we discuss two algorithms of this family.

Mean Shift. Mean shift [9,5] is based on an efficient rule to evolve the trajectories $y_i(t)$ when the window $k(x)$ can be written as $\psi(\|x\|_2^2)$ for a convex function $\psi(z)$ (for instance the Gaussian window has $\psi(z) \propto \exp(-z)$). The idea is to bound the window from below by the quadric $k(z') \geq k(z) + (\|z'\|_2^2 - \|z\|_2^2)\dot{\psi}(\|z\|_2^2)$. Substituting in (1) yields

$$P(y') \geq P(y) + \frac{1}{N} \sum_{j=1}^{N} (\|y' - x_j\|_2^2 - \|y - x_j\|_2^2)\dot{\psi}(\|y - x_j\|_2^2), \tag{2}$$

and maximizing this lower bound at $y = y_i(t)$ yields the mean-shift update rule

$$y_i(t+1) = \operatorname*{argmax}_y \frac{1}{N} \sum_{j=1}^{N} \|y - x_j\|_2^2 \dot{\psi}(\|y_i(t) - x_j\|_2^2) = \frac{\sum_{j=1}^{N} \dot{\psi}(\|y_i(t) - x_j\|_2^2)x_j}{\sum_{j=1}^{N} \dot{\psi}(\|y_i(t) - x_j\|_2^2)}. \tag{3}$$

---

[1] The term "kernel" is also used in the literature. Here we use the term "window" to avoid confusion with the kernels introduced in Sect. 3.

If the profile $\psi(z)$ is monotonically decreasing, then $P(y_i(t)) < P(y_i(t+1))$ at each step and the algorithm converges in the limit (since P is bounded [5]). The complexity is $O(dN^2T)$, where d is the dimensionality of the data space and T is the number of iterations. The behavior of the algorithm is illustrated in Fig. 1.

Medoid Shift. Medoid shift [20] is a modification of mean shift in which the trajectories $y_i(t)$ are constrained to pass through the points x_i, $i = 1, \ldots, N$. The advantage of medoid shift are: (i) only one step $y_i(1)$, $i = 1, \ldots, N$ has to be computed for each point x_i (because $y_i(t+1) = y_{y_i(t)}(1)$), (ii) there is no need for a stopping/merging heuristic (as these conditions are met exactly), and (iii) the data space $\mathcal{X}$ may be non-Euclidean (since to maximize (4) there is no need to compute derivatives). Eventually, points are linked by steps into a forest, with clusters corresponding to trees. The algorithm is illustrated in Fig. 1.

According to [20], the main drawback of medoid shift is speed. In fact, maximizing (3) restricted to the dataset amounts to calculating

$$y_i(1) = \operatorname*{argmax}_{y \in \{x_1, \ldots, x_N\}} \frac{1}{N} \sum_{j=1}^{N} \mathsf{d}^2(y, x_j) \dot{\phi}(\mathsf{d}^2(x_j, x_i)) \tag{4}$$

where $\mathsf{d}^2(x, y) = \|x - y\|_2^2$ in the Euclidean case. A basic implementation requires $O(N^3 + dN^2)$ operations, assuming $O(d)$ operations to evaluate $\mathsf{d}^2(x, y)$. However, by defining matrices $D_{kj} = \mathsf{d}^2(x_k, x_j)$ and $F_{ki} = \dot{\phi}(D_{ik})/N$, we can rewrite (4) as

$$y_i(1) = \operatorname*{argmax}_{k=1,\ldots,N} \sum_{j=1}^{N} D_{kj} F_{ji} = \operatorname*{argmax}_{k=1,\ldots,N} e_k^\top D F e_i \tag{5}$$

where e_i denotes the i-th element of the canonical basis.[2] As noted in [20], $O(N^{2.38})$ operations are sufficient by using the fastest matrix multiplication algorithm available. Unfortunately the hidden constant of this algorithm is too large to be practical (see [12], pag. 501). Thus, a realistic estimate of the time required is more pessimistic than what suggested by the asymptotic estimate $O(dN^2 + N^{2.38})$.

Here we note that a more delicate issue with medoid shift is that it may fail to properly identify the modes of the density $P(x)$. This is illustrated in Fig. 2, where medoid shift fails to cluster three real points $-1, +1$ and $+1/2$, finding two modes -1 and $+1$ instead of one. To overcome this problem, [20] applies medoid shift iteratively on the modes (in the example -1 and $+1$). However, this solution is not completely satisfactory because (i) the underlying model $P(x)$ is changed (similarly to *blurry* mean shift [9,3]) and (ii) the strategy does not work in all cases (for instance, in Fig. 2 points -1 and $+1$ still fail to converge to a single mode).

Finally, consider the interpretation of medoid shift. When $\mathcal{X}$ is a Hilbert space, medoid (and mean) shift follow approximately the gradient of the density $P(x)$

---

[2] For instance $e_2 = \begin{bmatrix} 0 & 1 & 0 & \ldots & 0 \end{bmatrix}^\top$.

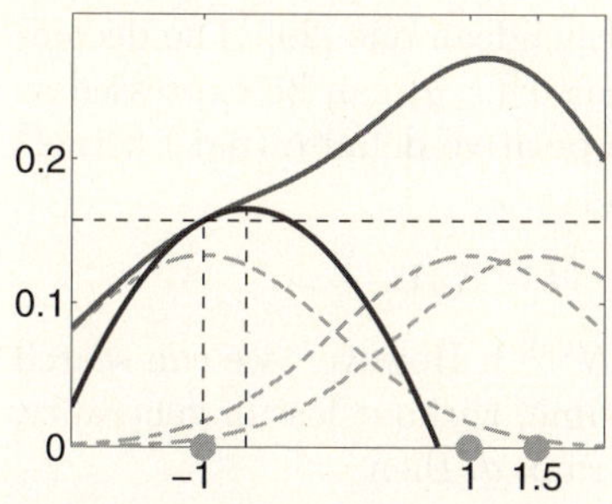 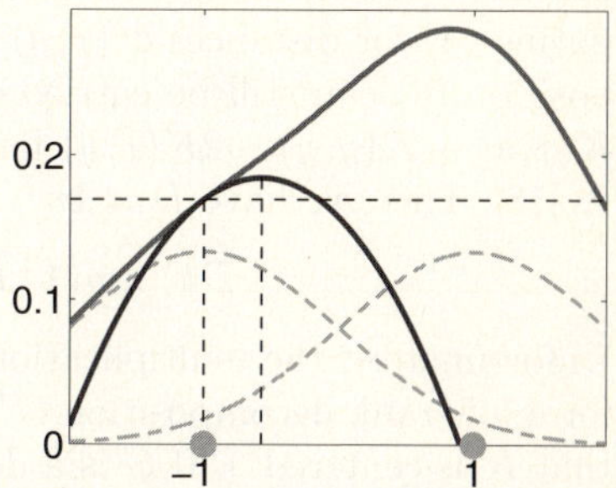

Fig. 2. Medoid shift over-fragmentation. Left. We apply medoid shift to cluster points $-1, +1, +1/2 \in \mathbb{R}$ using a Gaussian window of variance $\sigma^2 = 1$ (dashed green lines). The density $P(x)$ (red curve; Sect. 2) has a single mode, but medoid shift fails to move the point -1 towards the mode (i.e. $y_{-1}(1) = -1$). The reason is that the quadratic lower bound (2) (blue curve) is larger at -1 than it is at $+1$ or $+1/2$. Notice that mean shift would have moved -1 towards the mode by a small, but finite amount, eventually extracting the single mode. **Right.** The problem is not solved even if medoid shift is reiterated [20] on the two modes -1 and $+1$ (where $+1$ has double mass), even if the density $P(x)$ *does* become blurrier [2,20].

(by maximizing the lower bound (3)). The gradient depends crucially on the inner product and corresponding metric defined on $\mathcal{X}$, which encodes the cost of moving along each direction [22]. For general metric spaces $\mathcal{X}$, the gradient may not be defined, but the term $\mathbf{d}^2(x, y)$ in (4) has a similar direction-weighing effect. In later sections we will make this connection more explicit.

3   Fast Clustering

Faster Euclidean Medoid Shift. We show that the complexity of Euclidean medoid shift is only $O(dN^2)$ (with a small constant) instead of $O(dN^2 + N^{2.38})$ (with a large constant) [20]. Let $X = \begin{bmatrix} x_1 \ldots x_N \end{bmatrix}$ be the data matrix. Let $n = (X^\top \odot X^\top)\mathbf{1}$ be the vector of the squared norms of the data, where $\mathbf{1}$ denotes the vector of all ones and $\odot$ the Hadamard (component wise) matrix product. Then we have

$$D = \mathbf{1}n^\top + n\mathbf{1}^\top - 2X^\top X, \qquad DF = n(\mathbf{1}^\top F) + \mathbf{1}(n^\top F) - 2X^\top(XF). \quad (6)$$

The term $\mathbf{1}(n^\top F)$ has constant columns and is irrelevant to the maximization (5). Therefore, we need to compute

$$DF \propto n(\mathbf{1}^\top F) - 2X^\top(XF), \quad n = (X^\top \odot X^\top)\mathbf{1} = (I \odot X^\top X)\mathbf{1} \quad (7)$$

where I is the identity matrix.[3] It is now easy to check that each matrix product in (7) requires $O(dN^2)$ operations only.

---

[3] And we used the fact that $(I \odot AB)\mathbf{1} = (B^\top \odot A)\mathbf{1}$.

Kernel Medoid Shift. An advantage of medoid shift is the possibility of computing (4) for distances $\mathsf{d}^2(x,y)$ other than the Euclidean one [20]. The decomposition (6) can still be carried out if the distance $\mathsf{d}^2(x,y)$ can be expressed as $K(x,x) + K(y,y) - 2K(x,y)$ for an appropriate positive definite (p.d.) kernel[4] K [18]. Then we have $D = \mathbf{1}n^\top + n\mathbf{1}^\top - 2K$, and

$$DF \propto n(\mathbf{1}^\top F) - 2KF, \quad n = (I \odot K)\mathbf{1}.$$

Unfortunately, the multiplication KF is still $O(N^{2.38})$. However, we can search for a low-rank decomposition $G^\top G$ of K (we assume, without loss of generality, that K is centered[5]). If G is a decomposition of rank d, then

$$DF \propto n(\mathbf{1}^\top F) - 2G(G^\top F), \quad n = (I \odot G^\top G)\mathbf{1} = (G^\top \odot G^\top)\mathbf{1}$$

can still be computed in $O(dN^2)$ operations. The cost of decomposing K is typically around $O(d^2 N)$ [8,1]. See Fig. 3 for a basic implementation.

Quick Shift. In order to seek the mode of the density $P(x)$, it is not necessary to use the gradient or the quadratic lower bound (2). Here we propose *quick shift*, which simply moves each point x_i to the nearest neighbor for which there is an increment of the density $P(x)$. In formulas,

$$y_i(1) = \underset{j:P_j > P_i}{\operatorname{argmin}} D_{ij}, \quad P_i = \frac{1}{N} \sum_{j=1}^{N} \phi(D_{ij}). \tag{8}$$

Quick shift has four advantages: (i) simplicity; (ii) speed ($O(dN^2)$ with a small constant); (iii) generality (the nature of D is irrelevant); (iv) a tuning parameter to trade off under- and over-fragmentation of the modes. The latter is obtained because there is no *a-priori* upper bound on the length D_{ij} of the shifts $y_i(0) \to y_i(1)$. In fact, the algorithm connects all the points into a single tree. Modes are then recovered by breaking the branches of the tree that are longer than a threshold τ. Searching τ amounts to performing model selection and balances under- and over-fragmentation of the modes. The algorithm is illustrated in Fig. 1.

Quick shift is related to the classic algorithm from [13]. In fact, we can rewrite (8) as

$$y_i(1) = \underset{j=1,\ldots,N}{\operatorname{argmax}} \frac{\operatorname{sign}(P_j - P_i)}{D_{ij}}, \quad \text{and compare it to } y_i(1) = \underset{j:d(x_j,x_i) < \tau}{\operatorname{argmax}} \frac{P_j - P_i}{D_{ij}} \tag{9}$$

---

[4] The kernel K should not be confused with the Parzen window $k(z)$ appearing in (1). In the literature, it is common to refer to the Parzen window as "kernel", but in most cases it has rather different mathematical properties than the kernel K we consider here. An exception is when the window is Gaussian, in which cases $k(\mathsf{d}^2(x,y))$ is a p.d. kernel. In this case, we point out an interesting interpretation of mean shift as a local optimization algorithm that, starting from each data point, searches for the pre-image of the global data average computed in kernel space. This explains the striking similarity of the mean shift update Eq. (3) and Eq. (18.22) of [19].

[5] K is *centered* if $K\mathbf{1} = 0$. If this is not the case, we can replace K by $K' = HKH$, where $H = I - \mathbf{1}\mathbf{1}^\top/N$ is the so-called *centering matrix*. This operation translates the origin of the kernel space, but does not change the corresponding distance.

as given by [13]. Notice that $(P_j - P_i)/D_{ij}$ is a numerical approximation of the gradient of P in the direction $x_j - x_i$. The crucial difference is that maximizing the gradient approximation must be done in a neighborhood of each point defined *a-priori* by the choice of the parameter τ. Thus, model selection in [13] requires running the algorithm multiple times, one for each value of τ. In contrast, quick shift returns at once the solutions for all possible values of τ, making model selection much more efficient.

4   Cluster Refinement

In the previous section we introduced fast kernel medoid shift as an accelerated version of non-Euclidean medoid shift. Since medoid shift may over-fragment modes, quick shift was then proposed as a method to control under- and over-fragmentation by the choice of a parameter τ. No algorithm, however, guarantees the same accuracy of the slower mean shift.

It is then natural to ask whether mean shift could be extended to work in a non-Euclidean setting. [20] cites the problem of defining the mean as the major obstacle to this idea. [21] addresses this issue by defining mean shift vectors on the tangent space of a non-linear manifold, but no proof of convergence is given, and the applicability is limited by the fact that the data space needs to have a manifold structure known analytically.

A simple solution to this problem is to extend kernel medoid to a corresponding kernel mean shift procedure. Let $K(\cdot, \cdot)$ be a p.d. kernel on the data space $\mathcal{X}$. Then $K(x, \cdot)$ is an element of the so called *reproducing kernel Hilbert space* $\mathcal{H}$ [19], whose inner product is defined by letting $\langle K(x, \cdot), K(y, \cdot) \rangle_{\mathcal{H}} = K(x, y)$. Points $x \in \mathbb{R}^d$ are then identified with elements $K(x, \cdot)$ of the Hilbert space. Given this identification, we can write $\langle \cdot, x \rangle_{\mathcal{H}}$ for $\langle \cdot, K(x, \cdot) \rangle_{\mathcal{H}}$.

Kernel mean shift computes a "density[6]" on $\mathcal{H}$

$$P(y) = \frac{1}{N} \sum_{j=1}^{N} k(\mathsf{d}_{\mathcal{H}}^2(y, x_j)), \qquad y \in \mathcal{H} \tag{10}$$

where $\mathsf{d}_{\mathcal{H}}^2(x_j, y) = \langle y, y \rangle_{\mathcal{H}} + \langle x_j, x_j \rangle_{\mathcal{H}} - 2\langle y, x_j \rangle_{\mathcal{H}}$. Notice that $y \in \mathcal{H}$, unlike standard mean shift, does not belong necessarily to the data space $\mathcal{X}$ (up to the identification $x \equiv K(x, \cdot)$). However, if $k(z)$ is monotonically decreasing, then maximizing w.r.t. y can be restricted to the linear subspace $\mathrm{span}_{\mathcal{H}} X = \mathrm{span}_{\mathcal{H}}\{x_1, \ldots, x_n\} \subset \mathcal{H}$ (if not, the orthogonal projection of y onto that space decreases simultaneously all terms $\mathsf{d}_{\mathcal{H}}^2(x_j, y)$).

Therefore, we can express all calculations relative to $\mathrm{span}_{\mathcal{H}} X$. In particular, if $K_{ij} = K(x_i, x_j)$ is the kernel matrix, we have $\mathsf{d}_{\mathcal{H}}^2(x_j, y) = y^\top K y + e_j^\top K e_j - 2e_j^\top K y$ where e_j is the j-th vector of the canonical basis and y is a vector of N coefficients. As in standard mean shift, the shifts are obtained by maximizing the lower bound

---

⁶ The interpretation is discussed later.

(KERNEL) MEAN SHIFT	(KERNEL) MEDOID SHIFT
<pre>function Z = meanshift(G, sigma)

[d,N]   = size(G) ;
oN   = ones(N,1) ;
od   = ones(d,1) ;
n   = (G'.*G')*od ;

Z = G ;
T = 100 ;
for t=1:T
  m = (Z'.*Z')*od ;
  D = m*oN' + oN*n' - 2*(Z'*G) ;
  F = - exp(- .5 * D' / sigma^2) ;
  Y = F ./ (oN * (oN'*F)) ;
  Z = G*Y ;
end</pre> | <pre>function map = medoidshift(G, sigma)

[d,N]   = size(G) ;
oN   = ones(N,1) ;
od   = ones(d,1) ;
n   = (G'.*G')*od ;

D = n*oN' + oN*n' - 2*(G'*G) ;
F = - exp(- .5 * D' / sigma^2) ;
Q = n * (oN'*F) - 2 * G' * (G*F) ;

[drop,map] = max(Q) ;</pre> |

Fig. 3. Kernel mean and medoid shift algorithms. We show basic MATLAB implementations of two of the proposed algorithms. Here $K = G^\top G$ is a low-rank decomposition $G \in \mathbb{R}^{d \times N}$ of the (centered) kernel matrix and `sigma` is the (isotropic) standard deviation of the Gaussian Parzen window. Both algorithms are $O(dN^2)$ (for a fixed number of iterations of mean shift), reduce to their Euclidean equivalents by setting $G \equiv X$ and $Z \equiv Y$, and can be easily modified to use the full kernel matrix K rather than a decomposition $G^\top G$ (but the complexity grows to $O(N^3)$).

$$y_i(t+1) = \operatorname*{argmax}_{y \in \mathbb{R}^N} \sum_{j=1}^{N} (y^\top K y + e_j^\top K e_j - 2 e_j^\top K y) \dot\phi(\mathrm{d}_\mathcal{H}^2(x_j, y_i(t))).$$

Deriving w.r.t. y and setting to zero yields the update equation

$$y_i(t+1) = \frac{1}{\mathbf{1}^\top F e_i}(F e_i), \quad F_{ji} = \dot\phi(D_{ij}), \quad D_{ij} = \mathrm{d}_\mathcal{H}^2(y_i(t), x_j). \tag{11}$$

Low-rank approximation. Similarly to medoid shift, we can accelerate the algorithm by using a low-rank decomposition $K = G^\top G$ of the (centered) kernel matrix. It is useful to switch to matrix notation for all the quantities. Let $Y = [y_1, \ldots \ y_M]$ be the trajectory matrix and define $Z = GY$ the reduced coordinates.[7] The distance matrix D can be written compactly as

$$D = m\mathbf{1}^\top + \mathbf{1}n^\top - 2Y^\top K = m\mathbf{1}^\top + \mathbf{1}n^\top - 2Z^\top G;$$

where

$$m = (Y^\top \odot Y^\top K)\mathbf{1} = (Z^\top \odot Z^\top)\mathbf{1}, \quad n = (I \odot K)\mathbf{1} = (G^\top \odot G^\top)\mathbf{1}.$$

At each iteration D is calculated in $O(dN^2)$ operations. Then $F = \dot\phi(D^\top)/N$ is evaluated component-wise. Finally the trajectories Y (or equivalently Z) are updated by

$$Y \leftarrow F \operatorname{diag}(F\mathbf{1})^{-1}, \qquad Z \leftarrow GY.$$

---

[7] Similarly, the data matrix X has reduced coordinates equal to G.

in $O(N^2)$ operations. Notice that, by setting $G \equiv X$ and $Z \equiv Y$ in these equations, we obtain Euclidean mean shift back. See Fig. 3 for a basic implementation.

Interpretation, Regularization and Scaling. In Euclidean mean shift the function $P(x)$ is a non-parametric estimate of a probability density. Does the same interpretation hold in kernel space? For any fixed data set of size N, we can restrict our attention to the subspace space $\mathrm{span}_{\mathcal{H}} X \subset \mathcal{H}$ and interpret $P(x)$ as a probability density on this finite-dimensional space. Unfortunately, the number of dimensions of this space may be as large as the number of data points N, which makes the Parzen density estimate $P(x)$ inconsistent (in the sense that the variance does not converge to zero as $N \to \infty$). So how do we make sense of kernel mean shift? The idea is to use the fact that most of the dimensions of $\mathrm{span}_{\mathcal{H}} X$ are often unimportant. Formally, consider the eigen-decomposition $K = V \Sigma V^\top = G^\top G$, $G = \Sigma^{\frac{1}{2}} V^\top$ of the (centered) kernel matrix K. Assume $\Sigma^{\frac{1}{2}} = N \operatorname{diag}(\sigma_1, \ldots, \sigma_N)$, with $\sigma_1 \geq \sigma_2 \geq \cdots \geq \sigma_N$. According to this decomposition, vectors $x, y \in \mathrm{span}_{\mathcal{H}} X$ can be identified with their coordinates $g, z \in \mathbb{R}^N$ so that $\langle x, y \rangle_{\mathcal{H}} = \langle g, z \rangle$. Moreover the data matrix $G = \begin{bmatrix} g_1 \ldots g_n \end{bmatrix}$ has null mean[8] and covariance $GG^\top/N = \Sigma/N = \sigma_1^2 \operatorname{diag}(\lambda_1^2, \ldots, \lambda_N^2)$. If λ_i decay fast, the effective dimension of the data can be much smaller than N.

The simplest way to regularize the Parzen estimate is therefore to discard the dimensions above some index d (which also improves efficiency). Another option is to blur the coordinates z by adding a small Gaussian noise η of isotropic standard deviation ϵ, obtaining a regularized variable $z' = z + \eta$. The components of z with smaller variance are "washed out" by the noise, and we can obtain a consistent estimator of z' by using the regularized Parzen estimate $\sum_{i=1}^{N}(g_\epsilon * k)(z_i)$ (the same idea is implicitly used, for instance, in kernel Fisher discriminant analysis [15], where the covariance matrix computed in kernel space is regularized by the addition of $\epsilon^2 I$). This suggests that using a kernel with sufficient isotropic smoothing may be sufficient.

Finally, we note that, due to the different scalings $\lambda_1, \ldots, \lambda_N$ of the linear dimensions, it might be preferable to use an adapted Parzen window, which retains the same proportions [17]. This, combined with the regularization ϵ, suggests us to scale each axis of the kernel by $\sqrt{\sigma^2 \lambda_i^2 + \epsilon^2}$.[9]

---

[8] Because K is assumed to be centered, so that $\mathbf{1}^\top G^\top (G\mathbf{1}) = \mathbf{1}^\top K \mathbf{1} = 0$.

[9] So far we disregarded the normalization constant of the Parzen window $k(x)$ as it was irrelevant for our purposes. If, however, windows $k_\sigma(x)$ of variable width σ are used [6], then the relative weights of the windows become important. Recall that in the d dimensional Euclidean case one has $k_\sigma(0)/k_{\sigma'}(0) = (\sigma'/\sigma)^d$. In kernel space therefore one would have

$$\frac{k_\sigma(0)}{k_{\sigma'}(0)} = \sqrt{\prod_{i=1}^{N} \frac{\sigma'^2 \lambda_i^2 + \epsilon^2}{\sigma^2 \lambda_i^2 + \epsilon^2}}.$$

Fig. 4. Clustering on a manifold. By using kernel ISOMAP we can apply kernel mean and medoid shift to cluster points on a manifold. For the sake of illustration, we reproduce an example from [20]. From left to right: Kernel mean shift (7.8s), *non-iterated* kernel medoid shift (0.18s), iterated kernel medoid shift (0.48s), quick shift (0.12s). We project the kernel space to three dimensions $d = 3$ as the residual dimensions are irrelevant. All algorithms but non-iterated medoid shift segment the modes successfully. Compared to [20], medoid shift has complexity $O(dN^2)$, (with a small constant and $d = 3 \ll N$) instead of $O(N^3)$ (small constant) or $O(N^{2.38})$ (large constant).

5   Applications

5.1   Clustering on Manifolds

[20] applies medoid shift to cluster data on manifolds, based on the distance matrix D calculated by ISOMAP. If the kernel matrix $K = HDH'/2$, $H = I - \frac{1}{N}\mathbf{1}\mathbf{1}^\top$ is p.d., we can apply directly kernel mean or medoid shift to the same problem. If not, we can use the technique from [4] to regularize the estimate and enforce this property. In Fig. 4 this idea is used to compare kernel mean shift, kernel medoid shift and quick shift in a simple test case.

5.2   Image Segmentation

Image segmentation is a typical test case for mode seeking algorithms [5,16,20]. Usually mode seeking is applied to this task by clustering data $\{(p, f(p)), p \in \Omega\}$, where $p \in \Omega$ are the image pixels and $f(p)$ their color coordinates (we use the same color space of [5]).

As in [5], we apply mean shift to segment the image into super-pixels (mean shift variants can be used to obtain directly full segmentations [2,16,24]). We compare the speed and segmentation quality obtained by using mean shift, medoid shift, and quick shift (see Fig. 5 for further details).

Mean shift is equivalent to [5] and can be considered a reference to evaluate the other segmentations. Non-iterative medoid shift (first column) over-fragments significantly (see also Fig. 2), which in [20] is addressed by reiterating the algorithm. However, since our implementation is only $O(dN^2)$, medoid shift has at least the advantage of being much faster than mean shift, and can be used to speed up the latter. In Fig. 5 we compare the time required to run mean shift from scratch and from the modes found by medoid shift. We report the speedup (as the number of modes found by medoid shift over the number of pixels), the

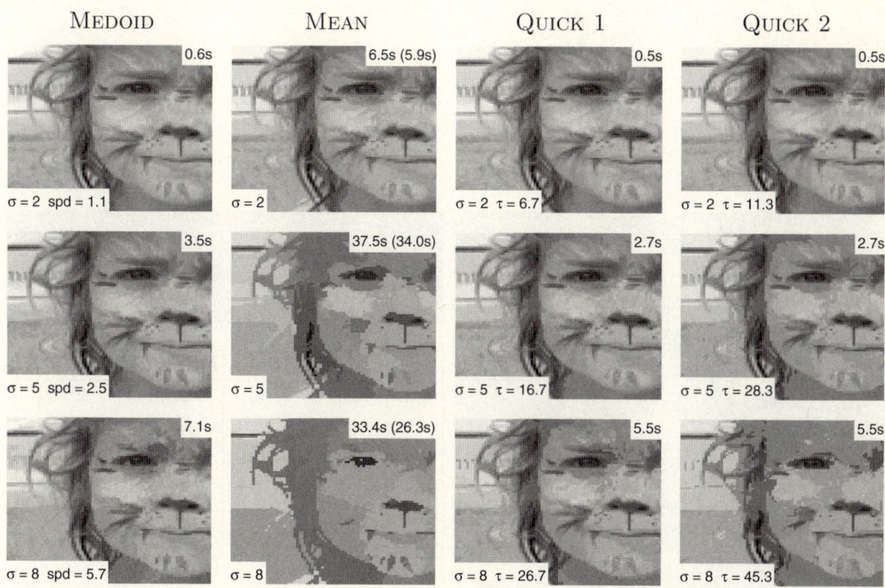

Fig. 5. Image segmentation. We compare different mode seeking techniques for segmenting an image (for clarity we show only a detail). We report the computation time in seconds (top-right corner of each figure). In order to better appreciate the intrinsic efficiency advantages of each method, we use comparable vanilla implementations of the algorithms (in practice, one could use heuristics and advanced approximation techniques [23] to significantly accelerate the computation). We use a Gaussian kernel of isotropic standard deviation σ in the spatial domain and use only one optimization: We approximate the support of the Gaussian window by a disk of radius 3σ (in the spatial domain) which results in a sparse matrix F. Therefore the computational effort increases with σ (top to bottom). The results are discussed in the text.

computation time of medoid+mean shift and, in brackets, the computation time of the mean shift part only. Interestingly, the efficiency increases for larger σ, so that the overall computation time actually *decreases* when σ is large enough.

Finally, we show the result of quick shift segmentation (last two columns) for increasing values of the regularization parameter τ. Notice that quick shift is run only once to get both segmentations (Sect. 3) and that the algorithm is in practice much faster than the other two, while still producing reasonable super-pixels.

5.3 Clustering Bag-of-Features

The interesting work [11] introduces a large family of positive definite kernels for probability measures which includes many of the popular metrics: χ^2 kernel, Hellinger's kernel, Kullback-Leibler kernel and l^1 kernel. Leveraging on these ideas, we can use kernel mean shift to cluster *probability measures*, and in

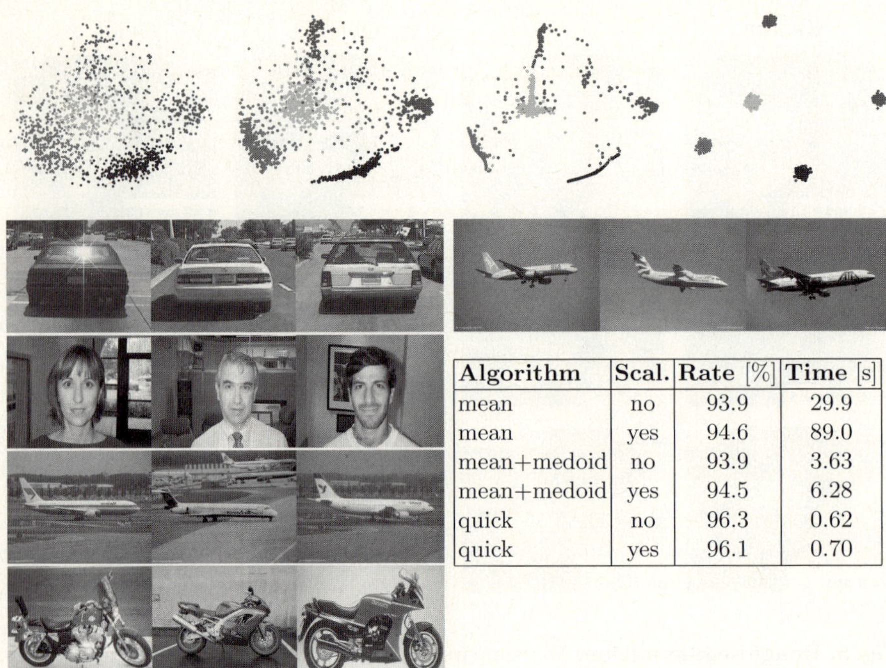

Algorithm	Scal.	Rate [%]	Time [s]
mean	no	93.9	29.9
mean	yes	94.6	89.0
mean+medoid	no	93.9	3.63
mean+medoid	yes	94.5	6.28
quick	no	96.3	0.62
quick	yes	96.1	0.70

Fig. 6. Automatic visual categorization. We use kernel mean shift to cluster bag-of-features image descriptors of 1600 images from Caltech-4 (four visual categories: airplanes, motorbikes, faces, cars). **Top.** From left to right, iterations of kernel mean shift on the bag-of-features signatures. We plot the first two dimensions of the rank-reduced kernel space (z vectors) and color the points based on the ground truth labels. In the rightmost panel the data converged to five points, but we artificially added random jitter to visualize the composition of the clusters. **Bottom.** Samples from the five clusters found (notice that airplane are divided in two categories). We also report the clustering quality, as the percentage of correct labels compared to the ground truth (we merge the two airplanes categories into one), and the execution time. We use basic implementations of the algorithms, although several optimizations are possible.

particular histograms, such as the ones arising in bag-of-features [7] or similar representations. In the rest of the section we experiment with the χ^2 kernel

$$K_{\chi^2}(x, y) = 2 \sum_{b=1}^{B} \frac{x_b y_b}{x_b + y_b}$$

where x and y are histograms of B bins.

Inspired by [10], we attempt to automatically infer the object categories of Caltech-4 in a completely unsupervised setting. We select at random 1600 images from the categories bike, airplanes, cars and faces. Instead of the more sophisticated representation of [10], we compute a basic bag-of-feature image representation as suggested by [25]: We extract multiscale Harris and DoG interest points (of fixed orientation; see [25] and ref. therein) and calculate SIFT descriptors [14], obtaining about 10^3 features per image. We then generate a vocabulary of 400 visual words by clustering a random selection of such descriptors by using k-means. For each image, we compute a bag-of-feature histogram x by counting the number of occurrences of each visual word in that image. Finally, we use the χ^2 kernel to generate the kernel matrix, that we feed to our clustering algorithms.

In Fig. 6 we compare kernel mean shift, kernel mean shift initialized by medoid shift, and quick shift. The problem we solve is considerably harder than [10], since in our case the number of clusters (categories) is unknown. All algorithms discover five (rather than four) categories (Fig. 6), but the result is quite reasonable since the category airplanes contains two distinct and visually quite different populations (grounded and airborne airplanes). Moreover, compared to [10] we do not try to explicitly separate an object from its background, but we use a simple holistic representation of each image.

The execution time of the algorithms (Fig. 6) is very different. Mean shift is relatively slow, at least in our simple implementation, and its speed greatly improves when we use medoid shift to initialize it. However, consistently with our image segmentation experiments, quick shift is much faster.

We also report the quality of the learned clusters (after manually merging the two airplane subcategories) as the percentage of correct labels. Our algorithm performs better than [10], that uses spectral clustering and reports 94% accuracy on selected prototypes and as low as 85% when all the data are considered; our accuracy in the latter case is at least 94%. We also study rescaling as proposed in Sect. 4, showing that it (marginally) improves the results of mean/medoid shift, but makes the convergence slower. Interestingly, however, the best performing algorithm (not to mention the fastest) is quick shift.

6   Conclusions

In this paper we exploited kernels to extend mean shift and other mode seeking algorithms to a non-Euclidean setting. This also clarifies issues of regularization and data scaling when complex spaces are considered. In this context, we showed how to derive a very efficient version of the recently introduced medoid shift algorithm, whose complexity is *lower* than mean shift. Unfortunately, we also showed that medoid shift often results in over-fragmented clusters. Therefore, we proposed to use medoid shift to initialize mean shift, yielding a clustering algorithm which is both efficient and accurate.

We also introduced quick shift, which can balance under- and over-fragmentation of the clusters by the choice of a real parameter. We showed that, in practice, this algorithm is very competitive, resulting in good (and sometimes better) segmentations compared to mean shift, at a fraction of the computation time.

Acknowledgment. Supported by AFOSR FA9550-06-1-0138 and ONR N00014-08-1-0414.

References

1. Bach, F.R., Jordan, M.I.: Kernel independent componet analysis. Journal of Machine Learninig Research 3(1) (2002)
2. Carreira-Perpiñán, M.: Fast nonparametric clustering with gaussian blurring mean-shift. In: Proc. ICML (2006)
3. Cheng, Y.: Mean shift, mode seeking, and clustering. PAMI 17(8) (1995)
4. Choi, H., Choi, S.: Robust kernel isomap. Pattern Recognition (2006)
5. Comaniciu, D., Meer, P.: Mean shift: A robust approach toward feature space analysis. PAMI 24(5) (2002)
6. Comaniciu, D., Ramesh, V., Meer, P.: The variable bandwidth mean shift and data-driven scale selection. In: Proc. ICCV (2001)
7. Csurka, G., Dance, C.R., Dan, L., Willamowski, J., Bray, C.: Visual categorization with bags of keypoints. In: Proc. ECCV (2004)
8. Fine, S., Scheinberg, K.: Efficient SVM training using low-rank kernel representations. Journal of Machine Learninig Research (2001)
9. Fukunaga, K., Hostler, L.D.: The estimation of the gradient of a density function, with applications in pattern recognition. IEEE Trans. 21(1) (1975)
10. Grauman, K., Darrell, T.: Unsupervised learning of categories from sets of partially matching image features. In: Proc. CVPR (2006)
11. Hein, M., Bousquet, O.: Hilbertian metrics and positive definite kernels on probability measures. In: Proc. AISTAT (2005)
12. Knuth, D.: The Art of Computer Programming: Seminumerical Algorithms, 3rd edn., vol. 2 (1998)
13. Koontz, W.L.G., Narendra, P., Fukunaga, K.: A graph-theoretic approach to nonparametric cluster analyisis. IEEE Trans. on Computers c-25(9) (1976)
14. Lowe, D.: Implementation of the scale invariant feature transform (2007), `http://www.cs.ubc.ca/~lowe/keypoints/`
15. Mika, S., Rätsch, G., Weston, J., Schölkopf, B., Müller, K.-R.: Fisher discriminant analysis with kernels. In: Proc. IEEE Neural Networks for Signal Processing Workshop (1999)
16. Paris, S., Durand, F.: A topological approach to hierarchical segmentation using mean shift. In: Proc. CVPR (2007)
17. Sain, S.R.: Multivariate locally adaptive density estimation. Comp. Stat. and Data Analysis, 39 (2002)
18. Schölkopf, B.: The kernel trick for distances. In: Proc. NIPS (2001)
19. Schölkopf, B., Smola, A.J.: Learning with Kernels. MIT Press, Cambridge (2002)
20. Sheikh, Y.A., Khan, E.A., Kanade, T.: Mode-seeking by medoidshifts. In: Proc. CVPR (2007)
21. Subbarao, R., Meer, P.: Nonlinear mean shift for clustering over analytic manifolds. In: Proc. CVPR (2006)
22. Sundaramoorthy, G., Yezzi, A., Mennucci, A.: Sobolev active contours. Int. J. Comput. Vision 73(3) (2007)
23. Yang, C., Duraiswami, R., Gumerov, N.A., Davis, L.: Improved fast Gauss transform and efficient kernel density estimation. In: Proc. ICCV (2003)
24. Yuan, X., Li, S.Z.: Half quadric analysis for mean shift: with extension to a sequential data mode-seeking method. In: Proc. CVPR (2007)
25. Zhang, J., Marszalek, M., Lazebnik, S., Schmid, C.: Local features and kernels for classification of texture and object categories: A comprehensive study IJCV (2006)

A Fast Algorithm for Creating a Compact and Discriminative Visual Codebook

Lei Wang[1], Luping Zhou[1], and Chunhua Shen[2]

[1] RSISE, The Australian National University, Canberra ACT 0200, Australia
[2] National ICT Australia (NICTA)*, Canberra ACT 2601, Australia

Abstract. In patch-based object recognition, using a compact visual codebook can boost computational efficiency and reduce memory cost. Nevertheless, compared with a large-sized codebook, it also risks the loss of discriminative power. Moreover, creating a compact visual codebook can be very time-consuming, especially when the number of initial visual words is large. In this paper, to minimize its loss of discriminative power, we propose an approach to build a compact visual codebook by maximally preserving the separability of the object classes. Furthermore, a fast algorithm is designed to accomplish this task effortlessly, which can hierarchically merge 10,000 visual words down to 2 in ninety seconds. Experimental study shows that the compact visual codebook created in this way can achieve excellent classification performance even after a considerable reduction in size.

1 Introduction

Recently, patch-based object recognition has attracted particular attention and demonstrated promising recognition performance [1,2,3,4]. Typically, a visual codebook is created as follows. After extracting a large number of local patch descriptors from a set of training images, k-means or hierarchical clustering is often used to group these descriptors into n clusters, where n is a predefined number. The center of each cluster is called "visual word", and a list of them forms a "visual codebook". By labelling each descriptor of an image with the most similar visual word, this image is characterized by an n-dimensional histogram counting the number of occurrences of each word. The visual codebook can have critical impact on recognition performance. In the literature, the size of a codebook can be up to 10^3 or 10^4, resulting in a very high-dimensional histogram.

A compact visual codebook has advantages in both computational efficiency and memory usage. For example, when linear or nonlinear SVMs are used, the complexity of computing the kernel matrix, testing a new image, or storing the support vectors is all proportional to the codebook size, n. Also, many algorithms working well in a low dimensional space will encounter difficulties such

---

* National ICT Australia is funded by the Australian Government's *Backing Australia's Ability* initiative, in part through the Australian Research Council. The authors thank Richard I. Hartley for many insightful discussions.

D. Forsyth, P. Torr, and A. Zisserman (Eds.): ECCV 2008, Part IV, LNCS 5305, pp. 719–732, 2008.

as singularity or unreliable parameter estimate when the dimensions increase. This is often called the "curse of dimensionality". A compact visual codebook provides a lower-dimensional representation and can effectively avoid these difficulties. Moreover, in patch-based object recognition, the histogram used to represent an image is essentially a discrete approximation of the distribution of visual words in that image. A large-sized visual codebook may overfit this distribution, as pointed out in [5]. Pioneering work of creating a compact and discriminative visual codebook has been seen recently in [4], which hierarchically merges the visual words in a large-sized initial codebook. To minimize the loss of discriminative ability, the work in [4] requires the new histograms to maximize the conditional probability of the true labels of training images (or image regions in their work). This is a rigorous but complicated criterion that involves non-trivial computation after each merging operation. Moreover, at each level of the hierarchy, the optimal pair of words to be merged are sought by an exhaustive search. These lead to a heavy computational load when dealing with large-sized initial codebooks.

Creating a compact codebook is essentially a dimensionality reduction problem. To preserve the discriminative power, any classification performance related criterion may be adopted, for example, the rigorous Bayes error rate, error bounds or distances, class separability measure, or that used in [4]. We pay particular interest to the class separability measure because of its simplicity and efficiency. By using this measure, we build a compact visual codebook that maximally preserves the separability of the object classes. More importantly, we propose a fast algorithm to accomplish this task effortlessly. By this algorithm, the class separability measure can be immediately evaluated once two visual words are merged. Also, searching for the optimal pair of words to be merged is cast as a 2D geometry problem and testing a small number of pairs is sufficient to find the optimal pair. Given an initial codebook of 10,000 visual words, the proposed fast algorithm can hierarchically merge them down to 2 words in ninety seconds. As experimentally demonstrated, our algorithm can produce a compact codebook which is comparable to or even better than that obtained by [4], but our algorithm needs much less computational overhead, especially when the size of the initial codebook is large.

2   The Scatter-Matrix Based Class Separability Measure

This measure involves the *Within-class scatter matrix* ($\mathbf{H}$), the *Between-class scatter matrix* ($\mathbf{B}$), and the *Total scatter matrix* ($\mathbf{T}$). Let $(\mathbf{x}, y) \in (\mathbb{R}^n \times \mathcal{Y})$ denote a training sample, where $\mathbb{R}^n$ stands for an n-dimensional input space, and $\mathcal{Y} = \{1, 2, \cdots, c\}$ is the set of c class labels. The number of samples in the i-th class is denoted by l_i. Let $\mathbf{m}_i$ be the mean vector of the i-th class and $\mathbf{m}$ be the mean vector of all classes. The scatter matrices are defined as

$$
\begin{aligned}
\mathbf{H} &= \sum_{i=1}^{c} \left[\sum_{j=1}^{l_i} (\mathbf{x}_{ij} - \mathbf{m}_i)(\mathbf{x}_{ij} - \mathbf{m}_i)^{\top} \right] \\
\mathbf{B} &= \sum_{i=1}^{c} l_i (\mathbf{m}_i - \mathbf{m})(\mathbf{m}_i - \mathbf{m})^{\top} \\
\mathbf{T} &= \sum_{i=1}^{c} \left[\sum_{j=1}^{l_i} (\mathbf{x}_{ij} - \mathbf{m})(\mathbf{x}_{ij} - \mathbf{m})^{\top} \right] = \mathbf{H} + \mathbf{B} \, .
\end{aligned}
\tag{1}
$$

A large class separability means small within-class scattering but large between-class scattering. A combination of two of them can be used as a measure, for example, $\mathrm{tr}(\mathbf{B})/\mathrm{tr}(\mathbf{T})$ or $|\mathbf{B}|/|\mathbf{H}|$, where $\mathrm{tr}(\cdot)$ and $|\cdot|$ denote the trace and determinant of a matrix, respectively. In these measures the scattering of data is evaluated through the mean and variance, which implicitly assumes a Gaussian distribution for each class. This drawback is overcome by incorporating the kernel trick and it makes the scatter-matrix based measure quite useful, as demonstrated in Kernel based Fisher Discriminate Analysis (KFDA) [6].

3   The Formulation of Our Problem

Given an initial codebook of n visual words, we aim to obtain a codebook consisting of m $(m \ll n)$ visual words in the sense that when represented with these m visual words, the c object classes can have maximal separability.

Recall that with a set of visual words, a training image can be represented by a histogram which contains the number of occurrences of each word in this image. Let $\mathbf{x}^n$ $(\mathbf{x}^n \in \mathbb{R}^n)$ and $\mathbf{x}^m$ $(\mathbf{x}^m \in \mathbb{R}^m)$ denote the histograms when n and m visual words are used, respectively. In the following, we first discuss an ideal way of solving our problem, and show that such a way is impractical for patch-based object recognition. This motivates us to propose the fast algorithm in this paper.

Inferring m visual words from the n initial ones is essentially a dimensionality reduction problem. It can be represented by a linear transform as

$$\mathbf{x}^m = \mathbf{W}^\top \mathbf{x}^n \tag{2}$$

where $\mathbf{W}$ $(\mathbf{W} \in \mathbb{R}^{n \times m})$ is an $n \times m$ matrix. Let $\mathbf{B}^n$ and $\mathbf{T}^n$ denote the *between*-class and *total*-class scatter matrices when the training images are represented by $\mathbf{x}^n$. The optimal linear transform, $\mathbf{W}^\star$, can be expressed as

$$\mathbf{W}^\star = \arg \max_{\mathbf{W} \in \mathbb{R}^{n \times m}} \frac{\mathrm{tr}(\mathbf{W}^\top \mathbf{B}^n \mathbf{W})}{\mathrm{tr}(\mathbf{W}^\top \mathbf{T}^n \mathbf{W})}. \tag{3}$$

Note that the determinant-based measure is not adopted because n is often much larger than the number of training images, making $|\mathbf{B}^n|$ and $|\mathbf{T}^n|$ zero. The problem in (3) has been studied in [7] recently[1]. The optimal $\mathbf{W}$ is located by solving a series of Semi-Definite Programming (SDP) problems. Nevertheless, this SDP-based approach quickly becomes intractable when n exceeds 100, which is far less than the number encountered in practical object recognition. Moreover, the $\mathbf{W}$ in patch-based object recognition may have the following constraints:

1. $\mathbf{W}_{ij} \in \{0, 1\}$ if requiring the m new visual words to have meaningful and determined content;[2]

---

[1] Note that this problem is not simply the Fisher Discriminant Analysis problem. Please see [7] for the details.

[2] For example, when discriminating *motorbikes* from *airplanes*, the content of a visual word will be "handle bar" and/or "windows" rather than 31% handle bar, 27% windows, and 42% something else.

2. $\sum_{j=1}^{m} \mathbf{W}_{ij} = 1$ if requiring that each of the n visual words *only be* assigned to one of the m visual words.
3. If no words are to be discarded, the constraint of $\sum_{i=1}^{n} \mathbf{W}_{ij} \geq 1$ will be imposed because each of the n visual words *must be* assigned to one of the m visual words;

This results in a large-scale integer programming problem. Efficiently and optimally solving it may be difficult for the state-of-the-art optimization techniques. In this paper, we adopt a suboptimal approach that hierarchically merges two words while maximally maintaining the class separability at each level.

4   A Fast Algorithm of Hierarchically Merging Visual Words

To make the hierarchical merging approach efficient, we need: i) Once two visual words are merged, the resulting class separability can be quickly evaluated; ii) In searching for the best pair of words to merge, the search scope has to be as small as possible. In the following, we show how these requirements are achieved with the scatter-matrix based class separability measure.

4.1   Fast Evaluation of Class Separability

Let $\mathbf{x}_i^t = [x_{i1}^t, \cdots, x_{it}^t]$ $(i = 1, \cdots, l)$ be the i-th training image when t visual words are used, where t $(t = n, n-1, \cdots, m)$ indicates the current level in the hierarchy. Let $\mathbf{K}^t$ be the Gram matrix defined by $\{\mathbf{K}^t\}_{ij} = \langle \mathbf{x}_i^t, \mathbf{x}_j^t \rangle$. Let $\mathbf{K}_{rs}^{t-1}$ be the resulting Gram matrix after merging the r-th and s-th words at level t. Their relationship is derived as

$$
\begin{aligned}
\{\mathbf{K}_{rs}^{t-1}\}_{ij} &= \langle \mathbf{x}_i^{t-1}, \mathbf{x}_j^{t-1} \rangle = \sum_{k=1}^{t-1} x_{ik}^{t-1} x_{jk}^{t-1} \\
&= \sum_{k=1}^{t} x_{ik}^t x_{jk}^t - x_{ir}^t x_{jr}^t - x_{is}^t x_{js}^t + (x_{ir}^t + x_{is}^t)(x_{jr}^t + x_{js}^t) \\
&= \sum_{k=1}^{t} x_{ik}^t x_{jk}^t + x_{ir}^t x_{js}^t + x_{is}^t x_{jr}^t \\
&= \{\mathbf{K}^t\}_{ij} + \{\mathbf{A}_{rs}^t\}_{ij} + \{\mathbf{A}_{rs}^t\}_{ji}
\end{aligned}
\tag{4}
$$

where $\mathbf{A}_{rs}^t$ is a matrix defined as $\mathbf{A}_{rs}^t = \mathbf{X}_r^t (\mathbf{X}_s^t)^\top$, where $\mathbf{X}_r^t$ is $[x_{1r}^t, \cdots, x_{lr}^t]$. Hence, it can be obtained that

$$
\mathbf{K}_{rs}^{t-1} = \mathbf{K}^t + \mathbf{A}_{rs}^t + (\mathbf{A}_{rs}^t)^\top.
\tag{5}
$$

A similar relationship exists between the class separability measures at t and $t-1$ levels. Let $\mathbf{B}^{t-1}$ and $\mathbf{T}^{t-1}$ be the matrices $\mathbf{B}$ and $\mathbf{T}$ computed with $\mathbf{x}^{t-1}$. It can be proven (the proof is omitted) that for a c-class problem,

$$
\text{tr}(\mathbf{B}_{rs}^{t-1}) = \sum_{i=1}^{c} \frac{\mathbf{1}^\top \mathbf{K}_{rs,i}^{t-1} \mathbf{1}}{l_i} - \frac{\mathbf{1}^\top \mathbf{K}_{rs}^{t-1} \mathbf{1}}{l}; \quad \text{tr}(\mathbf{T}_{rs}^{t-1}) = \text{tr}(\mathbf{K}_{rs}^{t-1}) - \frac{\mathbf{1}^\top \mathbf{K}_{rs}^{t-1} \mathbf{1}}{l}
\tag{6}
$$

where $\mathbf{K}_{rs,i}^{t-1}$ is computed by the training images from class i. It can be verified that $\mathbf{K}_{rs,i}^{t-1} = \mathbf{K}_i^t + \mathbf{A}_{rs,i}^t + (\mathbf{A}_{rs,i}^t)^\top$. The l_i is the number of training images from class i, and l is the total number. Note that $\mathbf{1}^\top \mathbf{A}_{rs}^t \mathbf{1} = \mathbf{1}^\top (\mathbf{A}_{rs}^t)^\top \mathbf{1}$, where $\mathbf{1}$ is a vector consisting of "1". By combining (5) and (6), we obtain that

$$
\begin{aligned}
\operatorname{tr}(\mathbf{B}_{rs}^{t-1}) &= \left(\sum_{i=1}^c \frac{\mathbf{1}^\top \mathbf{K}_i^t \mathbf{1}}{l_i} - \frac{\mathbf{1}^\top \mathbf{K}^t \mathbf{1}}{l} \right) + 2 \left(\sum_{i=1}^c \frac{\mathbf{1}^\top \mathbf{A}_{rs,i}^t \mathbf{1}}{l_i} - \frac{\mathbf{1}^\top \mathbf{A}_{rs}^t \mathbf{1}}{l} \right) \\
&= \operatorname{tr}(\mathbf{B}^t) + 2 \left(\sum_{i=1}^c \frac{\mathbf{1}^\top \mathbf{A}_{rs,i}^t \mathbf{1}}{l_i} - \frac{\mathbf{1}^\top \mathbf{A}_{rs}^t \mathbf{1}}{l} \right) \\
&\triangleq \operatorname{tr}(\mathbf{B}^t) + f(\mathbf{X}_r^t, \mathbf{X}_s^t),
\end{aligned}
\tag{7}
$$

where $f(\mathbf{X}_r^t, \mathbf{X}_s^t)$ denotes the second term in the previous step. Similarly,

$$
\begin{aligned}
\operatorname{tr}(\mathbf{T}_{rs}^{t-1}) &= \left(\operatorname{tr}(\mathbf{K}^t) - \frac{\mathbf{1}^\top \mathbf{K}^t \mathbf{1}}{l} \right) + 2 \left(\operatorname{tr}(\mathbf{A}_{rs}^t) + \frac{\mathbf{1}^\top \mathbf{A}_{rs}^t \mathbf{1}}{l} \right) \\
&= \operatorname{tr}(\mathbf{T}^t) + 2 \left(\operatorname{tr}(\mathbf{A}_{rs}^t) - \frac{\mathbf{1}^\top \mathbf{A}_{rs}^t \mathbf{1}}{l} \right) \\
&\triangleq \operatorname{tr}(\mathbf{T}^t) + g(\mathbf{X}_r^t, \mathbf{X}_s^t) \ .
\end{aligned}
\tag{8}
$$

Since both $\operatorname{tr}(\mathbf{B}^t)$ and $\operatorname{tr}(\mathbf{T}^t)$ have been computed at level t before any merging operation, the above results indicate that to evaluate the class separability after merging two words, only $f(\mathbf{X}_r^t, \mathbf{X}_s^t)$ and $g(\mathbf{X}_r^t, \mathbf{X}_s^t)$ need to be calculated.

In the following, we further show that at any level t ($m \le t < n$), $f(\mathbf{X}_r^t, \mathbf{X}_s^t)$ and $g(\mathbf{X}_r^t, \mathbf{X}_s^t)$ can be worked out with little computation. Three cases are discussed in turn.

i) *Neither the r-th nor the s-th visual word is newly generated at level t.*
   This means that both of them are directly inherited from level $t+1$. Assuming that they are numbered as p and q at level $t+1$, it can be known that

$$
f(\mathbf{X}_r^t, \mathbf{X}_s^t) = f(\mathbf{X}_p^{t+1}, \mathbf{X}_q^{t+1});
\tag{9}
$$

ii) *Just one of the r-th and the s-th visual words is newly generated at level t.*
   Assume that the r-th visual word is newly generated by merging the u-th and the v-th words at level $t+1$, that is, $\mathbf{X}_r^t = \mathbf{X}_u^{t+1} + \mathbf{X}_v^{t+1}$. Furthermore, assume that $\mathbf{X}_s^t$ is numbered as q at level $t+1$. It can be shown that

$$
\begin{aligned}
\mathbf{A}_{rs}^t &= \mathbf{X}_r^t (\mathbf{X}_s^t)^\top = (\mathbf{X}_u^{t+1} + \mathbf{X}_v^{t+1})(\mathbf{X}_q^{t+1})^\top \\
&= \mathbf{X}_u^{t+1}(\mathbf{X}_q^{t+1})^\top + \mathbf{X}_v^{t+1}(\mathbf{X}_q^{t+1})^\top \\
&= \mathbf{A}_{uq}^{t+1} + \mathbf{A}_{vq}^{t+1} .
\end{aligned}
\tag{10}
$$

In this way, it can be obtained that

$$
\begin{aligned}
f(\mathbf{X}_r^t, \mathbf{X}_s^t) &= 2 \left(\sum_{i=1}^c \frac{\mathbf{1}^\top \mathbf{A}_{rs,i}^t \mathbf{1}}{l_i} - \frac{\mathbf{1}^\top \mathbf{A}_{rs}^t \mathbf{1}}{l} \right) \\
&= 2 \left(\sum_{i=1}^c \frac{\mathbf{1}^\top \mathbf{A}_{uq,i}^{t+1} \mathbf{1}}{l_i} - \frac{\mathbf{1}^\top \mathbf{A}_{uq}^{t+1} \mathbf{1}}{l} \right) + 2 \left(\sum_{i=1}^c \frac{\mathbf{1}^\top \mathbf{A}_{vq,i}^{t+1} \mathbf{1}}{l_i} - \frac{\mathbf{1}^\top \mathbf{A}_{vq}^{t+1} \mathbf{1}}{l} \right) \\
&= f(\mathbf{X}_u^{t+1}, \mathbf{X}_q^{t+1}) + f(\mathbf{X}_v^{t+1}, \mathbf{X}_q^{t+1});
\end{aligned}
\tag{11}
$$

iii) *Both the r-th and the s-th visual words are newly generated at level t.*
   This case does not exist because only one visual word can be newly generated at each level of a hierarchical clustering.

The above analysis shows that $f(\mathbf{X}_r^t, \mathbf{X}_s^t)$ can be obtained either by directly copying from level $t+1$ or by a single addition operation. All of the analysis applies to $g(\mathbf{X}_r^t, \mathbf{X}_s^t)$. Hence, once the r-th and the s-th visual words are merged, the class separability measure, $\mathrm{tr}(\mathbf{B}_{rs}^{t-1})/\mathrm{tr}(\mathbf{T}_{rs}^{t-1})$, can be immediately obtained by two addition and one division operations.

Computational complexity. The time complexity of calculating $f(\mathbf{X}_i^n, \mathbf{X}_j^n)$ or $g(\mathbf{X}_i^n, \mathbf{X}_j^n)$ is analyzed. There are $n(n-1)/2$ values to be computed in total, each of which involves computing the matrix $\mathbf{A}_{ij}^n$ which needs l^2 multiplications. Both terms of $\mathbf{1}^\top \mathbf{A}_{ij,k}^n \mathbf{1}$ ($k = 1, 2, \cdots, c$) and $\mathbf{1}^\top \mathbf{A}_{ij}^n \mathbf{1}$ can be obtained by l^2 additions. Finally, $\sum_{i=1}^c (\frac{1}{l_i}) \mathbf{1}^\top \mathbf{A}_{ij,k}^n \mathbf{1} + (-\frac{1}{l}) \mathbf{1}^\top \mathbf{A}_{ij}^n \mathbf{1}$ can be worked out in $c+1$ multiplications and c additions. Hence, computing all $f(\mathbf{X}_i^n, \mathbf{X}_j^n)$ or $g(\mathbf{X}_i^n, \mathbf{X}_j^n)$ needs

$$\frac{n(n-1)}{2} \left[(l^2 + c + 1) \text{ multiplications} + (l^2 + c) \text{ additions} \right],$$

resulting in the complexity of $\mathcal{O}(n^2 l^2)$. In practice, the load of computing $\mathbf{A}_{ij}^n$ can be lower because the histogram $\mathbf{x}^n$ is often sparse. Also, $f(\mathbf{X}_i^n, \mathbf{X}_j^n)$ and $g(\mathbf{X}_i^n, \mathbf{X}_j^n)$ share the same $\mathbf{A}_{ij}^n$. The memory cost for storing all of the $f(\mathbf{X}_i^n, \mathbf{X}_j^n)$ and $g(\mathbf{X}_i^n, \mathbf{X}_j^n)$ in double precision format is $n(n-1) \times 8$ Bytes, leading to space complexity of $\mathcal{O}(n^2)$. When n equals $10,000$ (this is believed to be a reasonably large size for an initial visual codebook used in patch-based object recognition), the memory cost will be about 800 MByte, which is bearable for a desktop computer today. Moreover, the memory cost decreases quadratically with respect to the level because the total number of f or g is $t(t-1)/2$ at a given level t.

4.2 Fast Search for the Optimal Pair of Words to Merge

Although the class separability can now be quickly evaluated once a pair of words are merged, there are $\frac{t(t-1)}{2}$ possible pairs at level t from which we need to find the optimal pair to merge. If an exhaustive search is used to identify this optimal pair, the total number of pairs that are tested in the hierarchical merging process will be $\sum_{t=m+1}^n \frac{t(t-1)}{2}$. For $n = 10,000$ and $m = 2$, this number is as large as 1.67×10^{11}. Using an exhaustive search will significantly prolong the merging process. In the following, we propose a more efficient search strategy by making use of the properties of the scatter-matrix based class separability measure, which allows us to convert the search problem to a simple 2D geometry problem. Denote $f(\mathbf{X}_r^t, \mathbf{X}_s^t)$ and $g(\mathbf{X}_r^t, \mathbf{X}_s^t)$ by f^t and g^t in short, respectively. Recall that the class separability measure after merging two visual words is

$$\mathcal{J} = \frac{\mathrm{tr}(\mathbf{B}^{t-1})}{\mathrm{tr}(\mathbf{T}^{t-1})} = \frac{\mathrm{tr}(\mathbf{B}^t) + f^t}{\mathrm{tr}(\mathbf{T}^t) + g^t} = \frac{f^t - (-\mathrm{tr}(\mathbf{B}^t))}{g^t - (-\mathrm{tr}(\mathbf{T}^t))}$$

As illustrated in Fig. 1, geometrically, the value of $\mathcal{J}$ equals the slope of the line $\overline{AB}$ through $A(-\mathrm{tr}(\mathbf{T}^t), -\mathrm{tr}(\mathbf{B}^t))$ and $B(g^t, f^t)$.

The coordinates of A and B are restricted by the following properties of the scatter matrices:

i) From the definition in (1), it is known that

$$\mathrm{tr}(\mathbf{H}^t) \geq 0; \ \ \mathrm{tr}(\mathbf{B}^t) \geq 0; \ \ \mathrm{tr}(\mathbf{T}^t) = \mathrm{tr}(\mathbf{H}^t) + \mathrm{tr}(\mathbf{B}^t) \geq \mathrm{tr}(\mathbf{B}^t)$$

As a result, the point A must lie within the third quadrant of the Cartesian coordinate system gOf and above the line of $f - g = 0$. The domain of A is marked as a hatched region in Fig. 1.

ii) The coordinator of $B(g^t, f^t)$ must satisfy the following constraints:

$$\mathrm{tr}(\mathbf{B}^{t-1}) \geq 0 \Longrightarrow \mathrm{tr}(\mathbf{B}^t) + f^t \geq 0 \Longrightarrow f^t \geq -\mathrm{tr}(\mathbf{B}^t)$$
$$\mathrm{tr}(\mathbf{T}^{t-1}) \geq 0 \Longrightarrow \mathrm{tr}(\mathbf{T}^t) + g^t \geq 0 \Longrightarrow g^t \geq -\mathrm{tr}(\mathbf{T}^t)$$
$$\mathrm{tr}(\mathbf{T}^{t-1}) \geq \mathrm{tr}(\mathbf{B}^{t-1}) \Longrightarrow \mathrm{tr}(\mathbf{T}^t) + g^t \geq \mathrm{tr}(\mathbf{B}^t) + f^t$$
$$\Longrightarrow f^t - g^t - (\mathrm{tr}(\mathbf{T}^t) - \mathrm{tr}(\mathbf{B}^t)) \leq 0$$

They define three half-planes in the coordinate system gOf and the point $B(g^t, f^t)$ must lie within the intersection, the blue-colored region in Fig. 1.

Therefore, finding the optimal pair of words whose combination produces the largest class separability becomes finding the optimal point $B^\star$ which maximizes the slope of the line $\overline{AB}$, where the coordinate of A is fixed at a given level t.

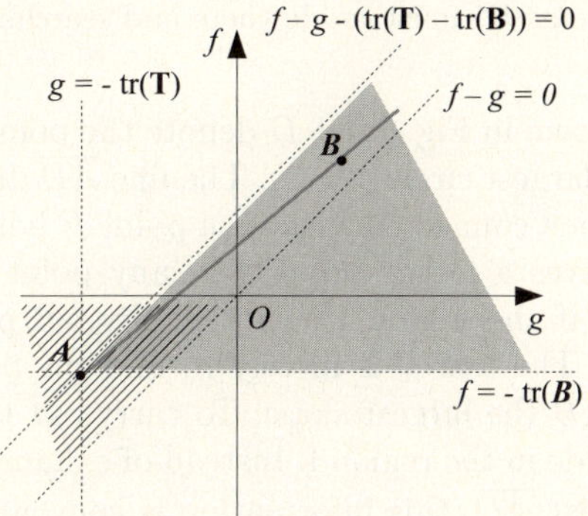

Fig. 1. Illustration of the region where $A(-\mathrm{tr}(\mathbf{T}^t), -\mathrm{tr}(\mathbf{B}^t))$ and $B(g^t, f^t)$ reside

Indexing structure. To realize the fast search, a polar coordinate based indexing structure is used to index the $t(t - 1)/2$ points of $B(g, f)$ at level t, as illustrated in Fig. 2. Each point B is assigned into a bin (i,j) according to its distance from the origin and its polar angle, where $i = 1, \cdots, K$ and $j = 1, \cdots, S$. The K is the number of bins with respect to the distance from the origin, whereas S is the number of bins with respect to the polar angle. In Fig. 2, this indexing structure is illustrated by K concentric circles, each of which is further divided into S segments. The total number of bins is KS. Through this indexing structure, we can know which points B reside in a given bin. In this paper, the number of circles K is set as 40, and their radius are arranged as $r_i = r_{i+1}/2$. The S is set as 36, which evenly divides $[0, 2\pi)$ into 36 bins.

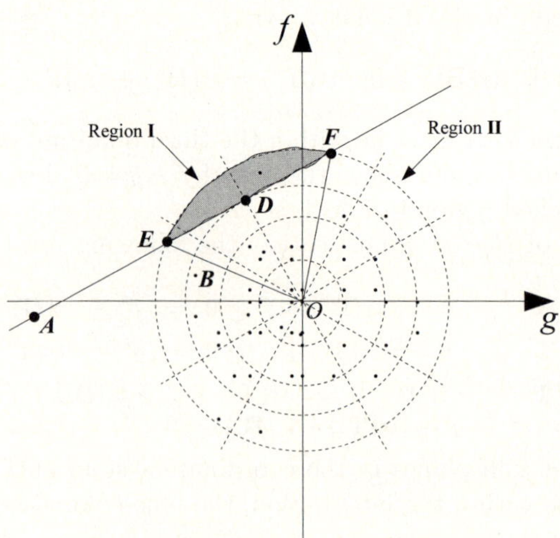

Fig. 2. The point A is fixed when searching for $B^\star$ which makes the line $\overline{AB}$ have the largest slope. The line $\overline{AD}$ is tangent to the second largest circle C_{K-1} at D, and it divides the largest circle C_K into two parts, region I and II. Clearly, a point B in region I always gives $\overline{AB}$ a larger slope than any point in region II. Therefore, if the region I is not empty, the best point $B^\star$ must reside there and searching region I is sufficient.

Search strategy. As shown in Fig. 2, let D denote the point where the line $\overline{AD}$ is tangent to the second largest circle, C_{K-1}. The line $\overline{AD}$ divides the largest circle C_K into two parts. When connected with A, a point B lying above $\overline{AD}$ (denoted by region I) always gives a larger slope than any point below it (denoted by region II). Therefore, if the region I is not empty, all points in the region II can be safely ignored. The search is merely to find the best point $B^\star$ from the region I which gives $\overline{AB}$ the largest slope. To carry out this search, we have to know which points reside in the region I. Instead of exhaustively checking each of the $\frac{t(t-1)}{2}$ points against $\overline{AD}$, this information is conveniently obtained via the above indexing structure. Let θ_E and θ_F be the polar angles of E and F where the line $\overline{AD}$ and C_K intersect. Denote the bins (with respect to the polar angle) into which they fall by S_1 and S_2, respectively. Thus, searching the region I can be accomplished by searching the bin (i, j) with $i = K$ and $j = S_1, \cdots, S_2$.[3] Clearly, the area of the searched region is much smaller than the area of C_K for moderate K and S. Therefore, the number of points $B(g, f)$ to be tested can be significantly reduced, especially when the point B distributes sparsely in the areas away from the origin. If the region I is empty, move the line $\overline{AD}$ to be tangent to the next circle, C_{K-2}, and repeat the above steps. After finding the optimal pair of words and merging them, all points $B(g, f)$ related to the two merged words will be removed. Meanwhile, new points related to the newly

---

[3] The region that is actually searched is slightly larger than the region I. Hence, the found best point $B^\star$ will be rejected if it is below the line $\overline{AD}$. This also means that the region I is actually empty.

generated word will be added and indexed. This process is conveniently realized in our algorithm by letting one word "absord" the other. Then, we finish the operation at level t and move to level $t-1$. Our algorithm is described in Table 1.

Before ending this section, it is worth noting that this search problem may be tackled by the dynamic convex hull [8] in computational geometry. Given the point A, the best point $B^\star$ must be a vertex of the convex hull of the points $B(g, f)$. At each level t, part of points $B(g, f)$ are updated, resulting in a dynamically changing convex hull. The technique of dynamic convex hull can be used to update the vertex set accordingly. This will be explored in future work.

Table 1. The fast algorithm for hierarchically merging visual words

---

Input: The l training images represented as $\{(\mathbf{x}_i, y_i)\}_{i=1}^{l}$ ($\mathbf{x}_i \in \mathbb{R}^n, y_i \in \{1, \cdots, c\}$). The n is the size of an initial visual codebook and y_i is the class label of $\mathbf{x}_i$
m: the size of the target visual codebook.
Output: The $n - m$ level merging hierarchy

Initialization:
  compute $f(\mathbf{X}_i^n, \mathbf{X}_j^n)$ and $g(\mathbf{X}_i^n, \mathbf{X}_j^n)$ $(1 \leq i < j \leq n)$ and store them in memory
  Index the $\frac{n(n-1)}{2}$ points of $B(g, f)$ with a polar coordinate quantized into bins
  Compute $A(-\mathrm{tr}(\mathbf{T}^n), -\mathrm{tr}(\mathbf{B}^n))$

Merging operation:
  for $t = n, n-1, \cdots, m$
    (1) **fast search** for the point $B(g^\star, f^\star)$ that gives the line $\overline{AB}$ the largest
       slope, where $f^\star = f^\star(\mathbf{X}_r^t, \mathbf{X}_s^t)$ and $g^\star = g^\star(\mathbf{X}_r^t, \mathbf{X}_s^t)$
    (2) **compute** $\mathrm{tr}(\mathbf{B}^{t-1})$ and $\mathrm{tr}(\mathbf{T}^{t-1})$ and update the point A:
       $\mathrm{tr}(\mathbf{B}^{t-1}) = \mathrm{tr}(\mathbf{B}^t) + f^\star(\mathbf{X}_r^t, \mathbf{X}_s^t);$ $\mathrm{tr}(\mathbf{T}^{t-1}) = \mathrm{tr}(\mathbf{T}^t) + g^\star(\mathbf{X}_r^t, \mathbf{X}_s^t)$
    (3) **update** $f(\mathbf{X}_r^t, \mathbf{X}_i^t)$ and $g(\mathbf{X}_r^t, \mathbf{X}_i^t)$
       $f(\mathbf{X}_r^t, \mathbf{X}_i^t) = f(\mathbf{X}_r^t, \mathbf{X}_i^t) + f(\mathbf{X}_s^t, \mathbf{X}_i^t);\ g(\mathbf{X}_r^t, \mathbf{X}_i^t) = g(\mathbf{X}_r^t, \mathbf{X}_i^t) + g(\mathbf{X}_s^t, \mathbf{X}_i^t)$
       remove $f(\mathbf{X}_s^t, \mathbf{X}_i^t)$ and $g(\mathbf{X}_s^t, \mathbf{X}_i^t)$
    (4) **re-index** $f(\mathbf{X}_r^t, \mathbf{X}_i^t)$ and $g(\mathbf{X}_r^t, \mathbf{X}_i^t)$
  end

---

5   Experimental Result

The proposed class separability measure based fast algorithm is tested on four classes of the Caltech-101 object database [9], including Motorbikes (798 images), Airplanes (800), Faces easy (435), and BACKGROUND_Google (520), as shown in Fig. 3. A Harris-Affine detector [10] is used to locate interest regions, which are then represented by the SIFT descriptor [11]. Other region detectors [12] and descriptors [13] can certainly be used because our algorithm has no restriction on this. The number of local descriptors extracted from the images of the four classes are about 134K, 84K, 57K, and 293K, respectively. Our algorithm is applicable to both binary and multi-class problems. This experiment focuses on the binary case, including both object categorization and object detection problems. To accumulate statistics, the images of the two object classes to be classified are randomly split as 10 pairs of training/test subsets. Restricted to the images in a training subset (those in a test subset are only used for test), their local descriptors are clustered to form the n initial visual words by using k-means clustering. Each image is then represented by a histogram containing the number of occurrences of each visual word.

Fig. 3. Example images of Motorbikes, Airplanes, Faces_easy, and BACK-GROUND_Google in [9] used in this experiment

Three algorithms are compared in creating a compact visual codebook, including k-means clustering (KMS in short), the algorithm proposed in [4] (PRO in short), and our class separability measure (CSM in short) based fast algorithm. In this experiment, the k-means clustering is used to cluster the local descriptors of the training images by gradually decreasing the value of k. Its result is used as a baseline. The CSM and PRO are applied to the initial n-dimensional histograms to hierarchically merge the visual words (or equally, the bins). For each algorithm, the obtained lower-dimensional histograms are used by a classifier to separate the two object classes. Linear and nonlinear SVM classifiers with a Gaussian RBF kernel are used. Their hyper-parameters are tuned via k-fold cross-validation. The three algorithms are compared in terms of: i) the time and memory cost with respect to the number of initial visual words; ii) the recognition performance achieved by the obtained compact visual codebooks. We aim to show that our proposed CSM-based fast algorithm can achieve the recognition performance comparable to or even better than the PRO algorithm but it is much faster in creating a compact codebook.

5.1   Result on Time and Memory Cost

The time and memory cost is independently evaluated on a synthetic data set. Fixing the number of training images at 100, the size of the initial visual codebook varies between 10 and 10,000. The number of occurrences of each visual

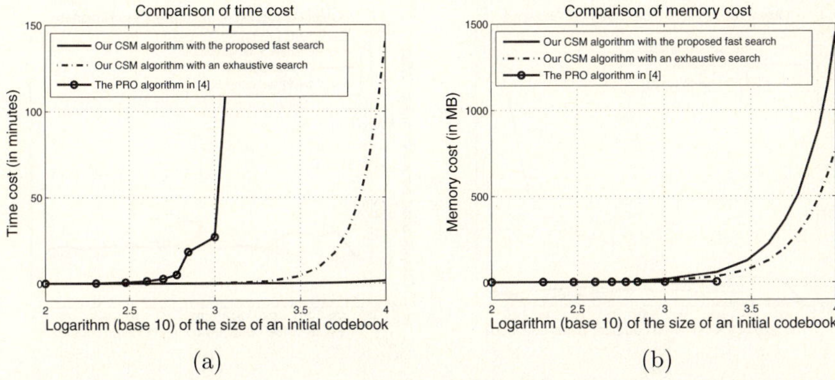

Fig. 4. Time and peak memory cost Comparison of our CSM algorithm (using the proposed fast search or an exhaustive search) and the PRO algorithm in [4]. The horizontal axis is the size (in logarithm) of an initial visual codebook, while the vertical axes are time and peak memory cost in (a) and (b), respectively. As shown, the CSM algorithm with the fast search significantly reduces the time cost for a large-sized visual codebook with acceptable memory usage.

word used in a histogram is randomly sampled from $\{0, 1, 2, \cdots, 99\}$. In this experiment, the CSM-based fast algorithm is compared with the PRO algorithm which uses an exhaustive search to find the optimal pair of words to merge. We implement the PRO algorithm according to [4], including a trick suggested to speed up the algorithm by only updating the terms related to the two words to be merged. Meanwhile, to explicitly show the efficiency of the fast search part in our algorithm, we purposely replace the fast search in the CSM-based algorithm with an exhaustive search to demonstrate the quick increase on time cost. A machine with 2.80GHz CPU and 4.0GB memory is used. The result is in Fig. 4. As seen in sub-figure(a), the time cost of the PRO algorithm goes up quickly with the increasing codebook size. It takes $1,624$ seconds to hierarchically cluster 1000 visual words to 2, whereas the CSM algorithm with an exhaustive search only uses 9 seconds to accomplish this. The less time cost is attributed to the simplicity of the CSM criterion and the fast evaluation method proposed in Section 4.1. The CSM algorithm with the fast search achieves the highest computational efficiency. It only takes 1.55 minutes to hierarchically merge 10,000 visual words to 2, and the time cost increases to 141.1 minutes when an exhaustive search is used. As shown in sub-figure(b), the price is that the fast search needs more memory (1.45GB for 10,000 visual words) to store the indexing structure. We believe that such memory usage is acceptable for a personal computer today. In the following experiments, the discriminative power of the obtained compact visual codebooks is investigated.

5.2   Motorbikes vs. Airplanes

This experiment discriminates the images of a motorbike from those containing an airplane. In each of the 10 pairs of training/test subsets, there are 959 training

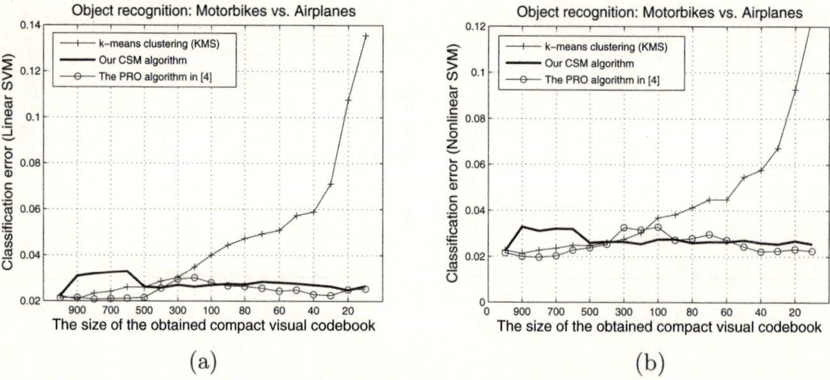

Fig. 5. Motorbikes vs. Airplanes Comparison of classification performance of the compact visual codebooks generated by k-means clustering (KMS), the PRO algorithm in [4], and our class separability measure (CSM) algorithm. Linear and nonlinear SVM classifiers are used in (a) and (b), respectively. The CSM-based algorithm still gives the excellent classification result when the codebook size has been considerably reduced.

images and 639 test images. An initial visual codebook of size $1,000$ is created by using k-means clustering. The CSM algorithm with the fast search hierarchically clusters them into 2 words in 6 seconds, whereas the PRO algorithm takes $6,164$ seconds to finish this. Based on the obtained compact visual codebook, a new histogram is created to represent each image. With the new histograms, a classifier is trained on a training subset and evaluated on the corresponding test subset. The average classification error rate is plotted in Fig. 5. The sub-figure (a) shows the result when a linear SVM classifier is used. As seen, the compact codebook generated by k-means clustering has poor discriminative power. Its classification error rate goes up with the decreasing size of the compact codebook. This is because k-means clustering uses the Euclidean distance between clusters as the merging criterion, which is not related to the classification performance. In contrast, the CSM and PRO algorithms achieve better classification performance, indicating that they well preserve the discriminative power in the obtained compact codebooks. For example, when the codebook size is reduced from 1000 to 20, these two algorithms still maintain excellent classification performance, with an increase of error rate less than 1%. Though the classification error rate of our CSM algorithm is a little bit higher (about 1.5%) at the initial stage, it soon drops to a level comparable to the error rate given by the PRO algorithm with the decreasing codebook size. Similar results can be observed from Fig. 5(b) where a nonlinear SVM classifier is employed.

5.3   Faces_Easy vs. Background_Google

This experiment aims to separate the images containing a face from the background images randomly collected from the Internet. In each training/test split, there are 100 training images and $1,498$ test images. The number of initial visual

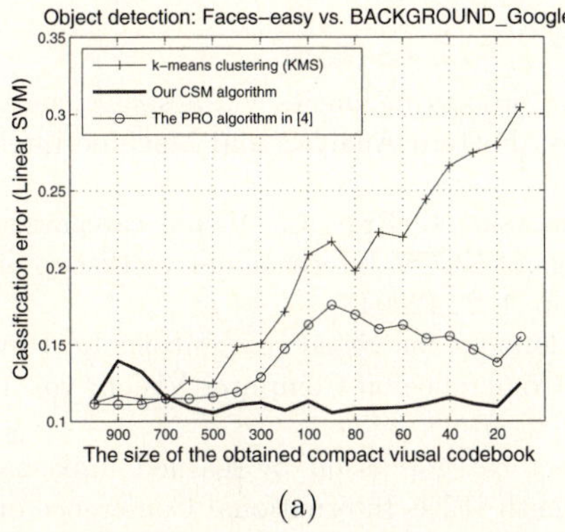

(a)

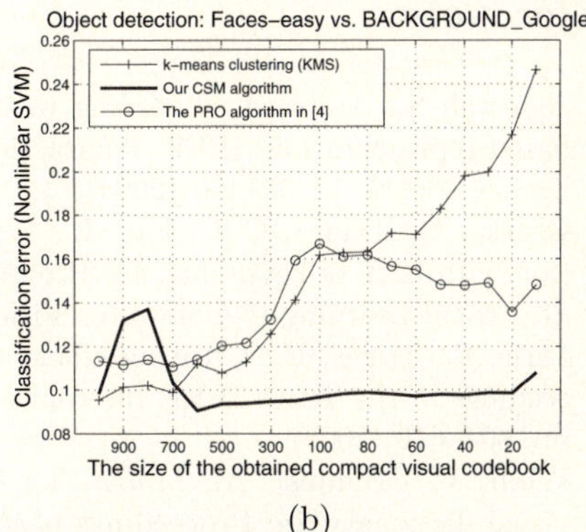

(b)

Fig. 6. Face-easy vs. Background_Google Comparison of classification performance of the small-sized visual codebooks generated by k-means clustering (KMS), the PRO algorithm in [4], and our proposed class separability measure (CSM). Linear and nonlinear SVM classifiers are used in (a) and (b), respectively. As shown, the CSM-based algorithm gives the best compact and discriminative codebooks.

words is $1,000$. They are hierarchically clustered into two words in 6 seconds by our CSM algorithm with the fast search and in $1,038$ seconds by the PRO algorithm. Again, with the newly obtained histograms, a classifier is trained and evaluated. The averaged classification error rates are presented in Fig. 6. In this experiment, the classification performance of the PRO algorithm is not as good as before. This might be caused by the hyper-parameters used in the PRO algorithm. Their values are preset according to [4] but may be task-dependent. In contrast, our CSM algorithm achieves the best classification performance. The small-sized compact codebooks consistently produce the error rate comparable to that of the initial visual codebook. This indicates that our algorithm effectively makes the compact codebooks preserve the discriminative power of the initial codebook. An additional advantage of our algorithm is that the CSM criterion is free of parameter setting. Meanwhile, a short "transition period" is observed on the CSM algorithm in Fig. 6, where the classification error rate goes up and then drops at the early stage. This interesting phenomenon will be looked into in future work.

6   Conclusion

To obtain a compact and discriminative visual codebook, this paper proposes using the separability of object classes to guide the hierarchical clustering of initial visual words. Moreover, a fast algorithm is designed to avoid a lengthy exhaustive search. As shown by the experimental study, our algorithm not only ensures the discriminative power of a compact codebook, but also makes the creation of a compact codebook very fast. This delivers an efficient tool for patch-based object recognition. In future work, more theoretical and experimental study will be conducted to analyze its performance.

References

1. Agarwal, S., Awan, A.: Learning to detect objects in images via a sparse, part-based representation. IEEE Transactions on Pattern Analysis and Machine Intelligence 26(11), 1475–1490 (2004)
2. Csurka, G., Dance, C.R., Fan, L., Willamowski, J., Bray, C.: Visual categorization with bags of keypoints. In: Proceedings of ECCV International Workshop on Statistical Learning in Computer Vision, pp. 1–22 (2004)
3. Jurie, F., Triggs, B.: Creating efficient codebooks for visual recognition. In: Proceedings of the Tenth IEEE International Conference on Computer Vision, vol. 1, pp. 604–610 (2005)
4. Winn, J., Criminisi, A., Minka, T.: Object categorization by learned universal visual dictionary. In: Proceedings of the Tenth IEEE International Conference on Computer Vision, vol. 2, pp. 1800–1807 (2005)
5. Varma, M., Zisserman, A.: A statistical approach to texture classification from single images. International Journal of Computer Vision 62(1-2), 61–81 (2005)
6. Mika, S., Rätsch, G., Weston, J., Schölkopf, B., Müller, K.R.: Fisher discriminant analysis with kernels. In: Hu, Y.H., Larsen, J., Wilson, E., Douglas, S. (eds.) Neural Networks for Signal Processing IX, pp. 41–48. IEEE, Los Alamitos (1999)
7. Shen, C., Li, H., Brooks, M.J.: A convex programming approach to the trace quotient problem. In: Yagi, Y., Kang, S.B., Kweon, I.S., Zha, H. (eds.) ACCV 2007, Part II. LNCS, vol. 4844, pp. 227–235. Springer, Heidelberg (2007)
8. Overmars, M.H., van Leeuwen, J.: Maintenance of configurations in the plane. Journal of Computer and System Sciences 23(2), 166–204 (1981)
9. Fei-Fei, L., Fergus, R., Perona, P.: Learning generative visual models from few training examples: an incremental bayesian approach tested on 101 object categories. In: Conference on Computer Vision and Pattern Recognition Workshop, vol. 12, pp. 178–178 (2004)
10. Mikolajczyk, K., Schmid, C.: Scale & affine invariant interest point detectors. International Journal of Computer Vision 60(1), 63–86 (2004)
11. Lowe, D.G.: Object recognition from local scale-invariant features. In: Proceedings of the Seventh IEEE International Conference on Computer Vision, vol. 2, pp. 1150–1157 (1999)
12. Mikolajczyk, K., Tuytelaars, T., Schmid, C., Zisserman, A., Matas, J., Schaffalitzky, F., Kadir, T., Gool, L.V.: A comparison of affine region detectors. International Journal of Computer Vision 65(1-2), 43–72 (2005)
13. Mikolajczyk, K., Schmid, C.: A performance evaluation of local descriptors. IEEE Transactions on Pattern Analysis and Machine Intelligence 27(10), 1615–1630 (2005)

A Dynamic Conditional Random Field Model for Joint Labeling of Object and Scene Classes

Christian Wojek and Bernt Schiele

Computer Science Department
TU Darmstadt
{wojek,schiele}@cs.tu-darmstadt.de

Abstract. Object detection and pixel-wise scene labeling have both been active research areas in recent years and impressive results have been reported for both tasks separately. The integration of these different types of approaches should boost performance for both tasks as object detection can profit from powerful scene labeling and also pixel-wise scene labeling can profit from powerful object detection. Consequently, first approaches have been proposed that aim to integrate both object detection and scene labeling in one framework. This paper proposes a novel approach based on conditional random field (CRF) models that extends existing work by 1) formulating the integration as a joint labeling problem of object and scene classes and 2) by systematically integrating dynamic information for the object detection task as well as for the scene labeling task. As a result, the approach is applicable to highly dynamic scenes including both fast camera and object movements. Experiments show the applicability of the novel approach to challenging real-world video sequences and systematically analyze the contribution of different system components to the overall performance.

1 Introduction

Today, object class detection methods are capable of achieving impressive results on challenging datasets (e.g. PASCAL challenges [1]). Often these methods combine powerful feature vectors such as SIFT or HOG with the power of discriminant classifiers such as SVMs and AdaBoost. At the same time several authors have argued that global scene context [2,3] is a valuable cue for object detection and therefore should be used to support object detection. This context-related work however has nearly exclusively dealt with static scenes. As this paper specifically deals with highly dynamic scenes we will also model object motion as an additional and important cue for detection.

Pixel-wise scene labeling has also been an active field of research recently. A common approach is to use Markov or conditional random field (CRF) models to improve performance by modeling neighborhood dependencies. Several authors have introduced the implicit notion of objects into CRF-models [4,5,6,7]. The interactions between object nodes and scene labels however are often limited to uni-directional information flow and therefore these models have not yet shown

D. Forsyth, P. Torr, and A. Zisserman (Eds.): ECCV 2008, Part IV, LNCS 5305, pp. 733–747, 2008.

the full potential of simultaneously reasoning about objects and scene. By formulating the problem as a *joint* labeling problem for object and scene classes, this paper introduces a more general notion of object-scene interaction enabling bidirectional information flow. Furthermore, as we are interested in dynamic scenes, we make use of the notion of dynamic CRFs [8], which we extend to deal with both moving camera and moving objects.

Therefore we propose a novel approach to jointly label objects and scene classes in highly dynamic scenes for which we introduce a new real-world dataset with pixel-wise annotations. Highly dynamic scenes are not only a scientific challenge but also an important problem, e.g. for applications such as autonomous driving or video indexing where both the camera and the objects are moving independently. Formulating the problem as a joint labeling problem allows 1) to model the dynamics of the scene and the objects separately which is of particular importance for the scenario of independently moving objects and camera, and 2) to enable bi-directional information flow between object and scene class labels.

The remainder of this paper is structured as follows. Section 2 reviews related work from the area of scene labeling and scene analysis in conjunction with object detection. Section 3 introduces our approach and discusses how object detection and scene labeling can be integrated as a joint labeling problem in a dynamic CRF formulation. Section 4 introduces the employed features, gives details on the experiments and shows experimental results. Finally, section 5 draws conclusions.

2   Related Work

In recent years, conditional random fields (CRFs) [9] have become a popular framework for image labeling and scene understanding. However, to the best of our knowledge, there is no work which explicitly models object entities in *dynamic* scenes. Here, we propose to model objects and scenes in a joint labeling approach on two different layers with different information granularity and different labels in a dynamic CRF [8].

Related work can roughly be divided into two parts. First, there is related work on CRF models for scene understanding, and second there are approaches aiming to integrate object detection with scene understanding.

In [10] Kumar&Hebert detect man-made structures in natural scenes using a single-layered CRF. Later they extend this work to handle multiple classes in a two-layered framework [5]. Kumar&Hebert also investigated object-context interaction and combined a simple boosted object detector for side-view cars with scene context of road and buildings on a single-scale database of static images. In particular, they are running inference separately on their two layers and each detector hypothesis is only modeled in a neighborhood relation with an entire region on the second layer. On the contrary, we integrate multi-scale objects in a CRF framework where inference is conducted jointly for objects and context. Additionally, we propose to model edge potentials in a consistent layout by exploiting the scale given by a state-of-the-art object detector [11]. Torralba *et al.* [7] use boosted classifiers to model unary and interaction potentials in order to jointly label object and scene classes. Both are represented by a dictionary of patches. However, the authors do not employ an object detector for entire objects. In our work we found a separate object detector to be essential for improved performance. Also Torralba *et al.* use separate layers for each object and scene class and thus inference is costly due to the high graph connectivity, and furthermore they also work on a single-scale database of static images. We introduce a sparse layer to represent object hypotheses and work on dynamic image sequences containing objects of multiple scales. Further work on simultaneous object recognition and scene labeling has been conducted by Shotton *et al.* [6]. Their confusion matrix shows, that in particular object classes where color and texture cues do not provide sufficient discriminative power on static images – such as boat, chair, bird, cow, sheep, dog – achieve poor results. While their Texton feature can exploit context information even from image pixels with a larger distance, the mentioned object classes remain problematic due to the unknown object scale. Furthermore, He *et al.* [4] present a multi-scale CRF which contains multiple layers relying on features of different scales. However, they do not model the explicit notion of objects and their higher level nodes rather serve as switches to different context and object co-occurrences. Similarly, Verbeek&Triggs [12] add information about class co-occurrences by means of a topic model. Finally, several authors proposed to adopt the CRF framework for object recognition as a standalone task [13,14,15] without any reasoning about the context and only report results on static single-scale image databases.

Dynamic CRFs are exploited by Wang&Ji [16] for the task of image segmentation with intensity and motion cues in mostly static image sequences. Similarly, Yin&Collins [17] propose a MRF with temporal neighborhoods for motion segmentation with a moving camera.

The second part of related work deals with scene understanding approaches from the observation of objects. Leibe *et al.* [18] employ a stereo camera system together with a structure-from-motion approach to detect pedestrians and cars in urban environments. However, they do not explicitly label the background classes which are still necessary for many applications even if all objects in the scene are known. Hoiem *et al.* [3] exploit the detected scales of pedestrians and cars together with a rough background labeling to infer the camera's viewpoint which in turn improves the object detections in a directed Bayesian network. Contrary to our work, object detections are refined by the background context but not the other way round. Also, only still images are handled while the presence of objects is assumed. Similarly, Torralba [2] exploits filter bank responses to obtain a scene prior for object detection.

3   Conditional Random Field Models

The following section successively introduces our model. It is divided into three parts: the first reviews single layer CRFs, the second additionally models objects in a separate layer and the last adds the scene's and objects' dynamics.

We denote the input image at time t with $\mathbf{x}^t$, the corresponding class labels at the grid cell level with $\mathbf{y}^t$ and the object labels with $\mathbf{o}^t$.

3.1 Plain CRF: Single Layer CRF Model for Scene-Class Labeling

In general a CRF models the conditional probability of all class labels $\mathbf{y}^t$ given an input image $\mathbf{x}^t$. Similar to others, we model the set of neighborhood relationships N_1 up to pairwise cliques to keep inference computationally tractable. Thus, we model

$$\log(P_{pCRF}(\mathbf{y}^t|\mathbf{x}^t, N_1, \Theta)) = \sum_i \Phi(y_i^t, \mathbf{x}^t; \Theta_\Phi) + \sum_{(i,j)\in N_1} \Psi(y_i^t, y_j^t, \mathbf{x}^t; \Theta_\Psi) - \log(Z^t)$$

(1)

Z^t denotes the so called partition function, which is used for normalization. N_1 is the set of all spatial pairwise neighborhoods. We refer to this model as *plain CRF*.

Unary Potentials. Our unary potentials model local features for all classes C including scene as well as object classes. We employ the joint boosting framework [19] to build a strong classifier $H(c, \mathbf{f}(x_i^t); \Theta_\Phi) = \sum_{m=1}^{M} h_m(c, \mathbf{f}(x_i^t); \Theta_\Phi)$. Here, $\mathbf{f}(x_i^t)$ denotes the features extracted from the input image for grid point i. M is the number of boosting rounds and c are the class labels. h_m are *weak learners* with parameters Θ_Φ and are shared among the classes for this approach. In order to interpret the boosting confidence as a probability we apply a softmax transform [5]. Thus, the potential becomes:

$$\Phi(y_i^t = k, \mathbf{x}^t; \Theta_\Phi) = \log \frac{\exp H(k, \mathbf{f}(x_i^t); \Theta_\Phi)}{\sum_c \exp H(c, \mathbf{f}(x_i^t); \Theta_\Phi)}$$

(2)

Edge Potentials. The edge potentials model the interaction between class labels at two neighboring sites y_i^t and y_j^t in a regular lattice. The interaction strength is modeled by a linear discriminative classifier with parameters $\Theta_\Psi = \mathbf{w}^T$ and depends on the difference of the node features $\mathbf{d}_{ij}^t := |\mathbf{f}(x_i^t) - \mathbf{f}(x_j^t)|$.

$$\Psi(y_i^t, y_j^t, \mathbf{x}^t; \Theta_\Psi) = \sum_{(k,l)\in C} \mathbf{w}^T \begin{pmatrix} 1 \\ \mathbf{d}_{ij}^t \end{pmatrix} \delta(y_i^t = k)\delta(y_j^t = l)$$

(3)

3.2 Object CRF: Two Layer Object CRF for Joint Object and Scene Labeling

Information that can be extracted from an image patch locally is rather limited and pairwise edge potentials are too weak to model long range interactions. Ideally, a complete dense layer of hidden variables would be added to encode possible locations and scales of objects, but since inference for such a model is computationally expensive we propose to inject single hidden variables $\mathbf{o}^t = \{o_1^t, \ldots, o_D^t\}$ (D being the number of detections) as depicted in figure 1(a). To instantiate those nodes any multi-scale object detector can be employed.

The additional nodes draw object appearance from a strong spatial model and are connected to the set of all corresponding hidden variables $\{\mathbf{y}^t\}_{o_n^t}$ whose

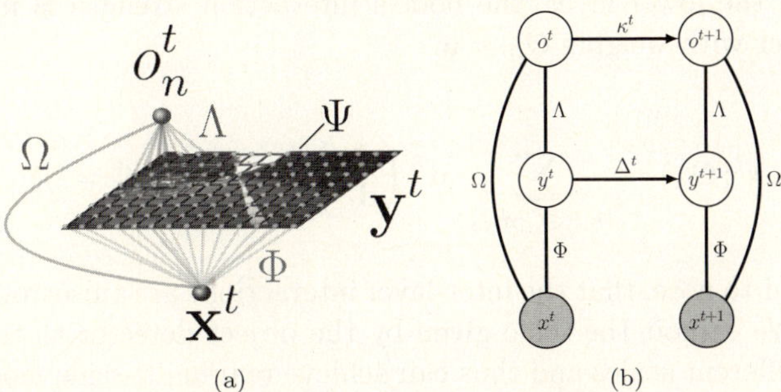

(a)   (b)

Fig. 1. (a) Graphical model for the *object CRF*; note that different edge colors denote different potentials; (b) Graphical model for our full *dynamic CRF*; observed nodes are grey, hidden variables are white, for the sake of readability we omit the spatial layout of $\mathbf{y}^t$ with the corresponding edge potential Ψ

evidence $\{\mathbf{x}^t\}_{o_n^t}$ support the object hypotheses. The new nodes' labels in this work are comprised of $O = \{object, background\}$; but the extension to multiple object classes is straight forward. Thus, we introduce two new potentials into the CRF model given in equation (1) and yield the *object CRF*:

$$\log(P_{oCRF}(\mathbf{y}^t, \mathbf{o}^t | \mathbf{x}^t, \Theta)) = \log(P_{pCRF}(\mathbf{y}^t | \mathbf{x}^t, N_2, \Theta)) + \tag{4}$$

$$\sum_n \Omega(o_n^t, \mathbf{x}^t; \Theta_\Omega) + \sum_{(i,j,n)\in N_3} \Lambda(y_i^t, y_j^t, o_n^t, \mathbf{x}^t; \Theta_\Psi)$$

Note that $N_2 \subset N_1$ denotes all neighborhoods where no object is present in the scene, whereas N_3 are all inter-layer neighborhoods with hypothesized object locations. Ω is the new unary object potential, whereas Λ is the inter-layer edge potential.

Unary Object Potentials. To define object potentials we use a state-of-the-art object detector. More specifically, we use a sliding window based multi-scale approach [11] where a window's features are defined by $\mathbf{g}(\{\mathbf{x}^t\}_{o_n^t})$ and classified with a linear SVM, the weights being $\mathbf{v}$ and b being the hyperplane's bias. To get a probabilistic interpretation for the classification margin, we adopt Platt's method [20] and fit a sigmoid with parameters s_1 and s_2 using cross validation.

$$\Omega(o_n^t, \mathbf{x}^t; \Theta_\Omega) = \log \frac{1}{1 + \exp(s_1 \cdot (\mathbf{v}^T \cdot \mathbf{g}(\{\mathbf{x}^t\}_{o_n^t}) + b) + s_2)} \tag{5}$$

Consequently, the parameters are determined as $\Theta_\Omega = \{\mathbf{v}, b, s_1, s_2\}$.

Inter-Layer Edge Potentials. For the inter-layer edge potentials we model the neighborhood relations in cliques consisting of two underlying first layer nodes y_i^t, y_j^t and the object hypothesis node o_n^t. Similar to the pairwise edge potentials on the lower layer, the node's interaction strength is modeled by a linear classifier with weights $\Theta_\Lambda = \mathbf{u}$.

$$\Lambda(y_i^t, y_j^t, o_n^t, \mathbf{x}^t; \Theta_\Lambda) = \sum_{(k,l)\in C; m\in O} \mathbf{u}^T \begin{pmatrix} 1 \\ \mathbf{d}_{ij}^t \end{pmatrix} \delta(y_i^t = k)\delta(y_j^t = l)\delta(o_n^t = m) \quad (6)$$

It is important to note, that the inter-layer interactions are anisotropic and scale-dependent. We exploit the scale given by the object detector to train different weights for different scales and thus can achieve real multi-scale modeling in the CRF framework. Furthermore, we use different sets of weights for different parts of the detected object enforcing an object and context consistent layout [15].

3.3 Dynamic CRF: Dynamic Two Layer CRF for Object and Scene Class Labeling

While the additional information from an object detector already improves the classification accuracy, temporal information is a further important cue. We propose two temporal extensions to the framework introduced so far. For highly dynamic scenes – such as the image sequences taken by a driving car, which we will use as an example application to our model, it is important to note that objects and the remaining scene have different dynamics and thus should be modeled differently. For objects we estimate their motion and track them with a temporal filter in 3D space. The dynamics for the remaining scene is mainly caused by the camera motion in our example scenario. Therefore, we use an estimate of the camera's ego motion to propagate the inferred scene labels at time t as a prior to time step $t + 1$.

Since both – object and scene dynamics – transfer information forward to future time steps, we employ directed links in the corresponding graphical model as depicted in figure 1(b). It would have also been possible to introduce undirected links, but those are computationally more demanding. Moreover, those might not be desirable from an application point of view, due to the backward flow of information in time when online processing is required.

Object Dynamics Model. In order to model the object dynamics we employ multiple extended Kalman filters [21] – one for each object. For the dynamic scenes dataset which we will use for the experimental section the camera calibration is known and the sequences are recorded from a driving car. Additionally, we assume the objects to reside on the ground plane. Consequently, Kalman filters are able to model the object position in 3D coordinates. Additionally, the state vector contains the objects' width and speed on the ground plane as well as the camera's tilt and all state variables' first derivative with respect to time.

For the motion model we employ linear motion dynamics with the acceleration being modeled as system noise which proved sufficient for the image sequences used below. The tracks' confidences are given by the last associated detection's score. Hence, we obtain the following integrated model:

$$\log(P_{tCRF}(\mathbf{y}^t, \mathbf{o}^t | \mathbf{x}^t, \Theta)) = \log(P_{pCRF}(\mathbf{y}^t | \mathbf{x}^t, N_2, \Theta)) + \tag{7}$$

$$\sum_n \kappa^t(o_n^t, \mathbf{o}^{t-1}, \mathbf{x}^t; \Theta_\kappa) + \sum_{(i,j,n) \in N_3} \Lambda(y_i^t, y_j^t, o_n^t, \mathbf{x}^t; \Theta_\Lambda)$$

where κ^t models the probability of an object hypothesis o_n^t at time t given the history of input images. It replaces the previously introduced potentials for objects Ω. The parameter vector consists of the detector's parameters and additionally of the Kalman filter's dynamics $\{A, W\}$ and measurement model $\{H_t, V_t\}$ and thus $\Theta_\kappa = \Theta_\Omega \cup \{A, W, H_t, V_t\}$.

Scene Dynamic Model. In the spirit of recursive Bayesian state estimation under the Markovian assumption, the posterior distribution of $\mathbf{y}^{t-1}$ is used as a prior to time step t. However, for dynamic scenes the image content needs to be transformed to associate the grid points with the right posterior distributions. In this work we estimate the projection Q from $\mathbf{y}^t$ to $\mathbf{y}^{t+1}$ given the camera's translation and calibration (Θ_{Δ^t}). Thus, we obtain an additional unary potential for $\mathbf{y}^t$.

$$\Delta^t(y_i^t, \mathbf{y}^{t-1}; \Theta_{\Delta^t}) = \log(P_{tCRF}(y_{Q^{-1}(i)}^{t-1} | \Theta)) \tag{8}$$

The complete *dynamic CRF* model including both object and scene dynamics as depicted in figure 1(b) then is

$$\log(P_{dCRF}(\mathbf{y}^t, \mathbf{o}^t, \mathbf{x}^t | \mathbf{y}^{t-1}, \mathbf{o}^{t-1}, \Theta)) = \log(P_{tCRF}(\mathbf{y}^t, \mathbf{o}^t | \mathbf{x}^t, \Theta)) +$$

$$\sum_i \Delta^t(y_i^t, \mathbf{y}^{t-1}; \Theta_{\Delta^t}) \tag{9}$$

3.4   Inference and Parameter Estimation

For inference in the undirected graphical models we employ sum-product loopy belief propagation with a parallel message update schedule. For parameter estimation we take a piecewise learning approach [22] by assuming the parameters of unary potentials to be conditionally independent of the edge potentials' parameters. While this no longer guarantees to find the optimal parameter setting for Θ, we can learn the model much faster as discussed by [22].

Thus, prior to learning the edge potential models we train parameters Θ_Φ, Θ_Ω for the unary potentials. The parameter set Θ_κ for the Kalman filter is set to reasonable values by hand.

Finally, the edge potentials' parameter sets Θ_Ψ and Θ_Λ are learned jointly in a maximum likelihood setting with stochastic meta descent [23]. As proposed by Vishwanathan *et al.* we assume a Gaussian prior with meta parameter σ on the linear weights to avoid overfitting.

4   Experiments

To evaluate our model's performance we conducted several experiments on two datasets. First, we describe our features which are used for texture and location based classification of scene labels on the scene label CRF layer. Then we introduce features employed for object detection on the object label CRF layer. Next, we briefly discuss the results obtained on the Sowerby database and finally we present results on image sequences on a new dynamic scenes dataset, which consist of car traffic image sequences recorded from a driving vehicle under challenging real-world conditions.

4.1   Features for Scene Labeling

Texture and Location Features. For the unary potential Φ at the lower level as well as for the edge potentials Ψ and inter-layer potentials Λ we employ texture and location features. The texture features are computed from the 16 first coefficients of the Walsh-Hadamard transform. This transformation is a discrete approximation of the cosine transform and can be computed efficiently [24,25] – even in real-time (e.g. on modern graphics hardware). The features are extracted at multiple scales from all channels of the input image in CIE *Lab* color space. As a preprocessing step, a and b channels are normalized by means of a gray world assumption to cope with varying color appearance. The L channel is mean-variance normalized to fit a Gaussian distribution with a fixed mean to cope with global lighting variations. We also found that normalizing the transformation's coefficients according to Varma&Zisserman [26] is beneficial. They propose to L_1-normalize each filter response first and then locally normalize the responses at each image pixel. Finally, we take the mean and variance of the normalized responses as feature for each node in the regular CRF lattice. Additionally, we use the grid point's coordinates within the image as a location cue. Therefore, we concatenate the pixel coordinates to the feature vector.

HOG. In the experiments described below we employ a HOG (Histogram of Oriented Gradients) detector [11] to generate object hypotheses. HOG is a sliding window approach where features are computed on a dense grid. First, histograms of gradient orientation are computed in *cells* performing interpolation with respect to the gradient's location and with respect to the magnitude. Next, sets of neighboring cells are grouped into overlapping *blocks*, which are normalized to achieve invariance to different illumination conditions. Our front and rear view car detector has a window size of 20×20 pixels. It is trained on a separate dataset of front and rear car views containing 1492 positive instances from the LabelMe database [27] and 178 negative images.

4.2   Results

Sowerby Dataset. The Sowerby dataset is a widely used benchmark for CRFs, which contains 7 outdoor rural landscape classes. The dataset comprises 104 images at a resolution of 96×64 pixels. Following the protocol of [5] we randomly selected 60 images for training and 44 images for testing. Some example images with inferred labels are shown in figure 2. However, this dataset does neither contain image sequences nor cars that can be detected with an object detector

Table 1. Comparison to previously reported results on the Sowerby dataset

	Pixel-wise accuracy	
	Unary classification	plain CRF model
He *et al.* [4]	82.4%	89.5%
Kumar&Hebert [5]	85.4%	89.3%
Shotton *et al.* [6]	85.6%	88.6%
This paper	84.5%	91.1%

and thus we can only compare our *plain CRF* model (equation 1) with previous work on this set.

The experiments show that our features and CRF parameter estimation is competitive to other state-of-the-art methods. Table 1 gives an overview of previously published results and how those compare to our model (see figure 3). While the more sophisticated Textons features [6] do better for unary classification, our CRF model can outperform those since our edge potentials are learned from training data. For this dataset we use a grid with one node for each input pixel, while the Gaussian prior σ was set to 1.25. The Walsh-Hadamard transform was run on the input images at the aperture size of 2, 4, 8 and 16 pixels. Moreover, we used a global set of weights for the isotropic linear classifiers of the edge potentials, but distinguish between north-south neighborhood relations and east-west neighborhood relations.

Input image   Unary classification   plain CRF result   Input image   Unary classification   plain CRF result

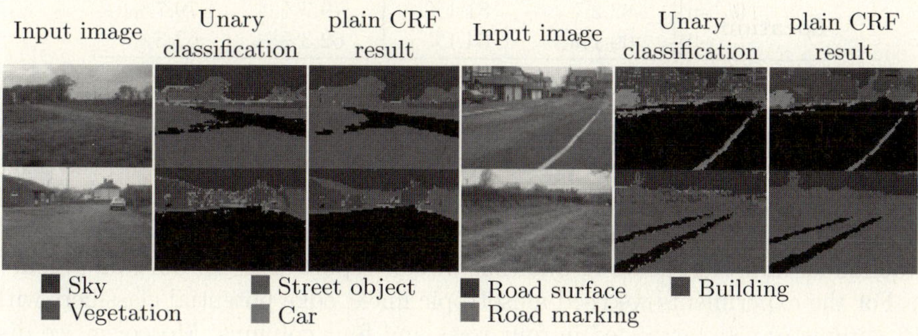

Fig. 2. *Sowerby* dataset example results

Dynamic Scenes Dataset. To evaluate our *object* and *dynamic CRF* we set up a new *dynamic scenes* dataset with image sequences consisting of overall 1936 images[1]. The images are taken from a camera inside a driving car and mainly show rural roads with high dynamics of driving vehicles at an image resolution of 752×480 pixels. Cars appear at all scales from as small as 15 pixels up to 200 pixels. The database consists of 176 sequences with 11 successive images each. It is split into equal size training and test sets of 968 images.

---

[1] The dataset is available at http://www.mis.informatik.tu-darmstadt.de.

To evaluate pixel level labeling accuracy the last frame of each sequence is labeled pixel-wise, while the remainder only contains bounding box annotations for the frontal and rear view car object class. Overall, the dataset contains the eight labels *void, sky, road, lane marking, building, trees & bushes, grass* and *car*. Figure 3 shows some sample scenes. For the following experiments we used 8×8 pixels for each CRF grid node and texture features were extracted at the aperture sizes of 8, 16 and 32 pixels.

We start with an evaluation of the unary classifier performance on the scene class layer. Table 2 lists the pixel-wise classification accuracy for different variations of the feature. As expected location is a valuable cue, since there is a huge variation in appearance due to different lighting conditions. Those range from bright and sunny illumination with cast shadows to overcast. Additionally, motion blur and weak contrast complicate the pure appearance-based classification. Further, we observe that normalization [26] as well as multi-scale features are helpful to improve the classification results.

Table 2. Evaluation of texture location features based on overall pixel-wise accuracy; Multi-scale includes feature scales of 8, 16 and 32 pixels, Single-scale is a feature scale of 8 pixels; note that these number do not include the CRF model – adding the *plain CRF* to the best configuration yields an overall accuracy of 88.3%

		Normalization			
		on		off	
		multi-scale	single-scale	multi-scale	single-scale
Location	on	82.2%	81.1%	79.7%	79.7%
	off	69.1%	64.1%	62.3%	62.3%

Next, we analyze the performance of the different proposed CRF models. On the one hand we report the overall pixel-wise accuracy. On the other hand the pixel-wise labeling performance on the car object class is of particular interest. Overall, car pixels cover 1.3% of the overall observed pixels. Yet, those are an important fraction for many applications and thus we also report those for our evaluation.

For the experiments we used anisotropic linear edge potential classifiers with 16 parameter sets, arranged in four rows and four columns. Moreover, we distinguish between north-south and east-west neighborhoods. For the inter-layer edge potentials we trained different weight sets depending on detection scale (discretized in 6 bins) and depending on the neighborhood location with respect to the object's center.

Table 3 shows recall and precision for the proposed models. Firstly, the employed detector has an equal error rate of 78.8% when the car detections are evaluated in terms of precision and recall. When evaluated on a pixel-wise basis the performance corresponds to 60.2% recall. The missing 39.8% are mostly due to the challenging dataset. It contains cars with weak contrast, cars at small scales and partially visible cars leaving the field of view. Precision for the detector evaluated on pixels is 37.7%. Wrongly classified pixels are mainly around the objects and on structured background on which the detector obtains false detections.

Table 3. Pixel-wise recall and precision for the pixels labeled as *Car* and overall accuracy on all classes

	No objects			With object layer			Including object dynamics		
	Recall	Precision	Acc.	Recall	Precision	Acc.	Recall	Precision	Acc.
CRF	50.1%	57.7%	88.3%	62.9%	52.3%	88.6%	70.4%	57.8%	88.7%
dyn. CRF	25.5%	44.8%	86.5%	75.7%	50.8%	87.1%	78.0%	51.0%	88.1%

Let us now turn to the performance of the different CRF models. Without higher level information from an object detector *plain CRFs* in combination with texture-location features achieve a recall of 50.1% with a precision of 57.7%. The recognition of cars in this setup is problematic since CRFs optimize a global energy function, while the car class only constitutes a minor fraction of the data. Thus, the result is mainly dominated by classes which occupy the largest regions such as sky, road and trees.

With higher level object information (*object CRF*) recall can be improved up to 62.9% with slightly lower precision resulting from the detector's false positive detections. However, when objects are additionally tracked with a Kalman filter, we achieve a recall of 70.4% with a precision of 57.8%. This proves that the object labeling for the car object class leverages from the object detector and additionally from the dynamic modeling by a Kalman filter.

Additionally, we observe an improvement of the overall labeling accuracy. While *plain CRFs* obtain an accuracy of 88.3%, the *object CRF* achieves 88.6% while also including object dynamics further improves the overall labeling accuracy to 88.7%. The relative number of 0.4% might appear low, but considering that the database overall only has 1.3% of car pixels, this is worth noting. Thus, we conclude that not only the labeling on the car class is improved but also the overall scene labeling quality.

When the scene dynamics are modeled additionally and posteriors are propagated over time (*dynamic CRF*), we again observe an improvement of the achieved recall from 25.5% to 75.7% with the additional object nodes. And also the objects' dynamic model can further improve the recall to 78.0% correctly labeled pixels. Thus, again we can conclude that the CRF model exploits both the information given by the object detector as well as the additional object dynamic to improve the labeling quality.

Finally, when the overall accuracy is analyzed while the scene dynamic is modeled we observe a minor drop compared to the static modeling. However, we again consistently observe that the object information and their dynamics allow to improve from 86.5% without object information to 87.1% with *object CRFs* and to 88.1% with the full model.

The consistently slightly worse precision and overall accuracy for the dynamic scene models need to be explained. Non-car pixels wrongly labeled as car are mainly located at the object boundary, which are mainly due to artifacts of the

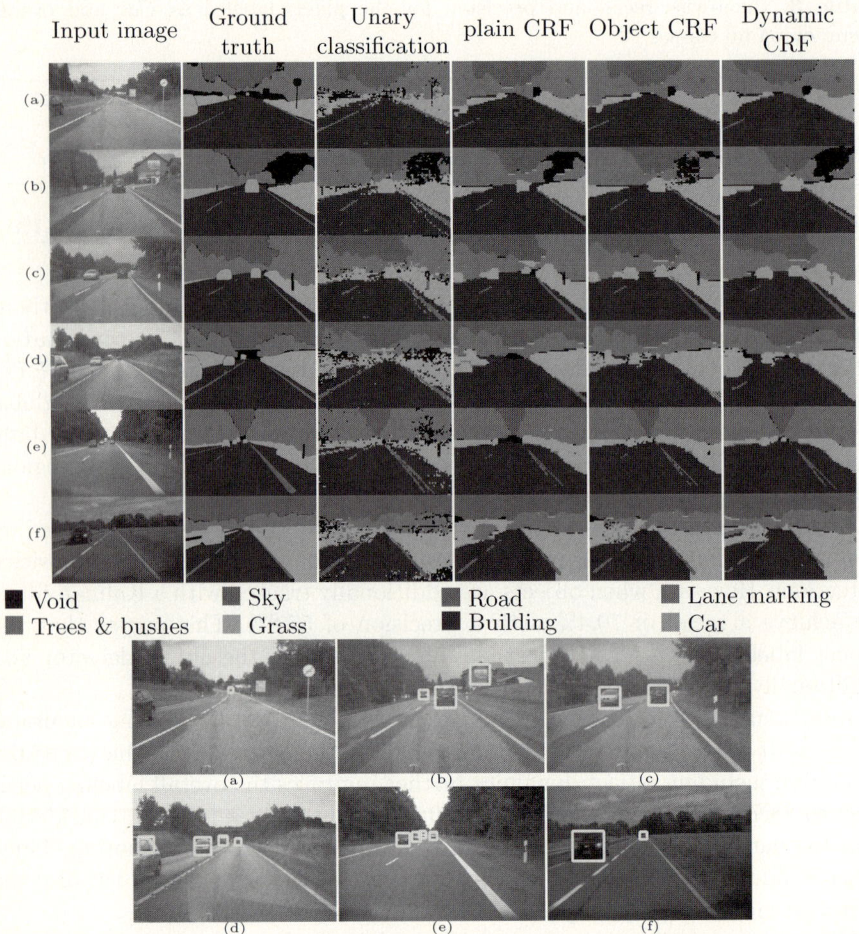

Fig. 3. *Dynamic scenes* dataset example result scene labels and corresponding detections in left-right order (best viewed in color); note that detections can be overruled by the texture location potentials and vice versa

scene label forward propagation. Those are introduced by the inaccuracies of the speedometer and due to the inaccuracies of the projection estimation.

A confusion matrix for all classes of the *dynamic scenes* database can be found in table 4. Figure 3 shows sample detections and scene labelings for the different CRF models to illustrate the impact of the different models and their improvements. In example (d) for instance the car which is leaving the field of view is mostly smoothed out by a *plain CRF* and *object CRF*, while the *dynamic CRF* is able to classify almost the entire area correctly. Additionally, the smaller cars which get smoothed out by a *plain CRF* are classified correctly by the *object* and *dynamic CRF*. Also note that false object detections as in example (c) do not result in a wrong labeling of the scene.

Table 4. Confusion matrix in percent for the *dynamic scenes* dataset; entries are row-normalized

True class	Fraction	Inferred Sky	Road	Lane marking	Trees & bushes	Grass	Building	Void	Car
Sky	10.4%	**91.0**	0.0	0.0	7.7	0.5	0.4	0.3	0.1
Road	42.1%	0.0	**95.7**	1.0	0.3	1.1	0.1	0.5	1.3
Lane marking	1.9%	0.0	36.3	**56.4**	0.8	2.9	0.2	1.8	1.6
Trees & bushes	29.2%	1.5	0.2	0.0	**91.5**	5.0	0.2	1.1	0.4
Grass	12.1%	0.4	5.7	0.5	13.4	**75.3**	0.3	3.5	0.9
Building	0.3%	1.6	0.2	0.1	37.8	4.4	**48.4**	6.3	1.2
Void	2.7%	6.4	15.9	4.1	27.7	29.1	1.4	**10.6**	4.8
Car	1.3%	0.3	3.9	0.2	8.2	4.9	2.1	2.4	**78.0**

5 Conclusions

In this work we have presented a unifying model for joint scene and object class labeling. While CRFs greatly improve unary pixel-wise classification of scenes they tend to smooth out smaller regions and objects such as cars in landscape scenes. This is particularly true when objects only comprise a minor part of the amount of overall pixels. We showed that adding higher level information from a state-of-the-art HOG object detector ameliorates this shortcoming. Further improvement – especially when objects are only partially visible – is achieved when object dynamics are properly modeled and when scene labeling information is propagated over time. The improvement obtained is bidirectional, on the one hand the labeling of object classes is improved, but on the other hand also the remaining scene classes benefit from the additional source of information.

For future work we would like to investigate how relations between different objects such as partial occlusion can be modeled when multiple object classes are detected. Additionally, we seek to improve the ego-motion estimation of the camera to further improve the performance. This will also allow us to employ motion features in the future. Finally, we assume that the integration of different sensors such as radar allow for a further improvement of the results.

Acknowledgements. This work has been funded, in part, by Continental Teves AG. Further we thank Joris Mooij for publically releasing libDAI and BAE Systems for the Sowerby Dataset.

References

1. Everingham, M., Zisserman, A., Williams, C., van Gool, L.: The pascal visual object classes challenge results. Technical report, PASCAL Network (2006)
2. Torralba, A.: Contextual priming for object detection. IJCV, 169–191 (2003)

3. Hoiem, D., Efros, A.A., Hebert, M.: Putting objects in perspective. In: CVPR (2006)
4. He, X., Zemel, R.S., Carreira-Perpiñán, M.Á.: Multiscale conditional random fields for image labeling. In: CVPR (2004)
5. Kumar, S., Hebert, M.: A hierarchical field framework for unified context-based classification. In: ICCV (2005)
6. Shotton, J., Winn, J., Rother, C., Criminisi, A.: Textonboost: Joint appearance, shape and context modeling for multi-class object recognition and segmentation. In: Leonardis, A., Bischof, H., Pinz, A. (eds.) ECCV 2006. LNCS, vol. 3951. Springer, Heidelberg (2006)
7. Torralba, A., Murphy, K.P., Freeman, W.T.: Contextual models for object detection using boosted random fields. In: NIPS (2004)
8. McCallum, A., Rohanimanesh, K., Sutton, C.: Dynamic conditional random fields for jointly labeling multiple sequences. In: NIPS Workshop on Syntax, Semantics (2003)
9. Lafferty, J.D., McCallum, A., Pereira, F.C.N.: Conditional random fields: Probabilistic models for segmenting and labeling sequence data. In: ICML (2001)
10. Kumar, S., Hebert, M.: Discriminative random fields: A discriminative framework for contextual interaction in classification. In: ICCV (2003)
11. Dalal, N., Triggs, B.: Histograms of oriented gradients for human detection. In: CVPR (2005)
12. Verbeek, J., Triggs, B.: Region classification with markov field aspect models. In: CVPR (2007)
13. Quattoni, A., Collins, M., Darrell, T.: Conditional random fields for object recognition. In: NIPS (2004)
14. Kapoor, A., Winn, J.: Located hidden random fields: Learning discriminative parts for object detection. In: Leonardis, A., Bischof, H., Pinz, A. (eds.) ECCV 2006. LNCS, vol. 3954. Springer, Heidelberg (2006)
15. Winn, J., Shotton, J.: The layout consistent random field for recognizing and segmenting partially occluded objects. In: CVPR (2006)
16. Wang, Y., Ji, Q.: A dynamic conditional random field model for object segmentation in image sequences. In: CVPR (2005)
17. Yin, Z., Collins, R.: Belief propagation in a 3D spatio-temporal MRF for moving object detection. In: CVPR (2007)
18. Leibe, B., Cornelis, N., Cornelis, K., Van Gool, L.: Dynamic 3D scene analysis from a moving vehicle. In: CVPR (2007)
19. Torralba, A., Murphy, K.P., Freeman, W.T.: Sharing features: Efficient boosting procedures for multiclass object detection. In: CVPR (2004)
20. Platt, J.: Probabilistic outputs for support vector machines and comparison to regularized likelihood methods. In: Smola, A.J., Bartlett, P., Schoelkopf, B., Schuurmans, D. (eds.) Advances in Large Margin Classifiers, pp. 61–74 (2000)
21. Kalman, R.E.: A new approach to linear filtering and prediction problems. Transactions of the ASME–Journal of Basic Engineering 82, 35–45 (1960)
22. Sutton, C., McCallum, A.: Piecewise training for undirected models. In: 21th Annual Conference on Uncertainty in Artificial Intelligence (UAI 2005) (2005)
23. Vishwanathan, S.V.N., Schraudolph, N.N., Schmidt, M.W., Murphy, K.P.: Accelerated training of conditional random fields with stochastic gradient methods. In: ICML (2006)
24. Hel-Or, Y., Hel-Or, H.: Real-time pattern matching using projection kernels. PAMI 27, 1430–1445 (2005)

25. Alon, Y., Ferencz, A., Shashua, A.: Off-road path following using region classification and geometric projection constraints. In: CVPR (2006)
26. Varma, M., Zisserman, A.: Classifying images of materials: Achieving viewpoint and illumination independence. In: Tistarelli, M., Bigun, J., Jain, A.K. (eds.) ECCV 2002. LNCS, vol. 2359. Springer, Heidelberg (2002)
27. Russell, B.C., Torralba, A., Murphy, K.P., Freeman, W.T.: LabelMe: A database and web-based tool for image annotation. IJCV 77, 157–173 (2008)

Local Regularization for Multiclass Classification Facing Significant Intraclass Variations

Lior Wolf and Yoni Donner

The School of Computer Science
Tel Aviv Univerisy
Tel Aviv, Israel

Abstract. We propose a new local learning scheme that is based on the principle of decisiveness: the learned classifier is expected to exhibit large variability in the direction of the test example. We show how this principle leads to optimization functions in which the regularization term is modified, rather than the empirical loss term as in most local learning schemes. We combine this local learning method with a Canonical Correlation Analysis based classification method, which is shown to be similar to multiclass LDA. Finally, we show that the classification function can be computed efficiently by reusing the results of previous computations. In a variety of experiments on new and existing data sets, we demonstrate the effectiveness of the CCA based classification method compared to SVM and Nearest Neighbor classifiers, and show that the newly proposed local learning method improves it even further, and outperforms conventional local learning schemes.

1   Introduction

Object recognition systems, viewed as learning systems, face three major challenges: First, they are often required to discern between many objects; second, images taken under uncontrolled settings display large intraclass variation; and third, the number of training images provided is often small.

Previous attempts to overcome these challenges use prior generic knowledge on variations within objects classes [1], employ large amounts of unlabeled data (e.g., [2]), or reuse previously learned visual features [3]. Here, we propose a more generic solution, that does not assume nor benefit from the existence of prior learning stages or of an additional set of training images.

To deal with the challenge of multiple classes, we propose a Canonical Correlation Analysis (CCA) based classifier, which is a regularized version of a recently proposed method [4], and is highly related to Fisher Discriminant Analysis (LDA/FDA). We treat the other two challenges as one since large intraclass variations and limited training data both result in a training set that does not capture well the distribution of the input space. To overcome this, we propose a new local learning scheme which is based on the principle of decisiveness.

In local learning schemes, some of the training is deferred to the prediction phase, and a new classifier is trained for each new (test) example. Such schemes

D. Forsyth, P. Torr, and A. Zisserman (Eds.): ECCV 2008, Part IV, LNCS 5305, pp. 748–759, 2008.

have been introduced by [5] and were recently advanced and shown to be effective for modern object recognition applications [6] (see references therein for additional references to local learning methods). One key difference between our method and the previous contribution in the field is that we do not select or directly weigh the training examples by their proximity to the test point. Instead, we modify the objective function of the learning algorithm to reward components in the resulting classifier that are parallel to the test example. Thus, we encourage the classification function (before thresholding takes place) to be separated from zero.

Runtime is a major concern for local learning schemes, since a new classifier needs to be trained or adjusted for every new test example. We show how the proposed classifier can be efficiently computed by several rank-one updates to precomputed eigenvectors and eigenvalues of constant matrices, with the resulting time complexity being significantly lower than that of a full eigen-decomposition. We conclude by showing the proposed methods to be effective on four varied datasets which exhibit large intraclass variations.

2   Multiclass Classification Via CCA

We examine the multiclass classification problem with k classes, where the goal is to construct a classifier given n training samples (x_i, y_i), with $x_i \in \mathbb{R}^m$ and $y_i \in \{1, 2, \ldots, k\}$. We assume $\sum_{i=1}^{n} x_i = 0$ (otherwise we center the data). Our approach is to find a transformation $T : \mathbb{R}^m \to \mathbb{R}^l$ and class vectors $v_j \in \mathbb{R}^l$ such that the transformed inputs $T(x_i)$ would be close to the class vector v_{y_i} corresponding to their class. Limiting the discussion at first to linear transformations, we represent T by a $m \times l$ matrix A such that $T(x) = A^\top x$. The formulation of the learning problem is therefore:

$$\min_{A, \{v_j\}_{j=1}^{k}} \sum_{i=1}^{n} \|A^\top x_i - v_{y_i}\|^2 \tag{1}$$

Define V to be the $k \times l$ matrix with v_j as its j'th row, so $v_j = V^\top e_j$. Also define $z_i = e_{y_i}$ where e_j is the j'th column of the identity $k \times k$ matrix I_k. Using these definitions, $v_{y_i} = V^\top z_i$ and Equation 1 becomes:

$$\min_{A, V} \sum_{i=1}^{n} \|A^\top x_i - V^\top z_i\|^2 \tag{2}$$

This expression can be further simplified by defining the matrices $X \in \mathbb{R}^{m \times n}$, $Z \in \mathbb{R}^{k \times n}$: $X = (x_1, x_2, \ldots, x_n)$, $Z = (z_1, z_2, \ldots, z_n)$. Equation 2 then becomes:

$$\min_{A, V} \operatorname{tr}(A^\top X X^\top A) + \operatorname{tr}(V^\top Z Z^\top V) - 2\operatorname{tr}(A^\top X Z^\top V) \tag{3}$$

This expression is not invariant to arbitrary scaling of A and Z. Furthermore, we require the l components of the transformed vectors $A^\top x_i$ and $V^\top z_i$ to be pairwise uncorrelated since there is nothing to be gained by correlations between them. Therefore, we add the constraints $A^\top X X^\top A = V^\top Z Z^\top V = I_l$, leading to the final problem formulation:

$$\max_{A,V} \quad \operatorname{tr}(A^\top X Z^\top V)$$

$$\text{subject to } A^\top X X^\top A = V^\top Z Z^\top V = I \tag{4}$$

This problem is solved through Canonical Correlation Analysis (CCA) [7]. A simple solution involves writing the corresponding Lagrangian and setting the partial derivatives to zero, yielding the following generalized eigenproblem:

$$\begin{pmatrix} 0 & X Z^\top \\ Z X^\top & 0 \end{pmatrix} \begin{pmatrix} a_i \\ v_i \end{pmatrix} = \lambda_i \begin{pmatrix} X X^\top & 0 \\ 0 & Z Z^\top \end{pmatrix} \begin{pmatrix} a_i \\ v_i \end{pmatrix} \tag{5}$$

where λ_i, $i = 1..l$ are the leading generalized eigenvalues, a_i are the columns of A, and v_i are, as defined above, the columns of V. To classifying a new sample x, it is first transformed to $A^\top x$, and then compared to the k class vectors, i.e., the predicted class is given by $\arg\min_{1 \leq j \leq k} \|A^\top x - v_j\|$.

This classification scheme is readily extendable to non-linear functions that satisfy Mercer's conditions by using Kernel CCA [8,9]. Kernel CCA is also equivalent to solving a generalized eigenproblem of the form of Equation 5, so although we refer directly to linear CCA throughout this paper, our conclusions are equally valid for Kernel CCA.

In Kernel CCA, or in the linear case when $m > n$, and in many other common scenarios, the problem is ill-conditioned and regularization techniques are required [10]. For linear regression, ridge regularization is often used, as is its equivalent in CCA and Kernel CCA [8]. This involves replacing $X X^\top$ and $Z Z^\top$ in Equation 5 with $X X^\top + \eta_X I$ and $Z Z^\top + \eta_Z I$, where η_X and η_Z are regularization parameters. In the CCA case presented here, for multiclass classification, since the number of training examples n is not smaller than the number of classes k, regularization need not be used for Z and we set $\eta_Z = 0$. Also, since the X regularization is relative to the scale of the matrix $X X^\top$, we scale the regularization parameter η_X as a fraction of the largest eigenvalue of $X X^\top$.

The multiclass classification scheme via CCA presented here is equivalent to Fisher Discriminant Analysis (LDA). We provide a brief proof of this equivalence. A previous lemma was proven by Yamada et al [4] for the unregularized case.

Lemma 1. *The multiclass CCA classification method learns the same linear transformation as multiclass LDA.*

Proof. The generalized eigenvalue problem in Equation 5, with added ridge regularization, can be represented by the following two coupled equations:

$$(X X^\top + \eta I_m)^{-1} X Z^\top v = \lambda a \tag{6}$$

$$(Z Z^\top)^{-1} Z X^\top a = \lambda v \tag{7}$$

Any solution (a, v, λ) to the above system satisfies:

$$(X X^\top + \eta I_m)^{-1} X Z^\top (Z Z^\top)^{-1} Z X^\top a = (X X^\top + \eta I_m)^{-1} X Z^\top \lambda v = \lambda^2 a \tag{8}$$

$$(Z Z^\top)^{-1} Z X^\top (X X^\top + \eta I_m)^{-1} X Z^\top v = (Z Z^\top)^{-1} Z X^\top \lambda a = \lambda^2 v \tag{9}$$

Thus the columns of the matrix A are the eigenvectors corresponding to the largest eigenvalues of $(XX^\top + \eta I_m)^{-1} X Z^\top (ZZ^\top)^{-1} Z X^\top$. Examine the product $ZZ^\top = \sum_{i=n} e_{y_i} e_{y_i}^\top$. It is a $k \times k$ diagonal matrix with the number of training samples in each class (denoted N_i) along its diagonal. Therefore, $(ZZ^\top)^{-1} = \mathrm{diag}(\frac{1}{N_1}, \frac{1}{N_2}, \ldots, \frac{1}{N_k})$. Now examine $XZ^\top$: $(XZ^\top)_{i,j} = \sum_{s=1}^{n} X_{i,s} Z_{j,s} = \sum_{s:y_s=j} X_{i,s}$. Hence, the j'th column is the sum of all training samples of the class j. Denote by $\bar{X}_j$ the mean of the training samples belonging to the class j, then the j'th column of $XZ^\top$ is $N_j \bar{X}_j$. It follows that

$$XZ^\top (ZZ^\top)^{-1} Z X^\top = \sum_{j=1}^{k} \frac{N_j^2}{N_j} \bar{X}_j \bar{X}_j^\top = \sum_{j=1}^{k} N_j \bar{X}_j \bar{X}_j^\top = S_B \tag{10}$$

Where S_B is the between-class scatter matrix defined in LDA [11]. Let $S_T = XX^\top$ be the total scatter matrix S_T. $S_T = S_W + S_B$ (where S_W is LDA's within-class scatter matrix), and using S_T in LDA is equivalent to using S_W. Hence, the multiclass CCA formulation is equivalent to the eigen-decomposition of $(S_W + \eta I)^{-1} S_B$, which is the formulation of regularized multiclass LDA.

Our analysis below uses the CCA formulation; the LDA case is equivalent, with some minor modifications to the way the classification is done after the linear transformation is applied.

3   Local Learning Via Regularization

The above formulation of the multiclass classification problem is independent of the test vector to be classified x. It may be the case that the learned classifier is "indifferent" to x, transforming it to a vector $A^\top x$ which has a low norm. Note that by the constraint $V^\top ZZ^\top V = I$, the norm of the class vectors v_j is $N_j^{-0.5}$ which is roughly constant for balanced data sets. This possible mismatch between the norm of the transformed example and the class vectors may significantly decrease the ability to accurately classify x. Furthermore, when the norm of $A^\top x$ is small, it is more sensitive to additive noise.

In local learning, the classifier may be different for each test sample and depends on it. In this work, we discourage classifiers that are indifferent to x, and have low $\|A^\top x\|^2$. Hence, to discourage indifference (increase decisiveness), we add a new term to the CCA problem:

$$\max_{A,V} \mathrm{tr}(A^\top X Z^\top V) + \bar{\alpha}\, \mathrm{tr}(A^\top x x^\top A)$$

$$\text{subject to} \quad A^\top XX^\top A = V^\top ZZ^\top V = I \tag{11}$$

$\mathrm{tr}(A^\top x x^\top A) = \|A^\top x\|^2$, and the added term reflects the principle of decisiveness. $\bar{\alpha}$ is a parameter corresponding to the trade-off between the correlation term and the decisiveness term. Adding ridge regularization as before to the solution of Equation 11, and setting $\alpha = \bar{\alpha}\lambda^{-1}$ gives the following generalized eigenproblem:

$$\begin{pmatrix} 0 & XZ^\top \\ ZX^\top & 0 \end{pmatrix} \begin{pmatrix} a \\ v \end{pmatrix} = \lambda \begin{pmatrix} XX^\top + \eta I - \alpha xx^\top & 0 \\ 0 & ZZ^\top \end{pmatrix} \begin{pmatrix} a \\ v \end{pmatrix} \tag{12}$$

Note that this form if similar to the CCA based multiclass classifier presented in Section 2 above, except that the ridge regularization matrix ηI is replaced by the local regularization matrix $\eta I - \alpha xx^\top$. We proceed to analyze the significance of this form of local regularization. In ridge regression, the influence of all eigenvectors is weakened uniformly by adding η to all eigenvalues before computation of the inverse. This form of regularization encourages smoothness in the learned transformation. In our version of local regularization, smoothness is still achieved by the addition of η to all eigenvalues. The smoothing effect is weakened, however, by α, in the component parallel to x. This can be seen by the representation $xx^\top = U_x \lambda_x U_x\top$ for $U_x^\top U_x = U_x U_x^\top = I$, with $\lambda_x = \mathrm{diag}(\|x\|^2, 0, \ldots, 0)$. Now $\eta I - \alpha xx^\top = U_x(\eta I - \alpha \lambda_x)U_x^\top$, and the eigenvalues of the regularization matrix are $(\eta - \alpha, \eta, \eta, \ldots, \eta)$. Hence, the component parallel to x is multiplied by $\eta - \alpha$ while all others are multiplied by η. Therefore, encouraging decisiveness by adding the term $\alpha\|A^\top x\|^2$ to the maximization goal is a form of regularization where the component parallel to x is smoothed less than the other components.

4   Efficient Implementation

In this section we analyze the computational complexity of our method, and propose an efficient update algorithm that allows it to be performed in time comparable to standard CCA with ridge regularization. Our algorithm avoids fully retraining the classifier for each testing example by training it once using standard CCA with uniform ridge regularization, and reusing the results in the computation of the local classifiers.

Efficient training of a uniformly regularized multiclass CCA classifier. In the non-local case, training a multiclass CCA classifier consists of solving Equations 6 and 7, or, equivalently, Equations 8 and 9. Let $r = \min(m, k)$, and note that we assume $m \leq n$, since the rank of the data matrix is at most n, and if $m > n$ we can change basis to a more compact representation. To solve Equations 8 and 9, it is enough to find the eigenvalues and eigenvectors of a $r \times r$ square matrix. Inverting $(XX^\top + \eta I_m)^{-1}$ and $(ZZ^\top)^{-1}$ and reconstructing the full classifier (A and V) given the eigenvalues and eigenvectors of the $r \times r$ matrix above can be done in $O(m^3 + k^3)$. While this may be a reasonable effort if done once, it may become prohibitive if done repeatedly for each new test example. This, however, as we show below, is not necessary.

Representing the local learning problem as a rank-one modification. We first show the problem to be equivalent to the Singular Value Decomposition (SVD) of a (non-symmetric) matrix, which is in turn equivalent to the eigendecomposition of two symmetric matrices. We then prove that one of these two matrices can be represented explicitly as a rank-one update to a constant (with regards to the new test example) matrix whose eigen-decomposition is computed only once. Finally, we show how to efficiently compute the eigen-decomposition of the modified matrix, how to derive the full solution using this decomposition and how to classify the new example in time complexity much lower than that of a full SVD.

Begin with a change of variables. Let $\bar{A} = (XX^\top + \eta I_m - \alpha xx^\top)^{\frac{1}{2}} A$ and $\bar{V} = (ZZ^\top)^{\frac{1}{2}} V$. By the constraints (Equation 11, with added ridge and local regularizations), $\bar{A}$ and $\bar{V}$ satisfy $\bar{A}^\top \bar{A} = A^\top (XX^\top + \eta I_m - \alpha xx^\top) A = I$ and $\bar{V}^\top \bar{V} = V^\top ZZ^\top V = I$. Hence, the new variables are orthonormal and the CCA problem formulation (Equation 4) with added ridge regularization becomes:

$$\max_{\bar{A}, \bar{V}} \operatorname{tr}(\bar{A}^\top (XX^\top + \eta I_m - \alpha xx^\top)^{-\frac{1}{2}} XZ^\top (ZZ^\top)^{-\frac{1}{2}} \bar{V})$$

$$\text{subject to} \qquad \bar{A}^\top \bar{A} = \bar{V}^\top \bar{V} = I \tag{13}$$

Define:

$$M_0 = (XX^\top + \eta I_m)^{-\frac{1}{2}} XZ^\top (ZZ^\top)^{-\frac{1}{2}} = U_0 \Sigma_0 R_0^\top \tag{14}$$

$$M = (XX^\top + \eta I_m - \alpha xx^\top)^{-\frac{1}{2}} XZ^\top (ZZ^\top)^{-\frac{1}{2}} = U\Sigma R^\top \tag{15}$$

where $U\Sigma R^\top$ is the Singular Value Decomposition (SVD) of M and similarly $U_0 \Sigma_0 R_0^\top$ for M_0. Then the maximization term of Equation 13 is $\bar{A}^\top U\Sigma R^\top \bar{V}$, which under the orthonormality constraints of Equation 13, and since we seek only l components, is maximized by $\bar{A} = U_{0|l}$ and $\bar{V} = R_{0|l}$, which are the l left and right singular vectors of M corresponding to the l largest singular values.

Since $M^\top M = R\Sigma^2 R^\top$, the right singular vectors can be found by the eigen-decomposition of the symmetric $M^\top M$. We proceed to show how $M^\top M$ can be represented explicitly as a rank-one update to $M_0^\top M_0$. Define $J_X = (XX^\top + \eta I_m)^{-1}$, then J_X is symmetric as the inverse of a symmetric matrix, and by the Sherman-Morrison formula [12],

$$(XX^\top + \eta_X I_m - \alpha xx^\top)^{-1} = (J_X - \alpha xx^\top)^{-1} = J_X + \frac{J_X \alpha xx^\top J_X}{1 - \alpha x^\top J_X x}$$

$$= J_X + \frac{\alpha}{1 - \alpha x^\top J_X x} (J_X x)(J_X x)^\top = (XX^\top + \eta_X I_m)^{-1} + \beta bb^\top \tag{16}$$

where $\beta = \frac{\alpha}{1 - \alpha x^\top J_X x}$ and $b = J_X x$. β and b can both be computed using $O(m^2)$ operations, since J_X is known after being computed once. Now,

$$M^\top M = (ZZ^\top)^{-\frac{1}{2}} ZX^\top (XX^\top + \eta I_m - \alpha xx^\top)^{-1} XZ^\top (ZZ^\top)^{-\frac{1}{2}}$$

$$= (ZZ^\top)^{-\frac{1}{2}} ZX^\top ((XX^\top + \eta I_m)^{-1} + \beta bb^\top) XZ^\top (ZZ^\top)^{-\frac{1}{2}}$$

$$= M_0^\top M_0 + \beta (ZZ^\top)^{-\frac{1}{2}} ZX^\top bb^\top XZ^\top (ZZ^\top)^{-\frac{1}{2}}$$

$$= M_0^\top M_0 + \beta cc^\top \tag{17}$$

$$\tag{18}$$

where $c = (ZZ^\top)^{-\frac{1}{2}}ZX^\top b$, and again c is easily computed from b in $O(km)$ operations. Now let $w = R_0^\top \frac{c}{\|c\|}$ (so $\|w\| = 1$) and $\gamma = \beta\|c\|^2$ to arrive at the representation

$$M^\top M = R_0(\Sigma_0^2 + \gamma ww^\top)R_0^\top \tag{19}$$

It is left to show how to efficiently compute the eigen-decomposition of a rank-one update to a symmetric matrix, whose eigen-decomposition is known. This problem has been investigated by Golub [13] and Bunch et al. [14]. We propose a simple and efficient algorithm that expands on their work. We briefly state their main results, without proofs, which can be found in the original papers.

The first stage in the algorithm described in Bunch et al. [14] is deflation, transforming the problem to equivalent (and no larger) problems $S + \rho zz^\top$ satisfying that all elements of z are nonzero, and all elements of S are distinct. Then, under the conditions guaranteed by the deflation stage, the new eigenvalues can be found. The eigenvalues of $S + \rho zz^\top$ satisfying that all elements of z are nonzero and all elements of S are distinct are the roots of $f(\lambda) = 1 + \rho \sum_{i=1}^{s} \frac{z_i^2}{d_i - \lambda}$, where s is the size of the deflated problem, z_i are the elements of z and d_i are the elements of the diagonal of S. [14] show an iterative algorithm with a quadratic rate of convergence, so all eigenvalues can be found using $O(s^2)$ operations, with a very small constant as shown in their experiments. Since the deflated problem is no larger than k, this stage requires $O(k^2)$ operations at most. Once the eigenvalues have been found, the eigenvectors of $\Sigma_0^2 + \gamma ww^\top$ can be computed by

$$\xi_i = \frac{(S - \lambda_i I)^{-1}z}{\|(S - \lambda_i I)^{-1}z\|} \tag{20}$$

using $O(k)$ operations for each eigenvector, and $O(k^2)$ in total to arrive at the representation

$$M^\top M = R_0 R_1 \Sigma_1 R_1^\top R_0 \tag{21}$$

Explicit evaluation of Equation 21 to find $\hat{V}$ requires multiplying $k \times k$, which should be avoided to keep the complexity $O(m^2 + k^2)$. The key observation is that we do not need to find V explicitly but only $A^\top x - v_i$ for $i = 1, 2, \ldots, k$, with v_i being the i'th class vector (Equation 1). The distances we seek are:

$$\|A^\top x - v_i\|^2 = \|A^\top x\|^2 + \|v_i\|^2 - 2v_i^\top A^\top x \tag{22}$$

with $\|v_i\|^2 = N_i$ (see Section 3). Hence, finding all exact distances can be done by computation of $x^\top AA^\top x - VA^\top x$, since $v_i^\top$ is the i'th row of V. Transforming back from $\bar{V}$ to V gives $V = (ZZ^\top)^{-\frac{1}{2}}\bar{V}$, where $(ZZ^\top)^{-\frac{1}{2}}$ needs to be computed only once. From Equations 6 and 21,

$$\begin{aligned}
A^\top x &= \Sigma_1^{-1}V^\top ZX^\top(XX^\top + \eta I_m - \alpha xx^\top)^{-1}x \\
&= \Sigma_1^{-1}R_1^\top R_0^\top(ZZ^\top)^{-\frac{1}{2}}ZX^\top\left((XX^\top + \eta I_m)^{-1} + \beta bb^\top\right)x
\end{aligned} \tag{23}$$

All the matrices in Equation 23 are known after the first $O(k^3 + m^3)$ computation and $O(k^2 + m^2)$ additional operations per test example, as we have shown

References

1. Fei-Fei, L., Fergus, R., Perona, P.: A bayesian approach to unsupervised one-shot learning of object categories. In: ICCV, Nice, France, pp. 1134–1141 (2003)
2. Belkin, M., Niyogi, P.: Semi-supervised learning on riemannian manifolds. Machine Learning 56, 209–239 (2004)
3. Bart, E., Ullman, S.: Cross-generalization: learning novel classes from a single example by feature replacement. In: CVPR (2005)
4. Yamada, M., Pezeshki, A., Azimi-Sadjadi, M.: Relation between kernel cca and kernel fda. In: IEEE International Joint Conference on Neural Networks (2005)
5. Bottou, L., Vapnik, V.: Local learning algorithms. Neural Computation 4 (1992)
6. Zhang, H., Berg, A.C., Maire, M., Malik, J.: Svm-knn: Discriminative nearest neighbor classification for visual category recognition. In: CVPR (2006)
7. Hotelling, H.: Relations between two sets of variates. Biometrika 28, 321–377 (1936)
8. Akaho, S.: A kernel method for canonical correlation analysis. In: International Meeting of Psychometric Society (2001)
9. Wolf, L., Shashua, A.: Learning over sets using kernel principal angles. J. Mach. Learn. Res. 4, 913–931 (2003)
10. Neumaier, A.: Solving ill-conditioned and singular linear systems: A tutorial on regularization (1998)
11. Hastie, T., Tibshirani, R., Friedman, J.: The elements of statistical learning: data mining, inference and prediction. Springer, Heidelberg (2001)
12. Sherman, J., Morrison, W.J.: Adjustment of an inverse matrix corresponding to changes in the elements of a given column or a given row of the original matrix. Annals of Mathematical Statistics 20, 621 (1949)
13. Golub, G.: Some modified eigenvalue problems. Technical report, Stanford (1971)
14. Bunch, J.R., Nielsen, C.P., Sorensen, D.C.: Rank-one modification of the symmetric eigenproblem. Numerische Mathematik 31, 31–48 (1978)
15. CalPhotos: A database of photos of plants, animals, habitats and other natural history subjects [web application], animal–mammals collection. bscit, University of California, Berkeley,
    `http://calphotos.berkeley.edu/cgi/`
    `img_query?query_src=photos_index&where-lifeform=Animal--Mammal`
16. Huang, G.B., Ramesh, M., Berg, T., Learned-Miller, E.: Labeled faces in the wild: A database for studying face recognition in unconstrained environments. University of Massachusetts, Amherst, Technical Report 07-49 (2007)
17. Rifkin, R., Klautau, A.: In defense of one-vs-all classification. Journal of Machine Learning Research 5 (2004)
18. Vedaldi, A.: Bag of features: A simple bag of features classifier (2007),
    `http://vision.ucla.edu/~vedaldi/`
19. Nister, D., Stewenius, H.: Scalable recognition with a vocabulary tree. In: CVPR (2006)
20. Nowak, E., Jurie, F., Triggs, B.: Sampling strategies for bag-of-features image classification. In: European Conference on Computer Vision. Springer, Heidelberg (2006)
21. Ojala, T., Pietikainen, M., Harwood, D.: A comparative-study of texture measures with classification based on feature distributions. Pattern Recognition 29 (1996)
22. Ahonen, T., Hadid, A., Pietikainen, M.: Face recognition with local binary patterns. In: Pajdla, T., Matas, J(G.) (eds.) ECCV 2004. LNCS, vol. 3024. Springer, Heidelberg (2004)
23. Huang, G.B., Jain, V., Learned-Miller, E.: Unsupervised joint alignment of complex images. ICCV (2007)

Saliency Based Opportunistic Search for Object Part Extraction and Labeling

Yang Wu[1,2], Qihui Zhu[2], Jianbo Shi[2], and Nanning Zheng[1]

[1] Institute of Artificial Intelligence and Robotics, Xi'an Jiaotong University
ywu@aiar.xjtu.edu.cn, nnzheng@mail.xjtu.edu.cn
[2] Department of Computer and Information Science, University of Pennsylvania
wuyang@seas.upenn.edu, qihuizhu@seas.upenn.edu,
jshi@cis.upenn.edu

Abstract. We study the task of object part extraction and labeling, which seeks to understand objects beyond simply identifiying their bounding boxes. We start from bottom-up segmentation of images and search for correspondences between object parts in a few shape models and segments in images. Segments comprising different object parts in the image are usually not equally salient due to uneven contrast, illumination conditions, clutter, occlusion and pose changes. Moreover, object parts may have different scales and some parts are only distinctive and recognizable in a large scale. Therefore, we utilize a multi-scale shape representation of objects and their parts, figural contextual information of the whole object and semantic contextual information for parts. Instead of searching over a large segmentation space, we present a saliency based opportunistic search framework to explore bottom-up segmentation by gradually expanding and bounding the search domain. We tested our approach on a challenging statue face dataset and 3 human face datasets. Results show that our approach significantly outperforms Active Shape Models using far fewer exemplars. Our framework can be applied to other object categories.

1   Introduction

We are interested in the problem of object detection with object part extraction and labeling. Accurately detecting objects and labeling their parts requires *going inside the object's bounding box* to reason about object part configurations. Extracting object parts with the right configuration is very helpful for recognizing object details. For example, extracting facial parts helps with recognizing faces and facial expressions, while understanding human activities requires knowing the pose of a person.

A common approach to solve this problem is to learn specific features for object parts [1][2]. We choose a different path which starts with bottom-up segmentation and aligns shape models to segments in test images. Our observation is that starting from salient segments, it is unlikely to accidentally align object parts to background edges. Therefore, we can search efficiently and avoid accidental alignment.

Our approach includes three key components: **correpondence**, **contextual information** and **saliency of segments**. There exist algorithms incorporating correspondence and contextual information such as pictorial structures [3] and contour context selection [4], both showing good performance on some object categories. The disadvantage

D. Forsyth, P. Torr, and A. Zisserman (Eds.): ECCV 2008, Part IV, LNCS 5305, pp. 760–774, 2008.

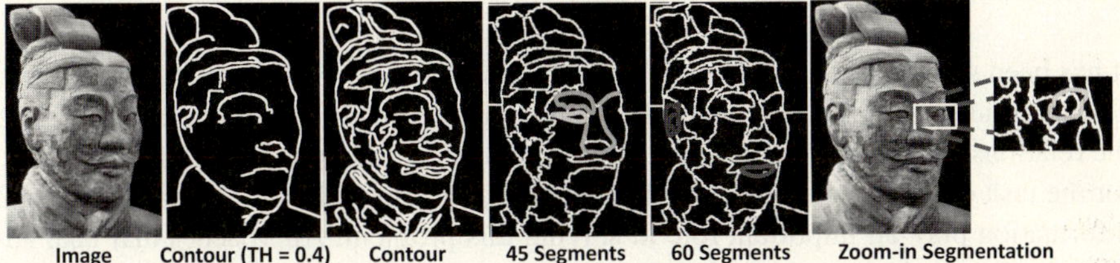

Fig. 1. Saliency of contours and segments. The second image is a group of salient contours from contour grouping [5] by setting a lower threshold to the average edge strength, while the third one contains all the contours from contour grouping. It shows that by thresholding the saliency of contour segments, we either get some foreground contours missing (under-segmented) or have a lot of clutter come in (over-segmented). The same thing happens to image segmentation. Segments comprising object parts pop out in different segmentation levels, representing different saliencies (cut costs). The last three images show such a case.

is that these methods ignore image saliency. Therefore, they cannot tell accidental alignment of faint segments in the background from salient object part segments. However, it is not easy to incorporate saliency. A naive way of using saliency is to find salient parts first, and search for less salient ones depending on these salient ones. The drawback is that a hard decision has to be made in the first step of labeling salient parts, and mistakes arising from this step cannot be recovered later. Moreover, object parts are not equally hard to find. Segments belonging to different object parts may pop out at different segmentation levels (with different numbers of segments), as shown in Figure 1. One could start with over-segmentation to cover all different levels. Unfortunately, by introducing many small segments at the same time, segment saliency will be lost, which defeats the purpose of image segmentation. Fake segmentation boundaries will also cause many false positives of accidentally aligned object parts.

We build two-level contexts and shape representations for objects and their parts, with the goal of high *distinctiveness* and *efficiency*. Some large object parts (*e.g.* facial silhouettes) are only recognizable as a whole in a large scale, rather than as a sum of the pieces comprising them. Moreover, hierarchical representation is more efficient for modeling contextual relationships among model parts than a single level representation which requires a large clique potential and long range connections. Two different levels of contextual information is explored: *figural context* and *semantic context*. The former captures the overall shape of the whole object, and the latter is formed by semantic object parts.

In this paper, we propose a novel approach called *Saliency Based Opportunistic Search* for object part extraction and labeling, with the following key contributions:

1. Different levels of context including both figural and semantic context are used.
2. Bottom-up image saliency is incorporated into the cost function.
3. We introduce an effective and efficient method of searching over different segmentation levels to extract object parts.

2   Related Work

It has been shown that humans recognize objects by their components [6] or parts [7]. The main idea is that object parts should be extracted and represented together with the relationships among them for matching to a model. This idea has been widely used for the task of recogize objects and their parts [8,9,3]. Figural and semantic contextual information play an important role in solving this problem. Approaches that take advantage of figural context include PCA and some template matching algorithms such as Active Shape Models (ASM) [10] and Active Appearance Models (AAM) [11]. Template matching methods like ASM usually use local features (points or key points) as searching cues, and constrain the search by local smoothness or acceptable variations of the whole shape. However, these methods require good initialization. They are sensitive to clutter and can be trapped in local minima. Another group of approaches are part-based models, which focus on semantic context. A typical case is pictorial structure [3]. Its cost function combines both the individual part matching cost and pair-wise inconsistency penalties. The drawback of this approach is that it has no figural context measured by the whole object. It may end up with many "OK" part matches without a global verification, especially when there are many faint object edges and occlusions in the image. Recently, a multiscale deformable part model was proposed to detect objects based on deformable parts [1], which is an example that uses both types of contextual information. However, it focuses on training deformable local gradient-based features for detecting objects, but not extracting object parts out of the images.

3   Saliency Based Opportunistic Search

Problem definition. The problem we are trying to solve is to extract and label object parts based on contextual information, given an image and its segmentations, as shown in Figure 1. Multiple models are used to cover some variations of the object (see Figure 2 for the models we have used on faces). Extracting and labeling object parts requires finding the best matched model. The problem can be formulated as follows:

Input

- Model: $M = \{M_1, M_2, \ldots, M_m\}$; each model M_k has a set of labeled parts $\{p_1^k, p_2^k, \ldots, p_n^k\}$. They are all shape models made of contours and line segments.
- Image: $S = \{s_1, s_2, \ldots, s_l\}$ is a set of region segments and contour segments coming from different segmentation levels from the image. For region segments, only boundaries are used for shape matching.

Output

- Best matched model M_k.
- Object part labels $L(S)$. $L(s_i) = j$, if s_i belongs to part p_j^k, or else $L(s_i) = 0$.

This can be formulated as a shape matching problem, which aims to find sets of segments whose shapes match to part models. However, the segments comprising the object parts are not equally hard to extract from the image, and grouping them to objects

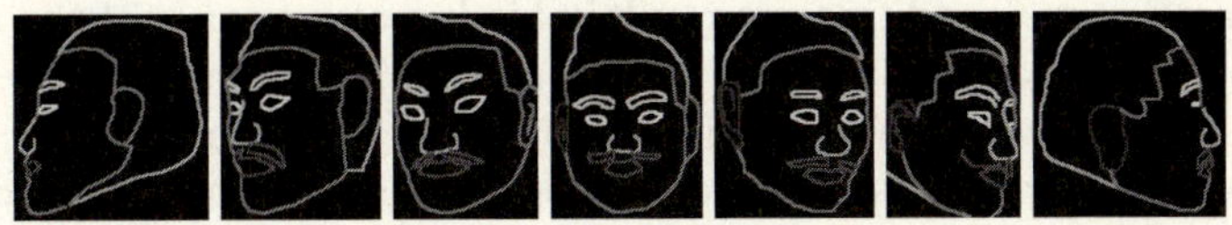

Fig. 2. Different models for faces. They are hand designed models obtained from 7 real images, each of them representing one pose. Facial features are labeled in different colors.

and object parts also requires checking the consistency among them. We call these efforts *"grouping cost"*, which is not measured by shape but can be helpful to differentiate segments belonging to object parts from those belonging to the background. Therefore, we combine these two into such a cost function:

$$C^{labeling} = C^{shape} + C^{grouping} \qquad (1)$$

C^{shape} measures the shape matching cost between shape models and labeled segments in the image, which relays much on *correspondence* and *context*. $C^{grouping}$ is the grouping cost, which can be measured in different ways, but in this paper it is mainly about the bottom-up *saliency* based editing cost.

The cost function above is based on the following three key issues.

1. **Correspondence** (u). *A way to measure the dissimilarity between a shape model and a test image.* The correspondence is defined on control points. Features computed on these control points represent the shape information and then the correspondences are used to measure the dissimilarity. Let $U^{\mathcal{M}} = \{a_1, a_2, \ldots, a_{N_a}\}$ be a set of control points on the model, and $U^{\mathcal{I}} = \{b_1, b_2, \ldots, b_{N_b}\}$ be the set on the image. We use u_{ij} to denote the correspondence between control points a_i and b_j where $u_{ij} = 1$ indicates they are matched, otherwise $u_{ij} = 0$. Note that this correspondence is different from the one between object parts and image segments.
2. **Context (x and y).** *The idea of using the context is to **choose the correct context** on both model and test image sides for shape matching invariant to clutter and occlusion.* **x** and **y** are used here to denote the context selection of either segments or parts on the model and the image, respectively.
3. **Saliency.** *A property of bottom-up segments which represents how difficult it is to separate the segment from the background.* Coarse-level segmentation tends to produce salient segments, while finer-level segmentation extracts less salient ones, but at the same time introduces background clutter. Local editing on the salient gap between two salient segments can help to get good segments out without bringing in clutter, but it needs contextual guidance.

Saliency based editing. Segmentation has problems when the image segments have different saliencies. Under-segmentation could end up with unexpected leakages, while over-segmentation may introduce clutter. A solution for this problem is to do some local editings. For example, adding a small virtual edge at the leakage place can make the segmentation much better without increasing the number of segments. *Zoom-in* in a small area is also a type of editing that can be effective and efficient, as presented in

Figure 1. Small costs for editing can result in big improvement on shape matching cost. This is based the shape integrity and the non-additive distance between shapes. However, editings need the contextual information from the model.

Suppose there are a set of possible editing operations $\mathbf{z}$ which might lead to better segmentation. $z_k = 1$ means that editing k is chosen, otherwise $z_k = 0$. Note that usually it is very hard to discover and precompute all the editings beforehand. Therefore, this editing index vector $\mathbf{z}$ is dynamic, and it appends on the fly. After doing some editings, some new segments/(part hypotheses) will come out, meanwhile we can still keep the original segments/parts. Therefore, a new variable $\mathbf{y}^{edit} = \mathbf{y}^{edit}(\mathbf{y}, \mathbf{z})$ is used to denote all the available elements which includes both the original ones in $\mathbf{y}$ and the new ones induced by editing $\mathbf{z}$. Let C_k^{edit} be the edit cost for editing k.

Our cost function (1) of object part labeling and extraction can be written as follows:

$$
\min_{\mathbf{x},\mathbf{y},\mathbf{z},\mathbf{u}} C^{labeling}(\mathbf{x},\mathbf{y},\mathbf{z},\mathbf{u}) = C^{shape}(\mathbf{x},\mathbf{y},\mathbf{z},\mathbf{u}) + C^{grouping}(\mathbf{z}) =
$$

$$
\sum_{i=1}^{N_a} [\beta \cdot \sum_{j=1}^{N_b} u_{ij} C_{ij}^{\mathcal{M}\leftrightarrow\mathcal{I}}(\mathbf{x}, \mathbf{y}^{edit}) + C_i^{\mathcal{F}\leftrightarrow\mathcal{M}}(\mathbf{x}, \mathbf{u})] + \sum_k C_k^{edit} z_k \quad (2)
$$

$$
s.t. \quad \sum_j u_{ij} \leq 1, \quad i = 1, ..., N_a
$$

$$
\begin{aligned}
\mathbf{x}: &\quad \textit{selection indicator of model segments/parts.} \\
\mathbf{y}: &\quad \textit{selection indicator of image segments/parts.} \\
\mathbf{z}: &\quad \textit{selection vector of editing operations.} \\
\mathbf{u}: &\quad \textit{correspondence of control points between the image and model.} \\
\mathbf{y}^{edit}(\mathbf{y},\mathbf{z}): &\quad \textit{selection indicator of image segments/parts edited by } \mathbf{z}.
\end{aligned}
$$

The three summations in equation (2) correspond to three different types of cost: *mismatch cost* $C^{\mathcal{M}\leftrightarrow\mathcal{I}}(\mathbf{x}, \mathbf{y}^{edit}, \mathbf{u})$, *miss cost* $C^{\mathcal{F}\leftrightarrow\mathcal{M}}(\mathbf{x}, \mathbf{u})$ and *edit cost* $C^{edit}(\mathbf{z})$. The mismatch cost, $C_{ij}^{\mathcal{M}\leftrightarrow\mathcal{I}}(\mathbf{x}, \mathbf{y}^{edit}) = \|f_i(\mathbf{x}) - f_j(\mathbf{y}^{edit})\|$ denotes the feature dissimilarity between two corresponding control points. To prevent the cost function from biasing to fewer matches, we add the miss cost $C_i^{\mathcal{F}\leftrightarrow\mathcal{M}}(\mathbf{x}) = \|f_i^{full} - (\sum_j u_{ij}) f_i(\mathbf{x})\|$ to denote how much of the model has not been matched by the image. It encourages more parts to be matched on the model side. There is a trade-off between $C_{ij}^{\mathcal{M}\leftrightarrow\mathcal{I}}$ and $C_i^{\mathcal{F}\leftrightarrow\mathcal{M}}$, where $\beta \geq 0$ is a controlling factor. Note that $\|\cdot\|$ can be any norm function[1].

The rest of this section focuses on the two parts of our cost function. Shape matching will be performed on two levels of contexts and saliency based editing will result in the opportunistic search approach.

3.1   Two-Level Context Based Shape Matching

We extend the shape matching method called contour context selection in [4] to two different contextual levels: "figural context selection" and "semantic context selection".

---

[1] In our shape matching we used L_1 norm.

Figural context selection. Figural context selection matches a segment-based holistic shape model to an object hypothesis represented by segments, which may have clutter and missing segments. We optimize the following cost function:

$$
\min_{\mathbf{x},\mathbf{y},\mathbf{u}} \ C^{figural}(\mathbf{x},\mathbf{y},\mathbf{u}) =
$$

$$
\sum_{i=1}^{N_a}[\,\beta\cdot\underbrace{\sum_{j=1}^{N_b} u_{ij}\,\|SC_i^{\mathcal{M}}(\mathbf{x})-SC_j^{\mathcal{I}}(\mathbf{y})\|}_{C_{ij}^{\mathcal{M}\leftrightarrow\mathcal{I}}(\mathbf{x},\mathbf{y}^{edit})}+\underbrace{\|SC_i^{\mathcal{F}}-(\textstyle\sum_j u_{ij})\cdot SC_i^{\mathcal{M}}(\mathbf{x})\|}_{C_i^{\mathcal{F}\leftrightarrow\mathcal{M}}(\mathbf{x},\mathbf{u})}\,)]
$$

$$
\text{s.t.} \ \sum_{i,j,i',j'} u_{ij}u_{i'j'}C_{i,j,i',j'}^{geo} \le C_{tol} \tag{3}
$$

where $SC_i^{\mathcal{M}}(\mathbf{x})$ and $SC_j^{\mathcal{I}}(\mathbf{y})$ is defined as the Shape Context centered at model control point a_i and image control point b_j. $C_{i,j,i',j'}^{geo}$ is the geometric inconsistent cost of correspondences $\mathbf{u}$. C_{tol} is the maximum tolerance of the geometric inconsistency. We use Shape Context [12] as our feature descriptor. Note that the size of Shape Context histogram is large enough to cover the whole object model, and this is a set-to-set matching problem. Details for this algorithm can be found in [4].

Semantic context selection. Similarly we explore semantic context to select consistent object part hypotheses. We first generate part hypotheses using almost the same context selection algorithm as the one presented above. The selection operates on parts instead of the whole object. Figure 3 shows an example of generating a part hypothesis.

In semantic context selection, we reason about semantic object parts. Hence we abstract each part (on either model or test image) as a point located at its center with its part label. We place control points on each one of the part centers.

Suppose C_j^{part} is the matching cost of part hypothesis j. We use $w_j^P = \dfrac{e^{\gamma C_j^{part}}}{e^\gamma} \in [\frac{1}{e^\gamma},1], \gamma \in [0,1]$ as its weight. Then the cost function for semantic context selection is:

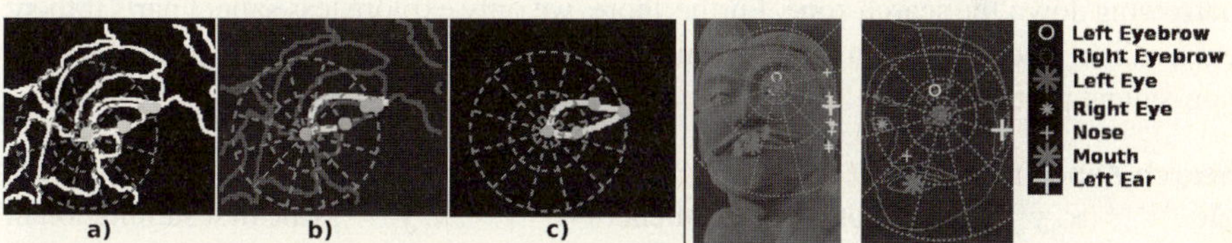

Fig. 3. Semantic context selection. Left: Part hypothesizing. a) A local part region around the eye in the image, with segments and control points. c) A model template of the eye with control points. Selection result on the image is shown in b). Right: Consistent part grouping. Semantic-level shape context centered on the left eye captures semantic contextual information of the image. A subset of those parts form a mutually consistent context and we group them by matching with the semantic-level shape context on the model shown in the middle.

$$\min_{\mathbf{x},\mathbf{y},\mathbf{u}} \quad C^{semantic}(\mathbf{x},\mathbf{y},\mathbf{u}) =$$

$$\sum_{i=1}^{N_a}[\,\beta \cdot \underbrace{\sum_{j=1}^{N_b} u_{ij} w_j^P \|SC_i^{\mathcal{M}}(\mathbf{x}) - SC_j^{\mathcal{I}}(\mathbf{y})\|}_{C_{ij}^{\mathcal{M}\leftrightarrow\mathcal{I}}(\mathbf{x},\mathbf{y}^{edit})} + \underbrace{\|SC_i^{\mathcal{F}} - (\textstyle\sum_j u_{ij}) \cdot SC_i^{\mathcal{M}}(\mathbf{x})\|}_{C_i^{\mathcal{F}\leftrightarrow\mathcal{M}}(\mathbf{x},\mathbf{u})}\,)]$$

$$(4)$$

The variable definitions are similar to figural context selection, except for two differences: 1) selection variables depend on the correspondences and 2) Shape Context no longer counts edge points, but object part labels.

The desired output of labeling $L(S)$ is implicitly given in the optimization variables. During part hypothesis generation, we put labels of candidate parts onto the segments. Then after semantic context selection, we confirm some labels and discard the others using the correspondence u_{ij} between part candidates and object part models.

3.2 Opportunistic Search

Labeling object parts using saliency based editing potentially requires searching over a very large state space. Matching object shape and its part configuration requires computing correspondences and non-local context. Both of them have exponentially many choices. On top of that, we need to find a sequence of editings, such that the resulting segments and parts produced by these editings are good enough for matching.

The key intuition of our saliency based opportunistic search is that we start from coarse segmentations which produce salient segments and parts to guarantee low saliency cost. We iteratively match configuration of salient parts to give a sequence of bounds to the *search zone* of the space which needs to be explored. The possible spatial extent of the missing parts is bounded by their shape matching cost and the edit cost (equally, saliency cost). Once the search space has been narrowed down, we "zoom-in" to the finer scale segmentation to rediscover missing parts (hence with lower saliency). Then we "zoom-out" to do semantic context selection on all the part hypotheses. Adding these new parts improves the bound on the possible spatial extent and might suggest new search zones. This opportunistic search allows both high *efficiency* and high *accuracy* of object part labeling. We avoid extensive computation by narrowing down the search zone. Furthermore, we only explore less salient parts if there exist salient ones supporting them, which avoids producing many false positives from non-salient parts.

Search Zone. In each step t of the search, given $(\mathbf{x}^{(t-1)}, \mathbf{y}^{(t-1)}, \mathbf{z}^{(t-1)}, \mathbf{u}^{(t-1)})$, we use $\Delta C^{\mathcal{M}\leftrightarrow\mathcal{I}}(\mathbf{x},\mathbf{y}^{edit})$ to denote the increment of $C^{\mathcal{M}\leftrightarrow\mathcal{I}}(\mathbf{x},\mathbf{y}^{edit})$ (the first summation in equation (2)). $\Delta C^{\mathcal{F}\leftrightarrow\mathcal{M}}(\mathbf{x},\mathbf{u})$ and $\Delta C^{edit}(\mathbf{z})$ are similarly defined. By finding missing parts, we seek to decrease the cost (2). Therefore, we introduce the following criterion for finding missing parts:

$$\beta\Delta C^{\mathcal{M}\leftrightarrow\mathcal{I}}(\mathbf{x},\mathbf{y},\mathbf{z}) + \Delta C^{\mathcal{F}\leftrightarrow\mathcal{M}}(\mathbf{x},\mathbf{u}) + \Delta C^{edit}(\mathbf{z}) \le 0 \qquad (5)$$

We write $C^{\mathcal{M}\leftrightarrow\mathcal{I}}(\mathbf{x},\mathbf{y},\mathbf{z}) = C^{\mathcal{M}\leftrightarrow\mathcal{I}}(\mathbf{x},\mathbf{y}^{edit})$ since $\mathbf{y}^{edit}$ depends on editing vector $\mathbf{z}$.

Algorithm 1. Saliency Based Opportunistic Search

1: Initialize using figural context selection. *For each part k, compute $\mathcal{Z}(k)$ based on u from figural context selection. Set $(\mathbf{x}^{(0)}, \mathbf{y}^{(0)}, \mathbf{z}^{(0)}, \mathbf{u}^{(0)})$ to zeros. Set $t = 1$.*

2: Compute search zones for all the missing parts. *Find all missing parts by thresholding the solution $\mathbf{x}^{(t-1)}$.*

       for each missing part p_k

            If $\mathcal{Z}(k) = \emptyset$, compute search zone set $\mathcal{Z}(k)$ by equation (9) and (10).

       end

3: Zoom-in search zone. *Update editing set $\mathbf{z}$.*

       for each $x_k^{(t-1)}$ where $\mathcal{Z}(k) \neq \emptyset$

            Perform Ncut segmentation for each zoom-in window indexed by elements in $\mathcal{Z}(k)$.

            Generate part hypotheses. Set $\mathcal{Z}(k) = \emptyset$.

            If no candidates can be found, go to the next missing part.

            Update $\mathbf{z}$ from part hypotheses.

       end

4: Evaluate configurations with re-discovered parts.

       Terminate if $\mathbf{z}$ does not change.

       Update $(\mathbf{x}^{(t)}, \mathbf{y}^{(t)}, \mathbf{z}^{(t)}, \mathbf{u}^{(t)})$ with the rediscovered parts using equation (4).

       Terminate if $C^{semantic}(\mathbf{x}, \mathbf{y}, \mathbf{u})$ does not improve.

       $t = t + 1$. Go to step 2.

The estimation of bounds is based on the intuition that if all the missing parts can be found, then no *miss* cost is needed to pay any more. Therefore, according to equation (4):

$$\Delta C^{\mathcal{F} \leftrightarrow \mathcal{M}}(\mathbf{x}) \geq -\sum_i C_i^{\mathcal{F} \leftrightarrow \mathcal{M}}(\mathbf{x}, \mathbf{u}). \tag{6}$$

This is the upper bound for the increment of either one of the other two items in equation (5) when any new object part is matched.

Suppose a new editing $\mathbf{z}_\alpha^{(t)} = 1|_{\mathbf{z}_\alpha^{(t-1)} = 0}$ matches a new object part a_k to a part hypothesis in the image b_ℓ. Let $k \leftrightarrow \ell$ indicate $u_{k\ell}^{(t)} = 1$ and $\sum_j u_{kj}^{(t-1)} = 0$. Then this editing at least has to pay the cost of matching a_k to b_ℓ (we do not know whether others will also match or not):

$$C|_{k \leftrightarrow \ell} = \beta \Delta C^{\mathcal{M} \leftrightarrow \mathcal{I}}(\mathbf{x}, \mathbf{y}, \mathbf{z})|_{k \leftrightarrow \ell} + C_\alpha^{edit}. \tag{7}$$

The first item on the right of equation (7) is the increment of *mismatch* $\Delta C^{\mathcal{M} \leftrightarrow \mathcal{I}}$ $(\mathbf{x}, \mathbf{y}^{edit})$ when a new object part a_k get matched to b_ℓ. It can be computed based on the last state of the variables $(\mathbf{x}^{(t-1)}, \mathbf{y}^{(t-1)}, \mathbf{z}^{(t-1)}, \mathbf{u}^{(t-1)})$. According to above equations, we get

$$\beta \Delta C^{\mathcal{M} \leftrightarrow \mathcal{I}}(\mathbf{x}, \mathbf{y}, \mathbf{z})|_{k \leftrightarrow \ell} + C_\alpha^{edit} - \sum_i C_i^{\mathcal{F} \leftrightarrow \mathcal{M}}(\mathbf{x}^{(t-1)}, \mathbf{u}^{(t-1)}) \leq 0 \tag{8}$$

Since we use Shape Context for representation and matching, the *mismatch* is non-decreasing. And also the editing cost is nonnegative, so we abtain the bounds for the new editing $\mathbf{z}_\alpha^{(t)} = 1|_{\mathbf{z}_\alpha^{(t-1)} = 0}$. Let $\mathcal{Z}(k)$ denote the search zone for object part k. Then we can compute two bounds for $\mathcal{Z}(k)$:

$$\textbf{(Supremum)} \quad \mathcal{Z}^{sup}(k) = \{\mathbf{z}_\alpha | \Delta C^{\mathcal{M} \leftrightarrow \mathcal{I}}(\mathbf{x}, \mathbf{y}, \mathbf{z})|_{k \leftrightarrow \ell} \leq \frac{1}{\beta} \sum_i C_i^{\mathcal{F} \leftrightarrow \mathcal{M}}(\mathbf{x}^{(t-1)}, \mathbf{u}^{(t-1)})\}$$

(9)

$$\textbf{(Infimium)} \quad \mathcal{Z}^{inf}(k) = \{\mathbf{z}_\alpha | C_\alpha^{edit} \leq \sum_i C_i^{\mathcal{F} \leftrightarrow \mathcal{M}}(\mathbf{x}^{(t-1)}, \mathbf{u}^{(t-1)})\} \qquad (10)$$

where $\mathcal{Z}^{sup}$ gives the supremum of the search zone, *i.e.* upper bound of zoom-in window size, and $\mathcal{Z}^{inf}$ gives the infimum of the search zone, *i.e.* lower bound of zoom-in window size. When the number of segments is fixed, the saliency of the segments decreases as the window size becomes smaller. $\mathcal{Z}^{sup}$ depends on *mismatch* and $\mathcal{Z}^{inf}$ depends on the *edit* cost (*i.e.* **saliency**). In practice, one can sample the space of the search zone, and check which ones fall into these two bounds.

Our opportunistic search is summarized in Algorithm 1.

4   Implementation

4.1   A Typical Example

We present more details on the opportunistic search using faces as an example in Figure 4. We found that usually the whole shape of the face is more salient than individual facial parts. Therefore, the procedure starts with figural context and then switchs to semantic context. We concretize our algorithm for this problem in the following steps. The same procedure can be applied to similar objects.

1. **Initialization: Object Detection.** Any object detection method can be used, but it is not a necessary step[2]. We used shape context voting [13] to do this task, which can handle different poses using a small set of positive training examples.
2. **Context Based Alignment.** First, use $C^{figural}$ in equation (3) to select the best matched model M_k and generate the correspondences $u^{figural}$ for rough alignment[3]. When the loop comes back again, update the alignment based on $u^{semantic}$. Estimate locations for other still missing parts.
3. **Part Hypotheses Generation.** Zoom in on these potential part locations by cropping the regions and do Ncut segmentation to get finer scale segmentation. Then match them to some predefined part models. The resulting matching score is used to prune out unlikely part hypotheses, according to the bound of the cost function.
4. **Part Hypotheses Grouping.** Optimize $C^{semantic}$ in equation (4). Note that the best scoring group may consist of only a subset of the actual object parts.
5. **Termination Checking.** If no better results can be obtained, then we go to the next step. Or else we update **semantic context** and go back to step 2.
6. **Extracting Facial Contours.** This is a special step for faces only. With the final set of facial parts, we optimize $C^{figural}$ again to extract the segments that correspond to the face silhouette, which can be viewed as a special part of the face.

---

[2] Figural context selection can also be used to do that [4].
[3] In practice, we kept best two model hypotheses.

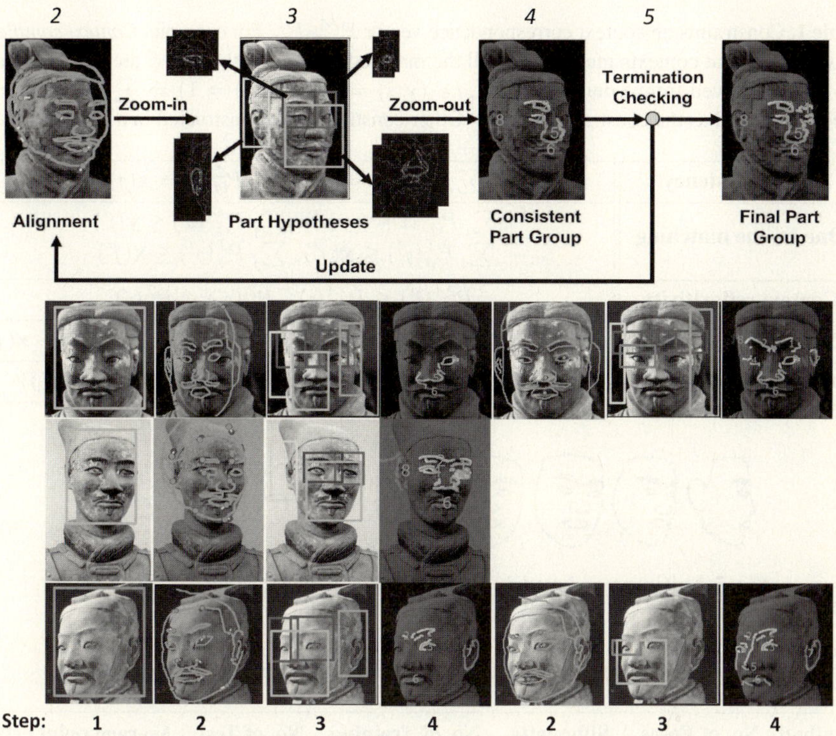

Fig. 4. Saliency based opportunistic search, using faces as an example. Top: the flowchart. Bottom: results of each step for 3 different examples. Typically the iteration converges after only one or two rounds. Rectangles with different colors indicate the zoom-in search zones for different parts. Note that when zoom-in is performed for the first time, two adjacent parts can be searched together for efficiency. This figure is best viewed in color.

4.2 Two-Level Context Selection

For simplification, we do not consider any editing in figural context selection. Then equation (3) is an integer programming problem, we relaxed the variables to solve it with LP. Details of this context selection algorithm can be found in [4].

For semantic context selection, we need to search for correspondences and part selection variables simultaneously because they are highly dependent, unlike the situation in figural context selection. Therefore, we introduce a *correspondence context vector* $P_{ij}^{\mathcal{M}} = u_{ij}\mathbf{x}$ to expand the selection space for model parts:

$$P_{ij}^{\mathcal{M}} \in \{0,1\}^{|U^{\mathcal{M}}|} : P_{ij}^{\mathcal{M}}(i') \Leftrightarrow u_{ij} = 1 \wedge \mathbf{x}(i') = 1 \tag{11}$$

Similarly, we define the *correspondence context vector* for image parts,

$$P_{ij}^{\mathcal{I}} \in \{0,1\}^{|U^{\mathcal{I}}|} : P_{ij}^{\mathcal{I}}(j') \Leftrightarrow u_{ij} = 1 \wedge \mathbf{y}(j') = 1 \tag{12}$$

In addition to the cost in equation (4), constraints on *context correspondence vector* $P^{\mathcal{M}}, P^{\mathcal{I}}$ are enforced such that the semantic context viewed by different parts are

Table 1. Constraints on context correspondence vector $P^{\mathcal{M}}, P^{\mathcal{I}}$. For example, *Context completeness* requires that contexts must include all the matched parts. If both i and i' are matched parts, the context viewed from i must include i', *i.e.* $(\mathbf{y}(i) = 1) \wedge (\mathbf{y}(i') = 1) \Rightarrow \sum_j P_{ij}^{\mathcal{M}}(i') = 1$, which is relaxed as the constraint in row 4. Other constraints are constructed in a similar way.

Self consistency	$\sum_j P_{ij}^{\mathcal{M}}(i) = \mathbf{y}(i),\ \sum_i P_{ij}^{\mathcal{I}}(j) = \mathbf{x}(j)$
One-to-one matching	$\sum_i P_{ij}^{\mathcal{M}}(i') \leq \mathbf{y}(i'),\ \sum_j P_{ij}^{\mathcal{M}}(i') \leq \mathbf{y}(i')$ $\sum_i P_{ij}^{\mathcal{I}}(j') \leq \mathbf{x}(j'),\ \sum_j P_{ij}^{\mathcal{I}}(j') \leq \mathbf{x}(j')$
Context reflexitivity	$P_{ij}^{\mathcal{M}}(i') \leq P_{ij}^{\mathcal{M}}(i),\ \ P_{ij}^{\mathcal{I}}(j') \leq P_{ij}^{\mathcal{I}}(j)$
Context completeness	$\mathbf{y}(i) - \sum_j P_{ij}^{\mathcal{M}}(i') \leq 1 - \mathbf{y}(i'),\ \ \mathbf{x}(j) - \sum_i P_{ij}^{\mathcal{I}}(j') \leq 1 - \mathbf{x}(j')$
Mutual context support	$\sum_j P_{ij}^{\mathcal{M}}(i') = \sum_{j'} P_{i'j'}^{\mathcal{M}}(i),\ \ \sum_i P_{ij}^{\mathcal{I}}(j') = \sum_{i'} P_{i'j'}^{\mathcal{I}}(j)$

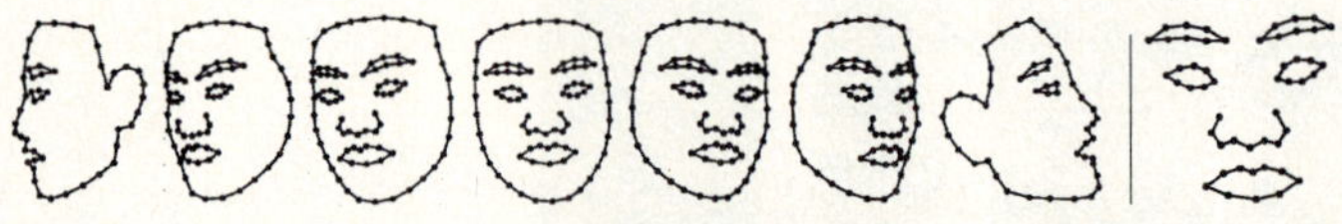

Fig. 5. Left: averaged models for ASM1. Right: averaged model for ASM3.

Table 2. Comparison of experimental details on Emperor-I dataset

Method	No. of Poses	Silhouette	No. of Training	No. of Test	Average point error
ASM1	7	w	138	86	0.2814
ASM2	5	w/o	127	81	0.2906
ASM3	3	w/o	102	70	0.3208
Ours	7	w	7+16	86	0.1503

Table 3. Average error, normalized by distance between eyes for ASM vs. our method

Method	Global	Eyebrows	Eyes	Nose	Mouth	Silhouette
ASM1	0.3042	0.2923	0.2951	0.2715	0.2524	0.3126
Ours	0.1547	0.2015	0.1142	0.1546	0.1243	0.1353

consistent with each other. These constraints are summarized by the table 1. The cost function and constraints are linear. We relaxed the variables and solved it with LP.

5   Experiments and Results

Datasets. We tested our approach on both statue faces from the Emperor-I dataset [14] and real faces from various widely used face databases (UMIST, Yale, and Caltech Faces). Quantitative comparison was done on the Emperor-I dataset and we also show some qualitative results on a sample set of all these datasets. The statue face dataset has

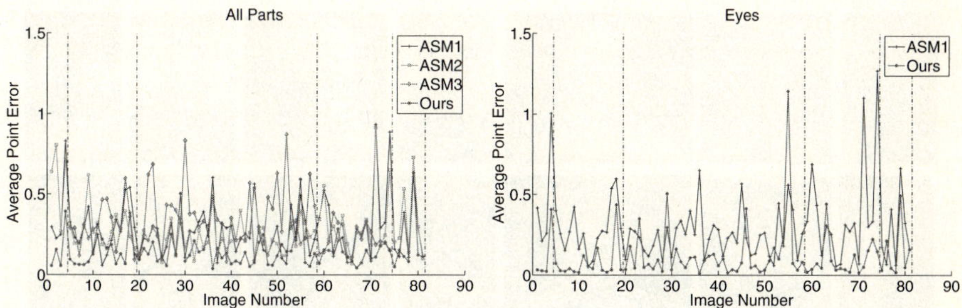

Fig. 6. Average point error vs. image number. All the values are normalized by the estimated distance of two eyes in each image. The vertical dot-dash lines separate images of different poses.

some difficulties that normal faces do not have: lack of color cue, low contrast, inner clutter, and great intra-subject variation.

Comparison measurement. The comparison is between Active Shape Models [10] and our approach. Since we extract facial parts by selecting contours, our desired result is that the extracted contours are all in the right places and correctly labeled. However, ASM generates point-wise alignment between the image and a holistic model. Due to the differences, we chose to use "normalized average point alignment error" measurement for alignment comparison.

Since our results are just labeled contours, we do not have point correspondences for computing the point alignment error. Therefore, we relaxed the measurement to the distance between each ground truth key point and its closest point on the contours belong to the same part. To make the comparison fair, we have exactly the same measurement for ASM by using spline interpolation to generate "contours" for its facial parts. We use 0.35 times the maximum height of the ground truth key points as an approximation of the distance between two eyes invariant to pose changes as the our normalizing factor.

Experiments. There are two aspects of our Emperor-I dataset that may introduce difficulties for ASM: few training examples with various poses and dramatic face silhouette changes. Therefore, we designed three variants of ASM to compensate for these challenges, denoted in our plots as "ASM1","ASM2","ASM3". Table 2 shows the differences. Basically, ASM2 and ASM3 disregard face silhouette and work on fewer poses that may have relatively more exemplars. Note that ASM3 even combined the training data of the three near-frontal poses as a whole. We used "leave-one-out" cross-validation for ASM. For our method, we picked up 7 images for different poses (one for each pose), labeled them and extracted the contours out to work as our holistic models. Moreover, we chose facial part models (usually combined by 2 or 3 contours) from a total of 23 images which also contained these 7 images. Our holistic models are shown in Figure 2 and Figure 5 shows those averaged ones for ASM.

In Figure 6, we show the alignment errors for all the facial parts together and also those only for the eyes. Other facial parts have similar results so we leave them out. Instead, we provide a summary in Table 3 and a comparison in the last column of Table 2, where each entry is the mean error across the test set or test set fold, as applicable. We

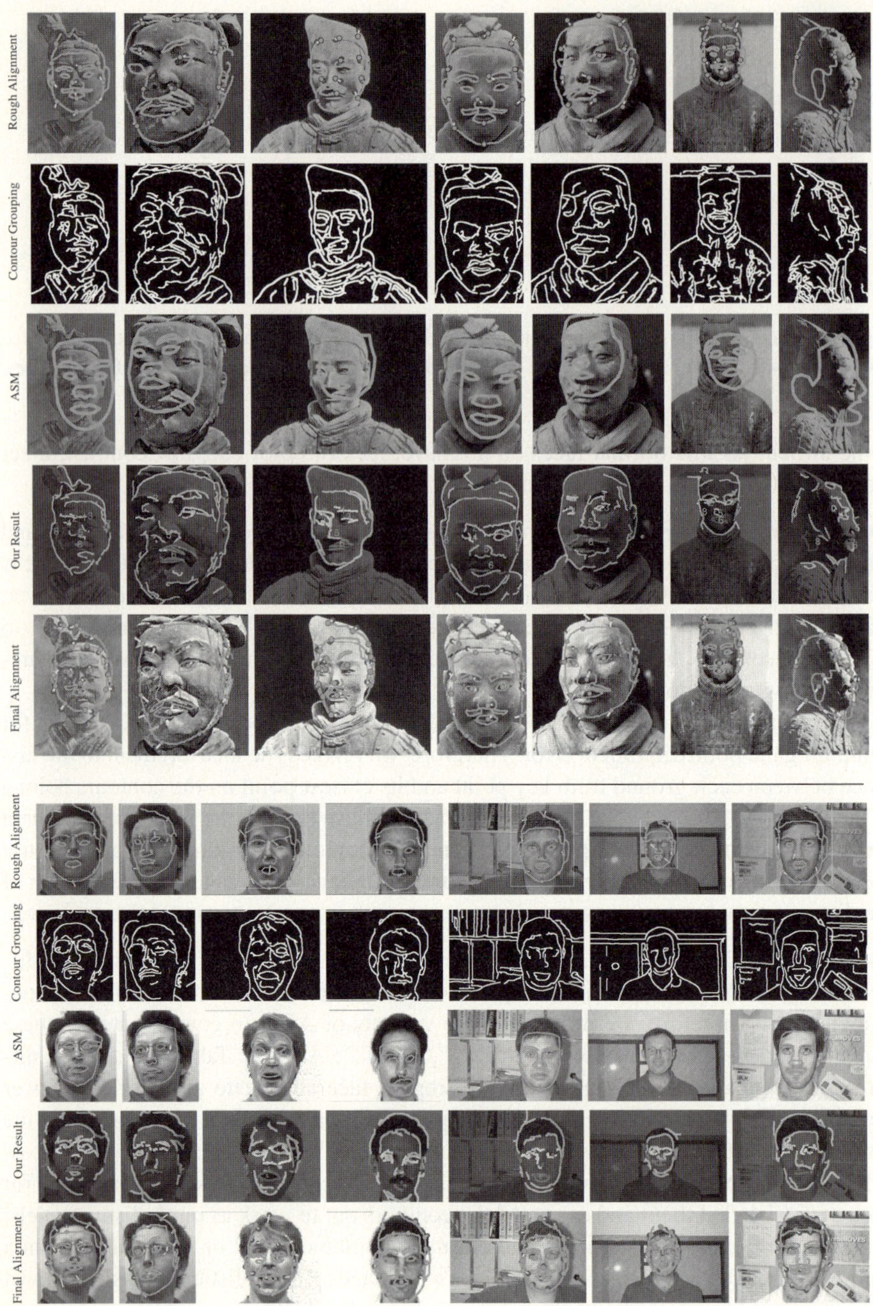

Fig. 7. A subset of the results. Upper group is on the Emperor-I dataset and the lower is for real faces from various face databases (1-2 from UMIST, 3-4 from Yale, and 5-7 from Caltech). Matched models, control points and labeled segments are superimposed on the images.

can see that our method performs significantly better than ASM on all facial parts with significantly fewer training examples. We provide a qualitative evaluation of the results in Figure 7, where we compare the result of ASM and our method on a variety of images containing both statue faces and real faces. These images show great variations, especially of those statue faces. Note that the models are only trained on statue faces.

6   Conclusion

We proposed an object part extraction and labeling framework which incorporates two-level contexts and saliency based opportunistic search. The combination of figural context on the whole object shape and semantic context on parts enables robustly search matching of object parts and image segments in cluttered images. Saliency further improves this search by gradually exploring salient bottom-up segmentations and bounding it via shape matching cost. Experimental results on several challenging face datasets demonstrate that our approach can accurately label object parts such as facial features and resist to accidental alignment.

Acknowledgment. This research was supported by China Scholarship Council, National Science Foundation (Grant NSF-IIS-04-47953(CAREER) and NSF-IIS-03-33036 (IDLP)), National Basic Research Program of China (Grant No. 2006CB708303 and No. 2007CB311005), and National High-Tech Research and Development Plan of China (Grant No. 2006AA01Z192). We would like to acknowledge the help from Praveen Srinivasan, and the discussions and technical help from Liming Wang. Special thanks are given to Geng Zhang for experimental help on ASM and the ground truth labeling.

References

1. Felzenszwalb, P., McAllester, D., Ramanan, D.: A discriminatively trained, multiscale, deformable part model. In: CVPR (2008)
2. Ferrari, V., Jurie, F., Schmid, C.: Accurate object detection with deformable shape models learnt from images. In: CVPR, pp. 1–8 (2007)
3. Felzenszwalb, P.F., Huttenlocher, D.P.: Pictorial structures for object recognition. IJCV 61(1), 55–79 (2005)
4. Zhu, Q., Wang, L., Wu, Y., Shi, J.: Contour context selection for object detection: A single exemplar suffices. In: ECCV (2008)
5. Zhu, Q., Shi, J.: Untangling cycles for contour grouping. In: ICCV (2007)
6. Biederman, I.: Recognition by components: A theory of human image understanding. PsychR 94(2), 115–147 (1987)
7. Pentland, A.: Recognition by parts. In: ICCV, pp. 612–620 (1987)
8. Amit, Y., Trouve, A.: Pop: Patchwork of parts models for object recognition. IJCV 75(2), 267–282 (2007)
9. Sudderth, E., Torralba, A., Freeman, W., Willsky, A.: Learning hierarchical models of scenes, objects, and parts. In: ICCV, pp. 1331–1338 (2005)
10. Cootes, T., Taylor, C., Cooper, D., Graham, J.: Active shape models: Their training and application. CVIU 61(1), 38–59 (1995)
11. Cootes, T., Edwards, G., Taylor, C.: Active appearance models. PAMI 23(6), 681–685 (2001)

12. Belongie, S., Malik, J., Puzicha, J.: Shape matching and object recognition using shape contexts. IEEE Trans. Pattern Anal. Mach. Intell. (2002)
13. Wang, L., Shi, J., Song, G., fan Shen, I.: Object detection combining recognition and segmentation. In: ACCV (1), pp. 189–199 (2007)
14. Chen, C.: The First Emperor of China. Voyager Company (1994)

Stereo Matching: An Outlier Confidence Approach

Li Xu and Jiaya Jia

Department of Computer Science and Engineering
The Chinese University of Hong Kong
{xuli,leojia}@cse.cuhk.edu.hk

Abstract. One of the major challenges in stereo matching is to handle partial occlusions. In this paper, we introduce the Outlier Confidence (OC) which dynamically measures how likely one pixel is occluded. Then the occlusion information is softly incorporated into our model. A global optimization is applied to robustly estimating the disparities for both the occluded and non-occluded pixels. Compared to color segmentation with plane fitting which globally partitions the image, our OC model locally infers the possible disparity values for the outlier pixels using a reliable color sample refinement scheme. Experiments on the Middlebury dataset show that the proposed two-frame stereo matching method performs satisfactorily on the stereo images.

1 Introduction

One useful technique to reduce the matching ambiguity for stereo images is to incorporate the color segmentation into optimization [1,2,3,4,5,6]. Global segmentations improve the disparity estimation in textureless regions; but most of them do not necessarily preserve accurate boundaries. We have experimented that, when taking the ground truth occlusion information into optimization, very accurate disparity estimation can be achieved. This shows that partial occlusion is one major source of matching errors. The main challenge of solving the stereo problems now is the appropriate outlier detection and handling.

In this paper, we propose a new stereo matching algorithm aiming to improve the disparity estimation. Our algorithm does not assign each pixel a binary visibility value indicating whether this pixel is partially occluded or not [7,4,8], but rather introduces soft Outlier Confidence (OC) values to reflect how confident we regard one pixel as an outlier. The OC values, in our method, are used as weights balancing two ways to infer the disparities. The final energy function is globally optimized using Belief Propagation (BP). Without directly labeling each pixel as "occlusion" or "non-occlusion", our model has considerable tolerance of errors produced in the occlusion detection process.

Another main contribution of our algorithm is the local disparity inference for outlier pixels, complementary to the global segmentation. Our method defines the disparity similarity according to the color distance between pixels and naturally transforms color sample selection to a general foreground or background color inference problem using image matting. It effectively reduces errors caused by inaccurate global color segmentation and gives rise to a reliable inference of the unknown disparity of the occluded pixels.

D. Forsyth, P. Torr, and A. Zisserman (Eds.): ECCV 2008, Part IV, LNCS 5305, pp. 775–787, 2008.
© Springer-Verlag Berlin Heidelberg 2008

We also enforce the inter-frame disparity consistency and use BP to simultaneously estimate the disparities of two views. Experimental results on the Middlebury dataset [9] show that our OC model effectively reduces the erroneous disparity estimate due to outliers.

2   Related Work

A comprehensive survey of the dense two-frame stereo matching algorithms was given in [10]. Evaluations of almost all stereo matching algorithms can be found in [9]. Here we review previous work dealing with outliers because, essentially, the difficulty of stereo matching is to handle the ambiguities.

Efforts of dealing with outliers are usually put in three stages in stereo matching – that is, the cost aggregation, the disparity optimization, and the disparity refinement. Most approaches use outlier truncation or other robust functions for cost computation in order to reduce the influence of outliers [2,11].

Window-based methods aggregate matching cost by summing the color differences over a support region. These methods [12,13] prevent depth estimation from aggregating information across different depth layers using the color information. Yoon and Kweon [14] adjusted the support-weight of a pixel in a given window based on the CIELab color similarity and its spatial distance to the center of the support window. Zitnick *et al.* [12] partitioned the input image and grouped the matching cost in each color segment. Lei *et al.* [15] used segmentation to form small regions in a region-tree for further optimization.

In disparity optimization, outliers are handled in two ways in general. One is to explicitly detect occlusions and model visibility [7,4,8]. Sun *et al.* [4] introduced the visibility constraint by penalizing the occlusions and breaking the smoothness between the occluded and non-occluded regions. In [8], Strecha *et al.* modeled the occlusion as a random outlier process and iteratively estimated the depth and visibility in an EM framework in multi-view stereo. Another kind of methods suppresses outliers using extra information, such as pixel colors, in optimization. In [16,6], a color weighted smoothness term was used to control the message passing in BP. Hirschmuller [17] took color difference as the weight to penalize large disparity differences and optimized the disparities using a semi-global approach.

Post-process was also introduced to handle the remaining outliers after the global or local optimization. Occluded pixels can be detected using a consistency check, which validates the disparity correspondences in two views [10,4,17,6]. Disparity interpolation [18] infers the disparities for the occluded pixels from the non-occluded ones by setting the disparities of the mis-matched pixels to that of the background. In [1,3,4,5,6], color segmentation was employed to partition images into segments, each of which is refined by fitting a 3D disparity plane. Optimization such as BP can be further applied after plane fitting [4,5,6] to reduce the possible errors.

Several disparity refinement schemes have been proposed for novel-view synthesis. Sub-pixel refinement [19] enhances details for synthesizing a new view. In [12] and [20], boundary matting for producing seamless view interpolation was introduced. These methods only aim to synthesize natural and seamless novel-views, and cannot be directly used in stereo matching to detect or suppress outliers.

3 Our Model

Denoting the input stereo images as I_l and I_r, and the corresponding disparity maps as $\mathcal{D}_l$ and $\mathcal{D}_r$ respectively, we define the matching energy as

$$E(\mathcal{D}_l, \mathcal{D}_r; I_l, I_r) = E_d(\mathcal{D}_l; I_l, I_r) + E_d(\mathcal{D}_r; I_l, I_r) + E_s(\mathcal{D}_l, \mathcal{D}_r), \qquad (1)$$

where $E_d(\mathcal{D}_l; I_l, I_r) + E_d(\mathcal{D}_r; I_l, I_r)$ is the data term and $E_s(\mathcal{D}_l, \mathcal{D}_r)$ defines the smoothness term that is constructed on the disparity maps. In our algorithm, we not only consider the spatial smoothness within one disparity map, but also model the consistency of disparities between frames.

As the occluded pixels influence the disparity estimation, they should not be used in stereo matching. In our algorithm, we do not distinguish between occlusion and image noise, but rather treat all problematic pixels as outliers. *Outlier Confidences* (OCs) are computed on these pixels, indicating how confident we regard one pixel as an outlier. The outlier confidence maps U_l and U_r are constructed on the input image pair. The confidence $U_l(x)$ or $U_r(x)$ on pixel x is a continuous variable with value between 0 and 1. Larger value indicates higher confidence that one pixel is an outlier, and vice versa.

Our model combines an initial disparity map and an OC map for two views. In the following, we first introduce our data and smoothness terms. The construction of the OC map will be described in Section 4.2.

3.1 Data Term

In the stereo configuration, pixel x in I_l corresponds to pixel $x - d_l$ in I_r by disparity d_l. Similarly, x in I_r corresponds to $x + d_r$ in I_l. All possible disparity values for d_l and d_r are uniformly denoted as set Ψ, containing integers between 0 and N, where N is the maximum positive disparity value. The color of pixel x in I_l (or I_r) is denoted as $I_l(x)$ (or $I_r(x)$). We define the data term $E_d(\mathcal{D}_l; I_l, I_r)$ on the left image as

$$E_d(\mathcal{D}_l; I_l, I_r) = \sum_x [(1 - U_l(x))(\frac{f_0(x, d_l; I_l, I_r)}{\alpha}) + U_l(x)(\frac{f_1(x, d_l; I_l)}{\beta})], \quad (2)$$

where α and β are weights. $f_0(x, d; I_l, I_r)$ denotes the color dissimilarity cost between two views. $f_1(x, d; I_l)$ is the term defined as the local color and disparity discontinuity cost in one view. $E_d(\mathcal{D}_r; I_l, I_r)$ on the right image can be defined in a similar way.

The above two terms, balanced by the outlier confidence $U_l(x)$, model respectively two types of processes in disparity computation. Compared to setting $U_l(x)$ as a binary value and assigning pixels to either outliers or inliers, our cost terms are softly combined, tolerating possible errors in pixel classification.

For result comparison, we give two definitions of $f_0(x, d_l; I_l, I_r)$ respectively corresponding to whether the segmentation is incorporated or not. The first is to use the color and distance weighted local window [14,6,19] to aggregate color difference between conjugate pixels:

$$f_0^{(1)}(x, d_l; I_l, I_r) = \min(g(\|I_l(x) - I_r(x - d_l)\|_1), \varphi), \qquad (3)$$

where $g(\cdot)$ is the aggregate function defined similarly to Equation (2) in [6]. We use the default parameter values (local window size 33×33, $\beta_{cw} = 10$ for normalizing color differences, $\gamma_{cw} = 21$ for normalizing spatial distances). φ determines the maximum cost for each pixel, whose value is set as the average intensity of pixels in the correlation volume.

The second definition is given by incorporating the segmentation information. Specifically, we use the Mean-shift color segmentation [21] with default parameters (spatial bandwidth 7, color bandwidth 6.5, minimum region size 20) to generate color segments. A plane fitting algorithm using RANSAC (similar to that in [6]) is then applied to producing the regularized disparity map d_{pf}. We define

$$f_0^{(2)}(x, d_l; I_l, I_r) = (1 - \kappa)f_0^{(1)}(x, d_l) + \kappa\alpha|d - d_{pf}|, \tag{4}$$

where κ is a weight balancing two terms.

$f_1(x, d_l; I_l)$ is defined as the cost of assigning local disparity when one pixel has chance to be an outlier.

$$f_1(x, d_l; I_l) = \sum_{i \in \Psi} \omega_i(x; I_l)\delta(d_l - i), \tag{5}$$

where $\delta(\cdot)$ is the Dirac function, Ψ denotes the set of all disparity values between 0 and N and $\omega_i(x; I_l)$ is a weight function for measuring how disparity d_l is likely to be i. We omit subscript l in the following discussion of $\omega_i(x; I_l)$ since both the left and right views can use the similar definitions.

For ease of explanation, we first give a general definition of weight $\omega_i'(x; I)$, which, in the following descriptions, will be slightly modified to handle two extreme situations with values 0 and 1. We define

$$\omega_i'(x; I) = 1 - \frac{\mathcal{L}(I(x), \mathbf{I}^i(W_x))}{\mathcal{L}(I(x), \mathbf{I}^i(W_x)) + \mathcal{L}(I(x), \mathbf{I}^{\neq i}(W_x))}, \tag{6}$$

where $I(x)$ denotes the color of pixel x and W_x is a window centered at x. Suppose after initialization, we have collected a set of pixels x' detected as inliers within each W_x (i.e., $U(x') = 0$), and have computed disparities for these inliers. We denote by $\mathbf{I}^i$ the set of inliers whose disparity values are computed as i. Similarly, $\mathbf{I}^{\neq i}$ are the inliers with the corresponding disparity values not equal to i. $\mathcal{L}$ is a metric measuring the color difference between $I(x)$ and its neighboring pixels $\mathbf{I}^i(W_x)$ and $\mathbf{I}^{\neq i}(W_x)$. One example is shown in Figure 1(a) where a window W_x is centered at an outlier pixel x. Within W_x, inlier pixels are clustered into $\mathbf{I}^1$ and $\mathbf{I}^{\neq 1}$. $\omega_1'(x; I)$ is computed according to the color similarity between x and other pixels in the two clusters.

(6) is a function to assign an outlier pixel x a disparity value, constrained by the color similarity between x and the clustered neighboring pixels. By and large, if the color distance between x and its inlier neighbors with disparity i is small enough compared to the color distance to other inliers, $\omega_i'(x; I)$ should have a large value, indicating high chance to let $d_l = i$ in (5).

Now the problem is on how to compute a metric $\mathcal{L}$ that appropriately measures the color distance between pixels. In our method, we abstract color sets $\mathbf{I}^i(W_x)$ and

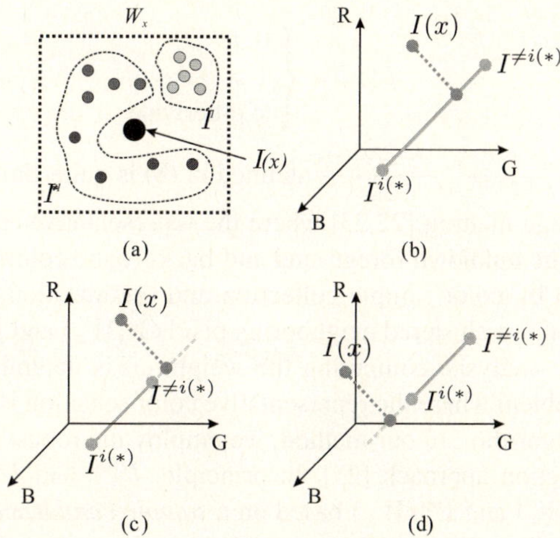

Fig. 1. Computing disparity weight ω'. (a) Within a neighborhood window W_x, inlier pixels are clustered into $\mathbf{I}^1$ and $\mathbf{I}^{\neq 1}$. (b)-(d) illustrate the color projection. (b) The projection of $I(x)$ on vector $I^{i(*)} - I^{\neq i(*)}$ is between two ends. (c-d) The projections of $I(x)$ are out of range, thereby are considered as extreme situations.

$\mathbf{I}^{\neq i}(W_x)$ by two representatives $I^{i(*)}$ and $I^{\neq i(*)}$ respectively. Then $\mathcal{L}$ is simplified to a color metric between pixels. We adopt the color projection distance along vector $I^{i(*)} - I^{\neq i(*)}$ and define

$$\mathcal{L}(I(x), c) = \|\langle I(x) - c, I^{i(*)} - I^{\neq i(*)}\rangle\|, \tag{7}$$

where $\langle \cdot, \cdot \rangle$ denotes the inner product of two color vectors and c can be either $I^{i(*)}$ or $I^{\neq i(*)}$. We regard $I^{i(*)} - I^{\neq i(*)}$ as a projection vector because it measures the absolute difference between two representative colors, or, equivalently, the distance between sets $\mathbf{I}^i(W_x)$ and $\mathbf{I}^{\neq i}(W_x)$.

Projecting $I(x)$ to vector $I^{i(*)} - I^{\neq i(*)}$ also makes the assignment of two extreme values 0 and 1 to $\omega_i(x; I)$ easy. Taking Figure 1 as an example, if the projection of $I(x)$ on vector $I^{i(*)} - I^{\neq i(*)}$ is between two ends, its value is obviously between 0 and 1, as shown in Figure 1 (b). If the projection of $I(x)$ is out of one end point, its value should be 0 if it is close to $I^{i(*)}$ or 1 otherwise (Figure 1 (c) and (d)). To handle the extreme cases, we define the final $\omega_i(x; I)$ as

$$\omega_i(x; I) = \begin{cases} 0 & \text{if } \langle I - I^{\neq i(*)}, I^{i(*)} - I^{\neq i(*)}\rangle < 0 \\ 1 & \text{if } \langle I^{i(*)} - I, I^{i(*)} - I^{\neq i(*)}\rangle < 0 \\ \omega_i'(x; I) & \text{Otherwise} \end{cases}$$

which is further expressed as

$$\omega_i = \mathcal{T}\left(\frac{(I - I^{\neq i(*)})^T(I^{i(*)} - I^{\neq i(*)})}{\|I^{i(*)} - I^{\neq i(*)}\|_2^2}\right), \tag{8}$$

where

$$T(x) = \begin{cases} 0 & x < 0 \\ 1 & x > 1 \\ x & \text{otherwise} \end{cases} \qquad (9)$$

Note that term $\frac{(I-I^{\neq i(*)})^T(I^{i(*)}-I^{\neq i(*)})}{\|I^{i(*)}-I^{\neq i(*)}\|_2^2}$ defined in (8) is quite similar to an alpha matte model used in image matting [22,23] where the representative colors $I^{i(*)}$ and $I^{\neq i(*)}$ are analogous to the unknown foreground and background colors. The image matting problem is solved by color sample collection and optimization. In our problem, the color samples are those clustered neighboring pixels $\mathbf{I}^i(W_x)$ and $\mathbf{I}^{\neq i}(W_x)$.

With the above analysis, computing the weight ω_i is naturally transformed to an image matting problem where the representative color selection is handled by applying an optimization algorithm. In our method, we employ the robust matting with optimal color sample selection approach [23]. In principle, $I^{i(*)}$ and $I^{\neq i(*)}$ are respectively selected from $\mathbf{I}^i(W_x)$ and $\mathbf{I}^{\neq i}(W_x)$ based on a *sample confidence* measure combining two criteria. First, either $I^{i(*)}$ or $I^{\neq i(*)}$ should be similar to the color of the outlier pixel I, which makes weight ω_i approach either 0 or 1 and the weight distribution hardly uniform. Second, I is also expected to be a linear combination of $I^{i(*)}$ and $I^{\neq i(*)}$. This is useful for modeling color blending since outlier pixels have chance to be the interpolation of color samples, especially for those on the region boundary.

Using the sample confidence definition, we get two weights and a neighborhood term, similar to those in [23]. Then we apply the Random Walk method [24] to compute weight ω_i. This process is repeated for all ω_i's, where $i = 0, \cdots, N$. The main benefit that we employ this matting method is that it provides an optimal way to select representative colors while maintaining spatial smoothness.

3.2   Smoothness Term

Term $E_s(\mathcal{D}_l, \mathcal{D}_r)$ contains two parts, representing intra-frame disparity smoothness and inter-frame disparity consistency:

$$E_s(\mathcal{D}_l, \mathcal{D}_r) = \sum_x [\sum_{x' \in \mathcal{N}_1(x)} (\frac{f_3(x, x', d_l, d_r)}{\lambda}) + \sum_{x' \in \mathcal{N}_2(x)} (\frac{f_2(x, x', d_l)}{\gamma}) +$$

$$\sum_{x' \in \mathcal{N}_1(x)} (\frac{f_3(x, x', d_r, d_l)}{\lambda}) + \sum_{x' \in \mathcal{N}_2(x)} (\frac{f_2(x, x', d_r)}{\gamma})], \qquad (10)$$

where $\mathcal{N}_1(x)$ represents the N possible corresponding pixels of x in the other view and $\mathcal{N}_2(x)$ denotes the 4-neighborhood of x in the image space. f_2 is defined as

$$f_2(x, x', d_i) = \min(|d_i(x) - d_i(x'))|, \tau), \quad i \in \{l, r\}, \qquad (11)$$

where τ is a threshold set as 2. To define (11), we have also experimented with using color weighted smoothness and observed that the results are not improved.

We define $f_3(\cdot)$ as the disparity correlations between two views:

$$f_3(x, x', d_l, d_r) = \min(|d_l(x) - d_r(x')|, \zeta) \text{ and}$$
$$f_3(x, x', d_r, d_l) = \min(|d_r(x) - d_l(x')|, \zeta) \quad , \qquad (12)$$

where ζ is a truncation threshold with value 1. We do not define a unique x' corresponding to x because x' is unknown in the beginning. The other reason is that both f_2 and f_3 are the costs for disparity smoothness. In f_2, all neighboring pixels are encoded in $\mathcal{N}_2$ though $d_i(x)$ is not necessarily similar to all $d_i(x')$. So we introduce f_3 with the similar thought for reducing the disparity noise in global optimization considering the inter-frame consistency.

4   Implementation

The overview of our framework is given in Algorithm 1, which consists of an initialization step and a global optimization step. In the first step, we initialize the disparity maps by minimizing an energy with the simplified data and smoothness terms. Then we compute the Outlier Confidence (OC) maps. In the second step, we globally refine the disparities by incorporating the OC maps.

Algorithm 1. Overview of our approach

---

1. **Initialization:**
   1.1 Initialize disparity map $\mathcal{D}$ by setting $U = 0$ for all pixels.
   1.2 Estimate Outlier Confidence map U.

2. **Global Optimization:**
   2.1 Compute data terms using the estimated outlier confidence maps.
   2.2 Global optimization using BP.

4.1   Disparity Initialization

To initialize disparities, we simply set all values in U_l and U_r to zeros and optimize the objective function combining (2) and (10):

$$(\sum_x \frac{f_0(x, d_l) + f_0(x, d_r)}{\alpha}) + E_s(\mathcal{D}_l, \mathcal{D}_r). \tag{13}$$

Because of introducing the inter-frame disparity consistency in (12), our Markov Random Field (MRF) based on the defined energy is slightly different from the regular-grid MRFs proposed in other stereo approaches [2,25]. In our two-frame configuration, the MRF is built on two images with $(4 + N)$ neighboring sites for each node. N is the total number of the disparity levels. One illustration is given in Figure 2 where a pixel x in I_l not only connects to its 4 neighbors in the image space, but also connects to all possible corresponding pixels in I_r.

We minimize the energy defined in (13) using Belief Propagation. The inter-frame consistency constraint makes the estimated disparity maps contain less noise in two frames. We show in Figure 3(a) the initialized disparity result using the standard 4-connected MRF without defining f_3 in (10). (b) shows the result using our $(4 + N)$-connected MRF. The background disparity noise is reduced.

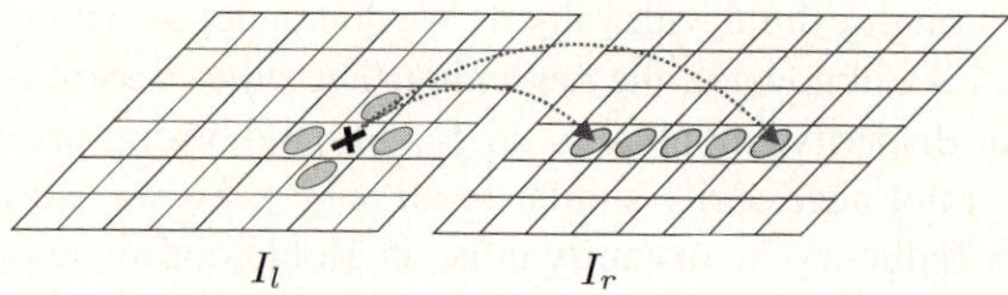

Fig. 2. In our dual view configuration, x (marked with the cross) is not only connected to 4 neighbors in one image, but also related to N possible corresponding pixels in the other image. The total number of neighbors of x is $4 + N$.

Depending on using $f_0^{(1)}$ in (3) or $f_0^{(2)}$ in (4) in the data term definition, we obtain two sets of initializations using and without using global color segmentation. We shall compare in the results how applying our OC models in the following global optimization improves both of the disparity maps.

4.2   Outlier Confidence Estimation

We estimate the outlier confidence map U on the initial disparity maps. Our following discussion focuses on estimating U_l on the left view. The right view can be handled in a similar way. The outlier confidences, in our algorithm, are defined as

$$U_l(x) = \begin{cases} 1 & |d_l(x) - d_r(x - d_l(x))| \geq 1 \\ \mathcal{T}(\frac{b_x(d^*) - b_{min}}{\|b_o - b_{min}\|}) & b_x(d^*) > t \wedge |d_l(x) - d_r(x - d_l(x))| = 0 \\ 0 & \text{Otherwise} \end{cases} \tag{14}$$

considering 2 cases.

Case 1: Our MRF enforces the disparity consistency between two views. After disparity initialization, the remaining pixels with inconsistent disparities are likely to be occlusions. So we first set the outlier confidence $U_l(x) = 1$ for pixel x if the inter-frame consistency is violated, i.e., $|d_l(x) - d_r(x - d_l(x))| \geq 1$.

Case 2: Besides the disparity inconsistency, pixel matching with large matching cost is also unreliable. In our method, since we use BP to initialize the disparity maps, the matching cost is embedded in the output disparity belief $b_x(d)$ for each pixel x. Here, we introduce some simple operations to manipulate it. First, we extract $b_x(d^*)$, i.e., the smallest belief, for each pixel x. If $b_x(d^*) < t$, where t is a threshold, the pixel should be regarded as an inlier given the small matching cost. Second, a variable b_o is computed as the average of the minimal beliefs regarding all occluded pixels detected in Case 1, i.e., $b_o = \sum_{U_l(x)=1} b_x(d^*)/K$ where K is the total number of the occluded pixels. Finally, we compute b_{min} as the average of top $n\%$ minimal beliefs among all pixels. n is set to 10 in our experiments.

Using the computed $b_x(d^*)$, b_o, and b_{min}, we estimate $U_l(\widetilde{x})$ for pixels neither detected as occlusions nor treated as inliers by setting

$$U_l(\widetilde{x}) = \mathcal{T}\left(\frac{b_{\widetilde{x}}(d^*) - b_{min}}{\|b_o - b_{min}\|}\right), \tag{15}$$

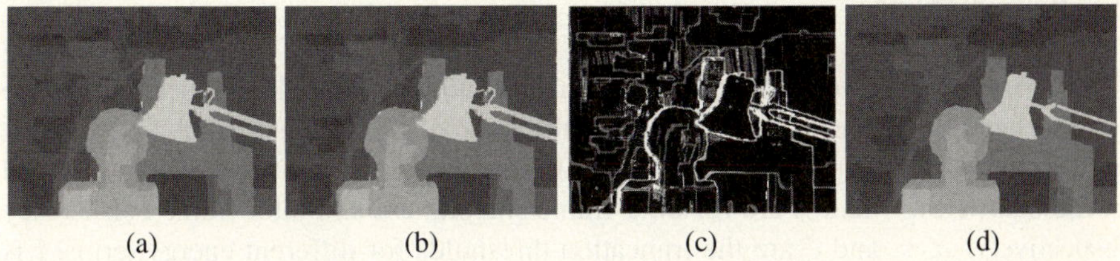

(a)   (b)   (c)   (d)

Fig. 3. Intermediate results for the "Tsukuba" example. (a) and (b) show our initial disparity maps by the 4-connected and $(4 + N)$-connected MRFs respectively without using segmentation. The disparity noise in (b) is reduced for the background. (c) Our estimated OC map. (d) A disparity map constructed by combining the inlier and outlier information. The disparities for the outlier pixels are set as the maximum weight ω_i. The inlier pixels are with initially computed disparity values.

where $\mathcal{T}$ is the function defined in (9), making the confidence value in range $[0, 1]$. (15) indicates if the smallest belief $b_x(d^*)$ of pixel x is equal to or larger than the average smallest belief of the occluded pixels detected in Case 1, the outlier confidence of x will be high, and vice versa.

Figure 3(c) shows the estimated outlier coefficient map for the "tsukuba" example. The pure black pixels represent inliers where $U_l(x) = 0$. Generally, the region consisting of pixels with $U_l(x) > 0$ is wider than the ground truth occluded region. This is allowed in our algorithm because $U_l(x)$ is only a weight balancing pixel matching and color smoothness. Even if pixel x is mistakenly labeled as an outlier, the disparity estimation in our algorithm will not be largely influenced because large $U_l(x)$ only makes the disparity estimation of x rely more on neighboring pixel information, by which $d(x)$ still has a large chance to be correctly inferred.

To illustrate the efficacy of our OC scheme, we show in Figure 3(d) a disparity map directly constructed with the following setting. Each inlier pixel is with initially computed disparity value and each outlier pixel is with the disparity i corresponding to the maximum weight ω_i among all ω_j's, where $j = 0, \cdots, N$. It can be observed that even without any further global optimization, this simple maximum-weight disparity calculation already makes the object boundary smooth and natural.

4.3   Global Optimization

With the estimated OC maps, we are ready to use global optimization to compute the final disparity maps combining costs (2) and (10) in (1). Two forms of $f_0(\cdot)$ ((3) and (4)) are independently applied in our experiments for result comparison.

The computation of $f_1(x, d; I)$ in (5) is based on the estimated OC maps and the initial disparities for the inlier pixels, which are obtained in the aforementioned steps. To compute ω_i for outlier pixel x with $U_l(x) > 0$, robust matting [23] is performed as described in Section 3.1 for each disparity level. The involved color sampling is performed in each local window with size 60×60. Finally, the smoothness terms are embedded in the message passing of BP. An acceleration using distance transform [25] is adopted to construct the messages.

5   Experiments

In experiments, we compare the results using and without using the Outlier Confidence maps. The performance is evaluated using the Middlebury dataset [10]. All parameters used in implementation are listed in Table 1 where α, β and κ are the weights defined in the data term. γ and λ are for intra-frame smoothness and inter-frame consistency respectively. φ, τ, and ζ are the truncation thresholds for different energy terms. t is the threshold for selecting possible outliers. As we normalize the messages after each message passing iteration by subtracting the mean of the messages, the belief b_{min} is negative, making $t = 0.9b_{min} > b_{min}$.

A comparison of the state-of-the-art stereo matching algorithms is shown in Table 2 extracted from the Middlebury website [9]. In the following, we give detailed explanations.

Table 1. The parameter values used in our experiments. N is the number of the disparity levels. $\bar{c}$ is the average of the correlation volume. b_{min} is introduced in (15).

Parameters	α	β	κ	γ	λ	φ	τ	ζ	t
value	φ	0.8	0.3	5.0	$5N$	$\bar{c}$	2.0	1.0	$0.9b_{min}$

Table 2. Algorithm evaluation on the Midellbury data set. Our method achieves overall rank 2 at the time of data submission.

Algorithm	Avg. Rank	Tsukuba nonocc	all	disc	Venus nonocc	all	disc	Teddy nonocc	all	disc	Cones nonocc	all	disc
Adap.BP [5]	2.3	1.11	1.37	5.79	**0.10**	**0.21**	**1.44**	4.22	7.06	11.8	**2.48**	7.92	7.32
Our method	3.6	**0.88**	1.43	**4.74**	0.18	0.26	2.40	5.01	9.12	12.8	2.78	8.57	**6.99**
DoubleBP [6]	3.7	0.88	**1.29**	4.76	0.14	0.60	2.00	3.55	8.71	**9.70**	2.90	9.24	7.80
SPDou.BP [19]	4.6	1.24	1.76	5.98	0.12	0.46	1.74	**3.45**	8.38	10.0	2.93	8.73	7.91
SymBP+occ [4]	8.8	0.97	1.75	5.09	0.16	0.33	2.19	6.47	10.7	17.0	4.79	10.7	10.9

Table 3. Result comparison on the Middlebury dataset using (1st and 3rd rows) and without using (2nd and 4th rows) OC Maps. The segmentation information has been incorporated for the last two rows.

Algorithm	Overall Rank	Tsukuba nonocc	all	disc	Venus nonocc	all	disc	Teddy nonocc	all	disc	Cones nonocc	all	disc
COLOR	16	1.12	3.29	5.92	0.49	1.48	6.78	10.5	16.9	21.1	3.42	12.1	8.26
COLOR+OC	5	**0.83**	1.41	**4.45**	0.25	0.31	3.22	10.1	14.6	19.9	3.22	9.82	7.40
SEG	4	0.97	1.75	5.23	0.30	0.70	3.98	5.56	9.99	13.6	3.04	8.90	7.60
SEG+OC	2	**0.88**	1.43	**4.74**	0.18	0.26	2.40	5.01	9.12	12.8	2.78	8.57	**6.99**

5.1   Results without Using Segmentation

In the first part of our experiments, we do not use the segmentation information. So data term $f_0^{(1)}$ defined in (3) is used in our depth estimation.

We iterate steps (1) and (2) until the step $t = T$ for which the change is below a small threshold. We then rank the similarity to the query x_1 with sim_T. Our experimental results in Section 6 demonstrate that the replacement of the original similarity measure sim with sim_T results in a significant increase in the retrieval rate.

The steps (1) and (2) are used in label propagation, which is described in Section 4. However, our goal and our setting are different. Although label propagation is an instance of semi-supervised learning, we stress that we remain in the unsupervised learning setting. In particular, we deal with the case of only one known class, which is the class of the query object. This means, in particular, that label propagation has a trivial solution in our case $\lim_{t \to \infty} f_t(x_i) = 1$ for all $i = 1, \ldots, n$, i.e., all objects will be assigned the class label of the query shape. Since our goal is ranking of the database objects according to their similarity to the query, we stop the computation after a suitable number of iterations $t = T$. As is the usual practice with iterative processes that are guaranteed to converge, the computation is halted if the difference $||f_{t+1} - f_t||$ becomes very slow, see Section 6 for details.

If the database of known objects is large, the computation with all n objects may become impractical. Therefore, in practice, we construct the matrix w using only the first $M < n$ most similar objects to the query x_1 sorted according to the original distance function sim.

4   Relation to Label Propagation

Label propagation is formulated as a form of propagation on a graph, where node's label propagates to neighboring nodes according to their proximity. In our approach we only have one labeled node, which is the query shape. The key idea is that its label propagates "faster" along a geodesic path on the manifold spanned by the set of known shapes than by direct connections. While following a geodesic path, the obtained new similarity measure learns to ignore irrelevant shape differences. Therefore, when learning is complete, it is able to focus on relevant shape differences. We review now the key steps of label propagation and relate them to the proposed method introduced in Section 3.

Let $\{(x_1, y_1) \ldots (x_l, y_l)\}$ be the labeled data, $y \in \{1 \ldots C\}$, and $\{x_{l+1} \ldots x_{l+u}\}$ the unlabeled data, usually $l \ll u$. Let $n = l + u$. We will often use L and U to denote labeled and unlabeled data respectively. The Label propagation supposes the number of classes C is known, and all classes are present in the labeled data[9]. A graph is created where the nodes are all the data points, the edge between nodes i, j represents their similarity $w_{i,j}$. Larger edge weights allow labels to travel through more easily. We define a $n \times n$ probabilistic transition matrix P as a row-wise normalized matrix w.

$$P_{ij} = \frac{w_{ij}}{\sum_{k=1}^{n} w_{ik}} \tag{4}$$

where P_{ij} is the probability of transit from node i to node j. Also define a $l \times C$ label matrix Y_L, whose ith row is an indicator vector for y_i, $i \in L$: $Y_{ic} = \delta(y_{i,c})$.

The label propagation computes soft labels f for nodes, where f is a $n \times C$ matrix whose rows can be interpreted as the probability distributions over labels. The initialization of f is not important. The label propagation algorithm is as follows:

1. Initially, set $f(x_i) = y_i$ for $i = 1, \ldots, l$ and $f(x_j)$ arbitrarily (e.g., 0) for $x_j \in X_u$

2. Repeat until convergence: Set $f(x_i) = \frac{\sum_{j=1}^{n} w_{ij} f(x_j)}{\sum_{j=1}^{n} w_{ij}}$, $\forall x_i \in X_u$ and set $f(x_i) = y_i$ for $i = 1, \ldots, l$ (the labeled objects should be fixed).

In step 1, all nodes propagate their labels to their neighbors for one step. Step 2 is critical, since it ensures persistent label sources from labeled data. Hence instead of letting the initial labels fade way, we fix the labeled data. This constant push from labeled nodes, helps to push the class boundaries through high density regions so that they can settle in low density gaps. If this structure of data fits the classification goal, then the algorithm can use unlabeled data to improve learning.

Let $f = \binom{f_L}{f_U}$. Since f_L is fixed to Y_L, we are solely interested in f_U. The matrix P is split into labeled and unlabeled sub-matrices

$$P = \begin{bmatrix} P_{LL} & P_{LU} \\ P_{UL} & P_{UU} \end{bmatrix} \tag{5}$$

As proven in [9] the label propagation converges, and the solution can be computed in closed form using matrix algebra:

$$f_U = (I - P_{UU})^{-1} P_{UL} Y_L \tag{6}$$

However, as the label propagation requires all classes be present in the labeled data, it is not suitable for shape retrieval. As mentioned in Section 3, for shape retrieval, the query shape is considered as the only labeled data and all other shapes are the unlabeled data. Moreover, the graph among all of the shapes is fully connected, which means the label could be propagated on the whole graph. If we iterate the label propagation infinite times, all of the data will have the same label, which is not our goal. Therefore, we stop the computation after a suitable number of iterations $t = T$.

5   The Affinity Matrix

In this section, we address the problem of the construction of the affinity matrix W. There are some methods that address this issue, such as local scaling [26], local liner approximation [17], and adaptive kernel size selection [27].

However, in the case of shape similarity retrieval, a distance function is usually defined, e.g., [1,3,4,5]. Let $D = (D_{ij})$ be a distance matrix computed by some shape distance function. Our goal is to convert it to a similarity measure in order to construct an affinity matrix W. Usually, this can be done by using a Gaussian kernel:

$$w_{ij} = \exp(-\frac{D_{ij}^2}{\sigma_{ij}^2}) \tag{7}$$

Previous research has shown that the propagation results highly depend on the kernel size σ_{ij} selection [17]. In [15], a method to learn the proper σ_{ij} for the kernel is introduced, which has excellent performance. However, it is not learnable in the case of few labeled data. In shape retrieval, since only the query shape has the label, the learning of σ_{ij} is not applicable. In our experiment, we use use an adaptive kernel size based on the mean distance to K-nearest neighborhoods [28]:

$$\sigma_{ij} = C \cdot \text{mean}(\{knnd(x_i), knnd(x_j)\}) \tag{8}$$

where $\text{mean}(\{knnd(x_i), knnd(x_j)\})$ represents the mean distance of the K-nearest neighbor distance of the sample x_i, x_j and C is an extra parameter. Both K and C are determined empirically.

6   Experimental Results

In this section, we show that the proposed approach can significantly improve retrieval rates of existing shape similarity methods.

6.1   Improving Inner Distance Shape Context

The IDSC [3] significantly improved the performance of shape context [1] by replacing the Euclidean distance with shortest paths inside the shapes, and obtained the retrieval rate of 85.40% on the MPEG-7 data set. The proposed distance learning method is able to improve the IDSC retrieval rate to **91.00%**. For reference, Table 1 lists some of the reported results on the MPEG-7 data set. The MPEG-7 data set consists of 1400 silhouette images grouped into 70 classes. Each class has 20 different shapes. The retrieval rate is measured by the so-called bull's eye score. Every shape in the database is compared to all other shapes, and the number of shapes from the same class among the 40 most similar shapes is reported. The bull's eye retrieval rate is the ratio of the total number of shapes from the same class to the highest possible number (which is 20×1400). Thus, the best possible rate is 100%.

In order to visualize the gain in retrieval rates by our method as compared to IDSC, we plot the percentage of correct results among the first k most similar shapes in Fig. 4(a), i.e., we plot the percentage of the shapes from the same class among the first k-nearest neighbors for $k = 1, \ldots, 40$. Recall that each class has 20 shapes, which is why the curve increases for $k > 20$. We observe that the proposed method not only increases the bull's eye score, but also the ranking of the shapes for all $k = 1, \ldots, 40$.

We use the following parameters to construct the affinity matrix: $C = 0.25$ and the neighborhood size is $K = 10$. As stated in Section 3, in order to increase computational efficiency, it is possible to construct the affinity matrix for only part of the database of known shapes. Hence, for each query shape, we first retrieve 300 the most similar shapes, and construct the affinity matrix W for only those shapes, i.e., W is of size 300×300 as opposed to a 1400×1400 matrix if we consider all MPEG-7 shapes. Then we calculate the new similarity measure

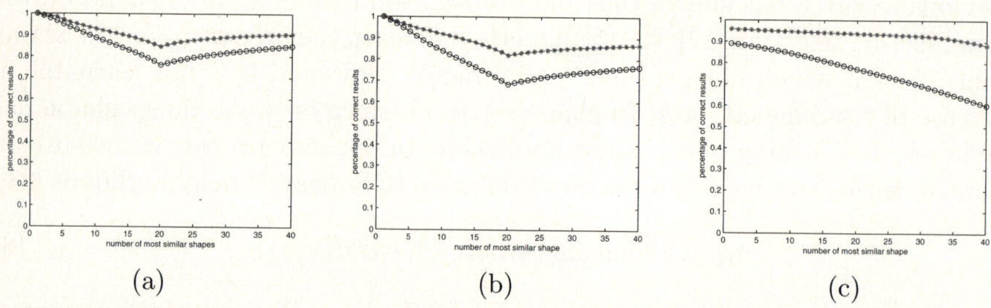

(a)   (b)   (c)

Fig. 4. (a) A comparison of retrieval rates between IDSC [3] (blue circles) and the proposed method (red stars) for MPEG-7. (b) A comparison of retrieval rates between visual parts in [4] (blue circles) and the proposed method (red stars) for MPEG-7. (c) Retrieval accuracy of DTW (blue circles) and the proposed method (red stars) for the *Face (all)* dataset.

Table 1. Retrieval rates (bull's eye) of different methods on the MPEG-7 data set

Alg.	CSS [29]	Vis. Parts [4]	SC +TPS [1]	IDSC +DP [3]	Hierarchical Procrustes [6]	Shape Tree [5]	IDSC+DP + our method
Score	75.44%	76.45%	76.51%	85.40%	86.35%	87.70%	**91.00%**

sim_T for only those 300 shapes. Here we assume that all relevant shapes will be among the 300 most similar shapes. Thus, by using a larger affinity matrix we can improve the retrieval rate but at the cost of computational efficiency.

In addition to the statistics presented in Fig. 4, Fig. 5 illustrates also that the proposed approach improves the performance of IDSC. A very interesting case is shown in the first row, where for IDSC only one result is correct for the query octopus. It instead retrieves nine apples as the most similar shapes. Since the query shape of the octopus is occluded, IDSC ranks it as more similar to an apple than to the octopus. In addition, since IDSC is invariant to rotation, it confuses the tentacles with the apple stem. Even in the case of only one correct shape, the proposed method learns that the difference between the apple stem is relevant, although the tentacles of the octopuses exhibit a significant variation in shape. We restate that this is possible because the new learned distances are induced by geodesic paths in the shape manifold spanned by the known shapes. Consequently, the learned distances retrieve nine correct shapes. The only wrong results is the elephant, where the nose and legs are similar to the tentacles of the octopus.

As shown in the third row, six of the top ten IDSC retrieval results of lizard are wrong. since IDSC cannot ignore the irrelevant differences between lizards and sea snakes. All retrieval results are correct for the new learned distances, since the proposed method is able to learn the irrelevant differences between lizards and the relevant differences between lizards and sea snakes. For the results of deer (fifth row), three of the top ten retrieval results of IDSC are horses. Compared

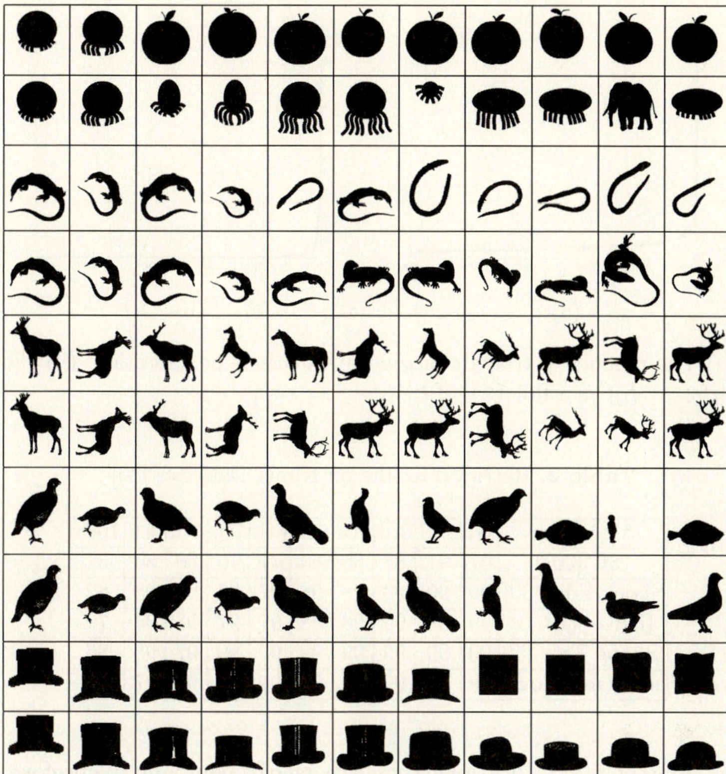

Fig. 5. The first column shows the query shape. The remaining 10 columns show the most similar shapes retrieved by IDSC (odd row numbers) and by our method (even row numbers).

to it, the proposed method (sixth row) eliminates all of the wrong results so that only deers are in the top ten results. It appears to us that our new method learned to ignore the irrelevant small shape details of the antlers. Therefore, the presence of the antlers became a relevant shape feature here. The situation is similar for the bird and hat, with three and four wrong retrieval results respectively for IDSC, which are eliminated by the proposed method.

An additional explanation of the learning mechanism of the proposed method is provided by examining the count of the number of violations of the triangle inequality that involve the query shape and the database shapes. In Fig. 6(a), the curve shows the number of triangle inequality violations after each iteration of our distance learning algorithm. The number of violations is reduced significantly after the first few hundred iterations. We cannot expect the number of violations to be reduced to zero, since cognitively motivated shape similarity may sometimes require triangle inequality violations [10]. Observe that the curve in Fig. 6(a) correlates with the plot of differences $||f_{t+1} - f_t||$ as a function of t shown in (b). In particular, both curves decrease very slow after about 1000

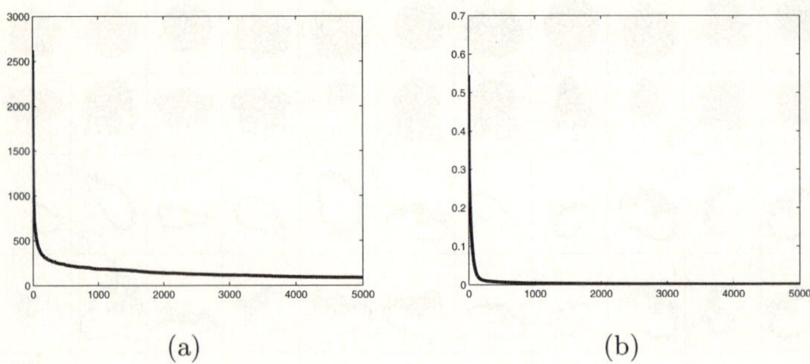

(a)   (b)

Fig. 6. (a) The number of triangle inequality violations per iteration. (b) Plot of differences $||f_{t+1} - f_t||$ as a function of t.

Table 2. Retrieval results on Kimia Data Set [30]

Algorithm	1st	2nd	3rd	4th	5th	6th	7th	8th	9th	10th
SC [30]	97	91	88	85	84	77	75	66	56	37
Shock Edit [30]	99	99	99	98	98	97	96	95	93	82
IDSC+DP [3]	99	99	99	98	98	97	97	98	94	79
Shape Tree [5]	99	99	99	99	99	99	99	97	93	86
our method	99	99	99	99	99	99	99	99	97	99

iterations, and at 5000 iterations they are nearly constant. Therefore, we selected $T = 5000$ as our stop condition. Since the situation is very similar in all our experiments, we always stop after $T = 5000$ iterations.

Besides MPEG-7, We also present experimental results on the Kimia Data Set [30]. The database contains 99 shapes grouped into nine classes. As the database only contains 99 shapes, we calculate the affinity matrix based on all of the shape in the database. The parameters used to calculate the affinity matrix are: $C = 0.25$ and the neighborhood size is $K = 4$. We changed the neighborhood size, since the data set is much smaller than the MPEG-7 data set. The retrieval results are summarized as the number of shapes from the same class among the first top 1 to 10 shapes (the best possible result for each of them is 99). Table 2 lists the numbers of correct matches of several methods. Again we observe that our approach could improve IDSC significantly, and it yields a nearly perfect retrieval rate.

6.2   Improving Visual Part Shape Matching

Besides the inner distance shape context [3], we also demonstrate that the proposed approach can improve the performance of visual parts shape similarity [4]. We select this method since it is based on very different approach than IDSC. In [4], in order to compute the similarity between shapes, first the best possible correspondence of visual parts is established (without explicitly computing the visual parts). Then, the similarity between corresponding parts is calculated and aggregated. The settings and parameters of our experiment are the same as for IDSC as reported in the previous section except we set $C = 0.4$. The accuracy of this method has been increased from 76.45% to 86.69% on the MPEG-7 data set, which is more than 10%. This makes the improved visual part method one of the top scoring methods in Table 1. A detailed comparison of the retrieval accuracy is given in Fig. 4(b).

6.3 Improving Face Retrieval

We used a face data set from [31], where it is called *Face (all)*. It addresses a face recognition problem based on the shape of head profiles. It contains several head profiles extracted from side view photos of 14 subjects. There exist large variations in the shape of the face profile of each subject, which is the main reason why we select this data set. Each subject is making different face expressions, e.g., talking, yawning, smiling, frowning, laughing, etc. When the pictures of subjects were taken, they were also encouraged to look a little to the left or right, randomly. At least two subjects had glasses that they put on for half of their samples.

The head profiles are converted to sequences of curvature values, and normalized to the length of 131 points, starting from the neck area. The data set has two parts, training with 560 profiles and testing with 1690 profiles. The training set contains 40 profiles for each of the 14 classes. As reported on [31], we calculated the retrieval accuracy by matching the 1690 test shapes to the 560 training shapes. We used a dynamic time warping (DTW) algorithm with warping window [32] to generate the distance matrix, and obtained the 1NN retrieval accuracy of 88.9% By applying our distance learning method we increased the 1NN retrieval accuracy to 95.04%. The best reported result on [31] has the first nearest neighbor (1NN) retrieval accuracy of 80.8%. The retrieval rate, which represents the percentage of the shapes from the same class (profiles of the same subject) among the first k-nearest neighbors, is shown in Fig. 4(c). The accuracy of the proposed approach is stable, although the accuracy of DTW decreases significantly when k increases. In particular, our retrieval rate for k=40 remains high, 88.20%, while the DTW rate dropped to 60.18%. Thus, the learned distance allowed us to increase the retrieval rate by nearly 30%. Similar to the above experiments, the parameters for the affinity matrix is $C = 0.4$ and $K = 5$.

7   Conclusion and Discussion

In this work, we adapted a graph transductive learning framework to learn new distances with the application to shape retrieval. The key idea is to replace the distances in the original distance space with distances induces by geodesic paths in the shape manifold. The merits of the proposed technique have been validated by significant performance gains over the experimental results. However, like semi-supervised learning, if there are too many outlier shapes in the shape database, the proposed approach cannot improve the results. Our future work will focus on addressing this problem. We also observe that our method is not limited to 2D shape similarity but can also be applied to 3D shape retrieval, which will also be part of our future work.

Acknowledgements

We would like to thank Eamonn Keogh for providing us the *Face (all)* dataset. This work was support in part by the NSF Grant No. IIS-0534929 and by the DOE Grant No. DE-FG52-06NA27508.

References

1. Belongie, S., Malik, J., Puzicha, J.: Shape matching and object recognition using shape contexts. IEEE Trans. PAMI 24, 705–522 (2002)
2. Tu, Z., Yuille, A.L.: Shape matching and recognition - using generative models and informative features. In: Pajdla, T., Matas, J(G.) (eds.) ECCV 2004. LNCS, vol. 3024, pp. 195–209. Springer, Heidelberg (2004)
3. Ling, H., Jacobs, D.: Shape classification using the inner-distance. IEEE Trans. PAMI 29, 286–299 (2007)
4. Latecki, L.J., Lakämper, R.: Shape similarity measure based on correspondence of visual parts. IEEE Trans. PAMI 22(10), 1185–1190 (2000)
5. Felzenszwalb, P.F., Schwartz, J.: Hierarchical matching of deformable shapes. In: CVPR (2007)
6. McNeill, G., Vijayakumar, S.: Hierarchical procrustes matching for shape retrieval. In: Proc. CVPR (2006)
7. Bai, X., Latecki, L.J.: Path similarity skeleton graph matching. IEEE Trans. PAMI 30, 1282–1292 (2008)
8. Srivastava, A., Joshi, S.H., Mio, W., Liu, X.: Statistic shape analysis: clustering, learning, and testing. IEEE Trans. PAMI 27, 590–602 (2005)
9. Zhu, X.: Semi-supervised learning with graphs. In: Doctoral Dissertation. Carnegie Mellon University, CMU–LTI–05–192 (2005)
10. Vleugels, J., Veltkamp, R.: Efficient image retrieval through vantage objects. Pattern Recognition 35(1), 69–80 (2002)
11. Latecki, L.J., Lakämper, R., Eckhardt, U.: Shape descriptors for non-rigid shapes with a single closed contour. In: CVPR, pp. 424–429 (2000)
12. Brefeld, U., Buscher, C., Scheffer, T.: Multiview dicriminative sequential learning. In: Gama, J., Camacho, R., Brazdil, P.B., Jorge, A.M., Torgo, L. (eds.) ECML 2005. LNCS (LNAI), vol. 3720. Springer, Heidelberg (2005)
13. Lawrence, N.D., Jordan, M.I.: Semi-supervised learning via gaussian processes. In: NIPS (2004)
14. Joachims, T.: Transductive inference for text classification using support vector machines. In: ICML, pp. 200–209 (1999)
15. Zhu, X., Ghahramani, Z., Lafferty., J.: Semi-supervised learning using gaussian fields and harmonic functions. In: ICML (2003)
16. Zhou, D., Bousquet, O., Lal, T.N., Weston, J., Scholkopf., B.: Learning with local and global consistency. In: NIPS (2003)
17. Wang, F., Wang, J., Zhang, C., Shen., H.: Semi-supervised classification using linear neighborhood propagation. In: CVPR (2006)

18. Zhou, D., Weston, J.: Ranking on data manifolds. In: NIPS (2003)
19. Roweis, S.T., Saul, L.K.: Nonlinear dimensionality reduction by locally linear embedding. Science 290, 2323–2326 (2000)
20. Fan, X., Qi, C., Liang, D., Huang, H.: Probabilistic contour extraction using hierarchical shape representation. In: Proc. ICCV, pp. 302–308 (2005)
21. Yu, J., Amores, J., Sebe, N., Radeva, P., Tian, Q.: Distance learning for similarity estimation. IEEE Trans. PAMI 30, 451–462 (2008)
22. Xing, E., Ng, A., Jordanand, M., Russell, S.: Distance metric learning with application to clustering with side-information. In: NIPS, pp. 505–512 (2003)
23. Bar-Hillel, A., Hertz, T., Shental, N., Weinshall, D.: Learning distance functions using equivalence relations. In: ICML, pp. 11–18 (2003)
24. Athitsos, V., Alon, J., Sclaroff, S., Kollios, G.: Bootmap: A method for efficient approximate similarity rankings. In: CVPR (2004)
25. Hertz, T., Bar-Hillel, A., Weinshall, D.: Learning distance functions for image retrieval. In: CVPR, pp. 570–577 (2004)
26. Zelnik-Manor, L., Perona, P.: Self-tuning spectral clustering. In: NIPS (2004)
27. Hein, M., Maier, M.: Manifold denoising. In: NIPS (2006)
28. Wang, J., Chang, S.F., Zhou, X., Wong, T.C.S.: Active microscopic cellular image annotation by superposable graph transduction with imbalanced labels. In: CVPR (2008)
29. Mokhtarian, F., Abbasi, F., Kittler, J.: Efficient and robust retrieval by shape content through curvature scale space. In: Smeulders, A.W.M., Jain, R. (eds.) Image Databases and Multi-Media Search, pp. 51–58 (1997)
30. Sebastian, T.B., Klein, P.N., Kimia, B.: Recognition of shapes by editing their shock graphs. IEEE Trans. PAMI 25, 116–125 (2004)
31. Keogh, E.: UCR time series classification/clustering page,
    http://www.cs.ucr.edu/~eamonn/time_series_data/
32. Ratanamahatana, C.A., Keogh, E.: Three myths about dynamic time warping. In: SDM, pp. 506–510 (2005)

Cat Head Detection - How to Effectively Exploit Shape and Texture Features

Weiwei Zhang[1], Jian Sun[1], and Xiaoou Tang[2]

[1] Microsoft Research Asia, Beijing, China
{weiweiz,jiansun}@microsoft.com
[2] Dept. of Information Engineering, The Chinese University of Hong Kong, Hong Kong
xtang@ie.cuhk.edu.hk

Abstract. In this paper, we focus on the problem of detecting the head of cat-like animals, adopting cat as a test case. We show that the performance depends crucially on how to effectively utilize the shape and texture features jointly. Specifically, we propose a two step approach for the cat head detection. In the first step, we train two individual detectors on two training sets. One training set is normalized to emphasize the shape features and the other is normalized to underscore the texture features. In the second step, we train a joint shape and texture fusion classifier to make the final decision. We demonstrate that a significant improvement can be obtained by our two step approach. In addition, we also propose a set of novel features based on oriented gradients, which outperforms existing leading features, e. g., Haar, HoG, and EoH. We evaluate our approach on a well labeled cat head data set with 10,000 images and PASCAL 2007 cat data.

1 Introduction

Automatic detection of all generic objects in a general scene is a long term goal in image understanding and remains to be an extremely challenging problem duo to large intra-class variation, varying pose, illumination change, partial occlusion, and cluttered background. However, researchers have recently made significant progresses on a particularly interesting subset of object detection problems, face [14,18] and human detection [1], achieving near 90% detection rate on the frontal face in real-time [18] using a boosting based approach. This inspires us to consider whether the approach can be extended to a broader set of object detection applications.

Obviously it is difficult to use the face detection approach on generic object detection such as tree, mountain, building, and sky detection, since they do not have a relatively fixed intra-class structure like human faces. To go one step at a time, we need to limit the objects to the ones that share somewhat similar properties as human face. If we can succeed on such objects, we can then consider to go beyond. Naturally, the closest thing to human face on this planet is animal head. Unfortunately, even for animal head, given the huge diversity of animal types, it is still too difficult to try on all animal heads. This is probably why we have seen few works on this fattempt.

In this paper, we choose to be conservative and limit our endeavor to only one type of animal head detection, cat head detection. This is of course not a random selection.

D. Forsyth, P. Torr, and A. Zisserman (Eds.): ECCV 2008, Part IV, LNCS 5305, pp. 802–816, 2008.

(a) cat-like animal   (b) cats

Fig. 1. Head images of animals of the cat family and cats

Our motivations are as follows. First, cat can represent a large category of cat-like animals, as shown in Figure 1 (a). These animals share similar face geometry and head shape; Second, people love cats. A large amount of cat images have been uploaded and shared on the web. For example, 2,594,329 cat images had been manually annotated in flickr.com by users. Cat photos are among the most popular animal photos on the internet. Also, cat as a popular pet often appears in family photos. So cat detection can find applications in both online image search and offline family photo annotation, two important research topics in pattern recognition. Third, given the popularity of cat photos, it is easy for us to get training data. The research community does need large and challenging data set to evaluate the advances of the object detection algorithm. In this paper, we provide 10,000, well labeled cat images. Finally and most importantly, the cat head detection poses new challenges for object detection algorithm. Although it shares some similar property with human face so we can utilize some existing techniques, the cat head do have much larger intra-class variation than the human face, as shown in Figure 1 (b), thus is more difficult to detect.

Directly applying the existing face detection approaches to detect the cat head has apparent difficulties. First, the cat face has larger appearance variations compared with the human face. The textures on the cat face are more complicated than those on the human face. It requires more discriminative features to capture the texture information. Second, the cat head has a globally similar, but locally variant shape or silhouette. How to effectively make use of both texture and shape information is a new challenging issue. It requires a different detection strategy.

To deal with the new challenges, we propose a joint shape and texture detection approach and a set of new features based on oriented gradients. Our approach is a two step approach. In the first step, we individually train a shape detector and a texture detector to exploit the shape and appearance information respectively. Figure 2 illustrates our basic idea. Figure 2 (a) and Figure 2 (c) are two mean cat head images over all training images: one aligned by ears to make the shape distinct; the other is aligned to reveal the texture structures. Correspondingly, the shape and texture detectors are trained on two differently normalized training sets. Each detector can make full use of most discriminative shape or texture features separately. Based on a detailed study of previous image and gradient features, e.g., Haar [18], HoG [1], EOH [7], we show that a new set of

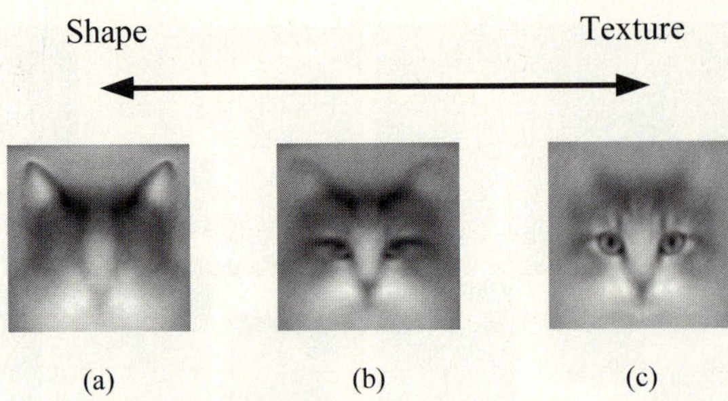

Fig. 2. Mean cat head images on all training data. (a) aligned by ears. More shape information is kept. (b) aligned by both eyes and ears using an optimal rotation+scale transformation. (c) aligned by eyes. More texture information is kept.

carefully designed Haar-like features on oriented gradients give the best performance in both shape and texture detectors.

In the second step, we train a joint shape and texture detector to fuse the outputs of the above two detectors. We experimentally demonstrate that the cat head detection performance can be substantially improved by carefully separating shape and texture information in the first step, and jointly training a fusion classifier in the second step.

1.1   Related Work

Since a comprehensive review of the related works on object detection is beyond the scope of the paper, we only review the most related works here.

Sliding window detection vs. parts based detection. To detect all possible objects in the image, two different searching strategies have been developed. The sliding window detection [14,12,18,1,17,15,20] sequentially scans all possible sub-windows in the image and makes a binary classification on each sub-window. Viola and Jones [18] presented the first highly accurate as well as real-time frontal face detector, where a cascade classifier is trained by AdaBoost algorithm on a set of Haar wavelet features. Dalal and Triggs [1] described an excellent human detection system through training a SVM classifier using HOG features. On the contrary, the parts based detection [5,13,9,6,3] detects multiple parts of the object and assembles the parts according to geometric constrains. For example, the human can be modeled as assemblies of parts [9,10] and the face can be detected using component detection [5].

In our work, we use two sliding windows to detect the "shape" part and "texture" part of the cat head. A fusion classifier is trained to produce the final decision.

Image features vs. gradient features. Low level features play a crucial role in the object detection. The image features are directly extracted from the image, such as intensity values [14], image patch [6], PCA coefficients [11], and wavelet coefficients [12,16,18]. Henry et al.[14] trained a neural network for human face detection using the image intensities in 20×20 sub-window. Haar wavelet features have become very popular since Viola and Jones [18] presented their real-time face detection system. The image features are suitable for small window and usually require a good photometric normalization. Contrarily, the gradient features are more robust to illumination changes. The gradient features are extracted from the edge map [4,3] or oriented gradients, which mainly include SIFT [8], EOH [7], HOG [1], covariance matrix[17], shapelet [15], and edgelet [19]. Tuzel et al. [17] demonstrated very good results on human detection using the covariance matrix of pixel's 1st and 2nd derivatives and pixel position as features. Shapelet [15] feature is a weighted combination of weak classifiers in a local region. It is trained specifically to distinguish between the two classes based on oriented gradients from the sub-window. We will give a detailed comparison of our proposed features with HOG and EOH features in Section 3.1.

2   Our Approach – Joint Shape and Texture Detection

The accuracy of a detector can be dramatically improved by first transforming the object into a canonical pose to reduce the variability. In face detection, all training samples are normalized by a rotation+scale transformation. The face is detected by scanning all sub-windows with different orientations and scales. Unfortunately, unlike the human face, the cat head cannot be well normalized by a rotation+scale transformation duo to the large intra-class variation.

In Figure 2, we show three mean cat head images over 5,000 training images by three normalization methods. In Figure 2 (a), we rotate and scale the cat head so that both eyes appear on a horizontal line and the distance between two ears is 36 pixels. As we can see, the shape or silhouette of the ears is visually distinct but the textures in the face region are blurred. In a similar way, we compute the mean image aligned by eyes, as shown in Figure 2 (c). The textures in the face region are visible but the shape of the head is blurred. In Figure 2 (b), we take a compromised method to compute an optimal rotation+scale transformation for both ears and eyes over the training data, in a least square sense. As expected, both ears and eyes are somewhat blurred.

Intuitively, using the optimal rotation+scale transformation may produce the best result because the image normalized by this method contains two kinds of information. However, the detector trained in this way does not show superior performance in our experiments. Both shape and texture information are lost to a certain degree. The discriminative power of shape features or texture features is hurt by this kind of compromised normalization.

2.1   Joint Shape and Texture Detection

In this paper, we propose a joint shape and texture detection approach to effectively exploit the shape and texture features. In the *training phase*, we train two individual detectors and a fusion classifier:

1. Train a shape detector, using the aligned training images by mainly keeping the shape information, as shown in Figure 2 (a); train a texture detector, using the aligned training image by mainly preserving the texture information, as shown in Figure 2 (c). Thus, each detector can capture most discriminative shape or texture features respectively.
2. Train a joint shape and texture fusion classifier to fuse the output of the shape and texture detectors.

In the *detection phase*, we first run the shape and texture detectors independently. Then, we apply the joint shape and texture fusion classifier to make the final decision. Specifically, we denote $\{c_s, c_t\}$ as output scores or confidences of the two detectors, and $\{f_s, f_t\}$ as extracted features in two detected sub-windows. The fusion classifier is trained on the concatenated features $\{c_s, c_t, f_s, f_t\}$.

Using two detectors, there are three kinds of detection results: both detectors report positive at roughly the same location, rotation, and scale; only the shape detector reports positive; and only the texture detector reports positive. For the first case, we directly construct the features $\{c_s, c_t, f_s, f_t\}$ for the joint fusion classifier. In the second case, we do not have $\{c_t, f_t\}$. To handle this problem, we scan the surrounding locations to pick a sub-window with the highest scores by the texture detector, as illustrated in Figure 3. Specifically, we denote the sub-window reported by the detector as $[x, y, w, h, s, \theta]$, where (x, y) is window's center, w, h are width and height, and s, θ are scale and rotation level. We search sub-windows for the texture/shape detector in the range $[x \pm w/4] \times [y \pm h/4] \times [s \pm 1] \times [\theta \pm 1]$. Note that we use real value score of the texture detector and do not make 0-1 decision. The score and features of the picked sub-window are used for the features $\{c_t, f_t\}$. For the last case, we compute $\{c_s, f_s\}$ in a similar way.

To train the fusion classifier, 2,000 cat head images in the validation set are used as the positive samples, and 4,000 negative samples are bootstrapped from 10,000 non-cat images. The positive samples are constructed as usual. The key is the construction of the negative samples which consist of all incorrectly detected samples by either the shape detector or the texture detector in the non-cat images. The co-occurrence relationship of the shape features and texture features are learned by this kind of joint training. The learned fusion classifier is able to effectively reject many false alarms by using both shape and texture information. We use support vector machine (SVM) as our fusion classifier and HOG descriptors as the representations of the features f_s and f_t.

The novelty of our approach is the discovery that we need to separate the shape and texture features and how to effectively separate them. The latter experimental results clearly validate the superiority of our joint shape and texture detection. Although the fusion method might be simple at a glance, this is exactly the strength of our approach: a simple fusion method already worked far better than previous non-fusion approaches.

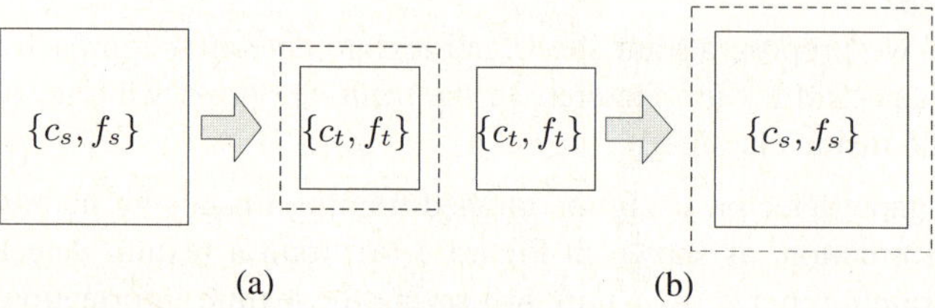

(a)   (b)

Fig. 3. Feature extraction for fusion. (a) given a detected sub-window (left) by the shape detector, we search a sub-window (right, solid line) with highest score by the texture detector in surrounding region (right, dashed line). The score and features $\{c_t, f_t\}$ are extracted for the fusion classifier. (b) similarly, we extract the score and features $\{c_s, f_s\}$ for the fusion.

3   Haar of Oriented Gradients

To effectively capture both shape and texture information, we propose a set of new features based on oriented gradients.

3.1   Oriented Gradients Features

Given the image I, the image gradient $\overrightarrow{g}(x) = \{g_h, g_v\}$ for the pixel x is computed as:

$$g_h(x) = G_h \otimes I(x), \quad g_v(x) = G_v \otimes I(x), \tag{1}$$

where G_h and G_v are horizontal and vertical filters, and $\otimes$ is convolution operator. A bank of oriented gradients $\{g_o^k\}_{k=1}^{K}$ are constructed by quantifying the gradient $\overrightarrow{g}(x)$ on a number of K orientation bins:

$$g_o^k(x) = \begin{cases} |\overrightarrow{g}(x)| & \theta(x) \in \text{bin}_k \\ 0 & \text{otherwise} \end{cases}, \tag{2}$$

where $\theta(x)$ is the orientation of the gradient $\overrightarrow{g}(x)$. We call the image g_o^k *oriented gradients channel*. Figure 4 shows the oriented gradients on a cat head image. In this example, we quantify the orientation into four directions. We also denote the sum of oriented gradients of a given rectangular region R as:

$$S^k(R) = \sum_{x \in R} g_o^k(x). \tag{3}$$

It can be very efficiently computed in a constant time using integral image technique [18].

Since the gradient information at an individual pixel is limited and sensitive to noise, most of previous works aggregate the gradient information in a rectangular region to form more informative, mid-level features. Here, we review two most successful features: HOG and EOH.

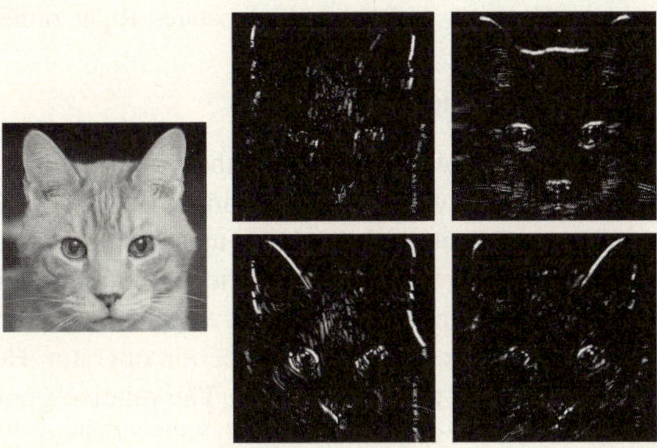

Fig. 4. Oriented gradients channels in four directions

HOG-cell. The basis unit in the HOG descriptor is the weighted orientation histogram of a "cell" which is a small spatial region, e.g., 8×8 pixels. It can be represented as:

$$\text{HOG-cell}(R) = [S^1(R), ..., S^k(R), ..., S^K(R)]. \tag{4}$$

The overlapped cells (e.g., 4×4) are grouped and normalized to form a larger spatial region called "block". The concatenated histograms form the HOG descriptor.

In Dalal and Triggs's human detection system [1], a linear SVM is used to classify a 64×128 detection window consisting of multiple overlapped 16×16 blocks. To achieve near real-time performance, Zhu et al. [21] used HOGs of variable-size blocks in the boosting framework .

EOH. Levi and Weiss [7] proposed three kinds of features on the oriented gradients:

$$\text{EOH}_1(R, k1, k2) = (S^{k_1}(R) + \epsilon)/(S^{k_2}(R) + \epsilon),$$

$$\text{EOH}_2(R, k) = (S^k(R) + \epsilon)/(\textstyle\sum_j (S^j(R) + \epsilon)),$$

$$\text{EOH}_3(R, \overline{R}, k) = (S^k(R) - S^k(\overline{R}))/sizeof(R),$$

where $\overline{R}$ is the symmetric region of R with respect to the vertical center of the detection window, and ϵ is a small value for smoothing. The first two features capture whether one direction is dominative or not, and the last feature is used to find symmetry or the absence of symmetry. Note that using EOH features only may be insufficient. In [7], good results are achieved by combining EOH features with Haar features on image intensity.

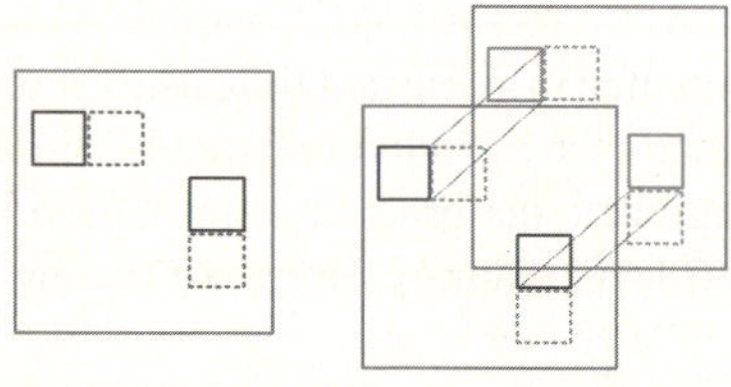

Fig. 5. Haar of Oriented Gradients. Left: in-channel features. Right: orthogonal features.

3.2   Our Features - Haar of Oriented Gradients

In face detection, the Haar features demonstrated their great ability to discover local patterns - intensity difference between two subregions. But it is difficult to find discriminative local patterns on the cat head which has more complex and subtle fine scale textures. On the contrary, the above oriented gradients features mainly consider the marginal statistics of gradients in a single region. It effectively captures fine scale texture orientation distribution by pixel level edge detection operator. However, it fails to capture local spatial patterns like the Haar feature. The relative gradient strength between neighboring regions is not captured either.

To capture both the fine scale texture and the local patterns, we need to develop a set of new features combining the advantage of both Haar and gradient features. Taking a close look at Figure 4, we may notice many local patterns in each oriented gradients channel which is sparser and clearer than the original image. We may consider that the gradient filter separates different orientation textures and pattern edges into several channels thus greatly simplified the pattern structure in each channel. Therefore, it is possible to extract Haar features from each channel to capture the local patterns. For example, in the horizontal gradient map in Figure 4, we see that the vertical textures between the two eyes are effectively filtered out so we can easily capture the two eye pattern using Haar features. Of course, in addition to capturing local patterns within a channel, we can also capture more local patterns across two different channels using Haar like operation. In this paper, we propose two kinds of features as follows:

In-channel features

$$\mathrm{HOOG}_1(R_1, R_2, k) = \frac{S^k(R_1) - S^k(R_2)}{S^k(R_1) + S^k(R_2)}. \tag{5}$$

These features measure the relative gradient strength between two regions R_1 and R_2 in the same orientation channel. The denominator plays a normalization role since we do not normalize $S^k(R)$.

Orthogonal-channel features

$$\mathrm{HOOG}_2(R1, R2, k, k^*) = \frac{S^k(R_1) - S^{k^*}(R_2)}{S^k(R_1) + S^{k^*}(R_2)}, \tag{6}$$

where k^* is the orthogonal orientation with respect to k, i.e., $k^* = k + K/2$. These features are similar to the in-channel features but operate on two orthogonal channels. In theory, we can define these features on any two orientations. But we decide to compute only the orthogonal-channel features based on two considerations: 1) orthogonal channels usually contain most complementary information. The information in two channels with similar orientations is mostly redundant; 2) we want to keep the size of feature pool small. The AbaBoost is a sequential, "greedy" algorithm for the feature selection. If the feature pool contains too many uninformative features, the overall performance may be hurt. In practice, all features have to be loaded into the main memory for efficient training. We must be very careful about enlarging the size of features.

Considering all combinations of R_1 and R_2 will be intractable. Based on the success of Haar features, we use Haar patterns for R_1 and R_2, as shown in Figure 5. We call the features defined in (5) and (6), Haar of Oriented Gradients (HOOG).

4   Experimental Results

4.1   Data Set and Evaluation Methodology

Our evaluation data set includes two parts, the first part is our own data, which includes 10,000 cat images mainly obtained from flickr.com; the second part is from PASCAL 2007 cat data, which includes 679 cat images. Most of our own cat data are near frontal view. Each cat head is manually labeled with 9 points, two for eyes, one for mouth, and six for ears, as shown in Figure 6. We randomly divide our own cat face images

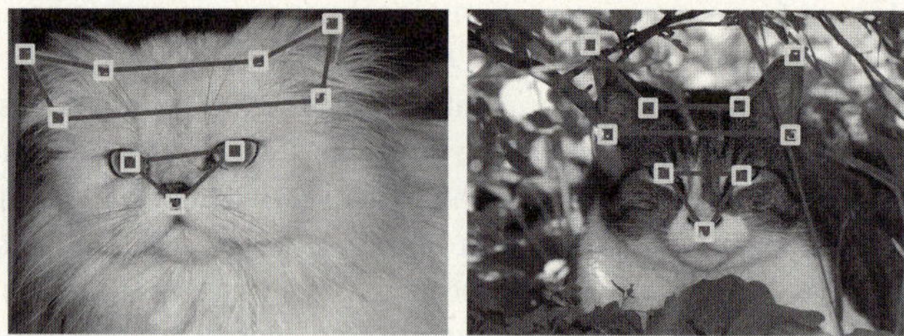

Fig. 6. The cat head image is manually labeled by 9 points

into three sets: 5,000 for training, 2000 for validation, and 3,000 for testing. We follow the PASCAL 2007 original separations of training, validation and testing set on the cat data. Our cat images can be downloaded from http://mmlab.ie.cuhk.edu.hk/ for research purposes.

We use the evaluation methodology similar to PASCAL challenge for object detection. Suppose the ground truth rectangle and the detected rectangle are r_g and r_d, and the area of those rectangles are A_g and A_d. We say we correctly detect a cat head only when the overlap of r_g and r_d is larger than 50%:

$$D(r_g, r_d) = \begin{cases} 1 & \text{if } \frac{(A_g \cap A_d)}{(A_g \cup A_d)} > 50\% , \\ 0 & \text{otherwise} \end{cases} , \tag{7}$$

where $D(r_g, r_d)$ is a function used to calculate detection rate and false alarm rate.

4.2   Implementation Details

Training samples. To train the shape detector, we align all cat head image with respect to ears. We rotate and scale the image so that two tips of ears appear on a horizontal line and the distance between two tips is 36 pixel. Then, we extract a 48×48 pixel region, centered 20 pixels below two tips. For the texture detector, a 32×32 pixel region is extracted. The distance between two eyes is 20 pixel. The region is centered 6 pixel below two eyes.

Features. We use 6 unsigned orientations to compute the oriented gradients features. We find the improvement is marginal when finer orientations are used. The horizontal and vertical filters are $[-1, 0, 1]$ and $[-1, 0, 1]^T$. No thresholding is applied on the computed gradients. For both shape and texture detector, we construct feature pools with 200,000 features by quantifying the size and location of the Haar templates.

4.3   Comparison of Features

First of all, we compare the proposed HOOG features with Haar, Haar + EOH, and HOG features on both shape detector and texture detector using our Flickr cat data set. For the Haar features, we use all four kinds of Haar templates. For the EOH features, we use default parameters suggested in [7]. For the HOG features, we use 4×4 cell size which produces the best results in our experiments.

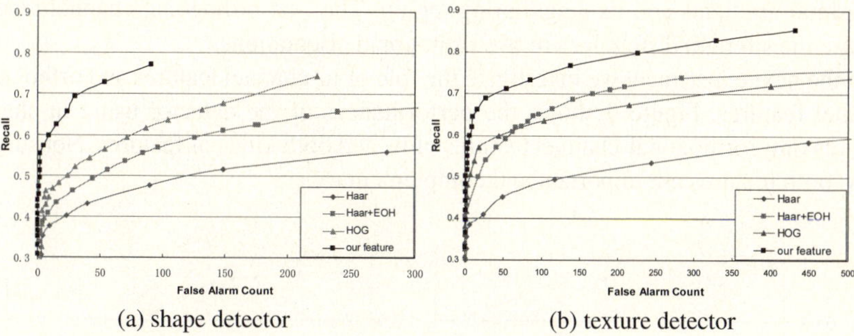

(a) shape detector   (b) texture detector

Fig. 7. Comparison of Haar, Haar+EOH, HOG, and our features

Figure 7 shows the performances of the four kinds of features. The Haar feature on intensity gives the poorest performance because of large shape and texture variations of the cat head. With the help of oriented gradient features, Haar + EOH improves the performance. As one can expect, the HOG features perform better on the shape detector than on the texture detector. Using both in-channel and orthogonal-channel information, the detectors based on our features produce the best results.

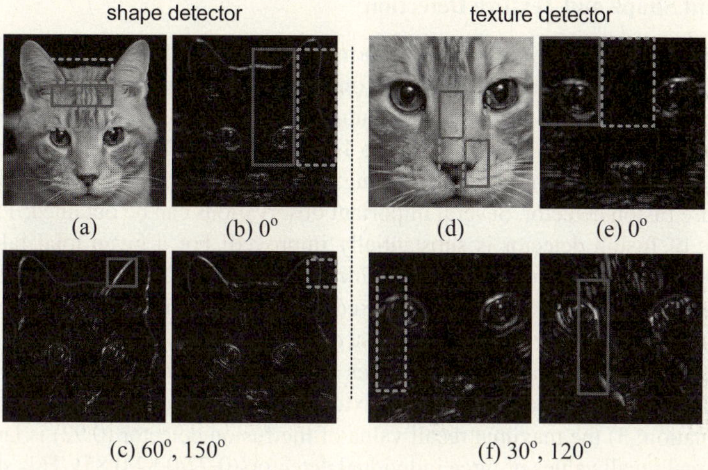

Fig. 8. Best features leaned by the AdaBoost. Left (shape detector): (a) best Haar feature on image intensity. (b) best in-channel feature. (c) best orthogonal feature on orientations 60° and 150°. Right (texture detector): (d) best Haar feature on image intensity. (e) best in-channel feature. (f) best orthogonal-channel feature on orientations 30° and 120°.

In Figure 8, we show the best in-channel features in (b) and (e), and the best orthogonal-channel features in (c) and (f), learned by two detectors. We also show the best Haar features on image intensity in Figure 8 (a) and (d). In both detectors, the best in-channel features capture the strength differences between a region with strongest horizontal gradients and its neighboring region. The best orthogonal-channel features capture the strength differences in two orthogonal orientations.

In the next experiment we investigate the role of in-channel features and orthogonal-channel features. Figure 9 shows the performances of the detector using in-channel features only, orthogonal-channel features only, and both kinds of features. Not surprisingly, both features are important and complementary.

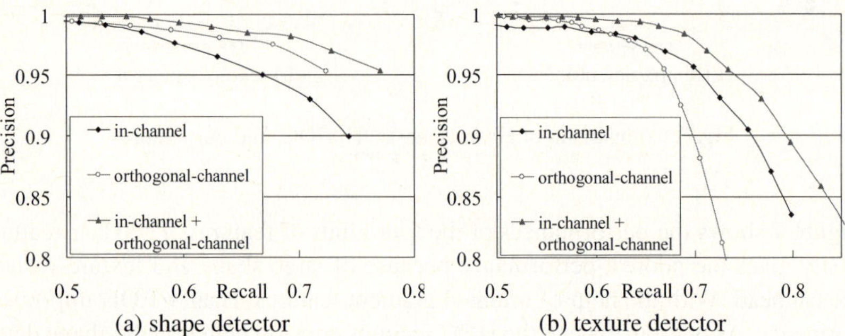

(a) shape detector   (b) texture detector

Fig. 9. The importance of in-channel features and orthogonal-channel features

4.4   Joint Shape and Texture Detection

In this sub-section, we evaluate the performance of the joint fusion on the Flickr cat data. To demonstrate the importance of decomposing shape and texture features, we also train a cat head detector using training samples aligned by an optimal rotation+scale transformation for the comparison. Figure 10 shows four ROC curves: a shape detector, a texture detector, a head detector using optimal transformation, and a joint shape and texture fusion detector. Several important observations can be obtained: 1) the performance of fusion detector is substantially improved! For a given total false alarm count 100, the recall is improved from 0.74/0.75/0.78 to 0.92. Or the total false alarm is reduced from 130/115/90 to 20, for a fixed recall 0.76. In image retrieval and search applications, it is a very nice property since high precision is preferred; 2) the head detector using optimal transformation does not show superior performance. The discriminative abilities of both shape and texture features are decreased by the optimal transformation; 3) the maximal recall value of the fusion detector (0.92) is larger than the maximal recall values of three individual detectors(0.77/0.82/0.85). This shows the complementary abilities of two detectors - one detector can find many cat heads which is difficult to the other detector; 4) note that the curve of fusion detector is very steep in the low false alarm region, which means the fusion detector can effectively improve the recall while maintain a very low false alarm rate.

The superior performance of our approach verifies a basic idea in object detection–context helps! The fusion detector finds surrounding evidence to verify the detection result. In our cat head detection, when the shape detector reports a cat, the fusion detector checks the surrounding shape information. If the texture detector says it may be a cat, we increase the probability to accept this cat. Otherwise, we decrease the probability to reject this cat.

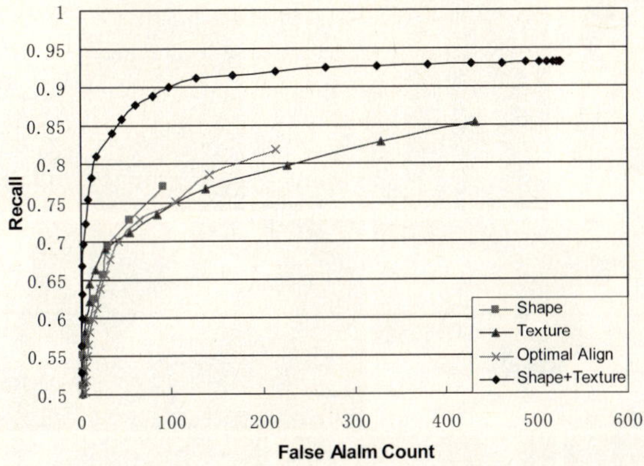

Fig. 10. Joint shape and texture detection

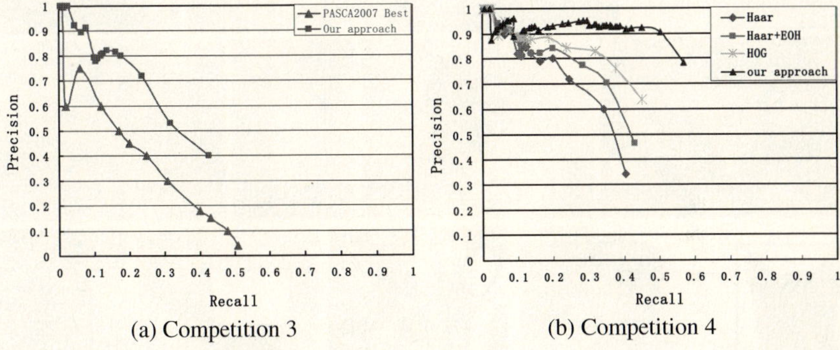

(a) Competition 3   (b) Competition 4

Fig. 11. Experiments on PASCAL 2007 cat data. (a) our approach and best reported method on Competition 3 (specified training data). (b) four detectors on Competition 4 (arbitrary training data).

Figure 12 gives some detection examples having variable appearance, head shape, illumination, and pose.

4.5   Experiment on the PASCAL 2007 Cat Data

We also evaluate the proposed approach on the PASCAL 2007 cat data [2]. There are two kinds of competitions for the detection task: 1) Competition 3 - using both training and testing data from PASCAL 2007; 2) Competition 4 - using arbitrary training data. Figure 11 (a) shows the precision-recall curves of our approach and the best reported method [2] on Competition 3. We compute the Average Precision (AP) as in [2] for a convenient comparison. The APs of our approach and the best reported method is 0.364 and 0.24, respectively. Figure 11(b) shows the precision-recall curves on Competition 4. Since there is no reported result on Competition 4, we compare our approach with the detectors using Haar, EOH, and HoG respectively. All detectors are trained on the

Fig. 12. Detection results. The bottom row shows some detected cats in PASCAL 2007 data.

same training data. The APs of four detectors (ours, HOG, Haar+EOH, Harr) are 0.632, 0.427, 0.401, and 0.357. Using larger training data, the detection performance is significantly improved. For example, the precision is improved from 0.40 to 0.91 for a fixed recall 0.4. Note that the PASCAL 2007 cat data treat the whole cat body as the object and only small fraction of the data contain near frontal cat face. However, our approach still achieves reasonable good results (AP=0.632) on this very challenging data (the best reported method's AP=0.24).

5   Conclusion and Discussion

In this paper, we have presented a cat head detection system. We achieved excellent results by decomposing texture and shape features firstly and fusing detection results secondly. The texture and shape detectors also greatly benefit from a set of new oriented gradient features. Although we focus on the cat head detection problem in this paper, our approach can be extended to detect other categories of animals. In the future, we are planing to extend our approach to multi-view cat head detection and more animal categories. We are also interest in exploiting other contextual information, such as the presence of animal body, to further improve the performance.

References

1. Dalal, N., Triggs, B.: Histograms of oriented gradients for human detection. In: CVPR, vol. 1, pp. 886–893 (2005)
2. Everingham, M., van Gool, L., Williams, C., Winn, J., Zisserman, A.: The PASCAL Visual Object Classes Challenge (VOC 2007) Results (2007),
   http://www.pascal-network.org/challenges/VOC/voc2007/
   workshop/index.html
3. Felzenszwalb, P.F.: Learning models for object recognition. In: CVPR, vol. 1, pp. 1056–1062 (2001)
4. Gavrila, D.M., Philomin, V.: Real-time object detection for smart vehicles. In: CVPR, vol. 1, pp. 87–93 (1999)
5. Heisele, B., Serre, T., Pontil, M., Poggio, T.: Component-based face detection. In: CVPR, vol. 1, pp. 657–662 (2001)
6. Leibe, B., Seemann, E., Schiele, B.: Pedestrian detection in crowded scenes. In: CVPR, vol. 1, pp. 878–885 (2005)
7. Levi, K., Weiss, Y.: Learning object detection from a small number of examples: the importance of good features. In: CVPR, vol. 2, pp. 53–60 (2004)
8. Lowe, D.G.: Object recognition from local scale-invariant features. In: ICCV, vol. 2, pp. 1150–1157 (1999)
9. Mikolajczyk, K., Schmid, C., Zisserman, A.: Human detection based on a probabilistic assembly of robust part detectors. In: Pajdla, T., Matas, J(G.) (eds.) ECCV 2004. LNCS, vol. 3021, pp. 69–82. Springer, Heidelberg (2004)
10. Mohan, A., Papageorgiou, C., Poggio, T.: Example-based object detection in images by components. IEEE Trans. Pattern Anal. Machine Intell. 23(4), 349–361 (2001)
11. Munder, S., Gavrila, D.M.: An experimental study on pedestrian classification. IEEE Trans. Pattern Anal. Machine Intell. 28(11), 1863–1868 (2006)
12. Papageorgiou, C., Poggio, T.: A trainable system for object detection. Intl. Journal of Computer Vision 38(1), 15–33 (2000)
13. Ronfard, R., Schmid, C., Triggs, B.: Learning to parse pictures of people. In: ECCV, vol. 4, pp. 700–714 (2004)
14. Rowley, H.A., Baluja, S., Kanade, T.: Neural network-based face detection. IEEE Trans. Pattern Anal. Machine Intell. 20(1), 23–38 (1998)
15. Sabzmeydani, P., Mori, G.: Detecting pedestrians by learning shapelet features. In: CVPR (2007)
16. Schneiderman, H., Kanade, T.: A statistical method for 3d object detection applied to faces and cars. In: CVPR, vol. 1, pp. 746–751 (2000)
17. Tuzel, O., Porikli, F., Meer, P.: Human detection via classification on riemannian manifolds. In: CVPR (2007)
18. Viola, P., Jones, M.J.: Robust real-time face detection. Intl. Journal of Computer Vision 57(2), 137–154 (2004)

19. Wu, B., Nevatia, R.: Detection of multiple, partially occluded humans in a single image by bayesian combination of edgelet part detectors. In: ICCV, vol. 1, pp. 90–97 (2005)
20. Xiao, R., Zhu, H., Sun, H., Tang, X.: Dynamic cascades for face detection. In: ICCV, vol. 1, pp. 1–8 (2007)
21. Zhu, Q., Avidan, S., Yeh, M.-C., Cheng, K.-T.: Fast human detection using a cascade of histograms of oriented gradients. In: CVPR, vol. 2, pp. 1491–1498 (2006)

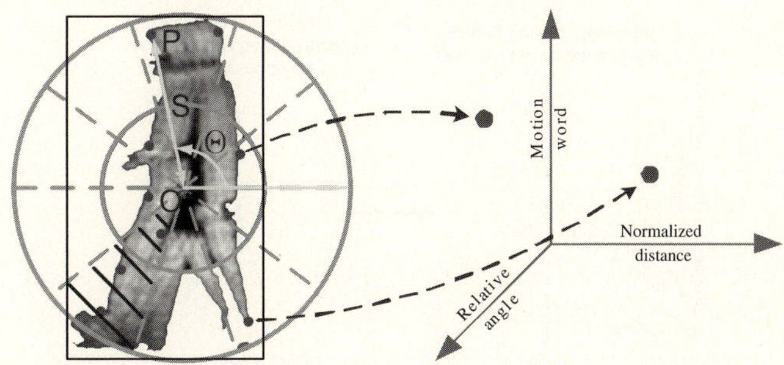

Fig. 4. Illustration of our MC representation (left) and its 3D descriptor (right). On the left, P denotes a MW at an interesting point, O denotes the reference point, Θ and S denote the relative angle and normalized distance between P and O in the support region (the black rectangle), respectively, and the shaded sector (blue) denotes the orientation of the whole representation. On the right, each MW is quantized into a point to generate a 3D MC descriptor. This figure is best viewed in color.

is inspired by Shape Context (SC) [20], which has been widely used in object recognition. The basic idea of SC is to locate the distribution of other shape points over relative positions in a region around a pre-defined reference point. Subsequently, 1D descriptors are generated to represent the shapes of objects.

In our representation, we utilize the polar coordinate system to capture the relative angles and distances between the MWs and the reference point (the pole of the polar coordinate system) for each action in the MIs, similar to SC. This reference point is defined as the geometric center of the human motion, and the relative distances are normalized by the maximum distance in the support region, which makes the MC insensitive to changes in scale of the action. Here, the support region is defined as the area which covers the human action in the MI. Fig.4 (left) illustrates our MC representation. Suppose that the angular coordinate is divided into M equal bins, the radial coordinate is divided into N equal bins and there are K MWs in the dictionary, then each MW can be put into one of the $M*N$ bins to generate a 3D MC descriptor for each MC representation, as illustrated in Fig.4 (right). To represent a human action in each video sequence, we sum up all the MC descriptors of this action to generate one 3D descriptor with the same dimensions.

When generating MC representations, another factor should also be considered, that is, the direction of the action, because the same action may occur in different directions. E.g. a person may be running in one direction or the opposite direction. In such cases, the distributions of the interesting points in the two corresponding MIs should be roughly symmetric about the y-axis. Combining the two distributions for the same action will reduce the discriminability of our representation. To avoid this, we define the orientation of each MC representation as the sector where most interesting points are detected, e.g. the shaded one (blue) in Fig. 4 (left). This sector can be considered to represent the main characteristics of the motion in one direction. For the same action but in the

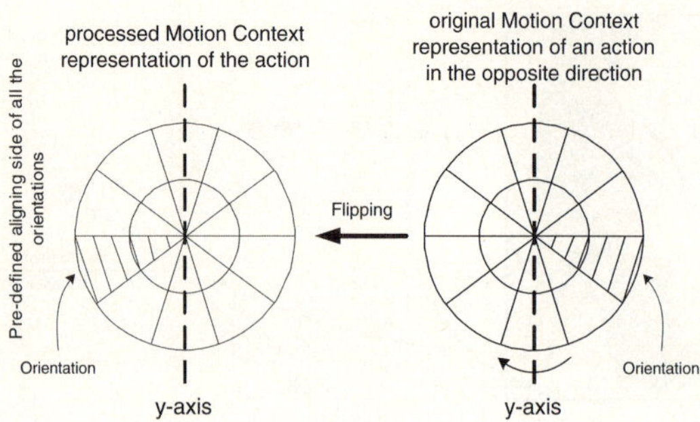

Fig. 5. Illustration of aligning an inconsistent MC representation of an action in the opposite direction. The pre-defined orientation of the actions is the left side of y-axis.

opposite direction, we then align all the orientations to the pre-defined side by flipping the MC representations horizontally around the y-axis. Thus our representation is symmetry-invariant. Fig.5 illustrates this process. Notice that this process is done automatically without the need to know the action direction.

The entire process of modeling human actions using the MC representation is summarized in Table 1.

Table 1. The main steps of modeling the human actions using the MC representation

---

Step 1 Obtain the MIs from the video sequences.
Step 2 Generate the MC representation for each human action in the MIs.
Step 3 Generate the 3D MC descriptor for each MC representation.
Step 4 Sum up all the 3D MC descriptors of an action to generate one 3D descriptor
to represent this action.

---

4   Action Recognition Approaches

We apply 3 different approaches to recognize the human actions based on the MWs or the 3D MC descriptors: pLSA, w^3-pLSA and SVM.

4.1   pLSA

pLSA aims to introduce an aspect model, which builds an association between documents and words through the latent aspects by probability. Here, we follow the terminology of text classification where pLSA was used first. The graphical model of pLSA is illustrated in Fig.6 (a).

Suppose $D = \{d_1, \ldots, d_I\}$, $W = \{w_1, \ldots, w_J\}$ and $Z = \{z_1, \ldots, z_K\}$ denote a document set, a word set and a latent topic set, respectively. pLSA models the joint probability of documents and words as:

$$P(d_i, w_j) = \sum_k P(d_i, w_j, z_k) = \sum_k P(w_j|z_k)P(z_k|d_i)P(d_i) \qquad (1)$$

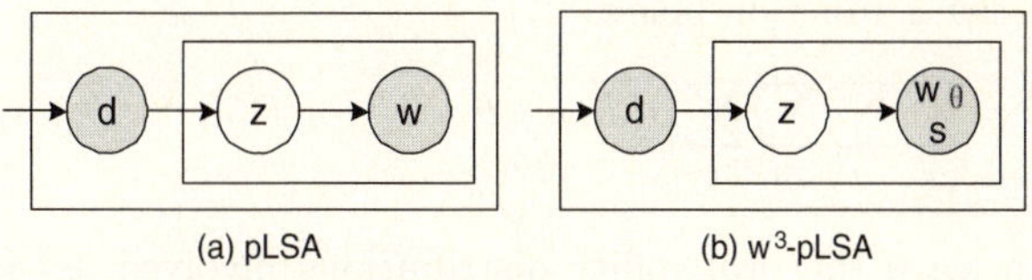

Fig. 6. Graphical models of pLSA (a) and our w^3-pLSA (b)

where $P(d_i, w_j, z_k)$ denotes the joint probability of document d_i, topic z_k and word w_j, $P(w_j|z_k)$ denotes the probability of w_j occurring in z_k, $P(z_k|d_i)$ denotes the probability of d_i classified into z_k, and $P(d_i)$ denotes the prior probability of d_i modeled as a multinomial distribution.

Furthermore, pLSA tries to maximize the $\mathcal{L}$ function below:

$$\mathcal{L} = \sum_i \sum_j n(d_i, w_j) \log P(d_i, w_j) \tag{2}$$

where $n(d_i, w_j)$ denotes the document-word co-occurrence table, where the number of co-occurrences of d_i and w_j is recorded in each cell.

To learn the probability distributions involved, pLSA employs the Expectation Maximization (EM) algorithm shown in Table 2 and records $P(w_j|z_k)$ for recognition, which is learned from the training data.

Table 2. The EM algorithm for pLSA

E-step:
$$P(z_k|d_i, w_j) \propto P(w_j|z_k)P(z_k|d_i)P(d_i)$$
M-step:
$$P(w_j|z_k) \propto \sum_i n(d_i, w_j)P(z_k|d_i, w_j)$$
$$P(z_k|d_i) \propto \sum_j n(d_i, w_j)P(z_k|d_i, w_j)$$
$$P(d_i) \propto \sum_j n(d_i, w_j)$$

4.2 w^3-pLSA

To bridge the gap between the human actions and our MC descriptors, we extend pLSA to develop a new graphical model, called w^3-pLSA. See Fig.6 (b), where d denotes human actions, z denotes latent topics, w, θ and s denote motion words, and the indexes in the angular and radial coordinates in the polar coordinate system, respectively.

Referring to pLSA, we model the joint probability of human actions, motion words and their corresponding indices in the angular and radial coordinates as

$$P(d_i, w_j, \theta_m, s_r) = \sum_k P(d_i, w_j, \theta_m, s_r, z_k) = \sum_k P(d_i)P(z_k|d_i)P(w_j, \theta_m, s_r|z_k) \tag{3}$$

and maximize the $\widehat{\mathcal{L}}$ function below.

$$\widehat{\mathcal{L}} = \sum_i \sum_j \sum_m \sum_r n(d_i, w_j, \theta_m, s_r) \log P(d_i, w_j, \theta_m, s_r) \tag{4}$$

Similarly, to learn the probability distributions involved, w^3-pLSA employs the Expectation Maximization (EM) algorithm shown in Table 3 and records $P(w_j, \theta_m, s_r | z_k)$ for recognition, which is learned from the training data.

Table 3. The EM algorithm for w^3-pLSA

E-step:
$$P(z_k | d_i, w_j, \theta_m, s_r) \propto P(w_j, \theta_m, s_r | z_k) P(z_k | d_i) P(d_i)$$
M-step:
$$P(w_j, \theta_m, s_r | z_k) \propto \sum_i n(d_i, w_j, \theta_m, s_r) P(z_k | d_i, w_j, \theta_m, s_r)$$
$$P(z_k | d_i) \propto \sum_{j,m,r} n(d_i, w_j, \theta_m, s_r) P(z_k | d_i, w_j, \theta_m, s_r)$$
$$P(d_i) \propto \sum_{j,m,r} n(d_i, w_j, \theta_m, s_r)$$

4.3   Support Vector Machine

A support vector machine (SVM) [8] is a powerful tool for binary classification tasks. First it maps the input vectors into a higher dimensional feature space, then it conducts a separating hyperplane to separate the input data, finally on each side of this hyperplane two parallel hyperplanes are conducted. SVM tries to find the separating hyperplane which maximizes the distance between the two parallel hyperplanes. Notice that in a SVM, there is an assumption that the larger the distance between the two parallel hyperplanes the smaller the generalization error of the classifier will be.

Specifically, suppose the input data is $\{(\mathbf{x_1}, y_1), (\mathbf{x_2}, y_2), \cdots, (\mathbf{x_n}, y_n)\}$ where $\mathbf{x_i}(i = 1, 2, \cdots, n)$ denotes the input vector and the corresponding $y_i(i = 1, 2, \cdots, n)$ denotes the class label (positive "1" and negative "-1"). Then the separating hyperplane is defined as $\mathbf{w} \cdot \mathbf{x} + b = 0$ and the two corresponding parallel hyperplanes are $\mathbf{w} \cdot \mathbf{x} + b = 1$ for the positive class and $\mathbf{w} \cdot \mathbf{x} + b = -1$ for the negative class, where $\mathbf{w}$ is the vector perpendicular to the separating hyperplane and b is a scalar. If a test vector $\mathbf{x_t}$ satisfies $\mathbf{w} \cdot \mathbf{x_t} + b > 0$, it will be classified as a positive instance. Otherwise, if it satisfies $\mathbf{w} \cdot \mathbf{x_t} + b < 0$, it will be classified as a negative instance. A SVM tries to find the optimal $\mathbf{w}$ and b to maximize the distance between the two parallel hyperplanes.

5   Experiments

Our approach has been tested on two human action video datasets from KTH [2] and Weizmann Institute of Science (WIS) [9]. The KTH dataset is one of the largest datasets for human action recognition containing six types of human actions: boxing, handclapping, handwaving, jogging, running, and walking. For each type, there are 99 or 100 video sequences of 25 different persons in 4 different scenarios: outdoors (S1), outdoors with scale variation (S2), outdoors with different clothes (S3) and indoors (S4), as illustrated in Fig.7 (left). In the WIS dataset, there are altogether 10 types of human actions: walk, run, jump, gallop sideways, bend, one-hand wave, two-hands wave, jump in place, jumping jack, and skip. For each type, there are 9 or 10 video sequences of 9 different persons with the similar background, as shown in Fig.7 (right).

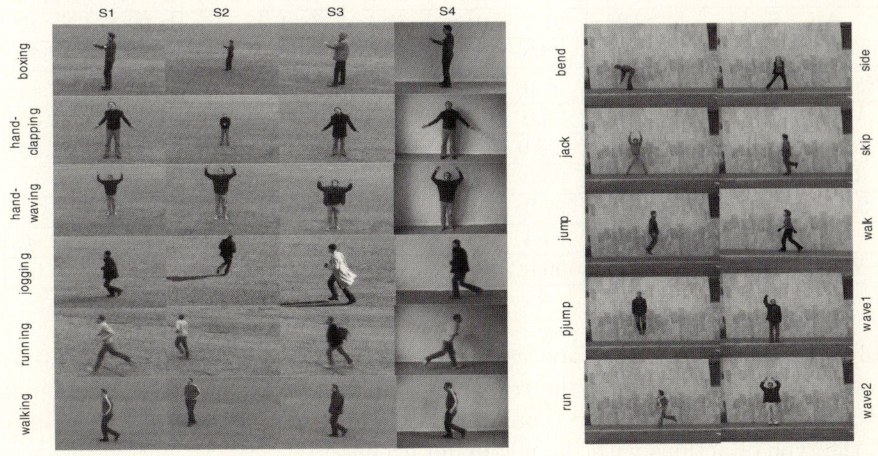

Fig. 7. Some sample frames from the KTH dataset (left) and the WIS dataset (right)

5.1   Implementation

To generate MC representations for human actions, we need to locate the reference points and the support regions first. Some techniques in body tracking (e.g. [21]) can be applied to locate the areas and the geometric centers of the human bodies in each frame group of a video sequence. The integration of the areas of a person can be defined as its support region and the mean of its centers can be defined as the reference point for this action in the MI. However, this issue is beyond the purpose of this paper. So considering that in our datasets each video sequence only contains one person, we simply assume that in each MI the support region of each human action covers the whole MI, and we adopted a simple method to roughly locate the reference points. First, we generated one MI from every 5-frame group of each video sequence empirically. Then a Gaussian filter was applied to denoise these MIs so that the motion information from the background was suppressed. Next, we used the Canny edge detector to locate the edges in each MI, and finally took the geometric center of the edge points as the reference point for the action.

After locating the reference points, we followed the steps in Table 1 to generate the MC representations for human actions. The detector and descriptor involved in Step 2 are the Harris-Hessian-Laplace detector [22] and the SIFT descriptor

Table 4. Comparison (%) between our approach and others on the KTH dataset

Rec.Con.	Tra.Str.	boxing	hand-c	hand-w	jogging	running	walking	average
MW+pLSA	SDE	85.2	91.9	91.7	71.2	73.6	82.1	82.62
	LOO	82.0	90.9	91.0	82.0	79.0	83.0	84.65
MW+SVM	SDE	90.4	84.8	82.8	65.1	76.1	82.0	80.20
	LOO	85.0	82.8	82.0	62.0	70.0	87.0	78.14
MC+w^3-pLSA	SDE	98.4	90.8	93.9	79.3	77.9	91.7	88.67
	LOO	95.0	97.0	93.0	88.0	84.0	91.0	***91.33***
MC+SVM	SDE	91.7	91.6	88.1	78.0	84.7	90.4	87.42
	LOO	88.0	93.9	91.0	77.0	85.0	90.0	87.49
Savarese et al. [14]	LOO	97.0	91.0	93.0	64.0	83.0	93.0	86.83
Wang et al. [10]	LOO	96.0	97.0	100.0	54.0	64.0	99.0	85.00
Niebles et al. [19]	LOO	100.0	77.0	93.0	52.0	88.0	79.0	81.50
Dollár et al. [3]	LOO	93.0	77.0	85.0	57.0	85.0	90.0	81.17
Schuldt et al. [2]	SDE	97.9	59.7	73.6	60.4	54.9	83.8	71.72
Ke et al. [24]	SDE	69.4	55.6	91.7	36.1	44.4	80.6	62.96
Wong et al. [25]	SDE	96.0	92.0	83.0	79.0	54.0	100.0	84.00

[13], and the clustering method used here is K-means. Then based on the MWs
and the MC descriptors of the training data, we trained pLSA, w^3-pLSA and
SVM for each type of actions separately, and a test video sequence was classified
to the type of actions with the maximum likelihood.

5.2   Experimental Results

To show the efficiency of our MC representation and the discriminability of the
MWs, we designed 4 different recognition configurations: MW+pLSA, MW+SVM,
MC+w^3-pLSA, and MC+SVM. Here we used libsvm [23] with the linear kernel. To
utilize the MWs, we employed the BOW model to represent each human action as
a histogram of the MWs without the $M*N$ spatial bins.

First, we tested our approach on the KTH dataset. We adopted two different
training strategies: split-data-equally (SDE) and leave-one-out (LOO). The SDE
strategy means that the video collection is divided into two equal sets randomly:
one as the training data (50 video sequences) and the other as the test data for
each type of actions, and we repeated this experiment for 15 times. In the LOO
strategy, for each type of actions, only the video sequences of one person are
selected as the test data and the rest as the training data, and when applying
this strategy to the KTH dataset, for each run we randomly selected one person
for each type of actions as the test data and repeated this experiment for 15
times. Empirically, in our model, the number of MWs is 100, and the numbers
of the quantization bins in the angular and radial dimensions are 10 and 2,
respectively. The number of latent topics in both graphical models is 40.

Table 4 shows our average recognition rate for each type of actions and the
comparison with others on the KTH dataset under different training strategies and recognition configurations. From this table, we can draw the following

Table 5. Comparison (%) between our approach and others on the WIS dataset. Notice that "✗" denotes that this type of actions was not involved in their experiments.

Rec.Con.	bend	jack	jump	pjump	run	side	skip	walk	wave1	wave2	ave.
MW+pLSA	77.8	100.0	88.9	88.9	70.0	100.0	60.0	100.0	66.7	88.9	84.1
MW+SVM	100.0	100.0	100.0	77.8	30.0	77.8	40.0	100.0	100.0	100.0	81.44
MC+w^3-pLSA	66.7	100.0	77.8	66.7	80.0	88.9	100.0	100.0	100.0	100.0	88.0
MC+SVM	100.0	100.0	100.0	88.9	80.0	100.0	80.0	80.0	100.0	100.0	*92.89*
Wang et al. [16]	100.0	100.0	89.0	100.0	100.0	100.0	89.0	100.0	89.0	100.0	96.7
Ali et al. [26]	100.0	100.0	55.6	100.0	88.9	88.9	✗	100.0	100.0	100.0	92.6
Scovanner [4]	100.0	100.0	67.0	100.0	80.0	100.0	50.0	89.0	78.0	78.0	84.2
Niebles et al. [6]	100.0	100.0	100.0	44.0	67.0	78.0	✗	56.0	56.0	56.0	72.8

conclusions: (1) MWs without any spatial information are not discriminative enough to recognize the actions. MW+pLSA returns the best performance (84.65%) using MWs, which is lower than the state of the art. (2) MC representation usually achieves better performances than MWs, which demonstrates that the distributions of the MWs are quite important for action recognition. MC+w^3-pLSA returns the best performance (91.33%) among all the approaches.

Unlike the KTH dataset, the WIS dataset only has 9 or 10 videos for each type of human actions, which may result in underfit when training the graphical models. To utilize this dataset sufficiently, we only used the LOO training strategy to learn the models for human actions and tested on all the video sequences. We compare our average recognition rates with others in Table 5. The experimental configuration of the MC representation is kept the same as that used on the KTH dataset, while the number of MWs used in the BOW model is modified empirically to 300. The number of latent topics is unchanged. From this table, we can see that MC+SVM still returns the best performance (92.89%) among the different configurations, which is comparable to other approaches and higher than the best performance (84.1%) using MW. These results demonstrate that our MC presentation can model the human actions properly with the distributions of the MWs.

6   Conclusion

We have demonstrated that our ***Motion Context*** (MC) representation, which is insensitive to changes in the scales and directions of the human actions, can model the human actions in the *motion images* (MIs) effectively by capturing the distribution of the *motion words* (MWs) over relative locations in a local region around the reference point and thus summarize the local motion information in a rich 3D descriptor. To evaluate this novel representation, we adopt two training strategies (split-data-equally (SDE) and leave-one-out (LOO)), design 4 different recognition configurations (MW+pLSA, MW+SVM, MC+w^3-pLSA, and MC+SVM) and test them on two human action video datasets from KTH and Weizmann Institute of Science (WIS). The performances are promising. For the KTH dataset, all configurations using MC outperform existing approaches where the best performances are obtained using w^3-pLSA (88.67% for SDE and 91.33% for LOO). For the WIS dataset, our MC+SVM returns the comparable performance (92.89%) using the LOO strategy.

References

1. Laptev, I., Lindeberg, T.: Space-time interest points. In: ICCV (2003)
2. Schuldt, C., Laptev, I., Caputo, B.: Recognizing human actions: a local svm approach. In: ICPR 2004, vol. III, pp. 32–36 (2004)
3. Dollár, P., Rabaud, V., Cottrell, G., Belongie, S.: Behavior recognition via sparse spatio-temporal features. In: VS-PETS (October 2005)
4. Scovanner, P., Ali, S., Shah, M.: A 3-dimensional sift descriptor and its application to action recognition. ACM Multimedia, 357–360 (2007)
5. Wang, Y., Loe, K.F., Tan, T.L., Wu, J.K.: Spatiotemporal video segmentation based on graphical models. Trans. IP 14, 937–947 (2005)
6. Niebles, J., Fei Fei, L.: A hierarchical model of shape and appearance for human action classification. In: CVPR 2007, pp. 1–8 (2007)
7. Hofmann, T.: Unsupervised learning by probabilistic latent semantic analysis. In: Mach. Learn., Hingham, MA, USA, vol. 42, pp. 177–196. Kluwer Academic Publishers, Dordrecht (2001)
8. Burges, C.J.C.: A tutorial on support vector machines for pattern recognition. In: Data Mining and Knowledge Discovery, vol. 2, pp. 121–167 (1998)
9. Blank, M., Gorelick, L., Shechtman, E., Irani, M., Basri, R.: Actions as space-time shapes. In: ICCV 2005, vol. II, pp. 1395–1402 (2005)
10. Wang, Y., Sabzmeydani, P., Mori, G.: Semi-latent dirichlet allocation: A hierarchical model for human action recognition. In: HUMO 2007, pp. 240–254 (2007)
11. Ikizler, N., Duygulu, P.: Human action recognition using distribution of oriented rectangular patches. In: HUMO 2007, pp. 271–284 (2007)
12. Efros, A., Berg, A., Mori, G., Malik, J.: Recognizing action at a distance. In: ICCV 2003, pp. 726–733 (2003)
13. Lowe, D.: Distinctive image features from scale-invariant keypoints. International Journal of Computer Vision 20, 91–110 (2003)
14. Savarese, S., Sel Pozo, A., Fei-Fei, J.N.L.: Spatial-temporal correlations for unsupervised action classification. In: IEEE Workshop on Motion and Video Computing, Copper Mountain, Colorado (2008)
15. Wang, Y., Tan, T., Loe, K.: Video segmentation based on graphical models. In: CVPR 2003, vol. II, pp. 335–342 (2003)
16. Wang, L., Suter, D.: Informative shape representations for human action recognition. In: ICPR 2006, vol. II, pp. 1266–1269 (2006)
17. Weinland, D., Ronfard, R., Boyer, E.: Free viewpoint action recognition using motion history volumes. Computer Vision and Image Understanding 104 (November/December 2006)
18. Bobick, A., Davis, J.: The recognition of human movement using temporal templates. PAMI 23(3), 257–267 (2001)
19. Niebles, J., Wang, H., Wang, H., Fei Fei, L.: Unsupervised learning of human action categories using spatial-temporal words. In: BMVC 2006, vol. III, p. 1249 (2006)

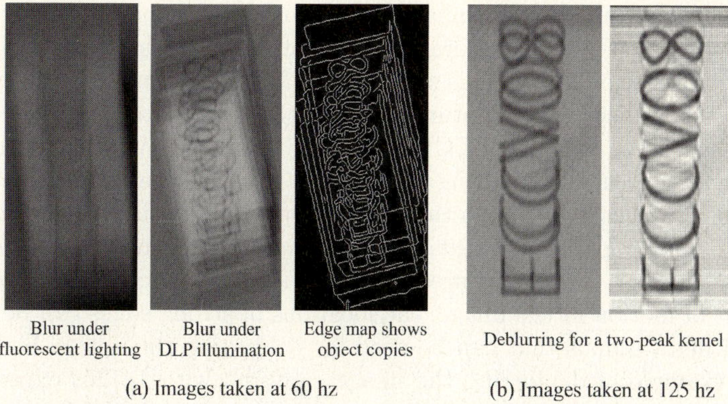

Fig. 9. Motion blurring under DLP illumination and fluorescent illumination: The scene consists of a heavy brick falling rapidly and an image is captured with exposures 1/60s (a) and 1/125s (b). Under fluorescent illumination, the motion blur appears as a smear across the image losing high frequencies. The temporal dithering in DLP projectors acts as a high frequency modulator that convolves with the moving object. The motion-blurred image still preserves some of the high spatial frequencies. Six copies of the text "ECCV08" in (a) and 2 copies in (b) are clearly visible.

component for each ball looks like the shading on a sphere (with dark edges) and the indirect component includes the interreflections between the balls (notice the bright edges). For the hand, the direct component is only due to reflection by the oils near the skin surface and is dark. The indirect component includes the effect of subsurface scattering and dominates the intensity. The checker pattern "flips" once in approximately 1/100s and hence we achieve separation at 100Hz. Due to finite resolution of the camera and the narrow depth of field of the projector, a 1-pixel blur is seen at the edges of the checker pattern. This results in the grid artifacts seen in the results.

6   Flutter Flash: Motion-Blur under DLP Illumination

Motion-blur occurs when the scene moves more than a pixel within the integration time of a camera. The blur is computed as the convolution of the scene motion with a box filter of width equal to the camera integration time. Thus, images captured of fast moving objects cause a smear across the pixels losing significant high frequencies. Deblurring images is a challenging task that many works have addressed with limited success. A recent approach by Raskar et al. [21] uses an electronically controlled shutter in front of the camera to modulate the incoming irradiance at speeds far greater than the motion of the object. In other words, the box filter is replaced by a series of short pulses of different widths. The new convolution between the object motion and the series of short pulses results in images that preserve more high frequencies as compared to the box filter. This "Flutter Shutter" approach helps in making the problem better conditioned. Our approach is similar in spirit to [21] with one difference: the fast shutter is simulated by the temporal dithering of the DLP illumination. Note that the DLP illumination dithering is significantly faster than mechanical shutters[1].

Figure 9 shows the images captured by with 1/60s exposure. The scene consists of a brick with the writing "ECCV08" falling vertically. When illuminated by a fluorescent source, the resulting motion-blur appears like a smear across the image. On the other hand, when the scene is illuminated using a DLP projector, we see 6 distinct copies of the text that are translated downward. A Canny edge detector is applied to the captured image to illustrate the copies. If we knew the extent of motion in the image, the locations of strong edges can be used as a train of delta signals that can be used for deblurring the image. In 9(b), we show an example of deblurring the image captured with 1/125s exposure. As in the deblurred images obtained the flutter shutter case, the DLP illumination preserves more high frequencies in the motion-blurred image.

7   Discussion

Speed vs. accuracy trade-off. One limitation of our approach is the requirement of a high speed camera. The acquisition speed of the camera and the effective speed of performance achieved depend on the task at hand and the signal-to-noise ratio of the captured images. For instance, the decision to use 10 frames for demultiplexing illumination or photometric stereo, or to use 20 frames for structured light, was mainly influenced by the noise characteristics of the camera. A more scientific exploration of this trade-off is required to better understand the benefits of our approach to each technique. A future avenue of research is to design 2D spatial intensity patterns that create temporal dithering codes that are optimal for the task at hand.

Issues in reverse engineering. The images shown in Figure 1 are dark for the input brightness range of 0 to 90. Despite the claim from manufacturers that the projector displays 8-bits of information, only about 160 patterns are usable for our experiments. To compensate for this, the projector performs spatial dithering in addition to temporal dithering in a few pixel blocks. This is an almost random effect that is not possible to reverse engineer without proprietary information from the manufacturers. We simply average a small neighborhood or discard such neighborhoods from our processing.

Other active vision techniques and illumination modulations. We believe that the temporal illumination dithering can be applied to a broader range of methods including pixel-wise optical flow estimation and tracking, projector defocus compensation and depth from defocus [12] and spectral de-multiplexing. While we exploit the temporal dithering already built-in to the projector, we do not have a way of controlling it explicitly. Better control is obtained by using a more expensive and special high speed MULE projector [17]. Finally, strobe lighting, fast LED [22] and flash modulation are also effective in temporally varying (not dithering) the illumination.

---

[1] Faster shutters can be realized by electronically triggering the camera.

Acknowledgements

This research was supported in parts by ONR grants N00014-08-1-0330 and DURIP N00014-06-1-0762, and NSF CAREER award IIS-0643628. The authors thank the anonymous reviewers for their useful comments.

References

1. Will, P.M., Pennington, K.S.: Grid coding: A preprocessing technique for robot and machine vision. AI 2 (1971)
2. Zhang, L., Curless, B., Seitz, S.M.: Rapid shape acquisition using color structured light and multi-pass dynamic programming. 3DPVT (2002)
3. Davis, J., Nehab, D., Ramamoothi, R., Rusinkiewicz, S.: Spacetime stereo: A unifying framework for depth from triangulation. In: IEEE CVPR (2003)
4. Curless, B., Levoy, M.: Better optical triangulation through spacetime analysis. In: ICCV (1995)
5. Young, M., Beeson, E., Davis, J., Rusinkiewicz, S., Ramamoorthi, R.: Viewpoint-coded structured light. In: IEEE CVPR (2007)
6. Scharstein, D., Szeliski, R.: High-accuracy stereo depth maps using structured light. In: CVPR (2003)
7. Zickler, T., Belhumeur, P., Kriegman, D.J.: Helmholtz stereopsis: Exploiting reciprocity for surface reconstruction. In: Heyden, A., Sparr, G., Nielsen, M., Johansen, P. (eds.) ECCV 2002. LNCS, vol. 2352, pp. 869–884. Springer, Heidelberg (2002)
8. Hertzmann, A., Seitz, S.M.: Shape and materials by example: A photometric stereo approach. In: IEEE CVPR (2003)
9. Wenger, A., Gardner, A., Tchou, C., Unger, J., Hawkins, T., Debevec, P.: Performance relighting and reflectance transformation with time-multiplexed illumination. ACM SIGGRAPH (2005)
10. Nayar, S.K., Krishnan, G., Grossberg, M.D., Raskar, R.: Fast separation of direct and global components of a scene using high frequency illumination. ACM SIGGRAPH (2006)
11. Sen, P., Chen, B., Garg, G., Marschner, S.R., Horowitz, M., Levoy, M., Lensch, H.P.A.: Dual photography. ACM SIGGRAPH (2005)
12. Zhang, L., Nayar, S.K.: Projection defocus analysis for scene capture and image display. ACM SIGGRAPH (2006)
13. Dudley, D., Duncan, W., Slaughter, J.: Emerging digital micromirror device (dmd) applications. In: Proc. of SPIE, vol. 4985 (2003)
14. Nayar, S.K., Branzoi, V., Boult, T.: Programmable imaging using a digital micromirror array. In: IEEE CVPR (2004)
15. Takhar, D., Laska, J., Wakin, M., Duarte, M., Baron, D., Sarvotham, S., Kelly, K., Baraniuk, R.: A new compressive imaging camera architecture using optical-domain compression. Computational Imaging IV at SPIE Electronic Imaging (2006)
16. Jones, A., McDowall, I., Yamada, H., Bolas, M., Debevec, P.: Rendering for an interactive 360 degree light field display. ACM SIGGRAPH (2007)
17. McDowall, I., Bolas, M.: Fast light for display, sensing and control applications. In: IEEE VR Workshop on Emerging Display Technologies (2005)
18. Raskar, R., Welch, G., Cutts, M., Lake, A., Stesin, L., Fuchs, H.: The office of the future: A unified approach to image-based modeling and spatially immersive displays. ACM SIGGRAPH (1998)

19. Cotting, D., Naef, M., Gross, M., Fuchs, H.: Embedding imperceptible patterns into projected images for simultaneous acquisition and display. In: ISMAR (2004)
20. Schechner, Y.Y., Nayar, S.K., Belhumeur, P.N.: A theory of multiplexed illumination. In: ICCV (2003)
21. Raskar, R., Agrawal, A., Tumblin, J.: Coded exposure photography: Motion deblurring using fluttered shutter. ACM SIGGRAPH (2006)
22. Nii, H., Sugimoto, M., Inami, M.: Smart light-ultra high speed projector for spatial multiplexing optical transmission. In: IEEE PROCAMS (2005)

term in the above equation is equal to 1. Equation (1) thus can be reduced to a linear projection of the light and the volume density,

$$I(y, z) = \int_x \rho(x, y, z) \cdot L(x, y) \, dx \ . \tag{2}$$

For media where the attenuation cannot be ignored, we present a simple, iterative method based on iterative relinearization (see §5.3).

5   Compressive Structured Light

In this section, we explain the idea of compressive structured light for recovering inhomogeneous participating media. For participating media, each camera pixel receives light from all points along the line of sight within the volume. Thus each camera pixel is an integral measurement of one row of the volume density. Whereas conventional structured light range finding methods seek to triangulate the position of a single point, compressed structured light seeks to reconstruct the 1D density "signal" from a few measured integrals of this signal.

This is clearly a more difficult problem. One way to avoid this problem is to break the integrals into pieces which can be measured directly. The price, however, is the deterioration of either spatial resolution or temporal resolution of the acquisition. Existing methods either illuminate a single slice at a time and scan the volume (see Fig. 2a and [4,3]), thus sacrificing temporal resolution, or they illuminate a single pixel per row and use interpolation to reconstruct the volume (e.g., Fig. 2b and [5]), sacrificing spatial resolution.

In contrast, the proposed compressive structured light method uses the light much more efficiently, projecting coded light patterns that yield "signatures," or integral measurements, of the unknown volume density function.

The didactic illustration in Fig. 1a depicts a simple lighting/viewpoint geometry under orthographic projection, with the camera viewpoint along the x-axis, and the projector emitting along the z-axis. Consider various coding strategies

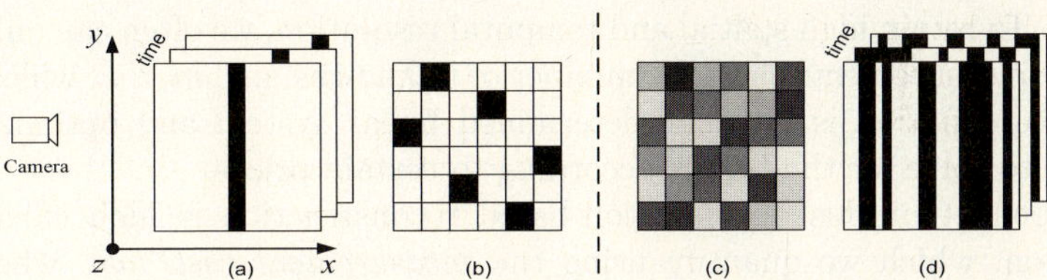

Fig. 2. Different coding strategies of the light $L(x,y)$ at time t for recovering inhomogeneous participating media: (a) scan (one stripe turned on) [4,3]; (b) laser-lines interpolation (one pixel turned on per one row) [5]; (c) Spatial coding of compressive structured light (all pixels are turned on with random values per time frame); (d) Temporal coding of compressive structured light (random binary stripes are turned on per time frame). Compressive structured light, shown in (c) and (d), recovers the volume by reconstructing the 1D signal along x-axis from a few integral measurements.

of the 3D light function $L(x, y, t)$: *Spatial* codes (Fig. 2c) recover the volume from a single image by trading spatial resolution along one dimension; *Temporal* codes (Fig. 2d) trade temporal resolution by emitting a sequence of vertical binary stripes (with no coding along y-axis), so that full spatial resolution is retained.[2]

We will see that these compressive structured light codes yield high efficiency both in acquisition time and illumination power; this comes at the cost of a more sophisticated reconstruction process, to which we now turn our attention.

5.1   Formulation

Consider first the case of spatial coding. Suppose we want to reconstruct a volume at the resolution $n \times n \times n$ (e.g., $n = 100$). The camera and the projector have the resolution of $M \times M$ pixels (e.g., $M = 1024$). Therefore, one row of voxels along the x-axis (refer to the red line in Fig. 1a) will receive light from $m = M/n$ (e.g., $m = 1024/100 \approx 10$) rows of the projector's pixels. The light scattered by these voxels in the viewing direction will then be measured, at each z-coordinate, by a vertical column of m camera pixels. Thus, using the fact that we have greater spatial projector/camera resolution than voxel resolution, we can have m measurements for each n unknowns. Similarly, we can also acquire these m measurements using temporal coding, i.e., changing the project light patterns at each of the m time frames.

Without loss of generality, we use $\mathbf{l}_1 = L(x, 1), \cdots, \mathbf{l}_m = L(x, m)$ to denote the m rows of pixels from the projector, and $b_1 = I(1, z), \cdots, b_m = I(m, z)$ to denote the image irradiance of the m pixels in the camera image. Let $\mathbf{x} = [\rho_1, \cdots, \rho_n]^T$ be the vector of the voxel densities along the row. Assuming no attenuation, the image irradiance for each of these m pixels is a linear projection of the light and the voxels' density from (2): $b_i = \mathbf{l}_i^T \mathbf{x}, i = 1, \cdots, m$. Rewriting these m equations in matrix form, we have: $\mathbf{Ax} = \mathbf{b}$, where $\mathbf{A} = [\mathbf{l}_1, \cdots, \mathbf{l}_m]^T$ is a $m \times n$ matrix, $\mathbf{b} = [b_1, \cdots, b_m]^T$ is a $m \times 1$ vector.

Thus, if attenuation is not considered, the problem of recovering the volume is formulated as the problem of reconstructing the 1D signal $\mathbf{x}$ given the constraints $\mathbf{Ax} = \mathbf{b}$. To retain high spatial and temporal resolution, we often can only afford far fewer measurements than the number of unknowns, i.e., $m < n$, which means the above equation is an underdetermined linear system and optimization is required to solve for the best $\mathbf{x}$ according to certain priors.

One benefit of this optimization-based reconstruction is high efficiency in acquisition, which we quantify using the *measurement cost*, m/n, where m is the number of the measurements and n is the number of unknowns (i.e., the dimension of the signal). For example, the measurement cost of the scanning method [4,3] is one. We show that by exploiting the sparsity of the signal, we can reconstruct the volume with much lower measurement cost (about $\frac{1}{8}$ to $\frac{1}{4}$).

---

[2] All of the 4 methods shown in Fig. 2 can be equally improved using color channels.

Table 1. Different norms used for reconstruction

Method	Optimization Functional	Constraints
Least Square (LS)	$\|\mathbf{Ax} - \mathbf{b}\|_2$	
Nonnegative Least Square (NLS)	$\|\mathbf{Ax} - \mathbf{b}\|_2$	$\mathbf{x} \geq 0$
CS-Value	$\|\mathbf{x}\|_1$	$\mathbf{Ax} = \mathbf{b}, \ \mathbf{x} \geq 0$
CS-Gradient	$\|\mathbf{x}'\|_1$	$\mathbf{Ax} = \mathbf{b}, \ \mathbf{x} \geq 0$
CS-Both	$\|\mathbf{x}\|_1 + \|\mathbf{x}'\|_1$	$\mathbf{Ax} = \mathbf{b}, \ \mathbf{x} \geq 0$

5.2 Reconstruction Via Optimization

Formulation. Solving the underdetermined linear system requires some prior (assumed) knowledge of the unknown signal, which can be represented as optimization functionals or constraints on the data. We consider several alternatives, as listed in Table 1. Besides the commonly-used Least Square (LS) and Nonnegative Least Square (NLS), we consider functionals using ℓ_1-norms, as these bias toward sparse representations:[3]

First, we observe that for many natural volumetric phenomena, often only a small portion of the entire volume is occupied by the participating media. For example, consider the beautiful ribbon patterns generated by smoke; similarly, sparsity was implicitly used to reconstruct (surface-like) flames [12]). This suggests the use of the ℓ_1-norm of the signal value (CS-Value).

Furthermore, the sparsity of *gradients* of natural images is well studied [20,21]. Related work in image restoration [22] uses nonlinear optimization to minimize "total variation," i.e., the sum of ℓ_2-norm of image gradient. In this vein, we consider the use of ℓ_1-norm on the signal's gradient (CS-Gradient).

Finally, consider a dynamic process, such as milk dissolving in water: here diffusion decreases the signal value's sparsity over time, but it increases the gradient sparsity. Motivated by this observation, we consider the sum of ℓ_1-norms of both the value and the gradient (CS-Both), so that the algorithm has the ability to "adapt" for the sparsity.

Analysis. Comparison of these reconstruction methods is first performed on 1D synthetic signals. These signals are randomly sampled rows from the volume density of smoke acquired in Hawkins et al. [3]. We restrict the measurement cost, m/n, to be 1/4. The measurement ensemble, $\mathbf{A}$, is generated in a way that each element is drawn independently from a normal distribution and each column is normalized to 1, which is effectively a white noise matrix and is known to be good for compressive sensing [7]. NRMSE (normalized root mean squared error) is used as the measure of error.

The reconstruction results are shown in Fig. 3. The commonly-used LS performs the worst, since it merely minimizes the errors without using any prior on the data. With the nonnegative constraint added, NLS has better performance. CS-Value and CS-Gradient are better than NLS given that both use one more

---

[3] LS and NLS are solved with SVD and Levenberg-Marquardt, respectively. The other functionals are formulated as Linear Programming (LP) and solved with GLPK [19].

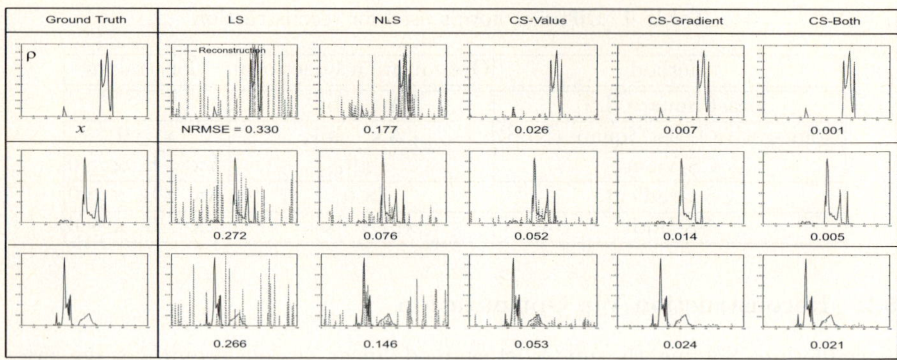

Fig. 3. Comparison of different reconstruction methods. The first column is the original signal. The remaining columns show reconstruction results (red dashed lines) for different methods, given the measurement cost, m/n, is equal to 1/4. The value below each plot is the NRMSE(normalized root mean squared error) of reconstruction.

prior—the sparsity on the signal value or on the signal gradient. The fact that CS-Gradient is better than CS-Value indicates that the sparsity on the signal gradient holds better than the sparsity on the signal value. Finally, as expected, CS-Both outperforms other methods due to its adaptive ability. In our trials, the favorable performance of CS-Both was not sensitive to changes of the relative weighting of the value and gradient terms. These observations carry over to the 3D setting (see Fig. 4), where we reconstruct a 128^3 volume; note that this requires 128×128 independent 1D reconstructions.

5.3   Iterative Attenuation Correction

Until now, we have not considered the attenuation in the image formation model in (1) yet. To take into account attenuation, we use a simple iterative relinearization algorithm as follows:

1. Assume no attenuation, solve the optimization problem with techniques from §5.2 to get the initial reconstruction of the volume density $\rho^{(0)}$.
2. At iteration k, assuming σ_t is known[4], compute the attenuated light as: $L^{(k)}(x, y, z) = \exp\left(- (\tau_1 + \tau_2)\right) \cdot L(x, y)$, where τ_1 and τ_2 are computed using $\rho^{(k-1)}$ as shown in §4.
3. With the attenuated light $L^{(k)}(x, y, z)$, (1) becomes a linear equation. We solve for $\rho^{(k)}$ and go to next iteration until it converges.[5]

Since our overall framework accommodates the scanning method [4,3] and the interpolation method [5] as special cases, the iterative algorithm could be directly applied to these prior methods as well.

---

[4] The attenuation coefficient, σ_t, of the participating medium can be obtained from literature, specified by a user, or be measured by a second camera taking the shadowgram of the volume.

[5] In practice, we found that the algorithm usually converges within 3-4 iterations.

6   Validation Via Simulation

To further validate our method, we perform simulations on a synthetic volume. The volume is generated from a triangular mesh of a horse and it is discretized into 128^3 voxels. For each voxel, if it is inside the mesh, the density is designed to be proportional to the distance from the center of the voxel to the center of the mesh, otherwise the density is 0. Fig. 4a shows the volume where blue corresponds to the lowest density while yellow corresponds to the highest density. A slice of the volume is shown in Fig. 4b.

Both spatial coding and temporal coding of compressive structured light are tested. The measurement cost, m/n, is fixed to 1/4. For spatial coding, we use a random color image with resolution of 1280×1280 as the coded light from the projector. This gives us $m = 1280/128 \times 3 = 30$ measurements to recover densities of 128 voxels on one row of the volume. Based on (1), a single image (shown in Fig. 4c) is generated from the camera view and used for reconstruction. For temporal coding, we use random binary stripes as illumination and generate 32 images for reconstruction. One of these images is shown in Fig. 4g. CS-Both is used to reconstruct the volume for both cases. As shown in Fig. 4, both methods accurately reconstruct the volume. Moreover, Fig. 4(right) shows the reconstruction errors and reconstructed slices at different iterations of attenuation correction, which demonstrates the effectiveness of the iterative algorithm.

We also evaluate different reconstruction methods at various measurement costs from 1/16 to 1. The results are shown as a table in Fig. 5. Conclusions similar to the ones from the previous 1D signal simulation (Fig. 3) can be drawn from these results: (1) As expected, all methods have improvements as the measurement cost increases. (2) Without using any prior of the data, LS is the worst for reconstruction with insufficient measurements. (3) CS-Gradient and CS-Both

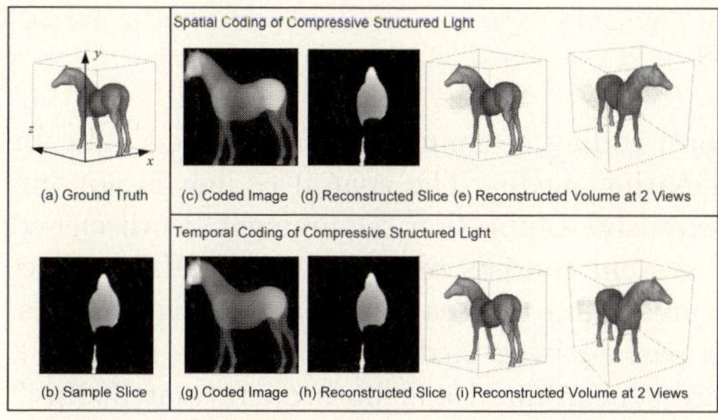

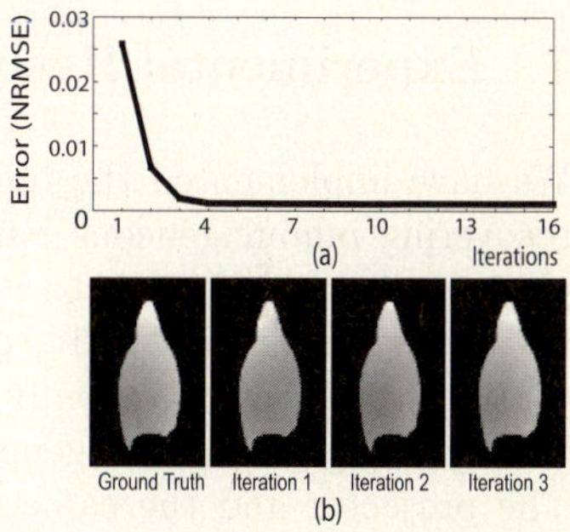

Fig. 4. Simulation results of volume reconstruction using compressive structured light. **LEFT**: (a) The original volume where blue means the lowest density and yellow means the highest density. (b) A slice of the volume. The top and the bottom row on the right shows the reconstruction results for spatial coding and temporal coding, respectively. For each row, from left to right are the coded image acquired by the camera, the reconstruction of the slice, and the reconstructed volume under two different views. **RIGHT**: (a) Reconstruction errors and (b) slices with iterative attenuation correction.

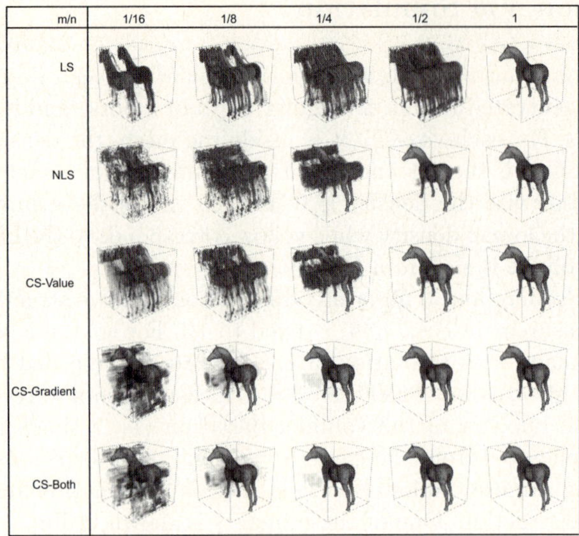

Fig. 5. Comparison of different reconstruction methods at different measurement costs, m/n. CS-Both outperforms other methods.

largely outperform other methods, especially for low measurement cost, which indicating strong sparsity in the signal's gradient. (4) CS-Both is better than CS-Gradient, especially at low measurement cost (e.g., as shown in Fig. 5 at $m/n = 1/16$). Based on these preliminary simulations, we chose to run our actual acquisition experiments with a measurement cost of 1/4 and the CS-Both optimization functional.

7   Experimental Results

We have implemented the temporal coding of compressive structured light for recovering inhomogeneous participating media. The spatial coding is not implemented currently due to its extensive calibration requirement, as discussed in §8. As shown in Fig. 1c, our system consists of a 1024×768 DLP projector and a 640×480 Dragonfly Express 8-bit camera, positioned at right angles, both viewing the inhomogeneous participating medium (milk drops in water). The projector and the camera are synchronized and both operate at 360fps.[6] Using 24 coded light patterns, we are able to recover a 128^3 volume at 15fps. These light patterns consist of 128 vertical stripes. Each stripe is assigned 0 or 1 randomly with the probability of 0.5. In this way, about half amount of the light is turned on for each measurement. We also tried alternative light patterns such as Hadamard codes, and found the random binary codes have better performance.

---

[6] The camera's resolution is set to 320×140 in order to achieve 360fps.

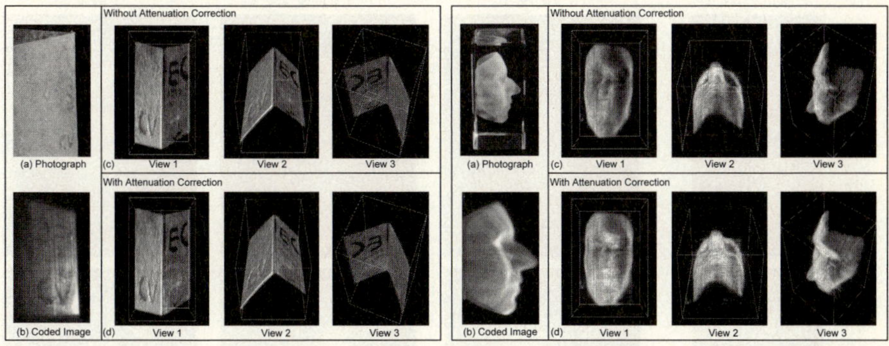

Fig. 6. Reconstruction results of **LEFT**: an object consisting of two glass slabs with powder where the letters "EC" are on the back slab and "CV" on the front slab, and **RIGHT**: point cloud of a face etched in a glass cube. Both examples show: (a) a photograph of the objects, (b) one of the 24 images captured by the camera, and reconstructed volumes at different views with (c) and without (d) attenuation correction.

We used this system to recover several types of inhomogeneous participating media, including, multiple translucent layers, a 3D point cloud of a face etched in a glass cube, and the dynamic process of milk mixing with water. The reconstructed volumes are visualized with the ray casting algorithm [23] in which the opacity function is set to the volume density.

We first perform reconstruction on static volumes. Fig. 6(left) shows the results of an object consisting of two glass slabs with powder on both. The letters "EC" are drawn manually on the back plane and "CV" on the front plane by removing the powder. Thus we create a volume in which only two planes have non-zero density. A photograph of the object is shown in Fig. 6a. We then reconstruct the volume using the proposed method. Fig. 6 shows one of the 24 captured images as well as the reconstructed volume at different views with and without attenuation correction. It shows that attenuation correction improves the results by increasing the density on the back plane.

Similarly, Fig. 6(right) show the reconstruction for a 3D point cloud of a face etched in a glass cube. As shown, our method also achieved good reconstruction of the volume. In this example, multiple scattering and attenuation within the point cloud are much stronger than the previous example. Thus in the reconstructed volume, the half of the face not directly visible to the camera has a lower estimated density (e.g., the relative darker area of the right eye in Fig. 6).

Finally, we use our system to reconstruct time-varying volumes. We take the dynamic process of milk drops dissolving in water as an example. We use a syringe to drip milk drops into a water tank as shown in the adjacent figure. With the proposed method, we are able to reconstruct time-varying volumes with high spatial resolution ($128 \times 128 \times 250$) at 15fps, which recovers the interesting patterns of the dynamic process (see Fig. 7).

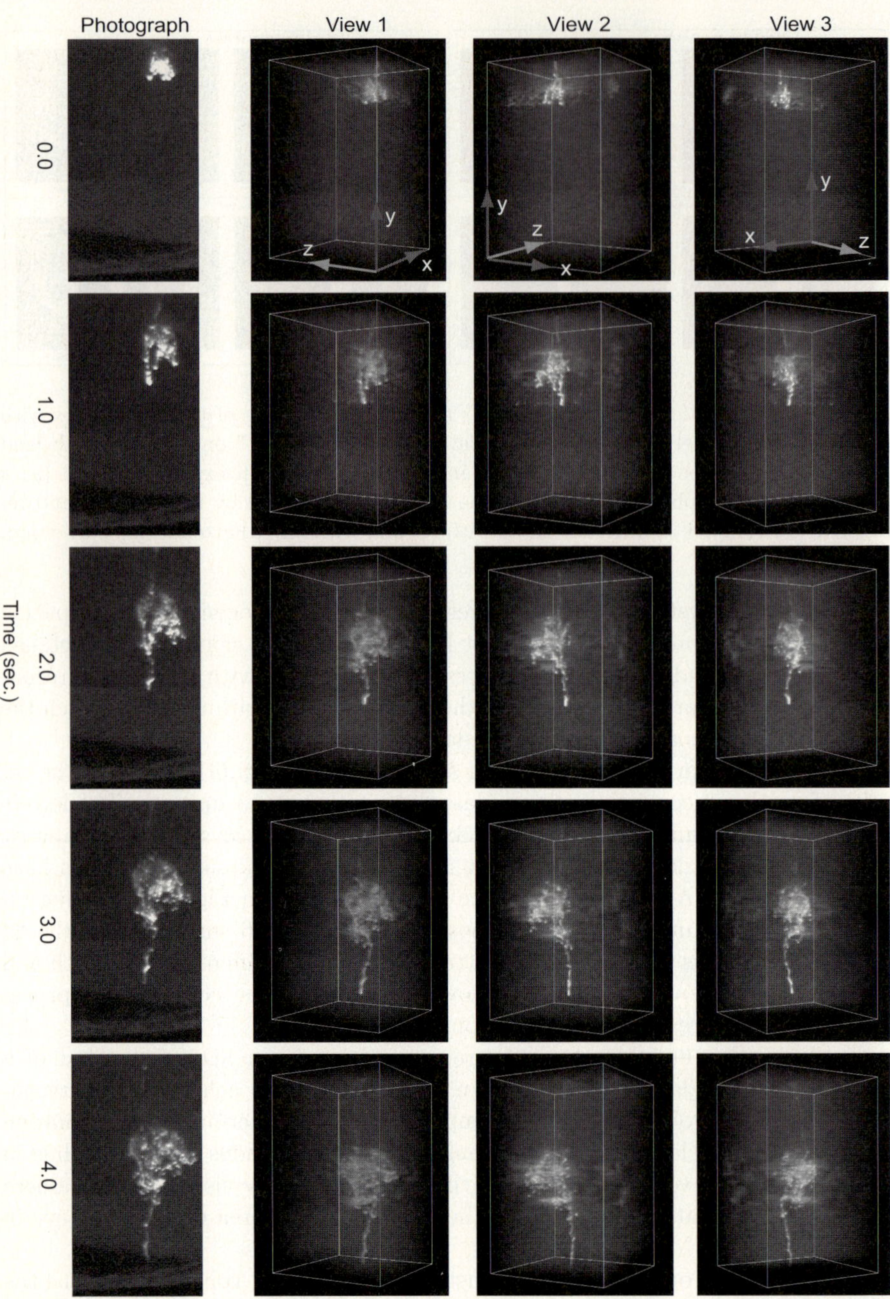

Fig. 7. Reconstruction results of milk drops dissolving in water. 24 images are used to reconstruct the volume at $128 \times 128 \times 250$ at 15fps. The reconstructed volumes are shown in three different views. Each row corresponds to one instance in time. The leftmost column shows the corresponding photograph (i.e., all projector pixels emit white) of the dynamic process.

These camera-based systems significantly reduce measurement time, but they also require special-purpose hardware and precise lighting control.

Passive methods for reflectometry that require only natural lighting provide an attractive alternative. In the computer graphics community, the inference of reflectance from natural images has been studied under the banner of 'inverse rendering'. Ramamoorthi et al. [7] derive an elegant framework for inverse rendering by interpreting the rendering equation as a convolution. This yields an important theoretical tool that, among other things, enables the recovery of reflectance through de-convolution. Unfortunately, this approach can only yield general isotropic BRDFs when the full 4D output light field is observed. More typically, one has access to a small number of images; and when this is the case, de-convolution can only yield radially-symmetric BRDFs[1], which are incapable of representing off-specular peaks and important grazing-angle effects [2,3].

Inverse rendering can also be formulated, as it is here, directly in the angular domain. Many approaches exist, and almost all of them rely on low-parameter BRDF models (Phong, Cook-Torrance, etc.) to make the problem tractable. Low-parameter BRDF models impose strong constraints on reflectance, and as a result, one can exploit them to recover more than just reflectance information from a set of input images. For example, there are methods for handling global illumination effects and anisotropic reflectance [8,9], spatial reflectance variation, and the simultaneous recovery of illumination and/or shape (e.g., [10,11]). (Patow et al. [12] provide a review.) Every parametric approach suffers from limited accuracy, however, because the expressiveness of existing low-parameter BRDF models is quite restricted [2,3]. This situation is unlikely to improve in the short term. Given the diversity of the world's materials, designing 'general purpose' low-parameter models that are simultaneously accurate, flexible and amenable to tractable analysis has proven to be a very difficult problem.

Unlike these existing approaches, our goal is to recover *general* reflectance information without the restrictions of radial symmetry or low-parameter models. By avoiding these restrictions, we can handle a broader class of materials. To maintain this generality, we find it necessary to assume isotropic reflectance, ignore global illumination effects, and require that shape and illumination be known *a priori*. While the tools we develop can likely be applied to other inverse rendering problems (see discussion in Sect. 5), we leave this for future work.

2   A Bivariate BRDF for Reflectometry

Passive reflectometry is not well-posed without some constraints on the BRDF. Indeed, a BRDF is a function of four (angular) dimensions, while an input image is a function of two. What we require is a way to constrain the BRDF without surrendering our ability to represent important phenomena. Here, we present an approach based on a bivariate representation for isotropic surface reflectance.

---

[1] A radially-symmetric BRDF is one that, like the Phong model, is radially symmetric about the reflection vector. It's angular domain has dimension one.

For many materials, the dimension of the BRDF domain can be reduced without incurring a significant loss of detail. The domain can be folded in half, for example, because reciprocity ensures that BRDFs are symmetric about the directions of incidence and reflection: $f(\mathbf{u}, \mathbf{v}) = f(\mathbf{v}, \mathbf{u})$. In many cases, the domain $(\theta_u, \phi_u, \theta_v, \phi_v)$ can be further 'projected' onto the 3D domain $(\theta_u, \theta_v, \phi_u - \phi_v)$ and then folded onto $(\theta_u, \theta_v, |\phi_u - \phi_v|)$. The projection is acceptable whenever a BRDF exhibits little change for rotations of the input and output directions (as a fixed pair) about the surface normal; and additional folding is acceptable whenever there is little change when reflecting the output direction about the incident plane. Materials that satisfy these two criteria—for some definition of 'little change'—are said to satisfy *isotropy* and *bilateral symmetry*, respectively. (It is also common to use the term *isotropy* to mean both.)

It is convenient to parameterize the BRDF domain in terms of halfway and difference angles [13]. Accordingly, the complete 4D domain is written in terms of the spherical coordinates of the halfway vector $\mathbf{h} = (\mathbf{u} + \mathbf{v})/\|\mathbf{u} - \mathbf{v}\|$ and those of the input direction with respect to the halfway vector: $(\theta_h, \phi_h, \theta_d, \phi_d)$. See Fig. 2. In this parameterization, the folding due to reciprocity corresponds to $\phi_d \to \phi_d + \pi$, and the projection due to isotropy (without bilateral symmetry) is one onto $(\theta_h, \theta_d, \phi_d)$ [13]. While undocumented in the literature, it is straightforward to show that bilateral symmetry enables the additional folding $\phi_d \to \phi_d + \pi/2$ which gives the 3D domain $(\theta_h, \theta_d, \phi_d) \subset [0, \pi/2]^3$.

Here, we consider an additional projection of the BRDF domain, one that reduces it from three dimensions down to two. In particular, we project $(\theta_h, \theta_d, \phi_d) \subset [0, \pi/2]^3$ to $(\theta_h, \theta_d) \in [0, \pi/2]^2$. A physical interpretation is depicted in Fig. 2, from which it is clear that the projection is acceptable whenever a BRDF exhibits little change for rotations of the input and output directions (as a fixed pair) about the halfway vector. This is a direct generalization of isotropy, bilateral symmetry and reciprocity, which already restrict the BRDF to be $\frac{\pi}{2}$-periodic for the same rotations. We refer to materials that satisfy this requirement (again, for some definition of 'little change') as being *bivariate*. The accuracy of bivariate representations of the materials in the MERL BRDF

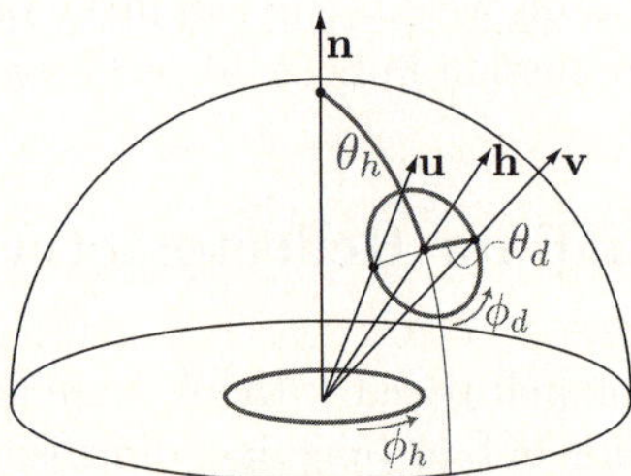

Fig. 2. Domain reduction for reciprocal, isotropic, bilaterally-symmetric, and bivariate BRDFs. Isotropic BRDFs are unchanged by rotations about the surface normal (i.e., changes in ϕ_h), while reciprocity and bilateral symmetry impose periodicity for rotations about the halfway vector (i.e., changes in ϕ_d). Here we consider *bivariate* BRDFs, which are constant functions of ϕ_d.

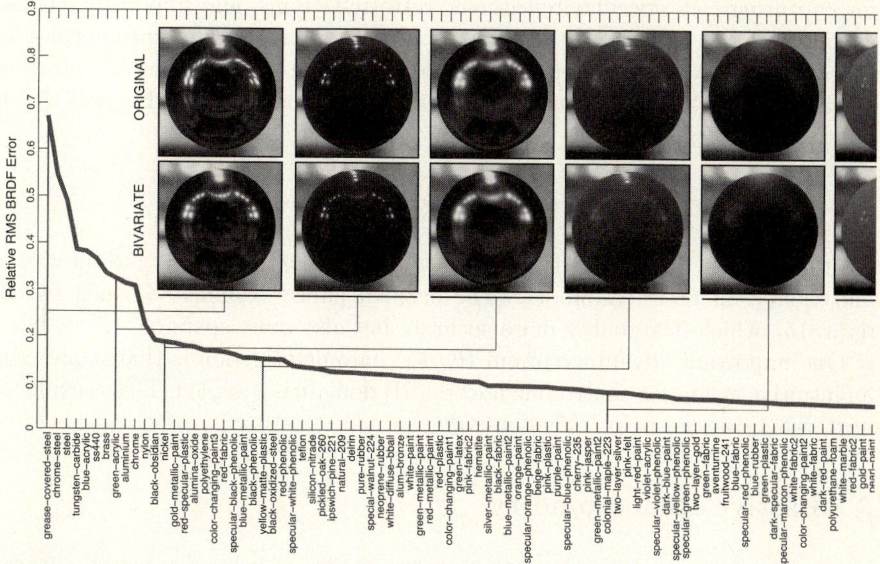

Fig. 3. Accuracy of bivariate representations of materials in the MERL BRDF database. Materials are in order of increasing accuracy, and representative renderings are shown for comparison. Most materials in the database are well-represented by a bivariate function. (Images embedded at high resolution; please zoom in.)

database [14] are shown in Fig. 3, where they are sorted by relative RMS BRDF error:

$$E_{\mathrm{rms}} = \left(\sum_{\theta_h,\theta_d,\phi_d} \frac{(f(\theta_h,\theta_d,\phi_d) - \bar{f}(\theta_h,\theta_d))^2}{(f(\theta_h,\theta_d,\phi_d))^2} \right)^{\frac{1}{2}}, \tag{1}$$

with

$$\bar{f}(\theta_h,\theta_d) = \frac{1}{|\Phi(\theta_h,\theta_d)|} \sum_{\Phi(\theta_h,\theta_d)} f(\theta_h,\theta_d,\phi_d).$$

Here, $\Phi(\theta_h,\theta_d)$ is the set of valid ϕ_d values given fixed values of θ_h and θ_d. The figure also shows synthetic images of materials that are more and less well-represented by a bivariate BRDF. Overall, our tests suggest that the overwhelming majority of the materials in the database are reasonably well-represented by bivariate functions. We even find that the bivariate reduction has positive effects in some cases. For example, the original green-acrylic BRDF has lens flare artifacts embedded in its measurements[2], and these are removed by the bivariate reduction (see Fig. 3).

Motivation for a bivariate representation is provided by the work of Stark et al. [2] who show empirically that a carefully-selected 2D domain is often sufficient

---

[2] W. Matusik, personal communication.

for capturing (off-)specular reflections, retro-reflections, and important Fresnel effects. The 2D domain (θ_h, θ_d) that is introduced above is homeomorphic to that of Stark et al., which is why it posesses these same properties. Stark et al. propose the '$\alpha\sigma$-parameterization' for two-dimensional BRDFs, and this is related to (θ_h, θ_d) by

$$\alpha = \sin^2 \theta_d, \quad \sigma = \frac{1}{2}\left(1 + \cos 2\theta_d\right)\sin^2 \theta_h.$$

For this reason, Figs. 2 and 3 can be seen as providing a new interpretation and validation for their model. (The original paper examined Cornell BRDF data [15], which is arguably more accurate but also quite sparse.)

One important advantage of our (θ_h, θ_d) parameterization is that it provides an intuitive means for controlling how the 2D domain is sampled. This is explored next, where we use it for reflectometry.

3   Passive Reflectometry

We assume that we are given one or more images of a known curved surface, and that these images are acquired under known distant lighting, such as that measured by an illumination probe. In this case, each pixel in the images provides a linear constraint on the BRDF, and our goal is to infer the reflectance function from these constraints. While the constraints from a single image are not sufficient to recover a general 3D isotropic BRDF [7], we show that they often *are* sufficient to recover plausible bivariate reflectance.

To efficiently represent specular highlights, retro-reflections and Fresnel effects, we can benefit from a non-uniform sampling of the 2D domain. While 'good' sampling patterns can be learned from training data [16], this approach may limit our ability to generalize to new materials. Instead, we choose to manually design a sampling scheme that is informed by common observations of reflectance phenomena. This is implemented by defining continuous functions $s(\theta_h, \theta_d)$ and $t(\theta_h, \theta_d)$ and sampling uniformly in (s, t). Here we use $s = 2\theta_d/\pi$, $t = \sqrt{2\theta_h/\pi}$ which increases the sampling density near specular reflections $(\theta_h \approx 0)$. With this in mind, we write the rendering equation as

$$I(\mathbf{v}, \mathbf{n}) = \int_{\Omega} L(R_n^{-1}\mathbf{u}) f(s(\mathbf{u}, R_n\mathbf{v}), t(\mathbf{u}, R_n\mathbf{v})) \cos \theta_u d\mathbf{u}, \tag{2}$$

where $\mathbf{v}$ is the view direction, $\mathbf{n}$ is the surface normal corresponding to a given pixel, and R_n is the rotation that sends the surface normal to the z-axis and the view direction to the xz-plane. We use overloaded notation for s and t, which depend on the incident and reflected directions indirectly through (θ_h, θ_d).

At each pixel, this integral is computed over the visible hemisphere of light directions Ω. Our use of a bivariate BRDF induces a 'folding' of this hemisphere because light directions $\mathbf{u}$ and $\mathbf{u}'$ that are symmetric about the view/normal plane correspond to the same point in our 2D BRDF domain. When the lighting and surface shape are known, we obtain a constraint from each pixel, and each

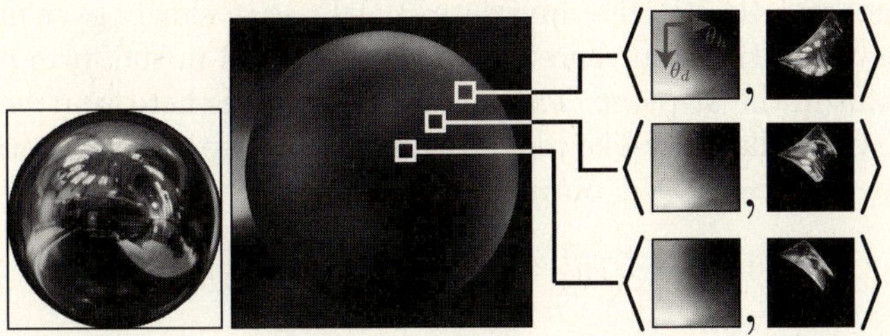

Fig. 4. Constraints on bivariate reflectance from natural lighting. Each pixel of an input image (*middle*) captured under distant illumination (*left*) gives a linear constraint that can be interpreted as an inner product of the 2D BRDF (*right*, first argument) and a visible hemisphere of lighting that is weighted, warped and folded across the local view/normal plane (*right*, second argument).

constraint can be interpreted as an inner product between the unknown BRDF and a hemisphere of illumination that is weighted by $\cos\theta_u$, folded across the local view/normal plane, and warped onto the st-plane. See Fig. 4.

To infer the BRDF from these constraints, we create a uniform grid $\mathcal{S} = \{(s_i, t_i)\}$ in the BRDF domain and approximate the rendering equation by a sum over a discrete set Ω_d of lighting directions on the hemisphere:

$$I(\mathbf{v}, \mathbf{n}) \approx \frac{2\pi}{|\Omega_d|} \sum_{\mathbf{u}_k \in \Omega_d} \left(\sum_{s_i, t_j \in N_k} \alpha_{i,j}^k L(R_n^{-1}\mathbf{u}_k) f(s_i, t_j) \right) \cos\theta_{u_k}, \qquad (3)$$

where N_k is the set of the four BRDF grid points that are closest to $s(\mathbf{u}_k, R_n\mathbf{v})$, $t(\mathbf{u}_k, R_n\mathbf{v})$, and $\alpha_{i,j}^k$ is the coefficient of the bilinear interpolation associated with these coordinates and s_i, t_j. (We find a piecewise linear approximation of the BRDF to be adequate.) This equation can be rewritten as

$$I(\mathbf{v}, \mathbf{n}) \approx \frac{2\pi}{|\Omega_d|} \sum_{(s_i, t_j) \in \mathcal{S}} f(s_i, t_j) \sum_{\mathbf{u}_k \in \text{bin}_{ij}} \alpha_{i,j}^k L(R_n^{-1}\mathbf{u}_k) \cos\theta_{u_k}, \qquad (4)$$

to emphasize its interpretation as an inner product.

Observations of distinct normals $\mathbf{n}_1 \dots \mathbf{n}_N$ obtained from one or more images provide constraints that are combined into a system of equations

$$I = Lf \qquad (5)$$

where $I = [I(\mathbf{v}, \mathbf{n}_1), \dots, I(\mathbf{v}, \mathbf{n}_N)]$ and L is a lighting matrix whose rows are given by the non-BRDF terms in Eq. 4. The goal is then to find f such that these constraints are satisfied. While this may work well in the noiseless case, in practice we require regularization to handle noise caused by the sensor, the bivariate approximation, the discretization of the rendering equation, and errors in the assumed surface shape.

As with general 4D BRDFs, bivariate BRDFs vary slowly over much of their domain. Regularization can therefore be implemented in the form of a smoothness constraint in the st-plane. There are many choices here, and we have found spatially-varying Tikhonov-like regularization to be especially effective. According to this design choice, the optimization becomes

$$\underset{f}{\mathrm{argmin}}\, \|I - Lf\|_2^2 + \alpha \left(\left\|\Lambda_s^{-1} D_s f\right\|_2^2 + \left\|\Lambda_t^{-1} D_t f\right\|_2^2 \right) \qquad (6)$$

subject to $f \geq 0$,

where D_s and D_t are $|\mathcal{S}| \times |\mathcal{S}|$ derivative matrices, and α is a tunable scalar regularization parameter. The matrices Λ_s and Λ_t are diagonal $|\mathcal{S}| \times |\mathcal{S}|$ matrices that affect non-uniform regularization in the bivariate BRDF domain. Their diagonal entries are learned from the MERL database by setting each to the variance of the partial derivative at the corresponding st domain point, where the variance is computed across all materials in the database. Probabilistically, this approach can be interpreted as seeking the MAP estimate with independent, zero-mean Gaussian priors on the bivariate BRDF's partial derivatives.

There are many possible alternatives for regularization. For example, one could learn a joint distribution over the entire bivariate domain, perhaps by characterizing this distribution in terms of a small number of modes of variation. However, we have found that the simple approach in Eq. 6 provides reasonable results, does not severely 'over-fit' the MERL database, and is computationally quite efficient (it is a constrained linear least squares problem).

3.1　Adequate Illumination

There is a question of when an environment is adequate for reflectometry to be well-posed and well-conditioned. An algebraic condition is readily available; we simply require the rank of the illumination matrix L to be sufficiently large (i.e., to approach $|\mathcal{S}|$). More intuitively, we require sufficient observations of all portions of the BRDF domain, with regions corresponding to specular reflections ($\theta_h \approx 0$), retro-reflections ($\theta_d \approx 0$), and grazing angles ($\theta_d \approx \pi/2$) being particularly important. In particular, we do not expect good results from simple environments composed of a small number of isolated point sources. This is in agreement with perceptual studies showing that humans are also unable to infer reflectance under such simple and 'unrealistic' conditions [1].

It is interesting to compare our approach to the convolution framework of Ramamoorthi et al. [7]. That approach enables a frequency domain analysis and provides very clear conditions for adequacy. For radially-symmetric BRDFs, for example, we know that an environment is adequate only if its band-limit exceeds that of the BRDF [7]. A frequency domain analysis is difficult to apply in the present case, however, because Eq. 5 does not represent a convolution. While an analysis of the conditions for adequate illumination in the bivariate case may be for worthwhile direction of future work, we focus instead on an empirical investigation here. We show that while the quality of the result depends on the environment, accurate reflectometry is achievable in many cases.

4   Evaluation and Results

We begin with an evaluation that uses images synthesized with tabulated BRDF data from the MERL database [14], measured illumination[3], and a physically based renderer[4]. Using these tools, we can render images for input to our algorithm as well as images with the recovered BRDFs for direct comparison to ground truth. In all cases, we use complete 3D isotropic BRDF data to create the images for input and ground-truth comparison, since this is closest to a real-world setting. Also, we focus our attention on the minimal case of a single input image; with additional images, the performance can only improve. It is worth emphasizing that this data is not free of noise. Sources of error include the fact that the input image is rendered with a 3D BRDF as opposed to a bivariate one, that normals are computed from a mesh and are stored at single precision, and that a discrete approximation to the rendering equation is used.

Given a rendered input image of a defined shape (we use a sphere for simplicity), we harvest observations from 8,000 normals uniformly sampled on the visible hemisphere to create an observation vector I of length 8,000. We discard normals that are at an angle of more than 80° from the viewing direction, since the signal to noise ratio is very low at these points. The bivariate BRDF domain is represented using a regular 32×32 grid on the st-plane, and our observation matrix L is therefore $M \times 1024$, where M is the number of useable normals. The entries in L are computed using Eq. 4 with 32,000 points uniformly distributed on the illumination hemisphere. With I and L determined, we can solve for the unknown BRDF as described in the previous sections.

We find it beneficial to use a small variant of the optimization in Eq. 6: we solve the problem twice using two separate pairs of diagonal weight matrices (Λ_s, Λ_t). One pair gives preference to diffuse reflectance, while the other gives preference to gloss. This provides two solutions, and we choose the one with lowest residual. Using this procedure, we were able to use the same weight matrices and regularization parameter (α) for all results in this paper. In every case, the optimizations were initialized with a Lambertian BRDF.

Results are shown in Fig. 5. The two left columns show results using a single input image synthesized with the Grace Cathedral environment. The recovered bivariate BRDFs are compared to the (3D) ground truth by synthesizing images in another setting (St. Peter's Basilica). Close inspection reveals very little noticeable difference between the two images, and the recovered BRDF is visually quite accurate. There are numerical differences, however, and these have been scaled by 100 for visualization. Note that some of this error is simply due to the bivariate approximation (see Fig. 6). The next two columns similarly show the recovery of the yellow-matte-plastic and green-acrylic materials, this time using the Cafe environment and the St. Peter's Basilica environment (alternately) for input and comparison to ground truth.

---

[3] Light probe image gallery: `http://www.debevec.org/Probes/`
[4] PBRT: `http://www.pbrt.org/`

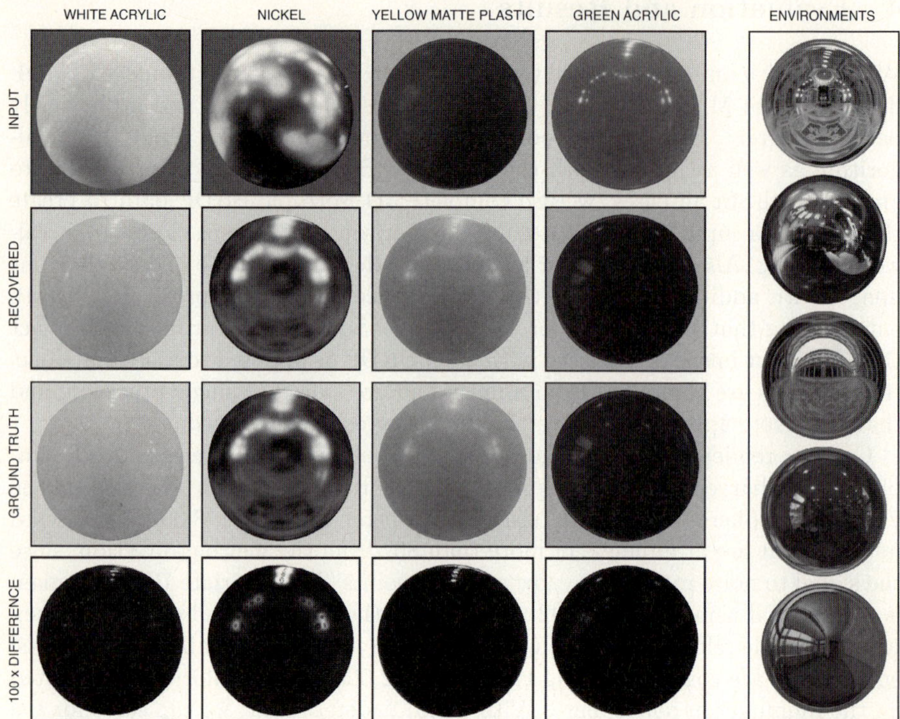

Fig. 5. Visual evaluation with MERL BRDF data. A bivariate BRDF is estimated from a single input image (*top*), and this estimate is used to render a new image under novel lighting (*second row*). Ground truth images for the novel environments are shown for comparison, along with difference images scaled by 100. Few noticeable differences exist. *Far right*: Environment maps used in the paper, top to bottom: St. Peter's Basilica, Grace Cathedral, Uffizi Gallery, Cafe and Corner Office.

In addition to these visual comparisons, we can also evaluate the recovered BRDFs quantitatively using scatter plots and RMS errors. The top of Fig. 6 shows incident-plane scatter plots for the red channels of three recovered BRDFs from Fig. 5, as well as the recovered colonial-maple BRDF from Fig. 1. While the scatter plots reveal clear deviations from ground truth, they suggest that the approach provides reasonable approximations for a variety of materials. This is true even though just a single image is used as input—many fewer than the 300 images that were used to collect the original data [14].

The bottom of the figure displays relative RMS errors for these four recovered BRDFs, along with corresponding results for all materials in the BRDF database. Shown is the accuracy (Eq. 1) of the bivariate BRDF for each material as estimated from one input image. This is done twice—once each using the Grace Cathedral and St. Peter's environments—and the curves are superimposed on the graph from Fig. 3, which shows the accuracy of the 'ground truth' bivariate reduction. (Note that the materials have been re-sorted for display

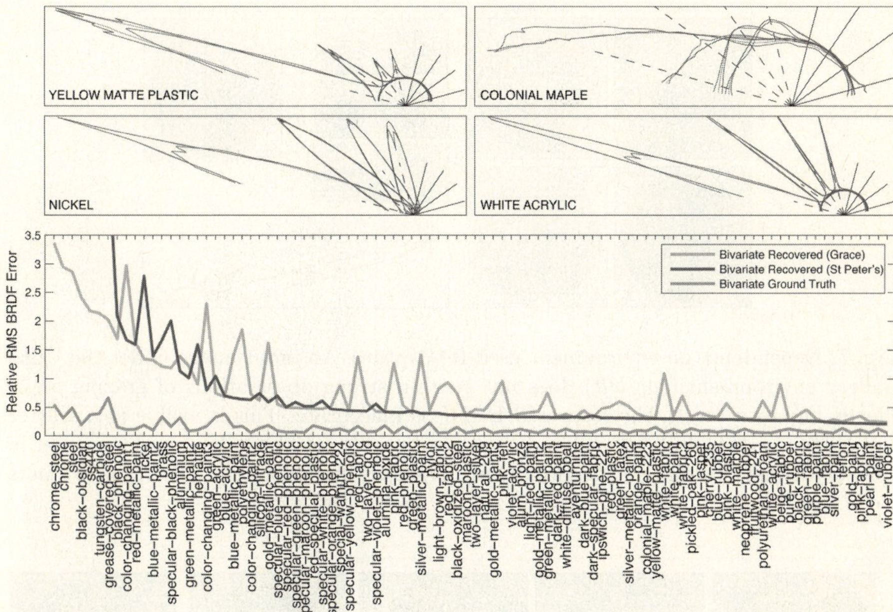

Fig. 6. Quantitative evaluation with MERL BRDF data. *Top*: Incident plane scatterplots for the four materials in Fig. 5, each showing: original 3D BRDF (*blue*); 'ground truth' bivariate BRDF (*green*); and BRDF recovered from one input image (*red*). *Bottom*: Relative RMS BRDF errors for all materials in the MERL database when each is recovered using a single image under the Grace Cathedral or St. Peter's environments. Vertical red lines match the scatterplots above.

purposes). The discrepancy between the results for the two different environments is expected in light of the discussion from Sect. 3.1. To further emphasize this environment-dependence, Fig. 7 compares estimates of yellow-matte-plastic using two different input images. The Uffizi Gallery environment (top left) does not provide strong observations of grazing angle effects, so this portion of the BRDF is not accurately estimated. This leads to noticeable artifacts near grazing angles when the recovered BRDF is used for rendering, and it is clearly visible in a scatter plot. When the Cafe environment is used as input, however, more accurate behavior near grazing angles is obtained.

4.1   Captured Data

The procedure outlined above was applied without change to captured data. Figure 8 shows the results for a number of materials. As before, each BRDF is recovered from a single input image (left), and the recovered BRDFs are used to render synthetic images of the same object from a novel viewpoint. The synthetic images are directly compared to real images captured in the same novel positions.

Captured data contains at least three significant sources of noise in addition to what exists in the rendered data above: 1) errors in the assumed surface geometry;

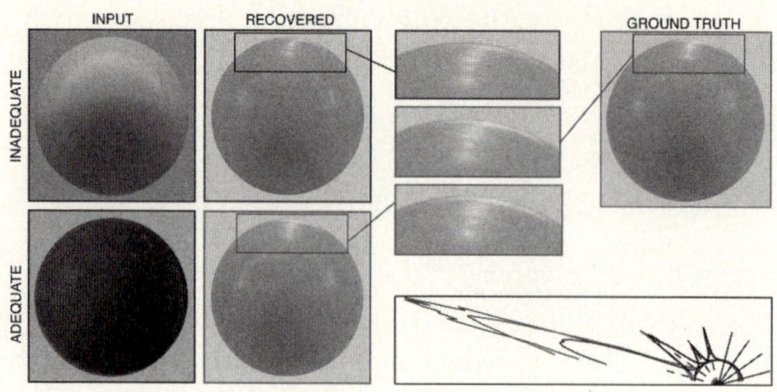

Fig. 7. Dependence on environment used for capture. An input image under the Uffizi Gallery environment (*top left*) does not contain strong observations of grazing angle effects, and as a result, the recovered BRDF is inaccurate. This is visible in a scatter plot (*bottom right*, black curves) and causes noticeable artifacts when used to render in a novel setting. If a different environment is used as input (*bottom left*) these artifacts are largely avoided.

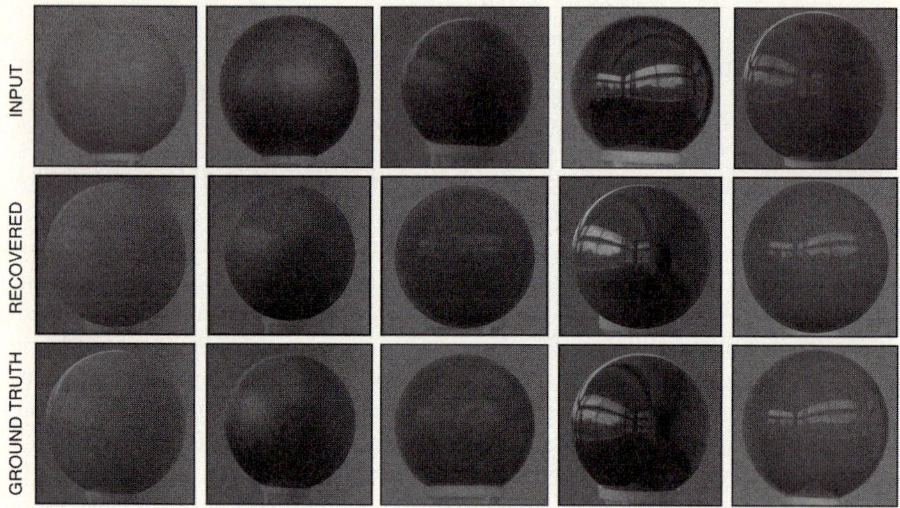

Fig. 8. Results using captured data. A BRDF is estimated from a single input image (*top*) under a known environment. This recovered BRDF is used to render a synthetic image for novel view within the same environment (*middle*). An actual image for the same novel position is shown for comparison (*bottom*). Despite the existence of non-idealities such as surface mesostructure and spatial inhomogeneity, plausible BRDFs are recovered.

2) surface mesostructure (e.g., the green sphere); and 3) spatial reflectance variations (e.g., the grey sphere). Presently, surface shape is computed by assuming the camera to be orthographic and estimating the center and radius of the sphere in the camera's coordinate system. Errors in this process, coupled with errors in

the alignment with the illumination probe, lead to structured measurement noise. Despite this, our results suggest that plausible BRDFs can be recovered for a diversity of materials.

5   Discussion

This paper presents a technique for 'light-weight' reflectometry that eliminates the need for active illumination and requires minimal infrastructure for acquisition. This is enabled by reducing the domain of isotropic bi-directional reflectance functions from three dimensions to two. We provide an empirical evaluation of this reduced representation that compliments recent work [2].

The proposed approach has clear advantages over existing inverse rendering techniques that recover reflectance from 2D images using de-convolution or low-parameter BRDF models. These existing methods recover reflectance functions that are one-dimensional (radially-symmetric) or zero-dimensional (parametric), respectively. In contrast, the method presented here recovers a two-dimensional reflectance function, and thereby matches the dimension of the output with that of the input. For this reason, it can be applied to a much broader class of surfaces.

One of the important things we give up in exchange for generality is the intuition provided by the convolution framework. It becomes difficult to characterize the necessary conditions for adequate illumination, and this suggests a direction for future work. In particular, it may be possible to clarify the role that 'environment foldings' (Fig. 4) play in reducing redundancy in L and 'enhancing the adequacy' of an environment.

There are a number of additional directions for future work. We presented one of many possible regularization schemes, and it is possible that others are more suitable. In exploring this possibility, one must be wary of 'overfitting' existing BRDF databases, since these may provide descriptions of only a fraction of the world's interesting materials. We have largely avoided this in our approach, but even so, we expect our method to be less successful for highly retro-reflective surfaces, which are not well represented in the MERL database.

Our focus in this work is the recovery of *general* reflectance functions, meaning those that are not necessarily well-represented by low-parameter models and those that are not radially-symmetric. For this reason, we considered the case in which the surface is homogeneous, its shape is known, and the illumination environment is also known. Relaxing these conditions is perhaps the most interesting direction for future work, and it is quite likely that the tools presented here will prove useful elsewhere (see [17] for a reconstruction application).

In this vein, the proposed framework provides an opportunity to explore the joint recovery of reflectance and illumination (f and L in Eq. 6), or at least the recovery of reflectance when lighting is unknown. Using our framework, this essentially becomes a blind de-convolution problem. It is possible that this line of research may eventually yield computational systems that can match the human ability to infer reflectance in uncontrolled conditions [1].

Acknowledgements

We thank Wojciech Matusik for helpful discussions regarding the MERL database. Support comes from an NSF CAREER award and a Sloan Foundation fellowship.

References

1. Fleming, R., Dror, R.O., Adelson, E.H.: Real-world illumination and the perception of surface reflectance properties. Journal of Vision 3 (2003)
2. Stark, M., Arvo, J., Smits, B.: Barycentric parameterizations for isotropic BRDFs. IEEE Transactions on Visualization and Computer Graphics 11, 126–138 (2005)
3. Ngan, A., Durand, F., Matusik, W.: Experimental analysis of brdf models. In: Eurographics Symposium on Rendering, pp. 117–126 (2005)
4. Ward, G.: Measuring and modeling anisotropic reflection. Computer Graphics (Proc. ACM SIGGRAPH) (1992)
5. Marschner, S., Westin, S., Lafortune, E., Torrance, K., Greenberg, D.: Image-based BRDF measurement including human skin. In: Proc. Eurographics Symposium on Rendering, pp. 139–152 (1999)
6. Ghosh, A., Achutha, S., Heidrich, W., O'Toole, M.: BRDF acquisition with basis illumination. In: Proc. IEEE Int. Conf. Computer Vision (2007)
7. Ramamoorthi, R., Hanrahan, P.: A signal-processing framework for inverse rendering. In: Proceedings of ACM SIGGRAPH, pp. 117–128 (2001)
8. Boivin, S., Gagalowicz, A.: Image-based rendering of diffuse, specular and glossy surfaces from a single image. In: Proceedings of ACM SIGGRAPH (2001)
9. Yu, Y., Debevec, P., Malik, J., Hawkins, T.: Inverse global illumination: recovering reflectance models of real scenes from photographs. In: Proceedings of ACM SIGGRAPH (1999)
10. Georghiades, A.: Incorporating the Torrance and Sparrow model of reflectance in uncalibrated photometric stereo. In: Proc. IEEE Int. Conf. Computer Vision, pp. 816–823 (2003)
11. Hara, K., Nishino, K., Ikeuchi, K.: Mixture of spherical distributions for single-view relighting. IEEE Trans. Pattern Analysis and Machine Intelligence 30, 25–35 (2008)
12. Patow, G., Pueyo, X.: A Survey of Inverse Rendering Problems. Computer Graphics Forum 22, 663–687 (2003)
13. Rusinkiewicz, S.: A new change of variables for efficient BRDF representation. In: Eurographics Rendering Workshop, vol. 98, pp. 11–22 (1998)
14. Matusik, W., Pfister, H., Brand, M., McMillan, L.: A data-driven reflectance model. ACM Transactions on Graphics (Proc. ACM SIGGRAPH) (2003)
15. Westin, S.: Measurement data, Cornell University Program of Computer Graphics (2003), http://www.graphics.cornell.edu/online/measurements/
16. Matusik, W., Pfister, H., Brand, M., McMillan, L.: Efficient isotropic BRDF measurement. In: Proc. Eurographics Workshop on Rendering, pp. 241–247 (2003)
17. Alldrin, N., Zickler, T., Kriegman, D.: Photometric stereo with non-parametric and spatially-varying reflectance. In: Proc. CVPR (2008)

Fusion of Feature- and Area-Based Information for Urban Buildings Modeling from Aerial Imagery*

Lukas Zebedin[1], Joachim Bauer[1], Konrad Karner[1], and Horst Bischof[2]

[1] Microsoft Photogrammetry
`irstname.lastname@microsoft.com`
[2] Graz University of Technology
`bischof@icg.tugraz.at`

Abstract. Accurate and realistic building models of urban environments are increasingly important for applications, like virtual tourism or city planning. Initiatives like Virtual Earth or Google Earth are aiming at offering virtual models of all major cities world wide. The prohibitively high costs of manual generation of such models explain the need for an automatic workflow.

This paper proposes an algorithm for fully automatic building reconstruction from aerial images. Sparse line features delineating height discontinuities and dense depth data providing the roof surface are combined in an innovative manner with a global optimization algorithm based on Graph Cuts. The fusion process exploits the advantages of both information sources and thus yields superior reconstruction results compared to the indiviual sources. The nature of the algorithm also allows to elegantly generate image driven levels of detail of the geometry.

The algorithm is applied to a number of real world data sets encompassing thousands of buildings. The results are analyzed in detail and extensively evaluated using ground truth data.

1   Introduction

Algorithms for the semi- or fully automatic generation of realistic 3D models of urban environments from aerial images are subject of research for many years. Such models were needed for urban planning purposes or for virtual tourist guides. Since the advent of web-based interactive applications like Virtual Earth and Google Earth and with the adoption of 3D content for mashups the demand for realistic models has significantly increased. The goal is to obtain realistic and detailed 3D models for entire cities.

This poses several requirements for the algorithm: First, it should not require any manual interaction because this would induce high costs. This restriction also dissuades the use of cadastral maps as they vary in accuracy, are not readily available everywhere and require careful registration towards the aerial data. Additionally such a dependency increases the cost at large scale deployment. Second, the algorithm should be flexible enough to generate accurate models for common urban roof structures without limiting itself to one specific type, like gabled roofs or rectangular outlines for example. This

---

* This work has been supported by the FFG project APAFA (813397) under the FIT-IT program.

also includes the requirement to be able to deal with complex compositions of roof shapes if those happen to be adjacent. Third, the algorithm should have a certain degree of efficiency as it is targeted at thousands of cities with millions of buildings in total. Last, the algorithm should be robust: the visual appearance should degrade gracefully under the presence of noise or bad input data quality.

In the following a survey and assessment of existing algorithms is given, which fail to meet one or more of the above mentioned requirements.

Among the early approaches are feature based modelling methods ([1,2,3,4,5]) which show very good results for suburban areas. The drawback of those methods is their reliance on sparse line features to describe the complete geometry of the building. The fusion of those sparse features is very fragile as there is no way to obtain the globally most consistent model.

The possibility of using additional data (cadastral maps and other GIS data in most cases) to help in the reconstruction task is apparent and already addressed in many publications ([6,7,8]). Such external data, however, is considered manual intervention in our work and thus not used.

A different group of algorithms concentrates on the analysis of dense altimetry data obtained from laser scans or dense stereo matching ([9,10]). Such segmentation approaches based solely on height information, however, are prone to failure if buildings are surrounded by trees and require a constrained model to overcome the smoothness of the data at height discontinuities. Guhno and Downman ([11]) combined the elevation data from a LIDAR scan with satellite imagery using rectilinear line cues. Their approach was, however, limited to determining the outline of a building. In our work we develop this approach further and embed it into a framework which overcomes the problems described above.

In [12] we have proposed a workflow to automatically derive the input data used in this paper. The typical aerial images used in the workflow have 80% along-strip overlap and 60% across-strip overlap. This highly redundant data is utilized in this paper. Similar approaches have been proposed by others ([13,14]), which demonstrate that it is possible to automatically derive a digital terrain model, digital elevation model, land use classification and orthographic image from aerial images. Figure 1 illustrates

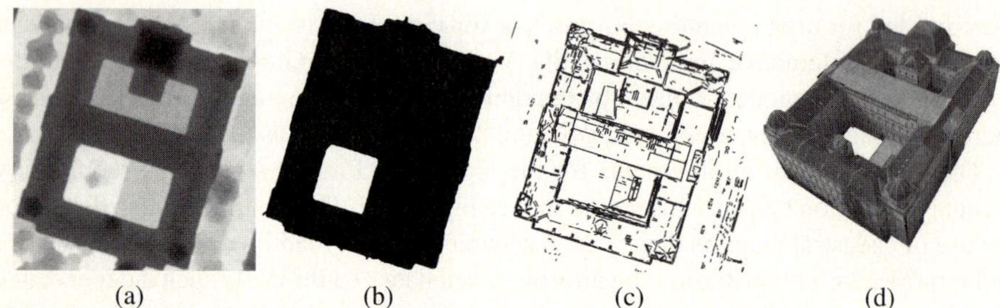

(a)   (b)   (c)   (d)

Fig. 1. These figures depict the data which is used for the reconstruction process: (a) height field, (b) building mask and (c) 3D line segments. Image (d) shows the obtained model by our algorithm.

the available data which is used for the reconstruction algorithm and also shows the result of the proposed algorithm.

Our proposed method does not need any manual intervention and uses only data derived from the original aerial imagery. It combines dense height data together with feature matching to overcome the problem of precise localization of height discontinuities. The nature of this fusion process separates discovery of geometric primitives from the generation of the building model in the spirit of the recover-and-select paradigm ([15]), thus lending robustness to the method as the global optimal configuration is chosen. The integration of the theory of instantaneous kinematics ([16]) allows to elegantly detect and estimate surfaces of revolution which describe a much broader family of roof shapes. A major feature of the proposed method is the possibility to generate various levels of geometric detail.

The rest of the paper is structured as follows: Chapter 2 gives a general overview of the method. In Chapter 3 we will describe the discovery of geometric primitives which are used to approximate the roof shape, whereas Chapter 4 discusses the building segmentation. Chapter 5 gives details about the fusion process which combines line features and dense image data. Results and experiments are outlined in Chapter 6. Finally, conclusions and further work are described in Chapter 7.

2   Overview of the Method

The workflow of the proposed method is outlined in Figure 2. Three types of information are necessary as input for the algorithm: Dense height data is generated by a dense image matching algorithm ([17]) (Figure 1a, represented as a height field) and gives a good estimate of the elevation, but suffers from oversmoothing at height discontinuities ([18]). Additionally a rough segmentation of the building is required (Figure 1b) which could be directly deduced from the height data for example. The third component are sparse 3D line segments (Figure 1c) which are obtained from line matching over multiple views ([1]).

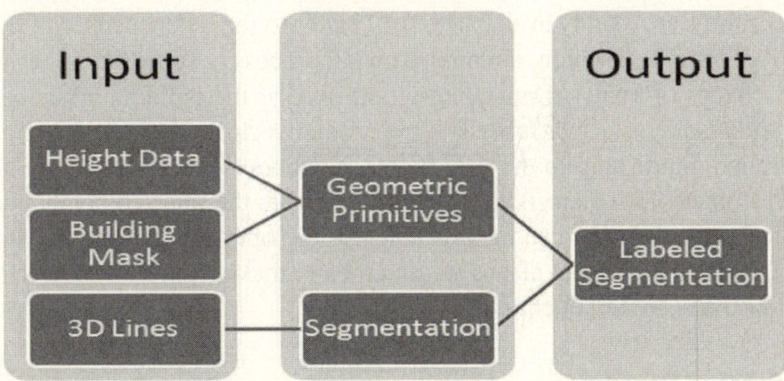

Fig. 2. Illustration of the single steps of the proposed method: height data and building mask are used to obtain a set of geometric primitives; In parallel the 3D lines are used to generate a segmentation of the building. Finally, a labeled segmentation is produced.

The building mask is combined with the dense height data, thus filtering out all 3D points which do not belong to the building. Afterwards the remaining points are grouped into geometric primitives. The geometric primitives are the basic building blocks for assembling the roof shape.

The 3D line segments are projected into the height field and used to obtain a line-based segmentation of the building. The 2D lines of the segmentation form polygons which are then assigned to one of the geometric primitives. Therefore, it is important that the 3D lines capture the location of the height discontinuities as each polygon is treated as one consistent entity which can be described by one geometric primitive. By extruding each of the 2D polygons to the assigned geometric primitive a 3D model of the building is generated.

Note that the algorithm presented in this paper makes no assumptions about the roof shape. Façades are modeled as vertical planes, because the oblique angle of the aerial images does not allow a precise reconstruction of any details.

3   Geometric Primitives

Geometric primitives form the basic building blocks which are used to describe the roof shape of a building. Currently two types of primitives, namely planes and surfaces of revolution, are used, but the method can be trivially extended to support other primitives. It is important to note, that the detection of geometric primitives is independent from the composition of the model. This means that an arbitrary amount of hypotheses can be collected and fed into later stages of the algorithm. As the order of discovery of the primitives is not important, weak and improbable hypotheses are also collected as they will be rejected later in the fusion step. If a primitive is missed, the algorithm selects another detected primitive instead which minimizes the incurred reconstruction error.

3.1   Planes

Efficiently detecting planes in point clouds for urban reconstruction is well studied and robust algorithms are readily available ([9]). Thanks to the independence of hypothesis discovery and model selection, a region growing process is sufficient in our workflow for the discovery of planes. Depending on the size of the building a number of random seed points are selected, for which the normal vector is estimated from the local neighbourhood. Starting from the seed points, neighbours are added which fit the initial plane estimate. This plane is regularly refined from the selected neighbours. Small regions are rejected to improve the efficiency of the optimization phase. Due to their frequency, close to horizontal planes are modified to make them exactly horizontal, the other oblique ones are left unchanged.

3.2   Surfaces of Revolution

Planar approximations of certain roof shapes (domes and spires for example) obtained from plane fitting algorithms, however, are not robust, visually displeasing and do not take the redundancy provided by the symmetrical shape into account. Therefore it is necessary to be able to deal with other shapes as well and combine them seamlessly to obtain a realistic model of the building.

Surfaces of revolution are a natural description of domes and spires and can be robustly detected. Mathematically such surfaces can be described by a 3D curve which moves in space according to an Euclidean motion. Instantaneus kinematics gives a relationship ([19]) between that Euclidean motion parameters and the corresponding velocity vector field. Using that connection it is possible to estimate the parameters of the Euclidean motion in a least squares sense given the normal vectors of the resulting surface.

The equation

$$v(x) = \bar{c} + c \times x \tag{1}$$

describes a velocity vector field with a constant rotation and constant translation defined by the two vectors $c, \bar{c} \in \mathbb{R}^3$. If a curve sweeps along that vector field, the normal vectors of all points on the resulting surface have to be perpendicular to the velocity vector at the associated point. Thus

$$n(x)v(x) = 0 \tag{2}$$

$$n(x)\,(\bar{c} + c \times x) = 0$$

holds, where $n(x)$ gives the normal vector at point x. With equation (2) it is possible to estimate the motion parameters given at least six point and normal vector pairs $(x, n(x))$ lying on the same surface generated by such a sweeping curve. In the case of point clouds describing an urban scene the parameter can be constrained by requiring the rotation axis to be vertical. This already reduces the degrees of freedom to two (assuming that z is vertical) and makes the problem easily solvable:

$$\bar{c} = (0, x, y)^T \qquad c = (0, 0, 1)^T$$

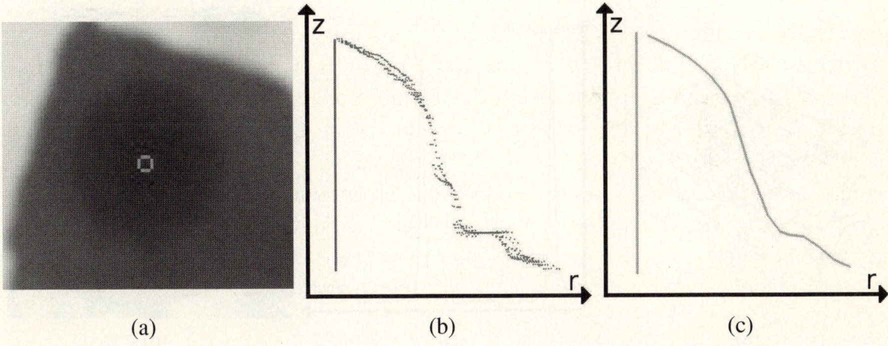

Fig. 3. Illustrations how starting with the dense height data the 3D curve is derived which generates the dome if it rotates around a vertical axis. (a) Raw height field with the detected axis, (b) all inliers are projected into the halfplane formed by axis and a radial vector, (c) the moving average algorithm produces a smooth curve.

where $\bar{c}$ gives the position of the axis and c denotes the vertical rotation axis. The remaining two unknown parameters are estimated by transforming each 3D point with the estimated normal vector $(x, n(x))$ into a Hough space ([20]). Local maxima in the accumulation space indicate axes for surfaces of revolution. For each axis all inliers are computed and projected into the halfplane spanned by the rotation axis and an arbitrary additional radial vector. The redundancy of the symmetrical configuration can be exploited by a moving average algorithm in order to estimate a smooth curve which generates the surface containing the inliers. Figure 3 illustrates those steps with a point cloud describing the shape of a spire.

4   Segmentation

The goal of the segmentation is to represent the general building structure - not only a rectangular shape - as a set of 2D polygons.

The approach of Schmid and Zisserman ([21]) is used for the generation of the 3D line set that is then used for the segmentation of the building into 2D polygons. A 3D line segment must have observations in at least four images in order to be a valid hypothesis. This strategy ensures that the reliability and geometric accuracy of the reported 3D line segments is sufficiently high. The presence of outliers is tolerable since the purpose of the 3D lines is to provide a possible segmentation of the building. Any 3D line that does not describe a depth discontinuity can be considered as an unwanted outlier which will contribute to the segmentation, but will be eliminated in the fusion stage.

The matched 3D line segments are used to obtain a 2D segmentation of the building into polygons by appying an orthographic projection. The 2D lines cannot be used directly to segment the building, however, as the matching algorithm often yields many short line segments describing the same height discontinuity. A grouping mechanism merges those lines to obtain longer and more robust lines. A weighted orientation

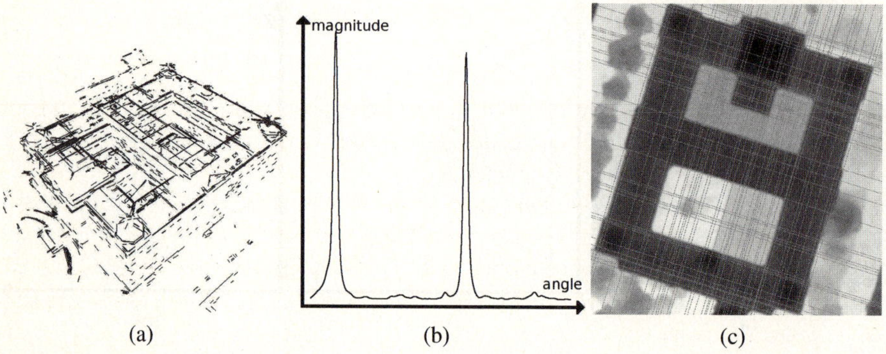

Fig. 4. Segmentation into polygons: (a) The matched 3D lines are projected into the $2\frac{1}{2}$D height field, (b) outliers are eliminated by a weighted orientation histogram which helps to detect principal directions of the building. (c) Along those directions lines are grouped, merged and extended to span the whole building.

histogram - the weights correspond to the length of each line - is created. The principal orientations are detected by finding local maxima in the histogram. Along those directions quasi parallel lines are grouped and merged thus refining their position.

Each grouped line is extended to span the whole building in order to simplify the segmentation process. The lines are splitting the area into a number of polygons. Each polygon is considered to be one consistent entity where the 3D points can be approximated by one geometric primitive.

Figure 4 illustrates this concept. The advantage of this approach is that no assumption or constraint of the shape, angles and connectivity of the building is necessary.

5   Information Fusion

Each polygon resulting from the segmentation is assigned to one geometric primitive (plane or surface of revolution, see Chapter 3). This labeling allows to create a piecewise planar reconstruction of the building - surfaces of rotation are approximated by a rotating polyline and therefore also yield piecewise planar surfaces in the polyhedral model.

The goal of the fusion step is to approximate the roof shape by the geometric primitives in order to fullfill an optimization criterion. In this paper we use the Graph Cuts algorithm with alpha-expansion moves ([22,23]), but other techniques like belief propagation are suited as well. The goal of this optimization is to select a geometric primitive for each polygon of the segmentation and to find an optimal trade-off between data fidelity and smoothness.

5.1   Graph Cuts Optimization

The Graph Cuts algorithm finds a very good approximation of the globally optimal solution for a broad range of tasks which can be stated as an energy minimization problem of the following form:

$$E(f) = \sum_{p \in P} D_p(f_p) + \lambda \cdot \sum_{\{p,q\} \in N} V_{p,q}(f_p, f_q) \qquad (3)$$

where $V_{p,q}(f_p, f_q)$ is called the smoothness term for the connected nodes p and q which are labeled f_p and f_q and $D_p(f_p)$ is called the data term which measures a data fidelity obtained by assigning the label f_p to node p.

In our approach the segmentation induces a set P of polygons, where each polygon represent a node of the graph. The neighboorhood relationship is reflected by the set N, which contains pairs of adjacent polygons, ie. polygons sharing an edge. The set of labels used in the optimization process represent the geometric primitives (planes and surfaces of revolution):

$$L = \{\text{plane}_1, \text{plane}_2, ..., \text{surface-of-revolution}_1, \text{surface-of-revolution}_2, ...\} \qquad (4)$$

Thus $f_p \in L$ reflects the label (current geometric primitve) assigned to node (polygon) $p \in P$.

The optimization using polygons is much faster than optimizing for each individual pixel because there are much fewer polygons than pixels. On the other hand it also exploits the redundancy of the height data because it is assumed that all pixels in one polygon belong to the same geometric primitive.

In our context the smoothness term measures the length of the border between two polygons and the data term measures the deviation between the observed surface (obtained from the dense image matching algorithm) and the fitted primitive. The following formulae are used to calculate those two terms:

$$D_p(f_p) = \sum_{x \in p} \left| height_{obs}(x) - height_{f_p}(x) \right| \tag{5}$$

$$V_{p,q}(f_p, f_q) = \begin{cases} length(border(p,q)) & \text{if } f_p \neq f_q \\ 0 & \text{if } f_p = f_q \end{cases} \tag{6}$$

where p and q denote two polygons and f_p is the current label of polygon p. The preset constant λ can be used to weight the two terms in the energy functional. The data term D_p calculates an approximation of the volume between the point cloud ($height_{obs}(x)$) and primitive f_p ($height_{f_p}(x)$) by sampling points x which lie within the polygon p. This sampling strategy allows to treat all geometric primitives similarly. because they are reduced to the incurred difference in volume and induced border to other polygons assigned to another geometric primitive. The smoothness term $V_{p,q}$ penalizes neighbouring polygons with different labels depending on their common border, thus favouring homogeneous regions.

The alpha-expansion move is used in order to efficiently optimize the labeling of all polygons with respect to all discovered primitives. The initial labeling can either be random or a labeling which minimizes only the data term for each individual polygon. After a few iterations (usually less than 5), the optimization converges and all 2D polygons can be extruded to the respective height of the assigned primitive to generate a polyhedral model of the building.

5.2   Levels of Detail

The second term in Equation (3) regularizes the problem and favors smooth solutions. Depending on the actual value of λ in Equation (3) different results are obtained. Higher values result in fewer and shorter borders at the cost of larger volumetric differences

Table 1. The impact of the smoothness parameter λ on the reconstructed model. The number of unique labels used after the Graph Cuts optimization iterations decreases as well as the number of triangles in the polygonal model. Δ Volume denotes the estimated difference in volume between the surface obtained by dense image matching and the reconstruced model (data term). The last column refers to the accumulated length of all borders in the final labeling (smoothness term).

λ	#Labels	#Triangles	Δ Volume $[m^3]$	Border Length $[m]$
5	7	79	1210.79	710.4
10	6	69	1677.19	349.4
20	4	42	1699.31	337.0
100	3	33	2293.36	290.4

between observed height values and reconstructed models. This feature can be used to generate different models with varying smoothness, trading data fidelity for geometric simplificationa as smaller details of the building are omitted. An example of such a simplification is shown in Figure 6. The relevant numbers for that building are given in Table 1.

6   Experiments

The first illustrative experiment was conducted on a test data set of a Graz. The ground sampling distance of the aerial imagery is 8cm. The examined building features four small cupolas at the corners. Additionally one façade is partially occluded by trees. Figure 5 shows the results of the reconstruction process. The texture of the façades is well aligned, implying that their orientation was accurately estimated by the 3D line matching. The domes are smoothly integrated into the otherwise planar reconstruction. Even the portion occluded by the tree has been straightened by the extension of the matched 3D lines.

The next example is taken from a data set of Manhattan, New York. This building shows that the reconstruction algorithm is not limited to façades perpendicular or parallel to each other. Figure 6 illustrates the effect of the smoothness term in the global optimization energy function. Various runs with different values for λ yield a reduced triangle count as the geometry is progressively simplified. Table 1 gives details about the solution for different values of λ. The Graph Cuts algorithm allows to find a globally optimal tradeoff between data fidelity and generalization. Those properties are expressed by the decreased length of borders and number of labels (which translate in general to fewer triangles) at the cost of an increase of the average difference between reconstructed and observed surface.

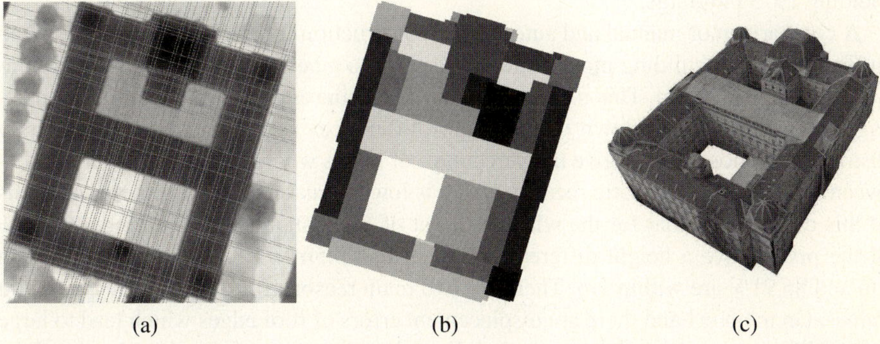

(a)   (b)   (c)

Fig. 5. The stages of the reconstruction are illustrated by means of the building of the Graz University of Technology: (a) Segmented height field, (b) labeled polygons after the Graph Cuts optimization, (c) screenshot of the reconstructed model ($\lambda = 5$)

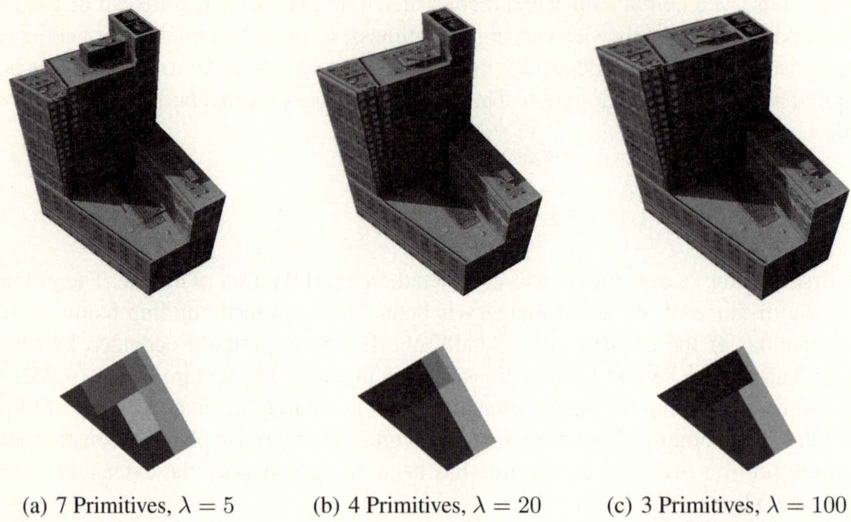

Fig. 6. Levels of Detail: The same building was reconstructed with different values for λ. The number of geometric primitives used to approximate the shape of the roof is decreasing with higher values for λ. In the upper row a screenshot of the reconstruction is depicted, below are illustrations of the matching labeling obtained by the Graph Cuts optimization.

Apart from judging the visual appearance of the resulting models, we assess the quality of the reconstructed models by comparing them to a ground truth which was obtained manually from the same imagery. For this purpose we use a stereoscopic device to trace the roof lines in 3D. Those roof lines are connected to form polygons and then extruded to the ground level. Those manually reconstructed models are considered ground truth data in this paper. Using this procedure the whole data set from Manhattan (consisting of 1419 aerial images at 15cm ground sampling distance) was processed yielding 1973 buildings.

A comparison of manual and automatic reconstruction for one building is illustrated in Figure 7. Both building models are converted into a height field with a ground sampling distance of 15cm. This makes it easy to determine and illustrate their differences. Figure 8 gives a break down of the height differences as a cummulative probabilty distribution. Those graphs give the percentage of pixels where the height difference between manual and automatic reconstruction is lower than a certain threshold. Analysis of this chart shows that for the whole data set of Manhattan (1973 buildings) 67.51% of the pixels have a height difference smaller than 0.5m, 72.85% differ by less than 1m and 86.91% are within 2m. There are two main reasons for discrepancies of height values: On the one hand there are displacement errors of roof edges which lead to large height differences, depending on the height of the adjacent roof. On the other hand the human operator is able to recognize small superstructurial details on the roofs like elevator shafts and air conditioning units which cause height differences usually below 2m. Those small features are sometimes missed by the automatic reconstruction.

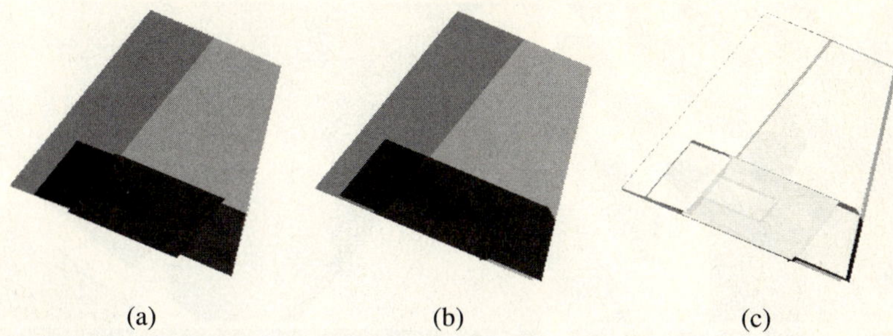

(a)   (b)   (c)

Fig. 7. Quality assessment with a manually generated ground truth: In (a) and (b) the height fields for the manually and automatically reconstructed building are shown, in (c) the height differences are shown. The largest difference in the placement of edges is about two pixels, which is about 30cm.

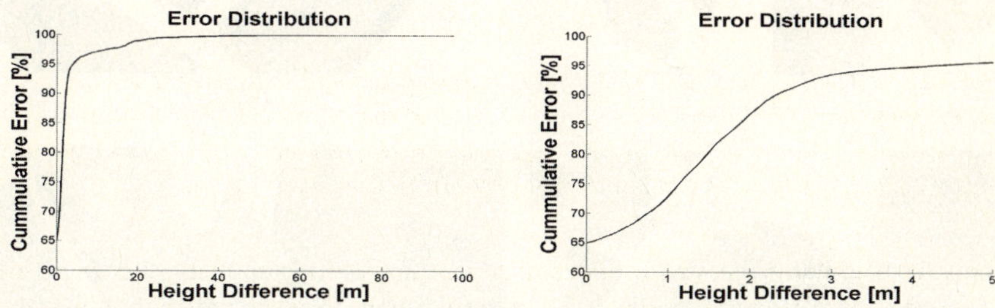

Fig. 8. The cummulative probabilty distribution of the height difference for manual and automatic reconstruction. The graph shows the error distribution for 1973 buildings from a data set of Manhattan, New York. The left image shows the graphs for height differences up to 100 meters; the right graph zooms on differences up to five meters.

Detailed views of typical results from the Manhattan data set are shown in Figure 9. The reconstruction of rectangular buildings is very successful, even though huge portions of their façades are occluded by trees. The integration of surfaces of revolution realistically models domes and spires (see 9b and 9d). It is important to note that for the purpose of visualization the surfaces of revolution are converted to triangle meshes by sampling them regularly (2m radially with 45 degrees of angular separation).

7 Conclusions and Future Work

In this paper we proposed a novell approach to reconstruct building models from aerial images by combining 3D line segments and dense image matching algorithms with a global optimization technique. The framework is able to use arbitrary basic geometric building blocks to describe the roof shape. The proposed surfaces of revolution elegantly describe domes and spires which are difficult to recover with an approach based on planes only. The combination of line based features and dense image matching

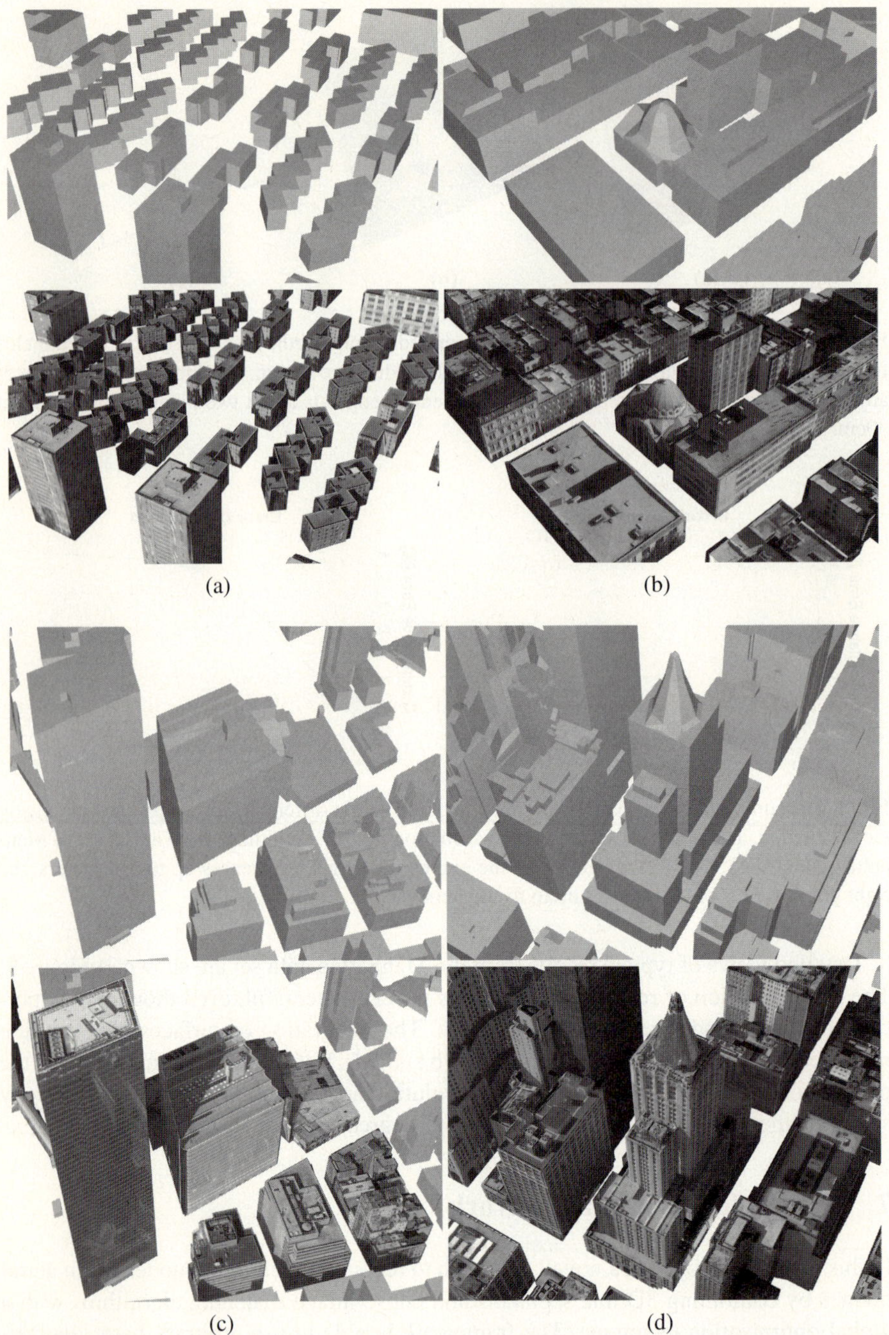

Fig. 9. Four detailed views of typical results for different types of buildings from the Manhattan data set: (a) rectangular buildings, (b) rectangular building with nicely integrated dome, (c) skyscrapers in downtown and (d) skyscraper with a spire

algorithms using a global optimization technique is very promising and is not restricted to the reconstruction of urban scenes from aerial imagery. Additionally it allows for the generation of different globally optimal levels of detail.

Future work will involve the investigation of other geometric primitives and methods to exploit symmetries encountered in common roof shapes like gabled roofs. Further research will be needed to evaluate the possibilities of this approach in other applications like streetside imagery.

References

1. Baillard, C., Zisserman, A.: Automatic Line Matching And 3D Reconstruction Of Buildings From Multiple Views. In: ISPRS Conference on Automatic Extraction of GIS Objects from Digital Imagery, vol. 32, pp. 69–80 (1999)
2. Bignone, F., Henricsson, O., Fua, P., Stricker, M.A.: Automatic Extraction of Generic House Roofs from High Resolution Aerial Imagery. In: European Conference on Computer Vision, Berlin, Germany, pp. 85–96 (1996)
3. Fischer, A., Kolbe, T., Lang, F.: Integration of 2D and 3D Reasoning for Building Reconstruction using a Generic Hierarchical Model. In: Workshop on Semantic Modeling for the Acquisition of Topographic Information, Munich, Germany, pp. 101–119 (1999)
4. Taillandier, F., Deriche, R.: Automatic Buildings Reconstruction from Aerial Images: a Generic Bayesian Framework. In: Proceedings of the XXth ISPRS Congress, Istanbul, Turkey (2004)
5. Vosselman, G.: Building Reconstruction Using Planar Faces in Very High Density Height Data. In: ISPRS Conference on Automatic Extraction of GIS Objects from Digital Imagery, Munich, vol. 32, pp. 87–92 (1999)
6. Baillard, C.: Production of DSM/DTM in Urban Areas: Role and Influence of 3D Vectors. In: ISPRS Congress, Instanbul, Turkey, vol. 35, p. 112 (2004)
7. Haala, N., Anders, K.H.: Fusion of 2D-GIS and Image Data for 3D Building Reconstruction. In: International Archives of Photogrammetry and Remote Sensing, vol. 31, pp. 289–290 (1996)
8. Suveg, I., Vosselman, G.: Reconstruction of 3D Building Models from Aerial Images and Maps. ISPRS Journal of Photogrammetry and Remote Sensing 58(3-4), 202–224 (2004)
9. Haala, N., Brenner, C.: Generation of 3D City Models from Airborne Laser Scanning Data. In: 3rd EARSEL Workshop on Lidar Remote Sensing on Land and Sea, Tallinn, Estonia, pp. 105–112 (1997)
10. Maas, H.G., Vosselman, G.: Two Algorithms for Extracting Building Models from Raw Laser Altimetry Data. In: ISPRS Journal of Photogrammetry and Remote Sensing, vol. 54, pp. 153–163 (1999)
11. Sohn, G., Dowman, I.: Data Fusion of High-Resolution Satellite Imagery and LIDAR Data for Automatic Building Extraction. ISPRS Journal of Photogrammetry and Remote Sensing 62(1), 43–63 (2007)
12. Zebedin, L., Klaus, A., Gruber-Geymayer, B., Karner, K.: Towards 3D Map Generation from Digital Aerial Images. ISPRS Journal of Photogrammetry and Remote Sensing 60(6), 413–427 (2006)
13. Chen, L.C., Teo, T.A., Shaoa, Y.C., Lai, Y.C., Rau, J.Y.: Fusion of Lidar Data and Optical Imagery for Building Modeling. In: International Archives of Photogrammetry and Remote Sensing, vol. 35(B4), pp. 732–737 (2004)
14. Hui, L.Y., Trinder, J., Kubik, K.: Automatic Building Extraction for 3D Terrain Reconstruction using Interpretation Techniques. In: ISPRS Workshop on High Resolution Mapping from Space, Hannover, Germany, p. 9 (2003)

15. Leonardis, A., Gupta, A., Bajcsy, R.: Segmentation of Range Images as the Search for Geometric Parametric Models. International Journal of Computer Vision 14(3), 253–277 (1995)
16. Pottmann, H., Leopoldseder, S., Hofer, M.: Registration without ICP, vol. 95, pp. 54–71 (2004)
17. Klaus, A., Sormann, M., Karner, K.: Segment-Based Stereo Matching Using Belief Propagation and a Self-Adapting Dissimilarity Measure. In: Proceedings of the 18th International Conference on Pattern Recognition, vol. 3, pp. 15–18. IEEE Computer Society Press, Washington (2006)
18. Scharstein, D., Szeliski, R.: A Taxonomy and Evaluation of Dense Two-Frame Stereo Correspondence Algorithms. International Journal of Computer Vision 47, 7–42 (2002)
19. Pottmann, H., Leopoldseder, S., Hofer, M.: Simultaneous Registration of Multiple Views of a 3D Object. In: Archives of the Photogrammetry, Remote Sensing and Spatial Information Sciences, vol. 34, Part 3A (2002)
20. Illingworth, J., Kittler, J.: A Survey of the Hough Transform. Computer Vision, Graphics and Image Processing 44(1) (1988)
21. Schmid, C., Zisserman, A.: Automatic Line Matching Across Views. In: IEEE Conference on Computer Vision and Pattern Recognition, pp. 666–671 (1997)
22. Boykov, Y., Veksler, O., Zabih, R.: Fast Approximate Energy Minimization Via Graph Cuts. In: International Conference on Computer Vision, Kerkyra, Corfu, vol. 1, pp. 377–384 (1999)
23. Kolmogorov, V., Zabih, R.: What Energy Functions Can Be Minimized Via Graph Cuts? In: European Conference on Computer Vision, Copenhagen, Denmark, vol. 3, pp. 65–81 (2002)

Printing: Mercedes-Druck, Berlin
Binding: Stein+Lehmann, Berlin